Computer Arithmetics for Nanoelectronics

Computer Arithmetics for Nanoelectronics

Vlad P. Shmerko
Svetlana N. Yanushkevich
Sergey Edward Lyshevski

CRC Press
Taylor & Francis Group
Boca Raton London New York

CRC Press is an imprint of the
Taylor & Francis Group, an **informa** business

CRC Press
Taylor & Francis Group
6000 Broken Sound Parkway NW, Suite 300
Boca Raton, FL 33487-2742

First issued in paperback 2017

© 2009 by Taylor & Francis Group, LLC
CRC Press is an imprint of Taylor & Francis Group, an Informa business

No claim to original U.S. Government works

ISBN 13: 978-1-138-11357-2 (pbk)
ISBN 13: 978-1-4200-6621-0 (hbk)

Library of Congress Cataloging-in-Publication Data

Shmerko, Vlad P.
    Computer arithmetics for nanoelectronics / authors, Vlad P. Shmerko, Svetlana N. Yanushkevich, Sergey Edward Lyshevski.
        p. cm.
    Includes bibliographical references and index.
    ISBN 978-1-4200-6621-0 (hardcover : alk. paper)
    1. Nanoelectronics--Mathematics. I. Yanushkevich, Svetlana N. II. Lyshevski, Sergey Edward. III. Title.

TK7874.84.S54 2009
621.381--dc22   2008052003

Visit the Taylor & Francis Web site at
http://www.taylorandfrancis.com

and the CRC Press Web site at
http://www.crcpress.com

This book is dedicated to the memory of
Claude Shannon

CONTENTS

18 Design for Testability 673

19 Error Detection and Error Correction 699

Preface

THE MOTIVATION: TOWARD PREDICTABLE TECHNOLOGIES

This book is motivated by the emerging device- and system-level solutions to implement various computing and processing platforms created with nano- and molecular technologies. This ultimately leads not only to the deployment of unique phenomena that which never have been utilized in conventional solid-state microelectronics devices but also to advanced system organizations and architectures. The three-dimensional (3D) device and system topologies, multiterminal multiple-valued devices, and other features are subproducts of those emerging solutions. For these new generations of computing devices and systems, the conventional design and analysis concepts and techniques are the valuable source of further developments.

There are many examples of the use of the fundamentals of computer structure design and computing in nanocomputing, including particular results, that have not been adopted in contemporary design. For example, the principle of programable logic arrays (PLA) has been adopted for the crossbar-based array design and classical fault-tolerant techniques modified for manufacturing of computing nanodevices and nanomemory. Computing paradigms based on cellular arrays are the focus of massive parallel nanoarrays. Hence, the computing paradigms of contemporary discrete devices play an essential role in the development of novel computing structures based on the technological advances in molecular electronics. This book addresses these problems.

These developments integrate theoretical computer science, computer engineering, electrical engineering, microelectronics engineering, and other areas. Researchers and students from computer and electrical engineering, chemistry, biology, mechanics, and physics departments were involved in this interdisciplinary study.

It should be noted that usage of the uniform terminology is another problem since the terms used in the area of nanocomputing are acquired from different disciplines. In particular, the term *molecular electronics* covers a broad range of topics, in particular, molecular computing devices, device physics, and biophysics of biomolecular devices. With a major emphasis on theoretical and applied computer science and engineering, in this book we focus on computing structure design based on computing primitives; the basic components such as switches, logic gates, 1-bit adders, and cells for computing arrays; and memory structures.

A wide variety of physical, biological and chemical phenomena, considered to be the candidates employed in processing, transforming, and storing the information at nano scale. These principles and phenomena are completely different compared to conventional solid-state devices. Various physical, biological, and chemical effects; phenomena; and transitions can be interpreted as computing events that are associated with the simplest logic operations, such as switching, OR, AND, and others. These logic operations correspond to the logic elements, which can be integrated and aggregated to perform complex functions.

This book provides the computing concepts, techniques, and tools in the form acceptable to specialists from various fields for the coherent and synergic design of computing structures utilizing the conventional and emerging (nano and molecular) processing devices. We also approach the problems of biomolecular processing typifying the biosystems' topologies, organizations, and potential principles of computing, such as multiple-valued, stochastic, parallel, reconfigurable, neuromorphical, etc. To this end, we elaborate the following two major aspects:

(*a*) The ability to interpret chemical, biological, and physical phenomena at the system level in terms of computing from the viewpoint of theoretical computer science and engineering, and

(*b*) The ability to utilize various chemical, biological, and physical phenomena to guarantee computing and processing at the device level.

This book integrates those developments by means of developing the fundamentals and theory of the information processing, computer structure and logic designs, stochastic computing, etc. Correspondingly, to reflect the aforementioned tasks and prospects, we title this book Computer Arithmetics for Nanoelectronics. In addition to conventional computing, fundamentals of biomolecular processing are introduced, in particular, new paradigms for delegation of computing properties into spatial nanostructures.

HOW THIS BOOK SATISFIES THE ESSENTIAL NEEDS OF INTERDISCIPLINARY COLLABORATION

With the advent of new technologies, a major shift has occurred in interdisciplinary studies. As a result, many engineers and scientists have found it necessary to understand the basic operation of digital systems, and how these systems can be designed if they carry out particular data-processing tasks. This trend has produced a need for basic knowledge in computing structure design to provide a unified overview of the interrelationship between fundamentals of digital system design, computer organization, architectures, and micro- and nanoelectronics.

To comprehend this book, no specialized knowledge of computer science, electrical circuits, electronics, physics, or chemistry is assumed. This book is written from interdisciplinary prospects for the development of a new generation of computer devices and systems.

There are 20 chapters in the book. Each chapter targets a specific tasks and covers particular problems. The key topic is computational data structures. We emphasize on choosing an appropriate data structure under the phenomenological criteria of technology. Computational properties of various data structures are examined and characterized. Our goal is to achieve balance in introducing the computational properties provided by data structures and the phenomenological properties of current and expected technologies. For example, logic networks and decision diagrams are different data structures, and each of them is characterized by a set of specific requirements in implementation. At the device level, processing and computational properties of various physical and chemical phenomena vary, and can be systemized with respect to the computational characteristics of data structures. The balancing of these characteristics and requirements can provide an appropriate implementation, and it can be illustrated as shown in Figure 0.1.

The main goal of this book is to elaborate a consensus between computational properties provided by data structures and phenomenological properties of the technology-driven nano and molecular devices. For example, for a given biomolecular phenomenon, its representation in terms of processing must be understood by specialists from related fields and efficiently applied in computing structure design. This book has been written with these objectives in mind.

This book contributes to design of computing structures with a focus on sound contemporary technologies and devices. It is up to date, comprehensive, and pragmatic in its approach. The book aims to provide balanced coverage of concepts, techniques, and practices. We found that five topics were essentially useful in our interdisciplinary studies:

Topic 1: Switch-based computing devices and molecular switches;

Topic 2: Multivalued data structures;

Topic 3: 3D computing structures and 3D molecular topologies;

Topic 4: Design for testability and imperfection of molecular structures; and

Topic 5: Natural computing (Figure 0.2).

The design cycle from task formulation to molecular-based implementation includes several phases, which are covered in the book's chapters. We start from a brief overview of the known computing devices reported in Chapter

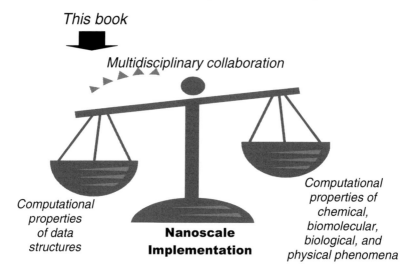

FIGURE 0.1
The main goal of this book is to contribute to the problem of the consensus between computational properties provided by data structures and phenomenological properties of nanoscale technology.

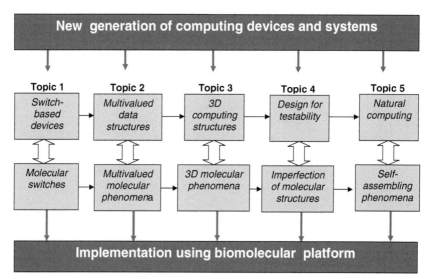

FIGURE 0.2
This book is based on seminars and discussions during the multidisciplinary project on development a new generation of computing devices.

2. The number systems, emphasizing binary arithmetic, are documented in Chapters 3 and 4. The residue number system is introduced as well, as it plays an essential role in many important applications such as encryption. Chapter 5 is a brief introduction to graphical data structures, which are useful computational models. The computational arithmetic must be embedded into the effects, phenomena, and transitions observed and exhibited by nano and molecular devices. The simplest computing devices are switches, and a trivial question is how computations can be mapped using the computing structures in which the operational nodes are implemented using switches covered in Chapters 6–10, 12, and 13. We introduced the theoretical base of various data structures in Chapter 6, their properties in Chapter 7, and techniques for their manipulation, optimization, and implementation in Chapters 8–10. This includes the polynomial forms of logic function, which are of the particular interest due to design tractability, simplicity and soundness as reported in Chapter 9. Polynomial forms are useful in many computational tasks and can be considered as appropriate candidates at the device-level implementation. Chapters 11–14 address the design techniques for the simplest logic networks, memory, and programmable devices. Molecular structures (aggregated devices) provide new opportunities for computing using multivalued logics. Techniques for representation and manipulation of multivalued signals and their implementation in discrete devices are reported in Chapter 11. The molecular devices are inherently 3D. We have proposed to delegate computational properties of the logic design data structures into 3D structures, as covered in Chapters 15 and 16. The information–theoretic measures and design aspects are given in Chapter 17. Chapters 18 and 19 report the testability problem. Natural computing based on various computing paradigms from nature is introduced in Chapter 20.

NEW CONCEPTS IN MULTIDISCIPLINARY STUDY

This book emphasizes the basic principles of computational arithmetic and computational structure design. Most of these principles are traditional in discrete devices design. However, the following key features distinguish this book from other known approaches in design and can be characterized as a major contribution:

▶ New design concepts and sound high-performance paradigms for data/signal processing and computing in 3D

▶ Extension of classical computing paradigms toward a 3D computing structures

▶ Reserving a central role for data structures as a key to applications; the relationships between various data structures and their manipulation through design represent the most important aspect in the design of a new generation of computing devices

▶ Fundamentals of the intermediate data structures; these data structures are the bridge between physical and chemical phenomena and techniques for computing structure design

▶ Techniques for natural computing, such as evolutionary strategy for arbitrary logic network design, and neural computing for the elementary switching functions

This book is dedicated to the memory of Claude Shannon. Claude Elwood Shannon's inventive genius, probably more than that of any other single person, has altered humankind's understanding of communication and digital systems. He was born in Petoskey, Michigan, in 1916. At age of 20 he graduated with degrees in mathematics and electrical engineering from the University of Michigan. During the summer of 1937, Shannon obtained an internship at Bell Laboratories, and returned to MIT to work on a Master's thesis "An Algebra for Theoretical Genetics." He graduated in 1940 with his Ph.D. in mathematics and S.M. degree in electrical engineering. Shannon's Master's thesis won him the Nobel Prize, along with fame and renown. His thesis has often been described as the greatest Master's thesis of all time. In his spare time, Shannon developed the chess machine and remote control mechanical mouse, the first to do so.

Shannon was the first to notice that the work of professor George Boole, done a century earlier, yielded the necessary mathematical framework for analyzing switching networks. He demonstrated that any logical statement could be implemented physically as a network of switches. Shannon's revolutionary paper on information theory still dominates the area of communication theory. It is likely that techniques based on Shannon's information theory will become some of the main design tools for nanosystems, computing systems for the age of nanotechnology.

Vlad P. Shmerko

Svetlana N. Yanushkevich

Sergey Edward Lyshevski

Calgary, Canada
Rochester, New York

1

Introduction

This book is aimed at introducing the reader to emerging computing based on various chemical, biochemical, and physical phenomena.

The main goal of this book is to elaborate a consensus between computational properties provided by data structures and phenomenological properties of the technology-driven nano devices:

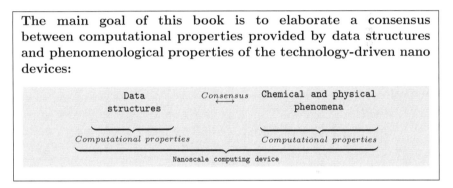

Though one may envision a departure from matured microelectronics, this may take a while due to limited compelling needs, infrastructure expenditure, necessity constraints, and the overall ability of microelectronics to fulfill the current and near-future needs. There is a compelling need to comprehend and examine data processing by molecular hardware from the interdisciplinary point of view in the basic sciences, engineering, technology, and medicine. Among various immense tasks, the authors focus mainly on theoretical computer science and engineering, examining how to accomplish data processing and computing in the envisioned molecular platforms.

1.1 Computational paradigms for nanocomputing structures

Nanocomputing, or computing by nano and/or molecular structures (primitives), adheres to the fundamental principles of logic design of discrete devices and various molecular phenomena. While these principles, being related to data processing, may be similar to the existing ones, the computing techniques should be refined, revisited or redefined to fit the technology requirements, device fundamentals, and system solutions. These problems can be addressed by applying contemporary logic design techniques to the design of computing structures. The computing paradigms and implementation principles define the form of computing structures, which embody the principles in the spatial and time solutions. This book provides a systematic view of the fundamentals of nano scale computing and introduces novel computing structures that satisfy the requirements of processing in three-dimensional (3D) space. Some particular cases of embodiment of contemporary design paradigms for discrete devices into nanocomputing structures are listed in Table 1.1.

Switch is the simplest computing device both in microelectronics and molecular (nano) technology. The simplest switch operates with one bit of information. For example, a switch can operate in two states, ON and OFF (connected–disconnected, open–closed, etc.). The other type of switch can direct a bit to one of two possible outputs (1-input 2-output switch). The switch is the basic element of implementation models in switching algebra.

Logic networks. These are the fundamental computing structures in logic design. Computing paradigm and design techniques of logic networks can be adopted from contemporary logic design of discrete systems, for example, two-level and multilevel logic network design and optimization. The elementary Boolean functions, such as AND, NAND, OR, NOR, and EXOR gate, can be computed using molecular switches. An arbitrary Boolean function can be computed by a logic network using the universal set (library) of logic gates.

Threshold networks. Logic networks that are designed using threshold gates are called *threshold networks*. An arbitrary Boolean function can be computed on a network of threshold gates. For example, two-input AND, OR, NOR, and NAND Boolean functions can be generated by a single threshold gate. Control over the type of Boolean function is exercised by the thresholds and weights.

Memory devices. Data storing can be accomplished by utilizing various phenomena exhibited, such as stable states, bistable states, m-stable states, and binding/unbinding. The implementation and utilization of these phenomena should be based on device physics applicable at the system level (logic networks, molecular aggregates, etc.).

Reprogrammable devices. These have provided significant performance improvements for many applications. FPGA (field programmable gate array)–based custom-computing systems can achieve high performance, providing near application-specific performance in an application-generic system. The simplest implementation of the field programmable method is a programmable logic device (PLD). PLDs are used primary for the *two-level* (sum-of-products) implementations of Boolean

TABLE 1.1

Nanocomputing structures adopt the computing paradigms, techniques, and data structures of contemporary computing

Contemporary computing	Transferring	Nano scale computing
Switch	$\Longrightarrow$	Molecular switch
Logic networks	$\Longrightarrow$	Logic networks over the library of molecular primitives
Threshold networks	$\Longrightarrow$	Logic networks over the library of molecular threshold gates
Memory devices	$\Longrightarrow$	Molecular memory devices
Reprogrammable logic arrays	$\Longrightarrow$	Crossbar-based computing arrays
Massive parallel computing	$\Longrightarrow$	Self-assembled molecular arrays
Self-reproducing automata	$\Longrightarrow$	Biomolecular self-assembly
Systolic arrays	$\Longrightarrow$	Systolic arrays over the library of molecular primitives
Stochastic computing	$\Longrightarrow$	Stochastic molecular computing
Fault-tolerant computing	$\Longrightarrow$	Molecular fault-tolerant computing
Neural networks	$\Longrightarrow$	Neuromorphic computing
Data structure	$\Longrightarrow$	Data structures for molecular computing

functions. PLDs have simple routing structures with predictable delay. PLDs and FPGAs are completely prefabricated devices. However, FPGAs are optimized for *multilevel* logic networks. This allows them to handle much more complex logic networks.

Example 1.1   *The most noticable feature of* **multi-FPGA systems** *is in the topology chosen to interconnect the blocks or chips. The most common topologies are* **mesh** *and* **crossbar** *topology. In a mesh, the chips in the system are connected to form a nearest neighbor pattern. This topology has the advantages of local interconnections, as well as expendability including to 3D, since meshes can be grown by adding resources to the edges of the array. Techniques of* **linear arrays**, *which are essentially 1D meshes, are common in contemporary design. Crossbar topologies separate the elements in the system into logic-bearing and routing-only chips. These crossbar topologies are not expendable. Another topology, which combines the expendablity of meshes and the simple routing of crossbars, is known as* **hierarchical crossbars**.

FPGA-like architectures can take advantage of architectural features from other computation domains and technology requirements, such as systolic arrays and digital signal processing, to provide a better resource structure than standard commercial general-purpose FPGA architectures.

Reconfiguration. The basic idea of reconfiguration is that the capability exists within a system to modify functionality after manufacturing. Reconfiguration is a widely recognized defect-and-fault-management technique in conventional design.

Example 1.2   *Examples of the* **reconfiguration** *are as follows: (a) ability to deactivate computing module within a module upon error diagnosis, (b) ability to switch to spare bits for single-cell failures in memory devices, and (c) bipass a memory device.*

Let us assume that a failure has been detected in a computing device. The failed unit must be bipassed, and a redundant resource must be activated and wired in. This is the key idea of fault avoidance through reconfiguration and redundancy.

Reconfiguration is often considered in conjunction with *redundancy* and *self-assembling*, and has been explored as a defect and fault mitigation approach for molecular-scale computing. the Reconfiguration can be obtained using so-called *polymorphic* principle of the computation. Polymorphic logic networks are those with multiple superimposed functionality; the output functions change due to changes in the operational point of the components.

Example 1.3 *Consider the **polymorphic** logic gates. Given the controlled supply voltages V_1 and V_2, the gate computes an AND and OR function on these voltages, respectively.*

Cellular arrays. The development process of biological organisms exploits essentially two mechanisms: (a) cellular differentiation (a cell copies its genetic material and splits into two identical cells), and (b) cellular division (the function a cell has to realize). In 1948, John Von Neumann introduced the model of *self-reproduction*. This model is known as a *cellular automata*. Self-reproduction is a prerequisite for any independent evolutionary process. A cellular automata is a finite or infinite network of identical deterministic finite state automata such that each cell has the same (finite) number and order of neighbors. Hence, a cellular automaton is a spatially distributed model of cells. The state of a cell at time $t+1$ depends on its own state and the states of neighboring cells at time t.

Cellular arrays are intrinsically parallel computing structures consisting of a regular latticework of identical logic elements (cells) that compute in lockstep while exchanging data with nearly cells. The properties of *local* computability, parallelism, and short-cascade chains make the performance of specific processing, such as matrix multiplication and vector operations, much higher. This is because the total length of interconnections from the inputs to the outputs decreases. Examples of cellular arrays are *crossbar-based* and *self-assembled* computing arrays.

(*a*) *Crossbar-based* computing arrays are based on the computing paradigm that a grid of the connected and addressable switches, sandwiched between the wires, have the property of storing and manipulation of data. Crossbar computing arrays are the computing paradigms that utilize a grid of the connected and addressable switches. These interfaced switches must ensure the specified functionality and capabilities from both device and circuitry prospectives.

Example 1.4 *OR, AND, and other elementary Boolean functions can be computed by the grid structure.*

This paradigm is attractive in realizing logic networks that implement memory and logic functions. The addressable molecular switches are set or reset by on–off switching, electrochemomechanical transitions, etc.

(b) *Self-assembled* computing arrays utilize and typify self-assembly, recognition, and other phenomena and mechanisms exhibited by biomolecules in living systems. There are no analog of this concept in microelectronics.

Biologically inspired computing is defined as a paradigm that typifies comprehended biological analogous of computing. Examples of biologically inspired computing are *artificial neural networks* and *evolutionary computing*.

(a) *Artificial neural networks* are an attempt to utilize natural-centric computing (with just moderate success to date) by implementing parallel processing capabilities and networked structures. The methods of contemporary logic design such as data structures design, their optimization, computer array structures, and others typify, to some extent, natural processing.

(b) *Evolutionary* computing centers on search and optimization, which are observed in living systems. Evolutionary algorithms have been used extensively in *evolvable hardware* [44]. Evolutionary algorithms have been shown to perform well in exploring large and complex design spaces, including logic network design.

Systolic arrays are special class of cellular arrays, that are based on the parallel-pipelined paradigm of computing. They are suitable for efficient computing if the parallel input/output is compatible with other devices in the system. *Linear* systolic arrays are a particular class of systolic arrays [47]. They do not require specific input/output interfaces when implemented as modules in computing systems. An arbitrary Boolean function can be computed on a linear systolic array.

Stochastic (probabilistic) computing is a fundamental concept in both contemporary design and nanocomputing. Envisioned nanotechnology- and molecular-based devices are expected to exhibit *deterministic* and *stochastic* electrochemomechanical phenomena that should be utilized to implement logic and memory functions. Stochastic computing promises to achieve fault tolerance by employing statistical principles in which deterministic logic signals are replaced with random variables.

Fault-tolerant computing can be based on various classical approaches. These developments can be applied for nanocomputing since the failure rate of nanodevices is expected to be high. For example, molecular devices are expected to have a high probability of failure due to synthesis (technological) complexity, uncertainties, nonideal processing functionality, varying characteristics, etc. There are many techniques in discrete devices design that can be useful in nanocomputing, including *masking logic*, and *noise-tolerant* and *failure-immune* concepts. *Error* *correction codes* and alternative number systems can be also used in the networked logic gates.

Neuromorphic networks are artificial neural networks that could be viewed to be biologically inspired or natural-centric ensure computing and memory storage. *Neuromorphic* networks utilize the principles of distributed computing, and the familiar examples of these models are the Hopfield network and the Boltzmann machine. At the gate level, this approach addresses artificial neural networks and is known as recursive stochastic models. They are based on the property of relaxation, that is, the ability to relax into a stable state in the absence of external excitations. *Relaxation* is referred to as the embedding of a correct solution into a network of logic nanocells called *artificial* neurons. There are no defined inputs and outputs in neuromorphic networks because they are distributed within the computing structure. Neuromorphic networks satisfy the criterion of fault-tolerant computing structures. One common model of neural networks is the *feedforward* network in which the processing unit is a linear threshold gate. A linear threshold gate computes an elementary Boolean function. It is wellknown that threshold gates are more computationally powerful than AND, OR, and NOT logic gates. Such a feedforward network can compute an arbitrary Boolean function.

Example 1.5   *The **Hopfield** networks can be used for modeling logic gates [9, 28]. Such a network is characterized by an energy function that has global minima only at the neuron states consistent with the truth table of the modeling gate. All other neuron states have higher energy. The energy function is uniquely specified by the weights and the thresholds of the neurons.*

Data structures provide various forms of representation of Boolean functions (algebraic, tabulated, and graphical forms). The theoretical base of data structures is Boolean and multivalued algebras. The special type of data structure, being used for the interpretation of molecular and physical phenomena, is called *intermediate* data. It is defined as *logic primitives*, which exhibit specified features of the molecular phenomenon and computing paradigm applying an appropriate encoding. Using specific assembling/interfacing/aggregating rules, logic primitives can be assembled into logic gates. The set of these gates (elementary Boolean functions) must be universal, that is, an arbitrary Boolean function can be computed using these specified library of gates. The role of intermediate data structures can be illustrated by the following scheme:

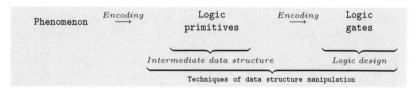

In this scheme, the encoding procedure plays a key role: all data must be specified via the assignments of the input and output variables, in particular, logical values 0s or 1s.

Example 1.6   *Using the symbolic quaternary alphabet $\mathcal{A} = \{A, B, C, D\}$, the* **intermediate data structure** *can be encoded (a) by the elements of this alphabet, (b) by specific orders of these elements in strings, and (c) by the operations with elements and strings, such as insertion, concatenation, deletion, and appending. These encoding schemes specify the set (library) of* **logic primitives** *and interactions between them. Such intermediate data structures are used in DNA computing (data are encoded by A, C, G, and T) and in quantum dot computing (data are encoding by dots).*

The analysis of Table 1.1 shows that the computing paradigms of contemporary discrete devices play an essential role in the development of novel computing structures based on technological advances of nanotechnology, molecular processing, and molecular (nano) electronics. This book addresses some major problems listed in Table 1.1.

1.2   Biological inspiration for computing

Processing principles of natural information are partially comprehended. The studies of natural data processing demonstrate that it is entirely different from the von Neumann computing paradigm.

1.2.1   Artificial neural networks

The human nervous system consists of small units called *neurons*. These neurons, when connected in tandem, form a nerve fiber. A biological neural net is a distributed collection of these nerve fibers. A neuron receives electrical signals from its neighboring neurons, processes those signals, and generates signals for other neighboring neurons attached to it. The operation of a biological neuron, which decides the nature of its output signal as a function of its input signals, is not clearly known to date. However, most biologists are of the opinion that a neuron, after receiving signals, estimates the weighted average of the input signals and limits the resulting amplitude of the processed signal by a nonlinear function.

Artificial neural networks, to some extent, can be considered to be a *biology-inspired* computational concept, that implements natural-centric data processing. The cell body in an artificial neural net is modeled by a linear activation function.

Example 1.7 *The number of neurons in the human brain is estimated to be 100 billions; mice and rats have 100 millions neurons, while honeybees and ants have 1 million neurons.*

Artificial neural networks can be designed and modeled to implement *massively parallel* architectures. A neural network is a network of nodes and links. The nodes are elementary computing units that could typify to some extent neurons assuming a conventional action potential premise. Each node has an activation level that corresponds to a neuron's rate of firing, while each link has a numeric weight that corresponds to the strength of a synapse. Such networks can be trained to recognize patterns and compute functions.

Neural networks seem to be attractive for nanoelectronics due to intrinsic fault tolerance. The degree of fault tolerance of neural network can be evaluated using the degree of its redundancy.

Example 1.8 *The human retina has 125 million rod cells and 6 million cone cells. An enormous data, among other tasks, is processed by the visual system and the brain in real-time. Real-time 3D image processing, ordinarily accomplished even by primitive vertebrates and insects that consume less than 1 μW energy to perform information processing, cannot be performed even by envisioned processors with trillions of transistors, device switching speed 1 THz, circuit speed 10 GHz, device switching energy 1×10^{-16} J, writing energy 1×10^{-16} J/bit, and read time 10 nsec.*

Example 1.9   *The information from the visual system, sensors and actuators is transmitted and processed within the nanoseconds range requiring μW of power. Performing enormous information processing tasks with immense performance, which are far beyond foreseen capabilities of envisioned parallel processors (which perform signal/data processing), the human brain consumes only 20 W. Only some of this power is required to accomplish information and signal/data processing.*

1.2.2   Evolutionary algorithms and evolvable hardware

Biological development is an example of a stunning mechanism that allows robust generation of complex organisms from a linear building plan, the DNA. *Evolutionary algorithms* are population-based stochastic search algorithms.

The evolution paradigm includes genetic algorithms, evolution strategies, evolutionary programming, and genetic programming. As an example of evolution paradigm, in this book, genetic algorithms are used for logic network design.

A genetic algorithm is an iterative procedure that consists of a finite-size population of individuals; each individual is represented by a string of symbols called the *genome*; a possible solution in a given problem space or search space is encoded in the genome. The algorithm sets out with an initial population of individuals that is generated at random. The individuals in the current population are decoded and evaluated according to an appropriate quality criterion called the *fitness function*. A new population is generated in the next generation. In this population, the individuals are selected according to their fitness and are transformed via genetically inspired operators, such as *crossover* and *mutation*. After some iterations, the genetic algorithm may find an acceptable solution.

Evolvable hardware is the application of evolutionary algorithms to the creation of electronic circuits. In evolutionary logic design, gate-level primitive components (gates, flip-flops, etc.) or graph-based abstraction are used to generate logic networks.

Example 1.10   *It was shown in many studies that various logic networks using FPGA of the configuration 10×10 cells can be created by a genetic algorithm with a population of size 50 and each string of size 2000 bits.*

1.2.3 Self-assembly

Molecular recognition, complementarity, and *aggregation* are well-established and sound principles. Molecular recognition implies the specific interaction between two or more molecules by means noncovalent (hydrogen bonding, metal coordination, hydrophobic forces, van der Waals forces, pi-pi interactions, electrostatic forces, etc.), and covalent bonds. For example, molecular recognition and molecular complementarity, exhibited by DNA, amino acids, and other biomolecules, can be significantly expanded utilizing organic and inorganic molecules. Stereochemistry studies the spatial arrangement of atoms, molecules, and molecular aggregates.

Most solid substances are crystalline in nature. Sometimes the particles of a sample of solid substance are themselves single crystals. Every crystal consists of atoms arranged in a 3D pattern that repeats itself regularly. The unit of this structure is the *unit cell*. For example, the unit cell of a cubic crystal is a cube. Crystals of many substances contain discrete groups of atoms, which are called *molecules*. The forces acting between atoms within a molecule are much stronger than those acting between molecules. At a low temperature, the molecules in a crystal lie rather quietly in their places. As the temperature increases, the molecules become more and more agitated. A molecule on the surface of the crystal is held to the crystal by the forces of attraction that its neighboring molecules exert on it. Forces of this kind are called *van der Waals* attractive forces.

Grouping means that larger objects are assembled out of smaller ones serving as their parts. It includes various procedures, such as clustering, class formation, and construction of strings. Typical example of grouping in biology is given below.

Example 1.11 *Cells form functional aggregated assemblies, which exhibit superior performance and capabilities. Living systems can be studied applying* **bottom-up** *hierarchies and* **top-down** *taxonomies.*

Specifically, many structural features of molecules and crystals that can be used in their interpretation in terms of computation are governed by *symmetry* and *grouping*. Symmetry operations, which leave one point in the space fixed, are called *point symmetry* operations. Point symmetry operations are *rotations* around an axis, *reflection* across a plane, and *inversion* through a point. The stacking of atoms or molecules side by side to build a crystal results in *translation* or *lattice* symmetry. The crystal lattice is the array of points at the corners of all of the unit cells in the crystal. The 3D array of symmetry elements itself is known as a *space group*.

Example 1.12 *Crystal growing is characterized by the grouping of molecules during solidification. A single crystal is the result of clusters gathered around a local center of crystallization. Crystals form their* **group hierarchies** *during crystal growth.*

An *aggregation* or *assembly* is formed an entity out of its parts. Each of the parts can also be obtained as a part of aggregation. *Self-assembly* is defined as the process by which an organized structure can be formed spontaneously from simple parts (molecules or various nanosize objects). It describes the assembly of natural structures such as crystals, DNA helices, and microtubules. The organization process is made into a desired structure via physical, chemical, or biochemical interactive processes involving, for example, electrostatic and surface forces. All these processes are very selective and reject defects so that the resulting structure is characterized by a high degree of perfection. Self-organization techniques are similar to the process of development of biological organisms.

Repeated duplication of a group of atoms by a screw axis produces a pattern called a *helix*. If the atoms are joined by chemical bonds in a continuous chain from one group to the next, the result is a helical molecule that extends the length of the crystal. It is possible to construct helical molecules by a symmetry operation similar to a screw axis, except that the angle of rotation from one group to the next is not an integral fraction of 360°. Some molecules of great biological importance have helical symmetry of this type, in particular, the α-helix of proteins and the helical backbone of the DNA molecule. Properties of helical symmetry can be encoded and interpreted in terms of computing Boolean functions.

The second law of thermodynamic states that in an isolated system, entropy can only decrease, not increase. Such system evolve to their state of maximum entropy, or thermodynamic equilibrium. The *thermodynamic* concept of entropy as the dissipation of heat is not very useful for computing systems. *Shannon entropy* is applicable to any system for which a state space can be defined. It expresses the degree of uncertainty about the state s of the system in terms of the probability distribution $P(s)$. In terms of Shannon entropy, the second law of thermodynamic can be expressed as "every system tends to its most probable state" [6]. At a molecular level, molecules are distributed homogeneously, and the most probable state of an isolated system is that of maximum entropy or thermodynamic equilibrium.

In molecular (nano) electronics, *self-assembly* is defined as a method of fabrication of the molecular computing structures that relies on chemicals, forming larger structures without centralized or external control. This is the spontaneous organization of molecules under thermodynamic equilibrium conditions into a structurally well-defined and rather stable arrangement. Self-assembly in this system is associated with *bottom-up* design.

The key engineering principle for molecular self-assembly is to design molecular building blocks that are able to undergo spontaneous stepwise interactions. In this design, the instructions are incorporated into the structural framework of each molecular component. The running of these instructions is based on the specific interaction patterns, environment, and the intermediate stages of the assembly.

One of the main issues encountered in logic design based on self-assembling paradigm is the ability to match the components (logic primitives) into combinations that result in the correct output of logic network. The complementary molecular primitives (^{CM}primitives) are defined as a set of simplest molecular structures. Each molecular primitive can be interpreted in terms of switches, the simplest computing operation. Molecular primitive are used for designing the molecular logic gates. Various subsets of the molecular primitive can be derived from the ^{CM}primitives. These subsets form various libraries for design molecular logic gates often called *multiterminal* and *multifunctional* molecular devices utilizing various molecules (aromatic, cyclic, and other), polypeptides, and side groups.

> **Example 1.13** **Multiterminal** *solid molecular devices can be engineered as cyclic molecules arranged from atoms ensuring functionality. These devices includes switches (two-terminal device), two-input logic gates (three-terminal devices), three-input logic gates (four-terminal devices), and various multiple-input multiple-output computing networks.*

Logic design using the molecular primitives is based on self-assembling and results in random or partially-ordered computing networks. The controllable self-assembling and robust binding/pairing can be implemented using, for example, the *templates*. *Fractal* assemblies that can be interpreted in terms of Boolean functions carry the information about these functions by the labeled topological structures. These fractals are acceptable as templates.

> **Example 1.14** *An example of a structure that can be constructed using algorithmic self-assembly is a* **Sierpinski triangle**. *This structure is also known as a Sierpinski gasket, which is a kind of fractal structure. A Sierpinski gasket is the* **Pascal triangle** *modulo two, i.e., the EXOR operation is used instead of arithmetic addition while forming the Pascal triangle. There are $(4^n + 2^n)/2$ elements in total in Pascal triangle.*

Fractal-like templates are useful for systematic design for molecular systems with logic processing capabilities. Methods of computer aided design (CAD) are classified into those for logic design based on logic primitives and those for designing molecular reactions.

1.3   Molecular computing devices

Synthetic chemistry allows one to synthesize a wide range of complex molecules from atoms linked by covalent bonds. Utilizing noncovalent and covalent intermolecular interactions, as well as precisely controlling spatial (structural) and temporal (dynamic) features, supramolecular chemistry provides methods to synthesize even more complex atomic aggregates.

Example 1.15   *The effective cell size of the envisioned microelectronic devices is projected to be 500×500 nm by the year 2025. Each of these devices will consist of billions of molecules. In contrast, molecular devices are expected to be synthesized from a couple of atoms or molecules.*

Molecular computing devices are comprised of

▶ Organic molecules
▶ Inorganic molecules
▶ Biomolecules

Molecular (nano) electronics focuses on fundamental/applied/experimental research and technology developments in the devising and implementation of novel, high-performance, enhanced-functionality, atomic/molecular devices, modules, and platforms (systems), as well as high-yield bottom-up fabrication. Molecular electronics centers on:

▶ Discovery of novel devices that are based on the new device physics
▶ Utilization of the exhibited unique phenomena, effects, and capabilities
▶ Devising of enabling topologies, organizations, and architectures
▶ Bottom-up, high-yield fabrication technologies

At the device level, the key differences between molecular and microelectronic devices are as follows:

The key differences between molecular and microelectronic devices
▶ Device physics and phenomena exhibited and utilized
▶ Performance, capabilities, and functionality achieved
▶ Topologies and organizations attained
▶ Fabrication processes, synthesis methods, and technologies used

The difference between microelectronic and molecular computing devices can be specified as follows. In microelectronic computing devices, individual molecules and atoms do not depict the overall device physics and do not define the device characteristics. In molecular computing devices, individual molecules and atoms explicitly define the overall device physics depicting the device performance, functionality, capabilities, and topologies.

There are fundamental differences at the system level. In particular, molecular electronics lead to novel organizations, advanced architectures, and the need for technology-centric, super-large-scale integration, novel interconnect, and interfacing. However, advanced techniques of logic design are the basis of the molecular electronics (Table 1.1).

Example 1.16 *In [42], a molecular computing structure called* **nanocell** *have been introduced. A nanonocell is a 2D network of self-assembled molecules that act as reprogrammable switches. An array of nanocells implement the concept of the FPGA; the nanocells can be programmed and reprogrammed after fabrication to perform a specific functions. Genetic algorithm can be used for designing the logic network of molecular switches in nanocells.*

Interconnect and interfacing

The characteristics of nano scale devices include, in particular, unreliable device performance, transfer function, interconnect limitations (the inability to provide global interconnections), thermal power generation, and regularity of layout. Nanoscale devices need to be interconnected locally and patterned into 2D or 3D arrays of cells of various topologies.

Quantum cellular automata architecture (QCA) is a typical example of a regular and locally interconnected array of cells interacting with its neighbors, but there are no wires in the signal paths. In contrast to conventional silicon-based designs, where information is transferred between devices by electrical current, in QCA information is transferred by Coulomb interaction, which passes the state of one cell to its neighbors.

Example 1.17 *The wire in QCAs is a chain of* **quantum cells**. *The QCA is characterized by the ability to cross wires in the plane; different wires carrying different binary values can cross each other at the same level without any interferance or crosstalk.*

In locally connected structures, the range of interaction and the connection complexity of each cell are independent of the number of cells. Therefore, these structure are scalable and massively parallel. Acceptable characteristics of reliability and robustness can be achieved by using special techniques. An example is a 2D cellular nonlinear network (CNN), which is an array of neuron-like cells [32, 39]. A cellular network can be mapped into a 3D topology [36].

Nanoscale design demand novel interconnect and interfacing solutions that ensure aggregation and assembly of molecular processing primitives in large-scale diverse modular modules. Molecular assemblies are comprised of functionalized aggregated molecules. In leaving organisms, biomolecules ultimately establish the biomolecular processing hardware. We focus on the *solid molecular devices*. In solid-state microelectronic devices, individual atoms and molecules have not been, and cannot be, utilized from the device physics prospective. The scaling down of microelectronic devices results in significant performance degradation due to quantum effects (interference, inelastic scattering, vortices, resonance, etc.), discrete impurities, and other features. In contrast, the molecular devices exhibit phenomena that can be uniquely utilized, ensuring device functionality and guaranteeing superior capabilities.

Organic synthesis is the collection of procedures for the preparation of specific molecules and molecular aggregates. In planning the syntheses of desired molecules, the *precursors* must be selected. One carries out the *retrosynthetic* analysis as

$$\text{Target molecule} \Longrightarrow \text{Precursor},$$

where the open arrow $\Longrightarrow$ denotes "is made from." Usually, more than one synthetic step is required. For example,

$$\text{Target molecule} \Longrightarrow \text{Precursor } 1 \Longrightarrow \cdots \Longrightarrow \text{Precursor Z} \Longrightarrow$$
$$\text{Starting molecule}$$

A linear synthesis, which is adequate for simple molecules, is a series of sequential steps to be performed, resulting in synthetic intermediates. For complex molecules, convergent or divergent synthesis is required. There are different procedures for synthesis of intermediates. For new synthetic intermediates, discovery, development, optimization, and implementation steps are needed.

In multiterminal solid molecular device, distinct quantum phenomena could be used to ensure the controlled characteristics. For example, quantum interaction, quantum interference, quantum transition, vibration, Coulomb effect, etc. The device physics, based on these and other phenomena and effects (electron spin, photon-electron-associated transitions, etc.), must be coherently complemented by the bottom-up synthesis of the molecular aggregates, that exhibit those phenomena.

Example 1.18   *Distinct solid molecular devices have been proposed,*
                   ranging from resistors to multiterminal devices [5,
                   10, 17, 26, 33, 40, 43]. These molecular devices are
                   comprised of organic, inorganic, and bio-molecules.
                   For example, Figure 1.1 shows, in particular, different
                   molecules that were functionalized in order to perform
                   acceptable characteristics for switching.

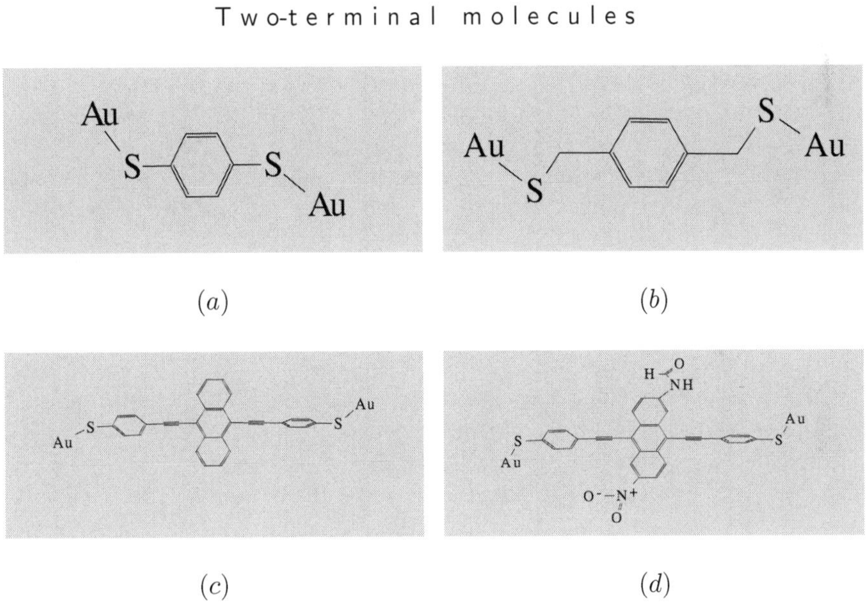

Two-terminal molecules

(a)   (b)

(c)   (d)

FIGURE 1.1
Molecules as potential two-terminal molecular devices: (a): 1,4-phenyledithiol
molecule and functionalized 1,4-phenyledithiol molecule; (b): 1,4-pheny-
lenedimethanethiol molecule; (c): 9,10-bis((2′-para-mercaptophenyl)-ethinyl)-
anthracene molecule; (d): 1,4-bis((2′-para-mercaptophenyl)-ethinyl)-2-acetyl-
amino-5-nitro-benzene molecule (Example 1.18).

Various problems associated with the devising, engineering, and analysis
of functional molecular devices are reported in [26]. In order to depart
from the symmetric organic molecular devices, asymmetric multiterminal
carbon-centered molecular devices were proposed. These molecular devices
are comprised from B, N, O, P, S, I and other atoms. To ensure synthesis
feasibility and practicality, these molecular devices are engineered from

cyclic molecules and their derivatives. The reported multiterminal molecules ensure the desired asymmetry of the voltage-current characteristics, while the saturation region or peaks-and-deeps should be examined.

By applying the voltage to the control terminal, one varies the potential, regulates the charge and electromagnetic field, and varies the interactions, as well as changes the tunneling affecting the electron transport. Hence, the input-output characteristics can be controlled.

The aggregation and interconnect of input/control/output terminals can be accomplished within the carbon framework. For example, (a) the electron transport is predefined or significantly affected by X_i and side groups; (b) atomic structures of side groups can exhibit transitions or interactions under the external electromagnetic excitations and thermal gradient; (c) side groups can be utilized as electron donating and electron withdrawing substituent groups, as well as interacting or interconnect groups.

The functional molecular switches operate on one bit. This simplest computing is associated with switching (from an OFF state to an ON state, and vice versa), using stimuli such as voltage pulses. In multiterminal solid molecular device, quantum effects could be used to ensure the controlled voltage-current characteristics. Figure 1.2a shows the two-terminal molecular devices of various complexity.

Three-terminal cyclic molecules are utilized as molecular devices. The three-terminal molecular device is based on the quantum interaction and controlled electron transport. The inputs signals V_A and V_B are supplied to the input terminals, while the output signal is V_{out}. These molecular AND and NAND gates are designed using cyclic molecules within the carbon interconnecting framework. By applying the voltage to the control terminal, one varies the potential, regulates the charge and electromagnetic field, varies the interactions, as well as changes the tunneling affecting the electron transport. Hence, the input-output characteristics can be controlled.

Example 1.19	*Three-terminal molecular devices can perform Boolean functions AND, NAND, OR, and NOR of two variables (two-input) logic primitives (Figure 1.2).*

Example 1.20	*Molecular and biomolecular devices can operate with the estimated transition energy 1×10^{-18} J, discrete energy levels (ensuring multiple-valued logics and memory) and femtosecond transition dynamics. These guarantee exceptional device transition (switching) speed, low losses, unique functionality, and other features ensuring superior overall performance.*

The term *molecular electronics* covers a broad range of topics, in particular,

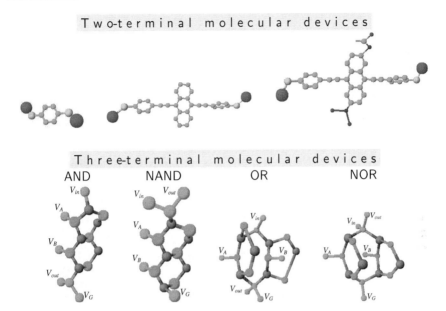

FIGURE 1.2
Molecular gates: AND, NAND, OR and NOR gates (Example 1.19).

molecular computing devices, device physics, and biophysics of biomolecular devices. Molecular materials for electronics deal with films or crystals that contain many trillions of molecules per functional unit. Molecular-scale electronics deal with one to few thousand molecules per device.

Molecular electronics involves the search for single molecules or small groups of molecules that can be used as the fundamental units for computing. The goal is to use these molecules to have specific properties and behaviors. Molecular devices for computing Boolean functions consist of (a) molecular terminals, and (b) molecular wires, which are materialized by means of interconnection phenomena.

Cross-talking

The computing units in molecular devices use different input signals; to assemble them into computing devices and systems of higher complexity, the standardization of the input/output signals are required. Most of the chemical reactions are of low specificity. That is, different reactions in one system may cause interference. This problem is called *cross-talking*. Cross-talking between the different reactions hinders assembling of chemical computing devices and systems.

Mixed silicon-based and molecular electronics

A hybrid between present silicon-based technology and technology based purely on molecular switches and molecular wires is a more viable path toward nano scale computing systems. A mixed paradigm presents opportunities for more tightly integrated mixed systems that utilize the advantage of the strong points of each technology.

Traditional digital logic design techniques can be used, and assemblies of switches and logic gates can be constructed. The resulting logic networks will operate in exactly the same manner as traditional silicon electronic-based circuits.

1.4   Fault tolerance

The stochastic nature of biomolecular systems can lead to random variation in the concentration of molecular species. Mutation or imperfect replication can alter the inserted gene sequences, possibly disabling them or causing them to operate in unforeseen ways. This type of computing is called *stochastic* computing. Stochastic computing achieves fault tolerance by employing statistical models in which deterministic logic signals are replaced with random variables. In this model, correct output signals are calculated with some probability. When a noise is accommodated, the Boolean function is replaced with a random function. The applications of various models for increasing the reliability of computing require small circuits or simple logic elements. Hence, efficient assembly and interconnects are critical for the implementation of stochastic models of computing. Conceptually, in stochastic computing, the problem of suppressing unwanted random effects is reformulated into the problem of efficient utilization of uncertainty.

The increase in complexity of the logic networks increases the probability of faults. Any architecture built from large numbers of these devices will contain a large number of defects, which fluctuate on time scale. The problem is to develop a computing structure, that is dynamically *defect tolerant*. That is, the problem is not only to test the correctness of logic networks but also to design networks resistant to faulty components. Biological systems are examples of complex fault-tolerant systems; these biological mechanisms can be used in the development of novel *self-test*, *self-repair*, and *self-replication* approaches. Self-replication means the capability of a machine to produce a copy of itself.

Fault detection is the analysis of errors to determine which components are faulty. Once the error is detected, the appropriate action must be taken. A property, required in all fault-tolerant computing techniques is that of *fault isolation*. Fault isolation aims to prevent a faulty unit from causing incorrect behavior in a nonfaulty unit.

As devices increase in complexity, defect, and contamination, control becomes even more important since defect tolerance is very low. Nanoscale devices will have a high probability of failure, that is, they are characterized by a high and dynamic failure process. These failures can be occur both during fabrication and at a steady rate after fabrication.

There are several approaches to deal with nanodevice failure rates: (a) design for testability; in this approach, the design cycle of logic network is considered under conditions of the testability in all design phases; (b) design based on redundant logic; the key assumption of this approach is that the redundance can be incorporated at various level of logic network over a set of possible faults; and (c) design based on probabilistic computing paradigm; the key idea of this approach is the adaptation to errors.

Assuming reliable computing in the presence of faults is called *fault tolerance*. For example, approaches to reliable computing in networks with faulty nodes and/or interconnects can utilize the error correction codes and residual number systems. These approaches state that it is possible to correct a class of faults if a library of *reliable* logic nanocells for implementing the correction is available. Error-reduction/correction in biocomputing systems becomes extremely important as the complexity of numerous connected biochemical reactions increases. This is needed for scalability and fault tolerance of the computing structures *Self-repairing* electronics are the areas of fault-tolerant computing. Traditional approaches to preserve electronics incorporate radiation shielding, insulation and redundancy at the expense of power and weight. The *self-adaptive* system can autonomously recover the lost functionality of a reconfigurable array.

Example 1.21 *NASA uses various approaches to prevent radiation, and extreme-temperature, hardened electronics, required by space missions to survive the harsh environments beyond Earth's atmosphere [21]. The* **self-adaptive** *systems are operating in extreme temperatures (from $120° C$ down to $-180° C$). The reasonable implementation of a self-adaptive principle is based on FPGAs. Another approach, called* **tuning reconfigurable** *electronics, is related to an inexpensive, navigation grade, miniaturized inertial measurement units, which surpass the current state-of-the art technology in performance, compactness (both size and mass), and power efficiency. This approach used by all NASA missions. The* **self-tuning** *techniques for reconfigurable micro-electro-mechanical systems (MEMS) gyroscopes based on* **evolutionary** *computation.*

Example 1.22   *Fault tolerant computing array are often based on the principle of* **reconfiguration**. *This principle is implemented by integration of redundant cells. Fault cells and their locations are detected, and the computer structure reconfigures around these faulty cells.*

The drawback of this approach is that the reliability of fault detection logic must be guaranteed.

1.5   Computing in 3D

It has been justified in a number of papers that 3D topologies of computer elements are optimal, physically realistic models for scalable computers. In the contemporary logic network design, the use of the third dimension is motivated by decreasing the interconnect topology. The third dimension is thought of as *layering*. For example, in chips, networks are typically assembled from layers, where logic cells occupy the bottom layers and their interconnections are routed in upper metal layers.

Example 1.23   *A widely accepted* **2D placement** *model for conventional microelectronics is a square grid with all logic network elements of unit square size with their input/output connections in the center. Network elements are then placed, in checkerboard fasion onto the grid. The logic network to be put in place is represented by a hypergraph, where network elemebts form the set of nodes. The goodness of the placement is measured by its total wire length. A* **3D placement** *of FPGA, for instance, provides a significant reduction of wire-lengths.*

In this book, the third dimension is considered not because of layers, but as a dimension that relates to electrochemomechanical phenomena [46]. It includes the space orientations and 3D relationship between molecules, as the key factor for achieving the desired functional logical properties. This is the motivation to search for adequate spatial computing models. A 3D directly interconnected molecular electronics concept utilizes a direct device-to-device aggregation based on molecular recognition, complementarity, and aggregation.

1.6 Multivalued processing

Multivalued algebra is a generalization of Boolean algebra, based upon a set of m elements $M = \{0, 1, 2, \ldots, m\}$, corresponding to multilevel signals and the corresponding operations. Multivalued logic has been proposed as the means to [7, 18, 19]

▶ Improve performance
▶ Increase the packing density of VLSI circuits
▶ Improve fault tolerance
▶ Reduce power and power dissipation
▶ Improve testability

In multivalued logic networks, more than two levels of a signal are used. There are many motivated examples for considering the processing of *multilevel* signals as biologically inspired processing.

Interconnections. One of the most promising approaches to solving the interconnection problems is the use of multivalued logic. The number of interconnections can be reduced with multilevel signal representation. The reduced complexity of interconnections makes the chip area and delay much smaller.

Packing density. The chip area can be evaluated using the interconnections. The number of interconnections for the networks that use m-level logic signals can be estimated as $1/\log_2 m$, compared with two-level ($m = 2$) logic signals, $1/\log_2 2$. In the case of 2D topology, the reduction becomes $1/(\log_2 m)^2$ [18]. The total area of interconnections can be calculated using the number of interconnections, topology of interconnections, and the length of interconnections.

Pads. The number of bonding pads in a chip can be reduced by using multilevel signals.

Testability can be improved because access to internal components of a logic network is available using extra pads.

Performance can be improved because of the decrease of the total length of interconnections and interconnections delay (the switching time of gates is much less than interconnection delay). Special encoding methods, such as radix-r signed-digit number system and residue symmetrical number system, as well as parallel hardware algorithms using multivalued logic, enable local computing. These are additional resources for increasing the performance.

Power dissipation can be decreased because the dynamic power dissipation is determined mainly by the interconnections.

Fault tolerance. Cross-talk noise is increased because of extremely small distances between wires. Special encoding methods, using multivalued logic enable reduction of cross-talk noise.

Sequential logic networks. Multivalued logic is the base for development of the racing-free asynchronous sequential elements and networks.

Special encoding methods for local computing. In signed-digit arithmetic, the carry propagation in arithmetic operations such as addition and substraction is localized by one digit position; that is, massive parallelism of the computation can be achieved. In the residue number system, addition and multiplication are inherently carry-free; this number system is suitable for massively parallel arithmetic operations. A signed-digit arithmetic and residue number system can be implemented by binary logic networks. However, multivalued logic provides them more efficiency compared with binary logic.

In practice, multivalued logic networks are modeled by a multilevel encoding of information.

1.7 Further study

[1] Adamatzky A. Computing with waves in chemical media: Massively parallel reaction-diffusion processors, *IEICE Trans. Electron.*, E87-C(11):1748-1756, 2004.

[2] Adleman LM. Molecular computation of solutions to combinatorial problems. *Science*, 226, November, pp. 1021–1024, 1994.

[3] Aguirre AH and Coello CAC. Evolutionary synthesis of logic circuits using information theory. In Yanushkevich SN, Ed., *Artificial Intelligence in Logic Design*, pp. 285–311, Kluwer, Dordrecht, 2004.

[4] Aoki T, Homma N, and Higuchi T. Evolutionary synthesis of arithmetic circuit structures. In Yanushkevich SN, Ed., *Artificial Intelligence in Logic Design*, pp. 39–72, Kluwer, Dordrecht, 2004.

[5] Aviram A and Ratner MA. Molecular rectifiers. *Chem. Phys. Letters*, 29:277–283, 1974.

[6] Beer S. *Decision and Control: The Meaning of Operational Research and Management Cybernetics*. John Wiley & Sons, New York, 1966.

[7] Butler JT. Multiple-valued logic. *IEEE Potentials*, 14(2):11–14, 1995.

[8] Carbone A and Seeman NC. Circuits and programmable self-assembling DNA structures. In *Proceedings of National Academy of Sciences*, 99(20):12577–12582, 2002.

[9] Chakradhar ST, Agrawal VD, and Bushnell ML. *Neural Models and Algorithms for Digital Testing.* Kluwer, Dordrecht, 1991.

[10] Chen J, Lee T, Su J, Wang W, Reed MA, Rawlett AM, Kozaki M, Yao Y, Jagessar RC, Dirk SM, Price DW, Tour JM, Grubisha DS, and Bennett DW. Molecular electronic devices, In Reed MA and Lee L, Eds., *Handbook of Molecular Nanoelectronics*, American Science Publishers, New York, 2003.

[11] Collier CP, Wong EW, Belohradsk M. Raymo FM, Stoddart JF, Kuekes PJ, Williams RS, and Heath JR. Electronically configurable molecular-based logic gates. *Science*, 285:391–394, July, 1999.

[12] Crawley D, Nikolić K, and Forshaw M, Eds., *3D Nanoelectronic Computer Architecture and Implementation.* Institute of Physics Publishing, UK, 2005.

[13] Das B and Abe S. Modeling molecular switches: A flexible molecule anchored to a surface. In Seminario JM, Ed., *Molecular and Nano Electronics: Analysis, Design and Simulation*, pp. 141–162, Elsevier, Amsterdam, 2007.

[14] de Castro LN. *Fundamentals of Natural Computing: Basic Concepts, Algorithms, and Applications.* Chapman & Hall/CRC Taylor & Francis Group, Boca Raton, FL, 2006.

[15] DeHon A. Array-based architecture for FET-based, nanoscale electronics. *IEEE Trans. Nanotechnology*, 2(1):23–32, 2003.

[16] Fredkin E and Toffoli T. Conservative logic. *Int. J. Theor. Phys.*, pp. 171–182, 2003.

[17] Ellenbogen JC and Love JC. Architectures for molecular electronic computers: Logic structures and an adder designed from molecular electronic diodes. *Proceedings IEEE*, 88(3):386–426, 2000.

[18] Hanyu T, Kameyama M, and Higuchi T. Prospects of multiple-valued VLSI processors. *IEICE Trans. Nanotechnology*, E76-C(3):383–392, 1993.

[19] Hanyu T. Challenge of a multiple-valued technology in recent deep-submicron VLSI. In *Proc. 31st IEEE Int. Symp. on Multiple-Valued Logic*, pp. 241–247, 2001.

[20] Karpovsky MG, Stanković RS, and Astola JT. *Spectral Logic and Its Applications for the Design of Digital Devices.* John Wiley & Sons, Hoboken, NJ, 2008.

[21] Keymeulen D. Self-repairing and tuning reconfigurable electronics: real world applications. In *Proc. NASA/ESA Conf. on Adaptive Hardware and Systems*, 2008.

[22] Kumar VKP and Tsai Y-C. Designing linear systolic arrays. *J. Parallel and Distributed Computing*, 7:441–463, 1989.

[23] Kung HT and Leiserson CE. Systolic arrays (for VLSI). In *Sparse Matrix Proceedings.* SIAM, Philadelphia, pp. 256–282, 1978.

[24] Kung SY. *VLSI Array Processors.* Prentice Hall, Englewood Cliffs, New York, 1988.

[25] Lent CS, Tougaw PD, Porod W, Bernstein GH. Quantum cellular automata. *Nanotechnology*, 4:49–57, 1993.

[26] Lyshevski SE. *Molecular Electronics, Circuits, and Processing Platforms.* CRC Press, Boca Raton, FL, 2007.

[27] Lyshevski SE. 3D multi-valued design in nanoscale integrated circuits. In *Proc. 35st IEEE Int. Symp. on Multiple-Valued Logic,* pp. 82–87, 2005.

[28] Macii E and Poncino M. An application of Hopfield networks to worst-case power analysis of RT-level VLSI systems. *Int. J. Sci.*, 35(8):783–792, 1997.

[29] Ma Y and Seminario JM. Analysis of programmable molecular electronic systems. In Seminario JM, Ed., *Molecular and Nano Electronics: Analysis, Design and Simulation*, pp. 96–140, Elsevier, Amsterdam, 2007.

[30] Negrini R and Sami MG. *Fault Tolerance Trough Reconfiguration in VLSI and WSI Arrays.* The MIT Press, Cambridge, MA, 1989.

[31] Peper F, Lee J, Abo F, Isokawa T, Adachi S, Matsui N, and Mashiko S. Fault-tolerance in nanocomputers: a cellular array approach. *IEEE Trans. Nanotechnology*, 3(1):187–201, 2004.

[32] Porod W, Lent CS, Toth G, Csurgay A, Huang Y-F, and Liu RW. *IEEE Abstracts*, p. 745, 1997.

[33] Reichert J, Ochs R, Beckmann D, Weber HB, Mayor M, and Lohneysen HV. Driving current through single organic molecules. *Physical Review Letters*, 88(17), 2002.

[34] Rosenblatt F. *Principles of Neurodynamics.* Spartan, New York, 1962.

[35] Rothemund PWK, Paradakis N, and Winfree E. Algorithmic self-assembly of DNA Sierpinski triangles. *PloS Biology* — www.plosbiology.org, 2(12,e424):2041–2053, 2004.

[36] Shmerko VP and Yanushkevich SN. Three-dimensional feedforward neural networks and their realization by nano-devices. *Artificial Intelligence Review, An Int. Science and Eng. J.* (UK), Special Issue on *Artificial Intelligence in Logic Design*, 20(3-4):473–494, 2004.

[37] Shukla SP and Bahar RI, Eds., *Nano, Quantum and Molecular Computing.* Kluwer, Dordrecht, 2004.

[38] De Silava AP, McClenaghan ND, and McCoy CP. Molecular logic systems. In Feringa BL, Ed., *Molecular Switches*, pp. 339–361, Wiley-VCH, Weinheim, Germany, 2001.

[39] Toth G, Lent CS, Tougaw PD, Brazhnik Y, Weng WW, Porod W, Liu RW, and Huang YF. Quantum cellular neural networks. *Superlattices and Microstructures*, 20,:473–478, 1996.

[40] Tour JM and James DK. Molecular electronic computing architectures. In Goddard WA, Brenner DW, Lyshevski SE, and Iafrate GJ, Eds., *Handbook of Nanoscience, Engineering and Technology*, pp. 4.1–4.28, CRC Press, Boca Raton, FL, 2003.

[41] Tour JM. *Molecular Electronics: Commercial Insights, Chemistry, Devices, Architecture and Programming.* World Scientific, Hackensack, NJ, 2003.

[42] Tour JM, Zandt WLV, Husband CP, Husband SM, Wilson LS, Franzon PD, and Nackashi DP. Nanocell logic gates for molecular computing. *IEEE Trans. Nanotechnology*, 1(2):100–109, 2002.

[43] Wang W, Lee T, Kretzschmar I, and Reed MA. Inelastic electron tunneling spectroscopy of an alkanedithiol self-assembled monolayer. *Nano Letters*, 4(4):643–646, 2004.

[44] Yao X and Higuchi T. Promises and challenges of evolutionary hardware. *IEEE Trans. Systems, Man, and Cybernetics, Part C: Applications and Reviews*, 29(1):87–97, 1999.

[45] Yanushkevich SN, Shmerko VP, and Lyshevski SE. *Logic Design of Nano ICs*, CRC Press, Boca Raton, FL, 2004.

[46] Yanushkevich SN. Logic design of computational nanostructures. *J. Computational and Theoretical Nanoscience*, 4(3):384–407, 2007.

[47] Yanushkevich SN. Spatial systolic array design for predictable nanotechnologies. *J. Computational and Theoretical Nanoscience*, 4(3):467–481, 2007.

[48] Yanushkevich SN, Shmerko VP, and Steinbach B. Spatial interconnect analysis for predictable nanotecnologies. *J. Computational and Theoretical Nanoscience*, 5(1):56–69, 2008.

[49] Yanushkevich SN, Miller DM, Shmerko VP, and Stanković RS. *Decision Diagram Techniques for Micro- and Nanoelectronic Design*, Taylor & Francis/CRC Press, Boca Raton, FL, 2006.

[50] Ziegler MM and Stan MR. CMOS/nano co-design for crossbar-based molecular electronic systems. *IEEE Trans. Nanotechnology*, 2(4):217–230, 2003.

2

Computational Nanostructures

2.1 Introduction

In this chapter, the basic principles for design of a computational structure are introduced. These principles, such as analysis and synthesis, assembling, top-down and bottom-up methodology, design styles, simulation, and modeling, can be adopted in nanodevices design, except for some particular cases when underlying physical and chemical phenomena address nontraditional approaches. This chapter also briefly introduces the computational models such as switch-based computing and homogeneous structures that are the focus of further chapters in this book.

The crucial point of computing nanostructure design is an approach that provides the delegation of computing abilities from initial data structure to the nanostructure. This approach is based on the specification of the form of graphical data structures, such as decision trees and diagrams. These data structures are characterized by computational properties that can be embodied in various spatial structures specified by nanotechnology. This chapter provides a brief introduction of this problem. Another aspect is that hybrid technologies can be used in computing structure design. This aspect requires using the unified approaches in design methodologies.

A *discrete system* is a combination of logic networks and discrete devices that is assembled to accomplish a desired result, such as the computing and transferring of data. Digital logic networks are used in all devices that process information in digital form. *Information* can be defined as recorded or communicated facts, or *data*. Information takes a variety of physical forms when being stored, communicated, or manipulated. Information on the nature of a physical phenomenon is conveyed by signals that assume a finite number of discrete values, that is, it is expressed as a finite sequence of symbols. A *signal* is defined as a function of one or more variables, and a *system* is defined as an entity that manipulates one or more signals to accomplish a function, thereby yielding new signals.

In this chapter, each signal is assumed to have only one of two values, denoted by the symbols 0 and 1. If the signals are constrained to only two values, the system is *binary*.

2.2   Theoretical background

Boolean algebra was introduced by George Boole in 1854 and applied by C. Shannon in 1938 to relay-contact networks, the first switching circuits. This theory is called *switching theory* and has been used ever since in the design of digital *logic networks* or *logic circuits*. Since then the technology has gone from relay-contacts through diode gates, transistor gates, and integrated circuits, to future nanotechnologies, and still Boolean algebra is its fundamental and unchanging basis. Modern logic design includes methods and techniques from various fields. In particular, digital signal processing is adopted in logic design for efficient manipulation of data; communication theory solves communication problems between computing components in logic networks; artificial intelligence methods and techniques are used for optimization at logic and physical levels of logic network design.

As far as predictable technologies are concerned, nontraditional computing paradigms that are based on various physical and chemical phenomena are studied. The assumed stochastic nature of many processes at nano scale implies that *random* signals must be used instead of *deterministic* signals. Random signals take on random values at any given moment in time and must be modeled probabilistically, while deterministic signals can be modeled as completely specified functions of time. The theoretical base of probabilistic logic signals is called *probabilistic logic*.

A unit for computing an elementary Boolean function (such as an AND, OR, NAND, NOR, EXOR, threshold function, or a special case of it, the majority function) at nano scale is defined as a *logic nanocell*. The elementary logic functions and arbitrary logic networks can be implemented by using only threshold (neuron) cells. A *switch* is the simplest logic operation that operates on a single bit. An arbitrary Boolean function (and, therefore, an arbitrary logic network) can be implemented using, for example, only NAND or only NOR nanocells, or only switches. If nanocells are affected by noise that influences their dynamic and steady-state behavior, they are referred to as *probabilistic nanocells*.

The classic computing paradigm implemented in most of today's computers is known as *von Neumann* architecture. In the computing paradigms based on the principle of *distributed* processing, information is encoded into the different components of a computing network. In *fault-tolerant* computing, the effects of faults are mitigated, and correct computations are guaranteed with a certain level of reliability. *Redundancy* is one of the fault-tolerant techniques that can be used if additional resources are available. Redundancy in distributed models cannot be achieved by integration of extra cell copies over the appropriate permutation and decision profile, but only by sharing of carriers of information between many components of the network. Assuming reliable computing in the presence of faults is called *fault-tolerance*; for example, approaches to reliable computing in networks with faulty nodes and/or interconnects, including error correction codes and residual number systems. These approaches state that it is possible to correct a class of faults if a library of *reliable* logic nanocells for implementing the correction is available.

2.3 Analysis and synthesis

The *specification* of a system is defined as a description of its function. The *implementation* of a system refers to the construction of a system. The *analysis* of a system has its objective to determine its specification from the implementation:

$$\underbrace{\texttt{Logic network} \longrightarrow \texttt{Specification}}_{Analysis}$$

Synthesis, or *design*, consists of obtaining an implementation that satisfies the specification of the system:

$$\underbrace{\texttt{Specification} \longrightarrow \texttt{Logic network}}_{Synthesis}$$

The central task in logic synthesis is to optimize the representation of a logic function with respect to various criteria.

Complex systems are specified at various levels of detail. At each level, the units of complexity are specified as *components* or a *subsystem* of the components, and a *basic element, or primitive* is defined as a component that has no internal structure.

A system can be examined at various levels of abstraction. Each such level is called a *design level* (Figure 2.1a). The following design levels are distinguished:

▶ The top design level is called the *architectural* or *system* level.
▶ The intermediate design level is called the *logic* level; this level is the subject of the present book.
▶ The bottom design level is called the *physical* level; this level is concerned with the details needed to manufacture or assemble the system.

At the physical level, a system is implemented by a complex interconnection of simplest elements (primitives). Because of high complexity, it is impractical to perform design and optimization at this level, motivating a move to the intermediate level of design. At the intermediate level, a modular structure provides a reasonable simplification of design. *Libraries* of standard modules significantly simplify the design of different systems. Assembling modules of increasing complexity into higher hierarchical blocks is achieved at the system level.

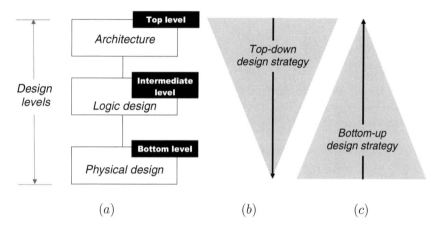

FIGURE 2.1
Design hierarchy (a), top-down (b), and bottom-up (c) design strategies.

2.3.1 Design hierarchy

A *hierarchical* approach to digital system design aims at reducing the cost of the design of a system, and improving the quality of the obtained solutions. The hierarchical approach to design makes a large system more manageable by reducing complexity and introducing a rational *partitioning* of the design processes. They can be designed, tested, and manufactured separately. This is the basis for *standardization*. The specified components can be mass produced at relatively low cost. In the design process, these components can be composed in standard *libraries* and reused with minor modifications. This facilitates the reduction of overall design time and cost. The *robustness* of the hierarchical approach provides many possibilities for avoiding design errors, design corrections, and repairs after manufacture.

Any design process includes a *design loop* that provides the possibility to carry out a *redesign* if errors are detected in simulation. This loop is repeated until the simulation indicates a successful design.

Example 2.1 *The use of standard cells is a classical example of how a restriction in the design space (a limited library of constrained cells) makes it possible to use intellectual and computational capabilities to find high-quality acceptable solutions through synthesis.*

2.3.2 Top-down design methodology

Design methodology is a system of ways of obtaining the implementation of a specified design.

A design that evolves from a generalized or abstract point of view and proceeds in steps to specific components is referred to as a *top-down design methodology* (Figure 2.1b):

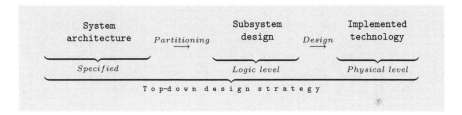

In this approach, the hierarchy tree is traversed from top to bottom. The system architecture is specified at the highest level first. The disadvantage of this approach is that no systematic procedure exists for optimization of the final implementation; that is, optimization at one particular level does not guarantee an optimal final solution. The success of the approach depends mainly on the experience and professional skills of the designer.

Example 2.2 *Given the architecture of a system, a designer may begin to detail subsystems in terms of small black boxes, each of which consists of an electronic circuit developed by "circuit designers." The designer has to know, in particular, (a) the logical function of each block, have a means be able of expressing those functions, and (b) the logic function the interconnected blocks provide.*

The top-down approach is currently used in the silicon industry, wherein small features such as transistors are etched into silicon using resisters and light. This hierarchy starts at the highest level of abstraction, the architecture level. Then it descends to the level of its component circuits, and finally to the level of the component switch and interconnect devices.

An example of an advanced top-down methodology is so-called *platform-based design*. A platform is defined as a family of the designs and not a single design. In this top-down process, the constraints that accompany the specification are mapped into constraints on the components of the platform.

Most artificial intelligence techniques based upon the top-down design paradigm are known as knowledge-based or expert systems.

2.3.3   Bottom-up design methodology

An alternative process to the top-down approach is a *bottom-up design methodology* (Figure 2.1c):

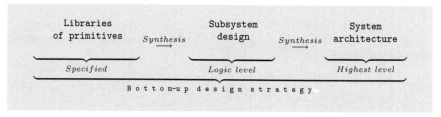

This is the reverse of the top-down design process. One starts with specific components in mind and proceeds by interconnecting these components into a generalized system. In a bottom-up design approach, the components at or near the lowest design level of the hierarchy tree are designed first. The architecture of the entire system is not specified until the top of the tree is reached. Unfortunately, in general, there is no systematic technique that results in correct system specification.

Bottom-up systems and models are those in which the global behavior emerges from the aggregative behavior of ensemble of relatively simple elements acting solely on the basis of local information.

> **Example 2.3**   *Natural computing inspired or based on biology is more rooted in* **bottom-up** *approaches. An example of bottom-up design is artificial neural networks. Their structure is not defined a priori; they are a result of the network interactions with the environment. For example, if the initial neural network has a single neuron, and more neurons are can be added until the network is capable of appropriately solving the problem.*

The bottom-up approach implies the construction of functionality into small features, such as molecules, with the opportunity to have the molecules further self-assemble into higher-ordered structural units such as transistors. Bottom-up methodologies are quite natural in that all systems in nature are constructed bottom-up.

Example 2.4 *Molecules with specific features assemble to form higher-order structures such as lipid bilayers.*

Molecular electronics proponents believe purposeful bottom-up design will be more efficient than the top-down method.

Example 2.5 *The device physics, based on quantum interaction, quantum interference, quantum transition, vibration, Coulomb effect, and other phenomena and effects (electron spin, photon-electron-assisted transitions, etc.), must be coherently complemented by the bottom-up synthesis of the molecular aggregates that exhibit those phenomena.*

However, in practice, there are many specific-area application devices and systems where top-down and bottom-up design methodologies can be efficiently used separately or in various combinations. This provides the possibility for designing simultaneously in several levels.

2.3.4 Design styles

In a general sense, the following design styles are distinguished:

Full custom design: This style provides freedom to the designer and is characterized by great flexibility; however, this style is not acceptable for the design of large systems.

Semi-custom design: Provides more possibilities for automation using, in particular, standardization; such as a *library* of standard cells.

Mixed design styles: Often provide an acceptable reduction to the flexibility of full-custom style, while opening up possibilities for the automation and optimization of semi-custom style.

Gate-array design meets the requirements of fabrication and simplifies the optimization problem; this style results in regular structures within the chip, that is, connected cells that are placed on a chip in a regular way.

These design styles are adopted in various technologies.

2.3.5 Modeling and simulation

A unit for computing elementary logic (Boolean) functions (such as an AND, OR, NAND, NOR, EXOR, threshold function, or the majority function) at nano scale is defined as a *logic nanocell*. The elementary logic functions and arbitrary logic networks can be implemented by using only threshold (neuron) cells. A *switch* is the simplest logic operation that operates on a single bit.

An arbitrary elementary logic function (and, therefore, an arbitrary logic network) can be implemented using, for example, only NAND or only NOR nanocells, or only switches. If nanocells are affected by noise that influence the dynamic and steady-state behavior, they are referred to as *probabilistic nanocells*.

The classic computing paradigm implemented in most of today's computers is known as *von Neumann* architecture. In the computing paradigms based on the principle of *distributed* processing, information is encoded into the different components of a computing network. In *fault-tolerant* computing paradigms, the effects of faults are mitigated, and correct computations are guaranteed with a certain level of reliability. *Redundancy* is one of the fault-tolerant techniques that can be used if additional resources are available. Redundancy in distributed models cannot be achieved by integration of extra cell copies over the appropriate permutation and decision profile, but only by sharing of carriers of information between many components of the network. Assuming reliable computing in the presence of faults is called *fault-tolerance*, which includes approaches to reliable computing in networks with faulty nodes and/or interconnects, error correction codes, and residual number systems, etc. These approaches state that it is possible to correct a class of faults if a library of *reliable* logic nanocells for implementing the correction is available.

Recursive stochastic models are based on the property of *relaxation* that is, an ability to relax into a stable state in the absence of external input. Relaxation is referred to as the *embedding* of a correct solution into a network of logic nanocells (neurons). In models based on *stochastic pulse stream encoding*, information is encoded in the average pulse rate or primary statistics of this stream.

The use of software simulation is an important part of any modern design process. The primary uses of a simulator are to check a design for functional correctness and to evaluate its performance. Simulators are key tools in determining whether design goals have been met and whether redesign is necessary.

2.4   Implementation technologies

The scaling of microelectronics down to nanoelectronics is the inevitable result of technological evolution (Figure 2.2). The most general classification of the trends in technology is based on grouping computers into *generations*. Using this criterion, five generations of computers are distinguished (Table 2.1). Each computer generation is 8 to 10 years in length.

The following can be compared against this scale:

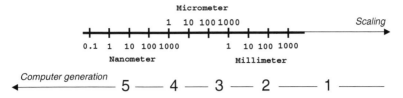

FIGURE 2.2
Progress from micro- to nanosize in computing devices.

TABLE 2.1
Computer generations are determined by the change in the dominant
technology

Generation	Dates	Technology
1	1950–1964	Vacuum tubes (zero-scale integration)
2	1965–1969	Transistors (small-scale integration)
3	1970–1979	Integrated circuits (medium-scale integration)
4	1980–2004	Large, very large, ultra large-scale integration
5	2005–	Nanotechnology (giga-scale integration)

The scaling of microelectronics down to nanoelectronics

- ▶ The size of an atom is approximately 10^{-10} m. Atoms are composed of subatomic particles; e.g., protons, neutrons and electrons. Protons and neutrons form the nucleus, with a diameter of approximately 10^{-15} m.
- ▶ 2D molecular assembly (1 nm).
- ▶ 3D functional nanoICs topology with doped carbon molecules ($2 \times 2 \times 2$ nm).
- ▶ 3D nanobioICs (10 nm).
- ▶ *E. coli* bacteria (2 mm) and ants (5 mm) have complex and high-performance integrated nanobiocircuitry.
- ▶ 1.5×1.5 cm 478-pin Intel® Pentium® processor with millions of transistors, and Intel 4004 Microprocessor (made in 1971 with 2,250 transistors).

Binary states are encountered in many different physical forms in various technologies. The standard approach is to use the digit symbols 0 and 1 to represent the two possible values of binary quantity. These symbols are referred to as *bits*.

Binary arithmetic has the following advantages:

(*a*) It can be implemented using on-off switches, the simplest binary devices.

(*b*) It provides for the simplest decision-making such as YES (1) and NO (0).

(*c*) Binary signals are more reliable than those formed by more than two quantization levels.

Significant evolutionary progress has been achieved in microelectronics. This progress (miniaturization, optimal design and technology enhancement) has been achieved by scaling down microdevices, approaching 45 nm sizing features for structures, while increasing the integration level (Figure 2.3). Complementary metal-oxide semiconductor (CMOS) technology is being enhanced, as nanolithography, advanced etching, enhanced deposition, novel materials, and modified processes are all currently used to fabricate ICs.

Example 2.6   *The channel length of metal-oxide-semiconductor field effect transistors (MOSFETs) has decreased from*

▶ *50 μm in 1960, to*

▶ *1 μm in 1990, and to*

▶ *130 nm in 2004, 65 nm in 2006, and 45 nm in 2007.*

Example 2.7   *The progress in miniaturization and integration can be observed, for example, on Intel processors:*

1971–1982: From Intel 4004 (1971, 2,250 transistors), to Intel 286 (1982, 120,000 transistors),

1993–2007: From Pentium (1993, 3,100,000 transistors), to Pentium 4 (2000, 42,000,000 transistors), to Itanium^{TM} 2 processor (2002), Pentium^® M processor (2003) with hundreds of millions of transistors, and Pentium Dual-Core in 2007.

The increases in packing density of the circuitry are achieved by shrinking the linewidths of the metal interconnects, by decreasing the size of other features, and by producing thinner layers in the multilevel device structures.

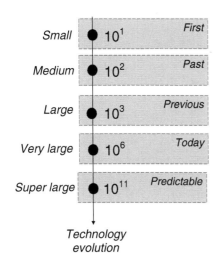

Small $\quad \bullet \; 10^1 \quad$ *First*

Medium $\quad \bullet \; 10^2 \quad$ *Past*

Large $\quad \bullet \; 10^3 \quad$ *Previous*

Very large $\quad \bullet \; 10^6 \quad$ *Today*

Super large $\quad \bullet \; 10^{11} \quad$ *Predictable*

Technology
evolution

Small-scale integration (SSI):
- ▶ *1960s, dozens of gates in a package*

Medium-scale integration (MSI):
- ▶ *1970s, hundreds of gates in a package*

Large-scale integration (LSI):
- ▶ *1980s, thousands of gates in a package*

Very large-scale integration (VLSI):
- ▶ *1980s and 1990s, hundreds of thousands of gates in a package*

Giga-scale integration:
- ▶ *Today, millions of gates in a package*

Tera-scale integration:
- ▶ *Expected, hundred millions of gates in a package*

FIGURE 2.3
Evolution of technologies: levels of integration of chips.

Example 2.8 *Commercial metal interconnect linewidths have decreased to 0.1 mm.*

Typical signal integrity effects include interconnect delay, cross-talk (in closely coupled lines the phenomenon of cross-talk can be observed), power supply integrity, and noise-on-delay effects. In the early days of very large-scale integration (VLSI) design, these effects were negligible because of relatively slow chip speed and low integration density. However, with the introduction of technological generations working at about 0.25 μm scale and below, there have been many significant changes in wiring and electrical performance.

As the number of computational and functional units on a single chip increases, the need for communication between those units also increases. Interconnection has started to become a dominant factor in chip performance. As chip speed continually increases, the increasing inductance of interconnections affects the signal parameters. The length of interconnect lines when measured in units of wire pitch increases dramatically.

Noise is a deviation of a signal from its intended or ideal value. Most noise in VLSI circuits is created by the system itself. Electromagnetic interference is an external noise source between subsystems. In deep submicron circuits, the noise created by parasitic components with digital switching exhibits the strongest effect on system performance. Noise has two deleterious effects on circuit performance: (a) when noise acts against a normal static signal, it can destroy the local information carried by the static node in the circuit and ultimately result in incorrect machine-state stored in a latch, and (b) when noise acts simultaneously with a switching node, this is manifest as a change in the timing of a transient.

2.5   Predictable technologies

The key to the predictable technologies of the future is changing the *computing paradigm*, that is, the model of computation and the physical effects for the realization of this model:

$$\underbrace{\text{Computing model}}_{Design} \xrightarrow{\textit{Implementation}} \underbrace{\text{Physical effects}}_{Technology}$$

Silicon loses its original band structure when it is restricted to very small sizes. The lithography techniques used to create the circuitry on the wafers also neared its technological limits.

There are fundamental technological differences among

▶ Nanoelectronic devices versus microelectronic ones (which can even be nanometers in size)

▶ Nanoelectronics versus microelectronics, for example, nanointegrated circuits versus integrated circuits

These enormous differences are due to differences in basic physics and other phenomena. The dimensions of nanodevices that have been made and characterized are a hundred times less than even newly designed microelectronic devices (including nanoFETs with 10 nm gate length). Nanoelectronics sizing leads to volume reduction by a factor of 1,000,000 in packaging, not to mention revolutionary enhanced functionality due to multiterminal and spatial features. For example, molecular electronics focuses on the development of electronic devices using small molecules with feature sizes on the order of a few nanometers.

Superconductive solid-state and molecular electronics focuses on the development of electronic devices using small particles and molecules with feature sizes on the order of a few nanometers. The ultimate goal of miniaturization is to use the minimum amount of atoms per electronic function.

Example 2.9   *A contemporary computer utilizes $\sim 10^{10}$ silicon-based devices, whereas the scaling factor gained from molecular-scale technology $\sim 10^{23}$ devices in a single beaker using routine chemical syntheses.*

An additional driving factor is the potential to utilize thermodynamically driven directed self-assembly of components such as chemically synthesized interconnects, active devices, and circuits.

In order to take advantage of the ultra-small size of molecules, one ideally needs an interconnect technology that (a) scales from the molecular dimensions, (b) can be structured to permit the formation of the molecular equivalent of large-scale diverse modular logic blocks as found in VLSI interconnect architectures, and (c) can be selectively connected to mesoscopically (100 nm scale) defined input/output (I/O).

Nanoscale devices must obey the fundamental constraints that the laws of physics impose on the manipulation of information. Some physical constraints, such as the fact that the speed at which information can travel optically through free space is upperbounded by a constant (the speed of light, 299 792 458 m/s), are still present in nanodimensions. The electrical transmission of signals along wires is slower than light, so the current propagation delays along dissipative wires are significant. The time of transmission is proportional to the square of the distance, unless the signals are periodically regenerated by additional buffering logic.

A wide variety of factors, such as voltage scaling and thermal noise, dramatically reduce the reliability of integrated systems. In nanotechnologies, process variations and quantum fluctuations occur in the operations of very-deep submicron transistors. Any computer with nano scale components will contain a significant number of defects, as well as massive numbers of wires and switches for communication purposes. It therefore makes sense to consider architectural issues and defect tolerance early in the development of a new paradigm.

Two main aspects are critical to the design of nanodevices. First is *the probabilistic behavior* of nanodevices (electrons, molecules), which means that a valid Boolean function can be calculated with some probability. Second is the *high defect rates* of nanodevices, meaning that because many of the fabricated devices have defects, their logic correctness is distorted.

There are two types of fault tolerances exhibited by a nanosystem: fault tolerance with respect to (a) data that are noisy, distorted, or incomplete, which results from the manner in which data are organized and represented in the nanosystem, and (b) physical degradation of the nanosystem itself. If certain nanodevices or parts of a network are destroyed, the network will continue to function properly. When the damage becomes extensive, the network will only affect the system's performance, as opposed to causing a complete failure. Self-assembling nanosystems are capable of this type of fault tolerance because they store information in a distributed (redundant) manner, in contrast to traditional storage of data in a specific memory location in which data can be lost in the case of a hardware fault.

The methods of stochastic computing provide another approach to overcoming the problem of the design of reliable computers from unreliable elements, i.e., nanodevices.

Example 2.10   *A signal may be represented by the probability that a logic level has a value of 1 or 0 at a clock pulse. In this way, random noise is deliberately introduced into the data. A quantity is represented by a clocked sequence of logic levels generated by a random process. Operations are performed via completely random data.*

2.6   Nanoelectronic networks

The specifications of nanoelectronic devices and networks are based on data structures adopted from classical logic design. This adaptation is done based on the premises of nano scale implementation structures.

2.6.1   CMOS-molecular electronics

The chemically directed self-assembly of molecular single-electron devices on prefabricated nanowires is considered as a promising way toward integrated circuits using, for example, single-electron tunneling.

2.6.2   Neuromorphic computing paradigm

A *Hopfield* network, also known as *neuromorphic* computing,[*] is based on distributed architecture principles, and can be implemented by various models [25]. Using a Hopfield network, an arbitrary logic function can be computed via a process similar to the gradual cooling process of metal. A value of a logic function given an assignment of Boolean variables is computed through the relaxation of neuron cells. Hopfield networks are capable of reliable computing, despite imperfect neuron cells and degradation of their performance. This is because of degraded neuron cells store information (in weights, thresholds, and topology) in a *distributed* (or redundant)

---

[*]The term "neuromorphic systems" was used for the first time by C. Mead [32] to define an interdisciplinary approach to the design of biologically inspired information processing systems.

manner. A single "portion" of information is not encoded in a single neuron cell, but is rather spread over many. This contrasts with the traditional computing paradigm (von Neumann architecture), in which data is stored in a specific memory location. The Hopfield computing paradigm is based on the concept of *minimization of energy* in a stochastic system. This concept is implemented using *McCulloch–Pitts* neuron cells for elementary logic functions, the *interconnections* between nanocells given a logic function, the *weights* assigned to the links between nanocells, and an *objective* function given the number of neuron cells, their thresholds, interconnections, and weights. The carriers of information in the Hopfield network are a particular topology of connections between neuron cells, the weights of the links, and the neuron thresholds.

2.6.3 Interconnect

An *interconnect* is considered with respect to various criteria, for example, number of switches, signal delays, total length, power dissipation, and cost. The traditional understanding of the interconnect as a wire in a voltage-controlled network does not work for certain nanotechnologies, especially for those where the carrier of charge (for example, single electron) is considered as a messenger that travels through the network of nanowires via multiplexing, or a signal propagated by the intractability of cells. In such a network, control signals (input variables) are separated from the messenger signal (function), and the problem of interconnection becomes the problem of compatibility of inputs and outputs. Below we consider switches on multiplexers as the most suitable candidates for interconnect nodes.

The components of biomolecular networks are not connected by physical wires that direct a signal to a precise location; molecules that are the inputs and outputs of these processes share a physical space and can commingle throughout the cell. It requires the special tools to isolate signals and prevent cross-talk, in which signals intended for one recipient are received by another. Unlike electronics, which typically rely on a clock to precisely synchronize signals, these biomolecular signals are asynchronous.

Nanoscale devices must obey the fundamental constraints that the laws of physics impose on the manipulation of information. Some physical constraints, such as the fact that the speed at which information can travel optically through free space is upperbounded by a constant (the speed of light, 299 792 458 m/s), are still present in nanodimensions. The electrical transmission of signals along wires is slower than light, so the current propagation delays along dissipative wires are significant. The time of transmission is proportional to the square of the distance, unless the signals are periodically regenerated by additional buffering logic.

The vulnerability of nanocircuits to defects and faults arising from instability and noise-proneness at nanometer scales [43] leads to unreliable and undesirable results of computation. Robustness to errors is an important design consideration for nanocomputers in the light of the noise and instabilities that affect the reliability of nanometer-scale devices.

2.6.4   Carbon-nanotube-based logic devices

Carbon nanotubes are sheets of graphite rolled into tubes. Depending on the direction in which the graphite sheet is rolled, the single-walled carbon nanotubes can be metallic or semiconductor. Semiconducting nanotubes are used for fabrication switches and transistors known as carbon-nanotube field-effect transistors (CNFETs). The CNFETs are considered as possible successors to silicon MOSFETs because CNFETs inherit current-voltage characteristics that are qualitatively to silicon MOSFETs. That is, MOS-based logic networks can be transferred to a CNFET-based design.

Storage elements are implemented as the molecular scale junction switches formed at the crosspoints of the wires. Addressing memory array (random access memory) can be designed based on the crosspoint array by using a decoder.

State-of-the art defect-tolerant techniques can be applied to the crossbar memory. The crossbar memory does not operate correctly if defects occurred during the fabrication. These defect can be located and identified. Using redundant rows and columns, the faults caused by the identified defects can be eliminated.

Example 2.11   *The following faults caused by the defects can be used in defect tolerance techniques: (a) broken nanowire, (b) stuck-at-low and stuck-at-high resistance state, (c) short at a crosspoint, and (d) open at a crosspoint.*

2.6.5   Crossbar-based computing structures

The *crossbar* structures can be designed using the bottom-up approach. Various physical phenomena can be utilized in crossbar structure design.

Example 2.12   *The following physical implementation can be utilized in crossbar structure design:*

   ▶ *Self-assembled nanowires*
   ▶ *Nano-imprinted wires*
   ▶ *Nanotubes*

Computing properties are delegated in crossbar structure using (a) crosspoint function and (b) topological properties such as distribution of this function in a rectangular array. The simplest function of the crosspoint is switch.

Example 2.13 *Functions of a crosspoint in a crossbar structure can be switches. The advantage of this array is that an arbitrary Boolean function can be directly assembled. The drawback is that the signal is degradated with respect to the number of active switches.*

The *crossbar* computing structures utilize the common paradigm for molecular electronics. The crossbar structure has a simple interconnection topology and is therefore attractive for fabrication. Different types of crossbar address different types of elements such as resistor, diode, or transistor. These elements can be combined to perform larger functions.

Example 2.14 *The basic crossbar structure consists of the combination of planes of parallel wires laid out in orthogonal directions.*

Each junction or crosspoint within the crossbar can be independently configured. After fabrication, the bistable junctions at the wire crossing can be customized, that is programmed to perform the designed Boolean function.

2.6.6   Noise

The term *noise* means undesirable variations of voltages and currents in a circuit. Noise is an unpredictable and random phenomenon. The noise become significant in a logic network when the noise signal causes incorrect logic functions. Noise in integrated circuits is the interference signal inducted by signals on neighboring wires or from the substrate. This noise is different from the intrinsic noise generated by active device components. Most CMOS digital circuits have a relatively high immunity to noise. However, as power supply levels have decreased, this property has diminished. Both the power supply distribution and signal distortion along coupled interconnect lines are interconnect-related problems. Interconnections in CMOS integrated circuits are multiconductor lines existing on different physical planes. The parasitic impedances of these conductor lines can be extracted from the geometric layout. The coupling capacitance is physically a fringing capacitance between neighboring interconnect lines, and strongly depends upon the physical structure of the adjacent interconnections.

Example 2.15   *For parallel metal lines on the same layers, the fringing capacitance will increase as the spacing between the interconnections and the thickness-to-width aspect ratio of the interconnection increases.*

Due to the screening effect of low-level interconnect, the metal-to-metal coupling capacitance among different layers also contributes to the total coupling capacitance. Coupling effects become more significant as the feature size is decreased. The induced noise voltage may cause extra power to be dissipated on the quiet line due to momentary glitches at the output of the logic gates.

Example 2.16   *At gigahertz operating frequencies and high integration densities, power dissipation densities are expected to approach $20W/cm^2$, a power density limit for an aircooled package devices. Such a power density is equivalent to 16.67 ampers of current for 1.2V power supply in a $0.1\mu m$ CMOS technology.*

2.7   Switch-based computing structures

Decision trees and diagrams are the graphical data structures, which correspond to the networks of switches. Therefore, they are called *switch-based computing structures*. This data structures are useful for two classes of problems: (a) for computing Boolean functions and (b) for embedding computational properties into spatial structures.

2.7.1   Switches

A *switch* is any device by means of which two or more conductors (electrical, electrochemical) can be conveniently connected or disconnected. The status of the contact, which can be opened or closed, can be represented by a variable x_i. A switch is the simplest computing device that operates on a single bit. Computations with n bits can be realized as compositions of such one-bit switching changes. Switches are the basic elements for construction of logic gates.

Switches have two states: *operated* or *active*, and *released* or or *inactive*. There are many different types of switches, but they fall into two general classes: *nonlocking* switches, which return to a normal state when released, and *locking* switches, which retain their state after the controlling level is released.

Example 2.17 *A NOT gate can be implemented as a single-bit switch.*

A switch whose level is operated by a molecular phenomenon is called a *molecular switch*. The state of the molecular switch "contacts" when the switch is not energized (activated), is called the *normal state* of the switch. Thus, a molecular switch can have both *normally open* and *normally closed* states.

Switches might be a clue to the resolving of the poor input–output gain in the device-centered (based on the library of logic gates) design considered above. Many nanocircuits, such as arrays of nanowires, charge state logic, and other devices, are nonuniliteral, i.e., they have no clearly distinguished input and output voltage. In terms of logic design, these circuits act as decision diagram-based pass-gate logic, or switch (simplest multiplexer)-based devices.

A Boolean function realized by a switch-based network has the value of 1 if there is closed path between two specified terminals of the logic network. The switch themselves are assumed to be operated by the binary variables of the Boolean function. A variable x_i and $\overline{x}_i$ may be associated with a switch that is closed if and only if $x_i = 1$ and $x_i = 0$, respectively. Switches associated with uncomplemented and complemented variables are assumed to be normally open ($x_i = 0$) and closed ($x_i = 1$), respectively.

Example 2.18 *When two switches are connected in series, the path through them is closed only when both of them are closed. This condition can be described by the equation $x_1 \times x_2 = 1$, where x_1 and x_2 are the binary variables associated with these two switches. In the parallel connection of switches, there is a closed path between the terminals if either switch is closed. Figure 2.4 shows the series and parallel connections of switches.*

Probabilistic switching paradigm is the base of models for the energy saving computing in contemporary logic design of discrete devices. Based on this model, probabilistic logic gates can be constructed. In these models, noise is considered as a resource to achieving the accurate energy and power characteristics of discrete devices.

A switch operates in a single bit. It is assumed that the minimum energy needed to compute a bit with error of $1 - p$ by idealized probabilistic switch is $kt \ln 2p$ Joules, where k is the Boltzmann's constant, and t is the temperature of the thermodynamic system.

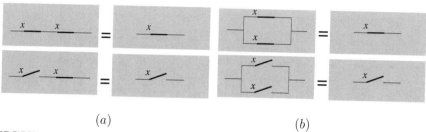

(a)   (b)

FIGURE 2.4

Two switches are connected in series; the closed path through them is closed only when both of them are closed (a). Two switches are connected in a parallel connection of switches; the path between the terminals is closed if either switch is closed (b) (Example 2.18).

2.7.2 Switch-based networks represented by decision diagrams

A decision diagram is a rooted directed acyclic graph consisting of

▶ The root node

▶ A set of terminal (constant) nodes

▶ A set of nonterminal nodes

connected with directed edges (links) (Figure 2.5). These components make up a *topology* of decision trees and diagrams. The topology is characterized by a set of topological parameters such as size, number of nonterminal nodes, number of levels, and number of links.

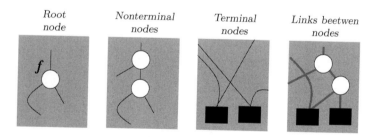

FIGURE 2.5

Components of a decision diagram.

Decision diagrams are graph-based data structures that represent algebraic ones. The algebraic data structure can be mapped into a decision tree or diagram, and vice versa.

Example 2.19 *Figure 2.6 shows the implementation of decision diagrams for the AND Boolean function using multiplexers.*

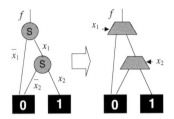

Decision diagram is a switch-based computing structure. Nonterminal nodes of the decision diagrams are replaced by multiplexers

FIGURE 2.6
Mapping of decision diagram into the multiplexer-based logic network (Example 2.19).

Manipulation of algebraic representations is based on the rules and axioms of Boolean algebra. These are aimed at simplification, factoring, composition, and decomposition. Manipulation of graphical data structures is based on the rules for reduction of nodes and links in decision trees and diagrams. There is a mapping between these representations. Algebraic expressions are relevant to the corresponding graphical data structures (hypercubes and hypercube-like structures, decision trees and diagrams, etc.)

Example 2.20 *Figure 2.7 shows the relationship between algebraic and graphical data structures for a 3-input NOR function. The algebraic form (a) is mapped into decision tree (b), which is reduced to the decision diagram (c).*

If the relationship between the basic rules of algebraic and graphical data structures established, an arbitrary logic function can be mapped into a corresponding decision diagram and vice versa. Because of the variety of algebraic structures and techniques for their synthesis, a wide variety of diagram types have been developed in the last four decades.

The implementation costs of decision trees and diagrams can be estimated directly from their topology. The other important properties, which can be evaluated on the diagram or tree, are a probability distribution that can be specified on the variables' spaces (which can be assumed uniform if not otherwise known) and information-theoretic measures that also use probability distributions. In Figure 2.8, the information-theoretic model of decision diagrams is given.

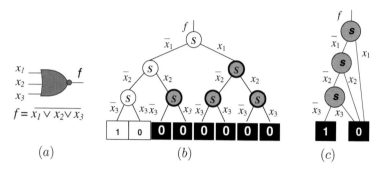

FIGURE 2.7

The 3-input NOR function (a), decision tree (b), and decision diagram (c) (Example 2.20).

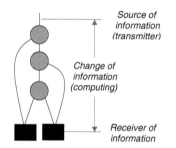

Source of information (transmitter)

Change of information (computing)

Receiver of information

The information-theoretic model includes:

▶ The source or transmitter of information, which corresponds to the input of the root node

▶ The channel where the information is transmitted. In a decision diagram, this corresponds to processing (changing) of information by nonterminal nodes

▶ The receiver of information, which corresponds to the terminal nodes of a decision diagram

FIGURE 2.8

Information-theoretic model of a decision diagram.

The topology of decision trees and diagrams can be changed by the manipulation of the algebraic representation or by direct manipulation of the graphical structure.

Example 2.21   *Manipulation of algebraic expression and graphical structure implies various topologies of decision diagrams (Figure 2.9). The topologies shown in Figure 2.9a,b,c are classified as planar topologies of decision diagrams (useful in the design of networks without crossings).*

Reduction of decision trees is aimed at simplification with respect to various criteria, in particular, reduction of the number of nonterminal nodes, satisfaction of topological constraints to meet the requirements of implementation, and reduction of the length of paths from root to terminal nodes.

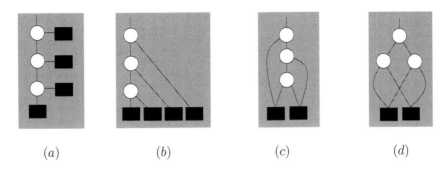

(a)   (b)   (c)   (d)

FIGURE 2.9
Various topologies of decision diagrams: (a) the topology is achieved by linearization of an algebraic expression of a Boolean function; (b) the topology is derived by linearization of an algebraic expression and manipulation (composition) of graphical elements; (c,d) the topologies are achieved by using various representations of Boolean functions and their symmetric properties (Example 2.21).

A path in the decision tree consists of the edges connecting a nonterminal node with a terminal node. A complete path connects the root node with a terminal node. Decision trees and diagrams are graphical representations of functional expressions. For instance, each complete path in a binary decision tree or diagram corresponds to a term in sum-of-products expression of a function f.

2.8   Spatial computational nanostructures

It has been justified in a number of papers that 3D topologies of computer elements are optimal, physically realistic models for scalable computers. In the contemporary logic network design, the use of the third dimension is motivated by decreasing of the interconnect topology. The third dimension is thought of as *layering*. For example, in chips, networks are typically assembled from about ten layers. In our study, the third dimension is considered not because of layers but as a dimension that relates to electrochemomechanical phenomena. It includes the space orientations, that is, the 3D relationship between molecules, as the key factor for achieving the desired functional logical properties.

2.8.1   Graph embedding problem

There are several documented approaches in computing in 3D. The first approach is based on the idea that 2D computational structures can be layered.   An example is the computer structures that are used in the contemporary logic network fabrication: logic network (for instance, a 2D silicon-based substrate) is layered and interconnected (between layers) to achieve the desired functionality. The interconnected layers form the third dimension. In this approach, the third dimension does not carry functional information about the implemented logic functions.   For example, this computational 3D model is not adequate to the natural 3D neural structure.

The second approach is based on the mapping of a logic function into a 3D system (a system with three coordinates) [3].   This approach requires complicated transformations of logic functions with respect to each dimension. The main drawback is that the known techniques of logic network design can not be used in the representation of logic functions in the 3D space. That is, each logic function requires a 3D expansion. This approach is nonacceptable in practice because it requires revision of the previously obtained techniques and tools. The third approach is related to the 3D cellular arrays and their particular case, 3D systolic arrays [29]. These structures are very complicated in design and control, and they are have not been completely implemented in practice. However, linear systolic arrays, which are reasonable for practice, can be used in the 3D computing structure design.

The following characteristics are critical in 3D computing: (a) embedding capabilities, (b) ability to extend the size of structure with minimal changes to the existing configuration, (c) ability to increase reliability and fault tolerance with minimal changes to the existing configuration, and (d) flexibility regarding technology-dependence and technology-independence.

There are two formulations of the embedding problem (Figure 2.10):

(*a*) Given a guest data structure, find an appropriate topology for its representation; this is a *direct* problem.

(*b*) Given a topological structure, find a corresponding data structure which is suitably represented by this topology; this is an *inverse* problem (using the properties of the host representation, specify possible data structures that satisfy certain criteria).

The direct and inverse problems of embedding address different problems of nanocomputing. In a direct problems, the topology is not specified, and the designer can map a representation of a logic function into spatial dimensions without strong topological limitations.   In this way, a high efficiency in embedding can be achieved. The inverse formulation of the problem assumes that the topology is constrained by technology. That is, the question is how to use a given topology in the efficient representation of logic functions.

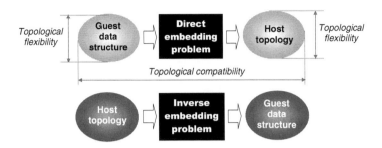

FIGURE 2.10
A formulation of direct and inverse embedding problems.

2.8.2   Embedding decision tree into spatial dimensions

There are a number of particular considerations for representing Boolean functions in nanospace:

(*a*) The Boolean functions have to be represented by a spatial data structure in which information about the function satisfies the requirements of the implementation technology.

(*b*) Information flow has to comply with the implementation topology.

(*c*) Data structure and information flow must be effectively embedded into the 3D topological structure.

In embedding problem, the initial (guest) structure must be specified with respect to computational abilities. A data structure becomes a computing network if the properties of data structures for the representation of Boolean functions are delegated. We utilize the fact that decision tree is topologically isomorphic to the H-tree, which can be embedded into a 3D hypercube. Hence, computational abilities are delegated to the H-tree. The 3D H-tree topology is constructed recursively from 2D elementary H-clusters. An H-tree embedded into a hypercube is called an N-hypercube topology.

The N-hypercube topology is the most suitable data structures for the representation of Boolean functions for the following reasons:

(*a*) They are 3D, and thus, meet the requirements of distributed spatial topology.

(*b*) They correspond to functional (Shannon expansion) and dataflow organization (information relation of variables and function values) requirements, as data structures such as decision trees can be embedded into hypercubes.

(*c*) They meet the requirements of certain nanotechnologies with local quantum effects and charge-state logic.

Example 2.22   *Given a Boolean function $f = \overline{x}_1 x_2 \vee x_1 \overline{x}_2 \vee x_1 x_2 x_3$. Its representation by a hypercube given in Figure 2.11a. Another representation is the hypercube-like structure obtained by embedding a decision tree of a function f into 3D H-tree (Figure 2.11(b)).*

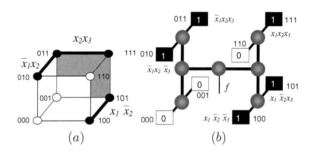

FIGURE 2.11

Representation of Boolean function $f = \overline{x}_1 x_2 \vee x_1 \overline{x}_2 \vee x_1 x_2 x_3$ by the classic hypercube (a) and hypercube-like structure (b) (Example 2.22).

2.9   Further study

Advanced topics of computational nanostructures

Topic 1: The fundamental theoretical sources are adopted from computer science, graph theory, artificial intelligence, and information theory. These techniques are the base of a 3D computing paradigm, in particular:

Graph theory provides techniques for delegation (embedding) of computing properties from 2D to 3D structures for the representation, manipulation, and optimization of logic functions using decision trees and diagrams. Graphical data structures become efficient tools in combination with techniques of logic function theory and information-theoretic approaches.

Logic function theory provides various techniques for manipulation of logic functions, including probabilistic Boolean and multivalued logic functions. These techniques address such problems as reliable computing using unreliable elements, fault-tolerant computing, and optimization techniques.

Information theory provides various techniques for manipulation and optimization of data structures using criteria of entropy; a topic of particular interest is the relationship of thermodynamic entropy (useful measure in molecular structures) and communication entropy (Shannon entropy).

Artificial intelligence provides the basis of neural network modeling.

Topic 2: The measure of computing structure. The five classic components of a computer structure are *input, output, memory, datapath,* and *control.* These are a technology-independent components. Description of this computer structure is based on a principle of abstraction, that is, each lower layer in the hierarchy of layers hiding details from the level above. The interface between the levels of abstraction is called the *instruction set.* The measure of computing structure is CPU execution time (in seconds):

$$\text{CPU execution time} = \underbrace{\frac{\text{Instructions}}{\text{Program}}}_{\text{Instruction count}} \times \underbrace{\frac{\text{Clock cycles}}{\text{Instruction}}}_{CPI} \times \underbrace{\frac{\text{Seconds}}{\text{Clock cycle}}}_{\text{Clock cycle time}}$$

where CPI is a clock cycles per instruction.

In computer structure, data are represented by bit patterns. These patterns can represent numbers (signed and unsigned integers, floating-point numbers), instructions, and other encoded parameters. Special instructions specify the meaning of the bit pattern.

Topic 3: Collision-based computing. The basic principle of collision-based computing is self-localizations [1, 56]. Truth values of logical variables are represented by the absence or presence of the traveling information quanta.

Further reading

[1] Adamatzky A. Computing with waves in chemical media: Massively parallel reaction-diffusion processors, *IEICE Trans. Electron.*, E87-C(11):1748-1756, 2004.

[2] Agraval V. An information theoretic approach to digital fault testing. *IEEE Trans. Comput.*, 30(8):582–587, 1981.

[3] Al-Rabady A and Perkowski M. Shannon and Davio sets of new lattice structures for logic synthesis in three-dimensional space. In *Proc. 5th Int. Workshop on Applications of the Reed–Muller Expansion in Circuit Design*, Mississippi State University, pp.165–184, 2001.

[4] Akers SB. Binary decision diagrams. *IEEE Trans. Comput.*, 27(6):509–516, 1978.

[5] Alexander MJ, Cohoon JP, Colflesh JL, Karro J, Peters EL, and Robins G. Placment and roting for three-dimensional FPGAs. In *Proc. 4th Canadian Workshop on Field-Programmable Devices*, pp. 11–18, 1996.

[6] Alspector J and Allen RB. A neuromorphic VLSI learning system. In Losleben P, Ed., *Advanced Research in VLSI: Proc. 1987 Stanford Conf.*, The MIT Press, Cambridge, MA, pp. 313–349, 1987.

[7] Ancona MG. Systolic processor design using single-electron digital circuits. *Superlattices and Microstructures*, 20(4):461–472, 1996.

[8] Asahi N, Akazawa M, and Amemiya Y. Single-electron logic device based on the binary decision diagram. *IEEE Trans. Electron Devices*, 44(7):1109–1116, 1997.

[9] Bachtold A, Hadley P, Nakanishi T, and Dekker C. Logic circuits with carbon nanotube transistors. *Science*, 294:1317–1320, 2001.

[10] Beiu V, Quintana JM, and Avedillo MJ. VLSI implementation of threshold logic – A comprehensive survay. *IEEE Neural Networks*, 14(5):1217–1243, 2003.

[11] Bryant RE. Graph-based algorithms for Boolean function manipulation. *IEEE Trans. Comput.*, 35(6):677–691, 1986.

[12] Brown BD and Card HC. Stochastic neural computing I: Computational elements. *IEEE Trans. Comput.*, 50(9):891–905, 2001.

[13] Chakradhar ST, Agrawal VD, and Bushnell ML. *Neural Models and Algorithms for Digital Testing*, Kluwer, Dordrecht, 1991.

[14] Crawley D, Nikolić K, Forshaw M, Ackermann J, Videlot C, Nguyen TN. Wang L, and Sarro PM. 3D molecular interconnection technology,   *J. Micromechanics and Microengineering*, 13(5):655–662, 2003.

[15] Ellenbogen JC and Love JC. Architectures for molecular electronic computers: Logic structures and an adder designed from molecular electronic diodes. *Proceedings IEEE*, 88(3):386–426, 2000.

[16] Evans WS and Schulman LJ. Signal propagation and noisy circuits. *IEEE Trans. Information Theory*, 45(7):2367–2373, 1999.

[17] Feringa BL, Ed., *Molecular Switches*, Wiley-VCH, Weinheim, Germany, 2001.

[18] Frank MP and Knight TF, Jr. Ultimate theoretical models of nanocomputers. *Nanotechnology*, 9:162–176, 1998.

[19] Frank MP. Approaching the physical limits of computing. In *Proc. 35th IEEE Int. Symp. Multiple-Valued Logic*, pp. 168–185, 2005.

[20] Gaines BR. Stochastic computing systems. In Tou JT, Ed., *Advances in Information Systems Science*, Plenum, New York, vol. 2, chap. 2, pp. 37–172, 1969.

[21] Gershenfeld N. Signal entropy and the thermodynamics of computation. *IBM Systems J.*, 35:577–586, 1996.

[22] Gerousis C, Goodnick SM, and Porod W. Toward nanoelectronic cellular neural networks. *Int. J. Circuits Theory Appl.* 28(6):523-535, 2000.

[23] Han J and Jonker P. A system archtecture solution for unreliable nanoelectronic devices. *IEEE Trans. Nanotechnology*, 1(4):201–208, 2002.

[24] Han J, Jonker P, Qi Y, and Fortes JAB. Toward hardware-redundant, fault-tolerant logic for nanoelectronics. *IEEE Design and Test Comput.*, July-August, pp. 328–339, 2005.

[25] Hopfield JJ. Neural networks and physical systems with emergent collective computational abilities. *Proceedings National Academy of Sciences*, 79:2554–2558, 1982.

[26] Hopfield JJ. Neurons with graded response have collective computational properties like those of two-state neurons. *Proceedings National Academy of Sciences*, 81(10):3088–3092, 1984.

[27] Hopfield JJ. Pattern recognition computation using action potential timing for stimulus representation. *Nature*, 376:3–36, 1995.

[28] Janer CL, Quero JM, Ortega JG, and Franquelo LG. Fully parallel stochastic computation architecture. *IEEE Trans. Signal Processing*, 44(8):2110–2117, 1996.

[29] Kung SY. *VLSI Array Processors*, Prentice Hall, Englewood Cliffs, NJ, 1988.

[30] McCulloch WS and Pitts WH. A logical calculus of ideas immanent in nervous activity. *Bulletin of Mathematical Biophysics*, 5:115–137, 1943.

[31] Majumdar A and Vrudhula SBK. Analysis of signal probability in logic circuits using stochastic models. *IEEE Trans. VLSI*, 1(3):365–379, 1993.

[32] Mead C. Neuromorphic electronic systems. *Proceedings IEEE*, 78(10):1629–1639, 1990.

[33] Minato S. *Binary Decision Diagrams and Applications for VLSI CAD*. Kluwer, Dordrecht, 1996.

[34] Moore EF and Shannon CE. Reliable circuits using less reliable relays. *J. Franklin Institute*, 262:191–208, Sept, 1956, and 262:281–297, Oct. 1956.

[35] Mullin AA. Stochastic combinational relay switching circuits and reliability. *IRE Trans. Circuit Theory*, 6(1):131–133, 1959.

[36] Negrini R and Sami MG. *Fault Tolerance Trough Reconfiguration in VLSI and WSI Arrays*. The MIT Press, Cambridge, MA, 1989.

[37] Palem KV. Energy aware computing through probabilistic switching: a study of limits. *IEEE Trans. Comput.*, 54(9):1123–1137, 2005.

[38] Parker KP and McCluskey EJ. Probabilistic treatment of general combinational networks. *IEEE Trans. Comput.*, 24(6):668–670, 1981.

[39] Peper F, Lee J, Abo F, Isokawa T, Adachi S, Matsui N, and Mashiko S. Fault-tolerance in nanocomputers: a cellular array approach. *IEEE Trans. Nanotechnology*, 3(1):187–201, 2004.

[40] Pierce WH. *Failure-Tolerant Computer Design*, Academic Press, San Diego, CA, 1965.

[41] Poole CP Jr. and Owens FJ. *Introduction to Nanotechnology*, John Wiley & Sons, New York, 2003.

[42] Reed MA and Tour JM. Computing with Molecules, *Scientific American*, pp. 86–93, June 2000.

[43] Sadek AS, Nikolić K, and Forshaw M. Parallel information and compuation with restriction for noise-tolerant nanoscale logic networks. *Nanotechnology*, 15:192–210, 2004.

[44] Shannon C. Reliable machines from unreliable components. *Notes by W. W. Peterson of Seminar at MIT*, March, 1956.

[45] Shmerko VP and Yanushkevich SN. Three-dimensional feedforward neural networks and their realization by nano-devices. *Artificial Intelligence Review*,

An Int. Science and Eng. J. (UK), Special Issue on *Artificial Intelligence in Logic Design*, 20(3-4):473–494, 2004.

[46] Siu KY, Roychowdhury VP, and Kailath T. Depth-size tradeoffs for neural computation. *IEEE Trans. Comput.*, 40(12):1402–1411, 1991.

[47] Shukla S and Bahar RI, Eds., *Nano, Quantum and Molecular Computing: Implications to High Level Design and Validation* Kluwer, Dordrecht, 2004.

[48] Tour JM, Zandt WLV, Husband CP, Husband SM, Wilson LS, Franzon PD, and Nackashi DP. Nanocell logic gates for molecular computing. *IEEE Trans. Nanotechnology*, 1(2):100–109, 2002.

[49] Tryon JG. Quadded logic. In [53], pp. 205–228.

[50] Von Neumann J. Probabilistic logics and the synthesis of reliable organisms from unreliable components. In Shannon CE and McCarthy J, Eds., *Automata Studies*, Princeton University Press, Princeton, NJ, pp. 43–98, 1956.

[51] Webster J, Ed., *Encyclopedia of Electrical and Electronics Engineering. Threshold Logic*, Vol. 22, John Wiley & Sons, New York, pp. 178–190, 1999.

[52] Wicker SW. *Error Control Systems for Digital Communication and Storage.* Prentice-Hall, New York, 1995.

[53] Wilcox RH and Mann WC, Eds., *Redundancy Techniques for Computing Systems.* Spartan Books, Washington, 1962.

[54] Winograd S and Cowan JD. *Reliable Computation in the Presence of Noise.* The MIT Press, Cambridge, MA, 1963.

[55] Yakovlev VV and Fedorov RF. *Stochastic Computing.* Engineering Industry Publishers, Moscow, 1974 (in Russian).

[56] Yamada K, Asai T, Hirose T, and Amemia Y. On digital LSI circuits exploiting collision-based fusion gates. *Int. J. of Unconventional Computing*, 4:45–59, 2007.

3

Binary Arithmetic

3.1 Introduction

The binary number system is the most important number system in digital design. This is because it is suited to the binary nature of the phenomena used in dominant microelectronic technology. Even in situations where the binary number system is not used as such, binary codes are employed to represent information at the signal level. However, other number systems can be useful in computing nanosystems. For example, 4,8,16, and 32 number systems can be used for memory devices. These systems address a multivalued logic. Multivalued logic values are often encoded using binary representations. However, humans prefer decimal numbers; that is, binary numbers must be converted into decimal numbers.

In this chapter, binary arithmetic is introduced. The binary number system is fundamental to computers and to digital electronics in general. This arithmetic operate with two values, 0s and 1s, and is the base of all computer devices. Binary arithmetic is used in computing nanodevices design because various chemical and physical phenomena can be encoded by 0s and 1s. Binary arithmetic can be also considered as a fundamental basis for hybrid computing devices design, in which the computing components are implemented using various technologies.

We introduce the techniques for the manipulation of binary numbers and show their relationship to other number systems such as decimal, hexadecimal, octal, and others. Binary data can be logically combined and computed by using theorems of Boolean algebra. Various number systems are examined that are used in digital data structures. These number systems, such as octal and hexadecimal, are used to simplify the manipulation of binary numbers.

3.2   Positional numbers

A number system is defined by its basic symbols, called *digits* or *numbers*, and the ways in which the digits can be combined to represent the full range of numbers we need. In a *positional* number system there is a finite set of digits. Each digit represents a nonnegative integer quantity. The number of distinct digits in the number system defines the *base* or *radix* of the number system. The numerical symbols are used to encode information. The encoding process can result in encoded information having desirable properties for the implementation.

3.2.1   The decimal system

The decimal number system is an example of a *positional* number system. The ten digits $0, 1, 2, \ldots, 9$ can be combined in various ways to represent any number. The fundamental method of constructing a number is to form a *sequence* or *string* of digits or *coefficients*

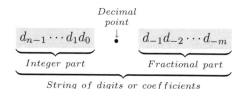

where integer and fractional parts are represented by n and m digits to the left and to the right of the *decimal point*, respectively. The subscript $i = -m, m-1, \ldots, 0, 1, \ldots, n$ gives the position of the digit. Depending on the position of digits in the string, each digit has an associated value of an integer raised to the power of 10 as follows:

$$
N = \underbrace{d_{n-1}d_{n-2}\cdots d_1 d_0 \ \overset{\substack{Decimal \\ point \\ \downarrow}}{\bullet} \ d_{-1}d_{-2}\cdots d_{-m}}_{String\ of\ coefficients}
$$

$$
= \underbrace{d_{n-1}\times 10^{n-1} + d_{n-2}\times 10^{n-2} + \cdots + d_1 \times 10^1 + d_0 \times 10^0}_{Computing\ the\ integer\ part}
$$

$$
= \underbrace{d_{-1}\times 10^{-1} + d_{-2}\times 10^{-2} + \cdots + d_{-m}\times 10^{-m}}_{Computing\ the\ fractional\ part} = \sum_{i=-m}^{n-1} d_i 10^i
$$

This method of representing numbers is called the *decimal system*. In the positional representation of digits: (a) each digit has a fixed value, or *weight*, determined by its position; (b) all the weights used in the decimal number system are powers of 10; (c) each decimal digit d_i ranges between 0 and 9; (d) the weighting of the digits is defined relative to the *decimal point*; this symbol means that digits to the left are weighted by positive powers of 10, giving integer values, while digits to the right are weighted by negative powers of 10, giving fractional values; and (e) fractions are denoted by sequences of digits whose weights are negative powers of 10.

Example 3.1   *The four digits in the number 2008 represent, from left to right, thousands (digit 2), hundreds (number 0), tens (number 0), and ones (number 8). Hence, this four-digit number can be represented in the following form:*

$$2008 = \sum_{i=0}^{3} 10^i \times d_i$$

$$= 2 \times 10^3 + 0 \times 10^2 + 0 \times 10^1 + 8 \times 10^0$$

The decimal number 747 in positional form is $747 = 7 \times 10^2 + 4 \times 10^1 + 7 \times 10^0.$

Example 3.2   *The decimal number 12.3456 consists of an integer part (12) and a fractional part (3456) separated by the decimal point. Thus, this number can be represented in the following form:*

$$12.3456 = \sum_{i=-4}^{1} 10^i \times d_i = \underbrace{1 \times 10^1 + 2 \times 10^0}_{Integer\ part}$$

$$+ \underbrace{3 \times 10^{-1} + 4 \times 10^{-2} + 5 \times 10^{-3} + 5 \times 10^{-4}}_{Fractional\ part}$$

The number 0.34_{10} *is represented as*

$$0.34_{10} = \sum_{i=-2}^{-1} 10^i \times d_i = 3 \times 10^{-1} + 4 \times 10^{-2} = {}^{34}/_{100}$$

3.2.2   Number radix

In general, an n-digit number in radix r consists of n digits, each taking one of r values: $\underbrace{0,\ 1,\ 2,\ \ldots,\ r-1.}_{Radix\ r\ system}$ A general number N in a positional number system is represented by the following formula:

$$\overbrace{\hspace{5cm}}^{n+m\ digits}$$

$$N = \underbrace{a_{n-1}a_{n-2}\cdots a_1 a_0}_{Integer\ part} \underset{\substack{\downarrow \\ Radix \\ point}}{\bullet} \underbrace{a_{-1}a_{-2}\cdots a_{-m}}_{Fractional\ part}$$

$$= \underbrace{a_{n-1} \times r^{n-1} + a_{n-2} \times r^{n-2} + \cdots + a_1 \times r^1 + a_0 \times r^0}_{Integer\ part}$$

$$+ \underbrace{a_{-1} \times r^{-1} + a_{-2} \times r^{-2} + \cdots + a_{-m} \times r^{-m}}_{Fractional\ part} = \sum_{i=-m}^{n-1} a_i r^i \qquad (3.1)$$

where a_i denotes a digit in the number system such that

$$0 \le a_i \le (r-1),$$

where r is the base of the number system, n is the number of digits in the integer part of N, and m is the number of digits in the fractional part of N. The integer part is separated from the fractional part by the *radix point*. The digits a_{n-1} and a_{-m} are referred to as the *most significant digits* (MSD) and the *least significant digits* (LSD) of the number N, respectively.

A number system is said to be of *base*, or *radix r*, because the digits are multiplied by powers of r, and this system uses r distinct digits.

Example 3.3   *The decimal number system is said to be of* base, *or radix, 10, because the digits are multiplied by powers of 10, and this system uses 10 distinct digits.*

To avoid possible confusion, the radix of the number system is often written as a decimal subscript appended to the number; that is, the subscript is placed after the LSD to indicate the radix of the number. When the context makes the radix obvious, it is not necessary to indicate the radix.

Example 3.4   *The radix of the number system is written as a decimal subscript as follows. The binary ($r = 2$) number 10110 can be written in the form 10110_2. The octal ($r = 8$) number 67344.25 is indicated in the form 67344.25_8. The decimal ($r = 10$) number 67390.845 is indicated in the form 67390.845_{10}.*

Equation 3.1 is used in number representation as follows: (a) choose the radix r of a number system; (b) choose the number of digits n in the integer part of N; (c) choose the number of digits m in the fractional part of N; (d) write the number N in the radix r number system.

Example 3.5 *In Table 3.1, the most useful number systems are listed. Observe that for those number systems of base less than 10, a subset of the digit symbols of the decimal number system is used. For example, maximal two digit numbers in various radix are as follows:* $11_2 = 3_{10}$, $22_3 = 8_{10}$, $33_4 = 15_{10}$, $77_8 = 63_{10}$, 99_{10}, $FF_{16} = 16 \times F + F$.

TABLE 3.1

The most important positional number systems used in data representation and computing (Example 3.5)

Base r	Number system	Digit symbols
2	Binary	0,1
3	Ternary	0,1,2
4	Quaternary	0,1,2,3
8	Octal	0,1,2,3,4,5,6,7
10	Decimal	0,1,2,3,4,5,6,7,8,9
16	Hexadecimal	0,1,2,3,4,5,6,7,8,9,A,B,C,D,E,F

Example 3.6 *Given* $n = 3$, *Equation 3.1 represents a number in hexadecimal, decimal, octal, and binary number systems by the following:*

$$N_{16} = \sum_{i=0}^{2} 16^i \times h_i = h_2 \times 16^2 + h_1 \times 16^1 + h_0 \times 16^0$$

$$N_{10} = \sum_{i=0}^{2} 10^i \times d_i = d_2 \times 10^2 + d_1 \times 10^1 + d_0 \times 10^0$$

$$N_8 = \sum_{i=0}^{2} 8^i \times o_i = o_2 \times 8^2 + o_1 \times 8^1 + o_0 \times 8^0$$

$$N_2 = \sum_{i=0}^{2} 2^i \times b_i = b_2 \times 2^2 + b_1 \times 2^1 + b_0 \times 2^0$$

The largest numbers that can be represented by three digits in these systems are $Max(N_{16}) = FFF_{16} = 4081_{10}$, $Max(N_{10}) = 999_{10}$, $Max(N_8) = 777_8 = 511_{10}$, *and* $Max(N_2) = 111_2 = 7_{10}$.

3.2.3 Fractional binary numbers

In the binary number positional representation, $B = \sum_{i=-n}^{m} 2^i \times b_i$, each binary digit or bit, b_i, is 0 or 1. The symbol "." becomes a *binary point*, separating bits on the left being weighted by positive powers of two, and those on the right being weighted by negative powers of two.

Example 3.7	*The binary number* 101.11_2 *is written as follows:*

$$101.11_2 = \sum_{i=-2}^{1} 2^i \times b_i$$

$$= \underbrace{1 \times 2^2 + 0 \times 2^1 + 1 \times 2^0}_{Integer\ part} + \underbrace{1 \times 2^{-1} + 1 \times 2^{-2}}_{Fractional\ part}$$

$$= 5\,{}^3/_4$$

Shifting the binary point one position to the left has the effect of dividing the number by two.

Example 3.8	*The binary number* 101.11_2 *represents the decimal number* $5\,{}^3/_4$. *Shifting the point one position to the left results in* 10.111_2, *that is,* $\overset{\leftarrow\,1}{101.11_2} = 10.111_2 = 2 + 0 + {}^1/_2 + {}^1/_4 + {}^1/_8 = 2\,{}^7/_8$.

Similarly, shifting the binary point one position to the right has the effect of multiplying the number by two. In computing, replacing arithmetic by shifts can occur when multiplying by constants.

Example 3.9	*The binary number* 101.11_2 *represents the decimal number* $5\,{}^3/_4$ *as follows:* $\overset{\rightarrow\,1}{101.11_2} = 1011.1_2 = (1 \times 2^2) + (0 \times 2^1) + (1 \times 2^0) + (1 \times 2^{-1}) + (1 \times 2^{-2}) = 5\,{}^3/_4$. *Shifting the point one position to the right results in* $1011.1_2 = 11\,{}^1/_2$, *that is,* $2^1) + (1 \times 2^0) + (1 \times 2^{-1}) = 8 + 0 + 2 + 1 + {}^1/_2 = 11\,{}^1/_2$.

Example 3.10	*Samples of integer and fractional parts of binary numbers and their corresponding decimal equivalents are given in Table 3.2.*

TABLE 3.2
Integer and fractional parts of binary numbers (Example 3.10)

Binary	Fraction (decimal)	Integer (decimal)
1.		$2^0 = 1$
10.		$2^1 = 2$
100.		$2^2 = 4$
1000.		$2^3 = 8$
10000.		$2^4 = 16$
100000.		$2^5 = 32$
0.1	$^1/_2$	$2^{-1} = 0.5$
0.01	$^1/_4$	$2^{-2} = 0.25$
0.001	$^1/_8$	$2^{-3} = 0.125$
0.0001	$^1/_{16}$	$2^{-4} = 0.0625$

3.2.4 Word size

In discrete systems, a *word* refers to a set of bits. *Word size* is defined as the number of bits in the binary numbers. Word size is typically a power of two and ranges from 8 bits, called a *byte*, to 32, 64, 128, or even 256 in some computers.

Example 3.11 *A 16-bit (two-byte) representation of a binary number with 8 bits for the integer part and 8 bits for the fractional part is shown below:*

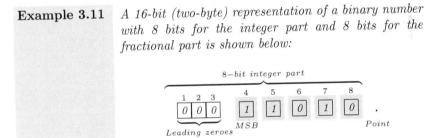

3.3 Counting in a positional number system

Counting is the fundamental operation in digital systems. A positional number system is well-suited for counting; that is, the counting is well automated and implemented in software and hardware.

Example 3.12 *In Table 3.3, the counting process is illustrated in various number systems in order to show regularities, which are useful for automation.*

TABLE 3.3
The first 16 integers in the binary, ternary, octal, decimal, and hexadecimal number systems (Example 3.12)

Binary	Ternary	Octal	Decimal	Hexadecimal
0	0	0	0	0
1	1	1	1	1
10	2	2	2	2
11	10	3	3	3
100	11	4	4	4
101	12	5	5	5
110	20	6	6	6
111	21	7	7	7
1000	22	10	8	8
1001	100	11	9	9
1010	101	12	10	A
1011	102	13	11	B
1100	110	14	12	C
1101	111	15	13	D
1110	112	16	14	E
1111	120	17	15	F

3.4 Basic arithmetic operations in various number systems

The four basic arithmetic operations: addition, subtraction, multiplication, and division can be performed in various positional number systems.

Example 3.13 *Addition and subtraction on the integer numbers in the binary, octal, decimal, and hexadecimal number systems are given in Table 3.4.*

3.5 Binary arithmetic

In the decimal system, the sign of a number is indicated by a special symbol, "+" or "−". In the binary system, the sign of a number is denoted by the

TABLE 3.4

Some arithmetic operations with unsigned numbers

Radix	Technique			
Binary ($r = 2$)	(15_{10})	$1\ 1\ 1\ 1_2$	(15_{10})	$1\ 1\ 1\ 1_2$
	(6_{10})	$+\ 0\ 1\ 1\ 0_2$	(6_{10})	$-\ 0\ 1\ 1\ 0_2$
	(21_{10})	$1\ 0\ 1\ 0\ 1_2$	(9_{10})	$1\ 0\ 0\ 1_2$
Octal ($r = 8$)	(15_{10})	$1\ 7_8$	(15_{10})	$1\ 7_8$
	(6_{10})	$+\ \ \ 6_8$	(6_{10})	$-\ \ \ 6_8$
	(21_{10})	$2\ 5_8$	(9_{10})	$1\ 1_8$
Decimal ($r = 10$)		$1\ 5_{10}$		$1\ 5_{10}$
	$+$	6_{10}	$-$	6_{10}
		$2\ 1_{10}$		9_{10}
Hexadecimal ($r = 16$)	(15_{10})	F_{16}	(15_{10})	F_{16}
	(6_{10})	$+\ 6_{16}$	(6_{10})	$-\ 6_{16}$
	(21_{10})	$1\ 5_{16}$	(9_{10})	9_{16}

left-most bit. Positive numbers are represented using the positional number representation. The so-called *unsigned* representation (magnitude only) is used to denote positive numbers.

Negative numbers can be represented in different ways. The most commonly used are sign-and-magnitude and complemented, which can be 1's complement or 2's complement notation (Figure 3.1).

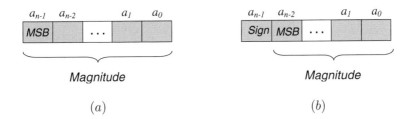

FIGURE 3.1

Unsigned number format (a) and sign-and-magnitude format (b).

Sign-and-magnitude is a two-point binary notation (one bit for sign and the rest for magnitude): NUMBER = <SIGN><MAGNITUDE> The *sign bit* is the leftmost bit, called the *most significant bit*, and it is equal to 0 for positive numbers and 1 for negative numbers.

While performing addition, the magnitudes are added, and the resulting sum is given the sign of the operands. If the operands have opposite signs, it is necessary to subtract the smaller number from the larger one (logic networks that compare and subtract numbers are needed). The range of signed integers is $-(2^{n-1}-1) \leq x \leq 2^{n-1}-1$. In such a system, zero has two representations: positive zero $00\ldots0$ and negative zero $10\ldots0$.

Example 3.14	*The 8-bit sign-and-magnitude representation of the number* 18_{10} *is*

$$18_{10} = \boxed{0} \overbrace{0010010}^{Sign \ Magnitude}$$

The 8-bit sign-and-magnitude form of the number -18_{10} *is*

$$-18_{10} = \boxed{1} \overbrace{0010010}^{Sign \ Magnitude}$$

Techniques for manipulation of binary numbers in sign-and-magnitude format are given in Table 3.5.

TABLE 3.5

Sign-and-magnitude techniques

Example		Technique
(3) 0 0 1 1 (2) + 0 0 1 0 ‾‾‾‾‾‾‾‾‾‾ (5) 0 1 0 1	(7) 0 1 1 1 (7) + 0 1 1 1 ‾‾‾‾‾‾‾‾‾‾ (−6) 1 1 1 0 Incorrect	If sign bits are both 0, perform addition of two binary numbers. If sum of the magnitudes is greater than $2^{n-1}-1$, the result is incorrect.
(+3) 0 0 1 1 (−2) + 1 0 1 0 ‾‾‾‾‾‾‾‾‾‾ (1) 0 0 0 1	(−3) 1 0 1 1 (+2) + 0 0 1 0 ‾‾‾‾‾‾‾‾‾‾ (−1) 1 0 0 1	If signs are different, compare the magnitudes. Subtract the smallest magnitude from the greater and assign to the result the sign of the greater magnitude.
(−3) 1 0 1 1 (−2) + 1 0 1 0 ‾‾‾‾‾‾‾‾‾‾ (−5) 1 1 0 1	(−7) 1 1 1 1 (−1) + 1 0 0 1 ‾‾‾‾‾‾‾‾‾‾ (0) 1 0 0 0 0 Incorrect	If sign bits are both 1, perform addition of both magnitudes; the sign bit of the result is 1. If sum of the magnitudes is greater than $2^{n-1}-1$, the result is incorrect.

3.6 Radix-complement representations

Consider the number D that consists of n digits d_i, $i = 1, 2, \ldots, n$ in the radix r number system. There are two types of *radix-complements* for representation of the number: the r radix-complement and the $r - 1$ radix-complement

$$\overline{D} = r^n - D \tag{3.2}$$

$$\overline{D} = (r^n - 1) - D \tag{3.3}$$

respectively.

> **Example 3.15** *Given the binary number $D = 101_2$ ($n = 3$), the 2's and 1's complements are $\overline{D} = 2^3 - 101 = 1000 - 101 = 011_2$ and $\overline{D} = (2^3 - 1) - 101 = 111 - 101 = 010_2$. Given the the decimal number $D = 125_{10}$ ($n = 3$), the 10's and 9's complements are $\overline{D} = 10^3 - 125 = 875_{10}$ and $\overline{D} = 10^3 - 1 - 125 = 874_{10}$.*

While the sign-and-magnitude system makes a number negative by changing its sign, a *complement number system* makes a number negative by taking its complement. Two numbers in a complement number system can be added or subtracted directly without the sign and magnitude checks required by the sign-and-magnitude system.

The *radix-complement* representation for decimal ($r = 10$) and binary ($r = 2$) number systems is shown in Figure 3.2.

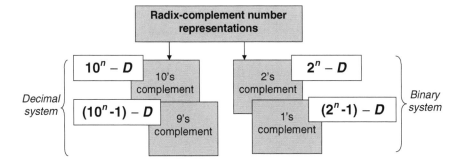

FIGURE 3.2
Radix-complement representations of decimal and binary numbers.

3.6.1   10's and 9's Complement systems

In 10's and 9's complement systems, positive numbers are represented using the same binary code as for unsigned numbers. Negative numbers are represented in a complement form.

10's Complement system

In the decimal number system, the radix-complement is called *10's complement*. The 10's complement of a decimal number is formed as follows:

Given: A decimal number $D = d_{n-1}d_n \ldots d_0$.
Step 1: Subtract each d_i from 9: $(9-d_{n-1}), (9-d_{n-2}), \ldots, (9-d_0)$.
Step 2: Add 1 to the resulting number.
Result: 10's complement representation.

Example 3.16   *Given the decimal number* $125 = d_2d_1d_0$, *its 10's complement is calculated as follows:*

$$\overline{125} = (9 - d_2)(9 - d_1)(9 - d_0) + 1$$
$$= (9 - 1)(9 - 2)(9 - 5) + 1 = 874 + 1 = 875,$$

which is the same result as the that obtained by the calculation in Example 3.15.

9's Complement

The 9's complement of a decimal number is formed as follows:

Given: A decimal number $D = d_{n-1}d_n \ldots d_0$.
Step 1: Subtract each d_i from 9: $(9-d_{n-1}), (9-d_{n-2}), \ldots, (9-d_0)$.
Result: 9's complement representation.

Example 3.17   *Given the decimal number 46, 125 and 5329, their 9's complements are calculated as follows:*

$$\overline{46} = (9 - d_1)(9 - d_0) = (9 - 4)(9 - 6) = 53$$
$$\overline{125} = (9 - d_2)(9 - d_1)(9 - d_0) = (9 - 1)(9 - 2)(9 - 5)$$
$$= 1874$$
$$\overline{5329} = (9 - d_3)(9 - d_2)(9 - d_1)(9 - d_0)$$
$$= (9 - 5)(9 - 3)(9 - 2)(9 - 9) = 4670$$

3.6.2 1's Complement system

In the 1's complement system, positive numbers are represented in the same way as unsigned numbers. Let a negative number $-P$ be given, where P is the magnitude of the number. An n-bit negative number K is obtained by subtracting the positive number P (magnitude) from $2^n - 1$:

$$K = (2^n - 1) - P \qquad (3.4)$$

An advantage of 1's complement representation is that a negative number is generated by complementing all bits of the corresponding positive number (magnitude). The addition of 1's complement numbers may require a correction, and the time needed to add two 1's complement numbers may be twice as long as the time needed to add two unsigned numbers. The 1's complement is formed as follows:

Given: A binary number $B = b_{n-1}b_n \ldots b_0$.
Step 1: Complement each digit of the binary number:

$$(1 - b_{n-1}), (1 - b_{n-2}), \ldots, (1 - b_0).$$

Step 2: Add 1 to the result.
Result: 1's complement representation.

Example 3.18 *An n-bit binary representation of numbers between $+7$ and -7 with 1's complement representation of negative numbers is given below.*

Binary number	Decimal number		Binary number	Decimal number
0111	+7		1111	−0
0110	+6		1110	−1
0101	+5		1101	−2
0100	+4		1100	−3
0011	+3		1011	−4
0010	+2		1010	−5
0001	+1		1001	−6
0000	+0		1000	−7

For example, the negative number -7_{10} is represented as a 4-bit number using the equation $K = (2^4 - 1) - 7 = 8_{10} = \boxed{1}\,000_2$. Alternatively, given $-7_{10} = 0111_2$, its complement is the desired number $-7_{10} = \overline{0111}_2 = \boxed{1}\,000_2$.

3.6.3   2's Complement

A code for representing an n-bit negative number K is obtained by subtracting its equivalent positive number P from 2^n, that is,

$$K = 2^n - P$$

An advantage of 2's complement representation is that when the numbers are added, the result is always correct. If there is a carry-out from the sign-bit position, it is simply ignored. The 2's complement is formed as follows:

> **Given:** The binary code for the magnitude of the negative
> number $B = b_{n-1}b_n \ldots b_0$.
> **Step 1:** Complement each digit of the code
>
> $$(1 - b_{n-1}), (1 - b_{n-2}), \ldots, (1 - b_0).$$
>
> **Result:** 2's complement representation.

Example 3.19   *Binary represen-tation of numbers between +7 and −7 with 2's complement representation of negative numbers is given below. Note that the binary codes of the positive numbers are represented exactly like the unsigned numbers.*

Binary number	Decimal number	Binary number	Decimal number
0111	+7	1111	−1
0110	+6	1110	−2
0101	+5	1101	−3
0100	+4	1100	−4
0011	+3	1011	−5
0010	+2	1010	−6
0001	+1	1001	−7
0000	0		

For example, the negative number -7_{10} is represented in 2's complement form as

$$-7_{10} = 2^4 - 7 = 9_{10} = 1001_2$$

Alternatively, $-7_{10} = \overline{0111}_2 + 1 = 1000_2 + 1 = 1001_2$.

Example 3.20   *Techniques for the addition of 2's complement binary numbers are demonstrated in Table 3.6.*

TABLE 3.6
Techniques for representation and addition of binary numbers:
unsigned, sign-and-magnitude, 1's and 2's complement

Techniques for computing binary numbers

Radix	Technique	
Unsigned	(3) 0 0 1 1 (2) + 0 0 1 0 (5) 0 1 0 1	(5) 0 1 0 1 (6) + 0 1 1 0 (11) 1 0 1 1
Sign-and-magnitude	(+7) 0 1 1 1 (−2) + 1 0 1 0 (+5) 0 1 0 1	(−5) 1 1 0 1 (−2) + 1 0 1 0 (−7) 1 1 1 1

1's complement
$K = (2^n - 1) - P$
$-5 = \underbrace{1111}_{2^4-1} - \underbrace{0101}_{P=5} = 1010$
$-9 = \underbrace{11111}_{2^5-1} - \underbrace{01001}_{P=9} = 10110$

(+5) 0 1 0 1
 +(−2) + 1 1 0 1
 (+3) $\boxed{1}$ 0 0 1 0
 + ⟶ 1
  0 0 1 1
	(−5) 1 0 1 0
	 +(−2) + 1 1 0 1
	 (−7) $\boxed{1}$ 0 1 1 1
	 + ⟶ 1
	  1 0 0 0

2's complement
$K = 2^n - P$

$-5 = \underbrace{1000}_{2^4} - \underbrace{0101}_{P=5} = 1011$
$-9 = \underbrace{10000}_{2^5} - \underbrace{01001}_{P=9} = 10111$

(+5) 0 1 0 1 +(−2) + 1 1 1 0 (+3) $\boxed{1}$ 0 0 1 1 ↑ *Ignore*	(−5) 1 0 1 1 +(−2) + 1 1 1 0 (−7) $\boxed{1}$ 1 0 0 1 ↑ *Ignore*

3.7   Conversion of numbers in various radices

Consider a radix r number that includes a radix point. First, the number must be separated into an integer part and a fractional part, since the parts must be converted differently. There are various algorithms for the conversion of numbers between systems.

Conversion of numbers from decimal to other radices

Given a decimal number, the conversion of this number to a radix r number is as follows. The conversion of a decimal integer to a radix r number is achieved by dividing the number and all successive quotients by r and accumulating the remainders.

The conversion of a fraction to a radix r fraction is accomplished by multiplying by radix r to give the integer and fraction; the new fraction is multiplied by r to give a new integer and a new fraction. This process is continued until the fractional part equals 0 or until there are enough digits to achieve sufficient accuracy.

Conversion of decimal integers to binary integers

For the conversion of decimal integers to binary integers, the following steps are needed:

> **Given:** A decimal integer.
> **Step 1:** Divide the decimal number by two to give a quotient and a remainder.
> **Step 2:** Divide the quotient by two to give a new quotient and a remainder.
> **Step 3:** Repeat division until the fractional part is 0 or until required number of digits is achieved.
> **Result:** The remainders are written bottom-up (from recent to the first obtained) of the desired binary number.

Example 3.21

$$50 : 2 = 25 + \boxed{0}$$
$$25 : 2 = 12 + \boxed{1}$$
$$12 : 2 = 6 + \boxed{0}$$
$$6 : 2 = 3 + \boxed{0}$$
$$3 : 2 = 1 + \boxed{1}$$
$$1 : 2 = 0 + \boxed{1}$$

Convert the decimal integer number 50_{10} into a binary number. Figure 3.3a illustrates the conversion. The sequence of remainders is 0,1,0,0,1,1. The result of conversion is obtained by reading the remainders in reverse order (from the bottom up in Figure 3.3): $50_{10} = 110010_2$.

Conversion of a decimal fraction to a binary fraction

For the conversion of a decimal fraction to a binary fraction, the following steps are needed:

> **Given:** A decimal fraction.
> **Step 1:** Multiply the fraction by 2.
> **Step 2:** Write down the obtained integer part. Multiply the fractional part of the obtained result by 2.
> **Step 3:** Repeat until 0 is obtained as a fractional part, or until the required accuracy is achieved.
> **Result:** A binary fraction.

Techniques for conversion of a decimal and binary numbers

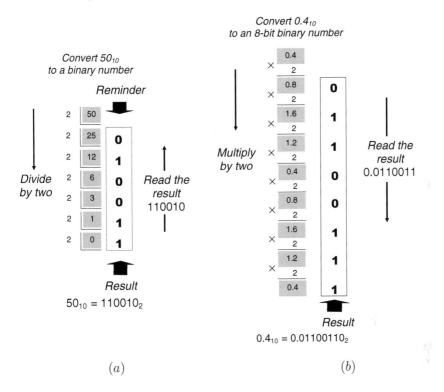

(a)

(b)

FIGURE 3.3

Conversion of a integer decimal number into a binary one (a) and conversion of a decimal fraction into a binary one (b) (Examples 3.21 and 3.22).

Example 3.22 *Convert the decimal fraction 0.4_{10} into a binary fraction. Represent the result using 8 bits. Figure 3.3b illustrates the conversion. The result of conversion is obtained by reading the remainders from the bottom-up, that is, $0.4_{10} = 01100110_2$.*

Conversion of decimal integers to octal integers

The following steps are needed to convert a decimal integer to an octal integer:

Conversion of decimal integer to octal integer

Given: A decimal integer.

Step 1: Divide the decimal number by 8 to give a quotient and a remainder.

Step 2: Divide the quotient by 8 to give new quotient and a remainder.

Step 3: Repeat division until the fractional part is 0 or until the required number of digits is achieved.

Result: The remainders written bottom-up (from recent to the first obtained) of the desired octal number.

Conversion of decimal integers to hexadecimal integers

The following steps are needed to convert decimal integer into hexadecimal integer:

Conversion of decimal integer to hexadecimal integer

Given: A decimal integer.

Step 1: Divide the decimal number by 16 to give a quotient and a remainder.

Step 2: Divide the quotient by 16 to give a new quotient and a remainder.

Step 3: Repeat division until the fractional part is 0 or until the required number of digits is achieved.

Result: The remainders written bottom-up (from most recent to first obtained) of the desired hexadecimal number.

A binary number can be converted to a decimal one via an intermediate system (octal or hexadecimal).

Techniques for conversion between binary, octal, and hexadecimal number systems are given in Table 3.7.

3.8 Overflow

The operations under consideration here are executed within the binary system as well other systems of restricted word length or number of digits. Below we will focus on the binary number system. Binary word size determines the range of the allowed number values of both operands and results.

TABLE 3.7
Techniques for conversion between binary, octal, and hexadecimal numbers systems

Techniques for conversion between numbers

Example	Comments
Binary to octal	
$1101001_2 = \underbrace{001}_{1}\ \underbrace{101}_{5}\ \underbrace{001}_{1} = 151_8$	Separate the bits into groups of three, starting from the right.
$10101.0101_2 = \underbrace{010}_{2}\ \underbrace{101}_{5}\ .\ \underbrace{010}_{2}\ \underbrace{100}_{4} = 25.24_8$	For the fractional part, start from the left.
Octal to binary	Replace each octal digit with a 3-bit string.
$457.24_8 = \underbrace{100}_{4}\ \underbrace{101}_{5}\ \underbrace{111}_{7}\ .\ \underbrace{010}_{2}\ \underbrace{100}_{4}$	For the fractional part, start from the left.
Binary to hexadecimal	
$101101101_2 = \underbrace{0001}_{1}\ \underbrace{0110}_{6}\ \underbrace{1101}_{D} = 16D_{16}$	Separate the bits into groups of four, starting from the right.
$10111.11_2 = \underbrace{0001}_{1}\ \underbrace{0111}_{7}\ .\ \underbrace{0011}_{3} = 17.3_{16}$	For the fractional part, start from the left.
Hexadecimal to binary	Replace each hexadecimal digit with the corresponding 4-bit string.
$5E4_{16} = \underbrace{0101}_{5}\ \underbrace{1110}_{E}\ \underbrace{0100}_{4}$	
$C2.A1_{16} = \underbrace{1100}_{C}\ \underbrace{0010}_{2}\ .\ \underbrace{1010}_{A}\ \underbrace{0001}_{1}$	For the fractional part, start replacing from the left.
Octal to hexadecimal	
$1352_8 = \underbrace{001}_{1}\ \underbrace{011}_{3}\ \underbrace{101}_{5}\ \underbrace{010}_{6}$	Separate the bits into groups of three, starting from the right, and then regroup into groups of four.
$= \underbrace{0010}_{2}\ \underbrace{1110}_{E}\ \underbrace{1010}_{A} = 2EA_{16}$	
Hexadecimal to octal	Replace each hexadecimal digit with a 4-bit binary string.
$2EA_{16} = \underbrace{0010}_{2}\ \underbrace{1110}_{E}\ \underbrace{1010}_{A}$	Separate the bits into groups of four, starting from the right, and then regroup into groups of three.
$= \underbrace{001}_{1}\ \underbrace{011}_{3}\ \underbrace{101}_{5}\ \underbrace{010}_{2} = 1352_8$	

Example 3.23   *Given a 4-bit word, the binary numbers that can be represented are varied: (a) from $-7_{10} = 1111_2$ to $+7_{10} = 0111_2$ for unsigned numbers; (b) from $-7_{10} = 1111_2$ to $+7_{10} = 0111_2$ in the sign-magnitude system; (c) from $-8_{10} = 1000_2$ to $+7_{10} = 0111_2$ for the 2's complement system.*

In particular, if an addition operation in the 2's complement system produces a result that does not fit the range -2^{n-1} to $2^{n-1} - 1$, then we say that *arithmetic overflow* has occurred. To ensure the correct operation, it is important to be able to detect the occurrence of overflow.

The following rules are used for detecting overflow:

Rule 1: An addition overflows if:

(a) The signs of the addends are the same and
(b) The sign of the sum is different from the addends' sign.

Rule 2: An addition overflows (Figure 3.4) if
the carry-in, C_{in}, and the carry-out of the most significant
bit in 2's complement representation are different, $C_{in} \neq C_{out}$.

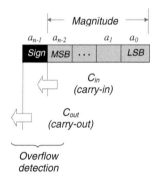

An addition overflows if the carry bit C_{in} into and carry bit C_{out} out of the sign position are different:

$$C_{in} \neq C_{out}$$

There is no overflow if

$$C_{in} = C_{out}$$

FIGURE 3.4
Overflow detection using carries.

Example 3.24   *Figure 3.5 illustrates four cases of overflow detection using carry bits C_{in} and C_{out} for the two 4-bit binary numbers. Note that in all cases the sign bit is included in computing the sum.*

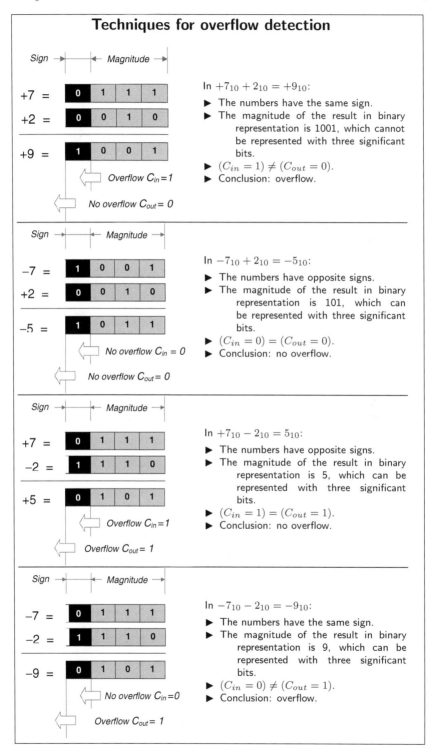

FIGURE 3.5

Overflow detection using carries C_{in} and C_{out}.

3.9   Implementation of binary arithmetic

Binary arithmetic is implemented by devices called *logic networks* for computing binary arithmetic, or arithmetic circuits. For example, *adder* and *multiplier* perform operations of addition and multiplication of binary numbers. Design of these devices is based on switching (Boolean) algebra.

Binary adders

The operands of addition are the *addend* and *augend*. The addend is added to the augend to form the sum. In most arithmetic circuits, the augmented operand (the augend) is replaced by the sum, whereas the addend is unchanged. An *n-bit binary adder* is a device that has two n-bit inputs $A = a_{n-1}, \ldots a_0$ and $B = b_{n-1}, \ldots b_0$ representing the operands A and B, respectively, and n-bit output $S = s_{n-1}, \ldots s_0$, and performs binary addition of the input operands. Additional input and output signals, *carry-in* C_i and *carry-out* C_{i+1} are used to implement module-based architecture, i.e., the design of larger adders. A device that performs addition at a single-bit position is the generic cell used not only to perform addition but also arithmetic multiplication and division.

Example 3.25   *In Figure 3.6, the problem of the 1-bit adder $(n = 1)$ design is introduced using symbolic representation (a), a truth table (b), and an algebraic description in terms of Boolean algebra (c). Such a device is referred to as a binary* **full-adder**. *A truth table defines an input–output relationship of inputs a_i, b_i, and c_i by listing all their combinations and corresponding outputs s_i and c_{i+1}. Each input combination is a separate row in the table. Algebraic description defines an this input–output relationship as a* **Boolean function**.

Details are given in further chapters.

3.10   Other binary codes

Specific binary codes are used in various tasks of logic design of discrete devices. Examples of such codes are the Gray code and binary-to-decimal code.

The implementation of 1-bit addition

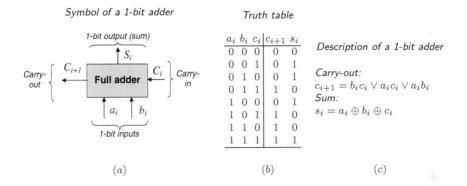

Symbol of a 1-bit adder

Truth table

a_i	b_i	c_i	c_{i+1}	s_i
0	0	0	0	0
0	0	1	0	1
0	1	0	0	1
0	1	1	1	0
1	0	0	0	1
1	0	1	1	0
1	1	0	1	0
1	1	1	1	1

Description of a 1-bit adder

Carry-out:
$c_{i+1} = b_i c_i \vee a_i c_i \vee a_i b_i$
Sum:
$s_i = a_i \oplus b_i \oplus c_i$

(a) (b) (c)

FIGURE 3.6
The 1-bit adder (full adder): symbolic representation (a), the truth table (b), and the formal description (Example 3.25).

3.10.1 Gray code

The *Gray code* is used for encoding the indexes of the nodes. There are several reasons to encode the indexes. The most important of them is to simplify analysis, synthesis, and embedding of topological structures. The Gray code is referred to as a *unit-distance* code. Let $b_n \ldots b_1 b_0$ be a binary representation of a positive integer number B and $g_n \ldots g_1 g_0$ be its Gray code. There is a relationship between the two codes: Binary code $b_n \ldots b_1 b_0 \iff$ Gray code $g_n \ldots g_1 g_0$ (Table 3.8).

TABLE 3.8
The relationship between binary and Gray code, $n = 3$

Binary code	Gray code	Gray code	Binary code
000	000	000	000
001	001	001	001
010	011	011	010
011	010	010	011
100	110	110	100
101	111	111	101
110	101	101	110
111	100	100	111

Suppose that $B = b_n \ldots b_1 b_0$ is given; then the i-th bit of the corresponding binary Gray code is generated as follows:

$$g_i = b_i \oplus b_{i+1} \tag{3.5}$$

where $b_{n+1} = 0$. Given the Gray code $G = g_n \ldots g_1 g_0$, the corresponding binary representation is derived as follows:

$$b_i = g_0 \oplus g_1 \oplus \ldots g_{n-i} = \bigoplus_{i=0}^{n-i} g_i \tag{3.6}$$

Table 3.8 illustrates the above transformation for $n = 3$.

Hamming distance

The *Hamming distance* is a useful measure in hypercube topology. The Hamming sum is defined as the bitwise operation of two codes, $g_{d-1} \ldots g_0$ and $g'_{d-1} \ldots g'_0$:

$$(g_{d-1} \ldots g_0) \oplus (g'_{d-1} \ldots g'_0) = (g_{d-1} \oplus g'_{d-1}), \ldots, (g_1 \oplus g'_1), (g_0 \oplus g'_0) \tag{3.7}$$

where $\oplus$ is an exclusive OR operation. If the sum is equal to 1, then the Hamming distance is said to be 1, that is, the codes G and G'.

In the hypercube, two nodes are connected by a link if and only if they have labels that differ by exactly one bit. The number of bits by which labels g_i and g_j differ is denoted by $h(g_i, g_j)$; this is the Hamming distance between the nodes.

Example 3.26   *The Hamming sum operation on two hypercubes for three-variable Boolean functions results in a 4D hypercube (Figure 3.7).*

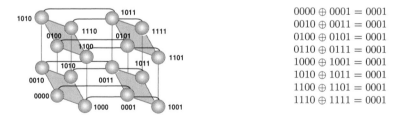

$$0000 \oplus 0001 = 0001$$
$$0010 \oplus 0011 = 0001$$
$$0100 \oplus 0101 = 0001$$
$$0110 \oplus 0111 = 0001$$
$$1000 \oplus 1001 = 0001$$
$$1010 \oplus 1011 = 0001$$
$$1100 \oplus 1101 = 0001$$
$$1110 \oplus 1111 = 0001$$

FIGURE 3.7

Hamming sum operations and corresponding products of variables of a 4D hypercube (Example 3.26).

3.10.2 Weighted codes

A 4-bit weighted code represents a decimal number as $w_3 a_3 + w_2 a_2 + w_1 a_1 + w_0 a_0$, where w_0, w_1, w_2, and w_3 are the weights of the code a_0, a_1, a_2, and a_3. In Table 3.9, several weighted binary codes for the decimal digits $0, 1, 2, \ldots 9$ are given.

TABLE 3.9
Weighted codes for decimal numbers

Decimal digit	8-4-2-1 code	6-3-1-1 code	Excess-3 code
0	0000	0000	0011
1	0001	0001	0100
2	0010	0011	0101
3	0011	0100	0110
4	0100	0101	0111
5	0101	0111	1000
6	0110	1000	1001
7	0111	1001	1010
8	1000	1011	1011
9	1001	1100	1100

Example 3.27 *Consider the codes $a_3 a_2 a_1 a_0 = 1000_{(8-4-2-1)}$ and $a_3 a_2 a_1 a_0 = 1011_{(6-3-1-1)}$ with weights $w_3 = 8, w_2 = 4, w_1 = 2, w_0 = 1$ and $w_3 = 6, w_2 = 3, w_1 = 1, w_0 = 1$, respectively. The corresponding decimal numbers are computed as follows:*

$$N_{(8-4-2-1)} = 8 \times a_3 + 4 \times a_2 + 2 \times a_1 + 1 \times a_0$$
$$= 8 \times 1 + 8 \times 0 + 8 \times 0 + 8 \times 0 = 8$$
$$N_{(6-3-1-1)} = 6 \times a_3 + 3 \times a_2 + 1 \times a_1 + 1 \times a_0$$
$$= 6 \times 1 + 3 \times 0 + 1 \times 1 + 1 \times 1 = 8$$

The Excess-3 code is obtained from 8-4-2-1 code by adding 3 (0011_2) to each group of four binary numbers of the corresponding 8-4-2-1 code.

Example 3.28   *Given the 8-4-2-1 code 1001 1000, its equivalent Excess-3 code is constructed as follows:*

$$
\begin{array}{ll}
1\ 0\ 0\ 1\ \ 1\ 0\ 0\ 0 & \longleftarrow\ 8-4-2-1\ code \\
+\ 0\ 0\ 1\ 1\ \ 0\ 0\ 1\ 1 & \longleftarrow\ adding\ 3 \\
\hline
1\ 1\ 0\ 0\ \ 1\ 0\ 1\ 1 & \longleftarrow\ Excess-3\ code
\end{array}
$$

Example 3.29   *Convert decimal numbers 28 into 8-4-2-1, 6-3-1-1, Excess-3, and Gray code:*

$$28_{(8-4-2-1)} = \boxed{0010}\ \boxed{1000}$$
$$\underbrace{}_{2}\ \underbrace{}_{8}$$

$$28_{(6-3-1-1)} = \boxed{0011}\ \boxed{1011}$$
$$\underbrace{}_{2}\ \underbrace{}_{8}$$

$$28_{(Excess-3)} = \boxed{0101}\ \boxed{1011}$$
$$\underbrace{}_{2}\ \underbrace{}_{8}$$

$$28_{(Gray)} = \boxed{0011}\ \boxed{1001}$$
$$\underbrace{}_{2}\ \underbrace{}_{8}$$

3.10.3   Binary-coded decimal

The 8-4-2-1 (BCD) code is example of weighted codes. Let two decimal digits (operands) are denoted by the binary codes $A = a_3a_2a_1a_0$ and $B = b_3b_2b_1b_0$. Note that 4-bit A and B cannot take values larger than 9 (e.g., 1010,1011,1100,1101,1110, and 1111). A carry-in and carry-out are denoted by C_0 and C_4. The data output (sum) is $S = s_3s_2s_1s_0$. This sum must be corrected. These cases are shown in Table in Figure 3.8, along with the circuit of BCD adder that consists of the binary adder and the correction network. The correction network implements addition of number 0110 SUM $Z = z_3z_2z_1z_0$ and carry-out c_{out}.

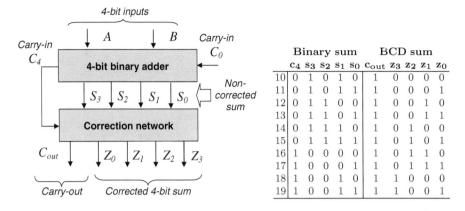

	Binary sum					BCD sum			
	c_4	s_3	s_2	s_1	s_0	c_{out}	z_3	z_2	z_1	z_0
10	0	1	0	1	0	1	0	0	0	0
11	0	1	0	1	1	1	0	0	0	1
12	0	1	1	0	0	1	0	0	1	0
13	0	1	1	0	1	1	0	0	1	1
14	0	1	1	1	0	1	0	1	0	0
15	0	1	1	1	1	1	0	1	0	1
16	1	0	0	0	0	1	0	1	1	0
17	1	0	0	0	1	1	0	1	1	1
18	1	0	0	1	0	1	1	0	0	0
19	1	0	0	1	1	1	1	0	0	1

FIGURE 3.8

Decade of BCD adder and the part of a truth table where correction is needed.

3.11   Further study

Advanced topics of number systems

Topic 1: Numbers and information. *Information* can be defined as recorded or communicated facts or data. Information takes various physical forms when being stored, communicated, or manipulated. Information in digital form is represented using a finite sequence of symbols. Processing digital information consists of forming a digit sequence or replacing the digits of a sequence with other digits. An *alphabet* is a finite collection of digits. Many alphabets can be used for representation of information. For example, an alphabet of ten digits is commonly employed to express numeric information. Because finite alphabets are used, a single digit can convey only a finite amount of information. Digits of a binary alphabet (0,1) convey a minimum of information.

Information theory provides a unit for measuring information and the mathematical means of computing the information content of a message. In this theory, the unit of information is defined as one *bit*. Each digit from a binary alphabet conveys exactly one bit of information. Each digit of an alphabet of k digits can convey as much as $\log_2 k$ bits of information. Sequences of digits are used to convey a greater amount of information that is provided by a single digit. The fundamental theorem of information theory is known as *Shannon's theorem*. This theorem predicts error-free data transmission in the presence of noise. Thus, the techniques of information theory are applied to the problems of the extraction of information from systems containing an element of randomness.

Topic 2: Floating-point number system is a numerical-representation system, defined by the IEEE 754 Standard. The main difficulty of fixed-point arithmetic (the word "fixed" refers to the fact that the radix point is placed at a fixed place in each number) is that the range of numbers that could be represented is limited. An alternative representation of numbers, known as the *floating-point format*, may be employed to eliminate the scaling factor problem. In this system, a string of digits (or bits) represents a real number. The radix point (decimal point, or binary point) can be placed anywhere relative to the digits within the string. This position is indicated separately in the internal representation. Numbers in float-point format consist of two parts: a *fraction* and an *exponent*. The advantage of floating-point representation is that it supports a much wider range of values than integers represented in computers using the *fixed point notation*. Unlike integers, which can represent exactly every number between the smallest and largest number, floating-point numbers are normally approximations for a numbers they cannot really represent.

The speed of performing floating point operations is used as performance measurement for computers. It is measured in "megaflops" (MFLOPs) (million floating-point operations per second), "gigaflops" (GFLOPs), "teraflops" (TFLOPs), etc. A floating-point operations are an addition, subtraction, multiplication, or division operation applied to a number in a single or double precision floating-point representation. Such data are used in scientific calculations. For example, the i860 processor (announced by Intel in 1989) was able to execute up to two floating-point operations and claimed to offer 100 MFLOPs.

Topic 3: The logarithmic number system is an arithmetic system used for representing real numbers. It was introduced as an alternative to the floating point number system. In the logarithmic number system, a number is represented by the logarithm of its absolute value.

Further reading

[1] Brown S and Vranesic Z. *Fundamentals of Digital Logic with VHDL Design*, McGraw-Hill, New York, 2000.

[2] Hayes JP. *Introduction to Digital Logic Design*, Addison-Wesley, Reading, MA, 1993.

[3] Koshy T. *Elementary Number Theory with Applications*. Harcourt/Academic Press, San Diego, CA, 2002.

[4] Marcovitz AB. *Introduction to Logic Design*, McGraw-Hill, Reading, MA, 2005.

[5] Muller JM, Scherbyna A, and Tisserand A. Semi-logarithmic number systems. *IEEE Trans. Comput.*, 47(2):145–151, February 1998.

[6] Shannon C. A Mathematical Theory of Communication. *Bell Systems Technical J.*, 27:379–423, 623–656, 1948.

[7] Swartzlander EE and Alexopoulos AG. The sign/logarithm number system. *IEEE Trans. Comput.*, 24:1238-1242, December 1975.

4

Residue Arithmetic

4.1 Introduction

The hardware implementation of an arithmetic algorithm is largely affected by a choice of a specific numbering system. In this chapter, the residue number system (RNS) is introduced. The attractive properties of RNS include carry-free, fault isolating, and modular characteristics, and are widely used, in particular, in high-speed digital signal processing. The most attractive property of RNS is that there is no carry propagation inside the set.

In RNS-based system, conversion procedures, from conventional binary representation to residue format and vise versa, are used:

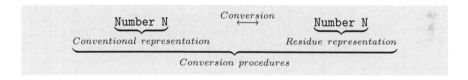

4.2 The basics of residue arithmetic

Residue arithmetic offers the alternative number format. An arithmetic operation performed on n-bit numbers may produce a result that is too long to be represented completely by n bits; that is, overflow occurs. The RNS has the property that i-th residue digit of sum, difference, and product is exclusively dependent on the i-th digits of the operands. This property determines that parallel operations can be performed on all residue digits.

In residue arithmetic, the results of all arithmetic operations are confined to some fixed set of m values such as $0, 1, \ldots, m-1$. Residue arithmetic ensures a finite word size in computing and is widely used in computing devices.

Figure 4.1 represents the correspondence between the ordinary and RNS computations.

A *residue* is defined as the remainder after a division. Given the

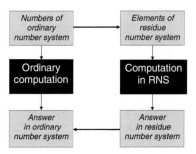

FIGURE 4.1

Correspondence between the ordinary and RNS computations.

representation of an integer N,

$$N = Im + r$$

where m is a check base and I is an integer, so that $0 \leq r \leq m$, and N is said to be equivalent to modulo m to r:

$$r \equiv N \ (mod \ m)$$

The numbers a and b are said to be *equivalent to modulo m* if the remainder obtained when a is divided by m is the same as the remainder that is obtained when b is divided by m:

$$a \equiv b \ (mod \ m)$$

(read as "a is congruent b modulo m").

Example 4.1 *If $a = 10$ and $b = 18$, then $10 \equiv 18 \ (mod \ 8)$ since*

$$10 = 1 \times 8 + \boxed{2} \qquad and \quad 18 = 2 \times 8 + \boxed{2}$$

(Remainder) ... *(Remainder)*

Example 4.2 *(a) $3 \equiv 17 \ (mod \ 7)$ because $3 - 17 = -14$ is divisible by 7. (b) $-2 \equiv 13 \ (mod \ 3)$ because $-2 - 13 = -15$ is divisible by 3. (c) $60 \equiv 10 \ (mod \ 25)$ because $60 - 10 = -50$ is divisible by 25. (d) $-4 \equiv -49 \ (mod \ 9)$ because $-4 - (-49) = 45$ is divisible by 9.*

It follows from these examples that an integer a is congruent to 0, $a \equiv 0 \ (mod \ m)$, if and only if it is divisible by m. Additional properties of congruence are:

▶ $a \equiv a \ (mod \ m)$ (reflexive property).

▶ If $a \equiv b \ (mod \ m)$, then $b \equiv a \ (mod \ m)$ (symmetry property).

▶ If $a \equiv b \ (mod \ m)$ and $b \equiv c \ (mod \ m)$, then $a \equiv c \ (mod \ m)$ (transitive property).

4.3 Addition in residue arithmetic

If $a \equiv b \ (mod \ m)$ and $c \equiv d \ (mod \ m)$, then (Figure 4.2)

$$a + c \equiv b + d \ (mod \ m)$$
$$a - c \equiv b - d \ (mod \ m)$$

Example 4.3 *Find the sum* $1017 + 2876 \ (mod \ 7)$. *There are two ways to calculate the sum. First approach:* $1017 + 2876 = 3893 \equiv 1 \ (mod \ 7)$. *Second approach:*

$$1017 \equiv 2 \ (mod \ 7) \quad and \quad 2876 \equiv 6 \ (mod \ 7)$$
$$1017 + 2876 \equiv 2 + 6 = 8 \equiv 1 \ (mod \ 7)$$

The second approach is preferable because it keeps the numbers involved small.
The sum of two congruences $13 \equiv 4 \ (mod \ 9)$ *and* $16 \equiv -2 \ (mod \ 9)$ *is computed as follows:* $13 + 16 \equiv 4 - 2 = 29 \equiv 2 \ (mod \ 9)$.

4.4 Multiplication in residue arithmetic

If $a \equiv b \ (mod \ m)$ and $c \equiv d \ (mod \ m)$, then (Figure 4.2)

$$ac \equiv bd \ (mod \ m)$$

Example 4.4 *Find the product* $1017 \times 2876 \ (mod \ 7)$. *Solution:* $1017 \equiv 2 \ (mod \ 7)$, $2876 \equiv 6 \ (mod \ 7)$, *and* $1017 \times 2876 \equiv 2 \times 6 = 12 \equiv 5 \ (mod \ 7)$. *The multiplication of two congruences* $13 \equiv 4 \ (mod \ 9)$ *and* $16 \equiv -2 \ (mod \ 9)$ *is computed as follows:* $13 \times 16 \equiv 4 \times (-2) = 208 \equiv -8 \ (mod \ 9)$.

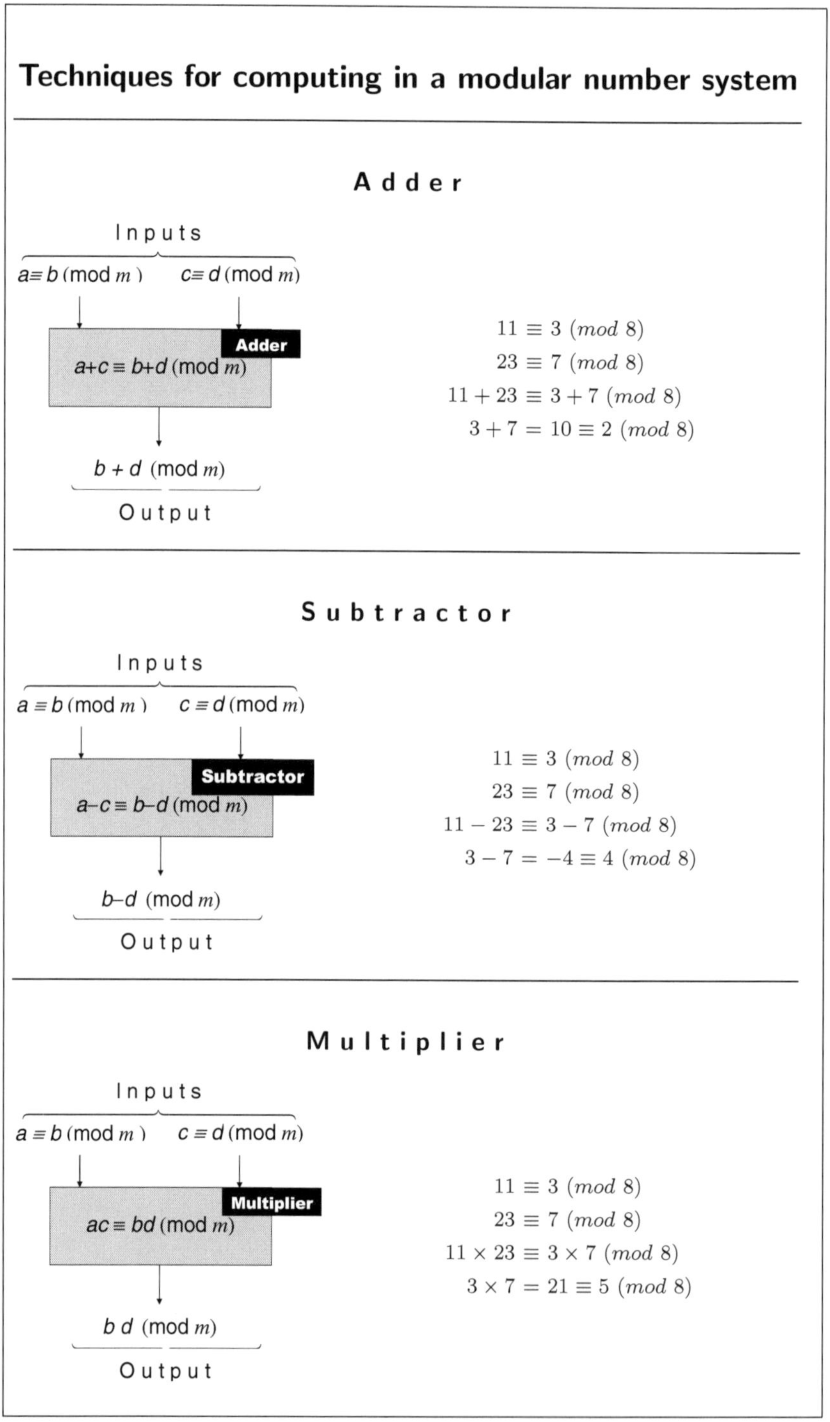

Techniques for computing in a modular number system

A d d e r

Inputs

$a \equiv b \,(\text{mod } m)$　　$c \equiv d \,(\text{mod } m)$

Adder

$a{+}c \equiv b{+}d \,(\text{mod } m)$

$b + d \,(\text{mod } m)$

Output

$$11 \equiv 3 \; (mod \; 8)$$
$$23 \equiv 7 \; (mod \; 8)$$
$$11 + 23 \equiv 3 + 7 \; (mod \; 8)$$
$$3 + 7 = 10 \equiv 2 \; (mod \; 8)$$

S u b t r a c t o r

Inputs

$a \equiv b \,(\text{mod } m)$　　$c \equiv d \,(\text{mod } m)$

Subtractor

$a{-}c \equiv b{-}d \,(\text{mod } m)$

$b{-}d \,(\text{mod } m)$

Output

$$11 \equiv 3 \; (mod \; 8)$$
$$23 \equiv 7 \; (mod \; 8)$$
$$11 - 23 \equiv 3 - 7 \; (mod \; 8)$$
$$3 - 7 = -4 \equiv 4 \; (mod \; 8)$$

M u l t i p l i e r

Inputs

$a \equiv b \,(\text{mod } m)$　　$c \equiv d \,(\text{mod } m)$

Multiplier

$ac \equiv bd \,(\text{mod } m)$

$b\,d \,(\text{mod } m)$

Output

$$11 \equiv 3 \; (mod \; 8)$$
$$23 \equiv 7 \; (mod \; 8)$$
$$11 \times 23 \equiv 3 \times 7 \; (mod \; 8)$$
$$3 \times 7 = 21 \equiv 5 \; (mod \; 8)$$

FIGURE 4.2
Modular adder, subtractor, and multiplier.

Example 4.5   *Addition and multiplication modulo 5 of two residues are given in the following tables:*

ADDITION   SUBTRACTION   MULTIPLICATION

$a + c \equiv b + d \pmod 5$ $a - c \equiv b - d \pmod 5$ $ac \equiv bd \pmod m$

+	0 1 2 3 4
0	0 1 2 3 4
1	1 2 3 4 0
2	2 3 4 0 1
3	3 4 0 1 2
4	4 0 1 2 3

−	0 1 2 3 4
0	0 4 3 2 1
1	1 0 4 3 2
2	2 1 0 4 3
3	3 2 1 0 4
4	4 3 2 1 0

×	0 1 2 3 4
0	0 0 0 0 0
1	0 1 2 3 4
2	0 2 4 1 3
3	0 3 1 4 2
4	0 4 3 2 1

For example, the residue 3 subtracted from the residue 1 modulo 5 is calculated as $1 - 3$ equals -2, and $-2 \equiv 3 \pmod 5$, so the difference is the residue 3. The product of the two residues 3 and 4 modulo 5 is the residue 2 because $3 \times 4 \equiv 2 \pmod 5$.

4.5   Computing powers in residue arithmetic

Computing powers of an integer modulo a natural number n is often just a mental rather than calculator exercise. The basic rule is as follows: if $a \equiv b \pmod m$, then $a^n \equiv b^n \pmod m$ for every positive integer n.

Example 4.6   *Find the products $1017^2 \pmod 7$, $1017^3 \pmod 7$, $1017^4 \pmod 7$, $1017^5 \pmod 7$, $1017^{12} \pmod 7$. Since $1017 \equiv 2 \pmod 7$,*

$$1017^2 \equiv 2^2 \pmod 7$$
$$1017^3 = 1017^2 \times 1017 \equiv 4 \times 2 = 8 \equiv 1 \pmod 7$$
$$1017^4 = 1017^3 \times 1017 \equiv 1 \times 2 = 2 \pmod 7$$
$$1017^5 = 1017^4 \times 1017 \equiv 2 \times 2 = 4 \pmod 7$$
$$1017^{12} = ((1017^4))^3 \equiv 2^3 = 8 \equiv 1 \pmod 7$$

The following property of a residue arithmetic is useful, for example, in electronic cash systems. The quadratic residues of a set modulo n are the elements that have a square root in the set.

Example 4.7   *For $n = 11$, the number 4 is a quadratic residue because 2^2 (mod 11) $= 4$. The number 5 is also a quadratic residue because 7^2 (mod 11) $= 5$. If n is prime, then there are $(n-1)/2$ quadratic residues. In the case of $n = 11$, the residues are:*

$$1^2 = 1 \ (mod \ 11) \qquad\qquad 6^2 = 3 \ (mod \ 11)$$
$$2^2 = 4 \ (mod \ 11) \qquad\qquad 7^2 = 5 \ (mod \ 11)$$
$$3^2 = 9 \ (mod \ 11) \qquad\qquad 8^2 = 9 \ (mod \ 11)$$
$$4^2 = 5 \ (mod \ 11) \qquad\qquad 9^2 = 4 \ (mod \ 11)$$
$$5^2 = 3 \ (mod \ 11) \qquad\qquad 10^2 = 1 \ (mod \ 11)$$

Each quadratic residue has two square roots if n is prime. One of them is smaller than $n/2$. The other is larger.

4.6   Solving modular equations

To solve a congruence or a system of congruences involving one or more unknowns means to find all possible values of the unknowns which make the congruence true.

Example 4.8   *Solution of the congruence $3x \equiv 1$ (mod 5) can be found by trying all possible values of x modulo 5:*

$$\text{IF } x = 0, \text{ THEN } \ 3x = 0 \equiv 0 \ (mod \ 5)$$
$$\text{IF } x = 1, \text{ THEN } \ 3x = 3 \equiv 3 \ (mod \ 5)$$
$$\text{IF } x = 2, \text{ THEN } \ 3x = 6 \equiv 1 \ (mod \ 5)$$
$$\text{IF } x = 3, \text{ THEN } \ 3x = 9 \equiv 4 \ (mod \ 5)$$
$$\text{IF } x = 4, \text{ THEN } \ 3x = 12 \equiv 2 \ (mod \ 5)$$

Since the modulus is 5, the integer x is in the range $0 \leq x \leq 5$. Thus, the only solution to the congruence is $x = 2$.

4.7   Complete residue systems

In residue arithmetic, an integer is represented as a set of residues with respect to a set of relatively prime integers called *moduli*. Residue arithmetic is defined in terms of a set of relatively prime moduli $\{r_1, r_2, \ldots, r_s\}$, where the greatest common divisor is equal to 1 for each pair of moduli. The set of integers $\{r_1, r_2, \ldots, r_s\}$ is called a *complete residue system modulo m* if $r_i \not\equiv r_j$ (mod m) whenever $i \neq j$, and for each integer n exists a corresponding r_i such that $n \equiv r_i$ (mod m).

Example 4.9 *The sets $\{1, 2, 3\}$, $\{-1, 0, 1\}$, and $\{1, 7, 9\}$ are all complete residue systems modulo 3. The set $\{0, 1, 2, 3, 4, 5\}$ is a complete residue system modulo 6. It should be noted that this set can be reduced to $\{1, 5\}$.*

While in ordinary arithmetic there is an infinite number of integers $0, 1, 2, \ldots$, in modular arithmetic there is essentially only a finite number of integers.

Example 4.10 *Given the arithmetic 8, there are only eight distinct integers $0, 1, 2, \ldots, 7$. Any other integer is congruent to one of these eight integers. For example, 8 is congruent to 0 (mod 8), $9 \equiv 1$ (mod 8), etc. Thus addition, subtraction, multiplication, and division may be used to yield*

$$\begin{aligned}
\text{ADDITION:} \quad & 3 + 7 \equiv 2 && (mod\ 8) \\
\text{SUBTRACTION:} \quad & 3 - 4 \equiv -1 \equiv 7 && (mod\ 8) \\
\text{MULTIPLICATION:} \quad & 3 \cdot 7 \equiv 5 && (mod\ 8) \\
\text{DIVISION:} \quad & 3 : 7 \equiv 5 && (mod\ 8)
\end{aligned}$$

The choice of moduli sets and the conversion of the residue to binary numbers are important issues in residue arithmetic. Intergers 2^r and $2^r \pm 1$ are commonly used as moduli for RNS, because the additions modulo 2^r or $2^r \pm 1$ can be implemented by r-bit binary adders.

An example of the application of a residue number system is public key cryptography. Two keys are used in the encryption and decryption of messages: one that must be kept secret and one that may be made public. These two keys are related mathematically by a so-called "one-way" function. A one-way function is easy to compute in one direction but very hard (computationally infeasible) to compute in the other direction. An example of a one-way function is multiplication versus factorization. It is simple to multiply two large prime numbers, but very hard to factor the result.

4.8 Further study

Advanced topics of number systems

Cryptography is the study of ways in which messages can be coded so that a third party, intercepting the code, will have great difficulty recovering the original text. *Cryptology* consists of *cryptography* and *cryptanalysis*. Cryptanalysis deals with breaking secret messages.

Advanced *encryption standards* such as RSA (Rivest–Shamir–Adleman) and AES (Advanced Encryption Standard) are based on number theory, encoding, and logic design techniques. Special hardware devices to implement these encryption algorithms have been manufactured.

The RSA encryption system relies upon the mathematics of modulo arithmetic. Both encryption and decryption are completed by raising to a power modulo a number which is the product of two large primes. The two primes are kept secret and the system can be broken if the two primes are recovered by factoring. The factoring is a process that has proven to be extremely difficult.

To encode a message using RSA, a user needs to create a *public* and *secret key* (Figure 4.3). They are chosen through several steps:

Step 1: Two large primes, p and q, are chosen (200–1000 bits). These numbers are chosen at random.

Step 2: The primes are multiplied together to yield $n = p \times q$. This is often at least 512 bit in practice.

Step 3: The secret key e is chosen. The greatest common denominator of e and $(p-1)(q-1)$ should be 1.

Step 4: The public key, d, is the inverse of $d \bmod (p-1)(q-1)$.

Step 5: The secret key is the pair of values n and e.

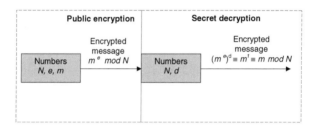

FIGURE 4.3

Using residue arithmetic in public key encryption.

Encryption is as follow. A message is converted into a number m less that n and decrypted by computing $(m^d \times mod\ n)^e \times mod\ n$. Specifically, this decryption works as follows:

$$(m^d \times mod\ n)^e \times mod\ n = m^{d^e} \times mod\ n$$
$$= m^{de} \times mod\ n = m \times mod\ n$$

Typical applications of cryptography is a *digital signature*. Digital signature is a message-dependent quantity that can be computed only by the sender of the message on the basis of some private information. It allows authentification of messages by guaranteeing that no one can forge the sender's signature and the sender cannot deny a message he sent. Applications requiring such signatures include business transactions, for example, between a bank and its customers. Digital signatures can also be used in environments where the interests of the parties involved are not necessary conflicting.

In [2], RNS floating-point arithmetic (addition, subtraction, multiplication, division, and square root) for an interval number is discussed with the goal to achieve reliable computation when hardware representations of numbers have inadequate precision. In [8], the residue-to-binary number converters for the RNS $\{2^n - 1, 2^n, 2^n + 1\}$ were designed using $2n$-bit or n-bit adders, that are twice as fast as generic ones, and achieve improvement in area and dynamic range as well. Particular modulo arithmetic involves Galois fields. Addition and multiplication in Galois fields, $GF(2^n)$ plays an important role in coding theory and is widely used in digital computers and data transmission or storage systems. The group theory is used to introduce algebraic system, called a *field*. A field is a set of elements in which we can do addition, subtraction, multiplication and division without leaving the set.

Further reading

[1] Harata Y, Nakamura Y, Nagase H, Takigawa M, and Takagi N. A High-speed multiplier using a redundant binary adder tree. *IEEE J. Solid States Circuits*, 22(1):28–34, 1987.

[2] Kinoshita E and Lee Ki-Ja. A residue arithmetic extension for reliable scientific computation. *IEEE Trans. Comput.*, 46(2):129-138, 1997.

[3] Koshy T. *Elementary Number Theory with Applications*. Harcourt/Academic Press, San Diego, CA, 2002.

[4] Menezes AJ, van Oorschot PC, and Vanstone SA. *Handbook of Applied Cryptography*. CRC Press, Boca Raton, FL, 1996.

[5] Paliouras V and Stouraitis T. Novel high-radix residue number system processors. *IEEE Trans. Circuits and Systems - Part II*, 47(10):1059–1073, October 2000.

[6] Shimabukuru K and Zukeran C. Reconfigurable current-mode multiple-valued residue arithmetic circuits. In *Proc. of the 28th IEEE Int. Sym. on Multiple-Valued Logic,*, pp. 282–287, 1998.

[7] Vinnakota B. and Rao VVB. Fast conversion technique for binary-residue number systems. *IEEE Trans. Circuits and Systems - I*, 41(12):927–929, December 1994.

[8] Wang Y. Residue-to-binary converters based on new Chinese remainder theorems. *IEEE Trans. Circuits and Systems - II*, 47(3):197–205, March 2000.

[9] Wei S and Shimizu K. Residue arithmetic circuits based on the signed-digit multiple-valued arithmetic circuits. In *Proc. of the 28th IEEE Int. Sym. on Multiple-Valued Logic*, pp. 276–281, 1998.

[10] Zimmermann R.,*Computer Arithmetic: Principles, Architectures, and VLSI Design*, Lecture notes, Integrated System Laboratory, ETH, Zürich, 1997.

5

Graph-Based Data Structures

5.1 Introduction

Theoretically, any problem associated with a finite graph is solvable. Graphical data structures are one of the ways for specification of logic functions. Graphs show the relationship between the variables and the logic function. The manipulation of the variables and terms of a function is replaced by the manipulation of graphical components (nodes, links, and paths).

5.2 Graphs in discrete device and system design

A logic network is modeled using an abstraction, which shows the relevant features without their associated details. In the design of digital devices, there are three levels of abstraction.

The *architectural* level, which represents a set of devices performing a set of operations. Graphs such as block diagrams are used at this level for viewing such features as parallelism of processing, data transfer, and position of devices.

The *logical* level shows how a network evaluates a set of logic functions. Graphical data structures such as directed graphs, or networks of nodes representing gates, are used at this level for representation, manipulation, and computing.

The *physical*, or *geometric*, level, which represents a network as a set of geometric entities. Graphical data structures such as placement and routing diagrams are used for a representation of components that is suitable for implementation of the available technology.

5.2.1 Graphs at the logical level

The use of graphs in logic design implies compliance with graph theory. Graph theory includes the classification of graphs, their properties, and techniques for their manipulation. The type of graph used in logic design depends on the specific problem; in particular: (a) if the direction of dataflow is important, *directed* graphs are used; (b) if data are transmitted using operations, bipartite graphs are appropriate; and (c) if a problem is represented by a decision table, trees and diagrams are used.

The latter graphical data structure - decision trees and diagrams, - can be used at almost any level of logic design. It is well suited to represent logic functions since any function that takes several values depending on its arguments (variables) can be represented by a graph called a *decision tree*.

Various algebraic forms of Boolean functions correspond to specific types of decision trees and diagrams. While the manipulation of Boolean functions in algebraic form is based on mathematical relations and theorems, their graphical equivalents are represented in the form of topological objects such as sets of nodes, levels, subtrees, and shapes, as well as in the form of functional characteristics such as distribution of nodes, weights, balance, and homogeneity. Various problems of logic analysis and synthesis have found an efficient solutions through the manipulation of these graphical objects.

Consider the general graph-based representation of a function. Information about the function is carried by the following graph components (Figure 5.1). *Intermediate* nodes operate on data, and *terminal* nodes indicate the values of computing functions. *Edges* or *links*, between nodes that are associated with data flow. *Topology* (local and global), which specifies the geometric properties and spatial relations of these nodes and edges, and, therefore, is unaffected by changes in the shape or size of the graph.

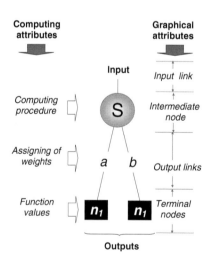

▶ Intermediate nodes indicate the computing procedures

▶ Links show the relationships between variables and computing procedures in intermediate nodes; links can be also labeled, for example, indices of variables or their weights

▶ Terminal nodes are assigned the resulting computed values

FIGURE 5.1

The elements of a graph representing a function.

5.2.2   Graphs at the physical design level

Physical design, or geometric-level synthesis, combines the specification of all geometric patterns of gates and their positions. It defines the physical layout of the chip. The layers of the layout correspond to the *masks* used for chip fabrication. The physical layout is the final target of a logic network.

The result of logic synthesis is a logic network, that is, a collection of interconnected parts. Each part is represented at the physical level by a shape and the area it occupies on the board, together with the location of one or more pins, where the electrical connections are made. Electrical connections are implemented by a tree or a routing graph.

The major steps in physical design are *placement* and *wiring*, also called *routing*. Placement means the positioning of components on the board in a manner that is feasible with respect to given technological limitations. Next, the connections between the pins of the components have to be routed onto the board. The main objective of placement is to minimize the total wire length of all connections.

5.3   Basic definitions

A *graph* consists of two sets V and E, where V is nonempty and each element of E is an unordered pair of distinct elements of V. A graph is defined as $G = (V, E)$. The following terminology is used in the representation of graphs. The elements of V are called *vertices* or *nodes*; the elements of E are called *edges* or *links*. The vertices v and w are said to be *incident* with the edge (vw). Two vertices are *adjacent* if they are the end vertices of an edge; two edges are *adjacent* if they have a vertex in common.

The terms *link*, *edge*, *connection*, and *interconnection* are used interchangeably. Note that the functional elements of a designed logic network correspond to the nodes, and the communication links correspond to the edges in the graph. In constructing graphs for the representation of computing structures, the following characteristics of graphs are used:

▶ The number of intermediate and terminal nodes.

▶ The *degree* of a node v, written $deg(v)$, is the number of edges incident with v.

▶ A link in E between the nodes v and u is specified as an unordered pair (v, u), and v and u are said to be *adjacent* to each other or are called *neighbors*.

▶ The *distance* between two nodes i and j in a graph G is the number of edges in G on the shortest path connecting i and j.

▶ The *diameter* of a graph G is the maximum distance between two nodes in G.

▶ A graph G is *connected* if a path exists between any pair of nodes i and j in G.

Example 5.1   *A graph and its components are represented in Figure 5.2.*

Graphical data structure

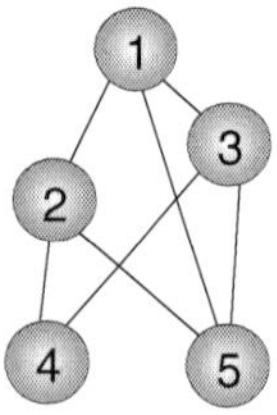

P r o p e r t i e s

▶ *The number of nodes:* 5
▶ *The degrees of the nodes v, $deg(v_i)$, are*

$$deg(1) = 3, \ deg(2) = 3, \ deg(3) = 3$$
$$deg(4) = 2, \ deg(5) = 3$$

▶ *The diameter is 2*
▶ *The graph G is connected (a path exists between any pair of nodes)*
▶ *All distances are 1, except for the distance 2 between nodes 1 and 4, 2 and 3, and 4 and 5*

FIGURE 5.2
A graph's components and characteristics (Example 5.1).

A graph may contain loops at vertices and/or multiple edges. A *loop* is an edge that is incident with only one vertex. *Multiple edges* are several edges incident with the same two vertices.

5.3.1   Directed graphs

A *directed* graph, or *digraph*, G consists of a finite set V of nodes and a set E of directed edges between the vertices, and is characterized by the following:

▶ The *indegree* of a node i is the number of edges in G leading to i. The *outdegree* of a node i is the number of edges outgoing from i.

▶ A node is called a *terminal node* if it has an outdegree of 0. If the outdegree of v is greater than 0, v is called an *internal* node. A node is called a *root* if it has an indegree of 0.

▶ The *adjacency* matrix $A = (a_{ij})$ is defined as a square matrix of size <NUMBER OF NODES> such that element $a_{ij} = 1$ if $(i, j) \in E$, and $a_{ij} = 0$ otherwise.

The adjacency matrix contains only two elements, 0 and 1. The graph and its adjacency matrix contain the same information; they are simply two alternative representations or data structures. A permutation of any two rows or columns in an adjacency matrix corresponds to relabeling the vertices and edges of the same graph.

Example 5.2 *The properties of the **directed** graph with four nodes and its adjacency matrix A are illustrated in Figure 5.3, where $in(v_i)$ and $out(v_i)$ are the indegrees and outdegrees of a node v_i.*

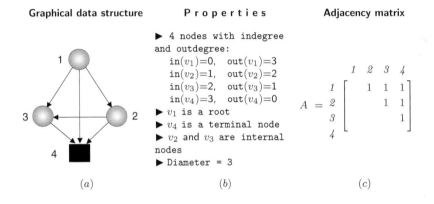

Graphical data structure

Properties

Adjacency matrix

▶ 4 nodes with indegree and outdegree:
$in(v_1)=0$, $out(v_1)=3$
$in(v_2)=1$, $out(v_2)=2$
$in(v_3)=2$, $out(v_3)=1$
$in(v_4)=3$, $out(v_4)=0$
▶ v_1 is a root
▶ v_4 is a terminal node
▶ v_2 and v_3 are internal nodes
▶ Diameter = 3

$$A = \begin{array}{c} \\ 1 \\ 2 \\ 3 \\ 4 \end{array} \begin{array}{cccc} 1 & 2 & 3 & 4 \\ \left[\begin{array}{cccc} & 1 & 1 & 1 \\ & & 1 & 1 \\ & & & 1 \\ & & & \end{array} \right] \end{array}$$

(a)   (b)   (c)

FIGURE 5.3
The directed graph (a), its properties (b), and an adjacency matrix (c) (Example 5.2).

5.3.2 Flow graphs

A *bipartite* graph is one whose vertices can be partitioned into two (disjoint) nonempty sets, S_1 and S_2, in such a way that every edge joints with a vertex in S_1 and a vertex in S_2. A *complete* bipartite graph is a bipartite graph in which every vertex in S_1 is joined to every vertex in S_2.

Typically, the edges in graphs are labeled by numbers, which are used in computing. A *weighted* graph is a graph in which each edge is assigned a nonnegative real number called a *weight*. A *flow* graph is a directed *bipartite* graph. A flow graph can be derived from matrix equations. In this case, the structure of a flow graph corresponds to the rule of matrix multiplication and the structure of the matrix (distribution of nonzero and zero elements).

Example 5.3   *Let a matrix transformation be represented by the equation* $\begin{bmatrix} 1 & 1 \\ 1 & 1 \end{bmatrix} \begin{bmatrix} a_1 \\ a_2 \end{bmatrix}$ *and* $\begin{bmatrix} 1 & 0 \\ 1 & 1 \end{bmatrix} \begin{bmatrix} a_1 \\ a_2 \end{bmatrix}$. *In Figures 5.4a and b, these equations (multiplications of a vector by matrix) are represented by flow graphs.*

$$\begin{bmatrix} 1 & 1 \\ 1 & 1 \end{bmatrix} \begin{bmatrix} a_1 \\ a_2 \end{bmatrix} = \begin{bmatrix} a_1 \times 1 + a_2 \times 1 \\ a_1 \times 1 + a_2 \times 1 \end{bmatrix} \qquad \begin{bmatrix} 1 & 0 \\ 1 & 1 \end{bmatrix} \begin{bmatrix} a_1 \\ a_2 \end{bmatrix} = \begin{bmatrix} a_1 \times 1 \\ a_1 \times 1 + a_2 \times 1 \end{bmatrix}$$

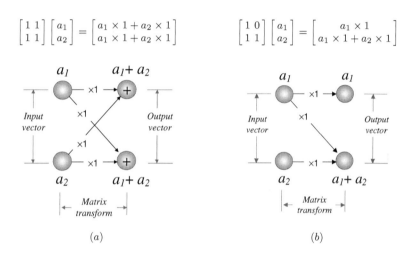

(a)   (b)

FIGURE 5.4

The structure of the flow graph corresponds to matrix multiplication: the matrix structure specifies the flow graph (Example 5.3).

5.3.3   Undirected graphs

In the case of undirected graphs, edges are considered unordered pairs, and therefore have no distinguishable direction. The *degree* of a node i in a graph G is the number of edges in G that are incident with i, that is, where the outdegree and the indegree coincide.

Example 5.4   *Figure 5.5 illustrates the properties of the **undirected** graph. Its adjacency matrix A is equal to the transposed matrix A, $A = A^T$.*

A *star graph* is a graph in which there is a particular node (the center) that connected to all the other nodes, while the rest of the nodes are connected only to the center (Figure 5.6).

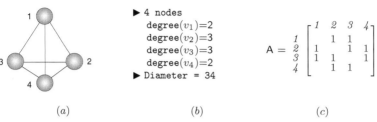

FIGURE 5.5

The undirected graph (a), its properties (b), and an adjacency matrix (c)
(Example 5.4).

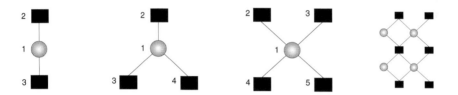

FIGURE 5.6

Star graphs.

5.3.4 A path in a graph

A *path* is a chain of edges between two vertices. A path can be measured by
a *length t*, which is the number of vertices in the path.

In the modeling of logic networks, the number associated with an edge is
some unit used in the problem to be solved, for example, a unit of time,
distance, or cost. The problem is formulated as finding the shortest path, or
minimizing the lengths of the paths. A *path* in a graph is a sequence of edges
and nodes leading from the root node to a terminal node. The *path length* in
a graph is the number of nonterminal nodes in the path.

> **Example 5.5** *A **decision tree** with 6 paths is given in Figure 5.7.
> For example, the paths $1 \rightarrow 2 \rightarrow 7$ and $1 \rightarrow 3 \rightarrow 5 \rightarrow 10$ are characterized by different features, such as the number of nodes and links.*

5.3.5 Isomorphism

Two graphs, $G_1(V_1, E_2)$ and $G_2(V_1, E_2)$, are isomorphic if there exists a
bijection $f : V_1 \rightarrow V_2$ such that $(u, v) \in E_1$ if and only if $(f(u), f(v)) \in E_2$.

Given the adjacency matrices A_1 and A_2 of the graphs G_1 and G_2,

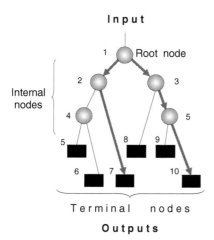

There are 6 paths from the root node to the terminal nodes. For example, the two highlighted paths are derived as follows:

▶ Path of length 2 from the root node through the node 2 to the terminal node 7

▶ Path of length 3 from the root node through the internal nodes 3 and 5 to the terminal node 10

FIGURE 5.7
A directed graph as a decision tree, with two paths highlighted (Example 5.5).

respectively, these graphs are isomorphic if and only if $A_2 = PA_1P^T$ for some permutation matrix P. That is, two graphs are isomorphic if their vertices can be labeled in such way that the corresponding adjacency matrices are equal. Two isomorphic graphs have

▶ The same number of nodes.
▶ The same number of nodes with a given degree.
▶ The same number of links.

Example 5.6   *Figure 5.8 shows two* **isomorphic** *graphs, where*

$$f(1) = D, f(2) = B, f(3) = C, f(4) = A, f(5) = E$$

Graph G_2 is obtained by relabeling the vertices of G_1, maintaining the corresponding edges in G_1 and G_2.

Example 5.7   *Two graphs representing 4D* **hypercubes** *are isomorphic (Figure 5.9), but they have different topologies.*

5.3.6   A subgraph and spanning tree

A graph $G_2(V_2, E_2)$ is a *subgraph* of $G_1(V_1, E_2)$ if $V_2 \subseteq V_1$ and $E_2 \subseteq E_1$.

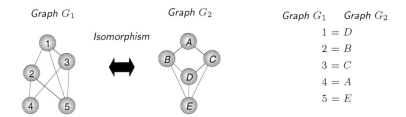

FIGURE 5.8
Examples of isomorphic graphs (Example 5.6).

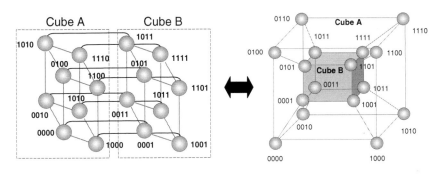

FIGURE 5.9
Isomorphic graphs (Example 5.7).

Example 5.8   *Figure 5.10 shows a graph (a) and a subgraph (b).*

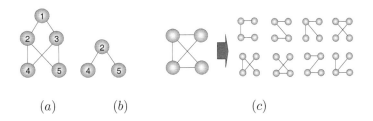

FIGURE 5.10
Examples of a subgraph (b) of a graph (a), and a graph and its eight spanning trees (c) (Examples 5.8, 5.9).

A *spanning tree* of a connected graph G is a subgraph that is a tree and that includes every vertex of G. These trees are used in the routing problem at the physical level of discrete device design.

Example 5.9   *Figure 5.10c shows a graph and all its corresponding* **spanning trees**.

5.3.7   Cartesian product

The *Cartesian product* of graphs provides a framework in which it is convenient to analyze as well as to construct new graphs:

$$\underbrace{\text{Graph } G_1 \quad \overset{\textit{Cartesian product}}{\times} \quad \text{Graph } G_1}_{\textit{New graph}}$$

Let $G_1 = (V_1, E_1)$ and $G_2 = (V_2, E_2)$ be two graphs. The product of G_1 and G_2, denoted $G_1 \times G_2 = (V_1 \times V_2, E)$, is a graph where the set of nodes is the product set

$$V_1 \times V_2 = \{x_1 x_2 | x_1 \in V_1, \ x_2 \in V_2\}$$
$$E = \{\langle x_1 x_2, y_1, y_2 \rangle | (x_1 = y_1, \ \langle x_2, y_2 \rangle \in E_2) \ or$$
$$(x_2 = y_2, \ \langle x_1, y_1 \rangle \in E_1)\}$$

It can be shown that a hypercube can be defined as the product of n copies of the complete graph with two vertices, K_2. That is,

$$H_n = H_{n-1} \times K_2$$

5.3.8   Planarity

Logic networks are modeled using graphs. In practical implementation, crossings are often expensive; they occupy space, require additional channels, and cause various other unwanted effects. Planar graph-based models can alleviate these difficulties.

A graph G is *planar* if it is isomorphic to a graph G' such that (*a*) the vertices and edges of G' are contained in the same plane, and (*b*) at most one vertex occupies, or one edge passes through any point on the plane. In other words, a graph is planar if it can be drawn on a plane with no two edges intersecting.

Example 5.10 *In Figure 5.11a, a **planar** graph is shown. There are no crossings between edges in this graph. The nonplanar graph G_2 in Figure 5.11b can be planarized, that is, converted to an isomorphic planar graph G_3 as depicted in Figure 5.11c.*

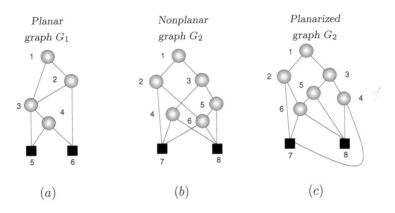

(a) (b) (c)

FIGURE 5.11

A planar graph (a); graph G_2 (b) can be planarized (c) (Example 5.10).

5.3.9 Operations on graphs

The *union* of two graphs $G_1 = (V_1, E_1)$ and $G_2 = (V_2, E_2)$ is another graph $G_3 = G_1 \cup G_2$, whose vertex set $V_3 = V_1 \cup V_2$ and edge set $E_3 = E_1 \cap E_2$:

$$\underbrace{\textbf{Graph } G_1 \quad \overset{Union}{\cup} \quad \textbf{Graph } G_1}_{New \ graph \ G_1 \cup G_2}$$

The *intersection* of two graphs $G_1 = (V_1, E_1)$ and $G_2 = (V_2, E_2)$ is another graph $G_4 = G_1 \cap G_2$, whose vertex set $V_4 = V_1 \cap V_2$ and edge set $E_4 = E_1 \cap E_2$, i.e., consisting only of those vertices and edges that are in both G_1 and G_2:

$$\underbrace{\textbf{Graph } G_1 \quad \overset{Intersection}{\cap} \quad \textbf{Graph } G_1}_{New \ graph \ G_1 \cap G_2}$$

The *ring sum* of the graphs G_1 and G_2, $G_1 \oplus G_2$, is a graph consisting of the vertex set $V_1 \cup V_2$ and of edges that are in either G_1 or G_2, but not in both:

$$
\underbrace{\text{Graph } G_1 \quad \overset{\text{Ring sum}}{\oplus} \quad \text{Graph } G_1}_{\text{New graph } G_1 \oplus G_2}
$$

These operations can be extended to include any finite number of graphs. A pair of vertices x, y in a graph G is said to be *merged* if the two vertices can be replaced by a single new vertex z, such that every edge that was incident on either x or y or on both is incident on the vertex z. This operation does not alter the number of edges, but reduces the number of vertices by one.

If the graphs G_1 and G_2 are edge disjoint, then

▶ $G_1 \cap G_2$ is a null graph.

▶ $G_1 \oplus G_2 = G_1 \cup G_2$.

If G_1 and G_2 are vertex disjoint, then $G_1 \cap G_2$ is empty. For any graph G,

▶ $G \cup G = G \cap G = G$.

▶ $G \oplus G$ is a null graph.

If g is a subgraph of G, then $G \oplus g$ is that subgraph of G which remains after all the edges in g have been removed from G; that is, $G \oplus g = G - g$.

Example 5.11   *Various operations on graphs are given in Table 5.1.*

5.3.10   Embedding

An embedding of a *guest* graph G into a *host* graph H is a one-to-one mapping $\varphi\colon V(G) \to V(H)$, along with a mapping α that maps an edge $(u, v) \in E(G)$ to a path between $\varphi(u)$ and $\varphi(v)$ in H:

$$
\underbrace{\underbrace{\text{Graph } G}_{\text{Guest}} \quad \overset{\text{Embedding}}{\longrightarrow} \quad \underbrace{\text{Graph } H}_{\text{Host}}}_{\text{Graph with new properties}}
$$

5.4   Tree-like graphs and decision trees

One class of graphical data structure, decision trees, is a convenient way to represent binary or multivalued logic functions. The number of possible values

TABLE 5.1
Operations on graphs (Example 5.11)

Techniques for computing: operations on graphs

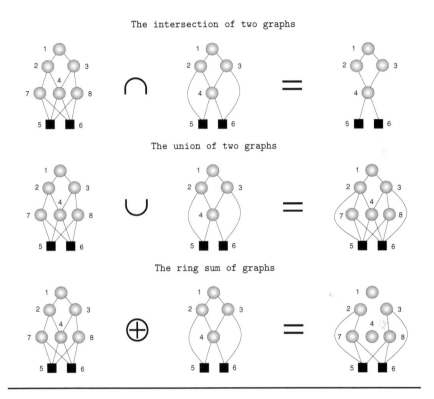

The intersection of two graphs

The union of two graphs

The ring sum of graphs

determines the tree's topology.

Example 5.12 *Figure 5.12 demonstrates a* **binary** *tree for the representation of Boolean functions (a), a* **ternary** *tree for the representation of ternary logic functions (b), and a* **mixed** *topology tree (c).*

5.4.1 Basic definitions

A *tree* is a rooted acyclic graph in which every node but the root has an indegree of 1. A special class of rooted trees is called *binary* trees. The *path length* of a tree can be defined as the sum of the path lengths from the root to all terminal vertices. Trees have several useful characteristics and properties.

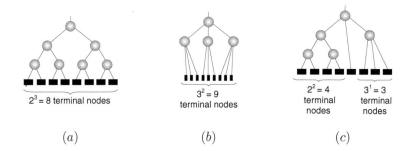

FIGURE 5.12
Binary (a), ternary (b), and mixed (c) trees (Example 5.12).

▶ For every node v there exists a unique path from the root to v. The length
of this path is called the *depth* or *level* of v.
▶ The *height* of a tree is equal to the greatest depth of any node in the tree.
▶ A node with no children is a *terminal* (*external*) node or *leaf*. A nonleaf
node is called an *internal* node.

A *complete* n-level p-tree is a tree with p^k nodes per level k for $k =
0, \ldots, n - 1$. A p^n-leaf complete tree has a level hierarchy (levels $0, 1, \ldots, n$);
the root is associated with the level zero, and its p children are on level one.
This link type describes the direction of data transmission between the child
and the parent, so that data are sent in only one direction at a time, up or
down.

Example 5.13   *Figure 5.13 shows how the parameters of a complete
3-level binary decision tree can be measured.*

5.4.2   Lattice topology of graphs

Certain classes of logic functions can be represented by trees with a lattice
topology. This allows for embedding trees into regular graphical structure
called a *lattice*.

Example 5.14   *Figure 5.14 demonstrates **lattice-like** trees and the
topology they can be embedded into.*

Example 5.15   *A decision diagram can be extended to a lattice-like
tree and then embedded into a lattice (Figure 10.30).*

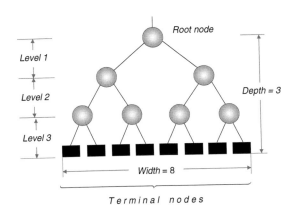

Root node

Level 1

Level 2

Level 3

Depth = 3

Width = 8

Terminal nodes

Parameters of the complete 3-level binary decision tree:

▶ Number of intermediate nodes is 7
▶ Function of the i-th intermediate node (f_i)
▶ Number of terminal nodes is 8
▶ Number of levels is 3
▶ Depth is 3
▶ Width is 8

FIGURE 5.13
Measurements of the parameters of a complete 3-level binary decision tree (Example 5.13).

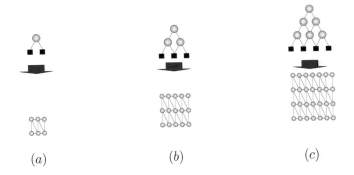

(a)   (b)   (c)

FIGURE 5.14
Embedding complete lattice-like trees into lattices (Example 5.14).

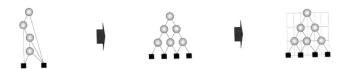

FIGURE 5.15
Embedding complete lattice-like trees into lattices (Example 10.42).

5.4.3 H-trees

An H-tree is a recursive topological construction of H_1-trees, where H_1 is defined as in Figure 5.16a. An H_{k+1}-tree can be constructed by replacing the leaves of H_k with H_1-trees. The number of terminal nodes in an H_k-tree is equal to 4^k. An H-tree makes optimal use of area and wire length. H-trees are the most common style for physical design of logic networks, in particular, clock routing.

Example 5.16   *In Figure 5.16, the $H_{k+1} = H_2$ tree is shown for $k = 1$.*

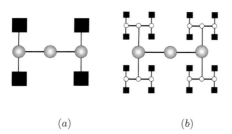

▶ *4 terminal nodes,*
▶ *3 nonterminal nodes*

The H_2-tree is constructed by replacing terminal nodes with H_1-trees; the H_1-tree includes:

▶ *$4 \times 4 = 16$ terminal nodes*
▶ *$4 \times 3 + 3 = 15$ nonterminal nodes*

(a)   (b)

FIGURE 5.16
H_1 (a) and H_2 (b) trees (Example 5.16).

5.4.4   Binary decision trees and functions

A *discrete* function f of n variables is defined as a function in which each variable takes exactly k values. A discrete function is *constant* if it assumes the same value wherever it is defined; it is *completely specified* if it is defined everywhere. A *decision tree* is a graphical model of the evaluation of a discrete function, wherein the value of a variable is determined and the next action is chosen accordingly:

$$\underbrace{\texttt{Discrete function } f \; \overset{Mapping}{\longrightarrow} \; \texttt{Graph } G}_{\textit{Graphical representation of } f}$$

In logic design, the focus is on the construction of decision trees from a function description, evaluating the efficiency of this representation, and the manipulation of the function represented by this tree.

There are other parameters of decision trees. For example, the distance $d(v_i, v_j)$ between two nodes v_i and v_j is the length of the shortest path between them, i.e., the number of edges in the shortest path. The *diameter* of a tree is defined as the length of the longest path in this tree. The notation of the *topology* of decision trees is used to distinguish their different shapes.

5.4.5 The relationship between decision trees and cube-like graphs

The terminal nodes of a complete decision tree correspond to the 2^n values of a Boolean function of n variables. On the other hand, these values can be represented by a hypercube. This implies one-to-one correspondence between the complete binary tree and the hypercube of a function.

Example 5.17 *Table 5.2 shows the complete decision trees and the corresponding hypercubes for $n = 1, 2, 3, 4$.*

The data structure, called *hypercube-like* topology, corresponds to the embedding of a complete binary tree into a multidimensional hypercube.

5.4.6 The simplification of graphs

The simplification of a graph means deleting some of its vertices or edges. When a vertex is removed from a graph, all edges incident with that vertex must also be removed. The result of deleting a vertex is a new graph or several new graphs. In logic design, the main task is the simplification of decision trees.

Example 5.18 *In Figure 5.17, the effect of deleting nodes is illustrated.*

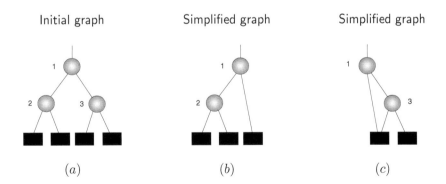

FIGURE 5.17

A complete decision tree (a) and its simplification by deleting the nodes labeled by 3 (b) and by 2 (c) (Example 5.18).

TABLE 5.2
The relationship between the complete binary decision trees and
hypercubes

Design example:
relationship between decision trees and hypercubes

Decision tree	Cube structure

A complete 1-level decision tree and 1D cube

A complete 2-level decision tree and 2D cube

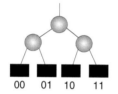

A complete 3-level decision tree and 3D cube

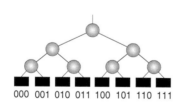 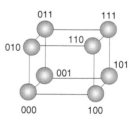

A complete 4-level decision tree and a 4D cube

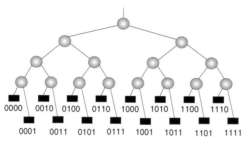 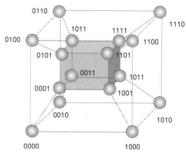

5.5 Voronoi diagrams

Given a set S of n points, called *sites*, the *Voronoi diagram* of S is a partition (decomposition) of space into regions (cells), such that each region consists of all points that are closer to one site than to any other. Voronoi diagrams are useful for modeling spatial structures, location optimization, and fractal generation. Most of the early work on Voronoi diagrams was motivated by crystallography. The objective in this respect was the study of regions arising from regularly placed sites. Molecular system consists of a number of distinct molecules. Equilibrium and other properties of the system depend on the spatial distribution of the sites, which can be conveniently represented by dividing the space between them according to the nearest-neighbor rule. The nearest-neighbor rule (the crystal growth model) forces the Voronoi regions to be convex polyhedra.

The elements of Voronoi diagrams (vertices, polygons or edges) can be assigned certain elements of computing data structures. For instance, geometric constraint solving consists of finding all possible placements of given geometrical primitives that satisfy a set of constraints between the primitives (Figure 5.18).

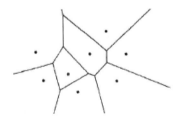

Voronoi diagram for eight points. The Voronoi diagram divides the plane according to the nearest-neighbor rule: each site is associated with the region of the plane closest to it. The Voronoi diagram maybe viewed as a data structure that organizes its defining sites in a prescribed manner.

FIGURE 5.18
Voronoi diagram.

A *Voronoi transform* represents an original topological structure in the form of a Voronoi diagram. *Computing* a Voronoi diagram is to transform an arbitrary topological data structure into a Voronoi diagram. This topological transformation can be implemented by various algorithms. Most of the algorithms for Voronoi diagram design explore particular properties of the input data structure. In the analysis and synthesis of topological structures, an appropriate metric should be chosen.

Point sets. A point set is simply a collection of points, called *sites*, in some space. The topology of the point set carries information about the configuration, such as the neighborhood of points, the nearness of two points, the connectivity of subsets of the point set, boundary points, and shape characteristics.

Euclidean space. Discrete topology can be defined in an m-dimensional Euclidean space $\mathbb{E}^m$. In a topological data representation $m = 1, 2$, or 3, there is a set of point operations (sum, subtraction, multiplication, etc.). Metric space is the only way of looking at distances between points in space. Formally, a metric space is a set S of arbitrary elements with a single valued, nonnegative real function d (called a *distance function*) that, for any two elements $x \in S$ and $y \in S$, gives the distance between them.

In the m-dimensional Euclidean space, $\mathbb{E}^m$, every point can be assigned m real numbers, their Cartesian coordinates $(x_1, x_2, \ldots, x_m)$. The distance between any two points $p = (x_1, x_2, \ldots, x_m) \in \mathbb{E}^m$ and $p' = (x'_1, x'_2, \ldots, x'_m) \in \mathbb{E}^m$ is given by the expression

$$< Distance >= d(p, p') = \sqrt{\sum_{i=1}^{m} (x_i - x'_i)^2}. \tag{5.1}$$

Other metrics. However, the distance is not limited to the Euclidean distance given by Equation 5.1. Any function can be chosen as long as certain conditions apply. These conditions come about as the generalization of the familiar properties of the Euclidean distance:

▶ Distance is non-negative, that is, $d(x, y) \geq 0$.

▶ Distance between two points is equal to zero if and only if the points coincide.

▶ Distance is symmetric, that is, $d(x, y) = d(y, x)$.

▶ Distance satisfies the triangle inequality, i.e., $d(x, y) \leq d(x, z) + d(y, z)$.

Data can be analyzed and synthesized in various metric spaces. There are a variety of different distance functions that can be used in the calculation of Voronoi diagrams (city-block, chessboard distance). The m-dimensional space $\mathbb{R}^m$ with city-block distance is an illustrative example of non-Euclidean metric space. Let $(x_1, x_2, \ldots, x_m)$ and $(x'_1, x'_2, \ldots, x'_m)$ be the Cartesian coordinates of points $p \in \mathbb{R}^m$ and $p' \in \mathbb{R}^m$, respectively. Then, the city-block distance is

$$< \text{city-block distance} >= d(p, p') = \sum_{i=1}^{m} |x_i - x'_i|.$$

The Voronoi region is the underlying topological component of a Voronoi data structure. In analysis, the Voronoi region carries useful information for verification of computing structures. In synthesis, a Voronoi region can be considered as a geometrical primitive for geometric constraint solving in the metric space $\mathbb{M}$.

Let $\mathbb{M}$ be a metric space with its associated distance function, $d(x, y)$, and P be a set of N distinct points in $\mathbb{M}$. These points are called *Voronoi sites* or simply *sites*.

Formally, the Voronoi region of the site $p_i \in P$ denoted by $V(p_i)$ is a subset of all points in $\mathbb{M}$ that are closer to the site p_i than to any other site $p_j \in P$, i.e.,

$$V(p_i) = \{x \in \mathbb{M} \; : \; d(x, p_i) \leqslant d(x, p_j), p_j \in P\}. \tag{5.2}$$

It follows from the Equation 5.2 that the Voronoi region carries the following information:

(a) The location of the site p_i with respect to the closer subset of sites.

(b) The local topology given by a subset of points.

A Voronoi diagram of the set P denoted by $V(P)$ is a partitioning of the metric space $\mathbb{M}$ into N Voronoi regions $V(p_i)$, $p_i \in P$, $i = 1, 2, \ldots, N$, such that each Voronoi region $V(p_i)$ contains points that are closer to the site p_i than to any other site in P.

Voronoi edge and Voronoi vertex. For some points, there are more than one closest generator. Points such as these act as an edge, or boundary, between adjacent Voronoi regions. A point is said to be on a Voronoi edge if every neighborhood of this point contains at least two points from two distinct Voronoi regions.

The Voronoi vertex can be defined by analogy with the Voronoi edge. A point is a Voronoi vertex if every neighborhood of this point contains at least three points from three distinct Voronoi regions. The set of all boundary points forms a network of lines called the *Voronoi skeleton*.

It follows from the above definitions that

▶ The Voronoi diagram is a partitioning of space into convex polygons, one polygon for each point $p_i \in P$.

▶ The Voronoi diagram of P is a collection of Voronoi regions $V(p_i)$. Each region $V(p_i)$ is precisely the region of the plane containing all points closer to the generator p_i than any other generator in P.

Curved boundaries between Voronoi regions. With points as generators, the Voronoi skeleton consists of straight lines. However, the ordinary Voronoi diagram defined for a set of point generators can be generalized in many ways, depending on the specific applications. For instance, more complex generators may allow us to generate curved boundaries between Voronoi regions. For the purpose of biometric data synthesis it is useful to introduce the Voronoi diagram of *closed connected* area primitives and therefore to extend the ordinary Voronoi diagram for geometrical objects such as line and curve segments, circles, polygons, etc.

An *area Voronoi diagram* can be constructed in a quite similar way to that of the ordinary Voronoi diagram. However, before proceeding further, the distance between a point and a set of points needs to be defined.

Given a set, S, of points in metric space $\mathbb{M}$, the distance, denoted by $d(x, S)$, from an arbitrary point $x \in \mathbb{M}$ to the set S can be defined as the greatest lower bound of distances between x and $x' \in S$ as x' ranges over all of S. In Euclidean space this is simply the shortest distance between x and the boundary of S.

Let $A = \{A_1, A_2, \ldots, A_N\} = \{A_i\}_{i=1}^{N}$ be a set of N *closed connected* areas in metric space $\mathbb{M}$. The areas in A are assumed not to overlap.

The area Voronoi region denoted by $V(A_i)$ and associated with the area (also called a generator) $A_i \in A$ is defined as a subset of all points in $\mathbb{M}$ that are closer to A_i than to any other area $A_j \in A$, i.e., $d(x, A_i) \leqslant d(x, A_j)$.

The area Voronoi diagram generated by the set $A = \{A_i\}_{i=1}^{N}$ and denoted by $V(A)$ is a partitioning of the metric space $\mathbb{M}$ into Voronoi regions $V(A_i)$, $i = 1, 2, \ldots, N$, such that each Voronoi region $V(A_i)$ contains points that are closer to the area A_i than to any other area in A.

Note that an area generator can be a polygonal figure as well as a line or curve segment, or a point. Therefore, the area Voronoi diagram can be interpreted as a generalization of the ordinary Voronoi diagram of point generators.

Curved boundaries between Voronoi regions are useful in topological data manipulation, and give, in particular, the possibility to generate more complex geometrical primitives and to define sophisticated constraints between them for geometric constraint solving.

Example 5.19   *Given a set of* **generators** $A = \{p,\ L\}$ *in* $\mathbb{E}^2$, *where p is a point and l is a horizontal line, the Voronoi diagram* $V(A)$ *consists of two regions separated by a parabolic arc (Figure 5.19).*

5.5.1   Direct and inverse Voronoi transform

Manipulation is often required to recover topological data. The Voronoi diagram is the result of performing a Voronoi transform. The problem can be formulated as recovering initial topology given a Voronoi diagram. The Voronoi diagram $V(P)$, can be considered as a *direct Voronoi transform*, denoted by **V**, and the Voronoi sites are said to be *generators* of the transformation. The concept of the *inverse Voronoi transform*, denoted by

The ordinary Voronoi generator can be extended from a point to a more complex geometrical object in order to obtain curved boundaries between Voronoi regions. A simple example of such a generalization is a set consisting of a point p and a straight line L.

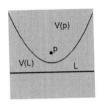

The generator set of the point p and the line L results in a Voronoi diagram consisting of two regions, $V(p)$ and $V(L)$, separated by a parabolic arc.

FIGURE 5.19

Example of the synthesis of topological structure with nonstraight lines (Example 5.19).

$\mathbf{V}^{-1}$, can be defined as a transform that restores the original distribution of generators P from their Voronoi diagram $V(P)$ (Figure 5.20).

The direct Voronoi transform $\mathbf{V} : P \mapsto V(P)$ maps a given set of Voronoi sites P into a Voronoi diagram $V(P)$ in such a way that with every Voronoi site $p_i \in P$ it associates a unique Voronoi region $V(p_i) \in V(P)$:
$<$ Generators $> \Rightarrow <$ Voronoi diagram $>$.

The inverse Voronoi transform $\mathbf{V}^{-1}$ can be defined as a transform that satisfies $P = \mathbf{V}^{-1}(\mathbf{V}(P))$ and therefore constitutes a map $\mathbf{V} : V(P) \mapsto P$ that associates with every Voronoi region $V(p_i) \in V(P)$ its Voronoi generator $p_i \in P$, that is, $<$ Voronoi diagram $> \Rightarrow <$ Generators $>$.

Example 5.20 *Given a set of generators in the Euclidean space $\mathbb{E}^2$ and the corresponding Voronoi diagram, denoted by*

$$< \text{Generators} > = P = \{p_1, p_2, p_3\} \text{ and}$$
$$< \text{Voronoi diagram} > = V(P) = \{V(p_1), V(p_2), V(p_3)\},$$

the two sets are in one-to-one correspondence, which can be represented by the direct $\mathbf{V}$ and inverse $\mathbf{V}^{-1}$ **Voronoi transforms** *(Figure 5.21):*

$$\mathbf{V} : p_1 \mapsto V(p_1), \ p_2 \mapsto V(p_2), \ p_3 \mapsto V(p_3),$$
$$\mathbf{V}^{-1} : V(p_1) \mapsto p_1, \ V(p_2) \mapsto p_2, \ V(p_3) \mapsto p_3.$$

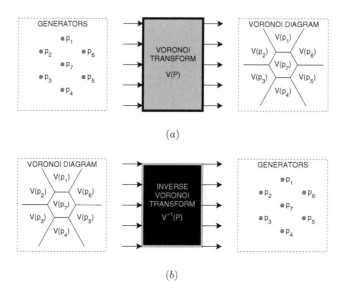

(a)

(b)

FIGURE 5.20
A direct Voronoi transform turns a set of generators P into a Voronoi diagram $V(P)$ (a), and inverse Voronoi transforms turns a Voronoi diagram $V(P)$ into a set of generators P (b).

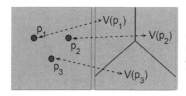

Direct Voronoi transform $\mathbf{V}(P)$, *which turns the set of generators* $P = \{p_1, p_2, p_3\}$ *into the Voronoi diagram* $V(P) = \{V(p_1), V(p_2), V(p_3)\}$, *defines a* **one-to-one** *correspondence between the sets* P *and* $V(P)$.

FIGURE 5.21
Voronoi transform $\mathbf{V} : P \mapsto V(P)$ (Example 5.20).

5.5.2   Distance mapping of feature points

A *distance map* is an extension of the distance function in metric space. Like the usual distance function that characterizes the separation of two points in space, the distance map is intended to characterize the spacial separation of a set of points. Furthermore, given a set of points, the analytical behavior of the distance map is closely related to the topological structure of its Voronoi diagram.

Metric space. A metric space $\mathbb{M}$ is a space with an associated distance function f which gives the distance between any two points in $\mathbb{M}$. A wide variety of alternate metrics can be used for the calculation of distances. The choice of a concrete distance function depends on the aims of analysis or synthesis as well as on the nature of a particular data set.

Example 5.21   *The results of calculation of the distance between two points, $p_1(0,0)$ and $p_2(1,1)$, in 2D space for different metrics:*

$$< Euclidean\ distance > = \sqrt{(x_2 - x_1)^2 + (y_2 - y_1)^2}$$
$$= \sqrt{(1 - 0)^2 + (1 - 0)^2} = \sqrt{2}$$
$$< Chessboard\ distance > = \max{(|x_2 - x_1|, |y_2 - y_1|)}$$
$$= \max{(|1 - 0|, |1 - 0|)} = 1$$
$$< City\text{-}block\ distance > = |x_2 - x_1| + |y_2 - y_1|$$
$$= |1 - 0| + |1 - 0| = 2$$

The distance function characterizes the separation of two points in metric space. However, an alternative point of view on the distance function can be more suitable: given a fixed point in a metric space, the distance function can be considered as a function of a single argument so that it gives the distance from any arbitrary point in the space to that fixed point.

5.5.3   Distance map

Let $\mathbb{M}$ be a metric space with its distance function $d(p, p')$. The distance from an arbitrary point p to a fixed point p' can be considered as a function of a single argument p. Such a function that assigns to each point in $\mathbb{M}$ a value equal to the distance from that point to the fixed point p' will be referred as a *distance map* of a *fixed point* in $\mathbb{M}$.

The distance map for a single fixed point can be easily extended to a *distance map* of a *set* $P = \{p_i\}_{i=1}^N$ of N distinct points in $\mathbb{M}$: the distance map, denoted by $d(p, P)$, assigns to each point p in $\mathbb{M}$ the minimal distance $d(p, p_i)$ as p_i ranges over all of P.

Euclidean distance map. Euclidean space $\mathbb{E}^2$ is an excellent metric space to study the analytical properties of the distance map, thanks to its nice distance function

$$d\ (p(x, y),\ p'(x', y')) = \sqrt{(x - x')^2 + (y - y')^2},$$

which is differentiable everywhere in $\mathbb{E}^2$. Therefore, the discussion will be restricted to $\mathbb{E}^2$ for a while, solely for the purposes of simplicity, and the necessary generalization will be made later.

The following example shows a contour plot of the distance map $d(p, P)$ for the set $P = \{p_1, p_2, p_3\}$ of three points in $\mathbb{E}^2$.

Example 5.22   *Figure 5.22 represents the contour plot of the distance.*

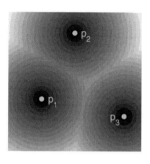

The contour plot of the distance map $d(p, P)$ for the set $P = \{p_1, p_2, p_3\}$ clearly demonstrates the points where the different circles in contour lines meet together. Given a set P of three points in $\mathbb{E}^2$: $P = \{p_1(-0.5, -0.25), p_2(0.0, 0.7), p_3(0.75, -0.5)\}$, the distance map $d(x, y) \equiv d(p(x, y), P)$ for the set P is

$$
\begin{aligned}
< \text{Distance map} > &= \min \; (\; d(p, p_1), \; d(p, p_2), \; d(p, p_3) \;) \\
&= \min \; (\; \sqrt{(x + 0.5)^2 + (y + 0.25)^2}, \\
&\qquad\quad \sqrt{x^2 + (y - 0.7)^2}, \\
&\qquad\quad \sqrt{(x - 0.75)^2 + (y + 0.5)^2} \;)
\end{aligned}
$$

FIGURE 5.22

Contour plot of the Euclidean distance function in $\mathbb{E}^2$ (Example 5.22).

The plot in Figure 5.22 from the cited example makes obvious a noteworthy topological behavior of the Euclidean distance map:

▶ In the areas close to the points p_1, p_2, and p_3, the contours surround each point as entire distinct circles. These actually are the contour lines of the individual distance functions: $d(p, p_1)$, $d(p, p_2)$, and $d(p, p_3)$, respectively. In the areas close to the points, the circles surrounding different points are well separated from each other.

▶ At the same time, the plot clearly demonstrates some contour lines that consist of circles meeting together at common points and lines. At these common points the distance map function becomes non-differentiable.

Thus, the Euclidean distance map has a set of *critical points*. Critical points are the points where the function has the gradient equal to zero or where one of its partial derivatives is not defined.

The relation between the critical points in the distance map and the Voronoi diagram can be expressed in the following statements, which are actually valid in any metric space associated with a monotonous distance function (Euclidean, chessboard, city-block and quasi-Euclidean):

▶ The points that lie on the Voronoi edges of the Voronoi diagram are the critical points of the distance map and these points represent the local maximums of the distance map.

▶ If critical points exist inside the Voronoi region, they are not the points of local maximums of the distance map.

Example 5.23 *Given the set P of three points in $\mathbb{E}^2$ and the **distance map** $d(x,y)$ from Example 5.22, Figure 5.23a represents a vector map of the distance map gradient and Figure 5.23b the points where the second derivatives, $\frac{\partial^2 d(x,y)}{\partial x^2}$ and $\frac{\partial^2 d(x,y)}{\partial y^2}$, of the distance map are negative.*

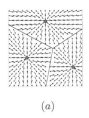

(a)

(b)

FIGURE 5.23
Analytical properties of the Euclidean distance function in $\mathbb{E}^2$: (a) the gradient plot of the Euclidean distance map generated by three points, and (b) the critical points where the second derivatives of the distance map are negative (Example 5.23).

The distance transform is used by many applications in order to map distances from feature points. The distance transform is also used in order to separate the adjoining features in biometric data and to identify primitives for the construction of new topological structures. The distance transform is normally applied to binary images and consists of calculating the distances between feature (colored) pixels and nonfeature (blank) pixels in the image; each pixel of the image is assigned a number that is the distance to the closest feature boundary.

Example 5.24 *Let the source image be the binary image in Figure 5.24. The **distance transform** assigns each pixel in the binary source image a number that is the distance to the closest feature (i.e., nonzero) pixel. The result is a grayscale distance map with real valued pixels.*

Initial structure   *Distance map*

FIGURE 5.24
Initial structure consisting of seven points and its distance map (Example 5.24).

A variety of alternate metrics such as the Euclidean, chessboard, city-block and quasi-Euclidean metrics can be used for the calculation of the distance transform.

Example 5.25   *Let the source binary image be a $[3 \times 3]$ matrix*

$$< Binary\ image >= I_{3\times 3} = \begin{bmatrix} I_{11} & I_{12} & I_{13} \\ I_{21} & I_{22} & I_{23} \\ I_{31} & I_{32} & I_{33} \end{bmatrix} = \begin{bmatrix} 0 & 0 & 0 \\ 0 & 1 & 0 \\ 0 & 0 & 0 \end{bmatrix},$$

where the only feature pixel has been placed at the center of the image. The distance transform replaces the zeros in the image by the distances indicated by the arrows:

$$\begin{bmatrix} 0 & 0 & 0 \\ 0 \leftarrow & 1 \rightarrow & 0 \\ 0 & 0 & 0 \end{bmatrix}$$

A distance transform based on different distance metrics is given in Table 5.3.

5.6   Further study

Advanced topics of graphical data structures

Topic 1: Graphical data structures in logic design. The following problems have received major consideration:

> *The Steiner tree problem* is formulated as follows: Given an edge-weighted graph and a nonempty subset of nodes called *terminal nodes T*, find

TABLE 5.3
Distance transform in Euclidean, chessboard, city-block, and quasi-Euclidean distance metric (Example 5.25)

Transform	Computing

E u c l i d e a n   m e t r i c

$$\begin{bmatrix} 0 & 0 & 0 \\ 0 & 1 & 0 \\ 0 & 0 & 0 \end{bmatrix} \Rightarrow \begin{bmatrix} I_{11} & I_{12} & I_{13} \\ I_{21} & I_{22} & I_{23} \\ I_{31} & I_{32} & I_{33} \end{bmatrix} = \begin{bmatrix} \sqrt{2} & 1 & \sqrt{2} \\ 1 & 0 & 1 \\ \sqrt{2} & 1 & \sqrt{2} \end{bmatrix}$$

$I_{11} = I_{13} = \sqrt{1^2 + 1^2} = \sqrt{2}$
$I_{31} = I_{33} = \sqrt{1^2 + 1^2} = \sqrt{2}$
$I_{12} = I_{32} = \sqrt{0^2 + 1^2} = 1$
$I_{21} = I_{23} = \sqrt{1^2 + 0} = 1$

C h e s s b o a r d   m e t r i c

$$\begin{bmatrix} 0 & 0 & 0 \\ 0 & 1 & 0 \\ 0 & 0 & 0 \end{bmatrix} \Rightarrow \begin{bmatrix} I_{11} & I_{12} & I_{13} \\ I_{21} & I_{22} & I_{23} \\ I_{31} & I_{32} & I_{33} \end{bmatrix} = \begin{bmatrix} 1 & 1 & 1 \\ 1 & 0 & 1 \\ 1 & 1 & 1 \end{bmatrix}$$

$I_{11} = I_{13} = \max(1, 1) = 1$
$I_{31} = I_{33} = \max(1, 1) = 1$
$I_{12} = I_{32} = \max(0, 1) = 1$
$I_{21} = I_{23} = \max(1, 0) = 1$

C i t y-b l o c k   m e t r i c

$$\begin{bmatrix} 0 & 0 & 0 \\ 0 & 1 & 0 \\ 0 & 0 & 0 \end{bmatrix} \Rightarrow \begin{bmatrix} I_{11} & I_{12} & I_{13} \\ I_{21} & I_{22} & I_{23} \\ I_{31} & I_{32} & I_{33} \end{bmatrix} = \begin{bmatrix} 2 & 1 & 2 \\ 1 & 0 & 1 \\ 2 & 1 & 2 \end{bmatrix}$$

$I_{11} = I_{13} = 1 + 1 = 2$
$I_{31} = I_{33} = 1 + 1 = 2$
$I_{12} = I_{32} = 0 + 1 = 1$
$I_{21} = I_{23} = 1 + 0 = 1$

a minimal weight-tree in G that spans T. The Steiner tree problem corresponds to placement in physical design; that is, positioning of the components on the board in a way that is feasible with respect to given technological limitations.

The bigraph-crossing problem. Another possible objective during placement is to minimize the number of wire crossings. Let $G = (V, E)$ be a bipartite graph with partitions V_1 and V_2. The solution is obtained by embedding G in the plane so that the nodes in V_i occupy distinct positions on the line $y = i$, and the edges are straight lines.

Recursive minimum-cut placement. The problem of recursive minimum-cut placement is stated by representing the logic network as a *hypergraph*. A hypergraph is like an undirected graph, but each *hyperedge*, rather than connecting two vertices, connects an arbitrary subset of vertices. The components correspond to the nodes, and the nets correspond to the hyperedges connecting these components (nodes). A decomposition of the logic network into two subnetworks is formulated as the problem of finding a balanced bipartition of the hypergraph with minimal cuts. This problem is also known as the *balanced hypergraph bipartition problem*.

Graph topology. The primary topic of interest regarding graphs that is used for physical or geometrical design is the manipulation of graph topology. The most often referred topologies include: hypercube topology, cube-connected cycles known as *CCC*-topology, pyramid topology, X-hypercube topology, hybrid topologies, and specific-purpose topologies (hyper-Peterson, hyperstar, Fibonacci cube, etc.).

(a) *Hypercube topology* has received considerable attention in classical logic design due mainly to its ability to interpret logic formulas and logic computations (Figure 15.2a). Hypercube-based structures are at the forefront of massively parallel computation because of the unique characteristics of hypercubes (fault tolerance, ability to efficiently permit the embedding of various topologies, such as lattices and trees).

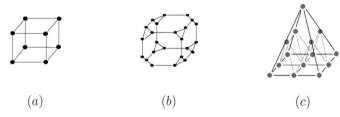

(a) (b) (c)

FIGURE 5.25
Spatial configurations: hypercube (a), CCC-hypercube (b), and pyramid (c).

(b) *Pyramid topology* is suitable for many computations based on the principle of hierarchical control, for example, decision trees and decision diagrams (Figure 15.2c). Some topologies are effective in particular cases, for example, for symmetric functions, partially specified functions, threshold functions, etc.

Topic 2: Computational paradigm based on embedding. Mapping abstract data structures into computing structures can be viewed as embedding some *guest* data structure into a given *host* data structure. Examples are matrix equations, flow graphs of algorithms, decision trees, and decision diagrams, which can be embedded into spatial dimensions using various techniques.

There are two formulations of the embedding problem:

(a) Given a guest data structure, find an appropriate topology for its representation; this is a *direct* problem.
(b) Given a topological structure, find the corresponding data structure that is suitably represented by this topology; this is an *inverse* problem. Using the properties of the host representation, specify possible data structures that satisfy it with respect to certain criteria.

The direct and inverse problems of embedding address different problems of computing. In direct problems, the topology is not specified, and the designer can map a representation of a logic function into spatial dimensions without strong topological limitations. The inverse formulation of the problem assumes that the topology is constrained by technology. That is, the question is how to use a given topology in the efficiently representation of logic functions.

Nanostructures are associated with a molecular/atomic physical platform. They have a truly 3D structure, instead of the 3D layout of silicon integrated circuits, composed of 2D layers with interconnections forming the third dimension. Figure 5.26 shows how a computing (guest) structure can be embedded into another (host) computing structure.

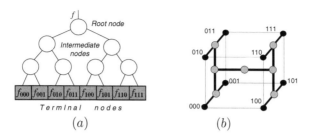

FIGURE 5.26
Embedding a decision tree into a hypercube-like structure.

Topic 3: Decision diagrams in logic design. *Minimization* of Boolean functions represented by their decision diagrams is accomplished through manipulation of the diagram. The key idea of a decision-diagram-based minimization is that by reducing the representation, Boolean manipulation becomes much simpler, computationally. Consequently, it provides a suitable data structure for the manipulation of Boolean functions.

Since decision diagram techniques are extended to various function manipulations, including calculation of Boolean differences, they can be employed in *sensitivity analysis*. Sensitivity functions, which include Boolean difference, specify the conditions under which the circuit output is sensitive to the changing of the value on a signal line. This is a component of automatic test generation since the sensitivity function is included in the equation that describes the set of all tests for each single fault (usually, "stuck-at" faults are considered).

Decision diagrams explicitly reflect the function's structure and contribution of the variables, they are the perfect tools for evaluation of the switching network activity.

Formal verification is accomplished in practical design by means of equivalence checking, which is conveniently implemented using decision diagrams. Decision diagrams can be applied directly to the task of testing the equivalence of two combinational circuits. This problem arises when comparing a circuit to a network derived from the system specification or when verifying that a performed optimization has not altered the circuit functionality.

At a logic level, decision diagrams are used mostly to represent a Boolean function to be implemented by a logic network. They are technology-independent models in this application; however, they are used to minimize the implementation cost. A minimum-cost representation can be derived from the decision diagram, and thus, achieve a network with a minimum number of AND and OR gates. This network can then be mapped into a transistor-level network, optimized in terms of transistors required in total for implementation of the gates.

At the implementation level, decision trees and diagrams can be directly mapped into a multiplexer-based tree. The resulting circuits have few interconnections and lend themselves well to large-scale integration (for instance, binary trees form an efficient interconnection pattern). Multiplexer networks can be used as universal logic modules, thereby reducing the number of basic components needed for logic design.

Example of the relationship between various data structures and tabulated form is given in Figure 5.27.

Design example: data structure transformations

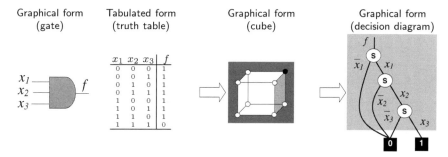

FIGURE 5.27
Relationships between various data structures.

Further reading

[1] Atallah MJ. *Algorithms and Theory of Computation.* Handbook, CRC Press, Boca Raton, FL, 1999.

[2] Berg C. *Hypergraphs. Combinatorics of Finite Sets.* North-Holland, 1989.

[3] Bhuyan N and Agrawal DP. Generalized hypercube and hyperbus structures for a computer network. *IEEE Trans. Comp.*, 33(1):323–333, 1984.

[4] Bowyer A. Computing Dirichlet tessellations. *Computer J.*, 24:162–166, 1981.

[5] Bryant RE. Graph-based algorithms for Boolean function manipulation. *IEEE Trans. Comp.*, 35(6):677–691, 1986.

[6] Butler JT and Sasao T. On the average path length in decision diagrams of multiple-valued functions. In *Proc. of the IEEE 33rd Int. Symp. on Multiple-Valued Logic*, pp. 383–390, 2003.

[7] Cormen TH, Leiserson CE, Rivest RL, and Stein C. *Introduction to Algorithms.* MIT Press, 2001.

[8] Cull P and Larson S. The mobius cubes. In *Proc. of the 6th Distributed Memory Computing Conf.*, pp. 699–702, 1991.

[9] Drechsler R and Becker B. *Binary Decision Diagrams. Theory and Implementation.* Kluwer, Dordrecht, 1998.

[10] Even S. *Graph Algorithms*. Computer Science Press, Rockville, Maryland, 1979.

[11] Fortune S. Voronoi diagrams and Delaunay triangulations. In Du DZ and Hwang F, Eds., *Computing in Euclidean Geometry*, pp. 193–233, World Scientific, 1992.

[12] Gibbons A. *Algorithmic Graph Theory*. Cambridge University Press, 1987.

[13] Goodaire EG and Parmenter MM. *Discrete Mathematics with Graph Theory*. Prentice-Hall, New York, 1998.

[14] Janković D, Stanković RS, and Drechsler R. Reduction of sizes of multi-valued decision diagrams by copy properties. In *Proc. of the IEEE 34th International Symp. on Multiple-Valued Logic*, pp. 223–228, 2004.

[15] Kumar VKP and Tsai Y-C. Designing linear systolic arrays. *J. Parallel and Distributed Computing*, 7:441–463, 1989.

[16] Lai YT and Wang PT. Hierarchical interconnection structures for field programmable gate arrays. *IEEE Trans. VLSI Systems*, 5(2):186–196, 1997.

[17] Leiserson CH. Fat-trees: universal networks for hardware-efficient supercomputing. *IEEE Trans. Comput.*, 34(10):892–901, 1985.

[18] Liotta G and Meijer H. Voronoi drawings of trees. *Computational Geometry*, 24(3):147–178, 2003.

[19] Mitchell T. Decision tree learning. In *Machine Learning*, pp. 52–78, McGraw-Hill, New York, 1997.

[20] Moret BME. Decision trees and diagrams. *Computing Surveys*, 14(4):593–623, 1982.

[21] Nagayama S and Sasao T. On the minimization of average path length for heterogeneous MDDs. In *Proc. of the IEEE 34th International Symp. on Multiple-Valued Logic*, pp. 216–222, 2004.

[22] Okabe A, Boots B, and Sugihara K. *Spatial Tessellations. Concept and Applications of Voronoi Diagrams*. Wiley, New York, NY, 1992.

[23] Perkowski MA, Chrzanowska-Jeske M, and Xu Y. Lattice diagrams using Reed-Muller logic. In *Proc. of the IFIP WG 10.5 Int. Workshop on Applications of the Reed-Muller Expansion in Circuit Design*. Japan, pp. 85–102, 1997.

[24] Öhring S and Das SK. Incomplete hypercubes: embeddings of tree-related networks. *J. Parallel and Distributed Computing*, 26:36–47, 1995.

[25] Saad Y and Schultz MH. Topological properties of hypercubes. *IEEE Trans. Comput.*, 37(7):867–872, 1988.

[26] Shen X, Hu Q, and Liang W. Embedding k-ary complete trees into hypercubes. *J. Parallel and Distributed Computing*, 24:100–106, 1995.

[27] Stanković RS. Simple theory of decision diagrams for representation of discrete functions. In *Proc. of the 4th Int. Workshop on Applications of the Reed-Muller Expansion in Circuit Design*. University of Victoria, BC, Canada, pp. 161–178, 1999.

[28] Stanković RS. Some remarks on basic characteristics of decision diagrams. In *Proc. of the 4th Int. Workshop on Applications of the Reed-Muller Expansion in Circuit Design.* University of Victoria, BC, Canada, pp. 139–148, 1999.

[29] Stanković RS. Unified view of decision diagrams for representation of discrete functions. *Int. J. Multi-Valued Logic and Soft Computing,* 8(2):237–283, 2002.

[30] Varadarajan R. Embedding shuffle networks into hypercubes. *J. Parallel and Distributed Computing,* 11:252–256, 1990.

[31] Wagner AS. Embedding the complete tree in hypercube. *J. Parallel and Distributed Computing,* 26:241–247, 1994.

[32] Wu AY. Embedding a tree network into hypercubes. *J. Parallel and Distributed Computing,* 2:238–249, 1985.

[33] Yanushkevich SN, Shmerko VP. and Lyshevski SE. *Logic Design of NanoICs.* CRC Press, Boca Raton, FL, 2005.

[34] Yanushkevich SN, Miller DM, Shmerko VP, and Stanković RS. *Decision Diagram Techniques for Micro- and Nanoelectronic Design.* Taylor & Francis/CRC Press, Boca Raton, FL, 2006.

6

Foundation of Boolean Data Structures

6.1 Introduction

Binary arithmetic provides a number possibilities for the manipulation of binary numbers. However, the design of devices for operation of binary data requires the detail, formal description of the manipulation. In 1854, about a century before the advent of modern computers, George Boole developed the theory of logical calculus, which has become known as *Boolean algebra*. He set forth a system of axioms and theorems on true–false relationships. If true is represented by 1 and false by 0, Boolean algebra can be applied to data representation and processing. In 1938, Claude Shannon applied Boolean algebra to Boolean circuit design. The basic element of these circuits was a switch that can be in one of two states: open or closed. Boole's theorems and Shannon's techniques help us design logic networks for calculations binary numbers. This particular form of Boolean algebra is called *switching theory*.

Boolean algebra is a universal data structure in the sense that (a) an arbitrary operations and manipulation of binary numbers can be represented in terms of Boolean algebra, and (b) this representation is technology independent, that is, it can be applied for description of any data structure independently of the underlying physical and chemical phenomena. However, various Boolean data structures based on the formalism of Boolean algebra can be developed with respect to the requirements of the technologies and fabrication. There are various algebras that are an extension or generalization of Boolean algebra, for example, probabilistic logic, fuzzy logic, and multivalued logic. In this chapter, we introduce the basic aspects of Boolean algebra that are used in Boolean data structure construction.

The theoretical foundations for digital design include

▶ Algebra over the two-element set $\{0, 1\}$

▶ Data structures for representing Boolean functions

▶ Techniques for the manipulation of these data structures, including transferring between different structures

6.2   Definition of algebra over the set $\{0,1\}$

A *universal* algebra consists of a set of elements and operations on this set. Boolean algebras are a particular case of universal algebra. There is an infinite number of different Boolean algebras. The basic requirements of any algebra employed to describe and manipulate Boolean functions include

▶ *Functional completeness* - The algebra must be capable of describing and manipulating an arbitrary Boolean function.

▶ *Flexibility* - The algebra must be amenable to the manipulation of Boolean functions so that design algorithms can be set up and employed with reasonable ease.

▶ *Implementability* - The basic connectives and any high-level functional operations should have simple and reliable physical logic circuit counterparts.

6.2.1   Boolean algebra over the set $\{0,1\}$

The simplest Boolean algebra is a two-valued Boolean algebra defined over the two-element set $B = \{0,1\}$. The algebra $\{B; \ \vee, \cdot; \ ^{-}; \ 0,1\}$ is called a *Boolean algebra*, or *switching algebra*. A two-valued Boolean algebra consists of (Figure 6.1)

▶ A set of two *elements*, $B = \{0,1\}$

▶ A set of *operations*: *binary* operations, Boolean sum $\vee$ and Boolean product $\cdot$ (also denoted $\wedge$), and the *unary* operation, complement, denoted by $^{-}$

▶ Two distinguished elements: a unique real number called a *zero element*, 0, and a *unity element*, 1

▶ A number of *axioms* or *postulates*

A binary operation is a rule that assigns to each pair of elements from a set of elements B a unique element from the same set B. Postulates are the basic assumptions from which it is possible to deduce the rules, theorems, and properties of the system.

6.2.2   Postulates

The *computation rules* of Boolean algebra, known as *Huntington's postulates*, or *laws*, are defined as follows (Figure 6.2):

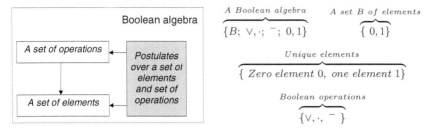

FIGURE 6.1
Boolean algebras are defined over sets of elements, operations, and postulates.

Computation rules of a Boolean algebra

▶ *Identities*

With respect to Boolean addition

$$x \vee 0 = x, \quad 0 \vee x = x, \quad x \vee 1 = 1$$
$$x \cdot 1 = x, \quad 1 \cdot x = x, \quad x \cdot 0 = 0$$

With respect to Boolean multiplication

▶ The *commutative* laws

With respect to Boolean addition

$$x \vee y = y \vee x \qquad \text{and} \qquad x \cdot y = y \cdot x$$

With respect to Boolean multiplication

▶ The *distributive* laws

Multiplication over addition

$$x \cdot (y \vee z) = (x \cdot y) \vee (x \cdot z)$$
$$x \vee (y \cdot z) = (x \vee y) \cdot (x \vee z)$$

Addition over multiplication

▶ *Inverse*: For every element $x \in B$, there exists an element $\bar{x} \in B$ called the *complement* of x such that

$$x \vee \bar{x} = 1 \text{ (additive inverse) and } x \cdot \bar{x} = 0 \text{ (multiplicative inverse)}$$

Techniques for proving some of the computational algebraic rules are given in Table 6.1.

6.2.3 The principle of duality

One can observe a consistent symmetry for each postulate: between identities with respect to the Boolean sum and product, between commutative laws with respect to the Boolean sum and product, and between distributive laws with respect to the Boolean sum and product. Symmetries can also be observed in the complement law. This fundamental interchangeability property of

The operations of Boolean algebra include the Boolean sum $\vee$, Boolean product $\cdot$, and complement $^{-}$

Operations with constants	$x \vee 0 = x$
	$x \cdot 1 = x$
Manipulation	$x \cdot (x \vee z) = (x \cdot y) \vee (x \cdot z)$
	$x \vee (x \cdot z) = (x \vee y) \cdot (x \vee z)$
Elimination of variables	$x \vee \overline{x} = 1$
	$x \cdot \overline{x} = 0$

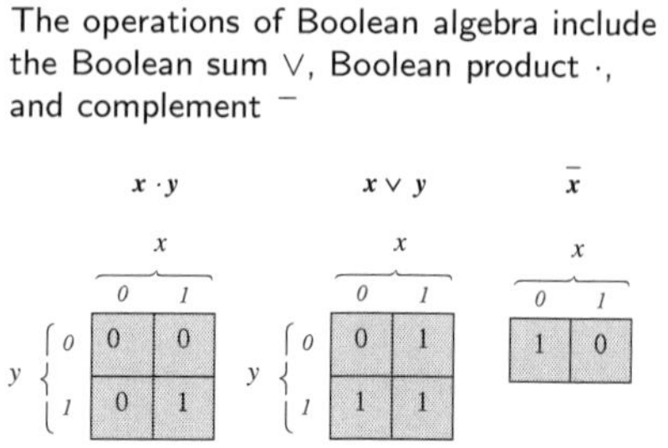

$x \cdot y$

	x	
	0	1
y 0	0	0
1	0	1

$x \vee y$

	x	
	0	1
y 0	0	1
1	1	1

$\overline{x}$

x	
0	1
1	0

FIGURE 6.2
Computation rules of Boolean algebra.

Boolean algebra is termed the *principle of duality*. Principle of duality states that each theorem of a Boolean algebra has a dual, which can be obtained as follows:

> **Given:** A Boolean function f.
> **Step 1:** Interchange the OR and AND operations in the expression.
> **Step 2:** Interchange the 0 and 1 elements of the expression.
> **Result:** The resulting function g is a dual of f.

Duality is used to prove that for each valid equation over a Boolean algebra, the dual equation is also valid. If we can prove, through a series of logical steps, that a given theorem is true, it immediately implies that the dual theorem is also true, since the dual of the logical steps that proves the original theorem, proves the dual theorem.

The principle of duality is the basis for deriving many of the properties of Boolean functions, utilized in efficient computing.

Example 6.1   *(a) The **dual** to the identity postulate, $x \vee 0 = x$, is the equation $x \cdot 1 = x$: $x \vee 0 = x \overset{Dual}{\Leftrightarrow} x \cdot 1 = x$. (b) The dual to $x(x \vee y) = x$ is $x \vee xy = x$.*

An arbitrary Boolean function f and its dual g are not equal, $g \neq f$.

Example 6.2   *Consider the Boolean function $f = \overline{x}_1 \overline{x}_3 \vee \overline{x}_1 x_2 \vee x_2 x_3$. The dual to this expression is*

$$g = (\overline{x}_1 \vee \overline{x}_3)(\overline{x}_1 \vee x_2)(x_2 \vee x_3) = \overline{x}_1 x_2 \vee \overline{x}_1 x_3 \vee x_2 \overline{x}_3$$

Note that $g \neq f$.

TABLE 6.1

Some of the computation rules

Computation rule	Formal description
Identity	$$x = \overbrace{x \vee 0}^{Identity} = \overbrace{x \vee (x \cdot \overline{x})}^{Inverse} = \overbrace{(x \vee x) \cdot (x \vee \overline{x})}^{Distributivity}$$ $$= \underbrace{(x \vee x)}_{Inverse} \cdot 1 = \underbrace{x}_{Identity} \vee x$$
Property of the element 1	$$x \vee 1 = \overbrace{(x \vee 1) \cdot 1}^{Identity} = \overbrace{(x \vee 1) \cdot (x \vee \overline{x})}^{Inverse} = \overbrace{x \vee (1 \cdot \overline{x})}^{Distributivity}$$ $$= x \vee \overline{x} = 1$$
Absorption	$$x \vee (x \cdot y) = \overbrace{(x \cdot 1) \vee (x \cdot y)}^{Identity} = \overbrace{x \cdot (1 \vee y)}^{Distributivity} = x \cdot 1 = x$$ $$x(x \vee y) = \overbrace{(x \vee 0) \cdot (x \vee y)}^{Identity} = \overbrace{x \vee (0 \cdot y)}^{Distributivity} = x \vee 0 = x$$
Adjacency	$$xy \vee x\overline{y} = x \overbrace{(y \vee \overline{y})}^{Inverse} = \overbrace{x \cdot 1}^{Identity} = x$$ $$(x \vee y) \cdot (x \vee \overline{y}) = x \overbrace{(y \vee \overline{y})}^{Inverse} = \overbrace{x \cdot 1}^{Identity} = x$$
Simplification	$$x \vee \overline{x}y = \overbrace{(x \vee \overline{x})(x \vee y)}^{Distributivity} = \overbrace{1 \cdot (x \vee y)}^{Identity} = x \vee y$$ $$x(\overline{x} \vee y) = \overbrace{x\overline{x} \vee xy}^{Distributivity} = \overbrace{0 \vee xy}^{Identity} = xy$$

Alternatively, a dual of a Boolean function can be found using the following algorithm:

> **Given:** A Boolean function f.
> **Step 1:** Find inverse of the Boolean function f: $\overline{f}$.
> **Step 2:** Complement the variables (change $\overline{x}$ to x and x to $\overline{x}$).
> **Result:** The resulting function g is a dual of f.

Example 6.3   *Given $f = x_1x_2 \vee x_3$, its complement is derived as follows: $\overline{f} = \overline{x_1x_2 \vee x_3} = \overline{x_1x_2}\,\overline{x_3} = (\overline{x}_1 \vee \overline{x}_2)\overline{x}_3 = \overline{x}_1\overline{x}_3 \vee \overline{x}_2\overline{x}_3$. Complement the variables of $\overline{f}$ and find the function $g = x_1x_3 \vee x_2x_3$. It is a dual to f, since g is the same function that can be obtained by finding a dual directly from f: $g = (x_1 \vee x_3)x_3 = x_1x_2 \vee x_2x_3$.*

6.2.4   Switch-based interpretation

Boolean postulates can be interpreted using switches. This interpretation is crucial for their implementation by various physical and chemical phenomena.

In Table 6.2, several of the computation rules of Boolean algebra are interpreted using switches. For example, the Boolean product is a series, and the Boolean sum is a parallel switch arrangement. Note that while $x = 0$ is represented by an open switch, the $\bar{x} = 1$ appearing in the same problem is represented by a closed switch.

6.2.5   Boolean algebra over Boolean vectors

A vector $\mathbf{a} = (a_1, a_2, \ldots, a_n)$ with n binary elements $a_i \in \{0, 1\}$, $i = 1, 2, \ldots, n$ is called an *n-dimensional Boolean vector*. We also denote the Boolean vector

as $\mathbf{a} = \begin{bmatrix} a_1 \\ a_2 \\ \cdot \\ \cdot \\ \cdot \\ a_n \end{bmatrix}$ or $\mathbf{a} = \begin{bmatrix} a_1 & a_2 & \ldots & a_n \end{bmatrix}^T$, where T means its transposition.

Example 6.4	*A 4D Boolean vector* $\mathbf{a} = (1, 0, 0, 1)$ *can be written as follows:* $\mathbf{a} = \begin{bmatrix} 1 \\ 0 \\ 0 \\ 1 \end{bmatrix} = \begin{bmatrix} 1 & 0 & 0 & 1 \end{bmatrix}^T.$

Let $\mathbf{B}^n = \{(a_1, a_2, \ldots, a_n)\}$, $a_i \in \{0, 1\}$, be the set of n-dimensional Boolean vectors. Consider two vectors, $\mathbf{a} = (a_1, a_2, \ldots, a_n)$ and $\mathbf{b} = (b_1, b_2, \ldots, b_n)$ in $\mathbf{B}^n$. The following operations with these vectors are specified: (a) Boolean sum ($\vee$): $\mathbf{a} \vee \mathbf{b} = (a_1 \vee b_1, \; a_2 \vee b_2, \ldots, a_n \vee b_n)$; (b) Boolean product ($\cdot$): $\mathbf{a} \cdot \mathbf{b} = (a_1 \cdot b_1, \; a_2 \vee b_2, \ldots, a_n \cdot b_n)$; and (c) complement: $\bar{\mathbf{a}} = (\bar{a}_1, \bar{a}_2, \ldots, \bar{a}_n)$.

If we specify the 0 and 1 elements in vector notation as $0 = \{(0, 0, \ldots, 0)\}$ and $1 = \{(1, 1, \ldots, 1)\}$, respectively, then the system $\langle \mathbf{B}^n, \vee, \cdot, 0, 1 \rangle$ is a *Boolean algebra over Boolean vectors*.

Example 6.5	*Consider the Boolean vectors* $\mathbf{a} = \begin{bmatrix} 0 & 1 & 1 & 0 \end{bmatrix}^T$ *and* $\mathbf{b} = \begin{bmatrix} 1 & 0 & 0 & 0 \end{bmatrix}^T$. *The following operations hold on these vectors:* **Boolean sum:** $\mathbf{a} \vee \mathbf{b} = \begin{bmatrix} 0 \vee 1 \\ 1 \vee 0 \\ 1 \vee 0 \\ 0 \vee 0 \end{bmatrix} = \begin{bmatrix} 1 \\ 1 \\ 1 \\ 0 \end{bmatrix}$; **Boolean product:** $\mathbf{a} \cdot \mathbf{b} = \begin{bmatrix} 0 \cdot 1 \\ 1 \cdot 0 \\ 1 \cdot 0 \\ 0 \cdot 0 \end{bmatrix} = \begin{bmatrix} 0 \\ 0 \\ 0 \\ 0 \end{bmatrix}$; **complement:** $\bar{\mathbf{a}} = \begin{bmatrix} \bar{0} \\ \bar{1} \\ \bar{1} \\ \bar{0} \end{bmatrix} = \begin{bmatrix} 1 \\ 0 \\ 0 \\ 1 \end{bmatrix}$; $\bar{\mathbf{b}} = \begin{bmatrix} \bar{1} \\ \bar{0} \\ \bar{0} \\ \bar{0} \end{bmatrix} = \begin{bmatrix} 0 \\ 1 \\ 1 \\ 1 \end{bmatrix}$.

TABLE 6.2
Computation rules in terms of switches

Computation rule	Switch-based interpretation
Idempotence $x \lor x = x$	
Idempotence $x \cdot x = x$	
Identity $x \lor 1 = 1$	
Identity $x \lor 0 = x$	
Identity $x \cdot 1 = x$	
Complement $\bar{x} \lor x = 1$	

6.2.6   DeMorgan's law

DeMorgan's law provides the possibility to manipulate complemented variables and equations. DeMorgan's law states:

$$\overline{x_1 \cdot x_2} = \overline{x}_1 \vee \overline{x}_2$$

$$\overline{x_1 \vee x_2} = \overline{x}_1 \cdot \overline{x}_2$$

These equations can be generalized for n variables $x_1, x_2, \ldots, x_n$ as follows: $\overline{x_1 \cdot x_2 \cdots x_n} = \overline{x}_1 \vee \overline{x}_2 \vee \cdots \vee \overline{x}_n$ and $\overline{x_1 \vee x_2 \vee \cdots \vee x_n} = \overline{x}_1 \cdot \overline{x}_2 \cdots \overline{x}_n$. Note that two successive complements cancel each other, that is, $\overline{\overline{x}} = x$. This rule is known as *involution*.

Example 6.6   *Examples of the manipulation of Boolean equations using* **DeMorgan's law** *are as follows:*

$$(a) \quad x_1 \cdot x_2 = \overbrace{\overline{\overline{x_1 \cdot x_2}}}^{Involution} = \overline{\overline{x}_1 \vee \overline{x}_2};$$

$$(b) \quad \overline{x_1 \cdot x_2} = \overline{x}_1 \vee \overline{x}_1;$$

$$(c) \quad \overline{x_1 \cdot x_2 \cdot x_3} = \overline{x}_1 \vee \overline{x}_2 \vee \overline{x}_3;$$

$$(d) \quad x_1 \vee x_2 = \overline{\overline{x_1 \vee x_2}} = \overline{\overline{x}_1 \cdot \overline{x}_2};$$

$$(e) \quad \overline{x_1 \vee x_2 \vee x_3} = \overline{x}_1 \cdot \overline{x}_2 \cdot \overline{x}_3;$$

$$(f) \quad \overline{(x_1 \vee x_2)(\overline{x}_3 \vee x_4)} = (\overline{x_1 \vee x_2}) \vee (\overline{\overline{x}_3 \vee x_4})$$
$$= \overline{x}_1 \overline{x}_2 \vee x_3 \overline{x}_4$$

6.3   Boolean functions

There are two distinguished forms of Boolean structures: *Boolean formulas*, and *Boolean functions*. Boolean formulas are a useful form of abstraction, but they are not acceptable for computation.

6.3.1   Boolean formulas

Boolean functions are particular functions that can be described in terms of *expressions* over Boolean algebra, called *Boolean formulas*. A Boolean formula of n variables is a *string of symbols* of $x_1, x_2, \ldots, x_n$, the binary operations of Boolean sum ($\vee$), Boolean product ($\cdot$), unary operation of the complement ($^-$), and brackets ( ). A Boolean formula is the Boolean function after specification of values, given assignments of variables.

Boolean formulas are useful for the study of various Boolean algebras and their relationships, but not acceptable for computing. In order to obtain Boolean expressions suitable for efficient computation, Boolean formulas must be transformed into Boolean functions:

$$\underbrace{\text{Boolean formulas}}_{Abstract\ data\ structure} \xrightarrow{Specification} \underbrace{\text{Boolean functions}}_{Computing\ data\ structure}$$

The relationship between Boolean formulas and Boolean functions is not one-to-one; many different formulas can represent the same Boolean function.

Example 6.7 *The **Boolean formulas** $f_1 = (x_1 \vee x_2) \cdot (x_3 \vee x_2 \cdot \overline{x}_4 \vee \overline{x}_1 \cdot x_2)$ and $f_2 = x_3 \cdot (x_1 \vee x_2) \vee x_2 \cdot (\overline{x}_1 \vee \overline{x}_4)$ represent the same **Boolean function** because*

$$\overbrace{(x_1 \vee x_2) \cdot (x_3 \vee x_2 \cdot \overline{x}_4 \vee \overline{x}_1 \cdot x_2)}^{f_1}$$
$$x_1(x_3 \vee x_2 \cdot \overline{x}_4 \vee \overline{x}_1 \cdot x_2) \vee x_2(x_3 \vee x_2 \cdot \overline{x}_4 \vee \overline{x}_1 \cdot x_2)$$
$$= x_1 \cdot x_3 \vee x_1 \cdot x_2 \cdot \overline{x}_4 \vee x_2 \cdot x_3 \vee x_2 \cdot \overline{x}_4 \vee \overline{x}_1 \cdot x_2$$
$$= x_1 \cdot x_3 \vee x_1 \cdot x_2 \cdot \overline{x}_4 \vee x_2 \cdot x_3 \vee x_2 \cdot \overline{x}_4 \vee \overline{x}_1 x_2$$
$$= x_1 \cdot x_3 \vee \cdot x_2 \cdot \overline{x}_4 (x_1 \vee 1) \vee x_2 \cdot x_3 \vee x_2 \cdot \overline{x}_4 \vee \overline{x}_1 x_2$$
$$= x_1 \cdot x_3 \vee x_2 \cdot x_3 \vee x_2 \cdot \overline{x}_4 \vee \overline{x}_1 \cdot x_2$$
$$= \underbrace{x_3 \cdot (x_1 \vee x_2) \vee x_2 \cdot (\overline{x}_1 \vee \overline{x}_4)}_{f_2}$$

6.3.2 Boolean functions

If a variable y depends on a variable x in such a way that each value of x determines exactly one value of y, then y is said to be a *function* of x. In terms of the inputs and outputs of some device, the function f is a rule that associates a unique output (exactly one) with each input. If the input is denoted by x, then the output is denoted by $f(x)$, or simply f. Thus, a function cannot assign two different outputs to the same input.

If $x_1, x_2, \ldots, x_n$ are variable inputs, there is a rule that assigns a unique value of f to given value of $x_1, x_2, \ldots, x_n$, and f is called a function of $x_1, x_2, \ldots, x_n$, $f = f(x_1, x_2, \ldots, x_n)$. The set of all possible inputs $x_1, x_2, \ldots, x_n$ is called the *domain* of f, and the set of outputs (resulting from varying $x_1, x_2, \ldots, x_n$ over the domain) is called the *range* of f.

Functions can be combined to define new functions. For example, the output of one function can be connected to the input of another. If the first function is f_1 and the second is f_2, then the composed function can be written as $f_1 \circ f_2$.

Functions can be described numerically by tables, geometrically by graphs, algebraically by formulas, or, in order to visualize the implementation of a Boolean function in terms of logic gates, logic diagrams. A logic diagram is a *visual syntax* for describing a logic network as an interconnection of logic gates, each of which emphasizes one particular input-to-output transformation of a logic signal.

A Boolean function described by an algebraic expression consists of

▶ Boolean (binary) variables

▶ The constants 0 and 1

▶ The logic operation symbols

Each of the variables appearing in the Boolean formula is assigned a value from the set $\{0, 1\}$ and evaluation is done using the operations of Boolean algebra. When a Boolean formula is evaluated for all possible *assignments* of values to the variables, the set of pairs (assignment, value) is a *Boolean function*. For a given value of the binary variables, the function can be equal to either 1 or 0.

> **Example 6.8**   *Boolean formulas can be evaluated using the following techniques:*
> *(a) Given the assignment $(x_1 = 1, x_2 = 0, x_3 = 1)$, the value of the Boolean formula $f = x_1 \vee \overline{x}_3 x_2 \vee x_2 \overline{x}_1$ is calculated as follows: $f = 1 \vee \overline{1} \cdot 0 \vee 0 \cdot \overline{1} = 1 \vee 0 \vee 0 = 1$.*
> *(b) The Boolean formula $f = x_1 \vee \overline{x}_2 x_3$ is equal to 1 if x_1 is equal to 1 or if both x_2 and x_3 are equal to 1. Formula f is equal to 0 otherwise.*

Boolean functions derived from formulas are called *algebraic* representations. Operations on Boolean variables such as Boolean sum, Boolean product, and complement can also be applied to the functions.

> **Example 6.9**   *If $f_1 = x_1 x_2$ and $f_2 = \overline{x}_1 \overline{x}_2$, then $f = f_1 \vee f_2 = x_1 x_2 \vee \overline{x}_1 \overline{x}_2$.*

6.4   Fundamentals of computing Boolean functions

There are many forms that allow one each Boolean function to be described using algebraic representations. The efficiency of computing Boolean functions depends on the form of their representation.

6.4.1 Literals and terms

The elementary groups of variables are Boolean terms:

The terms are composed of literals and are characterized as follows:

▶ A Boolean *literal* is either an uncomplemented or a complemented variable x_i. The i-th literal is denoted by $x_i^{c_i}$: where $x_i^0 = \overline{x}_i$ for $c_i = 0$, and $x_i^1 = x_i$ for $c_i = 1$.

▶ A Boolean *product term* or product, is either a single literal, or the AND (product) of literals, that is, literals connected by the AND operation: $x_1^{c_1} x_2^{c_2} \ldots x_k^{c_k}$.

▶ A Boolean *sum term*, or sum, is either a single literal, or the OR (sum) of literals: $x_1^{c_1} \vee x_2^{c_2} \vee \ldots \vee x_k^{c_k}$.

A Boolean function can be described using these groups of variables in various combinations. In determining the complexity of Boolean equations, one of the measures is the number of literals.

Example 6.10 *The Boolean function f of five variables is described by the two **terms**:* $f = \underbrace{(x_1 x_2 \overline{x}_3 x_4 x_5)}_{Product\ term} \underbrace{(\overline{x}_1 \vee x_2 \vee \overline{x}_3 \vee x_4)}_{Sum\ term}$. *The product and sum terms include five and four **literals**, respectively.*

6.4.2 Minterms and maxterms

The Boolean variables can be grouped into a *minterm* and a *maxterm*.

Minterms

A *minterm* of k variables $x_1, x_2, \ldots, x_k$ is a Boolean *product* of m literals, in which each variable appears exactly once in either true, x_i, or complemented form, $\overline{x}_i$, but not both. A minterm has the following properties: (a) each minterm has a value of 1 for exactly one combination of values of the variables, and (b) there are 2^k minterms of k variables $x_1, x_2, \ldots, x_k$.

Example 6.11   *(a) Four **minterms** can be constructed from two variables x_1 and x_2:*

$$x_1 x_2, \overline{x}_1 x_2, \ x_1 \overline{x}_2, \ and \ \overline{x}_1 \overline{x}_2.$$

(b) For a Boolean function of four variables x_1, x_2, x_3 and x_4, the products $x_1 \overline{x}_2 x_3 x_4$ and $x_1 \overline{x}_2 x_3 \overline{x}_4$ are minterms, but the product $x_1 \overline{x}_2 x_4$ is not a minterm since it lacks the variable x_3.

Maxterms

A maxterm of m variables $x_1, x_2, \ldots, x_m$ is a Boolean *sum* of m literals in which each variable appears exactly once in either true, x_i, or complemented form, $\overline{x}_i$, but not both. A maxterm has the following properties: (a) each maxterm has a value of 0 for exactly one combination of values of the variables, and (b) there are 2^m maxterms of m variables $x_1, x_2, \ldots, x_m$.

Example 6.12   *(a) Four **maxterms** can be constructed from the two variables x_1 and x_2:*

$$(x_1 \vee x_2), \ (\overline{x}_1 \vee x_2), \ (x_1 \vee \overline{x}_2), \ and \ (\overline{x}_1 \vee \overline{x}_2).$$

(b) For a Boolean function of four variables x_1, x_2, x_3 and x_4, the sums $\overline{x}_1 \vee \overline{x}_2 \vee x_3 \vee x_4$ and $x_1 \vee \overline{x}_2 \vee x_3 \vee \overline{x}_4$ are maxterms, but the sum $\overline{x}_1 \vee \overline{x}_2 \vee x_3$ is not a maxterm since it lacks the variable x_4.

6.4.3   Canonical SOP and POS expressions

Canonical, or *standard*, *sum-of-products* (SOP), respectively, and *product-of-sums* (POS) are expressions that consist of minterms and maxterms only.

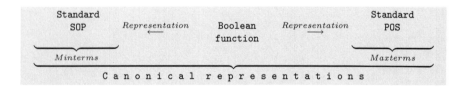

Given a Boolean function f of n variables, each product and sum term in these representations contains exactly n variables; the numbers of product and sum terms depends on the particular function.

Standard SOP representation of a Boolean function

Given: A Boolean function of n variables.

Step 1: Assume an assignment $c_1 c_2 ... c_n$ to the n variables, and find the corresponding value of the function. Repeat for all 2^n possible assignments of variables.

Step 2: Select those assignments for which the function assumes a value of 1, and derive the product of n variables according to the rule $x_1^{c_1} x_2^{c_2} ... x_n^{c_n}$, where $x_i^0 = \overline{x}_i$ for $c_i = 0$, and $x_i^1 = x_i$ for $c_i = 1$.

Step 3: Assemble the sum-of-products expression by combining the minterms using logical sum.

Result: The standard SOP expression.

Standard POS representation of a Boolean function

Given: A Boolean function of n variables

Step 1: Assume an assignment $c_1 c_2 ... c_n$ to the n variables, and find the corresponding value of the function. Repeat for all 2^n possible assignments of variables.

Step 2: Select those assignments for which the function assumes a value of 0, and derive the sum of n variables according to the rule $x_1^{c_1} \vee x_2^{c_2} \vee ... \vee x_k^{c_k}$, where $x_i^0 = \overline{x}_i$ for $c_i = 0$, and $x_i^1 = x_i$ for $c_i = 1$.

Step 3: Assemble the product-of-sums expression by combining the maxterms using logical product.

Result: The standard POS expression.

Example 6.13 *Given the Boolean function* $f = x_1 x_2 \vee x_3$, *the* standard SOP *expression is derived as follows:*

$$f = x_1 x_2 (\overline{x}_3 \vee x_3) \vee (\overline{x}_1 \vee x_1)(\overline{x}_2 \vee x_2) x_3$$
$$= x_1 x_2 \overline{x}_3 \vee x_1 x_2 x_3 \vee \overline{x}_1 \overline{x}_2 x_3 \vee \overline{x}_1 x_2 x_3 \vee x_1 \overline{x}_2 x_3 \vee x_1 x_2 x_3$$
$$= \overline{x}_1 \overline{x}_2 x_3 \vee x_1 \overline{x}_2 x_3 \vee \overline{x}_1 x_2 x_3 \vee x_1 x_2 \overline{x}_3 \vee x_1 x_2 x_3$$

The standard POS *expression is derived using the following manipulations:*

$$f = (x_1 \vee x_3)(x_2 \vee x_3) = (x_1 \vee \overline{x}_2 x_2 \vee x_3)(\overline{x}_1 x_1 \vee x_2 \vee x_3)$$
$$= (x_1 \vee \overline{x}_2 \vee x_3)(x_1 \vee x_2 \vee x_3)(\overline{x}_1 \vee x_2 \vee x_3)(x_1 \vee x_2 \vee x_3)$$
$$= (x_1 \vee \overline{x}_2 \vee x_3)(\overline{x}_1 \vee x_2 \vee x_3)(x_1 \vee x_2 \vee x_3)$$

The standard SOP and POS representations are dual:

$$\underbrace{\text{Standard SOP representation}}_{\text{Boolean function}} \overset{Duality}{\longleftrightarrow} \underbrace{\text{Standard POS representation}}_{\text{Boolean function}}$$

Example 6.14   *Given the standard POS expression $f = (x_1 \vee \overline{x}_2 \vee x_3)(\overline{x}_1 \vee x_2 \vee x_3)(x_1 \vee x_2 \vee x_3)$, its dual is a standard SOP $f = x_1\overline{x}_2x_3 \vee \overline{x}_1x_2x_3 \vee x_1x_2x_3$. Note that $f \neq g$.*

Any Boolean function can be expressed in standard SOP and POS form.

Example 6.15   *A Boolean function f, called the **majority** function, is expressed in the standard SOP and POS forms as follows:*

$$\overbrace{f = \overline{x}_1x_2x_3 \vee x_1\overline{x}_2x_3 \vee x_1x_2\overline{x}_3 \vee x_1x_2x_3}^{The\ SOP\ form}$$

$$\underbrace{= (x_1 \vee x_2 \vee x_3)\,(x_1 \vee x_2 \vee \overline{x}_3)\,(x_1 \vee \overline{x}_2 \vee x_3)\,(\overline{x}_1 \vee x_2 \vee x_3)}_{The\ POS\ form}$$

Let m_i and M_j be the i-th minterm and j-th maxterm, respectively, for a Boolean function f of n variables, where $i, j \in \{0, 1, 2, \ldots, 2^n - 1\}$. The short hand notation of an SOP form of the Boolean function is

$$\texttt{Minterm form} = \bigvee_i m(i)$$

where $\bigvee$ denotes the Boolean sum. The short hand notation of a POS form of the Boolean function is

$$\texttt{Maxterm form} = \prod_j M(j)$$

where $\prod$ denotes the Boolean product.

Example 6.16   *A Boolean function f of two variables x_1 and x_2 given the minterms $m_0 = \overline{x}_1\overline{x}_2$ and $m_3 = x_1x_2$, $i = 0, 3$, and maxterms $M_1 = x_1 \vee \overline{x}_2$ and $M_2 = \overline{x}_1 \vee x_2$, $j = 1, 2$. The minterm and maxterm expressions are as follows:*

$$f = \underbrace{\bigvee_{i=0,3} m(0,3) = \overline{x}_1\overline{x}_2 \vee x_1x_2}_{Minterm\ expression}$$

$$= \underbrace{\prod_{j=1,2} M(1,2) = \overbrace{(x_1 \vee \overline{x}_2)}^{M_1}\,\overbrace{(\overline{x}_1 \vee x_2)}^{M_2}}_{Maxterm\ expression}$$

6.4.4 Algebraic construction of standard SOP and POS forms

The standard SOP expression can be derived from the SOP form of a Boolean function by multiplying by the sum $x \vee \overline{x} = 1$.

Example 6.17 *A Boolean function of three variables is given in the SOP form $f = x_1 x_3 \vee x_2 x_3 \vee \overline{x}_1 \overline{x}_2 \overline{x}_3$. The missing variables in the first two product terms are added as follows: $f = x_1 \underbrace{(x_2 \vee \overline{x}_2)}_{Equal\ to\ 1} x_3 \vee \underbrace{(x_1 \vee \overline{x}_1)}_{Equal\ to\ 1} x_2 x_3 \vee \overline{x}_1 \overline{x}_2 \overline{x}_3$.*

This does not alter the equation because $x_1 \vee \overline{x}_1 = 1$ and $x_2 \vee \overline{x}_2 = 1$. The standard SOP expression is obtained by using the distributive law and by deleting any duplicated terms produced: $f = x_1 \overline{x}_2 x_3 \vee x_1 x_2 x_3 \vee \overline{x}_1 x_2 x_3 \vee \overline{x}_1 \overline{x}_2 \overline{x}_3$.

The standard POS expression can be derived from the POS form of a Boolean function by adding the product $x\overline{x} = 0$.

Example 6.18 *A Boolean function of three variables is given in the POS form $f = x_1(x_2 \vee \overline{x}_3)$. The missing variables x_2 and x_1 in the terms are added using the rule $x_2 \overline{x}_2 = 0$ and $x_1 \overline{x}_1 = 0$ to get the standard POS form:*

$$f = (x_1 \vee x_2 \overline{x}_2 \vee x_3 \overline{x}_3)(x_1 \overline{x}_1 \vee x_2 \vee x_3)$$
$$= (x_1 \vee x_2 \overline{x}_2 \vee x_3)(x_1 \vee x_2 \overline{x}_2 \vee \overline{x}_3)(x_1 \vee x_2 \vee x_3)(\overline{x}_1 \vee x_2 \vee x_3)$$
$$= \underbrace{(x_1 \vee x_2 \vee x_3)}_{Maxterm\ 0} \underbrace{(x_1 \vee \overline{x}_2 \vee x_3)}_{Maxterm\ 2} \underbrace{(x_1 \vee x_2 \vee \overline{x}_3)}_{Maxterm\ 1} \underbrace{(x_1 \vee \overline{x}_2 \vee \overline{x}_3)}_{Maxterm\ 3}$$
$$\underbrace{(\overline{x}_1 \vee x_2 \vee x_3)}_{Maxterm\ 4}$$

6.5 Proving the validity of Boolean equations

Proving the validity of Boolean data structures is widely used in logic design, in particular, for verification. The *verification* of a Boolean data structure (logic network, decision diagram, network of threshold elements) is the determination of whether or not this data structure implements its specific function. Verification plays a vital role in preventing incorrect logic network designs from being manufactured and used. There are many techniques for the verification of Boolean data structures. Most of these techniques are based on finding the Boolean equations and proving their validity.

The following approaches are used, in particular, to determine if a Boolean equation is valid (Figure 6.3): (*a*) construct a table of the values of the function (the truth table) and evaluate both sides of the equations for all combinations of values of the variables; (*b*) manipulate one side of the equation by applying various theorems until it is identical to the other side; and (*c*) reduce both sides of the equation independently to the same expression.

To prove that an equation is not valid, it is sufficient to show one combination of values of the variables for which the two sides of the equation have different values.

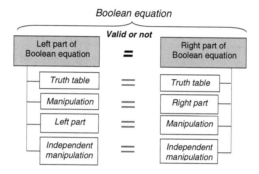

FIGURE 6.3

Techniques of proving the validity of Boolean equations.

Example 6.19

$$\overline{x}_1 x_2 \overline{x}_4 \vee x_2 x_3 x_4 \vee x_1 x_2 \overline{x}_3 \vee x_1 \overline{x}_2 x_4 = x_2 \overline{x}_3 \overline{x}_4 \vee x_1 x_4 \vee \overline{x}_1 x_2 x_3$$

$$\underbrace{\phantom{\overline{x}_1 x_2 \overline{x}_4 \vee x_2 x_3 x_4 \vee x_1 x_2 \overline{x}_3 \vee x_1 \overline{x}_2 x_4}}_{Left\ part} \quad \underbrace{\phantom{x_2 \overline{x}_3 \overline{x}_4 \vee x_1 x_4 \vee \overline{x}_1 x_2 x_3}}_{Right\ part}$$

One of the possible ways for proving this equation is to create truth tables for both parts. The left and right parts of the equation cover the minterms (correspond to the 1s in the table of the function's values of each part) as follows:

0100	0111	1100	1001	0100	0110
0110	1111	1101	1011	1100	0111

Right part: 1001, 1011, 1101, 1111

$$\underbrace{\overline{x}_1 x_2 \overline{x}_4}\ \underbrace{x_2 x_3 x_4}\ \underbrace{x_1 x_2 \overline{x}_3}\ \underbrace{x_1 \overline{x}_2 x_4} = \underbrace{x_2 \overline{x}_3 \overline{x}_4}\ \underbrace{x_1 x_4}\ \underbrace{\overline{x}_1 x_2 x_3}$$

These two sets of minterms correspond to the standard SOP expressions. These expressions are identical, and, thus, the truth tables for the left and right parts of the Boolean equation are identical.

6.6 Gates

The fundamental principle of logic design is the *principle of assembly* of an arbitrary computing network from simple computing elements called *gates*. A gate is a module implementing a simple Boolean function such as AND, OR, etc. These gates are the basic building blocks of combinational modules (logic networks).

6.6.1 Elementary Boolean functions

The simplest Boolean functions are called *elementary*. In Figure 6.4, all $2^{2^n} = 2^{2^2} = 16$ Boolean functions of two variables are given. The following two-variable functions among these 16 are special:

▶ The *constants* 0 ($f_0 = 0$) and 1 ($f_{15} = 1$) are the only two Boolean functions that do not possess any essential variables; in vector form, the Boolean constants 0 and 1 are $0 = [00 \ldots 0]^T$ and $1 = [11 \ldots 1]^T$, respectively.

▶ There are exactly four Boolean functions of a single variable: $f_3 = x_1$, $f_5 = x_2$, $f_{10} = \overline{x}_2$, and $f_{12} = \overline{x}_1$.

▶ The Boolean product is called the AND function, $f_1 = x_1 x_2$. The NAND function is the complemented AND function, $f_{14} = \overline{x_1 x_2}$.

▶ The Boolean sum is called the OR function, $f_7 = x_1 \vee x_2$. The NOR function is the complemented OR function, $f_8 = \overline{x_1 \vee x_2}$

▶ The exclusive OR, called the EXOR function, $f_6 = x_1 \oplus x_2$, and exclusive NOR, called the XNOR function, $f_9 = \overline{x_1 \oplus x_2}$.

In Table 6.3, the symbolic representations called *logic gates*, as well as switch models for the gates, are given.

6.6.2 Switch models for logic gates

In logic design, combinational logic networks are compounded of switches. Switch models represent many possible physical switch mechanisms, for example, the flow of electrons or other information carriers. A switch is represented by three connection points called *terminals* to which external signals may be applied or from which internal signals may be drawn. The lines attached to the terminals can represent wires, pipes, or other transmission media appropriate to the current technology. An *input terminal* allows a signal to enter the switch and change its state. An *output terminal* allows signals to leave the switch. The control variable x is applied to an input terminal of the switch. A terminal that can function both as an input and an output is said to be *bidirectional*. The data terminals of the switch models under consideration are assumed to be bidirectional.

TABLE 6.3
Elementary Boolean functions of two variables and their implementation using gates and switches

Techniques for representation of elementary Boolean functions

Function	Gate symbol	Truth table	Switch model
AND $f = x_1 x_2$		$\begin{array}{cc\|c} x_1 & x_2 & f \\ \hline 0 & 0 & 0 \\ 0 & 1 & 0 \\ 1 & 0 & 0 \\ 1 & 1 & 1 \end{array}$	
OR $f = x_1 \vee x_2$		$\begin{array}{cc\|c} x_1 & x_2 & f \\ \hline 0 & 0 & 0 \\ 0 & 1 & 1 \\ 1 & 0 & 1 \\ 1 & 1 & 1 \end{array}$	
NAND $f = \overline{x_1 x_2}$		$\begin{array}{cc\|c} x_1 & x_2 & f \\ \hline 0 & 0 & 1 \\ 0 & 1 & 1 \\ 1 & 0 & 1 \\ 1 & 1 & 0 \end{array}$	
NOR $f = \overline{x_1 \vee x_2}$		$\begin{array}{cc\|c} x_1 & x_2 & f \\ \hline 0 & 0 & 1 \\ 0 & 1 & 0 \\ 1 & 0 & 0 \\ 1 & 1 & 0 \end{array}$	
EXOR $f = x_1 \oplus x_2$		$\begin{array}{cc\|c} x_1 & x_2 & f \\ \hline 0 & 0 & 0 \\ 0 & 1 & 1 \\ 1 & 0 & 1 \\ 1 & 1 & 0 \end{array}$	
XNOR $f = \overline{x_1 \oplus x_2}$		$\begin{array}{cc\|c} x_1 & x_2 & f \\ \hline 0 & 0 & 1 \\ 0 & 1 & 0 \\ 1 & 0 & 0 \\ 1 & 1 & 1 \end{array}$	

Design example: Boolean functions of two variables

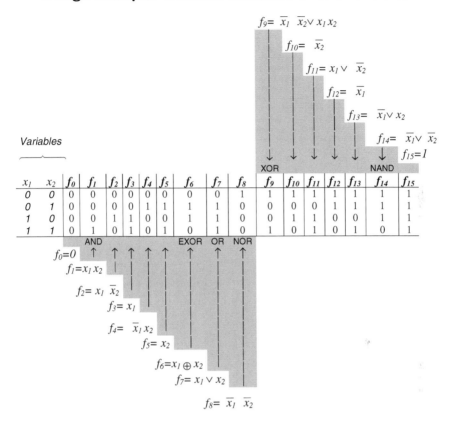

FIGURE 6.4
All 16 Boolean functions of two variables.

The effect of the switch on the output is determined by the *state* of the switch. In Figure 6.5a, the switch has two states:

▶ OPEN, or OFF. This state implies that there is no closed path through the switch connecting data terminals.

▶ CLOSED, or ON. This state implies that data terminals are connected via a path through the switch.

Using the switch shown in Figure 6.5a, switch-based models of elementary Boolean functions can be derived.

Example 6.20 *In Figure 6.5b, the switch-based model of the AND gate is shown.*

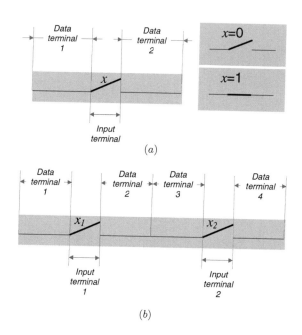

FIGURE 6.5
Model of the simplest switch (a) and switch-based models of the function
AND (b) (Example 6.20).

In Table 6.2, the computation rules of Boolean algebra are interpreted using
switches. These rules can be interpreted using the gate notation (Table 6.4).

6.6.3 Timing diagrams

The pulse has two edges: a *leading* edge and a *trailing* edge. For a positive-
going pulse, the leading edge is a *rising* edge, and the trailing edge is a *falling*
edge. The rising and falling edges are characterized by the rise and fall times,
respectively. The frequency f of a pulse waveform and the period T can be
calculated as $f = 1/T$. The basic timing waveform in computing device is
called the *clock*.

A *timing diagram* is a graphical representation of the relationship of two or
more waveforms with respect to each other on a time basis. Using a timing
diagram, it is possible to determine the logical level of a digital device at any
specified point in time.

Timing diagrams for the primary logic gates are given in Table 6.5.

TABLE 6.4
The computation rules in terms of gates

Techniques for manipulations of elementary Boolean functions

Computation rule	Gate-based interpretation
Idempotence $x \vee x = x$ $x \cdot x = x$	
Identity $x \vee 0 = x$ $x \vee 1 = 1$	
Identity $x \cdot 1 = x$ $x \cdot 0 = 0$	
DeMorgan's rule $\overline{x_1 \vee x_2} = \overline{x}_1 \overline{x}_2$	
Distributivity $x_1(x_2 \vee x_3)$ $= x_1 x_2 \vee x_1 x_3$	
Distributivity $x_1 \vee x_2 x_3$ $= (x_1 \vee x_2)(x_1 \vee x_3)$	

TABLE 6.5
Timing diagrams of the primary logic gates

Timing description of logic gates

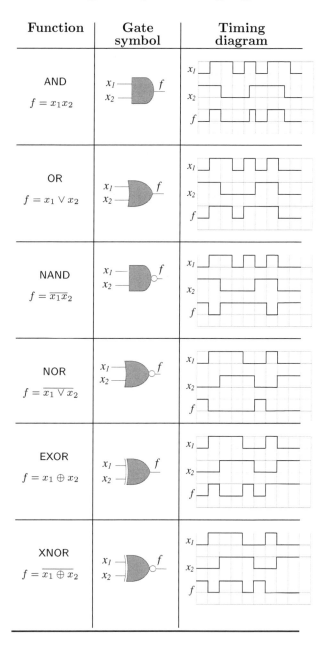

Function	Gate symbol	Timing diagram
AND $f = x_1 x_2$	x_1 — f, x_2	x_1, x_2, f
OR $f = x_1 \vee x_2$	x_1 — f, x_2	x_1, x_2, f
NAND $f = \overline{x_1 x_2}$	x_1 — f, x_2	x_1, x_2, f
NOR $f = \overline{x_1 \vee x_2}$	x_1 — f, x_2	x_1, x_2, f
EXOR $f = x_1 \oplus x_2$	x_1 — f, x_2	x_1, x_2, f
XNOR $f = \overline{x_1 \oplus x_2}$	x_1 — f, x_2	x_1, x_2, f

6.6.4 Performance parameters

The performance of a logic gate is the switching activity measured in terms of the propagation delay time and the power dissipation. The *propagation delay time* is a result of the limitation of *switching activity*. The shorter the propagation delay, the higher the speed of gate and higher the frequency at which it can operate. Propagation delay time of a logic gate is the time interval between the application of an input pulse and the occurrence of the output pulse.

The propagation delay time is used to evaluate the speed of the logic devices, and, therefore, the maximum operating frequency. The later cannot be exceeded since it can lead to violation of the timing for the surrounding logic and, therefore, incorrect operation of the devices.

6.7 Local transformations

A *local transformation* is defined as a set of rules for the simplification of a logic network. These rules are based on the theorems of Boolean algebra and polynomial algebra $GF(2)$ and applied locally. In contrast to the approaches that require the knowledge of the implemented Boolean function, the local transformations can be applied to the subfunctions; these local subfunctions are very simple. A local transformation is characterized as follows:

Property 1: *Technological requirements.* Using a local transformations, it is easy to satisfy the requirements given technology.

Property 2: *Automation.* In automation logic design tools, the consequences of a local transformation can not be predicted, but a functionally equivalent logic networks can be generated.

Property 3: *Flexibility of optimization.* The cost of the obtained network is evaluated, in particular, in terms of the number of logic gate or connections, and a solution with lowest cost can be selected as the final design.

Property 4: *Applicable to various data structures.* Local transformations can be applied to logic networks, decision trees, and diagrams.

Property 5: *Testability.* Local transformations can improve the conditions for testability.

Property 6: *Decomposability.* Local transformations can improve the conditions for decomposability.

The data structure that is obtained by applying the local transformations must be verified. For example, the functionality of logic network must be verified by comparison with an initial logic network (before application of local transformations). Local transformations for the logic networks using AND, OR, gates, and inverters, include the rules described below:

Rule 1: *Reduction of constants:*

- ▶ Remove a constant 1 that is connected to an AND gate.
- ▶ Remove a constant 0 that is connected to an OR gate.
- ▶ Replace an AND gate that has constant 0, with a constant 0.
- ▶ Replace an OR gate that has constant 1, with a constant 1.

Rule 2: *Reduction of duplicated gates:* If there two gates whose inputs and outputs are the same, remove one, and create a fan-out.

Rule 3: *Reduction of inverters:* Remove two inverters that are connected in series.

Rule 4: *Deletion of unused gates:* Remove gates whose outputs are not connected to other gates or terminals.

Rule 5: *Merging gates:* If two AND gates (OR gates) are connected in series, then merge them into one.

Example 6.21	*Figure 6.6 illustrates two types of the* **local transformations**. *Local transformation in the area A: the OR gate is replaced by the wire using the identity rule for variables and constants (Table 6.4)* $x_2 \vee 1 = 1$. *Local transformation in the area B: the AND gate is replaced by the wire based on the identity rule (Table 6.4)* $x_1 \cdot 1 = x_1$.

6.8   Properties of Boolean functions

There are several reasons for studying the properties of Boolean functions:

(a) The analysis of Boolean functions of many input variables is a task of exponential complexity that requires a great amount of time and memory resources. However, the complexity can be reduced if additional structural properties of the functions and related data structures be known.

(b) The properties of Boolean functions have different levels of significance in logic design (synthesis); this depends on their data structure and level of design abstraction. It is important to understand the resources available for the improvement of performance of design tools.

(c) Many techniques for the efficient manipulation of Boolean functions represented in various forms have been developed. These techniques utilize the properties of Boolean functions.

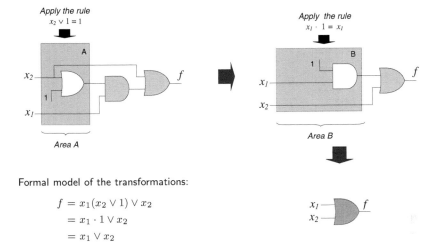

Formal model of the transformations:

$$f = x_1(x_2 \vee 1) \vee x_2$$
$$= x_1 \cdot 1 \vee x_2$$
$$= x_1 \vee x_2$$

FIGURE 6.6
Local transformations in logic network (Example 6.21).

In this chapter, the major properties of Boolean functions are introduced.

6.8.1   Self-dual Boolean functions

The property of Boolean algebra called the *duality principle* has been formulated in the form of *Huntington postulates*. These postulates state that given a valid identity, another valid identity can be obtained by

▶ Interchanging the operations OR and AND
▶ Interchanging the constants 0 and 1, that is, replacing 1s by 0s and 0s by 1s.

> **Example 6.22**   *Applying the duality principle to the identity $x_1 \cdot x_2 \vee \overline{x}_2 \cdot x_3 = x_1 \cdot x_2 \vee \overline{x}_2 \cdot x_3 \vee x_1 \cdot x_3$, we obtain another identity, $(x_1 \vee x_2) \cdot (\overline{x}_2 \vee x_3) = (x_1 \vee x_2) \cdot (\overline{x}_2 \vee x_3) \cdot (x_1 \vee x_3)$.*

A *self-dual* Boolean function f is the function that satisfies the following requirement:

$$f = \overline{f}(\overline{x}_1, \ldots, \overline{x}_n) \qquad (6.1)$$

The recognition of this and other properties can be accomplished using various data structures:

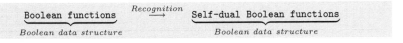

The examples below demonstrate the recognition of self-duality using algebraic data structure.

Example 6.23　　*The Boolean function $f = (x_1 \vee x_2 x_3)x_4$ is* **self-dual** *because its dual function is reduced to the initial one:*

$$\overline{x_1(\overline{x}_2 \vee \overline{x}_3) \vee \overline{x}_4} = (\overline{\overline{x}_1\overline{x}_2})(\overline{\overline{x}_1\overline{x}_3})x_4$$
$$= (x_1 \vee x_2)(x_1 \vee x_3)x_4$$
$$= (x_1 \vee x_1 x_3 \vee x_1 x_2 \vee x_2 x_3)x_4 = (x_1 \vee x_2 x_3)x_4 = f$$

There is an inverse symmetry in the truth tables of the self-dual functions: The upper half of the truth vector is equal to the complemented value of its lower part. This property can also be detected by inspection of the function's decision diagram.

Example 6.24　　*Figure 6.7a shows self-dual functions of two variables. This property can be used to generate the truth tables of the self-dual functions, given an arbitrary upper part of the truth vector (Figure 6.7b). Since there are four possible upper-part vectors (00, 01, 10, and 11), there are four self-dual truth vectors represented by the 2-variable decision tree as shown in Figure 6.7c.*

There are $2^{2^{n-1}}$ self-dual functions of n variables. For example, all Boolean functions of a single variable are self-dual, since, given $n = 1$, $2^{2^{1-1}} = 2$. Only $2^{2^{2-1}} = 4$ functions out of $2^{2^2} = 16$ are self-dual given two Boolean variables. A *self-antidual* function is the Boolean function such that

$$f = f(x_1, \ldots, x_n) = f(\overline{x}_1, \ldots, \overline{x}_n) \qquad (6.2)$$

Example 6.25　　*The Boolean function $f = x_1 \oplus x_2$ is* **self-antidual** *because $f(x_1, x_2) = (\overline{x}_1 \oplus 1) \oplus (\overline{x}_2 \oplus 1) = \overline{x}_1 \oplus \overline{x}_2 = f(\overline{x}_1, \overline{x}_2)$.*

The self-dual functions are used in *self-dual parity-checking*, a method of on-line error detection in logical networks.

6.8.2　Monotonic Boolean functions

A function f is *monotonically increasing*

$$\text{if } a \leq b \text{ implies } f(a) \leq f(b)$$

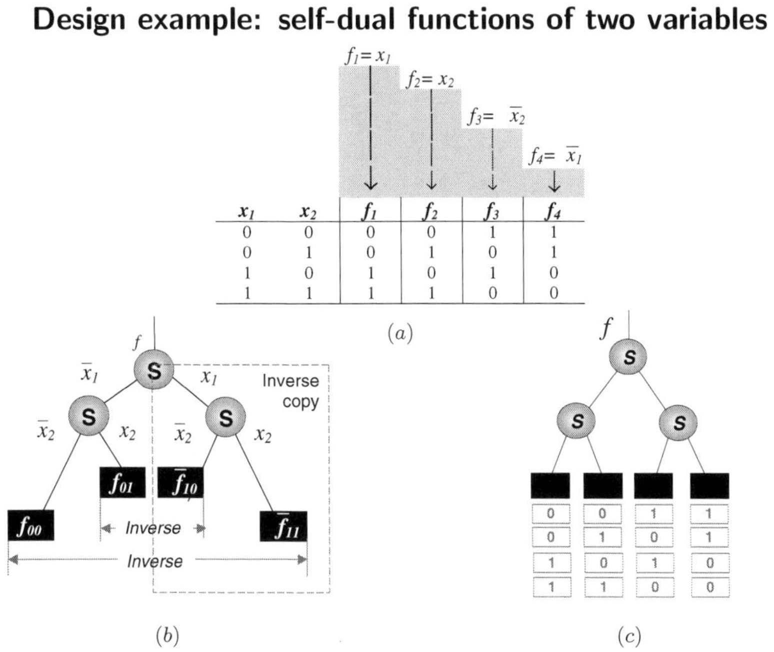

Design example: self-dual functions of two variables

x_1	x_2	f_1	f_2	f_3	f_4
0	0	0	0	1	1
0	1	0	1	0	1
1	0	1	0	1	0
1	1	1	1	0	0

(a)

(b)

(c)

FIGURE 6.7
Self-dual functions of two variables (a), detection (b), and generation (c) of
self-dual functions using decision trees (Example 6.24).

Similarly, it is *monotonically decreasing* function

$$\text{if } a \leq b \text{ implies } f(a) \geq f(b)$$

Recognition of the monotonicity property of a Boolean function has been
found useful in fault-tolerant circuit designs:

Monotonically increasing Boolean functions

Let $\mathbf{a} = (a_1, a_2, \ldots, a_n)$ and $\mathbf{b} = (b_1, b_2, \ldots, b_n)$ be Boolean vectors. Consider
the following relations between these vectors:

Relations between Boolean vectors

▶ If the elements of the vectors satisfy the condition $a_i \geq b_i$, $i = 1, 2, \ldots, n$, the
vector $\mathbf{a}$ is *equal to or greater than* $\mathbf{b}$. $\mathbf{a} \geq \mathbf{b}$;
▶ If $a_i \leq b_i$, the vector $\mathbf{a}$ is *equal to or less than* $\mathbf{b}$, then $\mathbf{a} \geq \mathbf{b}$.
▶ If $a_i \geq b_i$ and $a_i \leq b_i$, then the vectors $\mathbf{a}$ and $\mathbf{b}$ are *incomparable*.

For example, the vectors $\mathbf{a} = \begin{bmatrix} 0 \\ 1 \end{bmatrix}$ and $\mathbf{b} = \begin{bmatrix} 1 \\ 0 \end{bmatrix}$ are incomparable.

Example 6.26   *Consider two cases. <u>Case 1:</u> Let the Boolean vectors be* $\mathbf{a} = (a_1 a_2 a_3 a_4) = (1011)$ *and* $\mathbf{b} = (b_1 b_2 b_3 b_4) = (1001)$. *The vector* $\mathbf{a}$ *is greater than the vector* $\mathbf{b}$, *that is,* $f(\mathbf{a}) \geq f(\mathbf{b})$, *since* $a_1 = b_1$ $(1 = 1)$, $a_3 > b_3$ $(1 > 0)$, $a_2 = b_2$ $(0 = 0)$, *and* $a_4 = b_4$ $(1 = 1)$.
<u>Case 2:</u> Consider the Boolean vectors $\mathbf{a} = (a_1 a_2 a_3 a_4) = (1010)$ *and* $\mathbf{b} = (b_1 b_2 b_3 b_4) = (0111)$. *These vectors are* **incomparable** *because* $a_1 > b_1$ $(1 > 0)$, $a_3 = b_3$ $(1 = 1)$, $a_2 < b_2$ $(0 < 1)$, *and* $a_4 < b_4$ $(0 < 1)$.

If a Boolean function f meets the requirement

$$f(\mathbf{a}) \geq f(\mathbf{b}) \tag{6.3}$$

for any vectors $\mathbf{a}$ and $\mathbf{b}$ such that $\mathbf{a}$ is equal to or greater than $\mathbf{b}$, $\mathbf{a} \geq \mathbf{b}$, then f is a *monotonically increasing* function.

Example 6.27   *Figure 6.8 contains the* **monotonically increasing** *functions of two variables: 0, x_1, x_2, $x_1 x_2$, $x_1 \vee x_2$, and 1. Note that x_2 is a monotonically increasing function as well since $f(x_2 = 0) < f(x_2 = 1)$.*

A logic network that implements a monotonic function is called *a monotonic network*. The useful properties of monotonically increasing functions include the following:

Property 1: A logic product or sum of monotonically increasing function is a monotonically increasing function as well.

Property 2: A monotonically increasing function is described by an SOP *without complemented literals*. That is, they are implemented using AND and OR operations on noncomplemented variables and the constant functions 0 and 1 as input variables.

Example 6.28   *The logic network presented in Figure 6.9 implements a monotonically increasing function.*

Monotonically decreasing Boolean functions

Let $\mathbf{a}$ and $\mathbf{b}$ be Boolean vectors. If f satisfies the requirement $f(\mathbf{a}) \leq f(\mathbf{b})$ for any vectors $\mathbf{a}$ and $\mathbf{b}$ such that $\mathbf{a} \geq \mathbf{b}$, then f is a *monotonically decreasing* function.

Design example: monotonically increasing functions

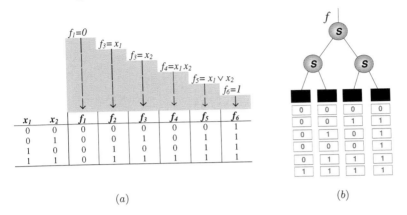

(a)   (b)

FIGURE 6.8
Monotonically increasing functions of two variables: truth table (a) and decision tree (b) (Example 6.27).

Design example: monotonically increasing functions

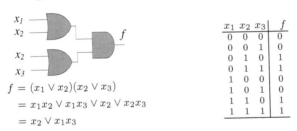

x_1	x_2	x_3	f
0	0	0	0
0	0	1	0
0	1	0	1
0	1	1	1
1	0	0	0
1	0	1	0
1	1	0	1
1	1	1	1

$f = (x_1 \lor x_2)(x_2 \lor x_3)$
$\;\;= x_1x_2 \lor x_1x_3 \lor x_2 \lor x_2x_3$
$\;\;= x_2 \lor x_1x_3$

FIGURE 6.9
Monotonically increasing functions of two variables (Example 6.28).

Example 6.29 *Figure 6.10 represents the* **monotonically decreasing** *functions of two variables:* $0, \; \overline{x}_1,$ $\overline{x}_2, \; \overline{x}_1\overline{x}_2, \; \overline{x}_1 \lor \overline{x}_2,$ *and 1.*

The useful properties of monotonically decreasing functions include the following:

Property 1: A logic sum or a product of the monotonically decreasing function is a monotonically decreasing function, too.

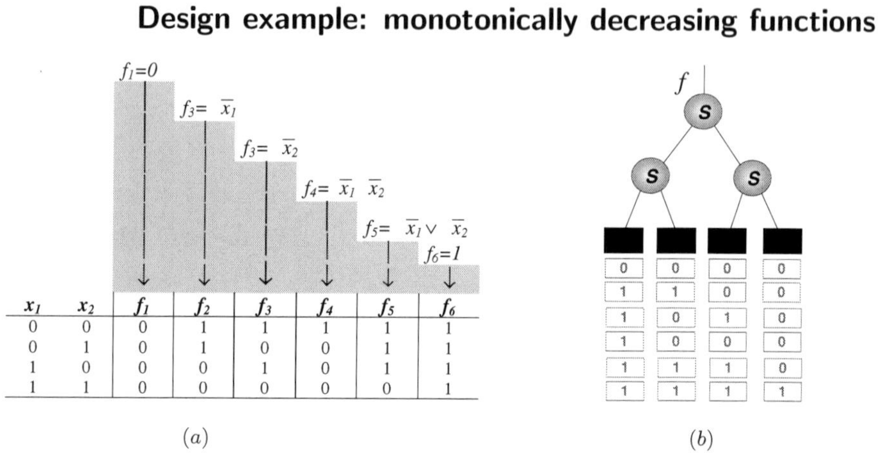

Design example: monotonically decreasing functions

x_1	x_2	f_1	f_2	f_3	f_4	f_5	f_6
0	0	0	1	1	1	1	1
0	1	0	1	0	0	1	1
1	0	0	0	1	0	1	1
1	1	0	0	0	0	0	1

$f_1=0$, $f_3=\overline{x}_1$, $f_3=\overline{x}_2$, $f_4=\overline{x}_1\,\overline{x}_2$, $f_5=\overline{x}_1\vee\overline{x}_2$, $f_6=1$

(a)   (b)

FIGURE 6.10
Monotonically decreasing functions of two variables: truth table (a) and decision tree (b) (Example 6.29).

Property 2: A monotonically decreasing function is described by an SOP *with complemented literals*. That is, they are implemented using AND and OR gates with complemented variables and the constant functions 0 and 1 as input variables.

Example 6.30   *Determine if the network given in Figure 6.11 (left) realizes a monotonically decreasing function. The output of the network $f = \overline{x}_1 \vee \overline{x}_2 \vee \overline{x}_3$ and the truth table reveal that this function is the logic sum of two monotonically decreasing functions, $\overline{x}_1\overline{x}_2$ and $\overline{x}_3$. Therefore, the function, implemented by the network, is monotonically decreasing.*

There are 2^{n+1} totally symmetric functions of n variables, $n + 2$ totally symmetric functions of n variables that are monotonically increasing and $n+2$ totally symmetric functions of n variables that are monotonically decreasing, $n + 2$ totally symmetric functions of n variables that are monotonically increasing and monotonically decreasing.

Unate Boolean functions

A Boolean function f that includes both monotonically increasing and decreasing functions is called a *unate* Boolean function. A Boolean function

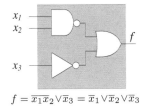

x_1	x_2	x_3	f
0	0	0	1
0	0	1	0
0	1	0	0
0	1	1	0
1	0	0	0
1	0	1	0
1	1	0	0
1	1	1	0

$$f = \overline{x_1}\overline{x_2} \vee \overline{x_3} = \overline{x_1} \vee \overline{x_2} \vee \overline{x_3}$$

FIGURE 6.11
A logic network that implements a monotonically decreasing Boolean function (Example 6.30).

is unate if and only if the minimal SOP expression contains either the literal x_i or $\overline{x}_i$ but not both.

Example 6.31 *The Boolean function $f = x_1x_2 \vee x_3x_4$ is* **unate**. *Function $f = x_2x_2x_3 \vee \overline{x}_2x_4$ is not unate since both x_2 and $\overline{x}_2$ appear in the minimal SOP expression.*

Example 6.32 *Determine if the logic network given in Figure 6.12 (left) realizes a unate function. The function $f = \overline{x}_1 \vee \overline{x}_2 \vee x_3$ is unate because the monotonically decreasing function $\overline{x_1}\overline{x_2}$ and monotonically increasing function x_3 are both contained in f.*

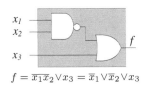

x_1	x_2	x_3	f
0	0	0	1
0	0	1	1
0	1	0	1
0	1	1	1
1	0	0	1
1	0	1	1
1	1	0	0
1	1	1	1

$$f = \overline{x_1}\overline{x_2} \vee x_3 = \overline{x}_1 \vee \overline{x}_2 \vee x_3$$

FIGURE 6.12
A logic network that implements a unate Boolean function (Example 6.32).

6.8.3 Linear functions

The property of linearity of a Boolean function is important for the cost-efficient implementations of logic networks; generally, linear functions require less gates if implemented using specific gates. Linearity also greatly simplifies the manipulation and transformation of Boolean functions:

$$\underbrace{\texttt{Boolean functions}}_{Boolean\ data\ structure} \quad \xrightarrow{Recognition} \quad \underbrace{\texttt{Linear Boolean functions}}_{Boolean\ data\ structure}$$

A *linear* Boolean function is represented by the equation

$$f = r_0 \oplus \bigoplus_{i=1}^{n} r_i x_i = r_0 \oplus r_1 x_1 \oplus \cdots \oplus r_n x_n \qquad (6.4)$$

where $r_i \in \{0,1\}$ is the i-th coefficient, $i = 1, 2, \ldots, n$. It should be noted that Expression 6.4 is a *parity* function if $r_1 = r_2 = \cdots = r_n = 1$.

Example 6.33   (a) *The Boolean function $f = x \vee y$ is* **nonlinear** *since its polynomial form $f = x \oplus y \oplus xy$ is not linear.*

                             (b) *The Boolean function $f = xy \vee \overline{x}\,\overline{y}$ is* **linear** *since its polynomial form is linear: $f = 1 \oplus x \oplus y$.*

Example 6.34   *There are eight linear Boolean functions $f(x,y)$ of two variables: $f_0 = 0$, $f_1 = 1$, $f_2 = x$, $f_3 = 1 \oplus x$, $f_4 = y$, $f_5 = 1 \oplus y$, $f_6 = x \oplus y$, and $f_7 = 1 \oplus x \oplus y$ (Figure 6.13).*

Linear expressions have several useful properties and implementations, in particular:

Property 1: There are 2^{n+1} linear expressions of n variables.

Property 2: A linear expression is either a self-dual or a self-antidual Boolean function:

$$f = \begin{cases} f(\overline{x}_1, \overline{x}_2, \ldots, \overline{x}_n) = \overline{f}(x_1, x_2, \ldots, x_n), & \text{if } \sum_{i=0}^{n+1} r_i = 1 \\[2mm] f(\overline{x}_1, \overline{x}_2, \ldots, \overline{x}_n) = f(x_1, x_2, \ldots, x_n), & \text{if } \sum_{i=0}^{n+1} r_i = 0 \end{cases}$$

Property 3: A Boolean function obtained by the linear composition of linear expressions is also a linear expression.

Design example: linear Boolean functions of two variables

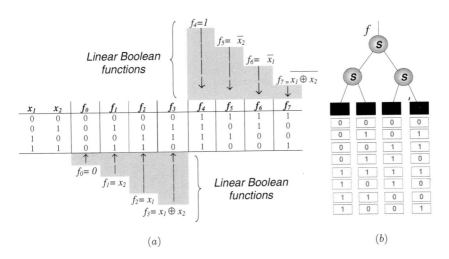

(a)   (b)

FIGURE 6.13

Linear functions of two variables (Example 6.34).

6.8.4 Universal set of functions

If an arbitrary Boolean function is completely represented by a set F of simple Boolean functions $\{f_1, f_2, \ldots, f_m\}$, this set of elementary functions is called *universal*, or *complete*. Five major classes of Boolean functions have been defined in the previous sections:

Five classes of Boolean functions
Class M_0, the set of 0-preserving functions
Class M_1, the set of 1-preserving functions
Class M_2, the set of self-dual functions
Class M_3, the set of monotonically increasing functions
Class M_4, the set of linear functions

In these classes, a *0-preserving function* is defined as $f(0, \ldots, 0) = 0$, and a *1-preserving function* is specified as $f(1, \ldots, 1) = 1$.

Example 6.35 *Functions $f_1 = x$, $f_2 = x_1 \vee x_2$, and $f_3 = x_1 x_2$ are 0- and 1-preserving functions because $f_1(0) = 0$, $f_1(1) = 1$, $f_2(0,0) = 0 \vee 0 = 0$, $f_2(1,1) = 1 \vee 1 = 1$, $f_3(0,0) = 0 \cdot 0 = 0$, and $f_3(1,1) = 1 \cdot 1 = 1$.*

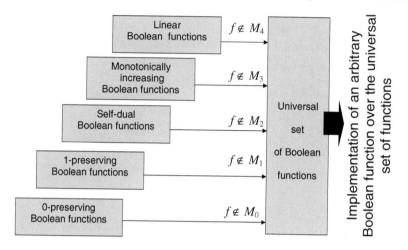

FIGURE 6.14

A universal set of Boolean functions is specified by five classes of Boolean functions.

Post theorem

A logic network can be considered as an interconnection of simple logic networks called *logic primitives*. The function of the network can be interpreted as performing a mathematical composition with operations corresponding to the logic primitives. The problem of completeness of logical primitives in a network can be transferred to the problem of the functionally completeness of the set of operations by which it is possible to implement an arbitrary Boolean function using a finite number of of logical primitives. The *Post theorem* provides a solution to this problem:

**Post theorem on the functionally complete set
of Boolean functions**

The set of Boolean functions F is *functionally complete*, or *universal*, if it includes (Figure 6.14)

► At least one function that is not monotonically increasing
► At least one function that is not self-dual
► At least one function that is not linear
► At least one function that is not 0-preserving
► At least one function that is not 1-preserving

Procedure to determine functional completeness

Given: A set of Boolean functions.
Step 1. Find at least one function that is not monotonically increasing.
Step 2. Find at least one function that is not self-dual.
Step 3. Find at least one function that is not linear.
Step 4. Find at least one function that is not 0-preserving.
Step 5. Find at least one function that is not 1-preserving.
Result: Identification of functional completeness if at least one of the above functions has been found in this set.

Design example: the minimal universal set

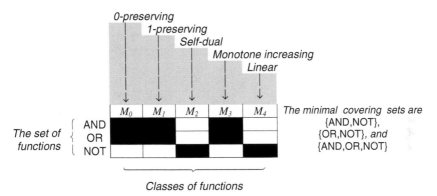

FIGURE 6.15

The minimal universal set for Boolean functions AND, OR, and NOT (Example 6.36).

Example 6.36 *The set of functions AND, OR, and NOT is a universal set because (Figure 6.15) the AND is a 0- and 1-preserving function, but neither a self-dual nor a linear function; the OR is a 0- and 1-preserving function, but neither a self-dual nor a linear Boolean function; the NOT is neither a 0- nor a 1-preserving function.*

6.9 Further study

Advanced topics of Boolean data structures

Topic 1: Multivalued logic. Boolean algebra is the simplest algebra. The motivation for its implementation is that two-valued Boolean algebra is suited to the two-state physical phenomena of today's technology. However, multivalued algebras have been the focus of interest for many years.

Multivalued data structures are the generalization of Boolean data structures for multivalued logic. Two-valued Boolean algebra is the dominating algebra, since it is suited to the two-state physical phenomena of today's technology. However, multivalued algebras were the focus of interest for many years. Multivalued algebra is based upon a set of m elements $M = \{0, 1, 2, \ldots, m\}$, corresponding to multivalued signals, and the corresponding operations. This means that multivalued logic networks operate with multivalued logic signals.

The primary advantage of multivalued signals is the ability to encode more information per variable than a binary system is capable of doing. Hence, less area is used for interconnections since each interconnection carries more information. Furthermore, the reliability of systems is also relevant to the number of connections because they are sources of wiring error and weak connections. For instance, the adoption of m-valued signals enables n pins to pass q^n combinations of values rather than just the 2^n limited by the binary representation.

There are many examples of the successful implementation of 8-, 16-, and 32-valued Boolean algebra in memory devices. In particular, a 32×32 bit multiplier based on the quaternary signed-digit number system consists of only three-stage signed-digit full adder using a binary-tree addition scheme. Another application of multivalued logic is residue arithmetic. Each residue digit can be represented by a multivalued code. For example, by this coding, *mod m* multiplication can be performed by a shift operator, and *mod m* addition can be executed using radix-arithmetic operators.

There are various universal (functionally complete) sets of operations for multivalued algebra: *Post* algebra, *Webb* algebra, *Bernstein* algebra, and *Allen and Givone* algebra.

Multivalued decision diagrams can be considered as an extension of decision diagram techniques for multivalued functions. Shannon expansion has been extended for multivalued logic as well.

It is expected that nanotechnology will provide new possibilities for implementing multivalued logic. Details can be found in the *Proceedings of the Annual IEEE Symposium on Multiple-Valued Logic.*

Topic 2: Probabilistic logic combines the capacity of probability theory to handle uncertainty with the capacity of classical predicate, or deductive, logic. Among the applications are probabilistic models for logic relations, where the truth values of sentences are probabilities.

Topic 3: Fuzzy logic. In the case of Boolean functions, a Boolean variable can take up only one of two possible values. In fuzzy logic, there are *membership functions* which can take up infinitely many values in the interval $[0, 1]$. That is, Boolean algebra can handle fuzzy logic (see the book *What Is Mathematics?* by R. Courant and H. Robbins, Oxford University Press, New York, 4th ed., 1947). Fuzzy logic devices provide the opportunity for adapting to variable input conditions (see, for example, the book *Fuzzy Logic with Engineering Applications* by T. J. Ross, second edition, Wiley, New York, 2004). Fuzzy logic operates with fuzzy variables that are different from logical variables.

Topic 4: Reversible logic. Contemporary computers are based on irreversible logic devices, which are being considered energy inefficient, due to the physical effects of energy dissipation (see, for example, "Irreversibility and Heat Generation in the Computing Process" by Rolf Landauer, *IBM Journal of Research and Development*, issue 5, pages 183–191, 1961). Further study by Charles H. Bennett ("Logical Reversibility of Computation," *IBM Journal of Research and Development*, issue 17, number 6, pages 525–532, 1973) targeted theoretical possibilities of reversible computing. With the discovery of a number of energy -recovering integrated circuit techniques, it became possible to exploit reversibility to reduce power consumption. An example of such circuits is the split-level charge recovery logic developed at MIT Artificial Intelligence Laboratory ("Asymptotically Zero Energy Split-Level Charge Recovery Logic" by S. G. Younis and T. F. Knight, Jr., *International Workshop on Low-Power Design*, pages 177–182, 1994). In 1982, E. F. Fredkin and T. Toffoli described a circuit implementation of a reversible element, later named the *Fredkin gate*, suggesting its possible realization on the Josephson junction-based systems ("Conservative Logic," *International Journal of Theoretical Physics*, volume 21, number 3, pages 219–253, 1982). The MIT group has further contributed to the design of asymptotically optimal computers, in particular, "adiabatic" reversible electronic logic technology for parallel reversible architectures (see, for example, "A Reversible Instruction Set Architecture and Algorithms" by J. Hall, *Physics and Computation*, pages 128–134, November 1994).

Further reading

[1] Brown S and Vranesic Z. *Fundamentals of Digital Logic with VHDL Design.* McGraw-Hill, 2000.

[2] Friedman AD and Menon PR. *Theory and Design of Switching Circuits.* Computer Science Press, Woodland Hills, CA, 1975.

[3] Givone DD. *Digital Principles and Design.* McGraw-Hill, New York, 2003.

[4] Hayes JP. *Introduction to Digital Logic Design.* Addison-Wesley, 1993.

[5] Karpovsky MG, Stanković RS, and Astola JT. *Spectral Logic and Its Applications for the Design of Digital Devices.* John Wiley & Sons, NJ, 2008.

[6] Kandel A and Lee SC. *Fuzzy Switching and Automata.* Crane, Russak, New York, 1979.

[7] Katz RH and Borriello G. *Contemporary Logic Design.* Pearson Prentice Hall, Upper Saddle River, NJ, 2005.

[8] Lee SC. *Modern Switching Theory and Digital Design.* Prentice-Hall, Englewood Cliffs, NJ, 1978.

[9] Mano MM and Kime CR. *Logic and Computer Design Fundamentals.* 3rd edition. Pearson Prentice Hall, Upper Saddle River, NJ, 2004.

[10] Marcovitz AB. *Introduction to Logic Design.* McGraw-Hill, 2005.

[11] Marinos PN. Fuzzy logic and its application to switching systems. *IEEE Trans. Comput.*, 18:343–348, 1969.

[12] Rine DC, Ed., *Computer Science and Multiple-Valued Logic: Theory and Applications.* Noth-Holland, Amsterdam, 1977.

[13] Sasao T. *Switching Theory for Logic Synthesis.* Kluwer, Dordrecht, 1999.

[14] Wakerly JF. *Digital Design. Principles and Practice.* 3rd edition. Prentice Hall, Upper Saddle River, NJ, 2001.

[15] Yanushkevich SN and Shmerko VP. *Introduction to Logic design.* Taylor & Francis/CRC Press, Boca Raton, FL, 2008.

7

Boolean Data Structures

7.1 Introduction

Switching theory is the underlying basis for development of Boolean data structures. Each data structure determines the efficiency of particular application od switching theory including manipulation of data, optimization, design of computing structures, and fabrication. Using various technological or fabrication criteria, an appropriate data structure can be chosen or developed.

Boolean data structures can be systemized with respect to various criteria. For example, data structure can be classified according to the form of the representation (algebraic, tabulated, matrix, or graphical form), to the embedding criteria (2D or 3D computing configurations), or the efficiency of encoding, such as DNA computing.

One of the main premises of computing structures design in nano scale is that computing abilities of basic structure must be delegated into another structure under the set of constrains. This delegation of computing properties addresses the problem of embedding of a guest structure into a host structure. Graph-based Boolean data structures are suitable for this problem.

Boolean *data structures* are of fundamental importance in logic design (Figure 7.1). No universal data structure exists that is efficient in all applications. Choosing an appropriate data structure is critical for the efficient representation, manipulation, and implementation of Boolean functions.

This chapter guides the reader through the process of the formalization, representation, and implementation of Boolean functions:

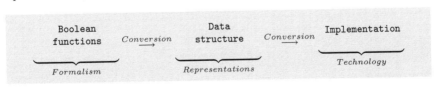

Boolean functions are implemented in computing devices using software and hardware tools. There are several steps for preparing a Boolean function for implementation, aiming at the simplification of the function and at choosing an appropriate form for implementation. No universal approach exists that

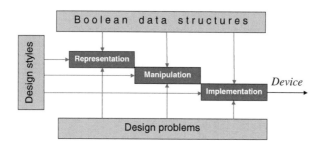

FIGURE 7.1
Design problems (representation, manipulation, and implementation) of
Boolean functions can be solved using different design styles.

is good for all situations. The effectiveness of preimplementation depends
on many factors; in particular, size (number of variables), decomposition
(partitioning into subfunctions), function properties, and type of data
structure.

A *data structure* representing a Boolean function is a mathematical model
of the function. A *data type* specifies the properties of this mathematical
model. There are, in particular, algebraic and graphical data types. Data
structure types for Boolean functions are the focus of this chapter.

7.2   Data structure types

A Boolean function f of n variables $x_1, x_2, \ldots x_n$ can be represented in
different mathematical *forms* or *descriptions*: (a) *graphical* forms, such as
logic networks, decision trees, decision and state diagrams, and hypercubes,
(b) *tabulated* forms such as truth tables and state tables, (c) *algebraic* and
matrix forms such as canonical descriptions and various transforms, and
(d) *mixed* graphical and tabulated forms such as K-maps All of these are
mathematical models or data structures of different data types.

7.3   Relationships between data structures

A Boolean function can be represented using different data structures. The
relationships between data structures are widely used in contemporary logic

design, particularly in the following:

```
  Truth table  ↔  Algebraic form
        K-map  ↔  Algebraic form
  Truth table  ↔  Decision tree
Decision tree  ↔  Decision diagram
Decision diagram  ↔  Logic network
Algebraic form  ↔  Decision tree
Algebraic form  ↔  Threshold gate-based network
```

The effectiveness of a given design depends on how efficiently these and other relationships between data structures are utilized.

> **Example 7.1**   *Tabulated forms such as truth tables and state tables can be used for small Boolean functions. K-maps, which combine the features of graphical and tabulated forms, are useful for the representation and optimization of Boolean functions in the so-called sum-of-product form, but are not applicable for polynomial forms. Graphical data structures, such as decision diagrams, are efficient not only for the representation and optimization of Boolean functions, but also for their implementation in hardware.*

7.4   The truth table

A Boolean function of n variables can be represented in a tabular form called a *truth table*:

7.4.1   Construction of the truth table

A truth table includes a list of combinations of 1s and 0s assigned to the binary variables, and a column that shows the value of the function for each binary combination. The number of rows in the truth table is 2^n, where n is the number of variables of the function. The binary combinations for the truth table are obtained from the binary numbers by counting from 0 through $2^n - 1$.

Example 7.2   *There are eight possible binary combinations for assigning bits to the three variables x_1, x_2, and x_3 of the Boolean function (Figure 7.2a) $f = x_1 \vee \overline{x}_2 x_3$. The column labeled f contains either 0 or 1 for each of these combinations. The table shows that the function is equal to 1 when $x_1 = 1$ or when $x_2 x_3 = 01$. It is equal to 0 otherwise.*

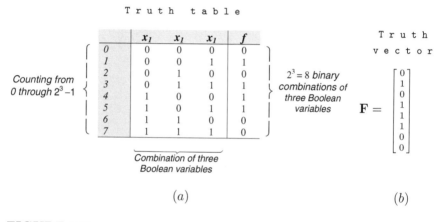

(a) (b)

FIGURE 7.2

Representation of a Boolean function in the form of a truth table (a) and truth vector (b) (Examples 7.2 and 7.4).

7.4.2   Truth tables for incompletely specified functions

Boolean algebra is flexible enough for the various situations that arise in design practice. This flexibility is demonstrated, in particular, by the utilization of Boolean algebra for the representation and manipulation of *incompletely specified* Boolean functions. As follows from the definition of Boolean algebra, Boolean formulas are not specified for assignments of variables. Boolean formulas become Boolean functions when their values are calculated. Suppose that some of these values cannot be computed; that is, they are unknown.

The $\times$ symbols in the truth table indicates that the value of the Boolean function is unknown; this is also called *don't care*; that is, it takes either the value 0 or 1 (Figure 7.3).

Functions that contain don't cares are called *incompletely specified* Boolean

	x_1	x_1	x_1	f	
0	0	0	0	0	
1	0	0	1	d	The value is **not specified** for the assignment 001
2	0	1	0	0	
3	0	1	1	1	
4	1	0	0	d	The value is **not specified** for the assignment 100
5	1	0	1	1	
6	1	1	0	d	The value is **not specified** for the assignment 110
7	1	1	1	0	

FIGURE 7.3
Truth table for the representation of an incompletely specified Boolean function f.

functions. Incompletely specified functions, if efficiently utilized, can bring valuable benefits. In particular, don't care values can provide additional opportunities for optimization by assuming that they take values of 0 or 1, depending on the optimization strategy.

Example 7.3   *The **truth table** of the incompletely specified Boolean function, given in Figure 7.3 can be drawn in a compact (reduced) form as follows:*

	$x_1 x_2 x_3$	f
0	000	0
2	010	0
3	011	1
5	101	1
7	111	0

7.4.3   Truth vector

A *truth vector* of a Boolean function is defined as a column f of the truth table,

$$\mathbf{F} = [f(0), f(1), \ldots, f(2^n - 1)]^T$$

The i-th element $f(i)$, $(i = 0, 1, 2, \ldots, 2^n - 1)$ of the truth vector is the value of a Boolean function given the assignment $i_1, i_2, \ldots, i_n$ (n-bit representation of i) of variables $x_1, x_2, \ldots, x_n$.

Example 7.4   *In Figure 7.2b, the **truth vector** $\mathbf{F}$ is derived by copying the column f of the truth table.*

7.4.4 Minterm and maxterm representations

Standard SOP and POS expressions are canonical representations of Boolean functions based on minterms and maxterms, respectively. Minterms and maxterms can be derived from a truth table. Minterms and maxterms correspond to the values 1 and 0 of the function, respectively.

> **Example 7.5**   *The Boolean function of three variables given in Table 7.1 is represented by four minterms and four maxterms.*

TABLE 7.1

Minterms and maxterms for a Boolean function of three variables

Assignment of variables x_1, x_2, x_3	Value of Boolean function	Minterm	Maxterm
000	1	$\overline{x}_1\overline{x}_2\overline{x}_3$	
001	1	$\overline{x}_1\overline{x}_2 x_3$	
010	0		$x_1 \vee \overline{x}_2 \vee x_3$
011	1	$\overline{x}_1 x_2 x_3$	
100	0		$\overline{x}_1 \vee x_2 \vee x_3$
101	0		$\overline{x}_1 \vee x_2 \vee \overline{x}_3$
110	0		$\overline{x}_1 \vee \overline{x}_2 \vee x_3$
111	1	$x_1 x_2 x_3$	

7.4.5 Reduction of truth tables

The following manipulations of truth tables are used for small Boolean functions: (a) reduction using shorthand notation, (b) reduction using particular properties of Boolean functions such as symmetry, and (c) reduction by replacing a group of rows by variables or subfunctions.

Reduction using shorthand notation

A Boolean function can be represented by its minterms only since the remaining values (0's) of the function can always be restored. This representation is called *shorthand notation*. Correspondingly, the truth table of the function can be reduced by removing the lines corresponding to maxterms.

Example 7.6   *Figure 7.4 shows the complete truth table and the reduced one. In a similar manner, a shorthand POS can be obtained, although the SOP form is used more often in practice.*

R e d u c t i o n   b a s e d   o n
s h o r t h a n d   S O P   n o t a t i o n

x_1	x_2	x_3	f
0	0	0	1
0	0	1	0
0	1	0	1
0	1	1	0
1	0	0	0
1	0	1	1
1	1	0	1
1	1	1	1

x_1	x_2	x_3	f
0	0	0	1
0	1	0	1
1	0	1	1
1	1	0	1
1	1	1	1

FIGURE 7.4
Truth table reduction (Example 7.6).

Reduction using variables and functions

Reduction of a truth table using variables and functions means partial replacement, together with reduction, of parts of the values of a function by its variables and subfunctions.

Example 7.7   *Figure 7.5 shows the results of the reduction of the complete truth tables.*

7.4.6   Properties of the truth table

Truth-table-based representations of Boolean functions are characterized by the following properties: (a) they allow for the observation of the values of functions and the corresponding assignments of variables; (b) they allow for observation of some of the properties of functions; and (c) they are ready for direct implementation. However, truth tables grow in size quickly and therefore can be useful only for functions with a small number of variables. In tabular representation, the function's values are explicitly shown, without taking into account their possible relationships.

Techniques for reduction of truth table based on grouping

Initial truth table

Reduced truth table
with respect to x_3

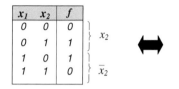

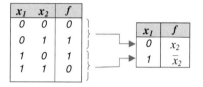

Reduced truth table
with respect to x_3

Reduced truth table
with respect to x_2 and x_3

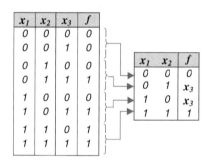

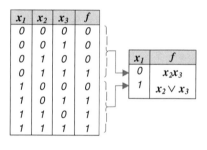

FIGURE 7.5
Truth table reduction (Example 7.7).

7.4.7 Deriving standard SOP and POS expressions from a truth table

In Figure 7.6, the procedures for deriving standard SOP and POS expressions from a truth table are given.

7.5 K-map

The *Karnaugh map* or *K-map* is a form of representation of a Boolean function that is an alternative to the truth table:

Deriving a standard SOP expression from a truth table

Given: A truth table of a Boolean function.
Step 1: Find the values of ⟦1⟧ in the truth table.
Step 2: Identify the corresponding (to the positions of the 1s) assignments of variables.
Step 3: Derive the minterms based on the rules:

(a) If the variable assignment is ⟦0⟧, then the variable is included in complemented form.

(b) If the variable assignment is ⟦1⟧, then the variable is included in the uncomplemented form.

Assemble variables into minterms using the AND operation
Step 4: Write the SOP form by combining the derived minterms using logic sum (∨).
Output: The canonical (standard) SOP.

Deriving a standard POS expression from a truth table

Given: A truth table of a Boolean function.
Step 1: Find the values of ⟦0⟧ in the truth table.
Step 2: Identify the corresponding (to the positions of the 1s) assignments of variables.
Step 3: Derive the maxterms based on the rules:

(a) If the variable assignment is ⟦0⟧, then the variable is included in uncomplemented form.

(b) If the variable assignment is ⟦1⟧, then the variable is included in the complemented form.

Assemble variables into minterms using the OR operation
Step 4: Write the POS form by combining the derived minterms using logic product (∧)
Output: The canonical (standard) POS

FIGURE 7.6
Procedures for deriving standard SOP and POS expressions from a truth table.

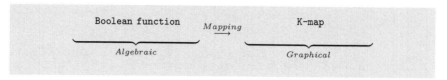

The K-map consists of *cells* that are labeled by variable assignments corresponding to the rows of the truth table (Figure 7.7). Each cell represents a minterm. In a K-map, each cell must differ from any adjacent cell by only one variable. For this, a K-map performs an algebraic factoring, such as $\bar{x}_1 x_2 \vee x_1 x_2 = \bar{x}_1 \vee x_1 x_2 = 1 \cdot (x_1 x_2) = x_2$. In this way, the map displays the factorable terms in a Boolean expression.

7.5.1 Representation of standard SOP and POS expressions using K-maps

In Figure 7.8, the procedures for the representation of standard SOP and POS expressions using K-maps are given.

7.5.2 A K-map for a Boolean function of two variables

A K-map is a 2D representation of the truth table of a Boolean function. Given a Boolean function of two variables, its four-row truth table can be regrouped into a K-map consisting of four cells. The columns of the map are labeled by the values of the variable x_1, and the rows are labeled by x_2. This label leads to the locations of minterms. The columns and rows of the K-map are encoded as follows:

$$
\text{2-variable K-map encoding} = \begin{cases} \overbrace{\begin{matrix} 0 & 1 \\ \bar{x}_1, & x_1, \end{matrix}}^{Encoding} \text{columns;} \\ \begin{matrix} 0 & 1 \\ \bar{x}_2, & x_2, \end{matrix} \text{rows.} \end{cases}
$$

7.5.3 A K-map for a Boolean function of three variables

A K-map for a Boolean function of three variables is constructed by placing two two-variable maps side by side. In this case, each valuation of variables x_1 and x_2 identifies a column in the map, while the value of x_3 distinguishes the two rows. The columns and rows of the K-map are encoded as follows:

$$
\text{3-variable K-map encoding} = \begin{cases} \overbrace{\begin{matrix} 00 & 01 & 11 & 10 \\ \bar{x}_1\bar{x}_2, & \bar{x}_1 x_2, & x_1 x_2, & x_1\bar{x}_2, \end{matrix}}^{Encoding} \text{columns;} \\ \begin{matrix} 0 & 1 \\ \bar{x}_3, & x_3, \end{matrix} \qquad\qquad\qquad \text{rows.} \end{cases}
$$

Techniques for Boolean function representation using K-maps

Boolean function of two variables

x_1	x_1	f
0	0	m_0
0	1	m_1
1	0	m_2
1	1	m_3

x_1

x_2		0	1
	0	m_0	m_2
	1	m_1	m_3

Boolean function of three variables

x_1	x_2	x_3	f
0	0	0	m_0
0	0	1	m_1
0	1	0	m_2
0	1	1	m_3
1	0	0	m_4
1	0	1	m_5
1	1	0	m_6
1	1	1	m_7

$x_1 x_2$

x_3		00	01	10	11
	0	m_0	m_2	m_6	m_4
	1	m_1	m_3	m_7	m_5

Boolean function of four variables

x_1	x_2	x_3	x_4	f	x_1	x_2	x_3	x_4	f
0	0	0	0	m_0	1	0	0	0	m_8
0	0	0	1	m_1	1	0	0	1	m_9
0	0	1	0	m_2	1	0	1	0	m_{10}
0	0	1	1	m_3	1	0	1	1	m_{11}
0	1	0	0	m_4	1	1	0	0	m_{12}
0	1	0	1	m_5	1	1	0	1	m_{13}
0	1	1	0	m_6	1	1	1	0	m_{14}
0	1	1	1	m_7	1	1	1	1	m_{15}

$x_1 x_2$

$x_3 x_4$		00	01	11	10
	00	m_0	m_4	m_{12}	m_8
	01	m_1	m_5	m_{13}	m_9
	11	m_3	m_7	m_{15}	m_{11}
	10	m_2	m_6	m_{14}	m_{10}

FIGURE 7.7
K-maps are alternatives to the truth tables of Boolean functions of two, three, and four variables.

7.5.4 A K-map for a Boolean function of four variables

A K-map for a Boolean function of four variables is constructed by placing two three-variable maps side by side. The columns and rows of the K-map are encoded as follows:

$$\text{4-variable K-map encoding} = \begin{cases} \overset{\text{Encoding}}{\overbrace{\begin{matrix} \overset{00}{\overline{x}_1\overline{x}_2,} & \overset{01}{\overline{x}_1 x_2,} & \overset{11}{x_1 x_2,} & \overset{10}{x_1\overline{x}_2,} \end{matrix}}} \text{ columns;} \\ \begin{matrix} \overset{00}{\overline{x}_3\overline{x}_4,} & \overset{01}{\overline{x}_3 x_4,} & \overset{11}{x_3 x_4,} & \overset{10}{x_3\overline{x}_4,} \end{matrix} \text{ rows.} \end{cases}$$

Representation of a standard SOP expression using a K-map

Given: A K-map.

Step 1: Identify the 1s in the K-map.

Step 2: Identify the assignments of variables corresponding to the
       row and line labels of the 1s cells.

Step 3: Derive the minterms using the identified assignments.

Output: Canonical (standard) SOP.

Representation of a standard POS expression using a K-map

Given: A K-map.

Step 1: Identify the 0s in the K-map.

Step 2: Identify assignments of variables corresponding to the row
       and line labels of the 0s cells.

Step 3: Derive the maxterms using the identified assignments

Output: Canonical (standard) POS.

FIGURE 7.8

Procedures for the representation of standard SOP and POS expressions using
K-maps.

Example 7.8   *K-maps for elementary Boolean functions of two
variables and their relationships with truth tables are
given in Table 7.2.*

7.5.5 A K-map for an incompletely specified Boolean function

An incompletely specified Boolean function can be represented by a K-map
in which the unspecified values are denoted using the symbol d.

Example 7.9   *Figure 7.9 illustrates the representation of the
incompletely specified Boolean function given the truth
vector $\mathbf{F} = [\, 0 \; \boxed{d} \; 0 \; 1 \; \boxed{d} \; 1 \; \boxed{d} \; 0 \,]^T$ by the K-map
and truth table, where $\boxed{d}$ is don't-care (unspecified)
value.*

TABLE 7.2
K-maps for elementary Boolean functions of two variables

Techniques for Boolean function representation using K-maps

Elementary Boolean function	Truth table	K-map
AND $f = x_1 x_2$ x_1 —⟩ f x_2 —	$\begin{array}{cc\|c} x_1 & x_2 & f \\ \hline 0 & 0 & 0 \\ 0 & 1 & 0 \\ 1 & 0 & 0 \\ 1 & 1 & 1 \end{array}$	x_1: 0, 1; x_2: 0, 1; cell ($x_1=1,x_2=1$) = 1
OR $f = x_1 \vee x_2$ x_1 —⟩ f x_2 —	$\begin{array}{cc\|c} x_1 & x_2 & f \\ \hline 0 & 0 & 0 \\ 0 & 1 & 1 \\ 1 & 0 & 1 \\ 1 & 1 & 1 \end{array}$	x_1: 0, 1; x_2: 0, 1; cells (0,1)=1, (1,0)=1, (1,1)=1
NAND $f = \overline{x_1 x_2}$ x_1 —⟩o f x_2 —	$\begin{array}{cc\|c} x_1 & x_2 & f \\ \hline 0 & 0 & 1 \\ 0 & 1 & 1 \\ 1 & 0 & 1 \\ 1 & 1 & 0 \end{array}$	x_1: 0, 1; x_2: 0, 1; cells (0,0)=1, (1,0)=1, (0,1)=1
NOR $f = \overline{x_1 \vee x_2}$ x_1 —⟩o f x_2 —	$\begin{array}{cc\|c} x_1 & x_2 & f \\ \hline 0 & 0 & 1 \\ 0 & 1 & 0 \\ 1 & 0 & 0 \\ 1 & 1 & 0 \end{array}$	x_1: 0, 1; x_2: 0, 1; cell (0,0)=1
EXOR $f = x_1 \oplus x_2$ x_1 —⟩ f x_2 —	$\begin{array}{cc\|c} x_1 & x_2 & f \\ \hline 0 & 0 & 0 \\ 0 & 1 & 1 \\ 1 & 0 & 1 \\ 1 & 1 & 0 \end{array}$	x_1: 0, 1; x_2: 0, 1; cells (1,0)=1, (0,1)=1
XNOR $f = \overline{x_1 \oplus x_2}$ x_1 —⟩o f x_2 —	$\begin{array}{cc\|c} x_1 & x_2 & f \\ \hline 0 & 0 & 1 \\ 0 & 1 & 0 \\ 1 & 0 & 0 \\ 1 & 1 & 1 \end{array}$	x_1: 0, 1; x_2: 0, 1; cells (0,0)=1, (1,1)=1

$x_1 x_2 x_3$	f	Comment
000	0	
001	d	$\overline{x}_1 \overline{x}_2 x_3$ or $x_1 \vee x_2 \vee \overline{x}_3$
010	0	
011	1	
100	d	$x_1 \overline{x}_2 \overline{x}_3$ or $\overline{x}_1 \vee x_2 \vee x_3$
101	1	
110	d	$x_1 x_2 \overline{x}_3$ or $\overline{x}_1 \vee \overline{x}_2 \vee x_3$
111	0	

$$x_1 x_2$$

	00	01	11	10
0	0	0	d	d
1	d	1	0	1

x_3

(a)   (b)

FIGURE 7.9
Representation of an incompletely specified Boolean function by a K-map and
truth table (Example 7.9).

7.6   Cube data structure

Boolean function can be represented as points in n-space. The collection of
2^n possible points is said to form of an n-*cube*, or a *Boolean hypercube*.

The reduced truth table introduced in Section 7.4.5 contained only
assignments of variables, corresponding to the minterms. These assignments
are called *primary cubes*.

Example 7.10   *Given a minterm $x_1 x_2 \overline{x}_3$, its assignment $[\,1\,1\,0\,]$ is a
primary* **cube**.

Primary cubes can be further reduced to cubes using Boolean *adjacency*
rules.

Example 7.11   *Consider the primary cubes $[\,1\,1\,0\,]$ and $[\,1\,1\,1\,]$.
These cubes are adjacent, since they are different in
one element only. This means that they can form a
new cube as $[\,1\,1\,0\,] \vee [\,1\,1\,1\,] = [\,1\,1\,\times\,]$, where
the symbol $\times$ is used to represent the missing variable.
This corresponds to the execution of the adjacency rule
on the two minterms: $x_1 x_2 \overline{x}_3 \vee x_1 x_2 x_3 = x_1 x_2 (\overline{x}_3 \vee
x_3) = x_1 x_2$. That is, the cube $[\,1\,1\,\times\,]$ corresponds to
the product term $x_1 x_2$ of the Boolean function of three
variables.*

The rules for the OR operation on cubes are presented in Figure 7.10a. The basic operations on a cube are sum, product, and complement.

Techniques for operations on the cubes

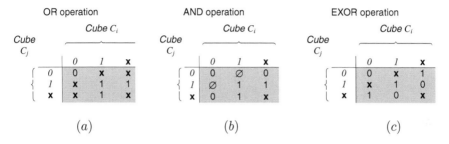

(a)   (b)   (c)

FIGURE 7.10
OR (a), AND (b), and EXOR (c) operations on the cubes.

An SOP of a Boolean function can be represented by a set of cubes, with OR operations between them, and the above rules can be used for reducing the SOP form.

Example 7.12   *Given a Boolean function $f = x_1 \lor x_2$, its cube representation is derived as follows:*

(a) *Using OR operations between the cubes: $f = x_1 \lor x_2 = [\,1 \times\,] \lor [\,\times\, 1\,]$.*

(b) *By reducing the truth table, using primary cubes and the rules from Figure 7.11:*

x_1	x_2	f
0	0	0
0	1	1
1	0	1
1	1	1

x_1	x_2	f
0	1	1
1	0	1
1	1	1

$$f = [\,0\,1\,] \lor [\,1\,0\,] \lor [\,1\,1\,]$$
$$= [\,1 \times \,] \lor [\,\times\, 1\,]$$

Note that
(a) the cube $[\,1 \times\,]$ is obtained by adjacing $[\,1\,0\,]$ and $[\,1\,1\,]$;
(b) the cube $[\,\times\, 1\,]$ is obtained by adjacing $[\,0\,1\,]$ and $[\,1\,1\,]$.

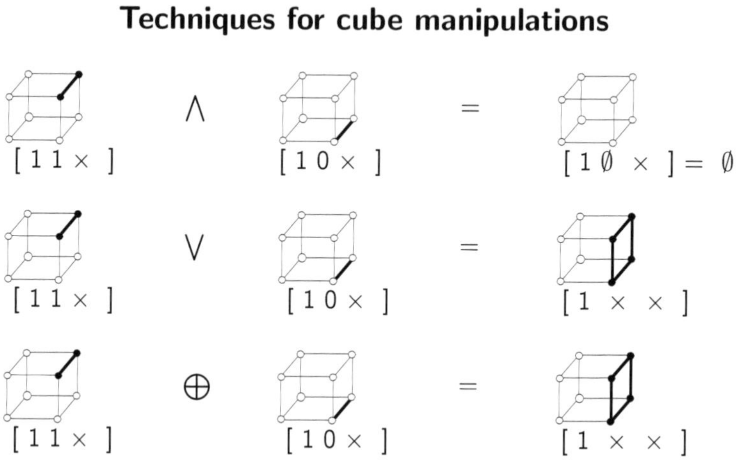

Techniques for cube manipulations

FIGURE 7.11
AND, OR, and EXOR operations on cubes (Example 7.12).

7.7 Graphical data structure for cube representation

An n-bit binary number can be represented by a point in n-space.

Example 7.13   *The set of 1-bit binary numbers 0 and 1 can be represented by two points in 1-space (a line). The set of 2-bit binary numbers 00,01,10 and 11 can be represented by four points in 2-space. A 3-bit binary number can be represented by eight points in 3-space (Figure 7.12).*

In general, n-bit number is represented by n-cube, or *Boolean hypercube*. A Boolean hypercube is a graphical data structure based on hypergraphs:

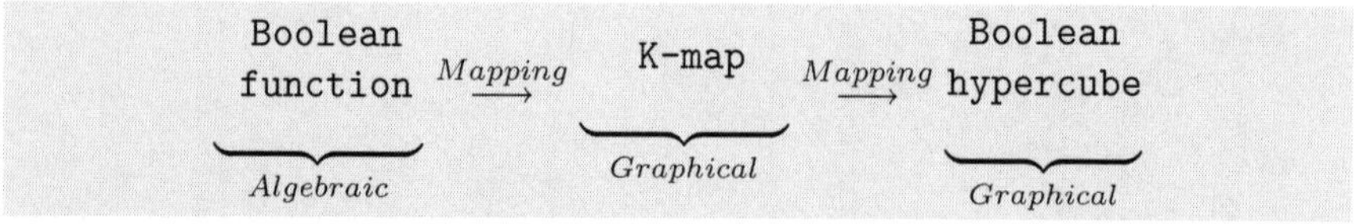

Operations on hypercubes are defined in a similar manner to operations on cubes. Thus, the manipulation of cubes involves OR, AND, and EXOR operations, applied to the appropriate literals, following the rules given in Figure 7.10. This data structure is useful for the representation, manipulation, and minimization of Boolean functions. A Boolean hypercube is defined as

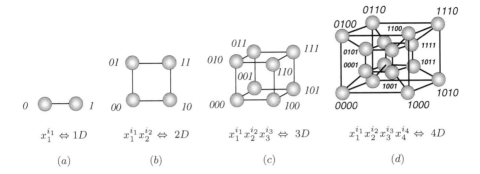

FIGURE 7.12
1D Boolean hypercube (a), 2D Boolean hypercube (b), 3D Boolean hypercube (c), and 4D Boolean hypercube (d) for representing one-, two-, three-, and four-variable Boolean functions, respectively (Example 7.13).

a collection of 2^m minterms $(m \leq n)$, product terms in which each of the n variables appears once:

$$\boxed{x_1^{i_1}} \ \boxed{x_2^{i_2}} \ \cdots \ \boxed{x_{n-1}^{i_{n-1}}} \ \boxed{x_n^{i_n}}$$

where $x_j^{i_j} = 1$ for $i_j = 0$, and $x_j^{i_j} = x_j$ for $i_j = 1$.

The Boolean hypercube encodes a Boolean function assigning codes to the 2^n vertices and $n \times 2^{n-1}$ edges. Figure 7.13 shows this assigning given a function of three variables $(n = 3)$. Operations between two Boolean hypercubes produce a new Boolean hypercube (product) that is useful in optimization problems. The following steps are applied for the representation of Boolean functions of n variables using Boolean hypercubes:

Given: A standard SOP form of a Boolean function.
Step 1. Determine the dimension of the Boolean hypercube.
Step 2. Construct the n axes corresponding to the variables. Along each axis, the variable x_i changes from 0 to 1.
Step 3. Assign the nodes of the Boolean hypercube with variable. assignments, so that neighboring nodes are encoded using the Gray code rule.
Output: Boolean hypercube.

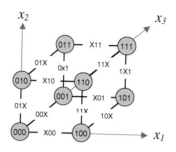

▶ *Each node is assigned to one of 8 codes (variable assignment)*

▶ *One edge out of 12 is assigned to a cube with one don't care ($\times$)*

▶ *One face out of 6 is assigned to a cube with two don't cares*

FIGURE 7.13

A Boolean hypercube data structure for representing and manipulating Boolean functions of three variables.

Example 7.14 *Various forms of representation of the Boolean function OR $f = x_1 \vee x_2$, and their relationships with each other are shown in Figure 7.14.*

Example 7.15 *Various forms of representation of a Boolean function $f = \overline{x}_3 \vee x_1\overline{x}_2$, and their relationships are given in Figure 7.15.*

Mapping the product into a Boolean hypercube

Let $x_j^{i_j}$ be a literal of a Boolean variable x_j such that $x_j^0 = \overline{x}_j$, $x_j^1 = x_j$ and $x_1^{i_1}x_2^{i_2}\ldots x_n^{i_n}$ is a product of literals. Topologically, this is a set of points on the plane numerated by $i = 0, 1, \ldots, n$. To map this set into a Boolean hypercube, the numbers must be encoded by the Gray code and represented by the corresponding graphs based on Hamming distance.

Example 7.16 *Figure 7.12 demonstrates the representation of Boolean functions using Boolean hypercubes. The product $x_1^{i_1}$ represents two variables, x_1 and $\overline{x}_1$. The product $x_1^{i_1}x_2^{i_2}$ corresponds to four minterms, x_1x_2, $\overline{x}_1x_2$, $x_1\overline{x}_2$, and $\overline{x}_1\overline{x}_2$. Products with four variables or more are represented by assembling 3D Boolean hypercubes.*

The 0D Boolean hypercube ($n = 0$) represents the constant 0. A line

Techniques for representation of the Boolean function

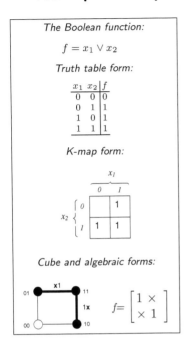

The Boolean function:

$$f = x_1 \vee x_2$$

Truth table form:

x_1	x_2	f
0	0	0
0	1	1
1	0	1
1	1	1

K-map form:

Cube and algebraic forms:

$$f = \begin{bmatrix} 1 & \times \\ \times & 1 \end{bmatrix}$$

▶ The 4 corners (vertices) correspond to the 4 rows of a truth table.

▶ Each vertex is identified by two coordinates.

▶ The horizontal coordinate is assumed to correspond to the variable x_1, and the vertical coordinate to x_2.

▶ The function f is equal to 1 for vertices 01, 10, and 11. The function f can be expressed as a set of vertices, $f = \{01, 10, 11\}$.

▶ The edge joins two vertices for which the labels differ in the value of only one variable.

▶ The letter x is used to denote the fact that the corresponding variable can be either 0 or 1.

▶ Vertices 01 and 11 are joined by the edge labeled x1.

▶ The edge 1x means a merger of the vertices 10 and 11.

▶ The term x_1 is the sum of minterms $x_1 \bar{x}_2$ and $x_1 x_2$. It follows that $x_1 \bar{x}_2 \vee x_1 x_2 = x_1$.

▶ The edges 1x and x1 define the function f in a unique way.

FIGURE 7.14

Representation of the Boolean function $f = x_1 \vee x_2$ in various forms (Example 7.14).

segment connects vertices 0 and 1, and these vertices are called the *face* of the 1D hypercube and denoted by ×. A 2D Boolean hypercube has four faces, denoted by the cubes 0×, 1×, ×0, and ×1. The total 2D Boolean hypercube can be denoted by ××.

Example 7.17 *Six faces of the Boolean hypercube described by the cubes ×× 0, ×× 1, 0××, 1××, ×1×, and ×0× (Figure 7.16), represent one-term products for a Boolean function of three variables.*

Example 7.18 *In Table 7.3, the Boolean hypercubes for elementary Boolean functions of three variables are represented.*

TABLE 7.3
The truth tables, K-maps, and Boolean hypercubes for elementary Boolean functions of three variables

Techniques for Boolean function representation

Elementary Boolean function	Truth table	K-map	Boolean hypercube
AND $f = x_1 x_2 x_3$ x_1 ── x_2 ── x_3 ── f	$\begin{array}{ccc|c} x_1 & x_2 & x_3 & f \\ \hline 0&0&0&0 \\ 0&0&1&0 \\ 0&1&0&0 \\ 0&1&1&0 \\ 1&0&0&0 \\ 1&0&1&0 \\ 1&1&0&0 \\ 1&1&1&1 \end{array}$	$x_3 \begin{cases} 0 \\ 1 \end{cases}$ with $x_1 x_2$ columns $00\ 01\ 11\ 10$; entry 1 at row 1, column 11	$f = [1\ 1\ 1]$
OR $f = x_1 \vee x_2 \vee x_3$ x_1 ── x_2 ── x_3 ── f	$\begin{array}{ccc|c} x_1 & x_2 & x_3 & f \\ \hline 0&0&0&0 \\ 0&0&1&1 \\ 0&1&0&1 \\ 0&1&1&1 \\ 1&0&0&1 \\ 1&0&1&1 \\ 1&1&0&1 \\ 1&1&1&1 \end{array}$	$x_3 \begin{cases} 0 \\ 1 \end{cases}$; row 0: $01\to1, 11\to1, 10\to1$; row 1: $00\to1,01\to1,11\to1,10\to1$	$f = \begin{bmatrix} \times & \times & 1 \\ \times & 1 & \times \\ 1 & \times & \times \end{bmatrix}$
NAND $f = \overline{x_1 x_2 x_3}$ x_1 ── x_2 ── x_3 ── f	$\begin{array}{ccc|c} x_1 & x_2 & x_3 & f \\ \hline 0&0&0&1 \\ 0&0&1&1 \\ 0&1&0&1 \\ 0&1&1&1 \\ 1&0&0&1 \\ 1&0&1&1 \\ 1&1&0&1 \\ 1&1&1&0 \end{array}$	$x_3 \begin{cases} 0 \\ 1 \end{cases}$; row 0: $00\to1,01\to1,11\to1,10\to1$; row 1: $00\to1,01\to1,10\to1$	$f = \begin{bmatrix} 0 & \times & \times \\ \times & 0 & \times \\ \times & \times & 0 \end{bmatrix}$
NOR $f = \overline{x_1 \vee x_2 \vee x_3}$ x_1 ── x_2 ── x_3 ── f	$\begin{array}{ccc|c} x_1 & x_2 & x_3 & f \\ \hline 0&0&0&1 \\ 0&0&1&0 \\ 0&1&0&0 \\ 0&1&1&0 \\ 1&0&0&0 \\ 1&0&1&0 \\ 1&1&0&0 \\ 1&1&1&0 \end{array}$	$x_3 \begin{cases} 0 \\ 1 \end{cases}$; entry 1 at row 0, column 00	$f = [0\ 0\ 0]$
EXOR $f = x_1 \oplus x_2 \oplus x_3$ x_1 ── x_2 ── x_3 ── f	$\begin{array}{ccc|c} x_1 & x_2 & x_3 & f \\ \hline 0&0&0&0 \\ 0&0&1&1 \\ 0&1&0&1 \\ 0&1&1&0 \\ 1&0&0&1 \\ 1&0&1&0 \\ 1&1&0&0 \\ 1&1&1&1 \end{array}$	$x_3 \begin{cases} 0 \\ 1 \end{cases}$; row 0: $01\to1,10\to1$; row 1: $00\to1,11\to1$	$f = \begin{bmatrix} 0 & 0 & 1 \\ 0 & 1 & 0 \\ 1 & 0 & 0 \\ 1 & 1 & 1 \end{bmatrix}$
XNOR $f = \overline{x_1 \oplus x_2 \oplus x_3}$ x_1 ── x_2 ── x_3 ── f	$\begin{array}{ccc|c} x_1 & x_2 & x_3 & f \\ \hline 0&0&0&1 \\ 0&0&1&0 \\ 0&1&0&0 \\ 0&1&1&1 \\ 1&0&0&0 \\ 1&0&1&1 \\ 1&1&0&1 \\ 1&1&1&0 \end{array}$	$x_3 \begin{cases} 0 \\ 1 \end{cases}$; row 0: $00\to1,11\to1$; row 1: $01\to1,10\to1$	$f = \begin{bmatrix} 0 & 0 & 0 \\ 0 & 1 & 1 \\ 1 & 0 & 1 \\ 1 & 1 & 0 \end{bmatrix}$

Techniques for representation of the Boolean function

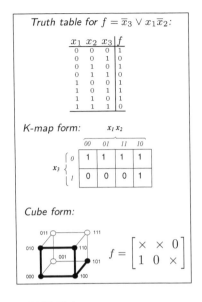

Truth table for $f = \overline{x}_3 \vee x_1\overline{x}_2$:

x_1	x_2	x_3	f
0	0	0	1
0	0	1	0
0	1	0	1
0	1	1	0
1	0	0	1
1	0	1	1
1	1	0	1
1	1	1	0

K-map form:

Cube form:

$$f = \begin{bmatrix} \times & \times & 0 \\ 1 & 0 & \times \end{bmatrix}$$

▶ There are five vertices that correspond to $f = 1$:

$$000, 010, 100, 101, 110$$

▶ Merging the vertex assignments yields

$$\mathbf{x}00, 0\mathbf{x}0, \mathbf{x}10, 1\mathbf{x}0, 10\mathbf{x}$$

▶ Four of these vertices can be merged to face, or term $\mathbf{x}\mathbf{x}0$. This term means that $f = 1$ if $x_3 = 0$, regardless of the values of x_1 and x_2.

▶ The function f can be represented in several ways. Some of the possibilities are

$$
\begin{aligned}
f &= \{000, 010, 100, 101, 110\} \\
&= \{0\mathbf{x}0, 1\mathbf{x}0, 101\} \\
&= \{\mathbf{x}00, \mathbf{x}10, 101\} \\
&= \{\mathbf{x}00, \mathbf{x}10, 10\mathbf{x}\} \\
&= \{\mathbf{x}\mathbf{x}0, 10\mathbf{x}\}
\end{aligned}
$$

FIGURE 7.15

Representation of the Boolean function $f = \overline{x}_3 \vee x_1\overline{x}_2$ by a Boolean hypercube: truth table and 3D Boolean hypercube representation (Example 7.15).

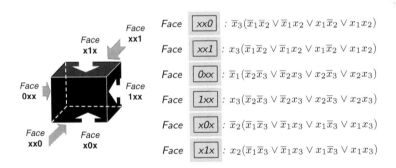

Face $\boxed{xx0}$: $\overline{x}_3(\overline{x}_1\overline{x}_2 \vee \overline{x}_1 x_2 \vee x_1\overline{x}_2 \vee x_1 x_2)$

Face $\boxed{xx1}$: $x_3(\overline{x}_1\overline{x}_2 \vee \overline{x}_1 x_2 \vee x_1\overline{x}_2 \vee x_1 x_2)$

Face $\boxed{0xx}$: $\overline{x}_1(\overline{x}_2\overline{x}_3 \vee \overline{x}_2 x_3 \vee x_2\overline{x}_3 \vee x_2 x_3)$

Face $\boxed{1xx}$: $x_3(\overline{x}_2\overline{x}_3 \vee \overline{x}_2 x_3 \vee x_2\overline{x}_3 \vee x_2 x_3)$

Face $\boxed{x0x}$: $\overline{x}_2(\overline{x}_1\overline{x}_3 \vee \overline{x}_1 x_3 \vee x_1\overline{x}_3 \vee x_1 x_3)$

Face $\boxed{x1x}$: $x_2(\overline{x}_1\overline{x}_3 \vee \overline{x}_1 x_3 \vee x_1\overline{x}_3 \vee x_1 x_3)$

FIGURE 7.16

Faces of the Boolean hypercube interpretation in the SOPs of a Boolean function of three variables (Example 7.17).

Hamming distance

Hamming distance is a useful measure of hypercube topology. The Hamming sum is defined as a bitwise operation:

$$(g_{d-1} \ldots g_0) \oplus (g'_{d-1} \ldots g'_0) = (g_{d-1} \oplus g'_{d-1}), \ldots, (g_1 \oplus g'_1), (g_0 \oplus g'_0) \quad (7.1)$$

where $\oplus$ is an exclusive OR operation. That is, the Hamming distance is the distance between two points on a Boolean hypercube with n vertexes defined as the number of coordinates (bit positions) at which the binary representations of the two points differ.

In the Boolean hypercube, two nodes are connected by a link (edge) if and only if they have labels that differ by exactly one bit. The number of bits by which the labels g_i and g_j differ is denoted by $h(g_i, g_j)$; this is the Hamming distance between the nodes.

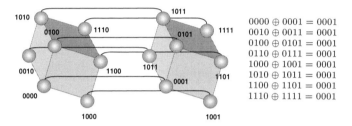

FIGURE 7.17
Hamming sum on Boolean hypercubes and the corresponding product of variables (Example 7.19).

Example 7.19	*Performing the Hamming sum on two Boolean hypercubes for a three-variable Boolean function is illustrated in Figure 7.17. The neighboring node's assignments differ by exactly one bit (the difference is shown using the Boolean EXOR operation)*

7.8   Logic networks

A logic network is a graphical data structure defined as an interconnection of gates that implements a combinational system. This graphical data structure can be described in various ways.

7.8.1   Design goals

A given Boolean function can be implemented in many ways using various gate types. The designer must choose the implementation that best meets the given design goals. A central problem in logic design is to obtain a logic network that realizes a given Boolean function using a given set of gates and has the lowest possible cost:

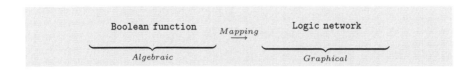

The cost of a logic network is conveniently measured in a technology-independent fashion by the total number of gates used. We assume that the logic gates and logic network can be of arbitrary size and can respond instantaneously to signal changes on their primary input lines. Physical constraints limit both the size and speed of real logic networks in important ways, of which the designer must be aware.

7.8.2 Basic components of a logic network

A logic network is composed of gates, external inputs and outputs, and connections. Connections are specified from external inputs and gate outputs to gate inputs and external outputs, and must meet the following requirements: (a) each gate input is connected to a constant 0 or 1, a network's external input, or a gate output; and (b) only one connection per gate input is allowed. The output load imposed on a gate output should not be greater than its fan-out factor. This load is computed as the sum of the input load factors of all the gate inputs that are connected to the gate output.

The transfer and processing of information requires the expenditure of energy, and a physical device can absorb or produce only a certain amount of energy without failing. This places an upper bound on the number of input sources that supply a gate, and another bound on the maximum number of output devices that may be supplied with signals from the gate's output. The number of inputs of gate is termed its *fan-in*; the number of connections, which are typically inputs to other gates that are connected to the gate's output, is termed its *fan-out*.

The composition of primitive components is accomplished by physically wiring the gates together. The tabulation of gate inputs and outputs and the nets to which they are connected is called a *netlist*. The *fan-in* of a gate is its number of inputs.

A logic network is *loop-free* or *cycle-free* if there exists only one path from any point of a network through the gates in the direction from input to output.

A logic network can be described as a *logic diagram*, a *netlist*, and a *set of language statements*. Throughout the design cycle, all these representations are used; that is, logic networks can be synthesized simultaneously using these forms.

Example 7.20   *Figure 7.18a shows the* **logic network** *implementing the Boolean function* $f = x_1 x_2 \vee x_2 x_3 \vee x_4$. *It can be represented by a list of gates and a list of connections (b), or by using a hardware language (c).*

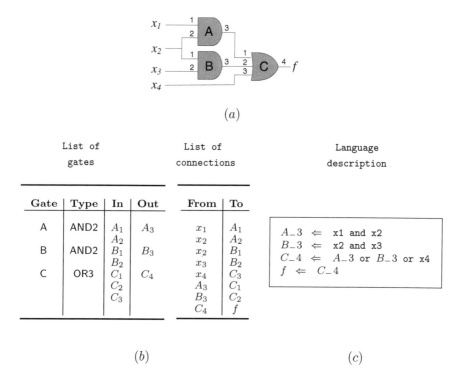

(a)

| List of gates | | | | | List of connections | | Language description |

Gate	Type	In	Out
A	AND2	A_1 A_2	A_3
B	AND2	B_1 B_2	B_3
C	OR3	C_1 C_2 C_3	C_4

From	To
x_1	A_1
x_2	A_2
x_2	B_1
x_3	B_2
x_4	C_3
A_3	C_1
B_3	C_2
C_4	f

$$A\_3 \;\Leftarrow\; \text{x1 and x2}$$
$$B\_3 \;\Leftarrow\; \text{x2 and x3}$$
$$C\_4 \;\Leftarrow\; A\_3 \text{ or } B\_3 \text{ or x4}$$
$$f \;\Leftarrow\; C\_4$$

(b)   (c)

FIGURE 7.18
Logic network description in the form of a logic diagram (a), a netlist (b), and a hardware description language (c) (Example 7.20).

A logic diagram

A logic diagram is a graphical representation of a logic network. This diagram consists of symbols of gates and connections that are drawn using the rules given above. The advantages of logic diagrams are as follows: (a) logic diagram is a directed graph, and therefore, its manipulation can be accomplished using graph-based techniques, and (b) logic diagram is convenient for network analysis, including visual analysis.

A netlist description

A netlist description is a reasonable form that is required in the simplification of the manipulation of a network. A netlist description consists of a *list of gates* and a *list of connections*. These lists can be considered as an analog of the adjustment matrix for representing directed graphs, taking into account the connection limitations.

Hardware description languages

Hardware description languages are text-based descriptions. The basic elements of these descriptions are statements aimed at translating the functional and structural properties of a logic networks into hardware equivalents.

7.8.3   Specification

The specification of a logic network is defined as the parameters and characteristics required for the representation of a network as a component of a larger logic network. According to the hierarchical design principle, this network must be suitable for assembling a large network from smaller logic networks. For this, the logic network must satisfy various requirements, including the *input load factor*, the *fan-out factor*, the *size*, the *propagation delay*, the *number of levels*, and the *dynamic* characteristics.

> **Example 7.21**   *Consider a logic network (Figure 7.19). The network has 5 gates, the fan-in for every gate is 2, and the fan-out for each gate is 1, except to the AND gate with inputs x_3 and x_4. The network characteristics are as follows: the input load factor is 6, fan-out factor is -4, and the number of levels is 2.*

7.8.4   Network verification

Each logic network design has to be verified with respect to various criteria before fabrication. In every design step, the exact input/output behavior of the system has to be specified, and functional behavior has to be realized within the bounds of the given technology. The description of system behavior is called *specification*. A realization of the specified system behavior is called an *implementation*.

   If it can be *proven* by any method that the implementation satisfies the given specifications, then the chip design is functionally correct. For this proof

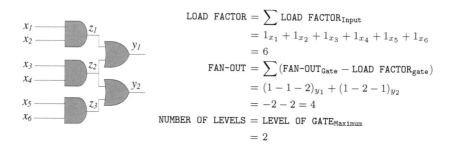

$$\text{LOAD FACTOR} = \sum \text{LOAD FACTOR}_{\text{Input}}$$
$$= 1_{x_1} + 1_{x_2} + 1_{x_3} + 1_{x_4} + 1_{x_5} + 1_{x_6}$$
$$= 6$$
$$\text{FAN-OUT} = \sum (\text{FAN-OUT}_{\text{Gate}} - \text{LOAD FACTOR}_{\text{gate}})$$
$$= (1 - 1 - 2)_{y_1} + (1 - 2 - 1)_{y_2}$$
$$= -2 - 2 = 4$$
$$\text{NUMBER OF LEVELS} = \text{LEVEL OF GATE}_{\text{Maximum}}$$
$$= 2$$

FIGURE 7.19
Logic network characteristics (Example 7.21).

of functional correctness, the following concepts are of practical interest: (a) *simulation*, (b) *formal* verification, and (c) *partial* verification.

Simulation means systematically evaluating the functional behavior of both the specification and the implementation for a large number of input vectors. If the outputs of the implementation and the specification agree for all these vectors, the design works correctly.

The aim of *formal verification* is a formal mathematical proof that the functional behaviors of the specification and the implementation coincides. An appropriate data structure must be chosen for applying this approach. Decision diagrams are suitable for the efficient formal verification of logic networks.

Partial verification is defined as a mathematical method for proving that an implementation satisfies at least the particular important properties of the specified system behavior. To guarantee that the functionality of a logic network has not been modified through the synthesis and optimization process, a verification task has to be solved.

Example 7.22 *In Figure 7.20, the initial network (left) is interpreted as a specification and implemented using NAND gates (right). It has to be proven that both the specification and the implementation are functionally equivalent; that is, both networks compute the same Boolean function:*

$$\overline{\overline{x_2 \overline{x}_3 x_1} x_4} = \overline{x_2 \overline{x}_3 x_1} \vee \overline{x}_4$$
$$= (\overline{x}_2 \vee \overline{x}_3) x_1 \vee \overline{x}_4$$
$$= x_1 \overline{x}_2 \vee x_1 \overline{x}_3 \vee \overline{x}_4 = f$$

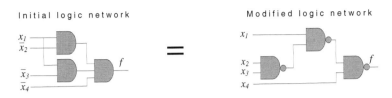

FIGURE 7.20
Logic network verification (Example 7.22).

7.9   Networks of threshold gates

An attractive property of threshold functions is that any elementary Boolean function can be implemented by a single threshold gate using certain control parameters. It follows from this property that a given threshold network can be used for computing different Boolean functions while varying the control parameters. The ability to keep the same topology in computing various Boolean functions is an important factor in implementation. Note that a logic network is designed for the implementation of a particular Boolean function; that is, the same logic network cannot be used for computing different Boolean functions.

7.9.1   Threshold functions

The class of threshold functions is of great importance for modeling biological neurons and for constructing artificial neural networks. The neuron's cell membrane is capable of sustaining a certain electric charge. When this charge reaches or exceeds a threshold k, the neuron "fires." This effect is modeled by means of threshold functions as proposed by McCulloch and Pitts.

A Boolean function f of n variables $x_1, x_2, \ldots, x_n$ is called a *threshold function* with weights $w_1, w_2, \ldots, w_n$ (real numbers) and *threshold k* if

$$f = 1 \text{ if and only if } \sum_{i=1}^{n} w_i \times x_i \geq k \qquad (7.2)$$

A threshold function is implemented by a threshold gate, as shown in Figure 7.21). This threshold gate can compute various elementary Boolean functions of two variables. For this, appropriate control parameters must be chosen.

Example 7.23   *The n-input OR and AND functions have thresholds of 1 and n, respectively. That is, in this **threshold gate**, increasing the number of active inputs results in the output 1.*

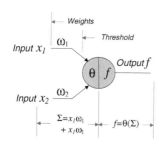

The input signal for $x_1, x_2 \in \{0, 1\}$ is

$$\Sigma = x_1 \times w_1 + x_2 \times w_2$$

Thresholding of Σ at the level Θ produces the output signal

$$f = \Theta(\Sigma)$$

FIGURE 7.21
Threshold gate.

There are several particular implementations of threshold functions. In particular, if $w_1 = w_2 = \ldots = w_n = 1$, $n = 2m+1$, $\Theta = m+1$, the threshold function implements the *majority* function. A majority function f is equal to 1 if the inputs have more 1s than 0s.

> **Example 7.24** *Let $m = 1$. Then $n = 2m + 1 = 2 \times 1 + 1 = 3$, and we have the **majority** function of three variables: $f = x_1 x_2 \vee x_2 x_3 \vee x_3 x_1$. This function is equal to 1 iff two or more inputs are equal to 1.*

7.9.2 McCulloch–Pitts models of Boolean functions

Two-input AND, OR, NOR, and NAND elementary logic functions can be computed by a single neural cell known as the *McCulloch–Pitts* model. Control over the type of logic function is exercised by the threshold θ and weights $w_i \in \{1, -1\}$ of the arithmetic sum $w_i \times x_1 + w_i \times x_2$, that is, the output is $f = w_1 \times x_1 + w_2 \times x_2 - \theta$. The Boolean function $f = x_1 \oplus x_2$ cannot be represented by one neuron cell. This limitation can be explained as follows. For the weights w_1, w_2 and a threshold of one neuron θ, the following inequalities hold: $0 \cdot w_1 + 1 \cdot w_2 \geq k$ because $f(0,1) = 1$; $1 \cdot w_1 + 0 \cdot w_2 \geq k$ because $f(1,0) = 1$; $1 \cdot w_1 + 1 \cdot w_2 < k$ because $f(1,1) = 0$.   The first two inequalities can be combined into $w_1 + w_1 \geq 2k$, which contradicts the third equation. Consequently, the function f cannot be represented in terms of a weighted threshold function. A solution to this problem is given in Table 7.4, where an EXOR function is implemented using three threshold gates.

> **Example 7.25** *In Table 7.4, the correspondence between the control parameters of the threshold gates and the implemented Boolean functions of two variables is illustrated.*

TABLE 7.4

The McCulloch–Pitts model for computing AND, OR, and NAND Boolean functions (Example 7.25)

AND $f = x_1 x_2$	OR $f = x_1 \vee x_2$	NAND $f = \overline{x_1 x_2}$

x_1 x_2	$\sum$	f
0 0	0	0
0 1	1	0
1 0	1	0
1 1	2	1

x_1 x_2	$\sum$	f
0 0	0	0
0 1	1	1
1 0	1	1
1 1	2	1

x_1 x_2	$\sum$	f
0 0	0	1
0 1	-1	1
1 0	-1	1
1 1	-2	0

7.9.3 Threshold networks

The *depth* of the logic network is defined to be the maximum path from the input to the output of this network. If all gates are grouped, the network can be arranged in *multilevel* network. The depth of a multilevel network is equal to the number of levels. A logic network can be defined as a *feedforward* network of logic gates (AND, NAND, OR, NOR, etc.) or threshold gates, that is, a network with no feedback.

The design of threshold networks aims in minimizing the depth, the size (number of threshold elements), the number of inputs of threshold elements, and the weight values of threshold elements.

Example 7.26 *A threshold network shown in Figure 7.22a is used for the implementation of AND-OR and OR-AND logic networks, which have the configurations of given threshold networks. The control parameters are taken from Table 7.4 as follows: (a) An AND-OR network (Figure 7.22b) is designed from a threshold network by specifications of the weights and thresholds for an AND function, that is, $w_{11} = w_{12} = \ldots = w_{43} = 1$, $\theta_1 = \theta_2 = \theta_3 = 2$, $\theta_4 = 1$. (b) An OR-AND network (Figure 7.22c) is designed by analogy: $w_{11} = w_{12} = \ldots = w_{43} = 1$, $\theta_1 = \theta_2 = \theta_3 = 1, \theta_4 = 2$.*

Threshold networks can be considered as an alternative to logic networks. In some cases, the networks of threshold elements more efficient compare with logic networks [4, 15]. This advantage is based on the following statements: (a) any symmetric function can be computed by the threshold network of depth

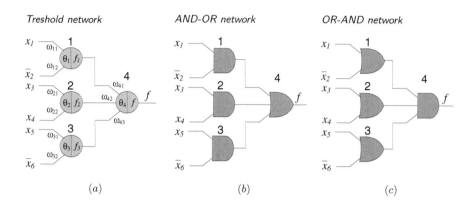

FIGURE 7.22
A network of threshold elements (a) and the corresponding AND-OR (b) and OR-AND (c) logic networks (Example 7.26).

two, and (b) an arbitrary Boolean function of n variables can be represented as a symmetric function of 2^n variables by repeating each variable x_j 2^{j-1} times.

7.10   Binary decision trees

A complete binary decision tree can be constructed from the truth table of a Boolean function. The procedure for constructing a complete binary decision tree is given in Figure 7.23. This procedure is based on the iterated application of a Shannon expansion:

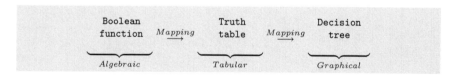

Example 7.27   *Figure 7.24 shows a complete* **binary decision tree** *(nodes implement Shannon expansion), derived for a Boolean function of three variables ($n = 3$), and order of variables $x_1 \longrightarrow x_2 \longrightarrow x_3$.*

Constructing a binary decision tree

Given: A Boolean function.

Step 1. Derive a truth table of the Boolean function.

Step 2. Derive the first node of the decision tree as follows:

(a) Choose a variable to assign to the root of the tree, x_i.

(b) Find the Shannon expansion $f = \overline{x}_i f_{x_i=0} \vee x_i f_{x_i=1}$.

(d) Assign $\overline{x}_i$ to the left outgoing branch and x_i to the right one.

Step 3. Choose another variable, x_j. Construct two nodes connected to the branches of the root node:

▶ $f_{x_i=0}$ is the input of the left node.

▶ $f_{x_i=1}$ is the input of the right node.

Find the Shannon expansion for both functions:

$$f_{x_i=0} = \overline{x}_j f_{\substack{x_i=0 \\ x_j=0}} \vee x_j f_{\substack{x_i=0 \\ x_j=1}} \quad \text{and} \quad f_{x_i=1} = \overline{x}_j f_{\substack{x_i=1 \\ x_j=0}} \vee x_j f_{\substack{x_i=1 \\ x_j=1}}$$

Step 4. Repeat Step 3 $n-2$ times, performing Shannon expansion of the factors with respect to the remaining $n-2$ variables.

Result: A complete binary decision tree.

Note: The order of variables can be fixed, or chosen using minimization criteria (for further reduction of the tree to a binary decision diagram).

FIGURE 7.23

Procedure for constructing a complete binary decision tree from the truth table of a Boolean function.

7.10.1 Representation of elementary Boolean functions using decision trees

Example 7.28 *Decision trees for two-input Boolean functions are given in Figure 7.25.*

7.10.2 Minterm and maxterm expression representations using decision trees

A minterm or maxterm expression can be mapped into a decision tree, and vice versa: Given a decision tree, the algebraic form of a standard SOP or POS expression can be recovered from this tree:

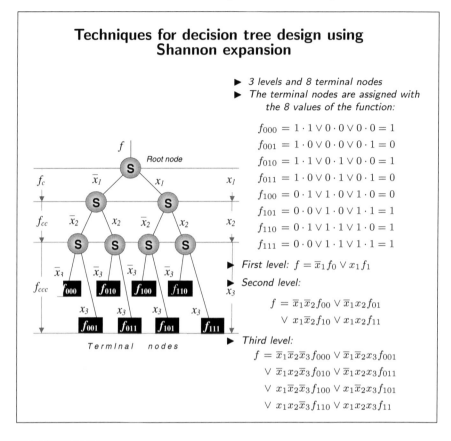

Techniques for decision tree design using Shannon expansion

▶ 3 levels and 8 terminal nodes
▶ The terminal nodes are assigned with the 8 values of the function:

$$f_{000} = 1 \cdot 1 \vee 0 \cdot 0 \vee 0 \cdot 0 = 1$$
$$f_{001} = 1 \cdot 0 \vee 0 \cdot 0 \vee 0 \cdot 1 = 0$$
$$f_{010} = 1 \cdot 1 \vee 0 \cdot 1 \vee 0 \cdot 0 = 1$$
$$f_{011} = 1 \cdot 0 \vee 0 \cdot 1 \vee 0 \cdot 1 = 0$$
$$f_{100} = 0 \cdot 1 \vee 1 \cdot 0 \vee 1 \cdot 0 = 0$$
$$f_{101} = 0 \cdot 0 \vee 1 \cdot 0 \vee 1 \cdot 1 = 1$$
$$f_{110} = 0 \cdot 1 \vee 1 \cdot 1 \vee 1 \cdot 0 = 1$$
$$f_{111} = 0 \cdot 0 \vee 1 \cdot 1 \vee 1 \cdot 1 = 1$$

▶ First level: $f = \overline{x}_1 f_0 \vee x_1 f_1$
▶ Second level:

$$f = \overline{x}_1 \overline{x}_2 f_{00} \vee \overline{x}_1 x_2 f_{01}$$
$$\vee \; x_1 \overline{x}_2 f_{10} \vee x_1 x_2 f_{11}$$

▶ Third level:

$$f = \overline{x}_1 \overline{x}_2 \overline{x}_3 f_{000} \vee \overline{x}_1 \overline{x}_2 x_3 f_{001}$$
$$\vee \; \overline{x}_1 x_2 \overline{x}_3 f_{010} \vee \overline{x}_1 x_2 x_3 f_{011}$$
$$\vee \; x_1 \overline{x}_2 \overline{x}_3 f_{100} \vee x_1 \overline{x}_2 x_3 f_{101}$$
$$\vee \; x_1 x_2 \overline{x}_3 f_{110} \vee x_1 x_2 x_3 f_{111}$$

FIGURE 7.24
The complete binary tree for representing a Boolean function f of three variables x_1, x_2, and x_3 (Example 7.27).

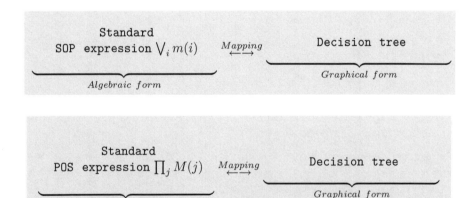

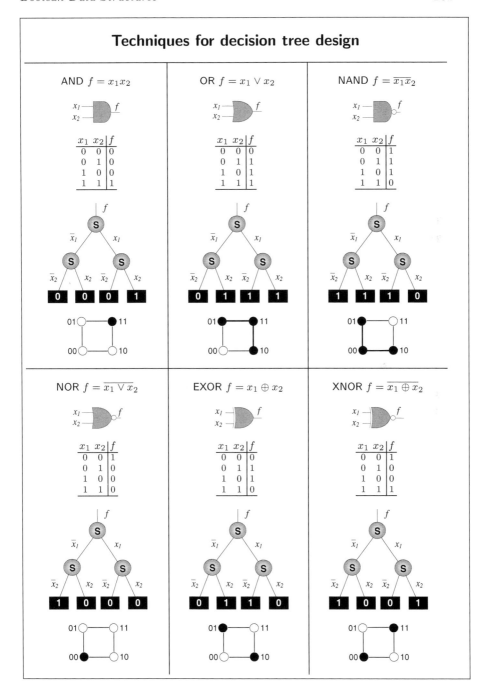

FIGURE 7.25
Complete decision trees for computing elementary Boolean functions compared to other data structures: algebraic, tabulated (truth table), and hypercubes (Example 7.28).

That is,

(*a*) The minterm in an SOP is mapped into a $\boxed{1}$ terminal node of the complete decision tree; a path in the tree from the root to a $\boxed{1}$ terminal node corresponds to a minterm in the SOP expression.

(*b*) The minterm in a POS is mapped into a $\boxed{0}$ terminal node of the complete decision tree; a path in the tree from the root to a $\boxed{0}$ terminal node corresponds to a maxterm in the POS expression.

Example 7.29   *Figure 7.26 shows the relationship of standard SOP and POS expressions and their representation using a decision trees for the Boolean function of two variables $f = x_1 \oplus x_2$.*

7.10.3   Representation of elementary Boolean functions by incomplete decision trees

In a complete decision tree, the number of nodes depends exponentially on the number of variables. Hence, this representation is not acceptable for a large number of variables. Algorithm for a decision tree reduction is as follows:

Example 7.30   *In Figure 7.27, reduced decision trees are used for representing four Boolean functions of two variables, AND, OR, NAND, and NOR. Each of these trees is constructed from the corresponding complete tree given in Figure 7.25 by deleting the node S, the linked terminal nodes, and the corresponding link reconstruction.*

One can observe from Example 7.30 that the reduced decision trees for AND, OR, NAND, and OR functions contain repeated terminal nodes. The complete binary trees do not correspond to the minimal Boolean expression that they realize. For example, an SOP restored from an OR function is $f = \overline{x}_1 x_2 \vee x_1$. However, this SOP is not minimal; further algebraic manipulations implies

$$f = \overline{x}_1 x_2 \vee x_1 = (x_1 \vee \overline{x}_1)(x_1 \vee x_2) = x_1 \vee x_2$$

The same can be observed for the NAND function: $f = \overline{x}_1 \vee x_1 \overline{x}_2 = \overline{x}_1 \vee \overline{x}_2 = \overline{x_1 x_2}$. Reduced decision trees contain fewer nodes than complete decision trees.

Techniques for decision tree design

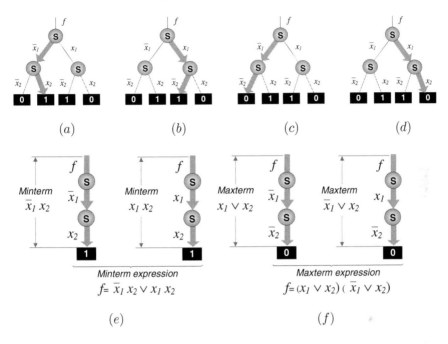

(a)   (b)   (c)   (d)

Minterm expression
$f = \overline{x}_1 x_2 \vee x_1 x_2$

Maxterm expression
$f = (x_1 \vee x_2)(\overline{x}_1 \vee x_2)$

(e)   (f)

FIGURE 7.26

Minterm and maxterm representation using decision trees: (a) minterm $\overline{x}_1 x_2$, (b) minterm $x_1 x_2$, (c) maxterm $x_1 \vee x_2$, (d) maxterm $\overline{x}_1 \vee x_2$, (e) the SOP expression $f = \overline{x}_1 x_2 \vee x_1 x_2$, and (f) the POS expression $f = (x_1 \vee x_2)(\overline{x}_1 \vee x_2)$ (Example 7.29).

Example 7.31 *Decision trees in Figure 7.28 implement the function $\overline{x}_1$. The reduction implies that $f = \overline{x}_1\overline{x}_2 \vee \overline{x}_1 x_2 = \overline{x}_1$ (Figure 7.28a). This effect is caused by the redundancy of the decision tree. Shannon expansion with respect to the variable x_2 is equal to 1 because $f = \underbrace{\overline{x}_2 f_0}_{\text{Equal to 1}} \vee \underbrace{x_2 f_1}_{\text{Equal to 1}} = 1$. Hence, the corresponding node can be deleted and replaced with a terminal node, labeled by 1 (Figure 7.28b).*

One can observe that reduction using the procedure described above still results in repeating terminal nodes in the reduced decision trees. Further reduction of these trees using this repetition leads to *decision diagrams*.

Binary decision tree reduction

Given: A complete binary tree for a Boolean function of n variables.

Step 1: Remove the nodes using the following rules:

    (a) Identify the nodes at the n-th level, which are connected to the terminal nodes of the same values.

    (b) Remove the nodes, and both terminal nodes, substituting them with a terminal node of the corresponding value.

Step 2: Repeat Step 1 for the remaining $n-1$ levels, starting from the bottom.

Result: A reduced binary decision tree.

7.11 Decision diagrams

The manipulation of graphical data structures is especially beneficial for large-size problems. Moreover, during the logic synthesis of circuits manufactured using certain technologies, it is not necessary to convert graph data structures into logic networks. This is because the topology of the obtained graph is itself a logic network.

A *path* in a decision diagram is a sequence of edges and nodes leading from the root node to a terminal node. A path carries local information for the algebraic representation of Boolean functions. A path can be measured by a *length t*, which is the number of vertices in the path. There is always a 0-length path from node u to node u'. The *path length* is the number of non-terminal nodes in the path. A binary decision diagram is a reduced binary decision tree:

The procedures given in Figure 7.29 result in a decision diagram for the representation of a Boolean function. Note that the reduced diagram does not necessary have the minimal number of nodes. The order of variables determines the size of the reduced decision diagram.

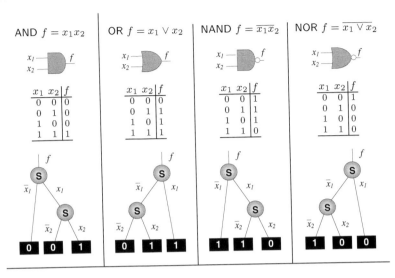

FIGURE 7.27
Representation of AND, OR, NAND, and NOR functions using reduced decision trees (Example 7.30).

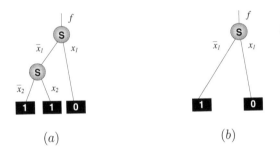

FIGURE 7.28
The redundant (a) and irredundant reduced (b) decision trees representing the function $f = \overline{x}_1$ (Example 7.31).

7.12   Further study

Advanced topics of data structures

Topic 1: A Hopfield computing paradigm is based on a *Hopfield network*, which is defined as a logic network of interconnected, binary-valued neuron cells that implement the distributed principle of encoding and processing of information. The result (a computed Boolean function) is obtained by decoding the neuron's states. Using a Hopfield network, an arbitrary logic

---

Algorithm 1: Deriving a binary decision diagram

Given: A Boolean function of n variables.
Step 1: Derive a complete binary decision tree.
Step 2: Reduce the decision tree.
Step 3: Merge repeating terminal nodes into two: `0` and `1`
Result: A binary decision diagram.

Algorithm 2: Deriving a binary decision diagram

Given: A Boolean function of n variables.
Step 1: Find a minimal SOP or POS.
Step 2: Derive a binary decision tree by deriving the path. that
     corresponds to the products.
Step 3: Reduce the decision tree to a decision diagram.
Result: A binary decision diagram.

FIGURE 7.29

Procedures for deriving binary decision diagrams from decision trees.

TABLE 7.5

Decision diagram parameters and characteristics

Characteristic	Definition
Size	The sum of the numbers of nodes at all levels, excluding the terminal nodes (always two), N_N
Depth, or number of levels	The number of levels, DEPTH; this is equivalent to the number of variables plus 1 if there are no redundant variables; that is, $n + 1$
Width	The maximum number of nodes at any level, WIDTH
Area	The area the decision diagram occupies is estimated as AREA=WIDTH×DEPTH
Path length	The number of edges on the path from the root to a terminal node
Weight	The number of ones in the terminal nodes
Homogeneity	The absolute difference between 1 and 0 terminal nodes

function can be computed via a process similar to the gradual cooling process of metal. Values of the Boolean function given an assignment of Boolean variables are computed through the relaxation of neuron cells.

Topic 2: Threshold logic theory relates neural network techniques to logic design, formulating logic functions in terms of inequalities. Many existing methods perform synthesis by representing each product term in an SOP expression of a function as a threshold gate or by converting each gate in a Boolean network into a threshold gate, and this requires advanced factorization algorithms. CMOS implementations of threshold gates were proposed in the 1980s. (see a survey of VLSI implementations of threshold logic in [1]). A threshold logic circuit design using new technologies, such as resonant-tunneling and quantum cellular arrays, have been considered just recently, upon development of efficient algorithms to factorize a multilevel network using algebraic or Boolean factorization.

Topic 3: Information-theoretic measures. A computing system can be seen as a process of communication between computer components. The classical concepts of information and entropy introduced by Shannon are the basis for this. The information-theoretic standpoint in computing is based on the following notations:

(a) *Source of information* is a stochastic process whereby an event occurs at time point i with probability p_i. That is, the source of information is defined in terms of the probability distribution for signals from this source. The problem is usually formulated in terms of sender and receiver of information and is used by analogy with communication problems.

(b) *Information engine* is the machine that deals with information.

(c) *Quantity of information* is a value of a function that occurs with a probability p; this quantity is equal to $(-\log_2 p)$.

(d) *Entropy*, $H(f)$, is the measure of the information content of X. The greater the uncertainty in the source output, the higher is its information content. A source with zero uncertainty would have zero information content and, therefore, its entropy would be equal to zero.

Let us assume that all combinations of values of variables occur with equal probability. A value of a function that occurs with the probability p carries a quantity of information equal to $<$ `Quantity of information` $> = -\log_2 p$ *bit*, where p is the probability of that value occurring. Note that information is measured in bits. The information carried by the value of a of a random variable x_i is equal to $I(x_i)_{|x_i=a} = -\log_2 p$ *bit*, where p is the quotient between the number of tuples whose i-th components equal a and the total number of tuples.

For example, the output of the AND function is equal to 0 with probability 0.25, and equal to 1 with probability 0.75. The entropy of the output signal is calculated as follows: $H_{out} = -0.25 \times \log_2 0.25 - 0.75 \times \log_2 0.75 = 0.81$ *bit/pattern* (see details in "An Information Theoretic Approach to Digital Fault Testing" by V. Agraval, *IEEE Transactions on Computers*, volume 30, number 8, pages 582–587, 1981).

Further reading

[1] Beiu V, Quintana JM, and Avedillo MJ. VLSI implementation of threshold logic – A comprehensive survey. *IEEE Neural Networks*, 14(5):1217–1243, 2003.

[2] Brown S. and Vranesic Z. *Fundamentals of Digital Logic with VHDL Design*, McGraw-Hill, New York, 2000.

[3] Bryant RE. Graph-based algorithms for Boolean function manipulation. *IEEE Trans. Comp.*, 35(6):677–691, 1986.

[4] Dertouzos ML. *Threshold Logic*. Wiley, New York, 1971.

[5] DeMicheli G. *Synthesis and Optimization of Digital Circuits*. McGraw Hill, New York, 1994.

[6] Givone DD. *Digital Principles and Design*. McGraw-Hill, New York, 2003.

[7] Hartmann CRP, Varshney PK, Mehrotra KG, and Gerberich CL. Application of information theory to the construction of efficient decision trees. *IEEE Trans. Information Theory*, 28(5):565–577, 1982.

[8] Hayes JP. *Introduction to Digital Logic Design*, Addison-Wesley, Reading, MA, 1993.

[9] Karpovsky MG, Stanković RS, and Astola JT. *Spectral Logic and Its Applications for the Design of Digital Devices*. John Wiley & Sons, Hoboken, NJ, 2008.

[10] Mano MM and Kime C. *Logic and Computer Design Fundamentals*. 3rd edition. Prentice Hall, Upper Saddle River, NJ, 2005.

[11] Marcovitz AB. *Introduction to Logic Design*, McGraw-Hill, Reading, MA, 2005.

[12] Meinel C and Theobald T. *Algorithms and Data Structures in VLSI Design*. Springer, Heidelberg, 1998.

[13] Minato S. *Binary Decision Diagrams and Applications for VLSI CAD*, Kluwer, Dordrecht, 1996.

[14] Rosenblatt F. *Principles of Neurodynamics*. Spartan, New York, 1962.

[15] Siu KY, Roychowdhury VP, and Kailath T. Depth-size tradeoffs for neural computation. *IEEE Trans. Comput.*, 40(12):1402–1411, 1991.

[16] Stanković RS. *Spectral Transform Decision Diagrams in Simple Questions and Simple Answers*. Nauka, Belgrade, 1998.

[17] Stanković RS and Astola JT. *Spectral Interpretation of Decision Diagrams*, Springer, Heidelberg, 2003.

[18] Yanushkevich SN and Shmerko VP. *Introduction to Logic design*, Taylor & Francis/CRC Press, Boca Raton, FL, 2008.

[19] Yanushkevich SN, Miller DM, Shmerko VP, and Stanković RS. *Decision Diagram Techniques for Micro- and Nanoelectronic Design*. Taylor & Francis/CRC Press, Boca Raton, FL, 2006.

8

Fundamental Expansions

8.1 Introduction

Fundamental expansions are the basis of various techniques for computing and data structure design. These techniques can be used for solving, in particular, the following problems:

Problem 1: Given a Boolean function f of n variables, find the efficient computation with respect to the variables; in this formulation, the particular properties of a Boolean function should be utilized in computing, that is, the order of the variables is important.

Problem 2: Given a symmetric Boolean function f of n variables, find the efficient computation with respect to the symmetric k variables; in this formulation, the efficient of computation depends on the efficiency of utilization of symmetric properties of function. However, if choice of the symmetric properties are taken into consideration, the number of steps can be reduced to steps.

Problem 3: Given a Boolean function f, analyze the sensitivity properties of this function; in this formulation, the behavior properties of Boolean function are the focus of interest.

Problem 4: Given a Boolean function f, represent this function in the polynomial form.

These and other problems can be efficiently solved using the fundamental expansions of Boolean algebra.

In this chapter, three fundamental expansions used in contemporary logic design are introduced (Figure 8.1):

▶ *Shannon expansion* for an arbitrary Boolean functions

▶ *Shannon expansion* for symmetric Boolean functions

▶ *Logic Taylor* expansion

The Shannon classical expansion provides the decomposition of an arbitrary Boolean function of n variables to a set of Boolean sub-functions of $m < n$ variables:

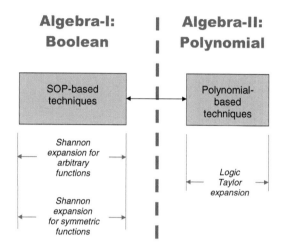

FIGURE 8.1
Fundamental expansions of Boolean functions.

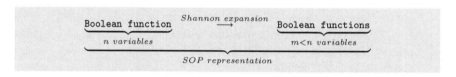

Many advanced techniques of logic analysis and synthesis are based on Shannon expansion. An arbitrary Boolean function can be expanded with respect to a single variable, a group of variables, and all variables. This is the primary fundamental expansion of logic design, introduced by Claude Shannon in 1937.

In practice, one of the advantages of knowing that a given Boolean function is symmetric is the potential implementation cost reduction offered by the implementation of such function using smaller number of gates. *Shannon expansion for symmetric function* is specific for the representation and realization of function with symmetric variables. This expansion is quite distinctive from the Shannon expansion for arbitrary Boolean function; this is the second fundamental expansion of logic design.

The third fundamental expansion is called the *logic Taylor expansion*; this is an analog of the classical Taylor expansion. This expansion describes the Boolean function in terms of *change* in function values, which is formulated using a *Boolean difference*:

Boolean function   *Logic Taylor expansion*   Polynomial   form
  SOP form   *In terms of differences*

The logic Taylor expansion provides various polynomial representations of Boolean functions using the logical operations EXOR and AND. The classical Taylor expansion is defined as a "polynomial for the function f about the point $x = x_i$, where each coefficient is defined in the terms of derivatives." In the logic Taylor expansion, "expansion about a point" is replaced by "polarity." Given a Boolean function of n variables, the logic Taylor expansion generates 2^n polynomials of different polarities. This expansion interpreted as a representation of a Boolean function of n variables by a set of polynomials of $m < n$ variables is called the *Davio expansion*:

$$\underbrace{\texttt{Polynomial form}}_{n\ variables} \xrightarrow{\textit{Davio expansion}} \underbrace{\texttt{Polynomial form}}_{m<n\ variables}$$

Davio expansion is a particular case of Shannon expansion.

8.2 Shannon expansion

There are Shannon expansions of an arbitrary Boolean function with respect to

- ▶ A single variable
- ▶ A group of variables
- ▶ All variables

Table 8.1 provides examples of Shannon expansion based on the above classification. In this table, Shannon expansion with respect to a single variable is defined as a *step*, or an *iteration* in the recursive procedure of Shannon expansion with respect to all variables.

8.2.1 Expansion with respect to a single variable

We denote a Boolean function f of n variables by $f = f(x_1, x_2, \ldots, x_n)$. The fact that variable x_i, $i \in \{1, 2, \ldots n\}$, takes the value 0 or 1 is represented in the form

$$f(x_1, \ldots, x_{i-1}, \overset{x_i}{\boxed{0}}, x_{i+1}, \ldots, x_n) \text{ and } f(x_1, \ldots, x_{i-1}, \overset{x_i}{\boxed{1}}, x_{i+1}, \ldots, x_n),$$

respectively.

Example 8.1   *Let $f = x_1 \vee x_2 \vee x_3$, then $f = f(x_1, x_2 = 0, x_3) = f(x_1, 0, x_3) = x_1 \vee 0 \vee x_3 = x_1 \vee x_3$ and $f = f(x_1, x_2 = 1, x_3) = f(x_1, 1, x_3) = x_1 \vee 1 \vee x_3 = 1$.*

TABLE 8.1
Types of Shannon expansion for an arbitrary Boolean function with respect to a single variable, a group of variables, and all variables

Type of expansion	Formal notation and example

W i t h   r e s p e c t   t o   a   s i n g l e   v a r i a b l e

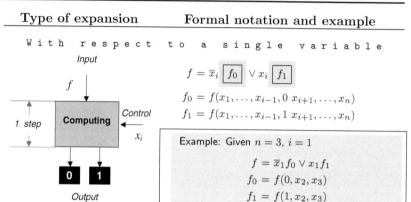

$$f = \overline{x}_i \boxed{f_0} \vee x_i \boxed{f_1}$$

$$f_0 = f(x_1, \ldots, x_{i-1}, 0\ x_{i+1}, \ldots, x_n)$$
$$f_1 = f(x_1, \ldots, x_{i-1}, 1\ x_{i+1}, \ldots, x_n)$$

Example: Given $n = 3$, $i = 1$

$$f = \overline{x}_1 f_0 \vee x_1 f_1$$
$$f_0 = f(0, x_2, x_3)$$
$$f_1 = f(1, x_2, x_3)$$

W i t h   r e s p e c t   t o   a   g r o u p   o f   v a r i a b l e s

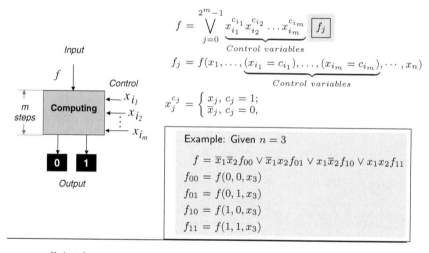

$$f = \bigvee_{j=0}^{2^m - 1} \underbrace{x_{i_1}^{c_{i_1}} x_{i_2}^{c_{i_2}} \ldots x_{i_m}^{c_{i_m}}}_{\text{Control variables}} \boxed{f_j}$$

$$f_j = f(x_1, \ldots, \underbrace{(x_{i_1} = c_{i_1}), \ldots, (x_{i_m} = c_{i_m})}_{\text{Control variables}}, \cdots, x_n)$$

$$x_j^{c_j} = \begin{cases} x_j, & c_j = 1; \\ \overline{x}_j, & c_j = 0, \end{cases}$$

Example: Given $n = 3$

$$f = \overline{x}_1 \overline{x}_2 f_{00} \vee \overline{x}_1 x_2 f_{01} \vee x_1 \overline{x}_2 f_{10} \vee x_1 x_2 f_{11}$$
$$f_{00} = f(0, 0, x_3)$$
$$f_{01} = f(0, 1, x_3)$$
$$f_{10} = f(1, 0, x_3)$$
$$f_{11} = f(1, 1, x_3)$$

W i t h   r e s p e c t   t o   a l l   v a r i a b l e s

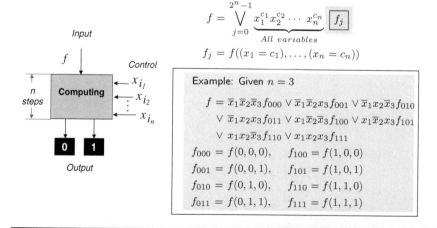

$$f = \bigvee_{j=0}^{2^n - 1} \underbrace{x_1^{c_1} x_2^{c_2} \cdots x_n^{c_n}}_{\text{All variables}} \boxed{f_j}$$

$$f_j = f((x_1 = c_1), \ldots, (x_n = c_n))$$

Example: Given $n = 3$

$$f = \overline{x}_1 \overline{x}_2 \overline{x}_3 f_{000} \vee \overline{x}_1 \overline{x}_2 x_3 f_{001} \vee \overline{x}_1 x_2 \overline{x}_3 f_{010}$$
$$\vee \overline{x}_1 x_2 x_3 f_{011} \vee x_1 \overline{x}_2 \overline{x}_3 f_{100} \vee x_1 \overline{x}_2 x_3 f_{101}$$
$$\vee x_1 x_2 \overline{x}_3 f_{110} \vee x_1 x_2 x_3 f_{111}$$

$f_{000} = f(0,0,0)$,	$f_{100} = f(1,0,0)$
$f_{001} = f(0,0,1)$,	$f_{101} = f(1,0,1)$
$f_{010} = f(0,1,0)$,	$f_{110} = f(1,1,0)$
$f_{011} = f(0,1,1)$,	$f_{111} = f(1,1,1)$

Any Boolean function $f = f(x_1, x_2, \ldots, x_n)$ of n variables can be expanded into two Boolean functions of $(n-1)$ variables:

$$f(\underbrace{x_1, x_2, \ldots, x_n}_{n\ variables}) = \overline{x}_i f(x_1, \ldots, x_{i-1}, \overset{\overline{x}_i}{\boxed{0}}, \underbrace{x_{i+1}, \ldots, x_n}_{})$$

$$\underset{\underset{OR}{\uparrow}}{\vee}, x_i f(x_1, \ldots, x_{i-1}, \overset{x_i}{\boxed{1}}, \underbrace{x_{i+1}, \ldots, x_n}_{n-1\ variables}) \qquad (8.1)$$

where $i \in \{1, 2, \ldots, n\}$. Equation 8.1 is called the *Shannon expansion*. Denoting

$$f_0 = f(x_1, \ldots, x_{i-1} \overset{\overline{x}_i}{\boxed{0}} x_{i+1}, \ldots, x_n) \text{ and } f_1 = f(x_1, \ldots, x_{i-1} \overset{x_i}{\boxed{1}} x_{i+1}, \ldots, x_n)$$

we can rewrite the Shannon expansion (Equation 8.1) in the form

$$f = \overline{x}_i f_0 \vee x_i f_1 \qquad (8.2)$$

Example 8.2 *Let $f = \overline{x}_1 x_2 \vee \overline{x}_3$ and $i = 2$. Then*

$$f_0 = f(\overline{x}_1, x_2 = 0, \overline{x}_3) = f(\overline{x}_1, 0, \overline{x}_3)$$
$$= \overline{x}_1 \cdot 0 \vee \overline{x}_3 = \overline{x}_3$$
$$f_1 = f(\overline{x}_1, x_2 = 1, \overline{x}_3) = f(\overline{x}_1, 1, \overline{x}_3)$$
$$= \overline{x}_1 \cdot 1 \vee \overline{x}_3 = \overline{x}_1 \vee \overline{x}_3$$
$$f = \overline{x}_2 f_0 \vee x_2 f_1 = \overline{x}_2 \boxed{(\overline{x}_3)} \vee x_2 \boxed{(\overline{x}_1 \vee \overline{x}_3)}$$

Equation 8.2 describes the expanding of any Boolean function of n variables with respect to n variables as an n-step process. Reduction of this equation can be achieved by eliminating a variable x_i via replacement of this variable by the possible values 0 and 1. From Equation 8.2 follows the recursive procedure for the expansion:

Step 1: Apply Shannon expansion with respect to a given variable.
Step 2: Repeat Step 1 to the next variable until the further application is not required or impossible.

Example 8.3   *Let $f = \overline{x}_1 x_2 \vee \overline{x}_3$. Find the expansion of this function: (a) with respect to variables x_1 and x_3, and (b) with respect to all variables $x_1 \longrightarrow x_2 \longrightarrow x_3$. Details of this computing are given in Figure 8.2.*

8.2.2   Expansion with respect to a group of variables

Shannon expansion can be done in terms of $m > 1$ variables. First, consider the case of two variables. Let the variables x_i and x_j, $i, j \in \{1, 2, \ldots n\}$, $i \neq j$ both take the value 0. Thus, the expansion cofactor f_{00} is represented as follows:

$$f_{00} = f(x_1, \ldots, x_{i-1}, \boxed{0}, x_{i+1}, \ldots, x_{j-1}, \boxed{0}, x_{j+1}, \ldots x_n)$$

Other cofactors are derived in a similar way:

$$f_{01} = f(x_1, \ldots, x_{i-1}, \boxed{0}, x_{i+1}, \ldots, x_{j-1}, \boxed{1}, x_{j+1}, \ldots x_n)$$

$$f_{10} = f(x_1, \ldots, x_{i-1}, \boxed{1}, x_{i+1}, \ldots, x_{j-1}, \boxed{0}, x_{j+1}, \ldots x_n)$$

$$f_{11} = f(x_1, \ldots, x_{i-1}, \boxed{1}, x_{i+1}, \ldots, x_{j-1}, \boxed{1}, x_{j+1}, \ldots x_n)$$

Example 8.4   *Given the Boolean function of three variables $f = \overline{x}_1 x_2 \vee \overline{x}_3$, the cofactors for expansion with respect to the variables x_1 and x_3 are formed as shown in Figure 8.2.*

Applying Expansion 8.1 recursively to the remaining $(n - m)$ variables $x_{i_1}, x_{i_2}, \ldots, x_{i_m}$, we obtain the formula

$$f = \bigvee_{j=0}^{2^m - 1} \underbrace{x_{i_1}^{c_{i_1}} x_{i_2}^{c_{i_2}} \ldots x_{i_m}^{c_{i_m}}}_{Control\ variables} f_j \qquad (8.3)$$

where cofactor f_j is defined as

$$f_j = f(x_1, x_2, \ldots, \underbrace{\boxed{x_{i_1} = c_{i_1}}, \boxed{x_{i_2} = c_{i_2}}, \ldots, \boxed{x_{i_m} = c_{i_m}}}_{Values\ of\ variables}, \ldots, x_{n-1}, x_n)$$

and

$$x_j^{c_j}, = \begin{cases} x_j, & c_j = 1 \\ \overline{x}_j, & c_j = 0 \end{cases}$$

Techniques for computing the Shannon expansion

Recursive expansion with respect to the variables x_1 and x_3

Step 1: Expand the function $\overline{x}_1 x_2 \vee \overline{x}_3$ with respect to the variable x_1:

$$f = \overline{x}_1 x_2 \vee \overline{x}_3$$
$$= \overline{x}_1 f_0 \vee x_1 f_1$$
$$= \overline{x}_1 (\overline{0} \cdot x_2 \vee \overline{x}_3) \vee x_1 (\overline{1} \cdot x_2 \vee \overline{x}_3)$$
$$= \overline{x}_1 \boxed{(x_2 \vee \overline{x}_3)} \vee x_1 \boxed{\overline{x}_3}$$

Step 2: Expand the result of Step 1 with respect to the variable x_3:

$$f = \overline{x}_3 f_{00} \vee x_3 f_{01}$$
$$= \overline{x}_3 (\overline{x}_1 (x_2 \vee \overline{0}) \vee x_1 (\overline{0})) \vee x_3 (\overline{x}_1 (x_2 \vee \overline{1}) \vee x_1 (\overline{1}))$$
$$= \overline{x}_3 (\overline{x}_1 \vee x_1) \vee x_3 (\overline{x}_1 x_2 \vee x_1 (0))$$
$$= \overline{x}_1 \overline{x}_3 \boxed{1} \vee x_1 \overline{x}_3 \boxed{1} \vee \overline{x}_1 x_3 \boxed{x_2} \vee x_1 x_3 \boxed{0}$$

Expansion with respect to a group of variables (x_1, x_3)

Step 1: Compute cofactors f_j, $j = 0, 1, 2, 3$, for the expansion

$$f = \overline{x}_1 x_2 \vee \overline{x}_3 = \bigvee_{j=0}^{2^2-1} x_1^{c_1} x_3^{c_3} f_j$$

as follows:

$$f_{00} = f(x_1 = 0, x_2, x_3 = 0) = f(0, x_2, 0) = \overline{0} x_2 \vee \overline{0} = \boxed{1}$$

$$f_{01} = f(x_1 = 0, x_2, x_3 = 1) = f(0, x_2, 1) = \overline{0} x_2 \vee \overline{1} = \boxed{x_2}$$

$$f_{10} = f(x_1 = 1, x_2, x_3 = 0) = f(1, x_2, 0) = \overline{1} x_2 \vee \overline{0} = \boxed{1}$$

$$f_{11} = f(x_1 = 1, x_2, x_3 = 1) = f(1, x_2, 1) = \overline{1} x_2 \vee \overline{1} = \boxed{0}$$

Step 2: Expand the function $\overline{x}_1 x_2 \vee \overline{x}_3$ with respect to a group of variables (x_1, x_3)

$$f = \overline{x}_1 \overline{x}_3 f_{00} \vee \overline{x}_1 x_3 f_{01} \vee x_1 \overline{x}_3 f_{10} \vee x_1 x_3 f_{11}$$
$$= \overline{x}_1 \overline{x}_3 (\overline{0} x_2 \vee \overline{0}) \vee \overline{x}_1 x_3 (\overline{0} x_2 \vee \overline{1}) \vee x_1 \overline{x}_3 (\overline{1} x_2 \vee \overline{0}) \vee x_1 x_3 (\overline{1} x_2 \vee \overline{1})$$
$$= \overline{x}_1 \overline{x}_3 \boxed{1} \vee \overline{x}_1 x_3 \boxed{x_2} \vee x_1 \overline{x}_3 \boxed{1} \vee x_1 x_3 \boxed{0}$$

FIGURE 8.2

Techniques for the Shannon expansion of the Boolean function $f = \overline{x}_1 x_2 \vee \overline{x}_3$ (Examples 8.3 and 8.4).

where $j = c_{i_1} c_{i_2} \ldots c_{i_m}$ is binary representation of j. Equation 8.3 describes computing the Shannon expansion with respect to m variables as an m-step process.

Example 8.5	*Using Equation 8.3, the cofactors of the Shannon expansion of the Boolean function $f = \overline{x}_1 \overline{x}_3 \vee x_1 x_2 \vee x_1 x_3$ of three variables ($n = 3$) with respect to the variables x_1 and x_2 ($m = 2$) are calculated as follows:*

$$\text{Co-FACTOR } f_{00} = 1 \cdot \overline{x}_3 \vee 0 \cdot 0 \vee 0 \cdot x_3 = \boxed{\overline{x}_3}$$

$$\text{Co-FACTOR } f_{01} = 1 \cdot \overline{x}_3 \vee 0 \cdot 1 \vee 0 \cdot x_3 = \boxed{\overline{x}_3}$$

$$\text{Co-FACTOR } f_{10} = 0 \cdot \overline{x}_3 \vee 1 \cdot 0 \vee 1 \cdot x_3 = \boxed{x_3}$$

$$\text{Co-FACTOR } f_{11} = 0 \cdot \overline{x}_3 \vee 1 \cdot 1 \vee 1 \cdot x_3 = \boxed{1}$$

$$f = \bigvee_{j=0}^{2^2-1} = \overline{x}_1 \overline{x}_2 \boxed{f_{00}} \vee \overline{x}_1 x_2 \boxed{f_{01}} \vee x_1 \overline{x}_2 \boxed{f_{10}} \vee x_1 x_2 \boxed{f_{11}}$$

$$= \overline{x}_1 \overline{x}_2 (\overline{x}_3) \vee \overline{x}_1 x_2 (\overline{x}_3) \vee x_1 \overline{x}_2 (x_3) \vee x_1 x_2 (1)$$

8.2.3   Expansion with respect to all variables

The boundary case of the Shannon expansion is encountered when setting $m = n$ in Equation 8.3:

$$f = \bigvee_{j=0}^{2^n-1} \underbrace{x_1^{c_1} x_2^{c_2} \cdots x_n^{c_n}}_{All \ variables} f_j \qquad (8.4)$$

where cofactor f_j is defined as

$$f_j = f(\underbrace{x_{i_1} = c_{i_1}, x_{i_2} = c_{i_2}, \ldots, x_{i_m} = c_{i_m}}_{Values \ of \ variables})$$

and

$$x_j^{c_j} = \begin{cases} x_j, & c_j = 1 \\ \overline{x}_j, & c_j = 0 \end{cases}$$

This equation describes the computing of any Boolean function of n variables by a single-step process.

Example 8.6   *Using Equation 8.4, the Shannon expansion of the Boolean function $f = \overline{x}_1\overline{x}_3 \vee x_1x_2 \vee x_1x_3$ with respect to all three variables $x_1, x_2,$ and x_3 is computed as shown in Figure 8.3.*

Techniques for computing the Shannon expansion

Expansion with respect to all variables x_1, x_2, and x_3

Step 1: Compute cofactors f_j, $j = 0, 1, \ldots, 7$, for the expansion

$$f = \overline{x}_1 x_2 \vee \overline{x}_3 = \bigvee_{j=0}^{2^3-1} x_1^{c_1} x_2^{c_2} x_3^{c_3} f_j$$

as follows:

$$f_{000} = f(000) = \overline{0} \cdot 0 \vee \overline{0} = \boxed{1}, \quad f_{001} = f(001) = \overline{0} \cdot 0 \vee \overline{1} = \boxed{0},$$

$$f_{010} = f(010) = \overline{0} \cdot 1 \vee \overline{0} = \boxed{1}, \quad f_{011} = f(011) = \overline{0} \cdot 1 \vee \overline{1} = \boxed{1},$$

$$f_{100} = f(100) = \overline{1} \cdot 0 \vee \overline{0} = \boxed{1}, \quad f_{101} = f(101) = \overline{1} \cdot 0 \vee \overline{1} = \boxed{0},$$

$$f_{110} = f(110) = \overline{1} \cdot 1 \vee \overline{0} = \boxed{0}, \quad f_{111} = f(111) = \overline{1} \cdot 1 \vee \overline{1} = \boxed{0}$$

Step 2: Expand the function $\overline{x}_1 x_2 \vee \overline{x}_3$ with respect to a group of variables $x_1, x_2,$ and x_3:

$$f = \overbrace{\overline{x}_1\overline{x}_2\overline{x}_3}^{000} f_{000} \vee \overbrace{\overline{x}_1\overline{x}_2 x_3}^{001} f_{001} \vee \overbrace{\overline{x}_1 x_2\overline{x}_3}^{010} f_{010} \vee \overbrace{\overline{x}_1 x_2 x_3}^{011} f_{011}$$

$$\vee \overbrace{x_1\overline{x}_2\overline{x}_3}^{100} f_{100} \vee \overbrace{x_1\overline{x}_2 x_3}^{101} f_{101} \vee \overbrace{x_1 x_2\overline{x}_3}^{110} f_{110} \vee \overbrace{x_1 x_2 x_3}^{111} f_{111}$$

$$= \overline{x}_1\overline{x}_2\overline{x}_3 \boxed{1} \vee \overline{x}_1\overline{x}_2 x_3 \boxed{0} \vee \overline{x}_1 x_2\overline{x}_3 \boxed{1} \vee \overline{x}_1 x_2 x_3 \boxed{1}$$

$$\vee x_1\overline{x}_2\overline{x}_3 \boxed{1} \vee x_1\overline{x}_2 x_3 \boxed{0} \vee x_1 x_2\overline{x}_3 \boxed{1} \vee x_1 x_2 x_3 \boxed{0}$$

FIGURE 8.3

Techniques for the Shannon expansion of the Boolean function $f = \overline{x}_1 x_2 \vee \overline{x}_3$ (Example 8.6).

8.2.4   Various forms of Shannon expansions

Shannon expansion provides a model for the multiplexing of Boolean functions, which enables representation of the Boolean function for further implementation as a logic network of universal gates such as multiplexers. Multiplexing can be implemented:

▶ Using an OR operation, such as in the Shannon expansion given in Equation 8.1

▶ Using an AND operation

▶ Using an EXOR operation

The dual of Shannon expansion

The dual of Shannon expansion (Equation 8.1) is the following expression:

$$
\overbrace{f\left(x_1, x_2, \ldots, x_n\right)}^{n \ variables} = \left(\overline{x}_i \vee \overbrace{f\left(x_1, \ldots, x_{i-1}\, \boxed{1}\, x_{i+1}, \ldots, x_n\right)}^{n-1 \ variables}\right)
$$

$$
\underset{\underset{AND}{\uparrow}}{\cdot}\ \left(x_i \vee \underbrace{f\left(x_1, \ldots, x_{i-1}\, \boxed{0}\, x_{i+1}, \ldots, x_n\right)}_{n-1 \ variables}\right) \quad (8.5)
$$

Example 8.7   *Let $f = x_1 \vee \overline{x}_2$. Its Shannon expansion (Equation 8.1) with respect to variable x_1 is represented as $f = \overline{x}_1(0 \vee \overline{x}_2) \vee x_1(1 \vee \overline{x}_2) = \overline{x}_1(\overline{x}_2) \vee x_1(1)$, and its dual (Equation 8.5) is $f = (\overline{x}_1 \vee (1 \vee \overline{x}_2))(x_1 \vee (0 \vee \overline{x}_2)) = (\overline{x}_1 \vee 1)(x_1 \vee \overline{x}_2)$. Both equations imply the initial function, since $f = \overline{x}_1\overline{x}_2 \vee x_1 = \overline{x}_2 \vee x_1$ and $f = (\overline{x}_1 \vee 1)(x_1 \vee \overline{x}_2) = x_1 \vee \overline{x}_2$.*

The EXOR form of Shannon expansion

Given a Shannon expansion in the form of Equation 8.1, the Boolean sum in this equation can be replaced by the EXOR function:

$$
\overbrace{f\left(x_1, x_2, \ldots, x_n\right)}^{n \ variables} = \overline{x}_i \overbrace{f\left(x_1, \ldots, x_{i-1}\, \boxed{0}\, x_{i+1}, \ldots, x_n\right)}^{n-1 \ variables}
$$

$$
\underset{\underset{EXOR}{\uparrow}}{\oplus}\ x_i \underbrace{f\left(x_1, \ldots, x_{i-1}\, \boxed{1}\, x_{i+1}, \ldots, x_n\right)}_{n-1 \ variables} \quad (8.6)
$$

It follows from this fact that

$$\overline{x}_i \, a \, \oplus x_i \, b = \overline{x}_i \, a \, \overline{x_i \, b} \vee \overline{\overline{x}_i \, a} \, x_i \, b = \overline{x}_i \, a \, (\overline{x}_i \vee \overline{b}) \vee (x_i \vee \overline{a})x_i \, b$$
$$= \overline{x}_i \, a \, \vee \overline{x}_i \, a \, \overline{b} \vee x_i \, b \vee x_i \, \overline{a} \, b = \overline{x}_i \, a \vee x_i \, b$$

Equation 8.6 is a polynomial form of the Shannon expansion. This expansion is used in the following form

$$f = \overline{x}_i f_0 \oplus x_i f_1 = (1 \oplus x_i)f_0 \oplus x_i f_1$$
$$= f_0 \oplus x_i f_0 \oplus x_i f_1 = f_0 \oplus x_i (f_0 \oplus f_1) \qquad (8.7)$$

Equation 8.7 is called *the positive Davio* expansion with respect to the variable x_i (details are given in Chapter 12).

Example 8.8   *Let $f = x_1 \vee \overline{x}_2$. Then the Shannon expansion in SOP and polynomial forms is derived as follows:*

SOP form: Polynomial form:

$$f = \overline{x}_1 f_0 \vee x_2 f_1 \qquad\qquad f = \overline{x}_1 f_0 \oplus x_1 f_1$$
$$= \overline{x}_1 (0 \vee \overline{x}_2) \vee x_1 (1 \vee \overline{x}_2) \quad = \overline{x}_1 (0 \vee \overline{x}_2) \oplus x_1 (1 \vee \overline{x}_2)$$
$$= \overline{x}_1 (\overline{x}_2) \vee x_1 (1) \qquad\qquad = \overline{x}_1 \overline{x}_2 \oplus x_1$$
$$= \overline{x}_1 \overline{x}_2 \vee x_1 = \overline{x}_2 \vee x_1$$

Both forms are equivalent:

$$f = \overline{x}_1 \overline{x}_2 \oplus x_1 \quad (polynomial \; form)$$
$$= \overline{\overline{x}_1 \overline{x}_2} x_1 \vee \overline{x}_1 \overline{x}_2 \overline{x}_1$$
$$= (x_1 \vee x_2)x_1 \vee \overline{x}_1 \overline{x}_2$$
$$= x_1 \vee x_1 x_2 \vee \overline{x}_1 \overline{x}_2$$
$$= x_1 \vee \overline{x}_1 \overline{x}_2 \vee \overline{x}_2 = x_1 \vee \overline{x}_2 \quad (SOP \; form)$$

8.3   Shannon expansion for symmetric Boolean functions

Symmetric functions are the special class of Boolean functions. These functions are characterized by properties of symmetry and can be efficiently used in design of networks with interchangeable wires of subfunctions. A typical problem of logic design is detection of symmetric properties, since knowledge of the symmetry properties of a Boolean function can be used for more efficient implementation of functions.

Shannon theorem provides a necessary and sufficient conditions for detection of symmetry in a Boolean function f of n variables, and for the representation of this function that utilizes this symmetry. Shannon expansion for symmetric functions is the third fundamental expansion of logic design (Figure 8.1). These theorems also provide a flexibility in logic design. In particular, Shannon theorems can be applied at the *global* and *local* levels of design; that is, detection of symmetries and expansion can be used for a total Boolean function (global) or its part (local); and these theorems can be used in various Boolean data structures.

8.3.1   Symmetric functions

There are two groups of symmetric Boolean functions:

(a) If two variables in a Boolean function f can be exchanged without changing the function itself, it is said that this function is *partially symmetric* with respect to these two variables.

(b) The boundary case of symmetry is *totally symmetric* functions: in totally symmetric function, all variables can be exchanged without changing the function itself.

The symmetry properties of Boolean functions can be recognized using detection algorithms on various data structures:

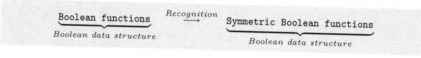

These include the algebraic forms of Boolean functions if they are available, decision diagrams, which are preferable for large-size functions, and logic networks.

Example 8.9   *The Boolean functions of three variables (a) $f = \overline{x}_1\overline{x}_2\overline{x}_3 \vee x_1x_2x_3$ and (b) $f = x_1\overline{x}_2x_3 \vee \overline{x}_1x_2x_3 \vee x_1x_2\overline{x}_3$ are **totally symmetric**, since permuting the variables $x_1, x_2,$ and x_3, respectively, leaves the function unchanged.*

Symmetric properties are specified in terms of Boolean data structure:

▶ In algebraic forms, symmetric properties are specified using permutation of variables in Boolean expressions

▶ In logic networks, symmetric properties are specified in terms of the permutation of inputs

▶ In decision trees and diagrams, symmetric properties are specified by the assigned values of the outgoing branches of a node

However, in all data structure types, the manipulation of their components must leaves the represented Boolean function unchanged.

Example 8.10 *The standard SOP form of the two-input OR function $f = x_1 \lor x_2$ is defined by three minterms: $x_1\overline{x}_2, \overline{x}_1 x_2$ and $x_1 x_2$. These minterms specify that the OR function is equal to 1 for the assignments of variables $x_1 x_2 = \{01, 10, 11\}$. However, the OR function is* **symmetric in variables** *x_1 and x_2. That is, f takes value 1 for assignments $x_1 x_2 = \{01, 10\}$. Observe that the number of 1s in these two assignments is the same (1).*

8.3.2 Partially symmetric Boolean functions

A Boolean function f is *partially symmetric* in k variables $x_1, x_2, ..., x_k,$ $k < n$, if and only if it is unchanged by any permutation of these k variables only. In other words, a Boolean function f is *partially symmetric* in an arbitrary pair of variables x_i and x_j if

$$f(x_1, x_2, ..., x_{i-1}, \boxed{0}, x_{i+1}, ..., x_{j-1}, \boxed{1}, x_{j+1}, ..., x_n)$$

$$= f(x_1, x_2, ..., x_{i-1}, \boxed{1}, x_{i+1}, ..., x_{j-1}, \boxed{0}, x_{j+1}, ..., x_n)$$

Hence, a Boolean function is partially symmetric symmetric if it is partially symmetric in any pair of variables x_i and x_j, for all i, j. There are $2^{3 \cdot 2^{n-2}}$ partially symmetric functions of n variables with respect to variables x_i and x_j.

Example 8.11 *The Boolean functions:*
(a) $f = x_2\overline{x}_3 \lor \overline{x}_2 x_3 \lor x_1 x_2 x_3$ is **partially symmetric** *with respect to variables x_2 and x_3 (Figure 8.4).*
(b) $f = x_1 x_2 \lor x_3$ is partially symmetric with respect to variables x_1 and x_2.
(b) $f = x_1\overline{x}_2 x_3\overline{x}_4 \lor \overline{x}_1 x_2\overline{x}_3 x_4$ is partially symmetric with respect to variables x_1 and x_3 and also with respect to the variables x_2 and x_4.
There are $2^{3 \cdot 2^{n-2}} = 2^{3 \cdot 2^{3-2}} = 64$ functions of three ($n = 3$) variables partially symmetric with respect to two variables.

8.3.3 Totally symmetric Boolean functions

A Boolean function f is *partially symmetric symmetric* if and only if it is unchanged by any permutation of variables. There are 2^{n+1} totally symmetric

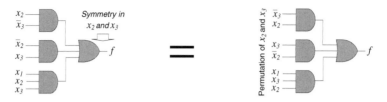

FIGURE 8.4

Logic networks for partially symmetric Boolean functions with respect to variables x_2 and x_3 (Example 8.11a).

functions of n variables. However, they are only small subclass of partially symmetric functions, that is,

$$\frac{\text{NUMBER OF PARTIALLY SYMMETRIC FUNCTIONS}}{\text{NUMBER OF TOTALLY SYMMETRIC FUNCTIONS}} = \frac{2^{3 \cdot 2^{n-2}}}{2^{n+1}} = 2^{3 \cdot 2^{n-2} - n + 1}$$

Example 8.12　　*There are $2^{n+1} = 2^{2+1} = 8$ totally symmetric Boolean functions of two variables (Figure 8.5).*

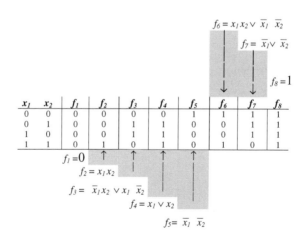

FIGURE 8.5

Totally symmetric Boolean functions of two variables (Example 8.12).

8.3.4 Detection of symmetric Boolean functions

The following theorem by Shannon provide conditions for detection and computing symmetric Boolean functions:

Theorem 8.1 *Necessary and sufficient condition that a Boolean function f of n variables be symmetric may be specified by starting a set of numbers $a_0, a_1, \ldots, a_k$ such that if exactly a_j, $j = 1, 2, \ldots a_k$, $0 \leq k \leq n$, of these variables of symmetry have the value 1, then the function has the value 1, and not otherwise.*

It follows from this Shannon theorem that

(a) The set of numbers $a_0, a_1, \ldots, a_k$ may be any set of positive integer numbers selected from the numbers zero to n, inclusive.

(b) The values of a symmetric function are determined **only by the number of literals which take the value 1** (the literal is a complemented or uncomplemented variable).

This theorem is the base for development of various techniques for symmetry detection.

8.3.5 Characteristic set

A set of integers numbers a_j, $j = 0, 1, \ldots k$, defined by Shannon theorem for detecting of symmetric function f of n variables, is called the *characteristic set*. The structure of the complete characteristic set can be represented in terms of minterms as follows:

$$A = \{ \boxed{a_0}, \boxed{a_1}, \boxed{a_2}, \ldots, \boxed{a_n} \} \tag{8.8}$$

where each number of the set corresponds to the group of minterms, except a_0 and a_n that correspond to a single minterms $M_{i_0} = M_0$ and $M_{i_n} = M_n$, respectively.

> **Example 8.13** *Techniques for computing of* **characteristic sets** *for symmetric Boolean functions are given in Table 8.2. Note that the first two functions are totally symmetric, and the last function is partially symmetric with respect to the variables x_1 and $\overline{x}_2$.*

TABLE 8.2

Technique for computing the characteristic set for symmetric Boolean functions

Symmetric function	Characteristic set

$f = x_1 \vee x_2$

x_1	x_2	f	MINTERM
0	0	0	
0	1	1	M_1
1	0	1	M_2
1	1	1	M_3

Standard SOP representation $f = M_1 \vee M_2 \vee M_3$.
Characteristic set

$$A = \left\{ \boxed{a_1}, \boxed{a_2} \right\} = \{1, 2\}$$

$a_1 \rightarrow \{M_1, M_2\} = M_{i_1}$ (Equation 8.8)

$a_2 \rightarrow \{M_3\} = M_{i_2}$

$f = x_1 \oplus x_2 \oplus x_3$

x_1	x_2	x_3	f	MINTERM
0	0	0	0	
0	0	1	1	M_1
0	1	0	1	M_2
0	1	1	0	
1	0	0	1	M_4
1	0	1	0	
1	1	0	0	
1	1	1	1	M_7

Standard SOP representation $f = M_1 \vee M_2 \vee M_4 \vee M_7$.
Characteristic set

$$A = \left\{ \boxed{a_1}, \boxed{a_3} \right\} = \{1, 3\}$$

$a_1 \rightarrow \{M_1, M_2, M_4\} = M_{i_1}$ (Equation 8.8)

$a_3 \rightarrow \{M_7\} = M_{i_3}$

$f = x_1 \bar{x}_2 \bar{x}_3 \vee \bar{x}_1 x_2 \bar{x}_3 \vee \bar{x}_1 \bar{x}_2 x_3$

x_1	x_2	x_3	f	MINTERM
0	0	0	0	
0	0	1	1	M_1
0	1	0	1	M_2
0	1	1	0	
1	0	0	1	M_4
1	0	1	0	
1	1	0	0	
1	1	1	1	

Standard SOP representation $f = M_1 \vee M_2 \vee M_4$.
Characteristic set

$$A = \left\{ \boxed{a_1} \right\} = \{1\}$$

$a_1 \rightarrow \{M_1, M_2, M_4\} = M_{i_1}$ (Equation 8.8)

8.3.6 Elementary symmetric functions

Using Shannon theorem for detecting symmetric functions, the *elementary symmetric functions* $S_0^n, S_1^n, \ldots, S_n^n$ of n variables can be defined as follows:

$$S_0^n = \overbrace{\bar{x}_1 \bar{x}_2 \cdots \bar{x}_{n-1} \bar{x}_n}^{0 \; inputs \; are \; equal \; to \; 1}$$

$$S_1^n = \overbrace{x_1 \bar{x}_2 \cdots \bar{x}_{n-1} \bar{x}_n}^{1 \; input \; is \; equal \; to \; 1} \vee \overbrace{\bar{x}_1 x_2 \bar{x}_3 \cdots \bar{x}_{n-1} \bar{x}_n}^{1 \; input \; is \; equal \; to \; 1} \vee \cdots \vee \overbrace{\bar{x}_1 \bar{x}_2 \cdots \bar{x}_{n-1} x_n}^{1 \; input \; is \; equal \; to \; 1}$$

$$\ldots \ldots \ldots \ldots \ldots$$

$$S_n^n = \overbrace{x_1 x_2 \ldots x_{n-1} x_n}^{n \; input \; are \; equal \; to \; 1}$$

The Boolean function $S_i^n = 1$ iff exactly i out of n inputs are equal to one.

That is, for all assignments of values to variables that have the same number of 1's, there is *exactly* one value of a symmetric function. The elementary symmetric functions $S_0^n, S_1^n, \ldots, S_n^n$ can be written in terms of minterms:

$$S_0^n = M_0, \ S_1^n = M_1 \vee M_2 \vee M_4 \vee \cdots \vee M_{2^t}, \ldots, S_n^n = M_n$$

In this notation, each elementary symmetric function is described by the group of minterms; the numbers of minterms can be described by the equation. For example, the elementary symmetric function S_1^n is described by the minterms with numbers $2^t, t = 0, 1, \ldots, T, \ T < 2^n$. Each minterm of this group consists of one uncomplemented variable in a product and corresponds element a_1 of the characteristic set.

Example 8.14 *There are 3 **elementary** symmetric Boolean functions of two variables (Table 8.3):*

$S_0^2 = M_0 = \bar{x}_1 \bar{x}_2$ – *exactly 0 out of 2 inputs are equal to 1.*
$S_1^2 = M_1 \vee M_2 = x_1 \oplus x_2$ – *exactly 1 out of 2 inputs are equal to 1.*
$S_2^2 = M_3 = x_1 x_2$ – *exactly 2 out of 2 inputs are equal to 1.*

TABLE 8.3
Elementary symmetric functions of two variables S_0^2, S_1^2, S_2^2, and their composition (Examples 8.14 and 8.16)

x_1	x_2	S_0^2	S_1^2	S_2^2	$S_{0,2}^2$	$S_{0,1}^2$	$S_{1,2}^2$	$S_{0,1,2}^2$
0	0	1	0	0	1	1	0	1
0	1	0	1	0	0	1	1	1
1	0	0	1	0	0	1	1	1
1	1	0	0	1	1	0	1	1

Elementary functions   *Composed symmetric functions*

8.3.7 Operations on elementary symmetric functions

Shannon theorem for detection of symmetric functions also provides for a considerable manipulation and simplification of symmetric Boolean functions in algebraic form. In manipulation, the characteristic sets are computed using the *set operation union* ∪, the *intersection* ∩, and *set difference* \ operations on sets can be used in the manipulation of characteristic sets.

Example 8.15   *Given symmetric Boolean functions f and g of three variables, and the characteristic sets of f and g, $A = \{1, 2, 4\}$ and $A = \{3, 5\}$, respectively. Operations on these sets are listed below: union: $A \cup B = \{1, 2, 3.4.5\}$, intersection: $A \cap B\{4\}$, and set difference: $\overline{A} = \{0, 1, 2, 3, 4, 5, 7\} \backslash \{1, 2, 4\} = \{0, 3, 5, 6, 7\}$ and $\overline{B} = \{0, 1, 2, 3, 4, 5, 7\} \backslash \{3, 5\} = \{0, 1, 2, 4, 6, 7\}$.*

Let the Boolean functions of n variables $f = S_A^n$ and $g = S_B^n$ be symmetric functions with respect to the set of variables A and B, $A, B \subseteq \{1, 2, \ldots n\}$, respectively; then the functions

LOGIC SUM: $f \vee g = S_{A \cup B}^n$

LOGIC MULTIPLICATION: $f \cdot g = S_{A \cap B}^n$

COMPLEMENT: $\overline{f} = S_{\overline{A}}^n$ and $\overline{g} = S_{\overline{B}}^n$

are also symmetric functions. New symmetric functions can be formed from the elementary symmetric Boolean functions.

Example 8.16   *In Table 8.3, symmetric functions $S_{0,2}^2, S_{0,1}^2, S_{1,2}^2$ and $S_{0,1,2}^2$ are formed as follows:*

(a) $S_0 \vee S_2 = S_{0 \cup 2} = S_{0,2}^2 = x_1 \oplus \overline{x}_2$ – *exactly 0 or 2 out of 2 inputs are equal to 1.*

(b) $S_0 \vee S_1 = S_{0 \cup 1} = S_{0,1}^2 = \overline{x}_1 \vee \overline{x}_2$ – *exactly 0 or 1 out of 2 inputs are equal to 1.*

(c) $S_1 \vee S_2 = S_{1 \cup 2} = S_{1,2}^2 = x_1 \vee x_2$ – *exactly 1 or 2 out of 2 inputs are equal to 1.*

(d) $S_0 \vee S_1 \vee S_2 = S_{0 \cup 1 \cup 2} = S_{0,1,2}^2 = 1$ – *exactly 0, 1, or 2 out of 2 inputs are equal to 1.*

Example 8.17   *The following operation can be used to generate new symmetric functions given symmetric functions of three variables:*

Logic sum: $S_{0,1,2,4}^4 \vee S_{2,3,4}^4 = S_{\{1,2,4\} \cup \{2,3,4\}}^4 = S_{0,1,2,3,4}^4$

Logic multiplication: $S_{0,1,2,4}^4 \cdot S_{2,3,4}^4 = S_{\{1,2,4\} \cap \{2,3,4\}}^4$

$= S_{2,4}^4$

Complement: $\overline{S}_{1,2,3,6}^7 = S_{\{0,1,2,3,4,5,6,7\} \backslash \{1,2,3,6\}}^7 = S_{0,4,5}^6$

8.3.8 Shannon expansion with respect to a group of symmetric variables

A convenient way of the representation of symmetric function is provided by *Shannon expansion for symmetric functions*: any Boolean function of n variables symmetric in the variables $x_1, x_2 \ldots, x_m$

$$f = f(\overbrace{x_1, x_2, \ldots, x_m}^{\substack{Symmetric \\ variables}}, \underbrace{x_{m+1}, \ldots, x_{n-1}, x_n}_{\substack{Nonsymmetric \\ variables}})$$

can be represented as

$$f = \bigvee_{i=0}^{m} \overbrace{S_i^m(x_1, x_2, \ldots, x_m)}^{\substack{Elementary \\ symmetric\ function}} f_i(\underbrace{x_{m+1}, \ldots, x_{n-1}x_n}_{\substack{Nonsymmetric \\ variables}}) \qquad (8.9)$$

where

$$f_i(x_{m+1}, \ldots, x_{n-1}x_n) = f(\overbrace{0, \ldots, 0,}^{i\ zeros}\ \overbrace{1, \ldots 1,}^{(m-i)\ ones}\ \underbrace{x_{m+1}, \ldots, x_{n-1}, x_n}_{\substack{Nonsymmetric \\ variables}})$$

Shannon expansion for symmetric function (Equation 8.9) assumes that the symmetry variables are already detected; that is, this expansion can be applied only to symmetric functions.

8.4 Techniques for computing symmetric functions

Techniques for computing partially and totally symmetric functions are based on the Shannon expansion for symmetric functions. The final goal is the representation of symmetric functions for efficient implementation, that is, efficient utilization of a symmetry property of Boolean function.

8.4.1 Computing partially symmetric functions

Techniques for computing partially symmetric functions are based on Shannon expansion (Equation 8.9).

Example 8.18   *The Boolean function* $f = x_1 x_2 x_3 x_4 \vee x_1 \bar{x}_2 x_3 \vee x_1 x_2 \bar{x}_3 \vee x_1 \bar{x}_2$ *is symmetric in variables* x_2 *and* x_3, *and can be represented in the form shown in Figure 8.6.*

8.4.2   Computing totally symmetric functions

Techniques for computing totally symmetric functions are based on Shannon expansion (Equation 8.9) for $m = n$: a totally symmetric function f can be uniquely represented by elementary symmetric functions:

$$f = \bigvee_{i \in A} S_i^n, \qquad A \subseteq 0, 1, 2, \ldots n \tag{8.10}$$

Example 8.19   *Table 8.4 shows two examples of expansion of totally symmetric functions.*

Example 8.20   *Given the totally symmetric Boolean functions of three variables, find its representation in terms of elementary symmetric functions:*
(a) $f = \bigvee m(1, 2, 4, 7)$, *its representation in terms of elementary symmetric functions as follows:*

$$f = \overbrace{\bar{x}_1 \bar{x}_2 x_3 \vee \bar{x}_1 x_2 \bar{x}_3 \vee x_1 \bar{x}_2 \bar{x}_3}^{S_1^3} \vee \overbrace{x_1 x_2 x_3}^{S_3^3} = S_1^3 \vee S_3^3 = S_{1,3}^3$$

(b) $f = (x_1 \vee x_2)(x_2 \vee x_3)(x_1 \vee x_3)$. *The solution is given in Figure 8.7. Note that this function is called a* **majority** *function.*

Example 8.21   *The n variable even parity check function (n is even) and odd parity check function are totally symmetric Boolean functions*

EVEN PARITY CHECK FUNCTION $= S_0^n \vee S_2^n \vee \cdots \vee S_n^n$
ODD PARITY CHECK FUNCTION $= S_1^n \vee S_3^n \vee \cdots \vee S_{n-1}^n$

respectively. In particular, given $n = 2$, *even and odd parity check functions are* $S_{0,2}^2 = S_0^2 \vee S_2^2 = \bar{x}_1 \bar{x}_2 \vee x_1 x_2$ *and* $S_1^2 = x_1 \bar{x}_2 \vee \bar{x}_1 x_2$, *respectively.*

Techniques for computing
partially symmetric Boolean functions

Given: Boolean function of three variables symmetric in variables x_2 and x_3

$$f = x_1 x_2 x_3 x_4 \vee x_1 \overline{x}_2 x_3 \vee x_1 x_2 \overline{x}_3 \vee x_1 \overline{x}_2$$

Find: The implementation

Step 1: Shannon expansion with respect to symmetric variables x_2 and x_3

$$f = \bigvee_{i=0}^{2} S_i^2 f_i$$

(a) **Assign** $x_2 x_3 = \{00\} \Rightarrow$ $\quad = S_0^2(x_2, x_3) f_0$ $\qquad \overbrace{\qquad\qquad}^{Equal\ to\ x_1 x_4}$
$$\left(x_1, \boxed{0}, \boxed{0}, x_4 \right)$$

(b) **Assign** $x_2 x_3 = \{10, 01\} \Rightarrow$ $\quad \vee\, S_1^2(x_2, x_3) f_1$ $\qquad \overbrace{\qquad\qquad}^{Equal\ to\ x_1}$
$$\left(\begin{array}{c} x_1, \boxed{1}, \boxed{0}, x_4 \\ x_1, \boxed{0}, \boxed{1}, x_4 \end{array} \right)$$

(c) **Assign** $x_2 x_3 = \{11\} \Rightarrow$ $\quad \vee\, S_2^2(x_2, x_3) f_2$ $\qquad \overbrace{\qquad\qquad}^{Equal\ to\ x_1}$
$$\left(x_1, \boxed{1}, \boxed{1}, x_4 \right)$$

$$= S_0^2(x_2, x_3) x_1 x_4$$
$$\vee\, S_1^2(x_2, x_3) x_1$$
$$\vee\, S_2^2(x_2, x_3) x_1$$

Step 2: Implement the expanded Boolean function

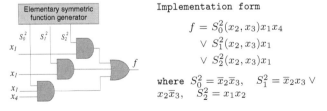

Elementary symmetric function generator

$S_0^2 \quad S_1^2 \quad S_2^2$

x_1

x_1

x_1
x_4

f

Implementation form

$$f = S_0^2(x_2, x_3) x_1 x_4$$
$$\vee\, S_1^2(x_2, x_3) x_1$$
$$\vee\, S_2^2(x_2, x_3) x_1$$

where $S_0^2 = \overline{x}_2 \overline{x}_3$, $\quad S_1^2 = \overline{x}_2 x_3 \vee x_2 \overline{x}_3$, $\quad S_2^2 = x_1 x_2$

FIGURE 8.6

Computing partially symmetric function based on Shannon expansion for symmetric functions (Example 8.18).

TABLE 8.4

Computing totally symmetric Boolean functions using Shannon expansion for symmetric functions

Techniques for computing totally symmetric Boolean functions

Symmetric function	Shannon expansion for symmetric functions

$$f = \bigvee_{i=0}^{2} S_i^2 f_i$$

$f = x_1 \vee x_2$

x_1	x_2	f
0	0	0
0	1	1
1	0	1
1	1	1

$$= S_0^2 f_0 \left(\overbrace{\begin{array}{cc} x_1 & x_2 \\ \downarrow & \downarrow \\ \boxed{0}, & \boxed{0} \end{array}}^{Equal\ to\ 0} \right) \vee S_1^2 f_1 \overbrace{\left(\begin{array}{cc} x_1 & x_2 \\ \downarrow & \downarrow \\ \boxed{0}, & \boxed{1} \\ \boxed{1}, & \boxed{0} \end{array} \right)}^{Equal\ to\ 1} \vee S_2^2 f_2 \overbrace{\left(\begin{array}{cc} x_1 & x_2 \\ \downarrow & \downarrow \\ \boxed{1}, & \boxed{1} \end{array} \right)}^{Equal\ to\ 1}$$

$$= S_1^2 \vee S_2^2 = S_{1,2}^2$$

where $S_1^2 = x_1 \vee x_2$ and $S_2^2 = x_1 x_2$

$$f = \bigvee_{i=0}^{3} S_i^3 f_i = S_0^3 f_0 \overbrace{\left(\begin{array}{ccc} x_1 & x_2 & x_3 \\ \downarrow & \downarrow & \downarrow \\ \boxed{0}, & \boxed{0}, & \boxed{0} \end{array} \right)}^{Equal\ to\ 0}$$

$$\vee S_1^3 f_1 \overbrace{\left(\begin{array}{ccc} x_1 & x_2 & x_3 \\ \downarrow & \downarrow & \downarrow \\ \boxed{0}, & \boxed{0}, & \boxed{1} \\ \boxed{0}, & \boxed{1}, & \boxed{0} \\ \boxed{1}, & \boxed{0}, & \boxed{0} \end{array} \right)}^{Equal\ to\ 1}$$

$f = x_1 \oplus x_2 \oplus x_3$

x_1	x_2	x_3	f
0	0	0	0
0	0	1	1
0	1	0	1
0	1	1	0
1	0	0	1
1	0	1	0
1	1	0	0
1	1	1	1

$$\vee S_2^3 f_1 \overbrace{\left(\begin{array}{ccc} x_1 & x_2 & x_3 \\ \downarrow & \downarrow & \downarrow \\ \boxed{0}, & \boxed{1}, & \boxed{1} \\ \boxed{1}, & \boxed{1}, & \boxed{0} \\ \boxed{1}, & \boxed{0}, & \boxed{1} \end{array} \right)}^{Equal\ to\ 0}$$

$$\vee S_3^3 f_2 \overbrace{\left(\begin{array}{ccc} x_2 & x_2 & x_3 \\ \downarrow & \downarrow & \downarrow \\ \boxed{1}, & \boxed{1}, & \boxed{1} \end{array} \right)}^{Equal\ to\ 1}$$

$$= S_1^3 \vee S_3^3 = S_{1,3}^3$$

where $S_1^3 = \overline{x}_1 \overline{x}_2 x_3 \vee \overline{x}_1 x_2 x_3$ and $S_3^3 = x_1 x_2 x_3$

Techniques for computing totally symmetric Boolean functions

Given: Totally symmetric Boolean function of three variables

$$f = (x_1 \vee x_2)(x_2 \vee x_3)(x_1 \vee x_3)$$

Find: The implementation

Step 1: Shannon expansion with respect to symmetric variables x_1, x_2, and x_3:

$$f = \bigvee_{i=0}^{3} S_i^2 f_i$$

(a) Assign $x_1 x_2 x_3 = \{000\} \Rightarrow$

$$= S_0^3 f_0 \overbrace{\left(\begin{array}{ccc} x_1 & x_2 & x_3 \\ \downarrow & \downarrow & \downarrow \\ \boxed{0}, & \boxed{0}, & \boxed{0} \end{array} \right)}^{Equal \ to \ 0}$$

(b) Assign $x_1 x_2 x_3 = \{001, 010, 100\} \Rightarrow$

$$\vee \ S_1^3 f_1 \overbrace{\left(\begin{array}{ccc} x_1 & x_2 & x_3 \\ \downarrow & \downarrow & \downarrow \\ \boxed{1}, & \boxed{0}, & \boxed{0} \\ \boxed{0}, & \boxed{1}, & \boxed{0} \\ \boxed{0}, & \boxed{0}, & \boxed{1} \end{array} \right)}^{Equal \ to \ 0}$$

(c) Assign $x_1 x_2 x_3 = \{011, 110, 101\} \Rightarrow$

$$\vee \ S_2^3 f_2 \left(\begin{array}{ccc} x_1 & x_2 & x_3 \\ \downarrow & \downarrow & \downarrow \\ \boxed{1}, & \boxed{1}, & \boxed{0} \\ \boxed{0}, & \boxed{1}, & \boxed{1} \\ \boxed{1}, & \boxed{0}, & \boxed{1} \end{array} \right)$$

(d) Assign $x_1 x_2 x_3 = \{111\} \Rightarrow$

$$\vee \ S_3^3 f_3 \overbrace{\left(\begin{array}{ccc} x_1 & x_2 & x_3 \\ \downarrow & \downarrow & \downarrow \\ \boxed{1}, & \boxed{1}, & \boxed{1} \end{array} \right)}^{Equal \ to \ 1}$$

Equal to 1 (under $S_2^3 f_2$ block)

$$= S_0^3 \cdot 0 \ \vee \ S_1^3 \cdot 0$$
$$\vee \ S_2^3 \cdot 1 \ \vee \ S_3^3 \cdot 1$$
$$= S_2^3 \vee S_3^3$$

Step 2: Implement the expanded Boolean function

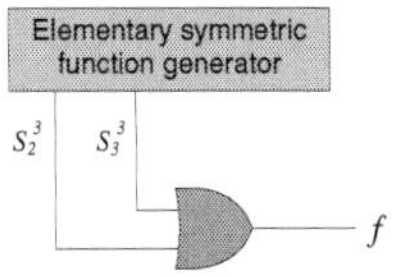

Implementation form

Elementary symmetric function generator

S_2^3 S_3^3

f

$$f = S_1^3(x_1, x_2, x_3) \vee S_3^3(x_1, x_2, x_3)$$

where

$$S_1^3(x_1, x_2, x_3) = x_1 x_2 \overline{x}_3 \vee x_1 \overline{x}_2 x_3 \vee \overline{x}_1 x_2 x_3$$
$$S_3^3(x_1, x_2, x_3) = x_1 x_2 x_3$$

FIGURE 8.7

Computing totally symmetric Boolean functions based on Shannon expansion for symmetric functions (Example 8.20).

8.4.3   Carrier vector

Characteristic sets A can be used for the compression of a truth vectors of symmetric functions. For a totally symmetric function f symmetric in n variables that is represented by a truth vector $\mathbf{F}$, this compression is significant and can be estimate as

$$\texttt{COMPRESSION} = \frac{\texttt{Size of the truth vector } \mathbf{F}}{\texttt{Size of the carrier vector } \mathbf{F}_c} = \frac{2^n}{n}$$

where $\mathbf{F}_c$ is the compressed truth vector called the *carrier vector* of a totally symmetric function. For example, for $n = 10$, $\texttt{COMPRESSION} = 2^{10}/10 = 102.4$.

Denote f_{a_j}, $0 \leq j \leq n$, the value of Boolean function f for assignments of the variables $x_1, x_2, \ldots, x_n$ that have j 1's. Then the carrier vector $\mathbf{F}_c$ can be specified as follows: a totally symmetric function $f(x_1, x_2, \ldots, x_n)$ is completely specified by a *carrier vector*

$$\mathbf{F}_c = [f_{a_0} \ f_{a_1} \cdots \ f_{a_n}]^T$$

such that if exactly a_j, $j = 0, 1, 2, \ldots n$ of the variables $x_1, x_2, \ldots, x_n$ have the value 1, then the function has the value f_{a_j} for all assignments of values to $x_1, x_2, \ldots, x_n$ that have j 1's, where $0 \leq j \leq n$.

> **Example 8.22**   *The **carrier vector** for the totally symmetric function of three variables is formed as shown in Figure 8.8.*

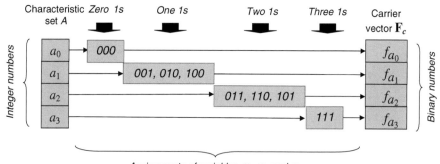

FIGURE 8.8
The carrier vector for the totally symmetric function of three variables (Example 8.22).

Example 8.23 *The totally symmetric function $f = \overline{x}_1\overline{x}_2\overline{x}_3 \vee x_1 x_2 x_3$ is specified by the carrier vector $\mathbf{F}_c = [1\ 0\ 0\ 1]^T$ since the function is 1 if and only if zero or three of the variables are 1 (Table 8.5).*

TABLE 8.5
Reducing the truth vector for a totally symmetric Boolean function (Examples 8.23 and 8.24)

x_1	x_2	x_3	Truth vector F	$\sum 1's$	Carrier vector $\mathbf{F}_c$
0	0	0	1	0	$0 \longrightarrow \boxed{1}$
0	0	1	0	1	$1 \longrightarrow \boxed{0}$
0	1	0	0	1	$2 \longrightarrow \boxed{0}$
0	1	1	0	2	$3 \longrightarrow \boxed{1}$
1	0	0	0	1	
1	0	1	0	2	
1	1	0	0	2	
1	1	1	1	3	

A carrier vector of a totally symmetric Boolean function can be represented in terms of an elementary symmetric functions.

Example 8.24 *The Boolean function f given by the carrier vector $\mathbf{F}_c = [1001]^T$ (Table 8.5) can be represented in terms of the elementary symmetric functions as $f = S_0^3 \vee S_3^3$.*

8.5 The logic Taylor expansion

The *logic Taylor expansion* is the third fundamental theorem for expanding Boolean functions (Figure 8.1). In contrast of the first two fundamental expansions that are used to the SOP representation of Boolean functions (Algebra I), the logic Taylor expansion is applied to the EXOR forms called the *polynomial* expressions (Algebra II).

8.5.1 Change in a digital system

Boolean difference is the formal model of change in a binary system. Simultaneous changes in several variables are described by Boolean difference of a higher order (second, third, and higher orders). In this chapter, Boolean differences are used for the expansion of Boolean functions using the so-called *logic Taylor expansion*.

A signal in a binary system is represented by two logical levels, 0 and 1. Let us formulate the task as detection of the change in this signal. The simplest solution is to deploy the EXOR operation (modulo 2 sum) of the signal s_{i-i} (before the "event") and the signal s_i (after the "event"), i.e., $s_{i-i} \oplus s_i$.

> **Example 8.25** *For the signal depicted in Figure 8.9, four possible combinations of the logical values or signals 0 and 1 are analyzed.*

It follows from this example that if not change itself but direction of change is the matter, then two logical values, 0 and 1, can characterize the behavior of the logic signal $s_i \in \{0, 1\}$ in terms of change, where 0 means any change of a signal, and 1 indicates that one of two possible changes has occurred, $0 \to 1$ or $1 \to 0$.

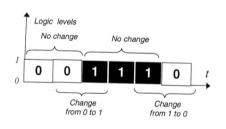
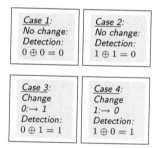

FIGURE 8.9
Change in a binary signal and its detection (Example 8.25).

8.5.2 Boolean difference

Let the i-th input of a Boolean function change from the value x_i to the opposite value, $\overline{x}_i$. This causes the circuit output to be changed from the initial value. Note that the values f_{x_i} and $f_{\overline{x}_i}$ are not necessarily different. The simplest way to recognize whether or not they are different is to find the difference between f_{x_i} and $f_{\overline{x}_i}$.

The Boolean difference of a Boolean function f of n variables with respect to a variable x_i is defined by the equation:

$$\frac{\partial f}{\partial x_i} = \underbrace{f(x_1, \ldots, \boxed{x_i}, \ldots, x_n)}_{Initial\ function} \oplus \underbrace{f(x_1, \ldots, \boxed{\overline{x}_i}, \ldots, x_n)}_{Function\ with\ complemented\ x_i} \quad (8.11)$$

It follows from the Shannon expansion that

$$f(x_1, \ldots, x_i, \ldots, x_n) = \overline{x}_i f_0 \oplus x_i f_1$$
$$f(x_1, \ldots, \overline{x}_i, \ldots, x_n) = x_i f_0 \oplus \overline{x}_i f_1$$

Hence, the Boolean difference (Equation 8.11) can be represented as follows

$$\begin{aligned}\frac{\partial f}{\partial x_i} &= (x_i f_0 \oplus \overline{x}_i f_1) \oplus (\overline{x}_i f_0 \oplus x_i f_1) \\ &= f_0(x_i \oplus \overline{x}_i) \oplus f_1(\overline{x}_i \oplus x_i) \\ &= f_0 \oplus f_1\end{aligned}$$

Therefore, the second definition of the Boolean difference (Equation 8.11) is

$$\frac{\partial f}{\partial x_i} = \underbrace{f(x_1, \ldots, \overset{x_i}{\boxed{0}}, \ldots, x_n)}_{x_i\ is\ replaced\ with\ 0} \oplus \underbrace{f(x_1, \ldots, \overset{x_i}{\boxed{1}}, \ldots, x_n)}_{x_i\ is\ replaced\ with\ 1}, \quad (8.12)$$

or

$$\frac{\partial f}{\partial x_i} = f_{x_i=0} \oplus f_{x_i=1} = f_0 \oplus f_1 \quad (8.13)$$

Thus, the simplest (but optimal) approach to calculating the Boolean difference includes two steps:

Computing the Boolean difference with respect to a variable

Given: A Boolean function.
Step 1: Replace x_i in the Boolean function with 0 to get a cofactor $f_{x_i=0}$; similarly, replacement of x_i with 1 yields $f_{x_i=1}$.
Step 2: Find the modulo 2 sum of the two cofactors.

Figure 8.10 gives an interpretation of the Boolean difference (Equation 8.11).

Example 8.26 *Given an OR function of two variables, find the* **Boolean difference** *with respect to the variable x_2. The result is given in Figure 18.7.*

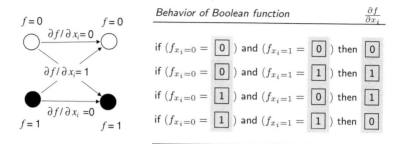

FIGURE 8.10
Formal description of change by Boolean difference.

Techniques for computing the Boolean difference

Boolean difference for a two-input OR gate with respect to the variable x_1	Boolean difference for a two-input OR gate with respect to the variable x_2
$$\frac{\partial f}{\partial x_1} = f_{x_1} \oplus f_{\overline{x}_1}$$ $$= (x_1 \vee x_2) \oplus (\overline{x}_1 \vee x_2)$$ $$= x_1 \oplus x_2 \oplus x_1 x_2 \oplus \overline{x}_1$$ $$\oplus x_2 \oplus \overline{x}_1 x_2$$ $$= 1 \oplus x_1 x_2 \oplus \overline{x}_1 x_2$$ $$= 1 \oplus x_2 = \boxed{\overline{x}_2}$$	$$\frac{\partial f}{\partial x_2} = f_{x_2} \oplus f_{\overline{x}_2}$$ $$= (x_1 \vee x_2) \oplus (x_1 \vee \overline{x}_2)$$ $$= x_1 \oplus x_2 \oplus x_1 x_2 \oplus x_1$$ $$\oplus \overline{x}_2 \oplus x_1 \overline{x}_2$$ $$= 1 \oplus x_1 x_2 \oplus x_1 \overline{x}_2$$ $$= 1 \oplus x_1 = \boxed{\overline{x}_1}$$
or	*or*
$$\frac{\partial f}{\partial x_1} = f_{x_1=0} \oplus f_{\overline{x}_1=1}$$ $$= (0 \vee x_2) \oplus (1 \vee x_2)$$ $$= x_2 \oplus 1 = \boxed{\overline{x}_2}$$	$$\frac{\partial f}{\partial x_2} = f_{x_2=0} \oplus f_{\overline{x}_2=1}$$ $$= (x_1 \vee 0) \oplus (x_1 \vee 1)$$ $$= x_1 \oplus 1 = \boxed{\overline{x}_1}$$

FIGURE 8.11
Computing the Boolean difference with respect to the variables x_1 and x_2 for a two-input OR gate (Example 18.8).

8.5.3 Boolean difference and Shannon expansion

It follows from the EXOR form of the Shannon expansion (Equation 8.6) that

$$f = \overline{x}_i f_0 \oplus x_i f_1 = (1 \oplus x_i) f_0 \oplus x_i f_1$$
$$= f_0 \oplus x_i f_0 \oplus x_i f_1 = f_0 \oplus x_i \underbrace{(f_0 \oplus f_1)}_{\frac{\partial f}{\partial x_i}}$$

Finally, the EXOR form of Shannon expansion

$$f = f_0 \oplus x_i \frac{\partial f}{\partial x_i} \qquad (8.14)$$

Therefore, Boolean difference is the component of an EXOR form of Shannon expansion. The EXOR form of Shannon expansion is called *positive Davio expansion.*

Example 8.27 *The Shannon expansion of the Boolean function $f = x_1 \oplus x_2$ with respect to the variable x_1 in terms of Boolean differences is calculated as follows: $f = f_0 \oplus x_1 \frac{\partial f}{\partial x_1} = x_2 \oplus x_1 \cdot 1$, where $\frac{\partial f}{\partial x_1} = f_0 \oplus f_1 = (0 \oplus x_2) \oplus (1 \oplus x_2) = x_2 \oplus \overline{x}_2 = 1$.*

8.5.4 Properties of Boolean difference

The Boolean difference (Equation 8.11) possesses the following properties:

Property 1: The Boolean difference is a Boolean function of $n-1$ variables $x_1, x_2, \ldots, x_{i-1}, x_{i+1}, \ldots, x_n$; i.e., it does not depend on the variable x_i. This also implies that the Boolean difference can be represented by a 2^{n-1} truth vector which is twice as small as the original truth vector.

Property 2: The value of the Boolean difference reflects the fact of change of the Boolean function f with respect to the i-th variable x_i: the Boolean difference is equal to 0 when such change occurs, and is equal to 1 otherwise.

Example 8.28 *The Boolean difference with respect to x_1 of the OR function $f = x_1 \vee x_2$ is equal to $\frac{\partial f}{\partial x_1} = \overline{x}_2$. The resulting function does not depend on the variable of differentiation, x_1. Similarly, the Boolean difference with respect to x_2 of the OR function is equal to $\frac{\partial f}{\partial x_2} = \overline{x}_1$, which does not depend on the variable x_2.*

8.5.5 The logic Taylor expansion

In the *classic* Taylor series, the coefficients are calculated as *derivatives* of the initial function at a *certain point*. By analogy, the coefficients of the *logic* Taylor series are *Boolean differences* (derivatives) with respect to a variable, and multiple Boolean differences at certain points (assignments of Boolean variables). A certain point in logic Taylor expansion is called the *polarity*. The polarity specifies which variables in Taylor expansion are complemented.

In particular, the logic Taylor series for a Boolean function f of two variables ($n = 2$) of the polarity $c \in \{0, 1, 2, 3\}$ (in binary form, $c = c_1 c_2 \in \{00, 01, 10, 11\}$) is defined by the equation

$$f = \overbrace{r_0^{(c)} \underbrace{(x_1 \oplus c_1)^0 (x_2 \oplus c_2)^0}_{Equal\ to\ 1}}^{0th\ product} \oplus \overbrace{r_1^{(c)} \underbrace{(x_1 \oplus c_1)^0 (x_2 \oplus c_2)^1}_{Equal\ to\ 1}}^{1st\ product}$$

$$\oplus \overbrace{r_2^{(c)} (x_1 \oplus c_1)^1 \underbrace{(x_2 \oplus c_2)^0}_{Equal\ to\ 1}}^{2nd\ product} \oplus \overbrace{r_3^{(c)} (x_1 \oplus c_1)^1 (x_2 \oplus c_2)^1}^{3rd\ product}$$

After simplification,

$$f = r_0^{(c)} \oplus r_1^{(c)} (x_2 \oplus c_2) \oplus r_2^{(c)} (x_1 \oplus c_1) \oplus r_3^{(c)} (x_1 \oplus c_1)(x_2 \oplus c_2)$$

The logic Taylor expansion for a Boolean function f of n variables of the polarity $c \in 0, 1, \ldots, 2^n - 1$, is defined by the equation

$$f = \bigoplus_{i=0}^{2^n - 1} r_i^{(c)} \overbrace{(x_1 \oplus c_1)}^{c_1\text{-polarity}}{}^{i_1} \ldots \overbrace{(x_n \oplus c_n)}^{c_n\text{-polarity}}{}^{i_n} \qquad (8.15)$$

$$\underbrace{\qquad\qquad\qquad\qquad}_{i\text{-th product}}$$

In Equation 8.15, $c_1, c_2, \ldots, c_n$ and $i_1, i_2, \ldots, i_n$ are the binary representations of the polarity c and index i, respectively. The value of $c_i \in \{0, 1\}$, $i = 0, 1, 2, \ldots, n$, specifies the x_i variable, that is, uncomplemented x_i if $c_i = 0$ or complemented $\bar{x}_i$ if $c_i = 1$. The i-th coefficient $r_i^{(c)}$, given polarity c, is calculated as follows:

$$r_i^{(c)} = \frac{\partial^{(j)} f(c)}{\partial x_1^{i_1} \partial x_2^{i_2} \ldots \partial x_n^{i_n}} \quad \text{and} \quad \partial x_i^{i_j} = \begin{cases} 1, & i_j = 0 \\ \partial x_j & i_j = 1 \end{cases} \qquad (8.16)$$

where parameter j is the order of Boolean differences, $j = \sum_{j=1}^{n} i_j$. The i-th coefficient $r_i^{(c)}$ is the value (0 or 1) of the j-order Boolean difference of the function $f(c)$, given polarity $x_1 = c_1, \ldots, x_n = c_n$.

Example 8.29 *Given polarity* $c = 2$, *compute the first coefficient* $(i = 1)$ $r_i^{(c)}$ *of the* **logic Taylor expansion** *for the Boolean function of three variables* $n = 3$. *Techniques for computing are introduced in Figure 8.12.*

Techniques for computing logic Taylor coefficients

Given: The polarity $c = 2$, the number of variables $n = 3$ of a Boolean function, the number of coefficient $i = 1$

Find: The coefficient $r_1^{(2)}$ of the logic Taylor expansion

Step 1: Write Equation 8.16 for $n = 3$, $i = 1$, and $c = 3$:

$$r_1^{(2)} = \frac{\partial^{(j)} f(2)}{\partial x_1^{i_1} \partial x_2^{i_2} \partial x_3^{i_3}}$$

Step 2: Compute the order j of Boolean difference for the first $(i = i_1 i_2 i_3 = 001)$ coefficient :

$$j = \sum_{j=1}^{3} i_j = 0 + 0 + 1 = 1$$

Step 3: Compute $\partial x_1^{i_1}, \partial x_2^{i_2}$, and $\partial x_3^{i_3}$ using Equation 8.16:

$$\partial x_1^{i_1} = \partial x_1^0 = 1; \quad \partial x_2^{i_2} = \partial x_2^0 = 1; \quad \partial x_1^{i_3} = \partial x_3^1 = \partial x_3$$

Step 4: Write the first coefficient of a logic Taylor expansion given second polarity for an arbitrary Boolean function of three variables using Equation 8.16 ($n = 3$, $i = 1$, $c = 3$, and $j = 1$):

$$r_1^{(2)} = \frac{\partial^{(1)} f(2)}{\partial x_1^{i_1} \partial x_2^{i_2} \partial x_3^{i_3}} = \frac{\partial^{(1)} f(2)}{\partial x_1^0 \partial x_2^0 \partial x_3^1} = \frac{\partial^{(1)} f(2)}{\partial x_3}$$

FIGURE 8.12

Computing the coefficients of logic Taylor expansion for an arbitrary Boolean function of three variables (Example 8.29)

It follows from Equation 8.15 that

▶ A variable x_j is 0-polarized if it enters into the expansion uncomplemented, and 1-polarized otherwise.

▶ Coefficients of the polynomial expression of a Boolean function are described in terms of a differences, formally defined as Boolean derivatives.

The logic Taylor expansion of an n-variable Boolean function f generates 2^n polynomial expressions of 2^n polarities.

Example 8.30 *There are $2^n = 2^2 = 4$ polynomial expressions of Boolean function of two variables $(n = 2)$ of polarities 0,1,2, and 3. Table 16.3 shows how each elementary function of two variables can be represented by four polynomial forms. For example, the OR gate given a **polarity** of $c = 3$ $(c_1 c_2 = 11)$ is represented by the two nonzero coefficients $r_0^{(3)}$ and $r_3^{(3)}$. That is, $f = r_0^{(3)} \oplus r_3^{(3)} \overline{x}_1 \overline{x}_2 = 1 \oplus \overline{x}_1 \overline{x}_2$, and corresponding truth vector $\mathbf{F} = [1001]^T$. This is an optimal representation of the OR function (optimal polarity).*

While the i-th coefficient $r_i^{(c)}(d)$ is described by a Boolean expression, it can be calculated in different ways; for example, symbolic or matrix transformations, cube-based technique, decision diagram technique, and probabilistic methods.

TABLE 8.6
Polynomial expressions as logic Taylor expansions of elementary Boolean functions (Example 8.30)

Techniques for computing logic Taylor expansion

Function	$r_0^{(c)}$	Vector form $\frac{\partial f}{\partial x_2}$	$\frac{\partial f}{\partial x_1}$	$\frac{\partial^2 f}{\partial x_1 \partial x_2}$	Algebraic form
x_1 —⟩ f x_2 — $f = x_1 \wedge x_2$	0 0 0 1	0 0 1 1	0 1 0 1	1 1 1 1	$f(c = 0) = x_1 x_2$ $f(c = 1) = x_1 \oplus x_1 \overline{x}_2$ $f(c = 2) = x_2 \oplus \overline{x}_1 x_2$ $f(c = 3) = 1 \oplus \overline{x}_2 \oplus \overline{x}_1 \oplus \overline{x}_1 \overline{x}_2$
x_1 —⟩ f x_2 — $f = x_1 \vee x_2$	0 1 1 1	1 1 0 0	1 0 1 0	1 1 1 1	$f(c = 0) = x_2 \oplus x_1 \oplus x_1 x_2$ $f(c = 1) = 1 \oplus \overline{x}_2 \oplus x_1 \overline{x}_2$ $f(c = 2) = 1 \oplus \overline{x}_1 \oplus \overline{x}_1 x_2$ $f(c = 3) = 1 \oplus \overline{x}_1 \overline{x}_2$
x_1 —⟩ f x_2 — $f = x_1 \oplus x_2$	0 1 1 0	1 1 1 1	1 1 1 1	0 0 0 0	$f(c = 0) = x_2 \oplus x_1$ $f(c = 1) = 1 \oplus \overline{x}_2 \oplus x_1$ $f(c = 2) = 1 \oplus x_2 \oplus \overline{x}_1$ $f(c = 3) = \overline{x}_2 \oplus \overline{x}_1$

The following example illustrates symbolic manipulation to derive algebraic polynomial forms.

Example 8.31 *Given $f = x_1 \lor x_2$, its EXOR form Shannon expansion with respect to both variables x_1 and x_2 is given in the first chart of Figure 8.13. To obtain an expression of polarity $c = 3$ ($c_1 c_2 = 11$), both x_1 and x_2 should be complemented), we have to make the substitution $a = \bar{a} \oplus 1$. On the other hand, the same result can be obtained using Boolean differences (the second chart in Figure 8.13), as follows from logic Taylor expansion.*

Techniques for computing polynomial expressions

The EXOR form Shannon expansion with respect to both variables x_1 and x_2 for a two-input OR gate	Logic Taylor expansion for the polarity $c = 3$ for a two-input OR gate
$f = \bar{x}_1 \bar{x}_2 f_{00} \oplus \bar{x}_1 x_2 f_{01}$ $\oplus x_1 \bar{x}_2 f_{10} \oplus x_1 x_2 f_{11}$ $= \bar{x}_1 \bar{x}_2 (0 \lor 0) \oplus \bar{x}_1 x_2 (0 \lor 1)$ $\oplus x_1 \bar{x}_2 (1 \lor 0) \oplus x_1 x_2 (1 \lor 1)$ $= \bar{x}_1 x_2 \oplus x_1 \bar{x}_2 \oplus x_1 x_2$	$f = f(3) \oplus \dfrac{\partial f(3)}{\partial x_2} \bar{x}_2 \oplus \dfrac{\partial f(3)}{\partial x_1} \bar{x}_1 \oplus \dfrac{\partial^2 f(3)}{\partial x_1 \partial x_2} \bar{x}_1 \bar{x}_2$ $= 1 \oplus 0 \cdot \bar{x}_2 \oplus 0 \cdot \bar{x}_1 \oplus 1 \cdot \bar{x}_1 \bar{x}_2$ $= 1 \oplus \bar{x}_1 \bar{x}_2$
Converting to the polarity $c = 3$ ($c_1 c_2 = 11$) $f = \bar{x}_1 (\bar{x}_2 \oplus 1) \oplus (\bar{x}_1 \oplus 1) \bar{x}_2$ $\oplus (\bar{x}_1 \oplus 1)(\bar{x}_2 \oplus 1)$ $= \bar{x}_1 \bar{x}_2 \oplus \bar{x}_1 \oplus \bar{x}_1 \bar{x}_2 \oplus \bar{x}_2$ $\oplus \bar{x}_1 \bar{x}_2 \oplus \bar{x}_1 \oplus \bar{x}_2 \oplus 1$ $= 1 \oplus \bar{x}_1 \bar{x}_2$	$f(3) = 1$ $\dfrac{\partial f(3)}{\partial x_2} = (x_1 \lor 0) \oplus (x_1 \lor 1)\big|_{x_1 = 1}$ $\qquad = (x_1 \oplus 1)\big|_{x_1 = 1} = \bar{x}_1\big|_{x_1 = 1 = \bar{1} = 0}$ $\dfrac{\partial f(3)}{\partial x_1} = (0 \lor x_2) \oplus (1 \lor x_2)\big|_{x_2 = 1}$ $\qquad = \bar{x}_2\big|_{x_2 = 1} = \bar{1} = 0$ $\dfrac{\partial f(3)^2}{\partial x_1 \partial x_2} = 1$

FIGURE 8.13
Techniques for construction of polynomial expressions of a given polarity using Shannon expansion and logic Taylor expansion (Example 8.31).

Example 8.32 *Given $f = x_1 \bar{x}_2 \lor x_3$, the logic Taylor expansion is derived for the polarity $c = 4$ ($c_1 c_2 c_3 = 100$), as shown in Figure 8.14.*

Techniques for computing a logic Taylor expansion

Given: A Boolean function $f = x_1 \overline{x}_2 \vee x_3$

Find: A logic Taylor expansion at the point $c = 4$ $(c_1 c_2 c_3 = 100)$

Step 1: Apply Equation 8.15 for $n = 3$:

$$f = \bigoplus_{i=0}^{2^3-1} \frac{\partial^3 f(c)}{\partial x_1^{i_1} \partial x_2^{i_2} \partial x_3^{i_3}} \overbrace{(x_1 \oplus c_1)}^{c_1-polarity}{}^{i_1} \overbrace{(x_2 \oplus c_2)}^{c_2-polarity}{}^{i_1} \overbrace{(x_3 \oplus c_3)}^{c_3-polarity}{}^{i_3}$$

Step 2: Substitute $c = 4$ $(c_1 = 1, c_2 = 0, c = 0)$ in the equation

$$f = \bigoplus_{i=0}^{2^3-1} \frac{\partial^3 f(4)}{\partial x_1^{i_1} \partial x_2^{i_2} \partial x_3^{i_3}} \overbrace{(x_1 \oplus 1)^{i_1} (x_2 \oplus 0)^{i_1} (x_3 \oplus 0)^{i_3}}^{Polarity\ 4}$$

Step 3: Expand the equation for $i = 0, 1, 2, \ldots, 7$:

$$f = \frac{\partial^3 f(4)}{\partial x_1^0 \partial x_2^0 \partial x_3^0} \overbrace{\overline{x}_1{}^0 x_2{}^0 x_3{}^0}^{1} \oplus \frac{\partial^3 f(4)}{\partial x_1^0 \partial x_2^0 \partial x_3^1} \overbrace{\overline{x}_1{}^0 x_2{}^0 x_3{}^1}^{x_3} \oplus \frac{\partial^3 f(4)}{\partial x_1^0 \partial x_2^1 \partial x_3^0} \overbrace{\overline{x}_1{}^0 x_2{}^1 x_3{}^0}^{x_2}$$

$$\underbrace{\qquad}_{1} \qquad \underbrace{\qquad}_{\partial x_3} \qquad \underbrace{\qquad}_{\partial x_2}$$

$$\oplus \frac{\partial^3 f(4)}{\partial x_1^0 \partial x_2^1 \partial x_3^1} \overbrace{\overline{x}_1{}^0 x_2{}^1 x_3{}^1}^{x_2 x_3} \oplus \frac{\partial^3 f(4)}{\partial x_1^1 \partial x_2^0 \partial x_3^0} \overbrace{\overline{x}_1{}^1 x_2{}^0 x_3{}^0}^{\overline{x}_1} \oplus \frac{\partial^3 f(4)}{\partial x_1^1 \partial x_2^0 \partial x_3^1} \overbrace{\overline{x}_1{}^1 x_2{}^0 x_3{}^1}^{\overline{x}_1 x_3}$$

$$\underbrace{\qquad}_{\partial x_2 \partial x_3} \qquad \underbrace{\qquad}_{\partial x_1} \qquad \underbrace{\qquad}_{\partial x_1 \partial x_3}$$

$$\oplus \frac{\partial^3 f(4)}{\partial x_1^1 \partial x_2^1 \partial x_3^0} \overbrace{\overline{x}_1{}^1 x_2{}^1 x_3{}^0}^{\overline{x}_1 x_2} \oplus \frac{\partial^3 f(4)}{\partial x_1^1 \partial x_2^1 \partial x_3^1} \overbrace{\overline{x}_1{}^1 x_2{}^1 x_3{}^1}^{\overline{x}_1 x_2 x_3}$$

$$\underbrace{\qquad}_{\partial x_1 \partial x_2} \qquad \underbrace{\qquad}_{\partial x_1 \partial x_2 \partial x_3}$$

Step 4: Compute the products $\partial x_1^{i_1} \partial x_2^{i_2} \partial x_3^{i_3}$ and $x_1^i x_2^i x_3^i$:

$$f = f(4) \oplus \frac{\partial f(4)}{\partial x_3} x_3 \oplus \frac{\partial f(4)}{\partial x_2} x_2 \oplus \frac{\partial^2 f(4)}{\partial x_2 \partial x_3} x_2 x_3 \oplus \frac{\partial f(4)}{\partial x_1} \overline{x}_1 \oplus \frac{\partial^2 f(4)}{\partial x_1 \partial x_3} \overline{x}_1 x_3$$

$$\oplus \frac{\partial^2 f(4)}{\partial x_1 \partial x_2} \overline{x}_1 x_2 \oplus \frac{\partial^3 f(4)}{\partial x_1 \partial x_2 \partial x_3} \overline{x}_1 x_2 x_3$$

FIGURE 8.14

Logic Taylor expansion of a Boolean function (Example 8.32).

Techniques for computing a logic Taylor expansion (continuation)

Step 5: Compute the values of Boolean differences:

$$f(4) = f(1,0,0) = 1 \cdot \overline{0} \vee 0 = \boxed{1}$$

$$\frac{\partial f(4)}{\partial x_3} = \overbrace{f(1,0,0) \oplus f(1,0,1)}^{With\ respect\ to\ x_3} = 1 \oplus (1 \cdot \overline{0} \vee 1) = \boxed{0}$$

$$\frac{\partial f(4)}{\partial x_2} = \overbrace{f(1,0,0) \oplus f(1,1,0)}^{With\ respect\ to\ x_2} = 1 \oplus (1 \cdot \overline{1} \vee 0) = \boxed{1}$$

$$\frac{\partial^2 f(4)}{\partial x_2 \partial x_3} = \frac{\partial}{\partial x_2}\left(\frac{\partial f(4)}{\partial x_3}\right) = \frac{\partial f(1,0,0)}{\partial x_3} \oplus \frac{\partial f(1,1,0)}{\partial x_3} = 0 \oplus 1 = \boxed{1}$$

$$\frac{\partial f(4)}{\partial x_1} = \overbrace{f(1,0,0) \oplus f(0,0,0)}^{With\ respect\ to\ x_1} = 1 \oplus (0 \cdot \overline{0} \vee 0) = \boxed{1}$$

$$\frac{\partial^2 f(4)}{\partial x_1 \partial x_3} = \frac{\partial}{\partial x_1}\left(\frac{\partial f(4)}{\partial x_3}\right) = \frac{\partial f(1,0,0)}{\partial x_3} \oplus \frac{\partial f(0,0,0)}{\partial x_3} = 0 \oplus 1 = \boxed{1}$$

$$\frac{\partial^2 f(4)}{\partial x_1 \partial x_2} = \frac{\partial}{\partial x_1}\left(\frac{\partial f(4)}{\partial x_2}\right) = \frac{\partial f(1,0,0)}{\partial x_2} \oplus \frac{\partial f(0,0,0)}{\partial x_2} = 1 \oplus 0 = \boxed{1}$$

$$\frac{\partial^3 f(4)}{\partial x_1 \partial x_2 \partial x_3} = \frac{\partial}{\partial x_1}\left(\frac{\partial^2 f(4)}{\partial x_2 \partial x_3}\right) = \frac{\partial^2 f(1,0,0)}{\partial x_2 \partial x_3} \oplus \frac{\partial^2 f(0,0,0)}{\partial x_2 \partial x_3} = 1 \oplus 0 = \boxed{1}$$

<u>Note:</u> The Boolean differences $\frac{\partial f(0,0,0)}{\partial x_2}$, $\frac{\partial f(0,0,0)}{\partial x_3}$, and $\frac{\partial f(0,0,0)}{\partial x_2 \partial x_3}$ are computed as follows:

$$\frac{\partial f(0,0,0)}{\partial x_2} = \overbrace{f(0,0,0) \oplus f(0,1,0)}^{With\ respect\ to\ x_2} = (0 \cdot \overline{0} \vee 0) \oplus (0 \cdot \overline{1} \vee 0) = 0$$

$$\frac{\partial f(0,0,0)}{\partial x_3} = \overbrace{f(0,0,0) \oplus f(0,0,1)}^{With\ respect\ to\ x_3} = (0 \cdot \overline{0} \vee 0) \oplus (0 \cdot \overline{0} \vee 1) = 1$$

$$\frac{\partial^2 f(0,0,0)}{\partial x_2 \partial x_3} = \frac{\partial f(0,0,0)}{\partial x_3} \oplus \frac{\partial f(0,1,0)}{\partial x_3} = 1 \oplus (f(0,1,0) \oplus f(0,1,1))$$

$$= 1 \oplus ((0 \cdot \overline{1} \vee 0) \oplus (0 \cdot \overline{1} \vee 1)) = 1 \oplus 1 = 0$$

Step 6: Write logic Taylor expansion using the computed Boolean differences:

$$f = \boxed{1} \oplus \boxed{0}\, x_3 \oplus \boxed{1}\, x_2 \oplus \boxed{1}\, x_2 x_3 \oplus \boxed{1}\, \overline{x}_1 \oplus \boxed{1}\, \overline{x}_1 x_3 \oplus \boxed{1}\, \overline{x}_1 x_2 \oplus \boxed{1}\, \overline{x}_1 x_2 x_3$$

FIGURE 8.15

Logic Taylor expansion of a Boolean function (continuation of Example 8.32).

8.6   Graphical representation of fundamental expansions

A useful feature of fundamental expansions is their relation to graphical data structures such as decision trees and decision diagrams. Equation 8.2 can be used for computing an arbitrary Boolean function in algebraic form. Another form of this equation is the graph. Figure 8.16a shows the representation of Equation 8.2 by a node of the graph.

8.6.1   Shannon expansion as a decision tree node function

Suppose that Shannon expansion, denoted by S, is implemented in this node, that is, the input is the function f, and outgoing links correspond to the parts of this equation. It is useful to interpret the function of the node as transfer-contact (Figure 8.16b).

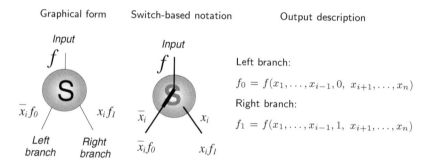

FIGURE 8.16

Graphical representation of a Shannon expansion with respect to a variable and its implementation as the transfer-contact (Equation 8.2).

Example 8.33   *A node of a decision tree that represents the Shannon expansion of the Boolean function $f = \overline{x}_1 x_3 \vee x_2 \overline{x}_3$ with respect to the variable x_1 is given in Figure 8.17.*

In the above examples, the Boolean function was replaced by two Boolean subfunctions with reduced variables. Shannon expansion can be applied to these subfunctions, too.

Figure 8.18 illustrates the computational details of the node. The node at the *intermediate* level results in two cofactors, f_0 and f_1.

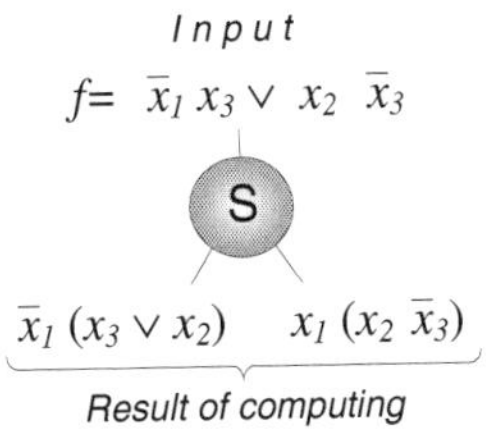

Input

$f = \overline{x}_1 x_3 \vee x_2 \ \overline{x}_3$

$$fo = 1 \cdot x_3 \vee x_2 \overline{x}_3 = x_3 \vee x_2$$
$$f_1 = 0 \cdot x_3 \vee x_2 \overline{x}_3 = x_2 \overline{x}_3$$
$$f = \overline{x}_1 f_0 \vee x_1 f_1$$
$$= \overline{x}_1 (x_3 \vee x_2) \vee x_1 (x_2 \overline{x}_3)$$

$\overline{x}_1 (x_3 \vee x_2)$ $x_1 (x_2 \ \overline{x}_3)$

Result of computing

FIGURE 8.17
Shannon expansion with respect to a variable x_1 of the Boolean function
$f = \overline{x}_1 x_3 \vee x_2 \overline{x}_3$ (Example 8.33).

Computing at the terminal level $f = \overline{x}_i f_0 \vee x_i f_1$

Output $f_0 = 0$, $f_1 = 0$: Node computes $f = \overline{x}_i \cdot 0 \vee x_i \cdot 0 = 0$
Output $f_0 = 0$, $f_1 = 1$: Node computes $f = \overline{x}_i \cdot 0 \vee x_i \cdot 1 = x_i$
Output $f_0 = 1$, $f_1 = 0$: Node computes $f = \overline{x}_i \cdot 1 \vee x_i \cdot 0 = \overline{x}_i$
Output $f_0 = 1$, $f_1 = 1$: Node computes $f = \overline{x}_i \cdot 1 \vee x_i \cdot 1 = 1$

A reduction of the decision tree is possible if $f_0 = 0$, $f_1 = 0$ (the node can be eliminated and the value $f = 0$ is assigned to the output), and if $f_0 = 1$, $f_1 = 1$ (the node can be eliminated and the value $f = 1$ is assigned to the output).

8.6.2 Matrix notation of the node function

A node in a decision tree of a Boolean function f corresponds to the Shannon expansion of a Boolean function with respect to a variable x_i: $f = \overline{x}_i f_0 \vee x_i f_1$, where $f_0 = f_{x_i=0}$ and $f_1 = f_{x_i=1}$. Here $f = f_{x_i=a}$ denotes the cofactor of f after assigning the constant a to the variable x_i.

The transformation assigned to a node of a decision tree can be represented in matrix notation. Given a function of a single variable x_i represented by the truth vector $\mathbf{F} = [\ f(0)\ f(1)\ ]^T$, its matrix notation is given as

$$f = [\ \overline{x}_i\ x_i\ ] \begin{bmatrix} 1 & 0 \\ 0 & 1 \end{bmatrix} \begin{bmatrix} f_0 \\ f_1 \end{bmatrix} = [\ \overline{x}_i\ x_i\ ] \begin{bmatrix} f_0 \\ f_1 \end{bmatrix} = \overline{x}_i f_0 \vee x_i f_1,$$

where $f_0 = f_{x_i=0}$, $f_1 = f_{x_i=1}$. Figure 8.19 shows the implementation of a node using a so-called *multiplexer* (details are given in Chapters 13 and 14).

8.6.3 Using Shannon expansion in decision trees

Recursive application of the Shannon expansion of a Boolean function with respect to all variables results in an n-level decision tree, called a *binary decision tree*.

Design example: a node in decision diagrams and trees

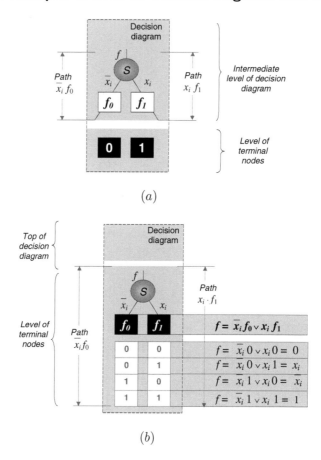

(a)

(b)

FIGURE 8.18
A node in decision diagrams and trees: computing at the intermediate level
(a) and computing at the terminal level (b).

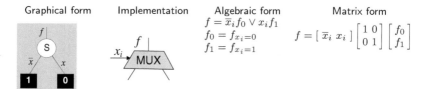

FIGURE 8.19
A node of a decision tree, its implementation by a multiplexer (MUX), and
algebraic and matrix descriptions.

Example 8.34 *Recursive application of the Shannon expansion with respect to variables x_1 and x_2 to the Boolean function $f = \overline{x}_1 x_2 \vee x_1 \overline{x}_2$ is illustrated in Figure 8.20.*

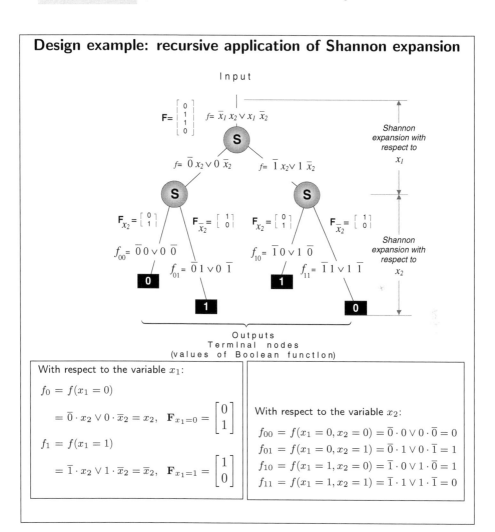

FIGURE 8.20

Recursive application of Shannon expansion to the Boolean function $f = \overline{x}_1 x_2 \vee x_1 \overline{x}_2$ using a decision tree (Example 8.34).

Example 8.35   *The symmetric property can be observed, for example, in decision diagrams, as illustrated in Figure 8.21.*

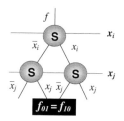

FIGURE 8.21

Symmetry conditions for adjacent variables of a decision tree (Example 8.35).

8.7   Further study

Advanced topics of Shannon and logic Taylor expansion

Topic 1: The arithmetic analog of the Shannon expansion of a Boolean function f with respect to a variable x_i. There exist:

▶ *The arithmetic analog of the positive Shannon* expansion $f = f_0 + x_i f_2$, where $f_0 = f_{x_i=0}$ and $f_2 = -f_{x_i=1} + f_{x_i=0}$.

▶ *The arithmetic analog of the negative Shannon* expansion $f = f_1 + \overline{x}_i f_2$, where $f_1 = f_{x_i=1}$.

For the Boolean variables x, x_1, and x_2, taking on the values of 0 or 1, the following is true, as introduced by the founder of Boolean algebra George Boole: $\overline{x} = 1 - x$, $x_1 \vee x_2 = x_1 + x_2 - x_1 x_2$, and $x_1 \wedge x_2 = x_1 x_2$. Other logic operations can be represented by arithmetic operations as well; for example, $x_1 \oplus x_2 = x_1 + x_2 - 2 x_1 x_2$. The right part of the equation is called an *arithmetic expression*.

Arithmetic representations of Boolean functions are known as *word-level* forms, and are a way to describe the parallel calculation of several Boolean functions at once. Another useful property of these arithmetic representations is linearization. A multioutput Boolean function can be represented by a linear word-level arithmetic polynomial and a linear word-level decision diagram.

Topic 2: Symmetric functions. There are many forms of symmetry in Boolean functions. In practice, one of the advantages of knowing that a given Boolean function is symmetric is the potential economy offered by the implementation of such function using smaller number of gates. In addition, in verification, the effectiveness of the *input matching* procedure can be increased if the symmetric input sets of the specification are known (the goal of verification is to check whether a given implementation follows the specification for which it was designed).

Let f be a Boolean function of n variables. Shannon expansion with respect to the variables x_i and x_j yields:

$$f = \overline{x}_i \overline{x}_j f_{\overline{x}_i \overline{x}_j} \vee \overline{x}_i x_j f_{\overline{x}_i x_j} \vee x_i \overline{x}_j f_{x_i \overline{x}_j} \vee x_i x_j f_{x_i x_j}$$

From this expansion follows that

(a) f is partially symmetric in variables (x_i, x_j) and $(\overline{x}_i, \overline{x}_j)$ if $f_{\overline{x}_i x_j} = f_{x_i \overline{x}_j}$;
(b) f is partially symmetric in variables $(x_i, \overline{x}_j)$ and $(\overline{x}_i, x_j)$ if $f_{\overline{x}_i \overline{x}_j} = f_{x_i x_j}$;
(c) f is multiform symmetric in variables $(x_i, x_j), (\overline{x}_i, \overline{x}_j)$ and $(x_i, \overline{x}_j), (\overline{x}_i, x_j)$ if $f_{x_i \overline{x}_j} = f_{\overline{x}_i x_j}$ and $f_{\overline{x}_i \overline{x}_j} = f_{x_i x_j}$, respectively.

For example, Shannon expansion of the Boolean function $f = \overline{x}_1 \overline{x}_2 \overline{x}_3 \vee x_1 x_2 \vee x_1 x_3 \vee x_2 x_3$ with respect to the variables x_1 and x_2 yields

$$f = \overline{x}_1 \overline{x}_2 \overbrace{(\overline{x}_3)}^{f_{\overline{x}_i \overline{x}_j}} \vee \overline{x}_1 x_2 \overbrace{(x_3)}^{f_{\overline{x}_i x_j}} \vee x_1 \overline{x}_2 \overbrace{(x_3)}^{f_{x_i \overline{x}_j}} \vee x_1 x_2 \overbrace{(x_3 \vee \overline{x}_3)}^{f_{x_i x_j}}$$

Hence, the Boolean function f is partially symmetric with respect to the variables (x_1, x_2) and $(\overline{x}_1, \overline{x}_2)$.

Shannon expansion for symmetric functions can be extended to *arbitrary* Boolean functions; that is, an arbitrary Boolean function can be represented with respect to the elementary symmetric functions (see, for example, the pioneered paper by R. C. Born and A. K. Scidmore "Transformation of Switching Functions to Completely Symmetric Switching Functions," *IEEE Transactions on Computers*, volume C-17, number 6, 1968, pages 596–599). The result of such symmetrization is a computing structure of a regular configuration. In the implementation, a regular structure is characterized by the local and short connections between computing elements, which is attractive for technology. In addition, logic design can be combined with layout, so that no special stage of placement and routing is necessary.

Symmetrization is the underlying principle of so-called *universal computing arrays*. For example, universal Akers array is a regular and planar layout of identical computing elements (multiplexers) that is able to compute an arbitrary Boolean function (S. B. Akers, "A Rectangular Logic Array," *IEEE Transactions on Computers*, volume C-21, number 8, 1972, pages 848–857). Another effective application of the symmetry is decision diagram techniques (see, for example, the paper by M. A. Perkowski, M. Chrzanowska-Jeske, and Y. Xu, "Lattice Diagrams Using Reed-Muller Logic," *Proceedings of the 3rd International Workshop on Applications of the Reed–Muller Expansion in Circuit Design*, Oxford University, 1997, pages 85–102).

The alternative data structure for symmetric functions, called *transeunt triangle* . Figure 8.22 shows how the *Pascal triangle* can be used for representations of symmetric Boolean functions. The Pascal triangle can be reduced to a transeunt triangle. This data structures for the function $f = \overline{x}_1 \overline{x}_2$

Data structures for symmetric Boolean functions

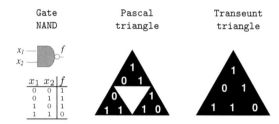

FIGURE 8.22

Representation of NAND gate using the Pascal triangle transeunt triangle.

is shown in Figure 8.22. The carrier vector $\mathbf{F}_c = [\,1\ 1\ 0\,]^T$ is formed from the initial truth vector $\mathbf{F} = [\,1\ 1\ 1\ 0\,]^T$.

Topic 3: The arithmetic analog of logic Taylor expansion represents a Boolean function in 2^n forms called *arithmetic fixed polarity forms*. An arithmetic analog of the logic Taylor expansion is expressed by the equation

$$f = \sum_{i=0}^{2^n-1} p_i^{(c)} \overbrace{\underbrace{\left(x_1 \oplus c_1\right)^{i_1} \ldots \overbrace{\left(x_n \oplus c_n\right)^{i_n}}^{c_n \; polarity}}_{i\text{-th product}}}^{c_1 \; polarity},$$

where $c_1 c_2 \ldots c_n$ and $i_1, i_2, \ldots, i_n$ are the binary representations of c (polarity) and i respectively; $p_i^{(c)}$ is a value of the arithmetic analog of a Boolean difference of f given c; i.e., $x_1 = c_1, \ldots, x_n = c_n$.

The arithmetic Taylor expansion produces 2^n arithmetic expressions, corresponding to 2^n polarities. Arithmetic Taylor expansions for several elementary Boolean functions are given in Table 8.7.

Topic 4: Spectral techniques are based on the fundamental expansion called *spectral transform expansion*. Many problems can be solved by transferring the problem in the original space isomorphically to some other space that reflects particular properties of the problem. Thus, in the new space, the problem is simpler or there are at least well known solution tools. Once the problem is solved, we have to be able to return to the original space. Spectral techniques offer alternative methods for solving complex tasks efficiently. The first applications of spectral techniques in logic design are related to the optimization problems and date back to the early 1950s.

Transferring a problem from the original domain into the spectral domain performs redistribution of the information content of a signal but does not reduce it. Spectral representations are canonical, i.e., a unique spectrum corresponds to a given function, and vice versa - the function can be reconstructed from the spectrum by the inverse transform. In many cases, due to particular properties a signal may express, the main portion of the information content of the signal is encoded in a (relatively) small number

TABLE 8.7
The arithmetic Taylor expansion of elementary Boolean functions

Function	F	$\frac{\partial f}{\partial x_2}$	$\frac{\partial f}{\partial x_1}$	$\frac{\partial^2 f}{\partial x_1 \partial x_2}$	Arithmetic expression
x_1 —⊐ f / x_2 / $y = x_1 \wedge x_2$	0	0	0	1	$x_1 x_2$
	0	0	1	-1	$x_1 - x_1 \bar{x}_2$
	0	1	0	-1	$x_2 - \bar{x}_1 x_2$
	1	-1	-1	1	$1 - \bar{x}_2 - \bar{x}_1 + \bar{x}_1 \bar{x}_2$
x_1 —⊐ f / x_2 / $y = x_1 \vee x_2$	0	1	1	-1	$x_2 + x_1 - x_1 x_2$
	1	-1	0	1	$1 - \bar{x}_2 + x_1 \bar{x}_2$
	1	0	-1	1	$1 - \bar{x}_1 + \bar{x}_1 x_2$
	1	0	0	-1	$1 - \bar{x}_1 \bar{x}_2$
x_1 —⊐ f / x_2 / $y = x_1 \oplus x_2$	0	1	1	-2	$x_2 + x_1 - 2x_1 x_2$
	1	-1	-1	2	$1 - \bar{x}_2 - x_1 + 2x_1 \bar{x}_2$
	1	-1	-1	2	$1 - x_2 - \bar{x}_1 + 2\bar{x}_1 x_2$
	0	1	1	-2	$\bar{x}_2 + \bar{x}_1 - 2\bar{x}_1 \bar{x}_2$

of spectral coefficients. For example, the AND function can be described in terms of so-called Walsh spectrum as follows

$$f = \frac{1}{4}(1 - (1 - 2x_2) - (1 - 2x_1) + (1 - 2x_1)(1 - 2x_2))$$

Decision trees are graphical representations of spectral transform expansions of Boolean function f with respect to a basis Q. In a decision tree, each path from the root node to the constant nodes corresponds to a basic function in Q. The constant nodes represent the Q-spectral coefficients.

Shannon [15] has shown that partially symmetric Boolean functions may be realized with considerably fewer components than most functions. McCluskey [11] studied the group invariance of Boolean functions. Epstein [8] studied the implementation of symmetric functions over the library of AND, OR, and NOT gates. Born and Scidmore [2] have shown that any Boolean function can be represented by a symmetric function. Arnold and Harrison [1] developed the concept of canonical symmetric functions. Lee SC and Lee ET [10] investigated symmetric properties of multivalued functions.

The size of a decision diagram can be reduced by using a variable order that places symmetric variables contiguously, for example, using sifting procedures. Symmetries can be utilized to improve the efficiency of functional equivalence checking. For example, Scholl et al. [14] used symmetries in decision diagram minimization.

Topic 5: Information notation of the Shannon expansion. Shannon expansion can be extended to the *integer* numbers and interpreted in terms of *Shannon information*. Let $A = \{a_1, a_2, \ldots, a_n\}$ be a complete set of events with the probability distribution $\{p(a_1), p(a_2), \ldots, p(a_n)\}$. The *entropy* of the finite field A is given by

$$H(A) = -\sum_{i=1}^{n} p(a_i) \cdot \log p(a_i)$$

where the logarithm is base 2. The entropy can never be negative; i.e., $\log p(a_i) \leq 0$, and thus $H(A) \geq 0$. The entropy is zero if and only if A contains one event only

Information notation of the Shannon expansion for a Boolean function f of n variables with respect to the variable x_i is represented by the equation

$$H^S(f|x_i) = p_{|x_i=0} \cdot H(f_{x_i=0}) + p_{|x_i=1} \cdot H(f_{x_i=1}).$$

Further reading

[1] Arnold RF and Harrison MA. Algebraic properties of symmetric and partially symmetric Boolean functions. *IEEE Trans. Electronic Circuits*, pp. 244–251, June, 1963.

[2] Born RC and Scidmore AK. Transformation of switching functions to completely symmetric switching functions. *IEEE Trans. Comput.*, 17(6):596–599, 1968.

[3] Butler JT, Dueck GW, Shmerko VP, and Yanushkevich SN. On the number of generators of transeunt triangles. *Discrete Applied Mathematics*, 108:309–316, 2001.

[4] Butler JT, Dueck GW, Shmerko VP, and Yanushkevich SN. Comments on SYMPATHY: fast exact minimization of fixed polarity Reed-Muller expansion for symmetric functions. *IEEE Trans. Computer-Aided Design of Integrated Circuits and Systems*, 19(11):1386–1388, 2000.

[5] Butler JT and Sasao T. On the properties of multiple-valued functions that are symmetric in both variables and labels. In *Proc. of the IEEE 28th Int. Symp. on Multiple-Valued Logic*, pp. 83–88, 1998.

[6] Butler JT, Herscovici D, Sasao T, and Barton R. Average and worst case number of nodes in binary decision diagrams of symmetric multiple-valued functions. *IEEE Trans. Comput.*, 46(4):491–494, 1997.

[7] Davio M, Deschamps JP, and Thayse A. *Discrete and Switching Functions*, McGraw-Hill, New York, 1978.

[8] Epstein G. Synthesis of electronic circuits for symmetric functions. *IRE Trans. Electronic Comput.*, pp. 57–60, March 1958.

[9] Hachtel GD and Somenzi F. *Logic Synthesis and Verification Algorithms*, Kluwer, Dordrecht, 1996.

[10] Lee SC and Lee ET. On multivalued symmetric functions. *IEEE Trans. Comput.*, pp. 312–317, March, 1972.

[11] McCluskey EJ. Yr. Detection of group invariance or total symmetry of a Boolean functions, *The Bell System Technical J.*, 35(6):1445–1453, 1956.

[12] Miller DM and Muranaka N.. Multiple-valued decision diagrams with symmetric variable nodes. In *Proc. IEEE Int. Symp. on Multiple-Valued Logic*, pp. 242–247, 1996.

[13] Pomeranz I and Reddy SM. On determining symmetries in inputs of logic circuits. *IEEE Trans. Computer-Aided Design of Integrated Circuits and Systems*, 13(11):1478–1433, 1994.

[14] Scholl C, Moller D, Molitor P, and Drechsler R. BDD minimization using symmetries. *IEEE Trans. Computer-Aided Design of Integrated Circuits and Systems*, 18(2):81–100, 1999.

[15] Shannon CE. A Symbolic analysis of relay and switching circuits, *Trans. American Institute of Electrical Engineers*, 57:713–723, Part B, 1938.

[16] Shannon CE and Weaver W. *The Mathematical Theory of Communication*, the University of Illinois Press, Urbana, Illinois, 1949.

[17] Stanković RS and Astola JT. *Spectral Interpretation of Decision Diagrams*, Springer, Dordrecht, 2003.

[18] Yanushkevich SN, Miller DM, Shmerko VP, and Stanković RS. *Decision Diagram Techniques for Micro- and Nanoelectronic Design*, Taylor & Francis/CRC Press, Boca Raton, FL, 2006.

9

Arithmetic of the Polynomials

9.1 Introduction

There are various chemical and physical phenomena that can be interpreted as the exclusive-or (EXOR) logic operation. Hence, the implementation of an EXOR cell can be followed directly from the chemical and physical phenomena instead of the EXOR representation using AND and OR operations. This is the motivation to consider polynomial arithmetic as a promising candidate for nanotechnology.

Arithmetic of the polynomials is the area of a logic network design based on the polynomial representation of Boolean functions. Polynomials are an algebraic form of data structure and can be considered as an extension of an SOP representation of Boolean functions. A Boolean function of n variables can be represented by n different polynomials. Hence, the problem of choosing the polynomial for a given Boolean function with respect to specified criteria addresses the optimization methods.

In the previous chapters, the SOP form and its dual equivalent, the POS form, of a Boolean function have been the focus of interest. The SOP forms were the data structures most often used for Boolean function representation during the 40s and 50s, when the first generation of computers was developed. Progress in computation methods dates from the beginning of the 70s, stimulated by the study of alternatives to the SOP forms (details are given in the section "Further study"). This study resulted in the development of so-called *polynomial* forms of Boolean functions. An arbitrary Boolean function can be represented in polynomial forms.

Polynomial forms can be represented using various canonical expansions such as logic Taylor expansion. A logic Taylor expansion at a point c is also called a *fixed polarity Reed–Muller expansion*. The value c is the so-called *fixed polarity* c. The binary representation of c, the n-tuple $(c_1, c_2, \ldots, c_n)$, determines the polarity of variables in the polynomial.

In operational domain domain, Boolean functions are computed using SOP representation. The EXOR analog of Shannon expansion, known as *Davio expansion*, operates in the *functional domain*. In a functional domain, Boolean functions are computed using polynomial representations.

Figure 9.11 shows the relationships between data structures in operational and functional domains that are used in decision diagram techniques. Techniques for decision diagram construction in an operational domain using Shannon expansion are based on algebraic and matrix notations of node functions. The same is true for a functional domain, where Davio expansion is used in the nodes. For example, a Boolean function in algebraic and matrix forms can be converted from an operational into a functional domain, and vice versa. This is the basis for decision diagram construction and computing.

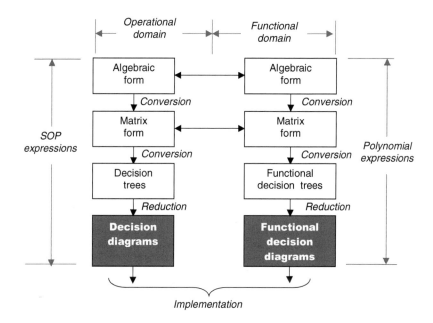

FIGURE 9.1
Data structures for the operational and functional domains of Boolean functions.

A decision tree using Davio expansion is called a *functional* decision tree. Note that in conversions of decision diagrams into functional decision diagrams and vice versa, algebraic and matrix forms are used as intermediate data structures:

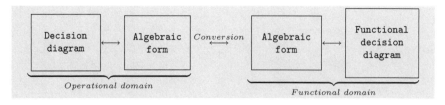

The differences between decision trees and functional decision trees are specified by the properties of SOP and polynomial expressions of Boolean functions. That is, decision trees using Shannon expansion in the nodes operate in SOP form, and functional decision trees using Davio expansion in the nodes operate in the polynomial form of a Boolean function.

After reduction, the decision tree becomes a decision diagram. By analogy, a functional decision tree is called a *functional decision diagram* after applying the reduction procedure. The reduction procedure for functional decision diagrams is different from the reduction of decision diagrams using Shannon expansion. This difference is specified by different techniques for simplification of SOP and polynomial expressions.

Similarly to decision diagrams based on Shannon expansion, functional decision diagrams are used for various design tasks such as representation, manipulation, optimization, and implementation of Boolean functions. The difference is that solutions of these tasks are in the functional domain.

The flexibility of polynomial forms

The motivation for studying polynomial forms of Boolean functions is as follows:

▶ Polynomial expressions provide additional flexibility in terms of choice of implementation technology. This property is efficiently utilized in logic design, especially in design of specific-area applications, in particular, encoding and encryption of information.

▶ There are various physical and molecular effects in predictable technology that can be interpreted as EXOR operations. Nanocomputing devices based on these effects can be used in logic network design and implementation.

▶ Polynomial forms are well suited to logic with more than two values - so-called *multivalued* logic. This fact is utilized in the design of some contemporary and next-generation devices.

In this chapter, only EXOR polynomial forms are considered. The corresponding algebra is called polynomial algebra in the field GF(2). Other polynomial forms are mentioned in the "Further study" section.

Example 9.1	*Figure 9.2 shows the EXOR gate and its AND–OR equivalent network. Using the criterion of the number of gates, the AND–NOR implementation is more complicated than EXOR-based networks.*

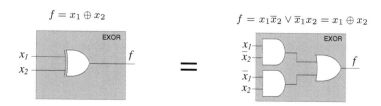

FIGURE 9.2
The EXOR function is represented by a two-level network of AND and OR gates (Example 9.1).

Similarities between SOP and polynomial forms

Given a Boolean function, the SOP expressions can be derived from the *standard* SOP form of this function using the simplification rules. Given a complete set of minterms, standard SOP expressions are formed using the correspondence of 1s in the truth vectors and minterms. By analogy, polynomial forms are derived from the correspondence of polarized minterms and nonzero coefficients of the vector of coefficients.

The standard, or canonical, SOP and polynomials forms are unique given a Boolean function. The number of terms in canonical SOP and polynomial expressions are equal to 2^n. Noncanonical SOP expressions can be derived from canonical SOP forms. Similarly, canonical and noncanonical polynomial expressions can be derived given a Boolean function. Figure 9.3 shows the structural similarity of standard SOP expressions and polynomial forms of Boolean functions:

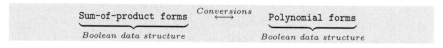

The completeness of the set of operations used in polynomial forms

The polynomial form is a representation of a Boolean function derived from the following universal set of operations over Boolean variables: the constant 1 and AND operations, and EXOR operations.

> **Example 9.2**   *(a) EXOR of single variables forms a so-called "linear polynomial"* $x_1 \oplus \cdots \oplus x_n$. *(b) EXOR and AND operations form a so-called a "nonlinear polynomial," such as, for example,* $x_1 \oplus x_1 x_2 \oplus x_1 x_2 x_3$. *(c) EXOR and AND operations and the constant 1 are used to implement an arbitrary Boolean function. For example, a complemented variable can be represented using EXOR and the constant 1:* $\overline{x} = x \oplus 1$.

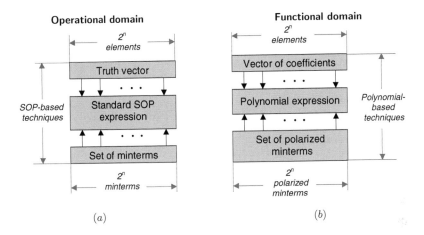

FIGURE 9.3
Structural similarity between standard SOP (a) and polynomial (b) forms.

Data structures for polynomial forms

Polynomial forms can be represented as follows (Figures 9.4 and 9.5):

▶ *Tabulated* forms, such as *functional tables* and *vectors of coefficients* (similar to truth tables and truth vectors for SOP forms)

▶ *Graphical* representations, such as *functional maps* and *functional cubes*, as well as *functional decision trees and diagrams* (similar to K-maps, cubes, and decision trees and diagrams for SOP forms)

▶ *Logic networks* of logic gates or threshold elements

Operational and functional domains

The relationship between *operational* and *functional* domains is the key to the synthesis and application of the polynomial forms of Boolean functions. All satellite Boolean data structures and the corresponding techniques are aimed at providing for representation, manipulation, optimization, and implementation of Boolean functions in the functional domain, namely (Figure 9.4):

(a) Each data structure has particular properties and characteristics and satisfies the requirements of specific tasks of the logic design cycle. There is no "universal" data structure that can be used in all phases of logic design.

(b) Each data structure plays a particular role in design, and is efficient only in solving particular tasks.

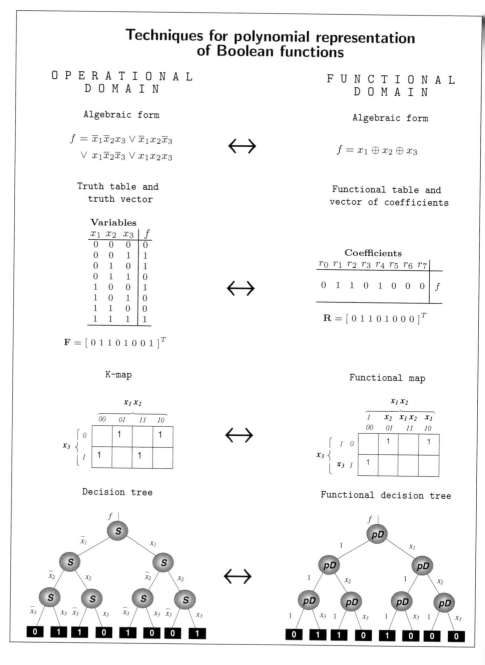

FIGURE 9.4

Data structures for representations of the EXOR function of two variables in the operational domain (SOP form) and the functional domain (polynomial form).

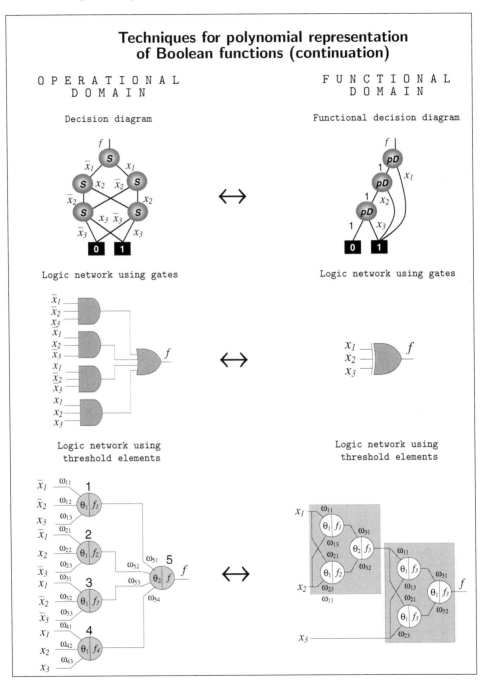

FIGURE 9.5

Data structures for representations of the EXOR function of two variables in the operational domain (SOP form) and the functional domain (polynomial form) (continuation of Figure 9.4).

(c) Each data structure can be converted into another one. These relationships between data structures are often used to achieve design goals.

Example 9.3   **Factoring** *techniques aim at a multilevel implementation of Boolean functions. For example, given the Boolean function* $f = x_1 x_2 \vee x_2 \overline{x}_3 x_4$*, it can be factored as follows:* $f = x_2(x_1 \vee \overline{x}_3 x_4)$*. The corresponding logic networks are shown in Figure 9.6.*

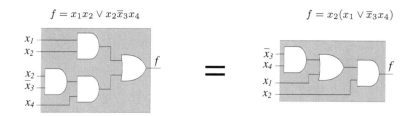

$$f = x_1 x_2 \vee x_2 \overline{x}_3 x_4 \qquad\qquad f = x_2(x_1 \vee \overline{x}_3 x_4)$$

FIGURE 9.6
Implementation of original and factored SOP expressions (Example 9.3).

Relationships between data structures

The relationships between various data structures in the operational and functional domains are used for the manipulation of polynomial expressions:

```
    Polynomial expression  ↔  Functional decision diagram
    Functional decision tree  ↔  Functional decision diagram
  Functional decision diagram  ↔  Matrix expression
            Algebraic form  ↔  Matrix form
       Polynomial expression  ↔  Truth table
```

Analysis and synthesis of EXOR networks

The combinational networks based on SOP representations are designed, or *synthesized* and *analyzed*, using Boolean algebra. Similarly, there are two classes of design problems based on polynomial descriptions: *analysis* and *synthesis*. The analysis problem is formulated as follows: Given an EXOR network, provide a tabular description of this network. The steps involved in synthesis of EXOR networks are basically the reverse of those involved in the analysis. Techniques for the analysis and synthesis of EXOR networks are based on various data structures: vector and algebraic forms, as well as functional decision trees and diagrams.

9.2 Algebra of the polynomial forms

In Chapter 1, Boolean algebra was defined as a set of elements, operations, and postulates. This algebraic structure is the formal basis of the SOP representation. The formal basis of the polynomial forms is the *finite field*. Finite fields are algebraic structures too, but they are characterized by the *elements*, *operations*, and *postulates* of a finite field. The theory of polynomial representations of Boolean functions has been adopted from related fields, such as digital signal processing. Details can be found in the "Further study" section.

9.2.1 Theoretical background

Fields

A *finite field* $\mathcal{F}$

▶ Is an algebraic structure defined as a set of elements, together with two binary operations, each having associative, commutative, and distributive properties, closure under addition and multiplication, inverse properties, and a unique element.

▶ The number of elements in the field is called the **order** of the field. A field with order m exists iff m is a **prime power**, i.e., $m = p^k$ for some integer k and a prime integer p. In this case, addition and multiplication are defined by a table composed such that the requirements for the field are true.

▶ In any finite field, the number of elements must be a power of a prime, p^k. This field is the Galois field.

▶ Every field with p^k elements is isomorphic to every other field with p^k elements. Some of these fields are useful in the representation, manipulation, analysis, and implementation of Boolean functions.

Binary operations are defined as *addition over a field* $\mathcal{F}$, and *multiplication over a field* $\mathcal{F}$.

Galois field

An example of a field $\mathcal{F}$ is a *Galois field* denoted as $GF(q)$:

(a) It consists of q elements $0, 1, 2, \ldots, q-1$.

(b) The number of elements in a Galois field must be equal to $q = p^n$, where p is a *prime* number and n is a positive integer (a natural number $p \geq 2$ is called prime if and only if the only natural numbers that divide p are 1 and p).

(c) In cases where $p = 2$, all 2^n elements are derived using a polynomial of degree n.

(d) Operations in $GF(q)$ are the modulo q sum and modulo q multiplication.

Example 9.4   *Galois field $GF(4)$ consists of four elements $0, 1, 2,$ and 3. Addition and multiplication in $GF(4)$ are defined as follows:*

Addition in $GF(4)$						Multiplication in $GF(4)$				
+	*0*	*1*	*2*	*3*		**·**	*0*	*1*	*2*	*3*
0	0	1	2	3		*0*	0	0	0	0
1	1	0	3	2		*1*	0	1	2	3
2	2	3	0	1		*2*	0	2	3	1
3	3	2	1	0		*3*	0	3	1	2

The field of integer numbers modulo a prime number k is a field.

Polynomials

A *polynomial* in the variable x is the representation of a function f as a sum over an algebraic field $\mathcal{F}$:

$$f = \sum_{i=0}^{N-1} a_i x^i \quad \text{over the field } \mathcal{F} \tag{9.1}$$

The values $a_0, a_1, \ldots, a_{N-1}$ are called *coefficients* of the polynomial. Expression 9.1 means that there exist many polynomials that are distinguished by the properties of the fields, namely, by the types of operation being addition and multiplication.

Example 9.5   *The following fields are used for the representation of Boolean functions: (a) Galois field of order 2, GF(2). This field consists of two elements, 0 and 1. In GF(2), the sum and multiplication correspond to EXOR and AND operations, respectively. (b) A set of integer numbers that includes only the elements 0 and 1. In this field, traditional sum and multiplication are used.*

In addition, the sets of rational and complex numbers, together with the arithmetic operations of sum and multiplication, can be used for various representations of Boolean functions.

9.2.2   Polynomials for Boolean functions

The polynomial in Equation 9.1 is defined for a single variable x. Boolean algebra operates with a set of variables $x_1, x_2, \ldots, x_n$. To apply this polynomial equation to Boolean functions, some restrictions are needed. These restrictions are based on the fundamental theorem of arithmetic, which states that every integer $i > 2$ can be written in the form

$$i = i_1 i_2 \ldots i_n \qquad (9.2)$$

for the unique *primes* $i_1 i_2 \ldots i_n$. This means that if any number is completely factored as in Expression 9.2, this expression is unique. Given a Boolean function of n variables $x_1, x_2, \ldots, x_n$,

$$f = \sum_{i=0}^{2^n - 1} a_i \overbrace{(x_1^{i_1} \cdots x_n^{i_n})}^{\text{Minterm over } \mathcal{F}} \qquad \text{over the field } \mathcal{F} \qquad (9.3)$$

where a_i is a coefficient, i_j is the j-th bit ($j = 1, 2, \ldots, n$) in the binary representation of the index $i = i_1 i_2 \ldots i_n$, and the *literal* $x_j^{i_j}$ is defined as

$$x_j^{i_j} = \begin{cases} 1, & \text{if } i_j = 0 \\ x_j, & \text{if } i_j = 1 \end{cases} \qquad \text{over the field } \mathcal{F} \qquad (9.4)$$

The group of variables $x_1^{i_1} x_2^{i_2} \cdots x_n^{i_n}$ is called a *minterm over the field* $\mathcal{F}$. While the values of Boolean functions are used in SOP (operational domain), the coefficients a_i are used in polynomial forms of Boolean functions (functional domain).

Example 9.6   *It follows from Equations 9.3 and 9.4 that*
$$f = a_0(x_1^{i_1} x_2^{i_2}) + a_1(x_1^{i_1} x_2^{i_2}) + a_2(x_1^{i_1} x_2^{i_2}) + a_3(x_1^{i_1} x_2^{i_2})$$
$$= a_0(x_1^0 x_2^0) + a_1(x_1^0 x_2^1) + a_2(x_1^1 x_2^0) + a_3(x_1^1 x_2^1)$$
$$= a_0 + a_1 x_2 + a_2 x_1 + a_3 x_1 x_2 \quad \text{over the field } \mathcal{F}$$
where "+" is a sum as defined in the field $\mathcal{F}$.

The polynomial form (Equation 9.3) is characterized as follows:

(*a*) The *operations* of sum and multiplication are specified by the properties of the field $\mathcal{F}$; that is, they are either *logical* or *arithmetic* operations.

(*b*) The *coefficients* a_i are computed for each Boolean function using the properties of the field $\mathcal{F}$.

(*c*) *Minterms* over the field $\mathcal{F}$ are specified by multiplication of *literals* $x_j^{i_j}$, $j = 1, 2, \ldots, n$, over $\mathcal{F}$ (Equation 9.4).

Example 9.7   *The following polynomials represent the Boolean function EXOR in different fields (algebras): $f_1 = x_1 \oplus x_2$ over GF(2) and $f_2 = x_1 + x_2 - 2x_1 x_2$ over the field of integers. In the polynomials f_1 and f_2, logical and arithmetical operations are used, respectively. In f_1, the coefficients are 0 or 1 because the field GF(2) consists of 0s and 1s only. The coefficients in the polynomial f_2 are integer numbers.*

Once the values of the Boolean variables x_1 and x_2 are assigned, the polynomials f_1 and f_2 assume the values of the initial function EXOR.

The above brief introduction to the basics of finite fields implies the following:

▶ The polynomial forms are more complicated compared to the SOP form because special techniques are required for computing the coefficients of polynomial forms.

▶ The polynomial forms are an extension of Boolean algebra, and as such they are different from SOP expressions.

9.3   GF(2) algebra

The term *polynomial forms over the field GF(2)* specifies the forms of Boolean functions in which minterms are combined using the EXOR operation. An arbitrary Boolean function can be represented by polynomial expression. The laws of the GF(2) algebra of polynomial forms are given in Table 9.1. In the example below, techniques for the manipulation of polynomial expressions using laws and identities are introduced.

Example 9.8 *Examples of applications of the laws of algebra of polynomial forms from Table 9.1 are as follows: (a) $x_1 \oplus x_2 \oplus x_1 x_2 = x_1 \oplus x_2(1 \oplus x_1) = x_1 \oplus x_2 \overline{x}_1$, (b) $1 \oplus x_1 x_2 \oplus x_1 x_3 = 1 \oplus x_1(x_2 \oplus x_3) = \overline{x_1(x_2 \oplus x_3)}$.*

TABLE 9.1
The GF(2) algebra and identities for manipulations

Techniques for computing the polynomial expressions

Laws and identities	Formal notation	Logic network
Associative law	$x_1 \oplus (x_2 \oplus x_3)$ $= (x_1 \oplus x_2) \oplus x_3$ $= x_1 \oplus x_2 \oplus x_3$ $x_1(x_2 x_3) = (x_1 x_2) x_3$ $= x_1 x_2 x_3$	
Distributive law	$x_1(x_2 \oplus x_3)$ $= x_1 x_2 \oplus x_1 x_3$	
Commutative law	$x_1 \oplus x_2 = x_2 \oplus x_1$ $x_1 x_2 = x_2 x_1$	
Identities for variables	$x \oplus x = 0, \quad x \cdot x = x$ $x \oplus \overline{x} = 1, \quad x \cdot \overline{x} = 0$	
Identities for variables and constants	$x \oplus 0 = x, \quad x \cdot 0 = 0$ $x \oplus 1 = \overline{x}, \quad x \cdot 1 = x$	

TABLE 9.2
The GF(2) algebra and identities for manipulations (continuation of Table 9.1)

Techniques for computing the polynomial expressions

Rules and identities	Formal notation	Logic network
Identities for constants	$0 \oplus 0 = 0, \quad 0 \cdot 0 = 0$ $1 \oplus 1 = 0, \quad 1 \cdot 1 = 1$ $1 \oplus 0 = 1, \quad 1 \cdot 0 = 0$	
DeMorgan's rules for polynomials	$x_1 \oplus \overline{x}_2 = \overline{x_1 \oplus x_2}$ $\overline{x}_1 \oplus x_2 = \overline{x_1 \oplus x_2}$ $\overline{x}_1 \oplus \overline{x}_2 = x_1 \oplus x_2$	
Relationships with SOP form	$x_1 \oplus x_2 = x_1 \overline{x}_2 \vee \overline{x}_1 x_2$ $\overline{x_1 \oplus x_2} = x_1 x_2 \vee \overline{x}_1 \overline{x}_2$	
Simplification rules	$x_1 x_2 \oplus \overline{x}_1 x_2 = x_2$ $\overline{x}_1 x_2 \oplus x_2 = x_1 x_2$ $x_1 x_2 \oplus x_2 = \overline{x}_1 x_2$	

Equivalent symbols for EXOR gates

$$x_1 \oplus x_2 = \overline{\overline{x}_1 \oplus x_2} = \overline{x_1 \oplus \overline{x}_2} = \overline{x}_1 \oplus \overline{x}_2$$

$$\overline{x_1 \oplus x_2} = \overline{x}_1 \oplus x_2 = x_1 \oplus \overline{x}_2 = \overline{\overline{x}_1 \oplus \overline{x}_2}$$

FIGURE 9.7
Equivalent symbols for EXOR and XNOR gates.

Using the theorems and rules from Table 9.1, equivalent symbols for the two input EXOR and XNOR gates can be derived (Figure 9.7). These logic symbols follow the rule that any pair of signals (inputs or outputs) of an EXOR or XNOR gate can be complemented without changing the EXOR or XNOR function, respectively.

9.3.1 Operational and functional domains

Forward and *inverse* transforms describe the relationship between *operational* and *functional* domains for Boolean data structures (Figure 9.8):

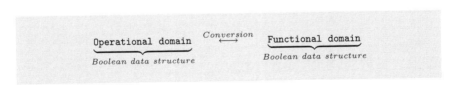

| Example 9.9 | *An SOP expression (operational domain) can be converted into polynomial form (functional domain) using algebraic, matrix, and cube forms.* |

The polarity of variables (complemented or uncomplemented) can be controlled for each transformation. Note that in SOP-based techniques, the polarities of variables can be changed using DeMorgan's rule. In the functional domain, this control can be implemented in matrix form.

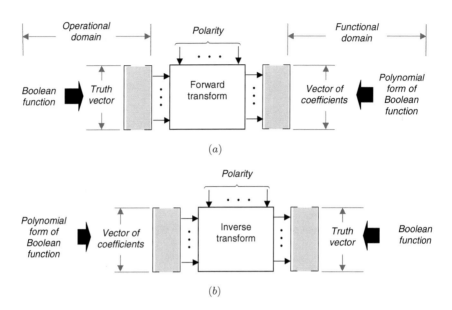

(a)

(b)

FIGURE 9.8
Forward and inverse transforms for the polynomial forms of Boolean functions.

9.3.2   The functional table

A Boolean function of n variables $f(x_i)$, $i = 1, 2, \ldots, n$, in the operational domain can be described in tabulated form using truth tables. In the functional domain, a Boolean function is represented in a polynomial form $f(x_i, a_j)$ $i, j \in \{1, 2, \ldots, n\}$, which is a function of variables x_i and coefficients a_j, and can be described by a *functional table*:

A functional table is a list of all combinations of 1s and 0s assigned to the binary coefficients $a_0, a_1, \ldots, a_{2^n-1}$ and corresponding polynomials. A functional table is characterized by the following properties:

▶ Each row corresponds to a combination of the 2^n coefficients $a_0, a_1, \ldots, a_{2^n-1}$ of the polynomial. The number of rows in the table is 2^{2^n}, where n is the number of variables of the Boolean function.

▶ The columns are the 2^{2^n} polynomial forms of the given Boolean functions. Each polynomial represents a Boolean function assuming various 2^n polarities of its n variables.

Example 9.10 *For a Boolean function f of two variables x_1 and x_2 ($n = 2$), the functional table over the field GF(2) is derived as follows. There are $2^n = 2^2 = 4$ coefficients a_0, a_1, a_2, and a_3, that specify $2^{2^n} = 2^{2^2} = 16$ polynomial forms of the Boolean function. Hence, the functional table contains 16 rows for all combinations of the coefficients a_0, a_1, a_2, a_3 and the column labeled f that contains 16 polynomials (Figure 9.9). For each polynomial in the functional table, a truth table can be derived. For example, the combination of coefficients $a_0 a_1 a_2 a_3 = 0011$ specifies the polynomial $x_1 \oplus x_1 x_2$.*

9.3.3   The functional map

The analog of the truth vector and K-map in the functional domain is the *vector of coefficients*. All 2^{2^n} possible vectors of coefficients for a Boolean function of n variables are represented by the *functional map*.

Example 9.11 *Let the Boolean function of three variables $f = \overline{x}_1 \overline{x}_2 \overline{x}_3 \vee \overline{x}_1 x_2 x_3$ be given in the form of a K-map (Figure 9.10). The functional map contains the coefficients of the polynomial in GF(2). The gates used for the implementation of the function in the operational domain are AND and OR, while the gates for polynomial implementation are AND and EXOR.*

Example 9.12 *In Figure 9.11, Boolean functions of two and three variables are represented in the operational and functional domains using a K-map and a functional map.*

9.3.4   Polarized minterms

The polynomial form of Equation 9.3 contains only uncomplemented variables. In order to achieve acceptable flexibility in a network design based on polynomial forms, so-called *polarized minterms* can be used. The polarized minterm is the product of *polarized literals*.

Techniques for computing
using functional and truth tables of Boolean functions

Functional table

a_0 a_1 a_2 a_3	Polynomial f		a_0 a_1 a_2 a_3	Polynomial f
0 0 0 0	0		1 0 0 0	1
0 0 0 1	$x_1 x_2$		1 0 0 1	$1 \oplus x_1 x_2$
0 0 1 0	x_1		1 0 1 0	$1 \oplus x_1$
0 0 1 1	$x_1 \oplus x_1 x_2$		1 0 1 1	$1 \oplus x_1 \oplus x_1 x_2$
0 1 0 0	x_2		1 1 0 0	$1 \oplus x_2$
0 1 0 1	$x_2 \oplus x_1 x_2$		1 1 0 1	$1 \oplus x_2 \oplus x_1 x_2$
0 1 1 0	$x_2 \oplus x_1$		1 1 1 0	$1 \oplus x_1 \oplus x_2$
0 1 1 1	$x_2 \oplus x_1 \oplus x_1 x_2$		1 1 1 1	$1 \oplus x_2 \oplus x_1 \oplus x_1 x_2$

Corresponding truth tables

$f = 0$

x_1 x_2	f
0 0	0
0 1	0
1 0	0
1 1	0

$f = x_1 x_2$

x_1 x_2	f
0 0	0
0 1	0
1 0	0
1 1	1

$f = x_1$

x_1 x_2	f
0 0	0
0 1	0
1 0	1
1 1	1

$f = x_1 \oplus x_1 x_2$
$= x_1 \overline{x}_2$

x_1 x_2	f
0 0	0
0 1	0
1 0	1
1 1	0

$f = x_2$

x_1 x_2	f
0 0	0
0 1	1
1 0	0
1 1	1

$f = x_2 \oplus x_1 x_2$
$= \overline{x}_1 x_2$

x_1 x_2	f
0 0	0
0 1	1
1 0	0
1 1	0

$f = x_2 \oplus x_1$
$= \overline{x}_1 x_2 \vee x_1 \overline{x}_2$

x_1 x_2	f
0 0	0
0 1	1
1 0	1
1 1	0

$f = x_2 \oplus x_1 \oplus x_1 x_2$
$= x_1 \vee x_2$

x_1 x_2	f
0 0	0
0 1	1
1 0	1
1 1	1

$f = 1$

x_1 x_2	f
0 0	1
0 1	1
1 0	1
1 1	1

$f = 1 \oplus x_1 x_2$
$= \overline{x_1 x_2}$

x_1 x_2	f
0 0	1
0 1	1
1 0	1
1 1	0

$f = 1 \oplus x_1$
$= \overline{x}_1$

x_1 x_2	f
0 0	1
0 1	1
1 0	0
1 1	0

$f = 1 \oplus x_1 \oplus x_1 x_2$
$= \overline{x_1 \overline{x}_2}$

x_1 x_2	f
0 0	1
0 1	1
1 0	0
1 1	1

$f = 1 \oplus x_2$
$= \overline{x}_2$

x_1 x_2	f
0 0	1
0 1	0
1 0	1
1 1	0

$f = 1 \oplus x_2 \oplus x_1 x_2$
$= \overline{\overline{x}_1 x_2}$

x_1 x_2	f
0 0	1
0 1	0
1 0	1
1 1	1

$f = 1 \oplus x_1 \oplus x_2$
$= \overline{x}_1 \overline{x}_2 \vee x_1 x_2$

x_1 x_2	f
0 0	1
0 1	0
1 0	0
1 1	1

$f = 1 \oplus x_2 \oplus x_1 \oplus x_1 x_2$
$= \overline{x}_1 \overline{x}_2$

x_1 x_2	f
0 0	1
0 1	0
1 0	0
1 1	0

FIGURE 9.9

Representation of Boolean functions of two variables in the form of a functional table (a) and the corresponding truth tables (b) (Example 9.10).

Design example: Operational and functional domains

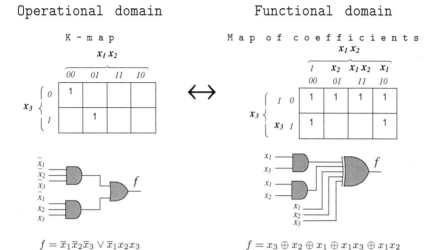

$$f = \overline{x}_1\overline{x}_2\overline{x}_3 \vee \overline{x}_1x_2x_3$$

$$f = x_3 \oplus x_2 \oplus x_1 \oplus x_1x_3 \oplus x_1x_2$$

FIGURE 9.10

Representation of the Boolean function $f = \overline{x}_1\overline{x}_2\overline{x}_3 \vee \overline{x}_1x_2x_3$ in the operational and functional domains (Example 9.11).

Literals and polarized literals

A literal is the representation of either an uncomplemented or a complemented variable:

$$\texttt{Literal} = x_j^{i_j} = \begin{cases} \overline{x}_j, & \text{if } i_j = 0 \\ x_j, & \text{if } i_j = 1 \end{cases} \qquad (9.5)$$

Literals in the form of Equation 9.5 are used in the standard SOP forms. The polarities of variables are specified by the particular Boolean functions. The polarity can be changed using DeMorgan's rule.

Example 9.13 *Let* $i = 0, 1, 2, 3$ $(i_1i_2 = 00, 01, 10, 11$ *for the representation of two Boolean variables). According to Equation 9.5, literals can be generated as follows:*

$$\{x_1^0x_2^0, \ x_1^0x_2^1, \ x_1^1x_2^0, \ x_1^1x_2^1\} = \{\overline{x}_1\overline{x}_2, \ \overline{x}_1x_2, \ x_1\overline{x}_2, \ x_1x_2\}$$

The *polarized* form of the literal provides an approach to *independent*

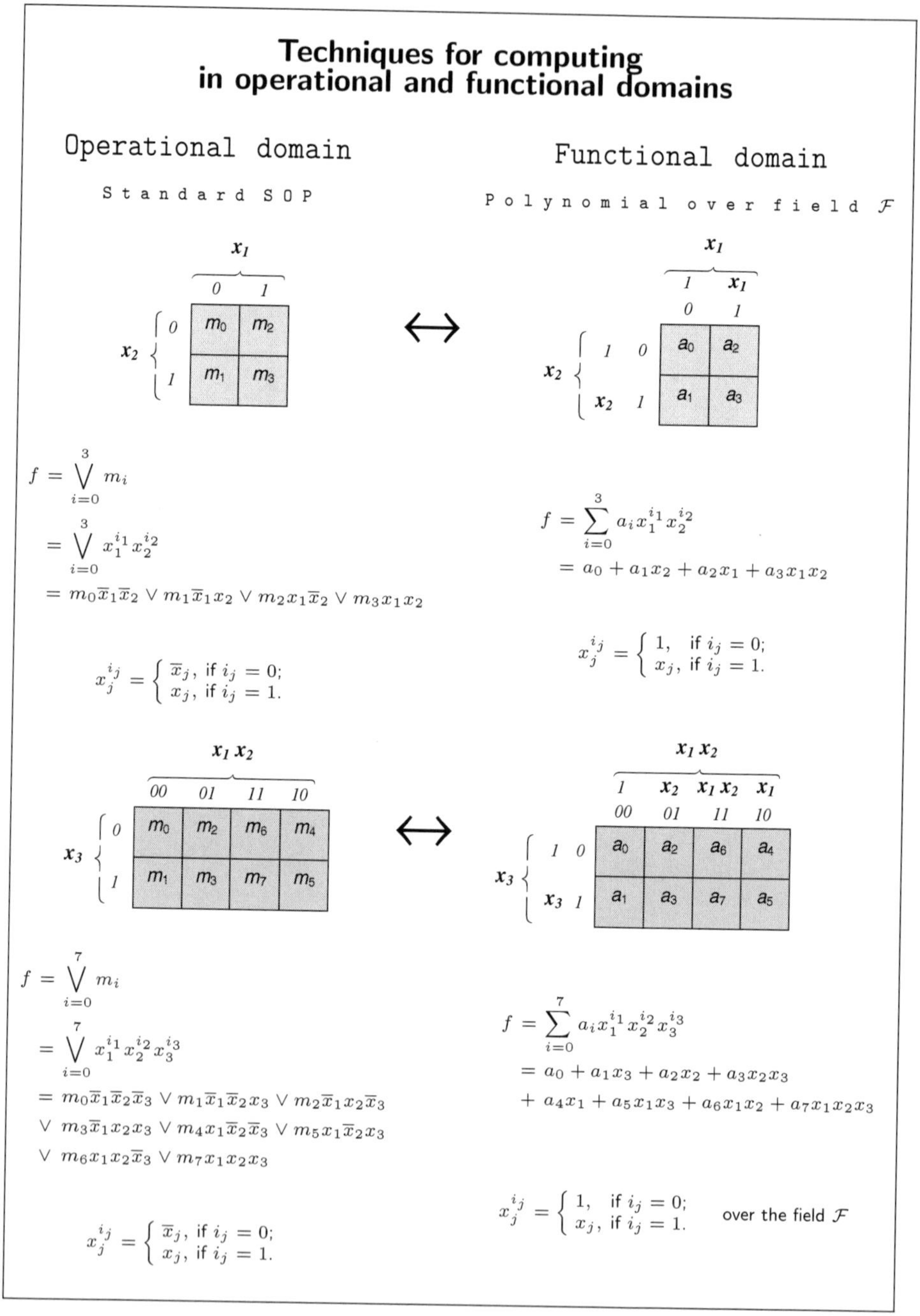

Techniques for computing in operational and functional domains

Operational domain

Standard SOP

x_1

	0	1
x_2 0	m_0	m_2
1	m_1	m_3

$\longleftrightarrow$

Functional domain

Polynomial over field $\mathcal{F}$

x_1

| | 1 | x_1 |
	0	1
x_2 0	a_0	a_2
1	a_1	a_3

$$f = \bigvee_{i=0}^{3} m_i$$

$$= \bigvee_{i=0}^{3} x_1^{i_1} x_2^{i_2}$$

$$= m_0 \overline{x}_1 \overline{x}_2 \vee m_1 \overline{x}_1 x_2 \vee m_2 x_1 \overline{x}_2 \vee m_3 x_1 x_2$$

$$x_j^{i_j} = \begin{cases} \overline{x}_j, & \text{if } i_j = 0; \\ x_j, & \text{if } i_j = 1. \end{cases}$$

$$f = \sum_{i=0}^{3} a_i x_1^{i_1} x_2^{i_2}$$

$$= a_0 + a_1 x_2 + a_2 x_1 + a_3 x_1 x_2$$

$$x_j^{i_j} = \begin{cases} 1, & \text{if } i_j = 0; \\ x_j, & \text{if } i_j = 1. \end{cases}$$

$x_1\,x_2$

	00	01	11	10
x_3 0	m_0	m_2	m_6	m_4
1	m_1	m_3	m_7	m_5

$\longleftrightarrow$

$x_1\,x_2$

| | 1 | x_2 | $x_1 x_2$ | x_1 |
	00	01	11	10
x_3 0	a_0	a_2	a_6	a_4
1	a_1	a_3	a_7	a_5

$$f = \bigvee_{i=0}^{7} m_i$$

$$= \bigvee_{i=0}^{7} x_1^{i_1} x_2^{i_2} x_3^{i_3}$$

$$= m_0 \overline{x}_1 \overline{x}_2 \overline{x}_3 \vee m_1 \overline{x}_1 \overline{x}_2 x_3 \vee m_2 \overline{x}_1 x_2 \overline{x}_3$$

$$\vee m_3 \overline{x}_1 x_2 x_3 \vee m_4 x_1 \overline{x}_2 \overline{x}_3 \vee m_5 x_1 \overline{x}_2 x_3$$

$$\vee m_6 x_1 x_2 \overline{x}_3 \vee m_7 x_1 x_2 x_3$$

$$x_j^{i_j} = \begin{cases} \overline{x}_j, & \text{if } i_j = 0; \\ x_j, & \text{if } i_j = 1. \end{cases}$$

$$f = \sum_{i=0}^{7} a_i x_1^{i_1} x_2^{i_2} x_3^{i_3}$$

$$= a_0 + a_1 x_3 + a_2 x_2 + a_3 x_2 x_3$$

$$+ a_4 x_1 + a_5 x_1 x_3 + a_6 x_1 x_2 + a_7 x_1 x_2 x_3$$

$$x_j^{i_j} = \begin{cases} 1, & \text{if } i_j = 0; \\ x_j, & \text{if } i_j = 1. \end{cases} \quad \text{over the field } \mathcal{F}$$

FIGURE 9.11

Techniques for the representation of Boolean functions in the operational and functional domains using K-maps and functional maps (Example 9.12).

control of the polarity of variables. A *polarized literal* is the representation of either an uncomplemented or a complemented variable specified by the control parameter called *polarity*, $c_1, c_2, \ldots, c_n$, $c_j \in \{0, 1\}$, $j = 1, 2, \ldots, n$:

$$
\begin{aligned}
\text{Polarized literal} &= (x_j \oplus c_j)^{i_j} \\
&= \begin{cases} 1, & \text{if } i_j = 0 \\ (x_j \oplus c_j), & \text{if } i_j = 1 \end{cases} \quad \text{over GF}(2)
\end{aligned} \tag{9.6}
$$

In Equation 9.6, the parameters i_j for the variable x_j are *separated* from the polarity c_j:

▶ Parameters i_j only specify the order of the minterms in the polynomial. They are an inherent property of a given form; that is, i_j are dependent parameters.

▶ Parameter c_j is an *independent* parameter.

The following example shows all possible combinations of the dependent and independent parameters of the literal.

Example 9.14 *For the polarity $c_j \in \{0, 1\}$ and parameter $i_j \in \{0, 1\}$, the complete set of polarized literals is generated as follows:*

$$
\text{Literal} \quad \overbrace{(x_j \oplus 0)^0}^{c_j = 0, \; i_j = 0} = x_j^0 = 1 \quad \text{over GF}(2)
$$

$$
\text{Literal} \quad \overbrace{(x_j \oplus 1)^0}^{c_j = 1, \; i_j = 0} = \overline{x}_j^0 = 1 \quad \text{over GF}(2)
$$

$$
\text{Literal} \quad \overbrace{(x_j \oplus 0)^1}^{c_j = 0, \; i_j = 1} = x_j^1 = x_j \quad \text{over GF}(2)
$$

$$
\text{Literal} \quad \overbrace{(x_j \oplus 1)^1}^{c_j = 1, \; i_j = 1} = \overline{x}_j^1 = \overline{x}_j \quad \text{over GF}(2)
$$

Minterm structure

The minterm is defined for the assignment $i_1, i_2, \ldots, i_n$ of Boolean variables $x_1, x_2, \ldots, x_n$ for which a Boolean function is equal to 1; that is, $x_1 = i_1, x_2 = i_2, \ldots, x_n = i_n$ if $f = 1$:

$$
\text{Minterm} = \overbrace{x_1^{i_1} x_2^{i_2} \cdots x_n^{i_n}}^{n \text{ literals}} \tag{9.7}
$$

where

$$x_j^{i_j} = \begin{cases} \overline{x}_j, & \text{if } i_j = 0 \\ x_j, & \text{if } i_j = 1 \end{cases} \tag{9.8}$$

These minterms are used in the standard SOP expressions. The simplest method for generating the minterms is to use the truth table of the Boolean function.

Polarized minterm structure

A *polarized minterm* is defined by the equation

Polarized minterm

n *polarized literals*

$$= \overbrace{(x_1 \oplus c_1)^{i_1} (x_2 \oplus c_2)^{i_2} \cdots (x_n \oplus c_n)^{i_n}}^{} \qquad \text{over GF(2)} \tag{9.9}$$

where

$$(x_j \oplus c_j)^{i_j} = \begin{cases} 1, & \text{if } i_j = 0 \\ (x_j \oplus c_j), & \text{if } i_j = 1 \end{cases} \qquad \text{over GF(2)}$$

In Equation 9.9, the polarities of the variables $x_1, x_2, \ldots, x_n$ are specified by the polarity parameters $c_1, c_2, \ldots, c_n$, respectively. An arbitrary polarity $c_i \in \{0, 1\}$ can be chosen for each Boolean variable x_i, $i = 1, 2, \ldots, n$.

Example 9.15   *Table 9.3 includes polarized minterms:*

Polarized minterm $= (x_1 \oplus c_1)^{i_1} (x_2 \oplus c_2)^{i_2}$ *over GF(2)*

For example, all four polarized minterms for the polarity $c = 2$ $(c_1 c_2 = 10)$ can be in four forms:

$$(x_1 \oplus 1)^0 (x_2 \oplus 0)^0 = 1; \quad (x_1 \oplus 1)^0 (x_2 \oplus 0)^1 = x_2;$$
$$(x_1 \oplus 1)^1 (x_2 \oplus 0)^0 = \overline{x}_1; \quad (x_1 \oplus 1)^1 (x_2 \oplus 0)^1 = \overline{x}_1 x_2$$

Figure 9.12 shows computing with the polarized minterms.

The polarity of each variable contributes to the polarity of a polynomial over the field GF(2).

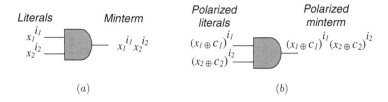

FIGURE 9.12
Computing of minterms (a) and polarized minterms (b) (Example 9.15).

TABLE 9.3
The polarized minterms $(x_1 \oplus c_1)^{i_1}(x_2 \oplus c_2)^{i_2}$ over GF(2) (Example 9.15)

Design example: polarized minterm computing

Polarity $c_1 c_2$	Polarized minterm $(x_1 \oplus c_1)^{i_1}(x_2 \oplus c_2)^{i_2}$	00	01	$i_1 i_2$ 10	11
00	$c_1=0,\ c_2=0$ $(x_1 \oplus 0)^{i_1}(x_2 \oplus 0)^{i_2}$	1	x_2	x_1	$x_1 x_2$
01	$c_1=0,\ c_2=1$ $(x_1 \oplus 0)^{i_1}(x_2 \oplus 1)^{i_2}$	1	$\overline{x}_2$	x_1	$x_1 \overline{x}_2$
10	$c_1=1,\ c_2=0$ $(x_1 \oplus 1)^{i_1}(x_2 \oplus 0)^{i_2}$	1	x_2	$\overline{x}_1$	$\overline{x}_1 x_2$
11	$c_1=1,\ c_2=1$ $(x_1 \oplus 1)^{i_1}(x_2 \oplus 1)^{i_2}$	1	$\overline{x}_2$	$\overline{x}_1$	$\overline{x}_1 \overline{x}_2$

9.4 Relationship between standard SOP and polynomial forms

Polynomial form over the field GF(2) of a Boolean function can be derived directly from the SOP expression of this function using the following algorithm:

Given: The standard SOP expression of a Boolean function
Find: The standard polynomial form
Procedure: Replace the OR operations with EXOR operation
Result: The standard polynomial form of the Boolean function

This procedure results in a nonoptimal polynomial representation, and additional manipulations are needed for its simplification.

Example 9.16   *The truth table of Boolean function is given in Table 9.4. A standard SOP of this function is $f = \overline{x}_1\overline{x}_2\overline{x}_3 \vee \overline{x}_1\overline{x}_2x_3 \vee \overline{x}_1x_2x_3 \vee x_1x_2x_3$. The polynomial expression is derived by replacing OR operations by the EXOR operations: $f = \overline{x}_1\overline{x}_2\overline{x}_3 \oplus \overline{x}_1\overline{x}_2x_3 \oplus \overline{x}_1x_2x_3 \oplus x_1x_2x_3$. Note that this polynomial expression consists of a minterms with variables of different polarities.*

TABLE 9.4
Relationship between a standard SOP and polynomial forms (Example 9.16)

Assignment of variables	Value of Boolean function	Minterm
000	1	$\overline{x}_1\overline{x}_2\overline{x}_3$
001	1	$\overline{x}_1\overline{x}_2x_3$
010	0	
011	1	$\overline{x}_1x_2x_3$
100	0	
101	0	
110	0	
111	1	$x_1x_2x_3$

9.5   Local transformations for EXOR expressions

In Chapter 4, a *local transformation* is defined as a set of rules for the simplification of a data structure. In this section, we consider a local transformation for a logic network, which consist of various types of logic gates, including EXOR gates. These transformations are based on the theorems of Boolean algebra and polynomial algebra GF(2), and are applied locally.

Local transformations for the logic networks using AND and OR gates and inverters were described in Chapter 4. The following rules can be applied to logic networks with EXOR gates.

Rule 1: *Replacing an EXOR gate with a constant*

> ▶ Replace an EXOR gate with a corresponding constant using the rules of identities for constants if the inputs of this gates are constants.
> ▶ Replace an EXOR gate with a corresponding constant using the rules of identities for variables if the inputs of this gates are literals of the same variable.

Rule 2: *Replacing an EXOR gate with a variable*

> ▶ An EXOR gate can be replaced with a variable using the rules of identity for variables and constants if one of the inputs is a constant

The rules for removing the duplicated gates, removing the unused gates, and merging the gates are similar to the ones for OR and AND gates.

Example 9.17 *Figure 9.13 illustrates two types of the local transformations:*

Local transformation in area A: the EXOR gate is replaced with the inverter using the identity rule for variables and constants (Table 9.1) $x_2 \oplus 1 = \overline{x}_2$.

Local transformation in area B: the inverter and EXOR gates are replaced by the wire using the simplification rule (Table 9.2) $x_1 \overline{x}_2 \oplus x_1 = x_1 (\overline{x}_2 \oplus 1) = x_1 x_2$.

9.6 Factorization of polynomials

Factoring of polynomial expressions is used, in particular, for dealing with the fan-in problem, and also if a logic network is designed using limited numbers of gate inputs. However, techniques for factoring SOP forms are not acceptable for polynomial forms. Factorization of polynomial expressions is based on the laws and identities given in Table 9.1. Similarly to SOP-based techniques, the factoring of polynomial expressions is based on the designer's experience and may be built into CAD tools to a limited extent. The application of various identities does not guarantee satisfactory results from factoring. In particular, an arbitrary Boolean variable can be replaced by its complement as follows:

$$x_1 \oplus x_2 = \overline{\overline{x}_1 \oplus x_2} = \overline{x_1 \oplus \overline{x}_2} = (1 \oplus x_1)x_2 \oplus 1 = x_1 \oplus (1 \oplus x_2) \oplus 1$$

Design example: local transformations

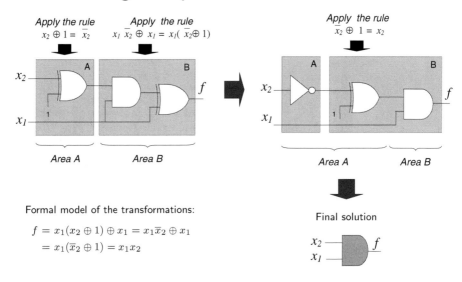

FIGURE 9.13
Local transformations in logic network (Example 9.17).

Extra variables can be included in an equation using the following properties: $x \oplus x \oplus x = x$ and $x \oplus x = 0$.

Example 9.18 *The polynomial expression*

$$f = \underbrace{1 \oplus x_4 \oplus x_3 \oplus x_2 \oplus x_2 x_3 \oplus x_1 x_3 \oplus x_1 x_2 \oplus x_1 x_2 x_3}_{\text{2-level logic network}}$$

can be directly implemented by the two-level logic network as shown in Figure 9.14a. The fan-in of the EXOR gate is equal to 7, and it is often not acceptable. Factoring results in the expression $f = \underbrace{1 \oplus x_4 \oplus (x_3 \oplus x_2 \oplus x_2 x_3) x_1}_{\text{4-level logic network}}$. *This polynomial expression is implemented by a four-level logic network (Figure 9.14b) using three-input EXOR gates.*

Design example : factoring

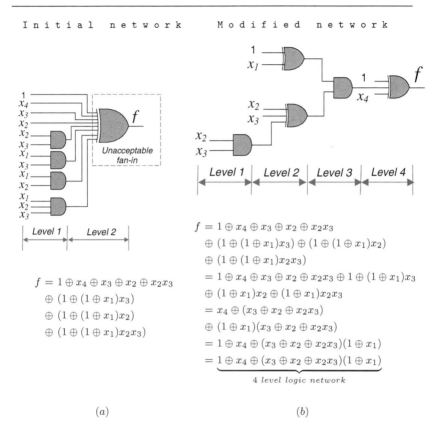

Initial network

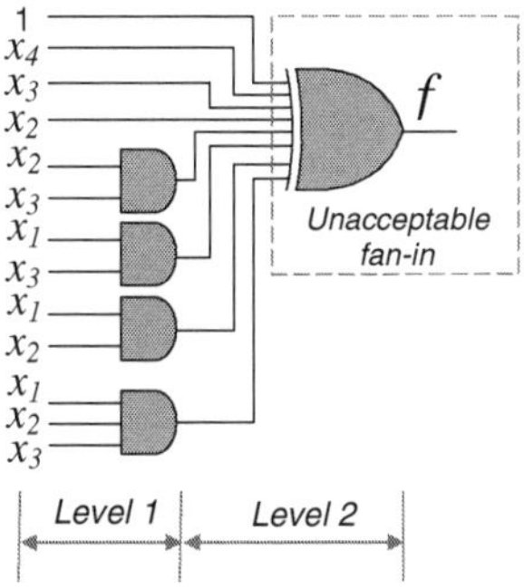

$$f = 1 \oplus x_4 \oplus x_3 \oplus x_2 \oplus x_2x_3 \oplus (1 \oplus (1 \oplus x_1)x_3) \oplus (1 \oplus (1 \oplus x_1)x_2) \oplus (1 \oplus (1 \oplus x_1)x_2x_3)$$

(a)

Modified network

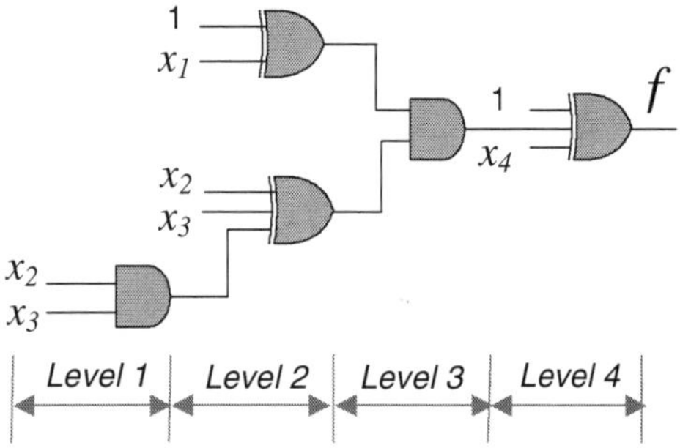

$$\begin{aligned} f &= 1 \oplus x_4 \oplus x_3 \oplus x_2 \oplus x_2x_3 \\ &\oplus (1 \oplus (1 \oplus x_1)x_3) \oplus (1 \oplus (1 \oplus x_1)x_2) \\ &\oplus (1 \oplus (1 \oplus x_1)x_2x_3) \\ &= 1 \oplus x_4 \oplus x_3 \oplus x_2 \oplus x_2x_3 \oplus 1 \oplus (1 \oplus x_1)x_3 \\ &\oplus (1 \oplus x_1)x_2 \oplus (1 \oplus x_1)x_2x_3 \\ &= x_4 \oplus (x_3 \oplus x_2 \oplus x_2x_3) \\ &\oplus (1 \oplus x_1)(x_3 \oplus x_2 \oplus x_2x_3) \\ &= 1 \oplus x_4 \oplus (x_3 \oplus x_2 \oplus x_2x_3)(1 \oplus x_1) \\ &= \underbrace{1 \oplus x_4 \oplus (x_3 \oplus x_2 \oplus x_2x_3)(1 \oplus x_1)}_{4\ level\ logic\ network} \end{aligned}$$

(b)

FIGURE 9.14
Two-level logic network implementation of a nonfactored polynomial expression (a) and four-level logic network implementation of a factored polynomial expression (Example 9.18).

9.7 Validity check for EXOR networks

Similarly to proving validity in SOP forms (Chapter 4), two Boolean functions in polynomial form can be said to be equivalent or nonequivalent. This check is based on algebra in GF(2).

The validity of two logic networks is referred to as an equivalence proof. The algorithm for the equivalence proof is given as follows:

Given: Two logic networks
Prove: The equivalence of these networks
Step 1: Describe the outputs for each network
Step 2: Form the logic equation for the outputs
Step 3: Prove this logic equation

Proving the validity of Boolean expressions that combine SOP and polynomial terms is performed by transforming both sides of the equation to one domain (operational or functional).

Example 9.19   *The equivalence of the logic networks given in Figure 9.15 can be determined by proving the validity of the Boolean equations:* $\underbrace{x_1 \oplus x_2}_{Left\ network}$ $= \underbrace{(x_1 \vee x_2) \oplus x_1 x_2}_{Right\ network}$.

Since $x_1 \vee x_2 = x_1 \oplus x_2 \oplus x_1 x_2$, *then* $x_1 \oplus x_2 = (x_1 \oplus x_2 \oplus x_1 x_2) \oplus x_1 x_2$ *and* $x_1 \oplus x_2 = x_1 \oplus x_2$.

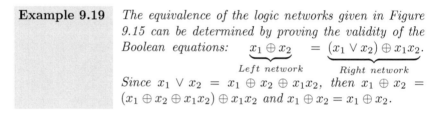

FIGURE 9.15
Proving the validity of two networks using manipulation of the Boolean functions in polynomial form (Example 9.19).

9.8   Fixed- and mixed-polarity polynomial forms

In terms of polynomial forms, two types of polarity are distinguished: *fixed* polarity and *mixed* polarity. In a fixed-polarity polynomial expression of a Boolean function f, every variable appears either complemented ($\overline{x}_i$) or uncomplemented (x_i), and never in both forms. There are 2^n fixed-polarity forms given a function of n variables. In a mixed-polarity form, a variable can appear in one or both polarities. There are 3^n mixed-polarity forms given a function f of n variables.

Example 9.20 *In Figure 9.16, the polynomial expressions*

$$\overbrace{\overline{x}_1 x_2 \overline{x}_3 \oplus \overline{x}_1 x_2 \oplus \overline{x}_1 \overline{x}_3}^{\textit{Fixed polarities of variables}} \text{ and } \overbrace{\overline{x}_1 x_2 \overline{x}_3 \oplus x_1 \overline{x}_2 \oplus \overline{x}_1 x_3}^{\textit{Mixed polarities of variables}}$$

are a fixed- and a mixed-polarity form, respectively.

Techniques for computing
the fixed- and mixed-polarity polynomial forms

Fixed-polarity polynomial expressions

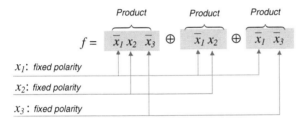

The variable x_2 appears uncomplemented; the variables x_1 and x_3 appear complemented only

Mixed-polarity polynomial expressions

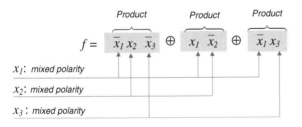

The variables $x_1, x_2,$ and x_3 appear in both uncomplemented and complemented forms

FIGURE 9.16
Illustration of fixed- and mixed-polarity polynomial forms (Example 9.20).

9.8.1 Fixed-polarity polynomial forms

A fixed-polarity polynomial expression of a Boolean function f of n variables is *unique*; that is, only one representation exists given a polarity $c\,(c_1, c_2, \ldots, c_n)$.

There are 2^n various polarities and two boundary cases among them: a *positive polarity* form in which all variables are uncomplemented, and a *negative polarity* form in which all variables are complemented (Figure 9.17).

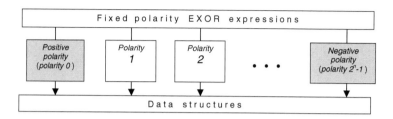

FIGURE 9.17
An arbitrary Boolean function can be represented in fixed-polarity polynomial form.

Example 9.21 *Polynomial expressions in positive polarity $c = 0$ ($c_1 c_2 = 00$) (uncomplemented variables) and the polarity $c = 2$, $c_1 c_2 = 10$ (only the variable x_1 is complemented) are as follows:*

$$\textbf{Polarity } c = 0 \ (c_1 = 0, c_2 = 0):$$

$$f = r_0(x_1 \oplus c_1)^{i_1}(x_2 \oplus c_2)^{i_2} \oplus r_1(x_1 \oplus c_1)^{i_1}(x_2 \oplus c_2)^{i_2}$$
$$\oplus r_2(x_1 \oplus c_1)^{i_1}(x_2 \oplus c_2)^{i_2} \oplus r_3(x_1 \oplus c_1)^{i_1}(x_2 \oplus c_2)^{i_2}$$
$$= r_0(x_1^0 x_2^0) \oplus r_1(x_1^0 x_2^1) \oplus r_2(x_1^1 x_2^0) \oplus r_3(x_1^1 x_2^1)$$
$$= r_0 \oplus r_1 x_2 \oplus r_2 x_1 \oplus r_3 x_1 x_2$$

$$\textbf{Polarity } c = 2 \ (c_1 = 1, c_2 = 0):$$

$$f = r_0(x_1 \oplus 1)^0(x_2 \oplus 0)^0 \oplus r_1(x_1 \oplus 1)^0(x_2 \oplus 0)^1$$
$$\oplus r_2(x_1 \oplus 1)^1(x_2 \oplus 0)^0 \oplus r_3(x_1 \oplus 1)^1(x_2 \oplus 0)^1$$
$$= r_0 \oplus r_1 x_2 \oplus r_2 \overline{x}_1 \oplus r_3 \overline{x}_1 x_2.$$

Let $f = x \vee y$. Then four fixed polarity polynomial expressions can be derived as shown in Figure 9.18.

Design example: polynomial forms of the OR gate

$f = x_1 \vee x_2$

0-polarity: $f = x_1 \oplus x_2 \oplus x_1 x_2$,	no complemented variables
1-polarity: $f = 1 \oplus \overline{x}_2 \oplus x_1 \overline{x}_2$,	x_2 is complemented
2-polarity: $f = 1 \oplus \overline{x}_1 \oplus \overline{x}_1 x_2$,	x_1 is complemented
3-polarity: $f = 1 \oplus \overline{x}_1 \overline{x}_2$,	x_1 and x_2 are complemented

FIGURE 9.18
Representation of the two-input OR gate by polynomial forms of $2^2 = 4$ different fixed polarities (Example 9.21).

Deriving a fixed polarity polynomial expansion given a Boolean function is a necessary step in the process of implementation of the function given a library of logic gates AND and EXOR together with a constant 1 signal. It forms a universal basis of operations for implementing an arbitrary Boolean function.

Example 9.22 *In Table 16.3, the expressions of the AND, OR, and EXOR Boolean functions of two variables are given in fixed polarities. For example, the polynomial in polarity $c = 3$ ($c_1 c_2 = 11$) for the OR function is represented by two nonzero coefficients $f = 1 \oplus \overline{x}_1 \overline{x}_2$. This is an optimal polynomial form of the OR function with respect to the criterion of the minimal number of literals.*

TABLE 9.5
Polynomial expressions of fixed polarities of elementary Boolean functions (Example 16.9)

Design example:
fixed polarity polynomial forms of the gates

	Coefficients				Polynomial expressions
	r_0	r_1	r_2	r_3	
x_1 —[ ]— f x_2 — $f = x_1 x_2$	0	0	0	1	$x_1 x_2$
	0	0	1	1	$x_1 \oplus x_1 \overline{x}_2$
	0	1	0	1	$x_2 \oplus \overline{x}_1 x_2$
	1	1	1	1	$1 \oplus \overline{x}_2 \oplus \overline{x}_1 \oplus \overline{x}_1 \overline{x}_2$
	r_0	r_1	r_2	r_3	
x_1 —[ ]— f x_2 — $f = x_1 \vee x_2$	0	1	1	1	$x_2 \oplus x_1 \oplus x_1 x_2$
	1	1	0	1	$1 \oplus \overline{x}_2 \oplus x_1 \overline{x}_2$
	1	0	1	1	$1 \oplus \overline{x}_1 \oplus \overline{x}_1 x_2$
	1	0	0	1	$1 \oplus \overline{x}_1 \overline{x}_2$
	r_0	r_1	r_2	r_3	
x_1 —[ ]— f x_2 — $f = x_1 \oplus x_2$	0	1	1	0	$x_2 \oplus x_1$
	1	1	1	0	$1 \oplus \overline{x}_2 \oplus x_1$
	1	1	1	0	$1 \oplus x_2 \oplus \overline{x}_1$
	0	1	1	0	$\overline{x}_2 \oplus \overline{x}_1$

9.8.2   Deriving polynomial expressions from SOP forms

Polynomial expressions can be derived from SOP forms using algebraic transformations. Given a canonical, or standard, SOP form, a polynomial form of mixed polarity is derived by replacing (a) the logical sum "$\vee$" with the EXOR operation "$\oplus$", and (b) the complement of a variable x with the $1 \oplus x$.

> **Example 9.23**   *A **0-polarity** polynomial form of the function $f = x_1 \vee x_2$ is derived as follows: $x_1 \vee x_2 = \overline{\overline{x_1 \vee x_2}} = \overline{\overline{x_1}\,\overline{x_2}} = (1 \oplus x_1)(1 \oplus x_2) \oplus 1 = x_1 \oplus x_2 \oplus x_1 x_2.$*

9.8.3   Conversion between polarities

Given one polarity of a polynomial expression of a Boolean function, one can convert it to another polarity expression by algebraic manipulations.

> **Example 9.24**   *A **mixed-polarity** polynomial expression can be transformed into a polynomial form of polarity $c = 2$ $(c_1 = 1, c_2 = 0)$:*
>
> $$f = \overbrace{\overline{x_1}\,\overline{x_2} \oplus \overline{x_1} x_2 \oplus x_1 x_2}^{Mixed\ polarity} = \overline{x_1}(1 \oplus x_2) \oplus \overline{x_1} x_2 \oplus \overbrace{(x_1 \oplus 1 \oplus 1)}^{\overline{x_1}} x_2$$
> $$= \overline{x_1} \oplus \overline{x_1} x_2 \oplus \overline{x_1} x_2 \oplus \overline{x_1} x_2 \oplus x_2 = \overline{x_1} \oplus \overline{x_1} x_2 \oplus x_2$$
>
> *Figure 9.19 illustrates the implementation of the initial SOP form (AND–OR logic network), the conversion of a mixed-polarity polynomial form (AND–EXOR logic network), and the fixed-polarity polynomial form, given $c = 2$. The polynomial form of the fixed-polarity $c = 1(c_1 = 0, c_2 = 1)$ is obtained as follows:*
>
> $$f = \overline{x_1}\,\overline{x_2} \oplus \overline{x_1} x_2 \oplus x_1 x_2$$
> $$= (x_1 \oplus 1)\overline{x_2} \oplus (x_1 \oplus 1)(\overline{x_2} \oplus 1) \oplus x_1(\overline{x_2} \oplus 1)$$
> $$= x_1 \overline{x_2} \oplus \overline{x_2} \oplus x_1 \overline{x_2} \oplus x_1 \oplus \overline{x_2} \oplus 1 \oplus x_1 \overline{x_2} \oplus x_1 = x_1 \overline{x_2} \oplus 1.$$

9.8.4   Deriving polynomial expressions from K-maps

K-maps are known as a graphical form for representing Boolean functions. Since canonical SOP forms can be derived directly from K-maps, the latter can be easily converted to mixed-polarity polynomial forms, which can be further simplified to EXOR expressions of other polarities:

Design example: converting logic network

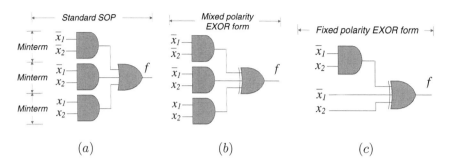

(a)   (b)   (c)

FIGURE 9.19
Interpretation of converting an SOP form into a mixed-polarity and then into
a fixed-polarity expression (Example 9.24).

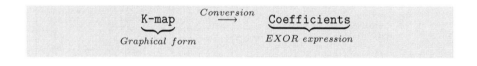

Example 9.25   *In Figure 9.20, mixed-polarity polynomial expressions
are derived from K-maps for two Boolean functions
of two and three variables. This is accomplished
by deriving the canonical SOP forms first and then
replacing "$\vee$" with "$\oplus$". The obtained polynomial
expressions of mixed polarity are not minimal and can
be further simplified to the expressions $x_1 \oplus x_2$ and
$x_1 \oplus x_2 \oplus x_3$, respectively.*

It should be noted that K-maps cannot be used for the minimization of
polynomial forms of Boolean functions. Special techniques are required for
minimizing polynomial expressions.

9.8.5   Simplification of polynomial expressions

Given a mixed-polarity polynomial form derived from a canonical SOP, further
algebraic transformations can lead to a minimized polynomial representation.
Minimization on a K-map cannot be directly applied to polynomial forms since
the rules for reducing are different in GF(2). Some rules are given below:

$$x \oplus \overline{x} = 1 \qquad x \oplus x = 0$$
$$x \oplus 1 = \overline{x} \qquad x \oplus 0 = x$$

Design example: deriving polynomial forms

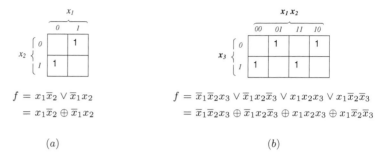

$f = x_1\overline{x}_2 \vee \overline{x}_1 x_2$

$\quad = x_1\overline{x}_2 \oplus \overline{x}_1 x_2$

$f = \overline{x}_1\overline{x}_2 x_3 \vee \overline{x}_1 x_2\overline{x}_3 \vee x_1 x_2 x_3 \vee x_1\overline{x}_2\overline{x}_3$

$\quad = \overline{x}_1\overline{x}_2 x_3 \oplus \overline{x}_1 x_2\overline{x}_3 \oplus x_1 x_2 x_3 \oplus x_1\overline{x}_2\overline{x}_3$

(a)   (b)

FIGURE 9.20
Deriving polynomial expressions from K-maps for Boolean functions of two variables (a) and three variables (b) (Example 9.25).

Since the EXOR operation is commutative and associative, the following rules hold true as well: $x_1 x_2 \oplus \overline{x}_1 x_2 = x_2, x_1\overline{x}_2 \oplus x_1 = x_1\overline{x}_2$ and $(x_1 \oplus x_2)x_1 = x_1\overline{x}_2$. The details of such minimizations are not the subject of this book. Exhaustive search techniques, that is, the generation of all possible mixed-polarity forms, can be applied to find the minimal one. Obviously, such techniques can only be applied to small functions, and for larger functions some heuristic approaches have been developed.

Example 9.26   *The expression $\overline{x}_1 \oplus \overline{x}_1 x_2 \oplus x_2$ obtained in Example 9.24 can be further simplified as follows:*

$$f = \overline{x}_1 \oplus \overline{x}_1 x_2 \oplus x_2 = \overline{x}_1(1 \oplus x_2) \oplus x_2$$
$$= (\overline{x}_1 \oplus x_2)(\overline{x}_2 \oplus x_2) = \overline{x}_1 \oplus x_2$$

The resulting expression is a minimal one.

9.9   Computing the coefficients of polynomial forms

The coefficients of polynomial forms can be calculated using matrix representations. In this section, the basics of matrix computations for deriving vectors of coefficients of fixed-polarity polynomial expressions are given. Matrix-based computations are well suited, in particular, for programming and fast transforms.

9.9.1 Matrix operations over GF(2)

In matrix operations in $GF(2)$, *binary* matrices consisting of 0s and 1s, are used.

Example 9.27 *A rectangular array of binary numbers,* $\mathbf{A} = \begin{bmatrix} 1 & 0 \\ 1 & 1 \end{bmatrix}$ *is a binary matrix with two columns and two rows. We thus say that* $\mathbf{A}$ *is a* 2×2 *(two by two) matrix. The first and second rows of* $\mathbf{A}$ *are* $[1\ 0]$ *and* $[1\ 1]$*, respectively.*

A matrix is a *square* matrix if it has the same number of rows and columns. A square matrix may be multiplied by itself. The *identity* matrix $\mathbf{I}$ is a matrix that has 1s along the diagonal and 0s everywhere else. If a matrix $\mathbf{A}$ has an inverse, then that inverse is denoted by $\mathbf{A}^{-1}$, and $\mathbf{A}\mathbf{A}^{-1} = \mathbf{A}^{-1}\mathbf{A} = \mathbf{I}$. The following matrix operations are used for deriving vectors representing the polynomial forms of Boolean functions:

Addition over GF(2): If a_{ij} and b_{ij} are the elements in the i-th row and j-th column of the binary matrices $\mathbf{A}$ and $\mathbf{B}$, respectively, then the matrix $\mathbf{C} = \mathbf{A} \oplus \mathbf{B}$ has elements $c_{ij} = a_{ij} \oplus b_{ij}$.

Multiplication over GF(2): If $a_{ik} \in \{0,1\}$ is the element in the i-th row and k-th column of a $2^n \times 2^m$ matrix $\mathbf{A}$, and $b_{kj} \in \{0,1\}$ is the element in the k-th row and j-th column of a $2^n \times 2^m$ matrix $\mathbf{B}$, then the $2^n \times 2^m$ matrix $\mathbf{C} = \mathbf{A}\mathbf{B}$ has elements $c_{ij} \in \{0,1\}$, $c_{ij} = \bigoplus_{k=1}^{2^n} a_{ik} b_{kj}$.

Inversion: The inverse of a $2^n \times 2^n$ matrix $\mathbf{A}$ is denoted as $\widehat{\mathbf{A}}$ or $\mathbf{A}^{-1}$ and satisfies $\mathbf{A}\mathbf{A}^{-1} = \mathbf{A}^{-1}\mathbf{A} = \mathbf{I}$, where $\mathbf{I}$ is the $2^n \times 2^n$ *identity* matrix containing unity entries on the diagonal and zeros elsewhere.

Example 9.28 *An example of multiplication of a binary matrix by a binary vector over GF(2) is given in Figure 9.21.*

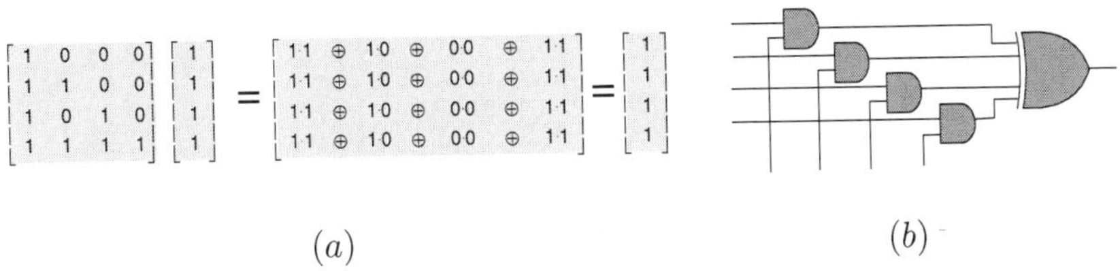

$$\begin{bmatrix} 1 & 0 & 0 & 0 \\ 1 & 1 & 0 & 0 \\ 1 & 0 & 1 & 0 \\ 1 & 1 & 1 & 1 \end{bmatrix} \begin{bmatrix} 1 \\ 1 \\ 1 \\ 1 \end{bmatrix} = \begin{bmatrix} 11 \oplus 10 \oplus 00 \oplus 11 \\ 11 \oplus 10 \oplus 00 \oplus 11 \\ 11 \oplus 10 \oplus 00 \oplus 11 \\ 11 \oplus 10 \oplus 00 \oplus 11 \end{bmatrix} = \begin{bmatrix} 1 \\ 1 \\ 1 \\ 1 \end{bmatrix}$$

(a)

(b)

FIGURE 9.21
Multiplication of a binary matrix by a binary vector over GF(2) (a) and corresponding logic network (b) (Example 9.28).

Example 9.29   *The following multiplications of $2^1 \times 2^1$ matrices result in identity matrices:*

$$(a) \quad \mathbf{A}\widehat{\mathbf{A}} = \begin{bmatrix} 1\ 0 \\ 1\ 1 \end{bmatrix} \begin{bmatrix} 1\ 0 \\ 1\ 1 \end{bmatrix} = \begin{bmatrix} 1\ 0 \\ 0\ 1 \end{bmatrix} = \mathbf{I} \quad over\ GF(2)$$

$$(b) \quad \mathbf{B}\widehat{\mathbf{B}} = \begin{bmatrix} 0\ 1 \\ 1\ 1 \end{bmatrix} \begin{bmatrix} 1\ 1 \\ 1\ 0 \end{bmatrix} = \begin{bmatrix} 1\ 0 \\ 0\ 1 \end{bmatrix} = \mathbf{I} \quad over\ GF(2)$$

$$(c) \quad \mathbf{C}\widehat{\mathbf{C}} = \begin{bmatrix} 1\ 0 \\ -1\ 1 \end{bmatrix} \begin{bmatrix} 1\ 0 \\ 1\ 1 \end{bmatrix} = \begin{bmatrix} 1\ 0 \\ 0\ 1 \end{bmatrix} = \mathbf{I}$$

$$(d) \quad \mathbf{D}\widehat{\mathbf{D}} = \begin{bmatrix} 0\ \ \ 1 \\ 1\ -1 \end{bmatrix} \begin{bmatrix} 1\ 1 \\ 1\ 0 \end{bmatrix} = \begin{bmatrix} 1\ 0 \\ 0\ 1 \end{bmatrix} = \mathbf{I}$$

Cases (a) and (b) concern operations over GF(2); in cases (c) and (d), identity matrices are obtained by operations over the field of integer numbers.

9.9.2   Polarized literals and minterms in matrix form

Polarized minterms are the basic components of the polynomial forms. Each polarized minterm is formed from n polarized literals, where n is the number of variables in a Boolean function.

In order to describe the polarized minterm in matrix form, its algebraic description must be converted into a description in matrix terms:

$$\underbrace{\text{Polarized literal}}_{Algebraic\ form} \overset{Conversion}{\longleftrightarrow} \underbrace{\text{Polarized literal}}_{Matrix\ form}$$

The function of the variable x_j, its polarity parameter c_j, and the parameter i_j that indicates the "presence" or "absence" of this variable in a logic expression is called the *polarized literal*; its algebraic form is as follows:

$$\text{Polarized literal} = (x_j \oplus c_j)^{i_j} = \begin{cases} 1, & \text{if } i_j = 0 \\ \\ (x_j \oplus c_j), & \text{if } i_j = 1 \end{cases} \quad \text{over GF(2)}$$

The matrix form of a polarized literal is based on the assumption that

▶ All operations are performed over GF(2)
▶ Multiplication of the *elementary* transform matrix and the truth vector of the variable x_j results in a vector of coefficients that corresponds to the simplest polynomial expressions, literals $x_j \oplus 0$ or $x_j \oplus 1$.

The elementary $2^1 \times 2^1$ transform matrix is denoted as $\mathbf{R}_{2^1}^{(c_j)}$ where $c_j \in \{0,1\}$ is the polarity of the variable x_j. The polarized literal corresponds to the elementary transform matrix for $c_j = 0$ and $c_j = 1$ as follows:

$$\text{Polarized literal} = \mathbf{R}_{2^1}^{(c_j)}$$

$$= \begin{cases} \mathbf{R}_{2^1}^{(0)} = \begin{bmatrix} 1 & 0 \\ 1 & 1 \end{bmatrix}, & \text{if } c_j = 0 \\ \\ \mathbf{R}_{2^1}^{(1)} = \begin{bmatrix} 1 & 0 \\ 1 & 1 \end{bmatrix}, & \text{if } c_j = 1 \end{cases} \quad \text{over GF(2)} \quad (9.10)$$

Computing the polarized literal means multiplying the elementary transform matrix $\mathbf{R}_{2^1}^{(c_j)}$ by the truth vector of a single variable x_j, $\mathbf{F} = [0\ 1]^T$.

Example 9.30   *In Figure 9.22, the polarized literal is computed by the multiplication of the elementary matrix and the truth vector for the variable x_j, $\mathbf{F} = \begin{bmatrix} 0 \\ 1 \end{bmatrix}$, over GF(2):*

$$\mathbf{R}_{2^1}^{(0)}\mathbf{F} = \begin{bmatrix} 1 & 0 \\ 1 & 1 \end{bmatrix} \begin{bmatrix} 0 \\ 1 \end{bmatrix} = \begin{bmatrix} 0 \\ 1 \end{bmatrix} \longrightarrow r_0 \oplus r_1 x_j = x_j$$

$$\mathbf{R}_{2^1}^{(1)}\mathbf{F} = \begin{bmatrix} 0 & 1 \\ 1 & 1 \end{bmatrix} \begin{bmatrix} 0 \\ 1 \end{bmatrix} = \begin{bmatrix} 1 \\ 1 \end{bmatrix} \longrightarrow r_0 \oplus r_1 x_j = 1 \oplus x_j$$

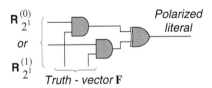

FIGURE 9.22
Computing a polarized literal in matrix form (Example 9.30).

Polarized minterms in matrix form

A polarized minterm can be described in a matrix equation as a conversion of its algebraic form into matrix form

$$
\underbrace{\text{Polarized minterm}}_{\text{Algebraic form}} \quad \overset{\textit{Conversion}}{\longleftrightarrow} \quad \underbrace{\text{Polarized minterm}}_{\text{Matrix form}}
$$

For this, a matrix description of polarized literals is used (Equation 9.10), that is, n polarized literals in algebraic form are replaced by their n matrix equivalents. In forming vectors and matrices of sizes $2^n \times 1$ and $2^n \times 2^n$, respectively, the operation of the Kronecker product must be used. Recall that the Kronecker product of the matrices

$$
\mathbf{A}_2 = \begin{bmatrix} a_{11} & a_{12} \\ a_{21} & a_{22} \end{bmatrix} \quad \text{and} \quad \mathbf{B}_2 = \begin{bmatrix} b_{11} & b_{12} \\ b_{21} & b_{22} \end{bmatrix}
$$

results in the 4×4 matrix

$$
\mathbf{A}_2 \otimes \mathbf{B}_2 = \begin{bmatrix} a_{11}\mathbf{B}_2 & a_{12}\mathbf{B}_2 \\ a_{21}\mathbf{B}_2 & a_{22}\mathbf{B}_2, \end{bmatrix} = \begin{bmatrix} a_{11}b_{11} & a_{11}b_{12} & a_{12}b_{13} & a_{12}b_{14} \\ a_{11}b_{21} & a_{11}b_{22} & a_{12}b_{23} & a_{12}b_{24} \\ a_{21}b_{31} & a_{21}b_{32} & a_{22}b_{33} & a_{22}b_{34} \\ a_{21}b_{41} & a_{21}b_{42} & a_{22}b_{43} & a_{22}b_{44} \end{bmatrix}
$$

The polarized minterm is formed using the Kronecker product between n elementary matrices $\mathbf{R}_{2^1}^{(c_j)}$ as follows:

$$
\mathbf{R}_{2^n}^{(c)} = \bigotimes_{j=1}^{n} \mathbf{R}_{2^1}^{(c_j)} \tag{9.11}
$$

where $\mathbf{R}_{2^1}^{(c_j)}$ is defined by Equation 9.10. The resulting $2^n \times 2^n$ matrix $\mathbf{R}_{2^n}^{(c)}$ represents a minterm of polarity $c = c_1 c_2 \dots c_n$.

Example 9.31 *For the polarity $c = 1$ ($c_1 = 1$, $c_2 = 0$) and parameter $i_j \in \{0, 1\}$, polarized literals are generated in matrix form as follows:*

$$
(x_j \oplus 0) \quad \longleftrightarrow \quad \mathbf{R}_{2^1}^{(0)} = \begin{bmatrix} 1 & 0 \\ 1 & 1 \end{bmatrix}
$$

$$
(x_j \oplus 1) \quad \longleftrightarrow \quad \mathbf{R}_{2^1}^{(1)} = \begin{bmatrix} 0 & 1 \\ 1 & 1 \end{bmatrix}
$$

$$
(x_j \oplus 0)(x_t \oplus 1) \quad \longleftrightarrow \quad \mathbf{R}_{2^1}^{(0)} \otimes \mathbf{R}_{2^1}^{(1)} = \begin{bmatrix} 1 & 0 \\ 1 & 1 \end{bmatrix} \otimes \begin{bmatrix} 0 & 1 \\ 1 & 1 \end{bmatrix}
$$

Example 9.32   *The Kronecker product of matrices* $\mathbf{R}_{21}^{(0)}$ *is computed as follows:*

$$\mathbf{R}_{22}^{(0)} = \mathbf{R}_{21}^{(0)} \otimes \mathbf{R}_{21}^{(0)} = \begin{bmatrix} 1 & 0 \\ 1 & 1 \end{bmatrix} \otimes \begin{bmatrix} 1 & 0 \\ 1 & 1 \end{bmatrix} = \begin{bmatrix} 1 & 0 & 0 & 0 \\ 1 & 1 & 0 & 0 \\ 1 & 0 & 1 & 0 \\ 1 & 1 & 1 & 1 \end{bmatrix}$$

Example 9.33   *A polarized minterm of fifth polarity* $(c = 5,\ n = 3)$ *is constructed in matrix form as shown in Figure 9.23.*

9.9.3   Computing the coefficients in fixed-polarity forms

Using Equation 9.11, the polarized minterms can be generated for a given polarity of a polynomial expression. In algebraic form, the polynomial expression is a sum of polarized minterms over $GF(2)$:

$$f = \bigoplus_{i=0}^{2^n-1} (x_1 \oplus c_1)^{i_1} \cdots (x_n \oplus c_1)^{i_n} \tag{9.12}$$

The particular properties of a Boolean function in algebraic polynomial representation (Equation 9.12) are specified by the index $i = i_1 i_2 \ldots i_n$ as follows:

$$(x_j \oplus c_j)^{i_j} = \begin{cases} 1, & \text{if } i_j = 0 \\ (x_j \oplus c_j), & \text{if } i_j = 1 \end{cases} \quad \text{over } GF(2)$$

In matrix form, this specification is realized by the multiplication of the matrix $\mathbf{R}_{2^n}^{(c_j)}$ and the truth vector $\mathbf{F}$ of the Boolean function f over $GF(2)$.

Forward transform

A forward transform is used for the representation of the truth vector of a Boolean function (operational domain) in the form of a vector of coefficients in polynomial form (functional domain):

$$\underbrace{\text{Truth vector}}_{Operational\ domain} \xrightarrow[\underset{over\ GF(2)}{}]{Transform} \underbrace{\text{Vector of coefficients}}_{Functional\ domain}$$

Specifically, given the truth vector $\mathbf{F} = [f(0)\ f(1) \ldots f(2^n - 1)]^T$, the vector of coefficients in polarity c, $\mathbf{R}^{(c)} = [r_0^{(c)}\ r_1^{(c)} \ldots r_{2^n-1}^{(c)}]^T$ is derived by the matrix equation as follows:

Techniques for deriving a polarized minterm in matrix form

Step 1: Find the corresponding elementary matrix for each literal:

$$\text{Algebraic form} \longrightarrow \underbrace{(x_1 \oplus 1)^{i_1}}_{\begin{bmatrix} 0 & 1 \\ 1 & 1 \end{bmatrix}} \underbrace{(x_2 \oplus 0)^{i_2}}_{\begin{bmatrix} 1 & 0 \\ 1 & 1 \end{bmatrix}} \underbrace{(x_3 \oplus 1)^{i_3}}_{\begin{bmatrix} 0 & 1 \\ 1 & 1 \end{bmatrix}}$$

$$\mathbf{R}_{2^1}^{(1)} \qquad \mathbf{R}_{2^1}^{(0)} \qquad \mathbf{R}_{2^1}^{(1)}$$

Step 2: Form the $2^3 \times 2^3$ transform matrix $\mathbf{R}_{2^3}^{(5)}$ for the fifth polarity as the Kronecker product of the elementary matrices:

$$\overbrace{\underbrace{\mathbf{R}_{2^1}^{(1)} \otimes \mathbf{R}_{2^1}^{(0)}}_{\text{The 1st step}}}^{\text{The 2nd step}} \otimes \overbrace{\mathbf{R}_{2^1}^{(1)}}^{\substack{\text{The 2nd step} \\ \text{The 1st step}}} = \begin{bmatrix} 0 & 1 \\ 1 & 1 \end{bmatrix} \otimes \begin{bmatrix} 1 & 0 \\ 1 & 1 \end{bmatrix} \otimes \begin{bmatrix} 0 & 1 \\ 1 & 1 \end{bmatrix}$$

$$= \begin{bmatrix} 0 & 1 \\ 1 & 1 \end{bmatrix} \otimes \overbrace{\begin{bmatrix} 0 & 1 & 0 & 0 \\ 1 & 1 & 0 & 0 \\ 0 & 1 & 0 & 1 \\ 1 & 1 & 1 & 1 \end{bmatrix}}^{\text{The 2nd step}} = \begin{bmatrix} 0 & 0 & 0 & 0 & 0 & 1 & 0 & 0 \\ 0 & 0 & 0 & 0 & 1 & 1 & 0 & 0 \\ 0 & 0 & 0 & 0 & 0 & 1 & 0 & 1 \\ 0 & 0 & 0 & 0 & 1 & 1 & 1 & 1 \\ 0 & 1 & 0 & 0 & 0 & 1 & 0 & 0 \\ 1 & 1 & 0 & 0 & 1 & 1 & 0 & 0 \\ 0 & 1 & 0 & 1 & 0 & 1 & 0 & 1 \\ 1 & 1 & 1 & 1 & 1 & 1 & 1 & 1 \end{bmatrix} = \mathbf{R}_{2^3}^{(5)}$$

Step 3: Use the matrix $\mathbf{R}_{2^3}^{(5)}$ for the matrix transform of the vector $\mathbf{F}$ to a vector of coefficients in the fifth polarity. In Boolean expressions, the variables x_1, x_2, and x_3 are used as $\overline{x}_1$, x_2, and $\overline{x}_3$, respectively; that is, the polarities of variables are fixed.

FIGURE 9.23
Deriving a polarized minterm in matrix form (Example 9.33).

$$\mathbf{R}^{(c)} = \mathbf{R}_{2^n}^{(c)} \cdot \mathbf{F} \quad \text{over GF(2)} \tag{9.13}$$

where the $2^n \times 2^n$ matrix $\mathbf{R}_{2^n}^{(c)}$ is generated by the Kronecker product:

$$\mathbf{R}_{2^n}^{(c)} = \bigotimes_{j=1}^{n} \mathbf{R}_{2^1}^{(c_j)}, \qquad \mathbf{R}_{2^1}^{(c)} = \begin{cases} \begin{bmatrix} 1 & 0 \\ 1 & 1 \end{bmatrix}, & c_j = 0 \\[2mm] \begin{bmatrix} 0 & 1 \\ 1 & 1 \end{bmatrix}, & c_j = 1 \end{cases}$$

Techniques for computing the polynomials

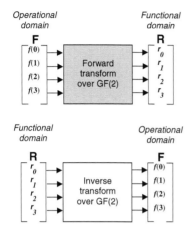

Operational domain

$$\mathbf{F} \begin{bmatrix} f(0) \\ f(1) \\ f(2) \\ f(3) \end{bmatrix}$$

Forward transform over GF(2)

Functional domain

$$\mathbf{R} \begin{bmatrix} r_0 \\ r_1 \\ r_2 \\ r_3 \end{bmatrix}$$

Functional domain

$$\mathbf{R} \begin{bmatrix} r_0 \\ r_1 \\ r_2 \\ r_3 \end{bmatrix}$$

Inverse transform over GF(2)

Operational domain

$$\mathbf{F} \begin{bmatrix} f(0) \\ f(1) \\ f(2) \\ f(3) \end{bmatrix}$$

The pair of forward and inverse transforms

$$\mathbf{R} = \mathbf{R}_{2^2} \cdot \mathbf{F} \quad \text{over GF(2)}$$
$$\mathbf{F} = \mathbf{R}_{2^2}^{-1} \cdot \mathbf{R} \quad \text{over GF(2)}$$

where

$$\mathbf{R}_{2^2} = \mathbf{R}_{2^2}^{-1}$$

$$\mathbf{R}_{2^2} = \bigotimes_{j=1}^{2} \mathbf{R}_{2^1}$$

$$\mathbf{R}_{2^1} = \begin{bmatrix} 1 & 0 \\ 1 & 1 \end{bmatrix}$$

FIGURE 9.24

Forward and inverse transforms for a Boolean function over GF(2) of two variables.

Example 9.34   *Given a Boolean function of two variables in the form of a truth vector* $\mathbf{F} = [1011]^T$, *the vector of coefficients is computed as follows:*

$$\mathbf{R}^{(2)} = \mathbf{R}_{2^2}^{(2)} \cdot \mathbf{F} = \begin{bmatrix} 0 & 0 & 0 & 1 \\ 0 & 0 & 1 & 1 \\ 0 & 1 & 0 & 1 \\ 1 & 1 & 1 & 1 \end{bmatrix} \begin{bmatrix} 1 \\ 0 \\ 1 \\ 1 \end{bmatrix} = \begin{bmatrix} 1 \\ 0 \\ 1 \\ 1 \end{bmatrix} \quad \text{over GF(2)}$$

where the matrix $\mathbf{R}_{2^2}^{(2)}$ *given* $c = 2$ *is generated using the Kronecker product* $\mathbf{R}_{2^2}^{(2)} = \mathbf{R}_{2^1}^{(1)} \otimes \mathbf{R}_{2^1}^{(0)} = \begin{bmatrix} 0 & 1 \\ 1 & 1 \end{bmatrix} \otimes \begin{bmatrix} 1 & 0 \\ 1 & 1 \end{bmatrix}$. *The vector of coefficients* $\mathbf{R}^{(2)} = [1\ 0\ 1\ 1]^T$ *corresponds to the expression* $f = 1 \oplus \overline{x}_1 \oplus \overline{x}_1 x_2$.

Example 9.35   *Given a logic network that implements a standard SOP expression, this logic network can be converted into an AND–EXOR network as shown in Figure 9.25.*

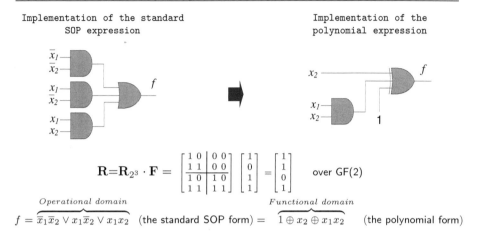

Design example:
conversion from operational to functional domain

Implementation of the standard SOP expression

Implementation of the polynomial expression

$$\mathbf{R} = \mathbf{R}_{2^3} \cdot \mathbf{F} = \begin{bmatrix} 1 & 0 & 0 & 0 \\ 1 & 1 & 0 & 0 \\ 1 & 0 & 1 & 0 \\ 1 & 1 & 1 & 1 \end{bmatrix} \begin{bmatrix} 1 \\ 0 \\ 1 \\ 1 \end{bmatrix} = \begin{bmatrix} 1 \\ 1 \\ 0 \\ 1 \end{bmatrix} \quad \text{over GF(2)}$$

Operational domain   *Functional domain*

$f = \overline{x}_1 \overline{x}_2 \vee x_1 \overline{x}_2 \vee x_1 x_2$ (the standard SOP form) $=$ $1 \oplus x_2 \oplus x_1 x_2$ (the polynomial form)

FIGURE 9.25
Conversion of the AND–OR logic network into the AND–EXOR network using forward transform (Example 9.35).

Inverse transform

The inverse transform is used for the conversion of a vector of coefficients of the polynomial form of a Boolean function (functional domain) into its truth vector (operational domain):

$$\underbrace{\textsf{Vector of coefficients}}_{Functional\ domain} \overset{Transform}{\underbrace{---\longrightarrow}_{over\ GF(2)}} \underbrace{\textsf{Truth vector}}_{Operational\ domain}$$

Given a vector of positive polarity polynomial coefficients $\mathbf{R} = [r_0 \ r_1 \dots r_{2^n - 1}]^T$, the truth vector $\mathbf{F} = [f(0) \ f(1) \dots f(2^n - 1)]^T$ of a Boolean function f is derived as follows (Figure 9.24):

$$\mathbf{F} = \mathbf{R}_{2^n}^{-1} \cdot \mathbf{R} \quad \text{over GF(2)} \tag{9.14}$$

where $\mathbf{R}_{2^1}^{-1} = \mathbf{R}_{2^1}$. Notice that the matrix $\mathbf{R}_{2^1}$ is a self-inverse matrix over AND and polynomial operations.

Example 9.36 *Given the an AND–EXOR network (Figure 9.26), this network can be converted into an AND–OR network as follows:*

Step 1: Compute the vector of coefficients $\mathbf{R} = [0\ 1\ 0\ 1]^T$.

Step 2: Compute the truth vector. Use the inverse transform given a certain polarity. Let the positive polarity be required: $\mathbf{F} = \mathbf{R}_{2^3}^{(-1)} \cdot \mathbf{R} = [0\ 1\ 0\ 0]^T$. *The truth vector corresponds to the standard SOP expression in algebraic form* $f = \overline{x}_1 x_2 \vee x_1 \overline{x}_2$.

Step 3: Design the AND–OR network.

Design example:
conversion from functional to operational domain

Implementation of the polynomial expression

Implementation of the standard SOP expression

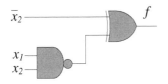

$$\mathbf{F} = \mathbf{R}_{2^3}^{(-1)} \cdot \mathbf{R} = \begin{bmatrix} 1 & 0 & 0 & 0 \\ 1 & 1 & 0 & 0 \\ 1 & 0 & 1 & 0 \\ 1 & 1 & 1 & 1 \end{bmatrix} \begin{bmatrix} 0 \\ 1 \\ 0 \\ 1 \end{bmatrix} = \begin{bmatrix} 0 \\ 1 \\ 0 \\ 0 \end{bmatrix} \qquad \text{over GF(2)}$$

$$f = \overline{x}_2 \oplus \overline{x_1 x_2} = x_2 \oplus 1 \oplus x_1 x_2 \oplus 1$$

$$\underbrace{}_{\substack{Functional \\ domain}} \qquad \underbrace{}_{\substack{Operational \\ domain}}$$

$$= \underbrace{x_2 \oplus x_1 x_2}_{} \quad \text{(the polynomial form)} = \underbrace{\overline{x}_1 x_2}_{} \quad \text{(the standard SOP form)}$$

FIGURE 9.26
Conversion of the AND–EXOR logic network into the AND-OR network using inverse transform (Example 9.36).

9.10 Decision diagrams

9.10.1 Function of the nodes

A node in a functional decision tree of a Boolean function f corresponds to the EXOR analog of Shannon expansion with respect to the variable x_i. It is called *Davio* expansion in honor of Professor *M. Davio*, who made pioneering contributions in the development of the theory of polynomial forms of Boolean functions.

In decision tree and diagram construction using an SOP form, only one type of node is used; that is, nodes that implement Shannon expansion.

Example 9.37 *Figure 9.27 shows the basic principle of mapping the term $1 \oplus x_2$ of a polynomial expression into a graph data structure.*

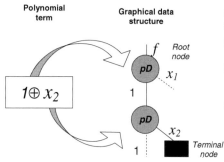

Polynomial term	Graphical data structure	Mapping the term $1 \oplus x_2$ of a polynomial expression into a graph data structure

▶ Two nodes (rooted and intermediate) and the terminal node are needed.

▶ Each node implements the positive Davio expansion, the terminal node can be marked by 0 or 1.

▶ If the terminal node is marked by 0, the graphical structure represents the term $1 \oplus x_2 = 0$; if the terminal node is marked by 1, the graphical structure represents the term $1 \oplus x_2 = 1$.

FIGURE 9.27
Mapping of the polynomial expression into the path of a decision diagram (Example 9.37).

There are two expansions in the functional domain: *positive Davio* expansion and *negative Davio* expansion. This is because the polynomial form of a Boolean function is characterized by polarity. These two expansions provide the construction of polynomial forms as follows:

Positive-polarity polynomials: If only the positive Davio expansion is applied, the resulting polynomial is of zero polarity; that is, all variables in the polynomial are uncomplemented.

Negative-polarity polynomials: If only the negative Davio expansion is applied, the resulting polynomial is of $2^n - 1$ polarity; that is, all variables in the polynomial are complemented.

Fixed-polarity polynomials: Application of both positive and negative Davio expansions results in fixed polarity; that is, polarity from 1 to $2^n - 2$ of the polynomial.

9.10.2 Algebraic form of the positive Davio expansions

Given a Boolean function f of n variables $x_1, x_2, \ldots, x_{i-1}, x_i, x_{i+1}, \ldots, x_n$,

$$f = f(x_1, x_2, \ldots, x_{i-1}, \boxed{x_i}, x_{i+1}, \ldots, x_n)$$

The positive Davio expansion with respect to the variable x_i is defined by the equation

$$f = f_0 \oplus x_i f_2 \qquad (9.15)$$

where $f_2 = f_0 \oplus f_1$.

Equation 9.15 is derived as follows. Shannon expansion of a Boolean function f with respect to the variable x_i results in the expression

$$f = \bar{x}_i f_0 \oplus x_i f_1 = (1 \oplus x_i) f_0 \oplus x_i f_1$$
$$= f_0 \oplus x_i f_0 \oplus x_i f_1 = f_0 \oplus x_i \underbrace{(f_0 \oplus f_1)}_{f_2}$$

Given $f_2 = f_0 \oplus f_1$, Equation 9.15 follows straightforwardly. From Equation 9.15 it follows that an arbitrary Boolean function f of n variables can be represented in expanded form with respect to the i-th variable x_i, $i \in 1, 2, \ldots, n$. Hence, positive Davio expansion given by Equation 9.15 is specified by the parameters f_0, f_1, and f_2:

Factor f_0: This is the function that is obtained from the function f by replacing the variable x_i by the logic value 0:

$$f_0 = f_{x_i=0} = f(x_1, \cdots, x_{i-1}, \boxed{x_i = 0}, x_{i+1}, \cdots, x_n)$$

Factor f_1: This is the function that is obtained from the function f by replacing the variable x_i by the logic value 1:

$$f_1 = f_{x_i=1} = f(x_1, \cdots, x_{i-1}, \boxed{x_i = 1}, x_{i+1}, \cdots, x_n)$$

Factor f_2: This is the function that is obtained by the EXOR sum of factors f_0 and f_1; that is,

$$f_2 = f_0 \oplus f_1 = f_{x_i=0} \oplus f_{x_i=1}$$

Factor $x_i f_2$: This is the function that is obtained by the AND multiplication of the variable x_i by the factor f_2, $x_i f_2$.

Computing the positive Davio expansion

▶ The node that implements the positive Davio expansion, denoted by pD, has two outputs:

    The left branch corresponds to the factor $1 \cdot f_0$

    The right branch corresponds to the factor $x_i \cdot f_2$

▶ Four possible combinations of the outputs f_0 and f_2 can be observed in computing:

    $\{f_0, f_2\} = \{0, 0\}$: Outputs of the left and right branches are both zero; hence, the input is $f = 0$.

    $\{f_0, f_2\} = \{0, 1\}$: The output of the right branch is 1; hence, the input is $f = x_i$.

    $\{f_0, f_2\} = \{1, 0\}$: Outputs of the left and right branches are both 1; hence, the input is $f = 1$.

    $\{f_0, f_2\} = \{1, 1\}$: The output of the left branch is 1; hence, the input is $f = \overline{x}_i$.

Example 9.38   *Let $f = x_1 \oplus x_2 \oplus x_1 x_3$. The positive Davio expansion of the Boolean function f with respect to the variable x_2 is defined as follows:*

$$f_0 = x_1 \oplus (x_2 = 0) \oplus x_1 x_3 = x_1 \oplus x_1 x_3$$

$$f_1 = x_1 \oplus (x_2 = 1) \oplus x_1 x_3 = 1 \oplus x_1 \oplus x_1 x_3$$

$$f_2 = f_0 \oplus f_1 = \underbrace{x_1 \oplus x_1 x_3}_{f_0} \oplus \underbrace{1 \oplus x_1 \oplus x_1 x_3}_{f_1} = 1$$

$$f = f_0 \oplus x_2 f_2 = x_1 \oplus x_1 x_3 \oplus x_2$$

Figure 9.28 illustrates the computational aspects of the positive Davio expansion.

9.10.3 Algebraic form of the negative Davio expansion

Given the Boolean function f of n variables $x_1, x_2, \ldots, x_{i-1}, x_i, x_{i+1}, \ldots, x_n$, the *negative Davio* expansion with respect to the variable x_i is expressed by the equation

$$f = f_1 \oplus \overline{x}_i f_2 \qquad (9.16)$$

By analogy with positive Davio expansion, negative Davio expansion is specified by the factors f_0, f_1, f_2, and $\overline{x}_i f_2$. Negative Davio expansion

Techniques for computing the positive Davio expansion

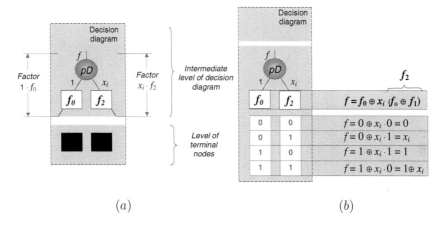

(a)   (b)

FIGURE 9.28
Nodes in functional decision diagrams and trees that implement the positive Davio expansion pD: function of the node (a) and computing of the node (b).

(Equation 9.16) with respect to the variable x_i is defined by analogy with positive Davio expansion:

$$
\begin{aligned}
f &= \overline{x}_i f_0 \oplus x_i f_1 \\
&= \overline{x}_i f_0 \oplus (1 \oplus \overline{x}_i) f_1 \\
&= \overline{x}_i f_0 \oplus f_1 \oplus \overline{x}_i f_1 \\
&= f_1 \oplus \overline{x}_i (f_0 \oplus f_1) \\
&= f_1 \oplus \overline{x}_i f_2
\end{aligned}
$$

Example 9.39 *Negative Davio expansion with respect to the variable x_1 is defined as follows:*

$$
\begin{aligned}
f &= f_1 \oplus x_2 f_2 \\
&= 1 \oplus x_1 \oplus x_1 x_3 \oplus \overline{x}_2
\end{aligned}
$$

Figure 9.29 illustrates the computational aspects of negative Davio expansion.

Computing negative Davio expansion

▶ The node that implements the negative Davio expansion, denoted by nD, has two outputs:

  The left branch corresponds to the factor $1 \cdot f_1$.
  The right branch corresponds to the factor $\overline{x}_i \cdot f_2$.

▶ Four possible combinations of the outputs f_1 and f_2 can be observed in computing:

  $\underline{\{f_1, f_2\} = \{0, 0\}}$: Outputs of the left and right branches are both zero; hence, the input is $f = 0$.

  $\underline{\{f_1, f_2\} = \{0, 1\}}$: The output of the right branch is 1; hence, the input is $f = \overline{x}_i$.

  $\underline{\{f_1, f_2\} = \{1, 0\}}$: Outputs of the left and right branches are both 1; hence, the input is $f = 1$.

  $\underline{\{f_1, f_2\} = \{1, 1\}}$: The output of the left branch is 1; hence, the input is $f = x_i$.

Techniques for computing the negative Davio expansion

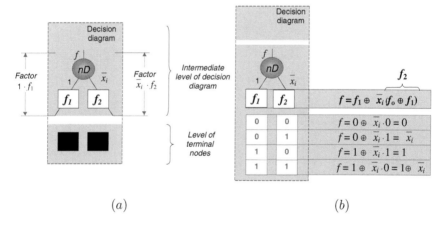

(a)   (b)

FIGURE 9.29

Nodes in the functional decision diagrams and trees that implement the negative Davio expansion nD: function of the node (a) and computing of the node (b).

9.10.4 Matrix forms of positive and negative Davio expansions

In matrix notation, the function f of the node is a Boolean function of a single variable x_i given by the truth vector $\mathbf{F} = [\, f(0) \; f(1) \,]^T$. Hence, the node can be described by the simplest matrix equation. The function of the node using a positive Davio expansion is as follows:

$$f = [\, 1 \; x_i \,] \begin{bmatrix} 1 & 0 \\ 1 & 1 \end{bmatrix} \begin{bmatrix} f_0 \\ f_1 \end{bmatrix} = [\, 1 \; x_i \,] \begin{bmatrix} f_0 \\ f_2 \end{bmatrix}$$
$$= f_0 \oplus x_i f_2 \tag{9.17}$$

Example 9.40   *Let the input variable be given by the truth vector* $\mathbf{F} = [\, f(0) \; f(1) \,]^T = [\, 1 \; 0 \,]^T$; *that is,* $f_0 = 1$ *and* $f_1 = 0$. *The positive Davio transform (Equation 9.17) is*

$$f = [\, 1 \; x_i \,] \begin{bmatrix} 1 & 0 \\ 1 & 1 \end{bmatrix} \begin{bmatrix} 1 \\ 0 \end{bmatrix} = [\, 1 \; x_i \,] \begin{bmatrix} 1 \\ 1 \end{bmatrix} = 1 \oplus x_i$$

In matrix notation, the function of a negative Davio node is as follows:

$$f = [\, 1 \; \overline{x}_i \,] \begin{bmatrix} 0 & 1 \\ 1 & 1 \end{bmatrix} \begin{bmatrix} f_0 \\ f_1 \end{bmatrix} = [\, 1 \; \overline{x}_i \,] \begin{bmatrix} f_1 \\ f_2 \end{bmatrix}$$
$$= f_1 \oplus \overline{x}_i f_2 \tag{9.18}$$

Example 9.41   *(Continuation of Example 9.40). The negative Davio transform (Equation 9.18) is*

$$f = [\, 1 \; \overline{x}_i \,] \begin{bmatrix} 0 & 1 \\ 1 & 1 \end{bmatrix} \begin{bmatrix} 1 \\ 0 \end{bmatrix} = [\, 1 \; \overline{x}_i \,] \begin{bmatrix} 1 \\ 1 \end{bmatrix} = 1 \oplus \overline{x}_i$$

9.10.5 Gate-level implementation of Shannon and Davio expansions

Consider the gate-level implementation of Shannon and Davio expansions. The logic networks are given in Figure 9.30 in comparison with a network for Shannon expansion.

Design example: Shannon and Davio expansions

OPERATIONAL DOMAIN

S h a n n o n
e x p a n s i o n
$f = \overline{x}_i f_0 \vee x_i f_1$

FUNCTIONAL DOMAIN

P o s i t i v e　　D a v i o
e x p a n s i o n
$f = f_0 \oplus x_i f_2$

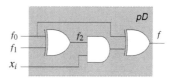

N e g a t i v e　　D a v i o
e x p a n s i o n
$f = f_1 \oplus \overline{x}_i f_2$

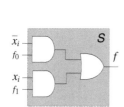

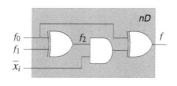

FIGURE 9.30
Gate-level representation of the nodes of decision trees and diagrams using Shannon and Davio expansions.

Example 9.42　*Given the Boolean function $f = x_1 \vee x_2$, its positive (pD) and negative (nD) Davio expansions with respect to the variable x_1 result in the polynomial expressions*

$$\textbf{pD: } f = f_0 \oplus x_1(f_0 \oplus f_1) = x_2 \oplus x_1 \underbrace{(x_2 \oplus 1)}_{f_0 \oplus f_1}$$
$$\underbrace{}_{f_0}$$

$$= x_2 \oplus x_1 \oplus x_1 x_2$$

$$\textbf{nD: } f = f_1 \oplus \overline{x}_1(f_0 \oplus f_1) = x_2 \oplus x_1 \underbrace{(x_2 \oplus 1)}_{f_0 \oplus f_1}$$
$$\underbrace{}_{f_1}$$

$$= 1 \oplus \overline{x}_1 \oplus \overline{x}_1 x_2$$

The logic networks for the Davio expansion given in Figure 9.30 can be used for computing by specification of the inputs; that is, $f_0 = x_2$ and $f_1 = 1$.

Table 9.6 summarizes the functions of the nodes for positive Davio and negative Davio expansions, labeled as pD and nD, respectively. For simplification, realization of the nodes is given using a single EXOR gate.

TABLE 9.6
The nodes of the functional decision tree that implement Davio expansion at the gate level, and their description in algebraic and matrix forms

Techniques for computing Davio expansions

Node	Realization	Algebraic form	Matrix form
Positive Davio node		$f = f_0 \oplus x_i f_2$ $f_0 = f\vert_{x_i=0}$ $f_2 = f\vert_{x_i=1} \oplus f\vert_{x_i=0}$	$f = [\,1\ x_i\,] \begin{bmatrix} 1 & 0 \\ 1 & 1 \end{bmatrix} \begin{bmatrix} f_0 \\ f_1 \end{bmatrix}$
Negative Davio node		$f = f_1 \oplus \bar{x}_1 f_2$ $f_1 = f\vert_{x_i=1}$ $f_2 = f\vert_{x_i=0} \oplus f\vert_{x_i=1}$	$f = [\,1\ \bar{x}_i\,] \begin{bmatrix} 0 & 1 \\ 1 & 1 \end{bmatrix} \begin{bmatrix} f_1 \\ f_2 \end{bmatrix}$

9.11 Techniques for functional decision tree construction

Techniques for functional decision tree construction consist of techniques for the reduction of functional decision trees, matrix-based designs, and manipulation of pD and nD nodes for conversion between polarities.

9.11.1 The structure of functional decision trees

The most important structural properties of the functional decision tree with positive Davio nodes are as follows:

▶ A Boolean function of n variables is represented by an n-level functional decision tree. The i-th level of the functional decision tree, $i = 1, \ldots, n$, includes 2^{i-1} nodes.

▶ Nodes at the n-th level are connected to 2^n terminal nodes, which take values 0 or 1. The nodes, corresponding to the i-th variable form the i-th level in the functional decision tree.

▶ In every path from the root node to a terminal node, the variables appear in a fixed order; the tree is thus said to be ordered.

▶ The values of constant nodes are the values of the coefficients of the polynomial expression for the Boolean function represented.

9.11.2 Design example: manipulation of pD and nD nodes

This design example introduces techniques for the design of functional decision diagrams for computing polynomial expressions of various polarities. This computing ability is provided by the distribution of pD and nD nodes in the levels of a decision tree. There are 2^n various combinations of the pD and nD nodes in the levels of a decision tree. Each combination corresponds to one polarity of a polynomial. There are two trivial cases in these 2^n combinations:

(a) The tree consisting of only pD nodes; it computes only the positive polarity polynomial (all variables are noncomplemented).

(b) The tree consisting of only nD nodes; it computes only the negative polarity polynomial (all variables are complemented).

Example 9.43 *Design functional decision trees for computing all positive fixed-polarity polynomial expressions of Boolean functions of two variables. Figure 9.31 shows all four possible decision trees. The functional decision tree that represents the polynomial of polarity $c = 1$ is shown in Figure 9.32. All possible coefficients of the polynomial expression are also shown.*

9.11.3 Design example: application of matrix transforms

This design example introduces techniques of functional decision tree construction using a matrix description of computing. Given a truth vector $\mathbf{F} = [f(0) \ f(1) \dots f(2^n - 1)]^T$ of the Boolean function f, the positive-polarity polynomial expression in algebraic form is defined as the following matrix transforms:

$$f = \widehat{\mathbf{X}} \ \mathbf{R}_{2^n} \ \mathbf{F} \quad \text{over GF}(2) \tag{9.19}$$

where the vector $\widehat{\mathbf{X}}$ and matrix $\mathbf{R}_{2^n}$ are constructed using the Kronecker product, denoted as $\otimes$:

$$\widehat{\mathbf{X}} = \bigotimes_{i=1}^{n} [\ 1 \ x_i\ ], \quad \mathbf{R}_{2^n} = \bigotimes_{i=1}^{n} \mathbf{R}_2,$$

and the elementary matrix $\mathbf{R}_2 = \begin{bmatrix} 1 & 0 \\ 1 & 1 \end{bmatrix}$.

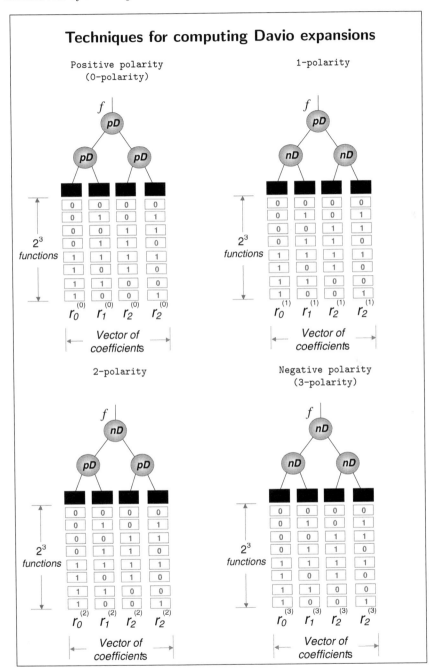

FIGURE 9.31

Functional decision trees for computing polynomial expressions in positive, negative, and fixed polarities of Boolean functions of two variables (Example 9.43).

Techniques for computing Davio expansions

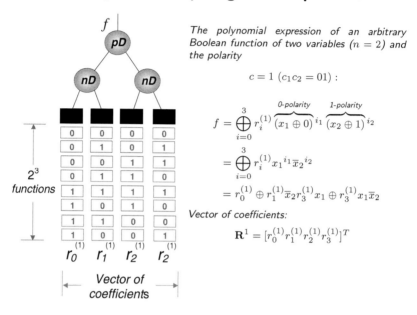

The polynomial expression of an arbitrary Boolean function of two variables $(n = 2)$ and the polarity

$$c = 1 \ (c_1 c_2 = 01):$$

$$f = \bigoplus_{i=0}^{3} r_i^{(1)} \overbrace{(x_1 \oplus 0)}^{0\text{-polarity}}{}^{i_1} \overbrace{(x_2 \oplus 1)}^{1\text{-polarity}}{}^{i_2}$$

$$= \bigoplus_{i=0}^{3} r_i^{(1)} x_1{}^{i_1} \overline{x}_2{}^{i_2}$$

$$= r_0^{(1)} \oplus r_1^{(1)} \overline{x}_2 r_3^{(1)} x_1 \oplus r_3^{(1)} x_1 \overline{x}_2$$

Vector of coefficients:

$$\mathbf{R}^1 = [r_0^{(1)} r_1^{(1)} r_2^{(1)} r_3^{(1)}]^T$$

FIGURE 9.32
The functional decision tree for a Boolean function of two variables (Example 9.43).

Example 9.44   *Given the Boolean functions of (a) a single variable x_i $(n = 1)$ and (b) two variables $x_1 \vee x_2$ $(n = 2)$, using Equation 9.19, the positive-polarity polynomial expressions are as follows:*

$$(a) \ n = 1: \ f = \widehat{\mathbf{X}} \, \mathbf{R}_{2^1} \, \mathbf{F} = [\, 1 \ x_i \,] \begin{bmatrix} 1 & 0 \\ 1 & 1 \end{bmatrix} \begin{bmatrix} 0 \\ 1 \end{bmatrix}$$

$$= [\, 1 \ x_i \,] \begin{bmatrix} 0 \\ 1 \end{bmatrix} = x_i$$

$$(b) \ n = 2: \ f = \widehat{\mathbf{X}} \, \mathbf{R}_{2^2} \, \mathbf{F}$$

$$= [\, 1 \ x_2 \ x_1 \ x_1 x_2 \,] \begin{bmatrix} 1 & 0 & 0 & 0 \\ 1 & 1 & 0 & 0 \\ 1 & 0 & 1 & 0 \\ 1 & 1 & 1 & 1 \end{bmatrix} \begin{bmatrix} 0 \\ 1 \\ 1 \\ 1 \end{bmatrix}$$

$$= [\, 1 \ x_2 \ x_1 \ x_1 x_2 \,] \begin{bmatrix} 0 \\ 1 \\ 1 \\ 1 \end{bmatrix} = x_2 \oplus x_1 \oplus x_1 x_2$$

The example below shows functional decision tree design using the truth vector of a Boolean function.

Example 9.45   *A Boolean function is given by the truth vector* $\mathbf{F} = [\,0\ 1\ 1\ 0\,]^T$ *(Figure 9.33).*

> **Step 1:** *Root pD node with input* $\mathbf{F} = [\,0\ 1\ 1\ 0\,]^T$. *Positive Davio expansion results in left branch,* $\mathbf{F}_{x_1=0} = [0\ 1]^T$, *and right branch,* $\mathbf{F}_{x_1=0} \oplus \mathbf{F}_{x_1=1} = [0\ 1]^T \oplus [1\ 0]^T = [\,1\ 1\,]^T$. *Both outputs results in functions and require further application of a Davio expansion.*
>
> **Step 2:** *Left node, left branch:* $\mathbf{F}_{x_1=0 \atop x_2=0} = [\,0\,]$. *Right branch:* $\mathbf{F}_{x_1=0 \atop x_2=0} \oplus \mathbf{F}_{x_1=0 \atop x_2=1} = [\,0\,] \oplus [\,1\,] = [\,1\,]$.
>
> **Step 3:** *Right node, left branch:* $\mathbf{F}_{x_1=1 \atop x_2=0} = [\,1\,]$. *Left branch:* $\mathbf{F}_{x_1=1 \atop x_2=0} \oplus \mathbf{F}_{x_1=1 \atop x_2=1} = [\,1\,] \oplus [\,1\,] = [\,0\,]$.

Example 9.46   *Let the Boolean function* $f = \overline{x}_1 \vee x_2$ *be given by its truth vector* $\mathbf{F} = [1\ 1\ 0\ 1]^T$. *The functional decision tree with all possible Davio expansion nodes is shown in Figure 9.34. The values of this tree's terminal nodes are used to derive the polynomial expression, in which the product terms are generated by the Kronecker product* $\widehat{\mathbf{X}}$. *The* 4×4 *transform matrix* $\mathbf{R}$ *is generated by the Kronecker product of the basic matrix* $\mathbf{R}_{21}$.

Example 9.47   *An arbitrary Boolean function* f *of three variables can be represented by the decision tree shown in Figure 9.35. Equation 9.15 is used as follows: (a) with respect to variable* x_1: $f = f_0 \oplus x_1 f_2$; *(b) with respect to variable* x_2: $f_0 = f_{00} \oplus x_2 f_{02}$, $f_1 = f_{10} \oplus x_2 f_{22}$; *(c) with respect to variable* x_3: $f_{00} = f_{000} \oplus x_3 f_{002}$, $f_{02} = f_{020} \oplus x_3 f_{022}$, $f_{20} = f_{200} \oplus x_3 f_{202}$, $f_{22} = f_{220} \oplus x_3 f_{222}$, *and (d) representation in polynomial form* $f = f_{000} = f_{002} x_3 \oplus f_{020} x_2 \oplus f_{022} x_2 x_3 \oplus f_{200} x_1 \oplus f_{202} x_1 x_3 \oplus f_{220} x_1 x_2 \oplus f_{222} x_1 x_2 x_3$.

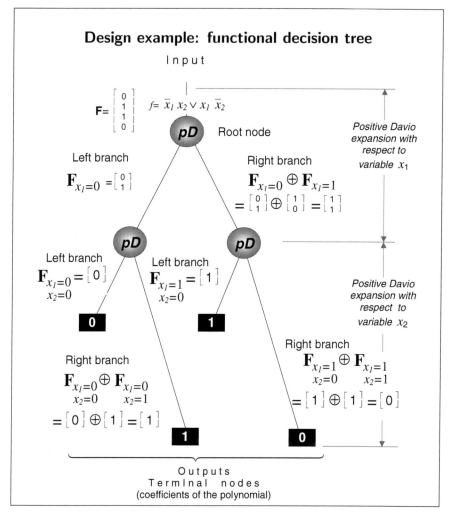

FIGURE 9.33
Functional decision tree design with positive Davio expansion in the nodes
using a truth vector of the Boolean function $f = \bar{x}_1 x_2 \vee x_2 \bar{x}_2$ (Example 9.45).

9.11.4 Design example: minterm computing

Example 9.48 *Figure 9.36 shows a representation of polarized
minterms using a decision tree for the Boolean
function of two variables.*

Design example: functional decision tree

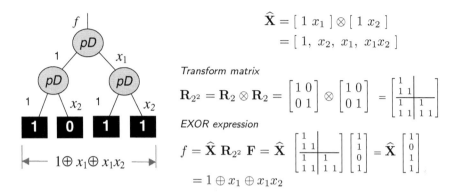

$$\widehat{\mathbf{X}} = [\, 1 \;\; x_1 \,] \otimes [\, 1 \;\; x_2 \,]$$
$$= [\, 1, \;\; x_2, \;\; x_1, \;\; x_1 x_2 \,]$$

Transform matrix

$$\mathbf{R}_{2^2} = \mathbf{R}_2 \otimes \mathbf{R}_2 = \begin{bmatrix} 1 & 0 \\ 0 & 1 \end{bmatrix} \otimes \begin{bmatrix} 1 & 0 \\ 0 & 1 \end{bmatrix} = \begin{bmatrix} 1 & & \\ 1 & 1 & \\ \hline 1 & & 1 \\ 1 & 1 & 1 & 1 \end{bmatrix}$$

EXOR expression

$$f = \widehat{\mathbf{X}} \, \mathbf{R}_{2^2} \, \mathbf{F} = \widehat{\mathbf{X}} \begin{bmatrix} 1 & & \\ 1 & 1 & \\ \hline 1 & & 1 \\ 1 & 1 & 1 & 1 \end{bmatrix} \begin{bmatrix} 1 \\ 1 \\ 0 \\ 1 \end{bmatrix} = \widehat{\mathbf{X}} \begin{bmatrix} 1 \\ 0 \\ 1 \\ 1 \end{bmatrix}$$

$$= 1 \oplus x_1 \oplus x_1 x_2$$

FIGURE 9.34

The functional decision tree for the Boolean function $f = \overline{x}_1 \vee x_2$ (Example 9.46).

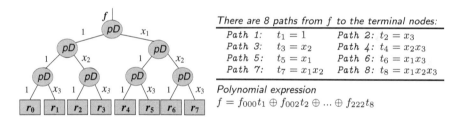

There are 8 paths from f to the terminal nodes:

Path 1:	$t_1 = 1$	*Path 2:*	$t_2 = x_3$
Path 3:	$t_3 = x_2$	*Path 4:*	$t_4 = x_2 x_3$
Path 5:	$t_5 = x_1$	*Path 6:*	$t_6 = x_1 x_3$
Path 7:	$t_7 = x_1 x_2$	*Path 8:*	$t_8 = x_1 x_2 x_3$

Polynomial expression
$$f = f_{000} t_1 \oplus f_{002} t_2 \oplus \ldots \oplus f_{222} t_8$$

FIGURE 9.35

Positive-polarity polynomial representation of a Boolean function of three variables by a functional decision tree (Example 9.47).

9.12 Functional decision tree reduction

The functional decision diagram for polynomial forms is derived from the functional decision tree by deleting redundant nodes and by sharing equivalent subgraphs. The rules below produce reduced functional decision diagrams.

9.12.1 Elimination rule

If the outgoing edge of a node labeled with x_i and $\overline{x}_i$ points to the constant zero, then delete the node and connect the edge to the other subgraph directly.

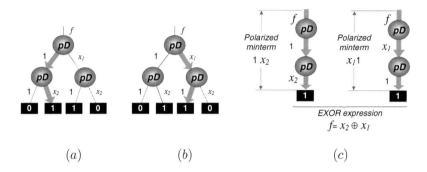

FIGURE 9.36
Polarized minterm representation using a decision tree: (a) polarized minterm
$1x_2$, (b) polarized minterm $x_2 1$, (c) polynomial expression $f = x_2 \oplus x_1$
(Example 9.48).

The formal basis of this rule is as follows (Figure 9.37): $\varphi = \varphi_0 \oplus x_i \varphi_2$. If
$\varphi_2 = 0$, then $\varphi = \varphi_0$.

9.12.2 Merging rule

In a tree, edges longer than one, that is, connecting nodes at non-successive
levels, can appear. For example, the length of an edge connecting a node at
the $(i-1)$-th level with a node at the $(i+1)$-th level is two.

Example 9.49 *Provide a formal explanation of the reduction of the
following functional decision trees:*

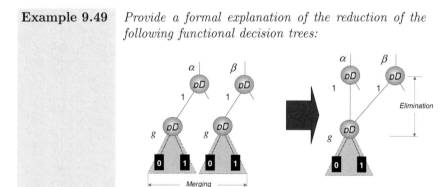

Example 9.50 *Application of reduction rules to the three-variable
NAND function is demonstrated in Figure 9.38.*

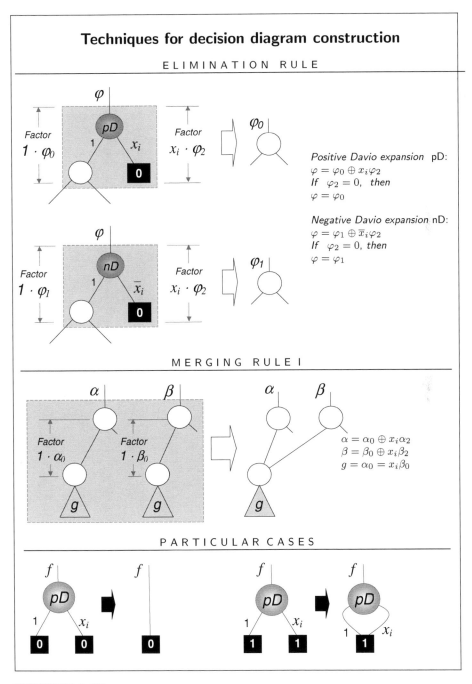

FIGURE 9.37

Reduction rules for functional decision diagram construction.

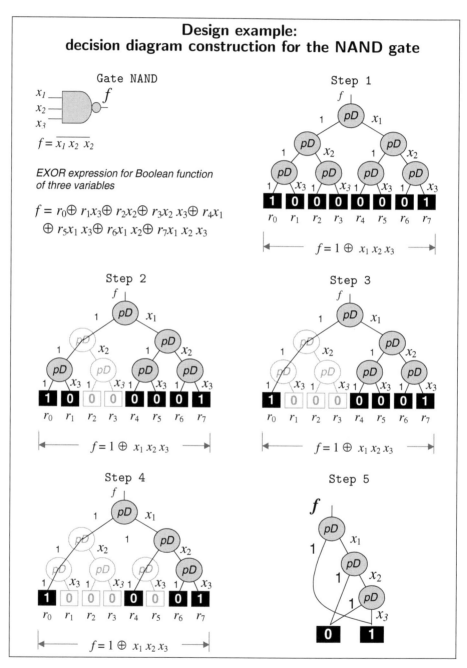

FIGURE 9.38

Functional decision diagram design using pD nodes for the three-variable NAND function (Example 9.50).

The functional decision diagram is derived from the functional decision tree by deleting redundant nodes and by sharing equivalent subgraphs. The rules below produce the reduced Davio diagram.

Example 9.51 *Figure 9.39 shows the derived functional decision diagrams for some Boolean functions of three variables.*

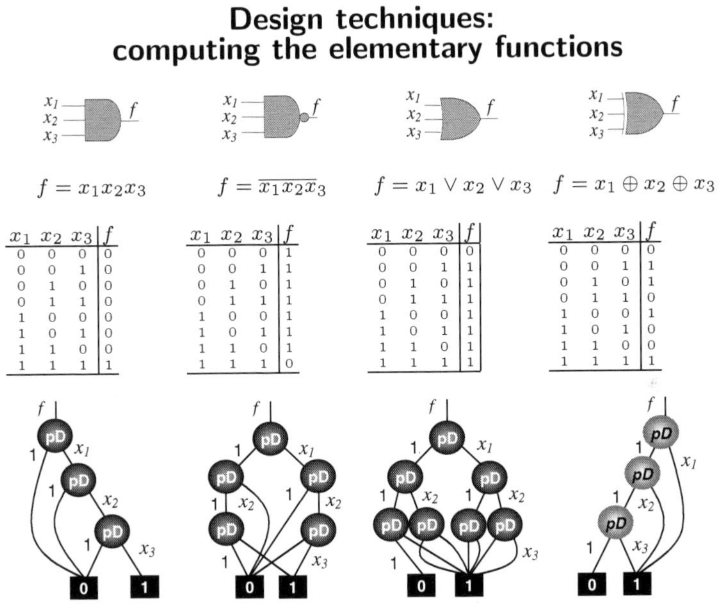

x_1	x_2	x_3	f
0	0	0	0
0	0	1	0
0	1	0	0
0	1	1	0
1	0	0	0
1	0	1	0
1	1	0	0
1	1	1	1

x_1	x_2	x_3	f
0	0	0	1
0	0	1	1
0	1	0	1
0	1	1	1
1	0	0	1
1	0	1	1
1	1	0	1
1	1	1	0

x_1	x_2	x_3	f
0	0	0	0
0	0	1	1
0	1	0	1
0	1	1	1
1	0	0	1
1	0	1	1
1	1	0	1
1	1	1	1

x_1	x_2	x_3	f
0	0	0	0
0	0	1	1
0	1	0	1
0	1	1	0
1	0	0	1
1	0	1	0
1	1	0	0
1	1	1	1

FIGURE 9.39
Functional decision diagrams for the Boolean functions AND, NAND, OR, and EXOR of three variables (Example 9.51).

9.13 Further study

Advanced topics of the functional decision diagrams

Topic 1: Computing polynomials using Pascal and transeunt triangles. The Pascal triangle for a Boolean function $f(x_1, x_2, ..., x_n)$ is a fractal structure formed by modulo 2 addition of 0s and 1s starting from the bottom row, which is the truth vector of f. The Pascal triangle for an n-variable Boolean function has a width of 2^n and a height of 2^n. Since the truth vector for functions with one or more variables has an even number of entries, it can be divided evenly into two parts. Each part produces, on its own, two subtriangles. In general,

---

▶ The bits along the triangle's left side are coefficients of zero polarity polynomial form RM_0.

▶ The bits along the right side of each subtriangle on the left side of the triangle are the coefficients of the polynomial form of polarity 1, RM_1.

▶ The bits along the left or right side of each following row of subtriangles, such that the bottom element is the j-th element of the truth vector $\mathbf{F}$, are the coefficients of polynomial forms of polarity j, RM_j, $j = 1, 2, \ldots, 2^n - 2$.

▶ The bits along the left side of the triangle are the coefficients of the polynomial form of polarity $2^n - 1$, RM_{2^n-1}.

Given the Boolean function $f = x_1 \bar{x}_2$ and its truth vector $\mathbf{F} = [0010]^T$, its Pascal triangles are shown in Figure 9.40. The location of coefficients of the polynomial expressions of polarities 0,1,2, and 3 are

$$RM(0) = [0011]^T, \quad RM(1) = [0001]^T, \quad RM(2) = [1111]^T, \quad RM(3) = [0101]^T$$

FIGURE 9.40
Encoding the Pascal triangles for the polynomial representation of the Boolean function $f = x_1 \bar{x}_2$.

The Pascal triangle for totally symmetric function called transeunt triangle is formed as follows (J. Butler, G. Dueck, V. Shmerko, S. Yanushkevich, "On the Number of Generators of Transeunt Triangles," *Discrete Applied Mathematics*, number 108, pages 309–316, 2001):

▶ The carrier vector is located at the base of the triangle.
▶ A vector of n 1s and 0s is formed by the exclusive OR of adjacent bits in the carrier vector.
▶ A vector of $(n-1)$ 1s and 0s is formed by the exclusive OR of adjacent bits in the previous vector, etc.
▶ At the apex of the triangle is a single 0 or 1.

The transeunt triangle can be generated from the carrier vector of coefficients in RM_0 or in RM_{2^n-1} by taking the exclusive OR of adjacent bits repeatedly until a single bit is obtained. This is because the exclusive OR is *self-invertible*. That is, given $A \oplus B$ and the value of A, we can find the value of B $(= A \oplus (A \oplus B))$. Figure 9.41 shows the transeunt triangle given the symmetric Boolean function $f = \overline{x}_1\overline{x}_2x_3 \vee x_1x_2x_3 \vee x_1\overline{x}_2\overline{x}_3 \vee \overline{x}_1x_2\overline{x}_3$. In this triangle, only the elements that correspond to a truncated truth vector of the symmetric function are given. The corresponding functional trees based on Davio expansion for $RM(0) = x_1 \oplus x_2 \oplus x_3$ and $RM(7) = 1 \oplus \overline{x}_1 \oplus \overline{x}_2 \oplus \overline{x}_3$, and the decision tree based on Shannon expansion, are given in Figure 9.41 as well.

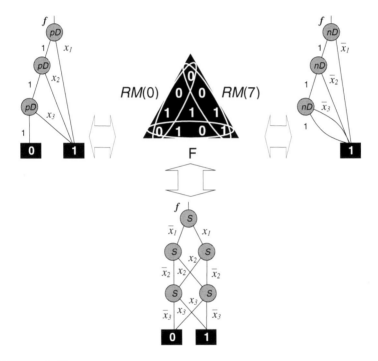

FIGURE 9.41

The relationship between the transeunt triangle and the corresponding decision diagrams for the totally symmetric Boolean function $f = x_1 \oplus x_2 \oplus x_3$.

Topic 2: Arithmetic polynomials. The fields over the numbers $\{0, 1\}$ and over the integer numbers $\{0, 1, 2, \ldots, N\}$ provide more possibilities for the representation of Boolean functions compared to SOP form.

Apparently, the first attempts to represent logic operations by arithmetical ones were taken by the founder of Boolean algebra, Boole (1854). He did not use the Boolean operators known today. Rather, he used arithmetic expressions. It is interesting to note that H. Aiken first found that arithmetic expressions can be useful in designing logic circuits and used them in the

Harvard MARK 3 and MARK 4 computers (H. H. Aiken, "Synthesis of Electronic Computing and Control Circuits," *Ann. Computation Laboratory of Harvard University*, XXVII, Harvard University, Cambridge, MA, 1951).

Arithmetic expressions are closely related to polynomial expressions over GF(2), but with variables and function values interpreted as the integers 0 and 1 instead of logic values. In this way, arithmetic expressions can be considered as integer counterparts of polynomial expressions over GF(2). For two Boolean variables x_1 and x_2, the following are true:

$$\overline{x} = 1 - x, \quad x_1 \vee x_2 = x_1 + x_2 - x_1x_2,$$
$$x_1 \wedge x_2 = x_1x_2, \quad x_1 \oplus x_2 = x_1 + x_2 - 2x_1x_2.$$

The right parts of the equations are called the *arithmetic expressions*. A Boolean function of n variables is the mapping $\{0,1\}^n \rightarrow \{0,1\}$, while an integer-valued function in arithmetical logic denotes the mapping $\{0,1\}^n \rightarrow \{0,1,\ldots,p-1\}$, where $p > 2$. For a Boolean function f of n variables, the arithmetic expression is given by

$$f = \sum_{i=0}^{2^n-1} a_i \cdot \underbrace{\left(x_1^{i_1} \cdots x_n^{i_n}\right)}_{i-th\ product}$$

where a_i is a coefficient (integer number); i_j is the j-th bit $1, 2, \ldots, n$, in the binary representation of the index $i = i_1 i_2 \ldots i_n$; and $x_j^{i_j}$ is defined as

$$x_j^{i_j} = \begin{cases} 1, & i_j = 0 \\ x_j, & i_j = 1 \end{cases}$$

Note that $\sum$ is arithmetic addition. For example, an arbitrary Boolean function of three variables is represented in the arithmetic expression

$$f = a_0 + a_1x_3 + a_2x_2 + a_3x_2x_3 + a_4x_1 + a_5x_1x_3 + a_6x_1x_2 + a_7x_1x_2x_3.$$

The coefficients a_i, $i = 0, 1, \ldots, 7$, are the integer numbers. These coefficients are derived from the truth table:

x_1 x_2 x_3	f		x_1 x_2 x_3	f
0 0 0	a_0		1 0 0	a_4
0 0 1	a_1		1 0 1	a_5
0 1 0	a_2		1 1 0	a_6
0 1 1	a_3		1 1 1	a_7

However, not all arbitrary chosen integer numbers specify a Boolean function. Special rules are used to determine the values of the polynomial coefficients. Arithmetic logic has many applications in contemporary logic design, for example, in the computation of signal probabilities for test generations, and switching activities for power and noise analysis.

Advanced topics of polynomial techniques, such as word-level representations, functional decision diagrams, probabilistic computing, and new computing paradigms can be found, in particular, in *Proceedings of the International Workshop on Applications of the Reed–Muller Expansion in Circuit Design*.

Topic 3: Decision diagrams for polynomial forms. Various decision diagrams for polynomial forms have been developed. They can be classified with respect to functions of nonterminal nodes, number of outgoing edges, decomposition rules, terminal nodes, and edges and their labels. In the case of the GF(p) field, the corresponding decision diagrams are called *bit-level* diagrams, allowing *p*-valued bits in the case of multivalued logic functions. Otherwise, decision diagrams are called *word-level* diagrams, since the value of a terminal node is represented by a word. The terminal nodes of decision diagrams can be fuzzy values.

Further reading

[1] Bryant RE. Graph-based algorithms for Boolean function manipulation. *IEEE Trans. Comp.*, 35(6):677–691, 1986.

[2] Butler JT and Sasao T. On the average path length in decision diagrams of multiple-valued functions. In *Proc. of the IEEE 33rd Int. Symp. on Multiple-Valued Logic*, pp. 383–390, 2003.

[3] Chen YA and Bryant RE. An efficient graph representation for arithmetic circuit verification. *IEEE Trans. Computer-Aided Design of Integrated Circuits and Systems*, 20(12):1443–1445, 2001.

[4] Davio M, Deschamps J-P, and Thayse A. *Discrete and Switching Functions*. McGraw–Hill, Maidenhead, 1978.

[5] Falkowski BJ. A note on the polynomial form of Boolean functions and related topics. *IEEE Trans. Comp.*, 48(8):860–864, 1999.

[6] Green D. *Modern Logic Design*, Addison-Wesley, 1986.

[7] Green DH. Families of Reed–Muller canonical forms, *Int. J. .Electronics*, 2:259–280, 1991.

[8] Jain J. Arithmetic transform of Boolean functions. In Sasao T and Fujita M, Eds., *Representations of Discrete Functions*, pp. 55–92, Kluwer, Dordrecht, 1996.

[9] Malyugin VD. *Paralleled Calculations by Means of Arithmetic Polynomials*. Physical and Mathematical Publishing Company, Russian Academy of Sciences, Moscow, 1997 (in Russian).

[10] Minato S. *Binary Decision Diagrams and Applications for VLSI CAD*, Kluwer, Dordrecht, 1996.

[11] Perkowski MA, Chrzanowska-Jeske M, and Xu Y. Lattice diagrams using Reed-Muller logic. In *Proc. IFIP WG 10.5 Int. Workshop on Applications of the Reed-Muller Expansion in Circuit Design*. Japan, pp. 85–102, 1997.

[12] Sasao T. Representation of logic functions using EXOR operators. In Sasao T and Fujita M, Eds., *Representations of Discrete Functions*, Kluwer, Dordrecht, pp. 29–54, 1996.

[13] Stanković RS. Simple theory of decision diagrams for representation of discrete functions. In *Proc. 4th Int. Workshop on Applications of the Reed-Muller Expansion in Circuit Design*. University of Victoria, BC, Canada, pp. 161–178, 1999.

[14] Stanković RS. Some remarks on basic characteristics of decision diagrams. In *Proc. 4th Int. Workshop on Applications of the Reed-Muller Expansion in Circuit Design*. University of Victoria, BC, Canada, pp. 139–148, 1999.

[15] Stanković RS. Unified view of decision diagrams for representation of discrete functions. *Int. J. Multi-Valued Logic and Soft Computing*, 8(2):237–283, 2002.

[16] Tsai CC and Marek-Sadowska M. Boolean functions classification via fixed polarity Reed–Muller forms. *IEEE Trans. Computers*, 46(2):173–186, 1997.

[17] Yanushkevich SN, Miller DM, Shmerko VP, and Stanković RS. *Decision Diagram Techniques for Micro- and Nanoelectronic Design*, Taylor & Francis/CRC Press, Boca Raton, FL, 2006.

10

Optimization of Computational Structures

10.1 Introduction

Optimization techniques can be applied to an arbitrary Boolean function described by the logical primitives such as AND, OR, NOR, NAND, EXOR, etc. Optimization results in representation with optimal or quasi-optimal parameters. These parameters are defined using the cost function and can include such parameters as a number of literals, and terms, and operations over the literal and terms. The goal of optimization is to remove the redundancy in the representation of Boolean functions. Because of various data structures for the representation of Boolean functions, there are different techniques for optimization.

Boolean functions can be considered in various forms called *Boolean data structures*, divided into *algebraic* and *graphical* representations. Algebraic representations include algebraic forms based on Boolean formulas, matrix forms, and tabulated forms. Graphical representations include K-maps, hypercubes, logic networks over the library of logic gates, networks of threshold elements, and decision diagrams.

Both algebraic and graphical forms are often not optimal; that is, they are redundant in their representation of a given Boolean function. The process of removing *redundancy* in Boolean data structures is called the *optimization* or *minimization* of Boolean functions.

Since optimization at various stages of design is critical for further implementation steps, the following tools are required: (a) efficient optimization algorithms and (b) efficient manipulation and conversion between data structures. In this chapter, approaches to the minimization of Boolean data structures given in SOP and POS forms are introduced. These minimization techniques result, respectively, in two-level AND–OR and OR–AND logic networks. Techniques based on K-maps, hypercubes, and decision diagrams are the focus of this chapter.

Decision trees, a graphical data structure for representing Boolean functions, were introduced in Chapter 2. In general, decision trees are

redundant data structures in which the redundancy of the representation
of Boolean functions is expressed in graphical form. Reducing decision
trees means removing redundancy in function representation. Hence, the
optimization of decision trees relates to the optimization of Boolean functions.
The resulting graphs are called *decision diagrams*. Besides the usage of
decision diagrams as an abstract data structure at almost every stage of logic
design, they are also suited for direct mapping into physical implementation
for some technologies.

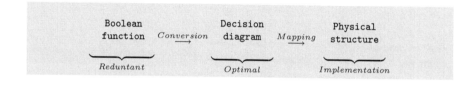

Example 10.1 *In Figure 10.1, the truth table of the three-input AND
gate is transferred into a cube and decision diagram.*

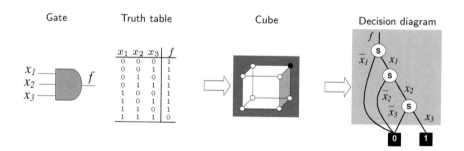

FIGURE 10.1
Relationships between truth tables, hypercubes, and decision diagrams for the
representation of Boolean function $f = x_1 x_2 x_3$ (Example 10.1).

10.2 Minterm and maxterm expansions

In Chapters 4 and 6, SOP and POS expansions were introduced as algebraic
data structures for the representation (description) of Boolean functions. In

the construction of these SOP and POS expansions, minterms and maxterms, respectively, were used.

A Boolean function f, written as a sum of minterms, is referred to as a *minterm expansion*, or a *standard sum-of-products (SOP)*. Let a Boolean function f of n variables be specified by its truth vector $\mathbf{F}$. The conversion of $\mathbf{F}$ into a minterm expansion means detecting 1s in the truth table and identifying the corresponding variable assignments.

Example 10.2 *Given the truth vector $\mathbf{F} = [0101]^T$, the shorthand notation of the SOP is $f = \bigvee m(1,3)$, since the truth vector $\mathbf{F}$ contains two ones, corresponding to the assignments $01_2 = 1_{10}$ and $11_2 = 3_{10}$ of the minterms $x_1^0 x_2^1 = \overline{x}_1 x_2$ and $x_1^1 x_2^1 = x_1 x_2$.*

A *maxterm expansion* or *standard product of sums (POS)* is the representation of a Boolean function as a product of maxterms. Let the Boolean function f of n variables be specified by its truth vector $\mathbf{F}$. To convert $\mathbf{F}$ into a maxterm expansion, the variable assignments corresponding to 0s in the truth vector must be identified, and the corresponding maxterms should be generated.

Example 10.3 *Given the truth vector $\mathbf{F} = [0101]^T$, the POS is $f = \prod M(0,2)$, since the truth vector $\mathbf{F}$ contains two zeros, corresponding to the assignments $00_2 = 0_{10}$ and $10_2 = 2_{10}$ of the maxterms $(x_1^0 \vee x_2^0) = (x_1 \vee x_2)$ and $(x_1^{\overline{1}} \vee x_2^{\overline{0}}) = (\overline{x}_1 \vee x_2)$.*

The minterm expression and maxterm expression can be directly converted into logic networks, as shown in Figure 10.2.

Example 10.4 *Given the standard SOP expression $f = x_1 \overline{x}_2 \vee \overline{x}_1 x_2$, the corresponding logic network is given in Figure 10.3.*

Noncanonical (nonstandard) SOP and POS expressions can be converted to two-level AND–OR and OR–AND networks, respectively, in the same way, except that the products and sums (which may not include all the function's variables) will be implemented instead of minterms and maxterms. This mapping is the most efficient, or least costly in terms of the number of logic gates, if the obtained SOP and POS are minimal forms. This will be discussed later in this chapter.

Two-level AND–OR and OR–AND logic networks can be directly converted into the corresponding standard SOP and POS expressions, respectively.

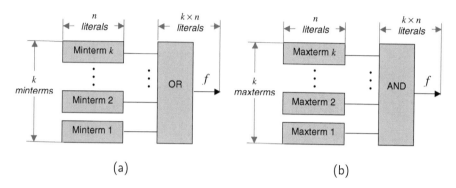

FIGURE 10.2
Converting minterm expressions (a) and maxterm expressions (b) into logic networks.

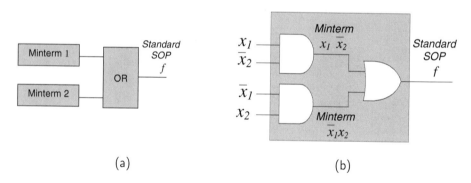

FIGURE 10.3
Mapping of a standard SOP expression into an AND–OR network (Example 10.4).

Example 10.5 *Figure 10.4a shows an OR–AND network and its conversion into a POS expression, which includes three sums, implemented by the two-input OR gates, combined by the products (three-input OR-gate): $f = (\overline{x}_1 \vee x_2)(\overline{x}_1 \vee x_3)(x_2 \vee x_3)$. Note that these sums are not maxterms since they do not include all three variables.*

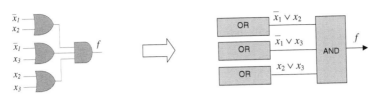

FIGURE 10.4
Converting an OR–AND logic network into a POS expression (Example 10.5).

10.3 Optimization of Boolean functions in algebraic form

Optimization of algebraic expressions of Boolean functions means their transformation into forms that better meet the implementation criteria (Figure 10.5). The optimization criteria of the SOP and POS forms relate, in particular, to the structure of the algebraic expression, the number of terms, the number of literals, and the fan-in.

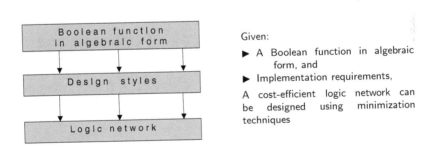

FIGURE 10.5
The conversion of Boolean functions into logic networks is based on various techniques.

The basic principle of minimization is *simplification*, or *reduction*. Techniques for the simplification of algebraic forms of Boolean functions are based on the following actions: (a) combining terms, (b) eliminating terms, (c) eliminating literals, and (d) adding redundant terms for the efficient elimination of terms and literals. Note that simplification does not guarantee that the resulting Boolean expression is minimal. Nevertheless, reduction is the first step to obtaining a minimal algebraic expression. The theoretical background of reduction is the consensus theorem.

10.3.1   The consensus theorem

The consensus theorem and its dual states that

$$XY \vee \overline{X}Z \vee \underbrace{YZ}_{\substack{Consensus \\ term}} = XY \vee \overline{X}Z \tag{10.1}$$

$$(X \vee Y)(\overline{X} \vee Z)\underbrace{(Y \vee Z)}_{\substack{Consensus \\ term}} = (X \vee Y)(\overline{X} \vee Z) \tag{10.2}$$

where the term XZ is called the *consensus term*. This term is redundant and is eliminated. Note that the Equation 10.2 is dual of the consensus theorem.

> **Example 10.6**   *The Boolean function* $f = x_1x_2x_3x_4 \vee \overline{x}_2x_3x_4x_5 \vee \overline{x}_1\overline{x}_2 \vee x_2x_3\overline{x}_5$ *can be simplified by adding the consensus term* $x_1x_3x_4x_5$ *and using the consensus theorem as follows:*
>
> $$f = x_1x_2x_3x_4 \vee \overline{x}_2x_3x_4x_5 \vee \overline{x}_1\overline{x}_2 \vee x_2x_3\overline{x}_5 \vee x_1x_3x_4x_5$$
> $$= \overline{x}_1\overline{x}_2 \vee x_2x_3\overline{x}_5 \vee x_1x_3x_4x_5$$

Generalization of the consensus theorem is as follows. Let xP_1 and $\overline{x}P_2$ be the Boolean terms where P_1 and P_2 are products of literals. Then the *operator* CONSENSUS is defined as

$$xP_1 \underbrace{\text{CONSENSUS}}_{Operator} \overline{x}P_2 = \underbrace{P_1P_2}_{Consensus \ term}$$

> **Example 10.7**   *The consensus operator exists for the following terms:*
>
> (a) $\boxed{x_1}\underbrace{\overline{x}_2x_3}_{P_1}$ CONSENSUS $\boxed{\overline{x}_1}\underset{P_2}{\underset{\uparrow}{x_4}} = \underbrace{\overline{x}_2x_3x_4}_{P_1P_2}$
>
> (b) $\overline{x}_1\boxed{\overline{x}_2}\overline{x}_3$ CONSENSUS $\overline{x}_1\boxed{x_2}\overline{x}_3 = \overline{x}_1\overline{x}_3$

The following property of the consensus operator is useful in reducing Boolean functions:

$$xP_1 \vee \overline{x}P_2 \vee P_1P_2 = xP_1 \vee \overline{x}P_2 \tag{10.3}$$
$$(x \vee P_1)(\overline{x} \vee P_2)(P_1 \vee P_2) = (x \vee P_1)(\overline{x} \vee P_2) \tag{10.4}$$

The consensus theorem (Equation 10.3) and its dual (Equation 10.4) states that the consensus terms P_1P_2 and $P_1 \vee P_2$, respectively, are redundant and can be removed from a Boolean expression.

10.3.2 Combining terms

Two terms can be combined using the *adjacency* rule: $XY \vee X\overline{Y} = X$. The adjacency rule can be used when

▶ Two product terms contain the same variables.
▶ Exactly one of the variables appears complemented in one term and not in the other.

When applied to an SOP expression, one product term may be duplicated, and each of the duplicates can be combined with two or more other terms. Techniques for combining terms are demonstrated in Table 10.1.

Example 10.8 *The Boolean function $f = x_1x_2x_3 \vee \overline{x}_1x_2x_3 \vee x_1\overline{x}_2x_3$ can be simplified by duplicating the product term $x_1x_2x_3$ twice and using the adjacency rule for each of the duplicates and the remaining terms: $f = x_1x_2x_3 \vee \overline{x}_1x_2x_3 \vee x_1x_2x_3 \vee x_1\overline{x}_2x_3 = x_2x_3 \vee x_1x_3$.*

10.3.3 Eliminating terms

Redundant terms can be eliminated using the *adjacency* rule $X \vee XY = X$ and the consensus theorem $XY \vee \overline{X}Z \vee YZ = XY \vee \overline{X}Z$. Techniques for eliminating terms are demonstrated in Table 10.1.

Example 10.9 *The Boolean function $f = \overline{x}_1x_2\overline{x}_3x_4 \vee \overline{x}_1x_2x_4 \vee x_1x_3x_4$ can be simplified by eliminating the product term $\overline{x}_1x_2\overline{x}_3x_4$: $f = (\overline{x}_1x_2\overline{x}_3x_4 \vee \overline{x}_1x_2x_4) \vee x_1x_3x_4 = \overline{x}_1x_2x_4 \vee x_1x_3x_4$.*

10.3.4 Eliminating literals

Redundant literals can be eliminated by the simplification rule: $X \vee \overline{X}Y = X \vee Y$. Techniques for eliminating literals are demonstrated in Table 10.1.

Example 10.10 *The Boolean function $f = x_1x_2 \vee x_1\overline{x}_2x_3$ can be simplified by eliminating the literal $\overline{x}_2$: $f = x_1x_2 \vee x_1\overline{x}_2x_3 = x_1x_2 \vee x_1x_3$.*

TABLE 10.1
Techniques of Boolean function simplification

The rule	Techniques
Combining terms	$(a)\ f = \overbrace{\underbrace{x_1x_2\overline{x}_3}_{X}\ \underbrace{\overline{x}_4}_{Y}}^{2\ terms} \vee \overbrace{\underbrace{x_1x_2\overline{x}_3}_{X}\ \underbrace{x_4}_{\overline{Y}}}^{1\ term} = x_1x_2\overline{x}_3$ $(b)\ f = \overbrace{x_1\overline{x}_2x_3 \vee x_1x_2x_3 \vee \overline{x}_1x_2x_3}^{3\ terms}$ $= x_1\overline{x}_2x_3 \vee \underbrace{x_1x_2x_3 \vee x_1x_2x_3}_{Duplicated} \vee \overline{x}_1x_2x_3$ $= \underbrace{x_1\overline{x}_2x_3 \vee x_1x_2x_3}_{Combining} \vee \underbrace{x_1x_2x_3 \vee \overline{x}_1x_2x_3}_{Combining} = \underbrace{x_1x_3 \vee x_2x_3}_{2\ terms}$
Eliminating terms	$(a)\ f = \overbrace{\underbrace{\overline{x}_1x_2}_{X} \vee \underbrace{\overline{x}_1x_2\ x_3}_{Y}}^{3\ terms} = \overbrace{\overline{x}_1x_2}^{2\ terms}$ $(b)\ f = \overbrace{\underbrace{\overline{x}_1x_2}_{Z}\ \underbrace{\overline{x}_3}_{\overline{X}} \vee \underbrace{x_2x_4}_{Y}\ \underbrace{x_3}_{X} \vee \underbrace{\overline{x}_1x_2}_{Z}\ x_4}^{3\ terms}$ $= \underbrace{\overline{x}_1x_2}_{Z}\ \underbrace{\overline{x}_3}_{\overline{X}} \vee \underbrace{x_2x_4}_{Y}\ \underbrace{x_3}_{X} \vee \underbrace{\overline{x}_1x_2}_{Z}\ \overbrace{x_2x_4}^{Y}$ $= \underbrace{\overline{x}_1x_2\overline{x}_3 \vee x_2x_3x_4}_{2\ terms}$
Eliminating literals	$f = \overbrace{\overline{x}_1x_2 \vee \overline{x}_1\overline{x}_2\overline{x}_3\overline{x}_4 \vee x_1x_2x_3\overline{x}_4}^{10\ literals}$ $= \overline{x}_1(x_2 \vee \overline{x}_2\overline{x}_3\overline{x}_4) \vee x_1x_2x_3\overline{x}_4$ $= \overline{x}_1(x_2 \vee \overline{x}_3\overline{x}_4) \vee x_1x_2x_3\overline{x}_4$ $= x_2(\overline{x}_1 \vee x_1x_3\overline{x}_4) \vee \overline{x}_1\overline{x}_3\overline{x}_4$ $= x_2(\overline{x}_1 \vee x_3\overline{x}_4) \vee \overline{x}_1\overline{x}_3\overline{x}_4$ $= \underbrace{\overline{x}_1x_2 \vee x_2x_3\overline{x}_4 \vee \overline{x}_1\overline{x}_3\overline{x}_4}_{8\ literals}$
Adding redundant terms	$f = \overbrace{x_1x_2 \vee x_2x_3 \vee \overline{x}_2\overline{x}_4 \vee x_1\overline{x}_3\overline{x}_4}^{4\ terms}$ $= x_1x_2 \vee x_2x_3 \vee \overline{x}_2\overline{x}_4 \vee x_1\overline{x}_3\overline{x}_4 \vee \underbrace{x_1\overline{x}_4}_{Adding}$ $= x_1x_2 \vee x_2x_3 \vee \overline{x}_2\overline{x}_4 \vee \underbrace{x_1\overline{x}_3\overline{x}_4 \vee x_1\overline{x}_4}_{Eliminating}$ $= \underbrace{x_1x_2 \vee x_2x_3 \vee \overline{x}_2\overline{x}_4}_{3\ terms}$

10.3.5 Adding redundant terms

Redundant terms can be useful for the efficient combining and elimination of terms and literals. Redundant terms do not change the value of a Boolean function. They can be incorporated into Boolean expressions in various ways; in particular, by adding $x\overline{x}$, and by multiplying by $x \vee \overline{x}$. Techniques for adding redundant terms are demonstrated in Table 10.1.

Example 10.11 *Adding x_2x_3 to the expression $x_1x_2 \vee \overline{x}_1x_3$ does not change the expression:*

$$x_1x_2 \vee \overline{x}_1x_3 \vee x_2x_3 = x_1x_2 \vee \overline{x}_1x_3 \vee (\overline{x}_1 \vee x_1)x_2x_3$$
$$= x_1x_2 \vee \overline{x}_1x_3 \vee \overline{x}_1x_2x_3 \vee x_1x_2x_3$$
$$= (x_1x_2 \vee x_1x_2x_3) \vee (\overline{x}_1x_3 \vee \overline{x}_1x_2x_3)$$
$$= x_1x_2 \vee \overline{x}_1x_3$$

Adding $x_1x_2x_3$ to the term x_1 does not change the term: $f = x_1 \vee x_1x_2x_3 = x_1 \cdot (1 \vee x_2x_3) = x_1 \cdot 1 = x_1$.

The formal basis of the verification is proving the validity of Boolean data structures (Chapter 4). In the example below, the techniques for proving validity of Boolean functions is extended by using consensus theorem and elimination rule.

Example 10.12 *Prove that*

$$\underbrace{\overline{x}_1x_2\overline{x}_4 \vee x_2x_3x_4 \vee x_1x_2\overline{x}_3 \vee x_1\overline{x}_2x_4}_{Left\ part}$$
$$= \underbrace{x_2\overline{x}_3\overline{x}_4 \vee x_1x_4 \vee \overline{x}_1x_2x_3}_{Right\ part}$$

Proving using algebraic manipulations include: (a) adding three consensus terms, (b) combining terms, and (c) elimination terms to the left part as follows:

$$\text{LEFT PART} \quad \vee \quad \underbrace{x_2\overline{x}_3\overline{x}_4 \vee \overline{x}_1x_2x_3 \vee x_1x_2x_4}_{Consensus\ terms}$$
$$= x_1x_4 \vee \underbrace{\overline{x}_1x_2\overline{x}_4 \vee x_2x_3x_4 \vee x_1x_2\overline{x}_3}_{Eliminated\ terms} \vee x_2\overline{x}_3\overline{x}_4 \vee \overline{x}_1x_2x_3$$
$$= x_2\overline{x}_3\overline{x}_4 \vee x_1x_4 \vee \overline{x}_1x_2x_3$$

10.4   Implementing SOP expressions using logic gates

Let a Boolean function be represented in SOP form, and let AND, OR, NOR, and NAND gates - a *library* of gates - be available for the design of logic networks. The problem is formulated as mapping, or implementing, a function in the form of a logic network using a library of gates. In Chapter 4, a set of logic operations was called *functionally complete* if any Boolean function could be expressed in terms of the operations from this set.

> **Example 10.13** *The set AND, OR, and NOT is functionally complete. Indeed, an SOP form of a Boolean function can be realized using only AND, OR, and NOT operations. This means that, given a Boolean function in its SOP form, the function can be implemented as a logic network of AND, OR and NOT gates.*

Some single Boolean operations, such as NAND or NOR, form a functionally complete set in themselves.

> **Example 10.14** *An arbitrary Boolean function can be implemented by using only NAND gates or only NOR gates.*

10.4.1   Two-level logic networks

A two-level logic network composed of AND and OR gates can be converted into a network composed of NAND gates or NOR gates, or combinations of AND, OR, NAND, and NOR gates. These conversions are carried out by manipulating Boolean functions using the property $x = \overline{\overline{x}}$ and by applying DeMorgan's rule:

$$x_1 \vee x_2 \vee \ldots \vee x_m = \overline{\overline{x_1 \vee x_2 \vee \ldots \vee x_m}} = \overline{\overline{x}_1 \overline{x}_2 \ldots \overline{x}_m} \qquad (10.5)$$

$$x_1 x_2 \ldots x_m = \overline{\overline{x_1 x_2 \ldots x_m}} = \overline{\overline{x}_1 \vee \overline{x}_2 \vee \ldots \vee \overline{x}_m}. \qquad (10.6)$$

> **Example 10.15** *Consider a SOP $f = x_1 \vee x_2 \overline{x}_2 x_3 \vee \overline{x}_2 x_3 x_4$ This expression can be implemented using a two-level AND–OR network. Manipulation of this expression results in another two-level expressions, corresponding to the networks shown in Table 10.2.*

The algorithm for designing a two-level NAND-only logic network is given below:

TABLE 10.2
Techniques for manipulating Boolean functions in two-level logic network design

Design example: two-level network design

Conversion	Two-level logic network

(a) AND-OR

$$f = x_1 \lor x_2\overline{x}_2x_3 \lor \overline{x}_2x_3x_4$$

(b) OR-AND

$$f = (x_1 \lor x_2 \lor x_3)(x_1 \lor \overline{x}_2 \lor \overline{x}_3)(x_1 \lor \overline{x}_3 \lor x_4)$$

AND-OR   OR-AND

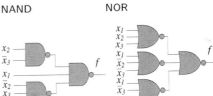

(c) NAND

$$f = x_1 \lor x_2\overline{x}_2x_3 \lor \overline{x}_2x_3x_4$$
$$= \overline{\overline{x_1 \lor x_2\overline{x}_2x_3 \lor \overline{x}_2x_3x_4}}$$
$$= \overline{\overline{x}_1 \overline{x_2\overline{x}_3} \overline{\overline{x}_2x_3x_4}}$$

(d) NOR

$$f = (x_1 \lor x_2 \lor x_3)(x_1 \lor \overline{x}_2 \lor \overline{x}_3)(x_1 \lor \overline{x}_3 \lor x_4)$$
$$= \overline{\overline{(x_1 \lor x_2 \lor x_3)(x_1 \lor \overline{x}_2 \lor \overline{x}_3)(x_1 \lor \overline{x}_3 \lor x_4)}}$$
$$= \overline{\overline{(x_1 \lor x_2 \lor x_3)} \lor \overline{(x_1 \lor \overline{x}_2 \lor \overline{x}_3)} \lor \overline{(x_1 \lor \overline{x}_3 \lor x_4)}}$$

NAND   NOR

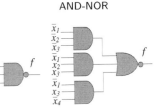

(e) OR-NAND

$$f = x_1 \lor x_2\overline{x}_2x_3 \lor \overline{x}_2x_3x_4$$
$$= \overline{\overline{x_1 \lor x_2\overline{x}_2x_3 \lor \overline{x}_2x_3x_4}}$$
$$= \overline{\overline{x}_1 \overline{x_2\overline{x}_3} \overline{\overline{x}_2x_3x_4}}$$
$$= \overline{\overline{x}_1(\overline{x}_2 \lor x_3)(x_2 \lor \overline{x}_3 \lor \overline{x}_4)}$$
$$= x_1 \lor \overline{(\overline{x}_2 \lor x_3)} \lor \overline{x_2 \lor \overline{x}_3 \lor \overline{x}_4}$$

(f) AND-NOR

$$f = \overline{(x_1 \lor x_2 \lor x_3)} \lor \overline{(x_1 \lor \overline{x}_2 \lor \overline{x}_3)} \lor \overline{(x_1 \lor \overline{x}_3 \lor x_4)}$$
$$= \overline{\overline{x}_1\overline{x}_2\overline{x}_3 \lor \overline{x}_1x_2x_3 \lor \overline{x}_1x_3\overline{x}_4}$$

OR-NAND   AND-NOR

(g) OR-NAND

$$f = x_1 \lor x_2\overline{x}_2x_3 \lor \overline{x}_2x_3x_4$$
$$= \overline{\overline{x_1 \lor x_2\overline{x}_2x_3 \lor \overline{x}_2x_3x_4}}$$
$$= \overline{\overline{x}_1 \overline{x_2\overline{x}_3} \overline{\overline{x}_2x_3x_4}}$$
$$= \overline{\overline{x}_1(\overline{x}_2 \lor x_3)(x_2 \lor \overline{x}_3 \lor \overline{x}_4)}$$

(h) NAND-AND

$$f = \overline{x}_1\overline{x}_2\overline{x}_3 \lor \overline{x}_1x_2x_3 \lor \overline{x}_1x_3\overline{x}_4$$
$$= \overline{\overline{x}_1\overline{x}_2\overline{x}_3} \, \overline{\overline{x}_1x_2x_3} \, \overline{\overline{x}_1x_3\overline{x}_4}$$

OR-NAND   NAND-AND

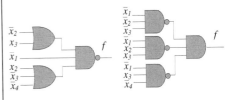

Step 1: Find a minimal SOP for the Boolean function f.
Step 2: Draw the corresponding two-level AND-OR logic network.
Step 3: Replace AND and OR gates with NAND gates leaving the gate interconnections unchanged.
Step 4: If the output gate has any single literals as inputs, complement these literals.

The algorithm for designing a two-level NOR-only logic network is as follows:

Step 1: Find a minimal POS for the Boolean function f.
Step 2: Draw the corresponding two-level OR-AND logic network.
Step 3: Replace OR and AND gates with NOR gates, leaving the gate interconnections unchanged.
Step 4: If the output gate has any single literals as inputs, complement these literals.

10.4.2 Multilevel logic networks

If the design criteria require conversion of a Boolean expression into a multilevel logic network, so-called *factorization* of the expression is used. Factoring is implemented using the SOP or POS expressions. If an SOP or POS expression is given, it has to be checked first for a degenerate form.

Degenerate SOP and POS forms

Algebraic expressions of a Boolean function f of n variables can include products and sums of one or more (up to n, in minterms and maxterms) variables. The following forms of SOP expressions are of particular interest in function manipulation:

▶ SOP expressions in which products are minterms (n variables in each minterm)

▶ SOP expressions in which products consist of two or more variables, but less than n

▶ SOP expressions in which products may consist of a single variable, are so-called *degenerate* SOPs

> **Example 10.16** *(a) The SOP form of the Boolean function f_1 of three variables, $f_1 = x_1 x_2 \overline{x}_3 \vee \overline{x}_1 x_2 x_3 \vee \overline{x}_1 \overline{x}_2 \overline{x}_3$, is the sum of minterms, or canonical SOP. The representation $f_2 = x_1 x_2 \overline{x}_3 \vee x_4$ of a function f_2 is a degenerate SOP expression.*

By analogy, in the POS form of a Boolean function, all sums are the sums of single variables. Three types of POS are distinguished:

▶ POS expressions in which sums are maxterms

▶ POS expressions in which sums may consist of two or more variables

▶ *Degenerate* POS expressions in which sums may consist of a single variable

Example 10.17 *The Boolean function f_1 of three variables, $f_1 = (x_1 \vee x_2 \vee \overline{x}_3)(\overline{x}_1 \vee x_2 \vee x_3)(\overline{x}_1 \vee \overline{x}_2 \vee \overline{x}_3)$, is represented by the product of maxterms. The representation of a Boolean function in the form $f_2 = (x_1 \vee x_2 \vee \overline{x}_3)x_4$ is the degenerate POS expression.*

The rationale for considering the generate form of an SOP expression is its usage for the simplification of implementation using AND and OR logic gates.

Factored expression

A *factored* expression of a Boolean function is defined as follows:

▶ A literal is a factored expression.

▶ A logical sum is a factored expression.

▶ A logical product of factored expressions is a factored expression.

It follows from this definition that in factored expressions, complements are permitted only for the literals. Factorization is the conversion of a degenerate SOP form into a degenerate POS form:

A Boolean expression is said to be

▶ *Totally factored* if it is in maxterm form; if a Boolean function is in fully factored form, the function cannot be factored any further.

▶ *Partially factored* if it is in POS form; if a Boolean function is in partially factored form, the function can be factored further.

The distributive laws considered in Chapter 4 can be used to factor Boolean expressions.

Example 10.18 *The terms $(\overline{x}_1 \vee x_2)(x_1 \vee x_3)x_2$ and $(x_1 \vee x_2)(x_3 \vee (\overline{x}_4(x_5 \vee \overline{x}_6)))$ are totally and partially factored expressions, respectively. The term $(\overline{x_1 \vee x_2})x_4$ is not a factored expression.*

Example 10.19 *Techniques for factoring using the distributive law are illustrated below for the Boolean functions* $f_1 = x_1 \vee \overline{x}_2 x_3 x_4$ *and* $f_2 = \overline{x}_3 x_4 \vee \overline{x}_3 \overline{x}_5 \vee \overline{x}_6 x_7$:

$$(a)\ f_1 = \underbrace{x_1 \vee \overline{x}_2 x_3 x_4}_{Distributive\ law} = (x_1 \vee \overline{x}_2)\ \underbrace{(x_1 \vee x_3 x_4)}_{Distributive\ law}$$

$$= \underbrace{(x_1 \vee \overline{x}_2)(x_1 \vee x_3)(x_1 \vee x_4)}_{Totally\ factored\ form}$$

$$(b)\ f_2 = \overline{x}_3 x_4 \vee \overline{x}_3 \overline{x}_5 \vee \overline{x}_6 x_7 = \overline{x}_3 \underbrace{(x_4 \vee \overline{x}_5)}_{Distributive\ law} \vee \overline{x}_6 x_7$$

$$= \underbrace{(\overline{x}_3 \vee \overline{x}_6 x_7)}_{Distributive\ law}\ \overbrace{(x_4 \vee \overline{x}_5 \vee \overline{x}_3 x_7)}$$

$$= \underbrace{(\overline{x}_3 \vee \overline{x}_6)(\overline{x}_3 \vee x_7)(x_4 \vee \overline{x}_5 \vee \overline{x}_6)(x_4 \vee \overline{x}_5 \vee x_7)}_{Totally\ factored\ form}$$

The factored expressions f_1 *and* f_2 *are in maxterm form; these expressions cannot be factored any further.*

10.4.3 Conversion of factored expressions into logic networks

Factored expressions can be converted into logic networks. Totally factored expressions (the Boolean function can not be factored any further) have several useful properties, in particular, with respect to (a) the number of literals and terms, (b) fan-out and fan-in of logic networks, and (c) the number of gates of logic networks.

Property 1: The sizes (in number of literals and terms) for totally factored expressions for a Boolean function f and its complement $\overline{f}$ are the same. This is because in a factored expression, the complement of f, $\overline{f}$, is obtained by replacing AND with OR operations, and by replacing the variables with their complements.

Property 2: A factored expression corresponds to a fan-out-free multilevel logic network.

Property 3: The amount of memory needed for factored expressions is not as large as that needed for storing SOPs or truth tables.

Property 4: Factoring:

(a) increases the number of gates, (b) decreases fan-in, and (c) reduces the number of literals in the expression.

An SOP expression can be implemented by one or more OR gates and by feeding a single AND gate at the logic network output. The example below illustrates how factoring can be used to deal with the fan-in problem.

Example 10.20 *Suppose that the available gates have a maximum fan-in of three, and implements the Boolean function $f = x_1\overline{x}_2\overline{x}_3\overline{x}_4 \vee x_1x_2\overline{x}_3x_4$. The direct implementation of the function f (Figure 10.6) does not meet the requirement of the maximum allowed fan-in. A solution is achieved by factoring the function $f = x_1\overline{x}_3(\overline{x}_2\overline{x}_4 \vee x_2x_4)$.*

SOP expression   Factored SOP expression

$f = x_1\overline{x}_2\overline{x}_3\overline{x}_4 \vee x_1x_2\overline{x}_3x_4$ $f = x_1\overline{x}_3(\overline{x}_2\overline{x}_4 \vee x_2x_4)$

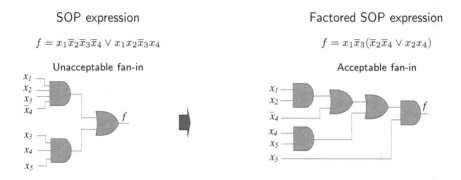

FIGURE 10.6
Implementation of the SOP form of a Boolean function and the factored form of the SOP expression (Example 10.20).

10.5 Minimization of Boolean functions using K-maps

In Chapter 2, the K-map was introduced as a useful data structure for the representation of Boolean functions. The K-map also provides a systematic way of performing optimization on Boolean functions:

To find a minimal SOP expression, the K-map is used for the representation of Boolean function in its standard (canonical) SOP form. Each 1 in the K-map corresponds to the minterm. Reduction process is based on the use of

the identities $x_1 x_2 \vee \overline{x}_1 x_2 = x_2$ and $\overline{x}_1 \vee \overline{x}_1 x_2 = x_1 \vee x_2$ to combine and simplify the product terms. Any two adjacent squares in a K-map correspond to minterms that differ in only one variable.

Minimization of Boolean functions of two variables

A two-variable K-map is a 2D representation of a two-variable truth table. It is used for minimizing SOP or POS expressions of a Boolean function of two variables (Figure 10.7).

Example 10.21 *Consider the K-map shown in Figure 10.7a.*

(a) *Let $m_1 = \overline{x}_1 x_2$ and $m_3 = x_1 x_2$ ($m_2 = m_4 = 0$). The two minterms m_1 and m_3 differ in that the variable x_1 appears in its complimented form in m_1 and its uncomplemented form in m_3. These minterms can be combined $\overline{x}_1 x_2 \vee x_1 x_2 = x_2$.*

(b) *Let $m_0 = \overline{x}_1 \overline{x}_2$, $m_1 = \overline{x}_1 x_2$, and $m_2 = x_1 \overline{x}_2$ ($m_3 = 0$). The minterm m_0 can be combined with both the minterm m_1 and m_2: $\overline{x}_1 \overline{x}_2 \vee \overline{x}_1 x_2 \vee x_1 \overline{x}_2 = \overline{x}_1 \vee x_1 \overline{x}_2 = \overline{x}_1 \vee x_2$.*

Example 10.22 *Techniques for minimization based on K-maps for Boolean functions of two variables are illustrated in Figure 10.8.*

Minimization of Boolean functions of three variables

The K-map of a function of three variables is represented as a 2×4 table, placed so that the neighbor cells are adjacent (the corresponding variable assignments of the two neighbor cells differ by exactly one bit).

Example 10.23 *Minimization of various Boolean functions of three variables using their K-maps is illustrated in Figure 10.9. For example, the minimal SOP expression in the case (d) corresponds to the following manipulation*

$$f = \overline{x}_1 \overline{x}_2 \overline{x}_3 \vee \overline{x}_1 \overline{x}_2 x_3 \vee x_1 \overline{x}_2 x_3$$
$$= \overline{x}_1 \overline{x}_2 \vee x_1 \overline{x}_2 = \overline{x}_2$$

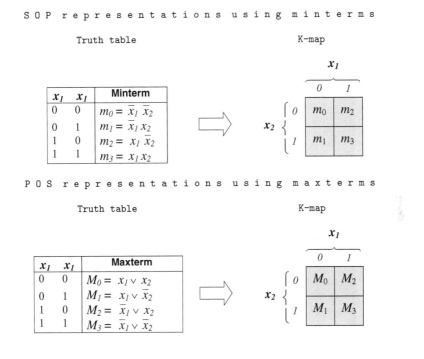

FIGURE 10.7
K-maps for the representation of Boolean functions of two variables in the standard SOP form (a) and standard POS form (b) (Example 10.21).

Minimization of Boolean functions of four variables

The K-map for an SOP or POS expression of a Boolean function of four variables is a 4×4 table.

> **Example 10.24** *Figure 10.10 contains various examples of K-maps of Boolean functions of four variables and the results of minimization.*

Don't-care conditions in minimization

The values of a Boolean function which are not specified are called *don't-care conditions*. Such a Boolean function is called *incompletely specified* function. In the K-map, don't-care conditions are marked by a d entry. In combining squares, a d entry is used whenever possible to form larger blocks of squares. A d-square can be neglected if it is not needed in minimization. Incompletely specified Boolean function is described in terms of the don't-care conditions d.

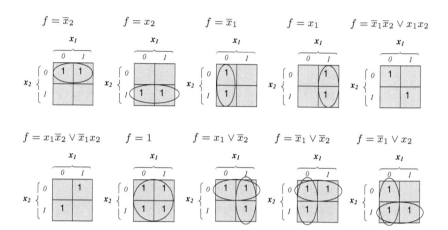

FIGURE 10.8
Technique for minimization of Boolean functions of two variables using K-maps (Example 10.22).

Example 10.25 *Consider K-map given in Figure 10.11. In the case (a), the don't-care condition d corresponding to the value of Boolean function $f(110)$ is not used in the minimization. In the case (b), the don't-care condition d corresponding to $f(010)$ and $f(110)$ are used.*

Minimal POS expressions

The minimal POS expression can be derived using K-map. In this case, 0 entries are considered in the K-map. Each 0 entry corresponds to the maxterm. That is, a K-map represents a standard (canonical) POS. The minimal POS expression is obtained in exactly the same manner as for SOP forms except that we try to enclose the largest number of 0s into groups.

Example 10.26 *The minimization of Boolean functions in a POS form is shown in Figure 10.12.*

Conversion of minterm expressions into maxterm expressions, and vice versa

The same K-map represents both minterms (corresponding to 1-cells in the K-map) and maxterms (corresponding to 0-cells). Minimization of the SOP form

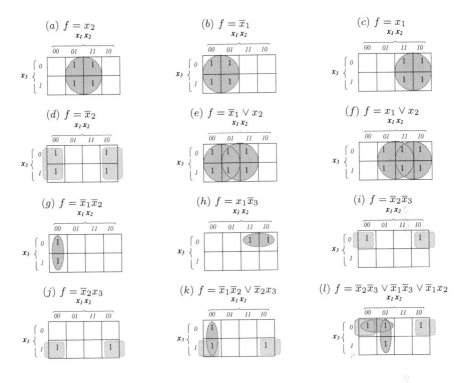

FIGURE 10.9

Techniques for the minimization of Boolean functions of three variables using K-maps (Example 10.23).

of the function is normally performed on the 1-cells by combining adjacent cells. To accomplish the minimization of a POS, 0-cells must be considered, and the adjacent cells must be combined when possible. This operation on the 0-cells of the same map results in obtaining the POS for the same function and can be considered as a conversion of the minimal SOP to a minimal POS form of the Boolean function.

Example 10.27 *Conversion of the minimal SOP expression $f = x_1 \vee x_2\overline{x}_3 \vee \overline{x}_2x_3x_4$ into the minimal POS expression using K-maps is shown in Figure 10.13).*

The consensus theorem (Equation 10.3) and its dual (Equation 10.4) can be used in various Boolean data structures: algebraic equations, truth tables, hypercubes, K-maps, logic networks, and decision diagrams. The example below illustrates the consensus terms in K-maps.

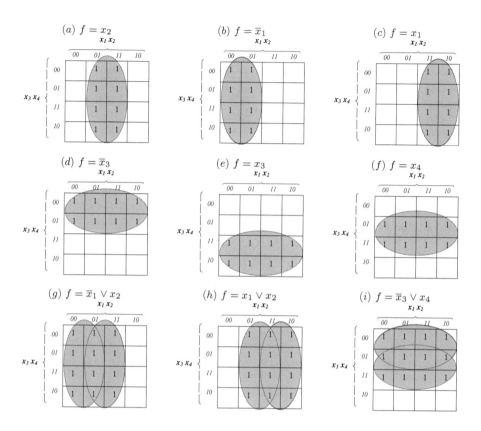

FIGURE 10.10
Techniques for the minimization of Boolean functions of four variables using K-maps (Example 10.24).

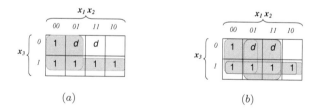

FIGURE 10.11
Don't-care conditions in minimization (Example 10.25).

Example 10.28 *Figure 10.14 shows the consensus terms for Boolean functions of three and four variables. In both functions, these terms can be removed.*

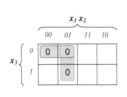

Standard POS expression:
$f = (x_1 \vee x_2 \vee x_3)(x_1 \vee \overline{x}_2 \vee x_3)(x_1 \vee \overline{x}_2 \vee \overline{x}_3)$.
The reduction rules: $(x \vee y)(x \vee \overline{y} = x)$ and
$(x \vee y)(x \vee \overline{y} \vee z) = (x \vee y)(x \vee z)$.
The minimization using K-map corresponds to the
following manipulation:
$f = (x_1 \vee x_2 \vee x_3)(x_1 \vee \overline{x}_2 \vee x_3)(x_1 \vee \overline{x}_2 \vee \overline{x}_3)$
$\quad = (x_1 \vee x_3)(x_1 \vee \overline{x}_2 \vee \overline{x}_3) = (x_1 \vee x_3)(x_1 \vee \overline{x}_2)$

FIGURE 10.12

Minimization of Boolean function given the standard POS expression
(Example 10.26).

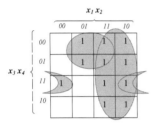

Minimumal SOP form
$f = x_1 \vee x_2\overline{x}_3 \vee \overline{x}_2 x_3 x_4$

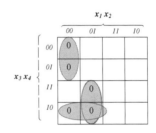

Minimal POS form
$f = (x_1 \vee x_2 \vee x_3)(x_1 \vee \overline{x}_2 \vee \overline{x}_3)(x_1 \vee \overline{x}_3 \vee x_4)$

FIGURE 10.13

Minimal SOP and POS expressions for the Boolean function (Example 10.27).

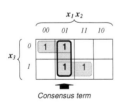

Consensus term

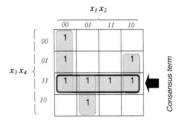

FIGURE 10.14

Consensus theorem in the form of K-maps (Example 10.28).

10.6 Optimization of Boolean functions using decision trees and decision diagrams

The conversion of a decision tree into a decision diagram is called the
optimization of this graphical structure. This approach to optimization can

be used for small- and medium-sized Boolean functions, since the number of nodes in the complete decision tree of a Boolean function of n variables is $2^{2n} - 1$; i.e., it depends exponentially on n. For instance, the complete decision tree of a function of one hundred variables has 2^{100} terminal nodes, and the total number of nodes is $2^{200} - 1$. In contemporary logic design, a decision diagram for a network is constructed and dynamically optimized directly from the circuit netlist. The disadvantage of this approach is that the structure obtained may not be canonical, since it depends heavily on the network structure and implementation. Decision trees do not have this flaw, since they are canonical, being constructed based on the truth tables of the functions. For medium-sized Boolean functions, reduced decision trees can be constructed using the properties and/or incompleteness of the data.

> **Example 10.29** *The complete decision trees and the reduced ordered decision diagrams for four Boolean functions are given in Figure 10.15.*

10.6.1 The formal basis for the reduction of decision trees and diagrams

The reduction procedures for decision trees and diagrams are based on the major theorems of Boolean algebra. Below, we give the graph-based interpretation of these rules.

> **Example 10.30** *The absorption rules (see Chapter 4), $xy \vee x\overline{y} = x$ and $(x \vee y)(x \vee \overline{y}) = x$, correspond to a node reduction in binary decision trees and diagrams (Figure 10.16).*

The consensus theorem states that $xy \vee \overline{x}z \vee yz = xy \vee \overline{x}z$. This theorem is used to eliminate the redundant terms in a Boolean expression.

> **Example 10.31** *An interpretation of the consensus theorem in terms of binary decision diagrams is given in Figure 10.17.*

Note that the resulting decision diagram is minimal with respect to the number of nodes and the order of variables. The Boolean equation, restored from the decision diagram, is not necessary minimal.

10.6.2 Decision tree reduction rules

A Boolean function can be represented by a decision tree that can be reduced and further converted into a decision diagram:

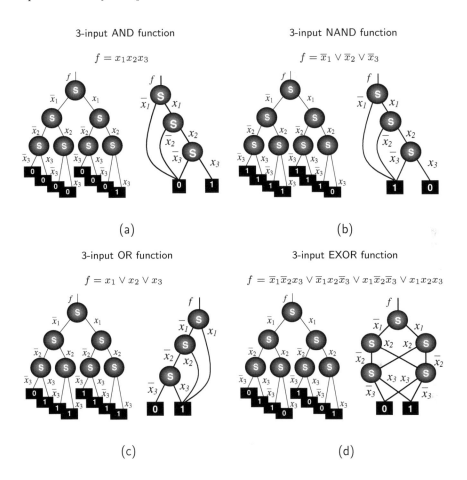

FIGURE 10.15
The complete decision trees and the reduced ordered decision diagrams for
the 3-input AND (a), NAND (b), OR (c) and EXOR (d) functions (Example
10.29).

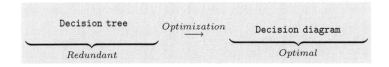

The initial binary tree or diagram has a given order of variables, that is, it
is an *ordered decision tree* or *diagram*. The resulting decision tree or diagram
is called a *reduced ordered binary decision tree* or *diagram (ROBDD)*.

Algebraic form

$$f = xy \vee x\overline{y}$$
$$= x(y \vee \overline{y}) = x$$

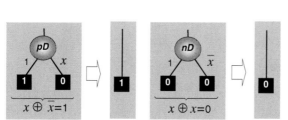

Graphical form

(a) *The terminal nodes*
     $1 \cdot x\overline{y}$ *and* $1 \cdot xy$ *are merged,*
(b) *Edges* $\overline{y}$ *and* y *are merged,*
(c) *The lower node* S *is eliminated*

FIGURE 10.16
Absorption rules of Boolean algebra in algebraic form and in terms of decision diagrams (Example 10.30).

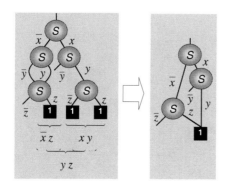

Algebraic form

$$f = xy \vee \overline{x}z \vee yz = xy \vee \overline{x}z$$

Graphical form

(a) *The terminal nodes* $xy\overline{z}$ *and* xyz
     are merged to $1 \cdot xy$, *and lower right* S *node is eliminated*
(b) *Lower left* S *node for variable* y *is eliminated*

FIGURE 10.17
Interpretation of the consensus theorem in terms of decision diagrams (Example 10.31).

Example 10.32 *An arbitrary Boolean function f of three variables can be represented by a complete decision tree, as shown in Figure 10.18. To design this tree, Shannon expansion is used as follows:*

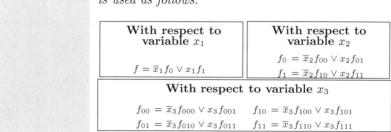

With respect to variable x_1	With respect to variable x_2
	$f_0 = \overline{x}_2 f_{00} \vee x_2 f_{01}$
$f = \overline{x}_1 f_0 \vee x_1 f_1$	$f_1 = \overline{x}_2 f_{10} \vee x_2 f_{11}$
With respect to variable x_3	
$f_{00} = \overline{x}_3 f_{000} \vee x_3 f_{001}$ $f_{10} = \overline{x}_3 f_{100} \vee x_3 f_{101}$	
$f_{01} = \overline{x}_3 f_{010} \vee x_3 f_{011}$ $f_{11} = \overline{x}_3 f_{110} \vee x_3 f_{111}$	

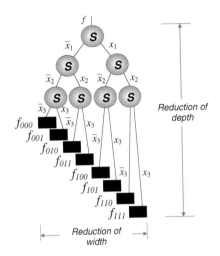

Boolean function of three variables:

$$f = f_{000}\overline{x}_1\overline{x}_2\overline{x}_3 \vee f_{001}\overline{x}_1\overline{x}_2x_3$$
$$\vee \; f_{010}\overline{x}_1x_2\overline{x}_3 \vee f_{011}\overline{x}_1x_2x_3$$
$$\vee \; f_{100}x_1\overline{x}_2\overline{x}_3 \vee f_{101}x_1\overline{x}_2x_3$$
$$\vee \; f_{110}x_1x_2\overline{x}_3 \vee f_{111}x_1x_2x_3$$

Characteristis of the decision tree:

▶ 3 levels
▶ 7 intermediate nodes
▶ 8 terminal nodes
▶ 2^k nodes in k-th level, $k = 0, 1, 2$

FIGURE 10.18
Implementation of the Shannon expansion using decision tree for a Boolean function of three variables (Example 10.32).

A decision tree is *reduced* if it does not contain any vertex v whose successors lead to the same node, and if it does not contain any distinct vertices v and v' such that the subgraphs rooted in v and v' are isomorphic.

The reduced decision tree is still a rooted acyclic graph in which every node but the root has an indegree 1. It can be further reduced to a decision diagram. In the decision diagram, the nodes can have an indegree higher than 1. A binary decision diagram is a directed acyclic graph that:

▶ Contains exactly one root.
▶ Has terminal nodes labeled by the constants 1 and 0 and has an indegree ≥ 2.
▶ Has internal nodes with an indegree ≥ 2 and an outdegree equal to 2.

In an *ordered* decision diagram, a linear variable order is placed on the input variables. The variables' occurrences on each path of this diagram have to be consistent with this order. A binary decision diagram represents a Boolean function f of n variables in the following way:

▶ Internal nodes represent a Shannon expansion and are labeled by S; each internal node except the root has an indegree of ≥ 1 and an outdegree equal to 2.
▶ Links between the nodes are labeled by a Boolean variable x_i, $i = 1, 2, ..., n$, and each internal node has an $\overline{x}_i$-link and x_i-link.
▶ Each assignment to the input variable x_i defines a unique path from the root to one of the terminal nodes.

▶ The label of the terminal node gives the value of Boolean function.

Example 10.33 *Figure 10.19 shows the correspondence of the path in the binary decision diagram and the Boolean product $\overline{x}_1 x_2$.*

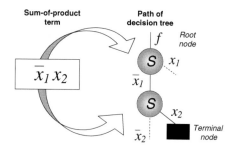

▶ The path is a chain of two nodes (root, intermediate node(s) and one terminal node) connected by links.
▶ If the terminal node is marked by 0, the decision diagram represents the sum term $(x_2 \vee \overline{x}_1) = 0$; if the terminal node is marked by 1, the decision diagram represents the product term $\overline{x}_1 x_2 = 1$.

FIGURE 10.19
A path in a decision diagram corresponds to the logical product or sum (Example 10.33).

The decision diagram for a given Boolean function f is derived from the decision tree for f by deleting redundant nodes, and by sharing equivalent subgraphs. The reduction rules are summarized in Figure 10.20 and considered in detail below.

Elimination rule

The elimination rule allows for the removal of redundant subtrees representing subfunctions. It states that if two descendent subtrees of a node are identical, then the node can be deleted, and the incoming edges of the deleted node can be connected to the corresponding successor. In terms of Shannon expansion, implemented by the node, this means that if $f_0 = f_1 = g$ then

$$f = \overline{x}_i f_0 \vee x_i f_0 = \overline{x}_i g \vee x_i g = (\overline{x}_i \vee x_i)g = g$$

Merging rules

Let the Shannon expansions for the Boolean functions α and β be

$$\alpha = \overline{x}_i \alpha_0 \vee x_i \alpha_1 \quad \text{and} \quad \beta = \overline{x}_i \beta_0 \vee x_i \beta_1$$

respectively.

Merging rule I

If $g = \alpha_0 = \beta_1$, then

$$\alpha = g \vee x_i \alpha_1 \quad \text{and} \quad \beta = \overline{x}_i \beta_0 \vee g$$

Merging rule II

If $\alpha_0 = \beta_0$ and $\alpha_1 = \beta_1$, then $\alpha = \beta$, and the nodes corresponding to α and β can be merged into one node with an indegree 2.

Particular cases of elimination of internal nodes connected to the terminal nodes are shown in Figure 10.21. Elimination of the node in Figures 10.21a and 10.21b corresponds to the equations

$$f = \overline{x}_i f_0 \vee x_i f_1 = \overline{x}_i 0 \vee x_i 0 = 0 \quad \text{and}$$
$$f = \overline{x}_i f_0 \vee x_i f_1 = \overline{x}_i 1 \vee x_i 1 = 1$$

The combination of the merging and eliminating rules is illustrated in Figure 10.21c for a particular case of g. Figure 10.21d shows the reduction of the logic network to one connection (wire) based on the elimination rule:

$$f = x_i g \vee \overline{x}_i g = g(x_i \vee \overline{x}_i) = g$$

Deriving a reduced decision tree from a complete decision tree

Given: A complete decision tree for a Boolean function.

Construct: A reduced decision tree.

Step 1: Apply elimination rule where possible to remove the redundant nodes (variables in the products) of isomorphic subtrees.

Step 2: Repeat Step 1 as necessary until no more nodes can be eliminated.

Techniques for reduction of decision trees

ELIMINATION RULE

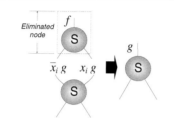

If $f_0 = f_1 = g$, then the Shannon expansion implies

$$f = \overline{x}_i g \vee x_i g = g$$

MERGING RULE I

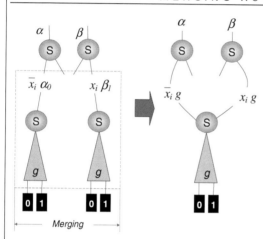

The Shannon expansions for the functions α and β are

$$\alpha = \overline{x}_i \alpha_0 \vee x_i \alpha_1$$
$$\beta = \overline{x}_i \beta_0 \vee x_i \beta_1$$

respectively. If

$$g = \alpha_0 = \beta_1$$

then

$$\alpha = g \vee x_i \alpha_1$$
$$\beta = \overline{x}_i \beta_0 \vee g$$

MERGING RULE II

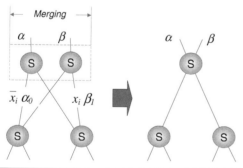

The Shannon expansions for the functions α and β are

$$\alpha = \overline{x}_i \alpha_0 \vee x_i \alpha_1$$
$$\beta = \overline{x}_i \beta_0 \vee x_i \beta_1$$

respectively. If

$$\alpha_0 = \beta_0, \quad \text{and} \quad \alpha_1 = \beta_1$$

then

$$\alpha = \beta$$

FIGURE 10.20

Reduction rules for the construction of a decision diagram from a decision tree.

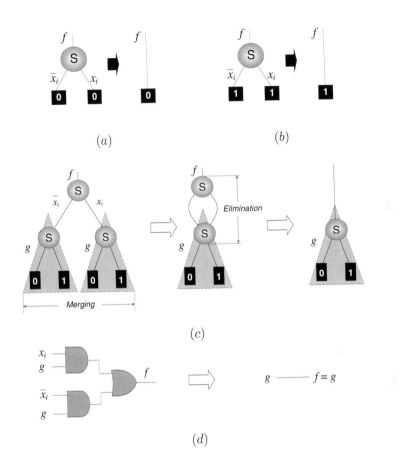

FIGURE 10.21
Application of the elimination rule (a), (b), combination of merging and elimination rules (c), and equivalent reduction using logic networks (d).

Example 10.34 *Figures 10.22 and 10.23 introduce various aspects of constructing reduced decision trees from complete ones. Note that the Boolean function $f = \overline{x}_1 \vee x_1 x_2$ derived from the tree (Figure 10.22a) is not minimal because $f = \overline{x}_1 \vee x_1 x_2 = (\overline{x}_1 \vee x_1)(\overline{x}_1 \vee x_2) = \overline{x}_1 \vee x_2$. However, the resulting decision tree is minimal with respect to the number of nodes and the order of variables. The same situation is demonstrated by the example given in Figure 10.22c where the function obtained is not minimal:*

$$f = \overline{x}_1 x_2 \vee x_1 x_2 x_3 \vee x_1 \overline{x}_2 = \overline{x}_1 x_2 \vee x_1 (x_2 x_3 \vee \overline{x}_2)$$
$$= \overline{x}_1 x_2 \vee x_1 (x_2 \vee \overline{x}_2)(x_3 \vee \overline{x}_2) = \overline{x}_1 x_2 \vee x_1 x_3 \vee x_1 \overline{x}_2$$

Techniques for minimization of Boolean functions

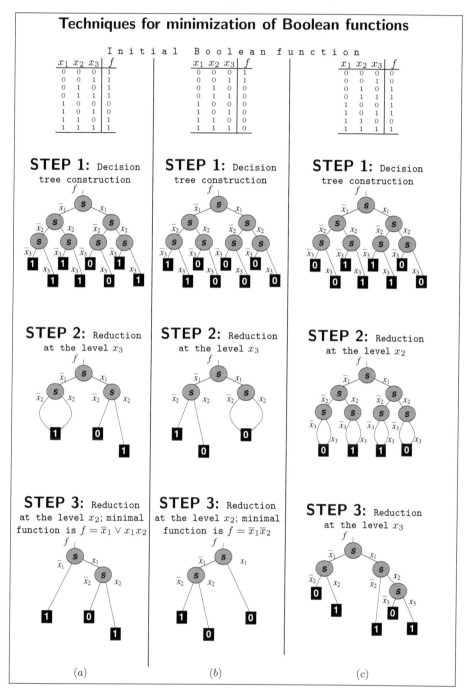

FIGURE 10.22

Techniques for minimization of Boolean functions of three variables using decision trees (Example 10.34).

The reduced decision trees are further transformed into decision diagrams. The complete algorithm for deriving the decision diagrams given a complete binary tree is as follows.

**Deriving a reduced decision tree from
a complete decision tree**

Given: A complete decision tree for a Boolean function.
Construct: A reduced decision tree.
Step 1: Apply elimination rule where possible, to remove the redundant nodes (variables in the products) of isomorphic subtrees.
Step 2: Apply merging rules I and II, to remove the redundant nodes and isomorphic subtrees.
Step 3: Repeat Steps 1 and 2 as necessary until no more nodes can be eliminated.
Result: Reduced decision tree.

Example 10.35 *Application of the reduction rules to the complete decision tree of the three-input NOR function is demonstrated in Figure 10.24.*

If an initial reduced decision tree is given, the decision diagram can be constructed from this tree using the same reduction rules.

Example 10.36 *A* **reduced** *decision tree that implements the Boolean function $f = x_1 \vee x_2 \overline{x}_2 x_3 \vee \overline{x}_2 x_3 x_4$ is shown in Figure 10.25. This tree can be reduced to a decision diagram, which is only different in the number of terminal nodes. The minimum SOP and POS can be derived from either the tree or the diagram.*

10.7 Decision diagrams for symmetric Boolean functions

The properties of Boolean functions considered in Chapter 7 can be applied for the reduction of decision trees. In particular, symmetries of the functions can be efficiently utilized in the reduction procedure. A Boolean function f is *symmetric* in variables x_i and x_j if

$$f(x_i = 1, x_j = 0) = f(x_i = 0, x_j = 1)$$

That is, a totally symmetric Boolean function is unchanged by any permutation of its variables.

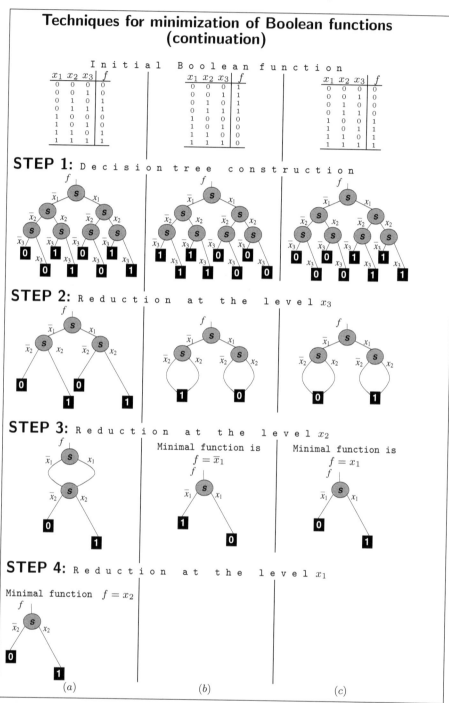

FIGURE 10.23

Techniques for minimization of a Boolean function of three variables using decision trees (continuation of Figure 10.22; Example 10.34).

Design example: decision diagram construction

GIVEN: 3-input NOR gate $f = \overline{x_1 \vee x_2 \vee x_3}$

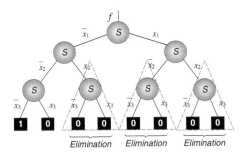

STEP 1: Initial complete decision tree; the order of variables $x_1 \to x_2 \to x_3$

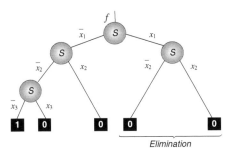

STEP 2: Reduced decision tree derived from Step 1 by using the elimination rule

STEP 3: Reduced ordered decision diagram derived from Step 2 by using the elimination rule

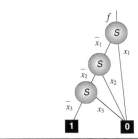

FIGURE 10.24

The three-input NOR function (a), the binary decision tree (b), and the binary decision diagram with the lexicographical order of variables (c) (Example 10.35).

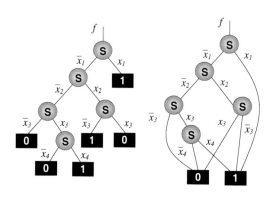

The path 1: $x_1 = 1$
The path 2: $\overline{x}_1\overline{x}_2\overline{x}_3 = 0$
The path 3: $\overline{x}_1x_2\overline{x}_3 = 1$
The path 4: $\overline{x}_1x_2x_3 = 0$
The path 5: $\overline{x}_1\overline{x}_2x_3\overline{x}_4 = 0$
The path 6: $\overline{x}_1\overline{x}_2x_3x_4 = 1$
A minimum SOP:

$$f = x_1 \vee \overline{x}_1x_2\overline{x}_3 \vee \overline{x}_1\overline{x}_2x_3x_4$$

A minimum POS:

$$f = (\overline{x}_1 \vee \overline{x}_2 \vee \overline{x}_3)$$
$$\cdot \; (\overline{x}_1 \vee \overline{x}_2x_3 \vee \overline{x}_4)$$
$$\cdot \; (\overline{x}_1 \vee \overline{x}_2x_3 \vee \overline{x}_4)$$

FIGURE 10.25
The implementation of the Boolean function $f = x_1 \vee x_2\overline{x}_2x_3 \vee \overline{x}_2x_3x_4$
(Example 10.36).

Example 10.37 *Using the above definition, total symmetry can be detected:*

(a) $f = x_1 \oplus x_2$ *and* $x_1x_2 \vee x_2x_3 \vee x_1x_3$ *are totally symmetric functions.*

(b) $f = x_1x_2 \vee x_3$ *is not a totally symmetric function. It is symmetric with respect to* x_1 *and* x_2.

Example 10.38 *In Figure 10.26, the symmetry in variables* x_i *and* x_j *is utilized in the decision diagram construction. The Shannon expansion results in*

$$f = \overline{x}_i\overline{x}_j f_{00} \vee \overbrace{\overline{x}_ix_j f_{01} \vee x_i\overline{x}_j f_{10}}^{Symmetry} \vee x_ix_j f_{11}$$
$$= \overline{x}_i\overline{x}_j f_{00} \vee (\overline{x}_ix_j \vee x_i\overline{x}_j)f^* \vee x_ix_j f_{11}$$

where $f_{01} = f_{10} = f^*$. *Instead of the four-element truth vector, this totally symmetric function can be described by the **reduced** three-element truth vector.*

The *reduced* truth vector of a totally symmetric Boolean function f is formed from a truth vector of f by removing the entries that are identical because of symmetry. It contains all of the information to completely specify

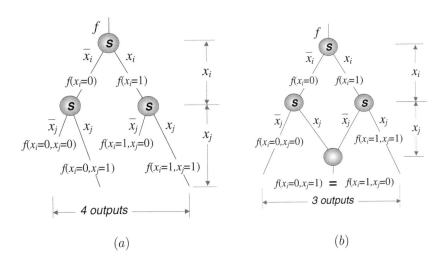

FIGURE 10.26
Fragment of a binary decision tree for a totally symmetric Boolean function
(a), and the decision diagram (b) (Example 10.38).

a symmetric function. For all assignments of values to variables that have the
same number of 1s, there is *exactly* one value of a symmetric function. For a
totally symmetric function of n variables, the reduced truth vector has $n + 1$
elements.

Example 10.39 *Let $a, b, c, d \in \{0, 1\}$. Then, the* **reduced truth
vectors** *for the totally symmetric Boolean functions of
two and three variables are formed as shown in Figure
10.27.*

Therefore, a symmetric function is completely specified by a reduced truth
vector $A = [a_0, a_1, \ldots, a_n]$, such that $f(x_1, x_2, \ldots, x_n)$ is a_i for all assignments
of values to $x_1, x_2, \ldots, x_n$ that have i 1's, where $0 \leq i \leq n$.

Example 10.40 *Figure 10.28 shows how to generate all totally symmetric
Boolean functions of two variables from a reduced
decision tree: $f = \overline{x}_1 \overline{x}_2 f_{00} \vee (\overline{x}_1 x_2 \vee x_1 \overline{x}_2) f^* \vee x_1 x_2 f_{11}$.
For example, given the assignments $f_{00} = f^* = f_{11} = 1$*

$$f = \overline{x}_1 \overline{x}_2 \cdot 1 \vee (\overline{x}_1 x_2 \vee x_1 \overline{x}_2) \cdot 1 \vee x_1 x_2 \cdot 1$$
$$= \overline{x}_1 (\overline{x}_2 \vee x_2) \vee x_1 (\overline{x}_2 \vee x_2) = \overline{x}_1 \vee x_1 = 1$$

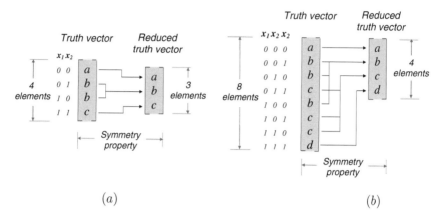

FIGURE 10.27
Forming the reduced truth vector for a totally symmetric Boolean function of
two (a) and three (b) variables (Example 10.39).

10.8   Measurement of the efficiency of decision diagrams

A decision diagram is characterized, similarly to a decision tree, by the *size,
depth, width, area,* and the *efficiency of reduction* of a decision tree or diagram
of size SIZE₁ to a tree or diagram of size SIZE₂:

$$100 \times \frac{\text{SIZE}_1}{\text{SIZE}_2}\%$$

Example 10.41 *In Figure 10.29, the* **reduction measures** *for the
decision tree and decision diagram are given.*

10.9   Embedding   decision   diagrams   into   lattice
structures

While decision diagrams often provide a compact representation of Boolean
functions, their layout is not much simpler than that of "traditionally"
designed logic networks, making placement and routing a difficult task. As
an alternative, *lattice diagrams* have been proposed. The number of nodes at

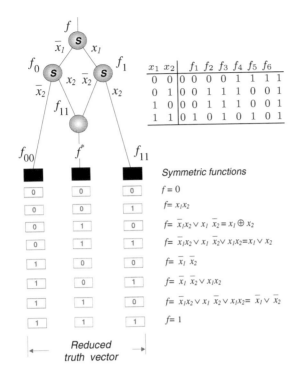

$x_1\,x_2$	f_1	f_2	f_3	f_4	f_5	f_6			
0 0	0	0	0	0	1	1	1	1	1
0 1	0	0	1	1	1	0	0	0	1
1 0	0	0	1	1	1	0	0	0	1
1 1	0	1	0	1	0	1	0	1	0

Symmetric functions

$f = 0$

$f = x_1 x_2$

$f = \bar{x}_1 x_2 \vee x_1\,\bar{x}_2 = x_1 \oplus x_2$

$f = \bar{x}_1 x_2 \vee x_1\,\bar{x}_2 \vee x_1 x_2 = x_1 \vee x_2$

$f = \bar{x}_1\,\bar{x}_2$

$f = \bar{x}_1\,\bar{x}_2 \vee x_1 x_2$

$f = \bar{x}_1 x_2 \vee x_1\,\bar{x}_2 \vee x_1 x_2 = \bar{x}_1 \vee \bar{x}_2$

$f = 1$

Reduced
truth vector

FIGURE 10.28
The decision tree for a totally symmetric Boolean function of two variables
(Example 10.40).

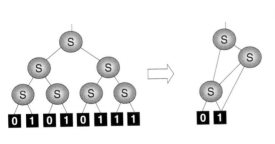

The complete decision tree:
▶ The size is 7
▶ The depth is 3
▶ The width is 8
▶ The area is 24

The decision diagram
▶ The size is 3
▶ The efficiency of reduction is

$$100 \times \frac{\text{SIZE}_1}{\text{SIZE}_2}\% = \frac{7}{3}\% = 233\%$$

FIGURE 10.29
The complete decision tree and the decision diagram for the Boolean function
$f_1 = x_1 x_2 \vee x_3$ (Example 10.41).

each level is linear, which makes the diagram fit onto a 2D structure suitable for resolution of the routing problem.

The embedding problem is formulated as follows: Embed a given decision diagram (the guest structure) into an appropriate regular lattice structure (the host structure). The particular solution to this problem is as follows: (a) the host structure is specified as the lattice structure and (b) the decision diagram must be transferred to a configuration that can be embedded into a lattice structure.

> **Example 10.42** *An example of embedding a decision tree into a* **lattice** *is given in Figure 10.30.*

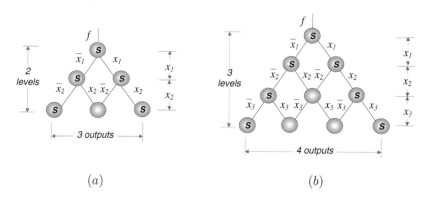

FIGURE 10.30
Embedding a decision tree into a lattice given a Boolean function of two (a) and three (b) variables (Example 10.42).

Note that an arbitrary Boolean function represented by a decision tree can be embedded into a lattice structure using the technique known as *symmetrization* based on *pseudo-symmetry*.

10.10 Further study

Advanced topics of logic optimization

Topic 1: Implementation-driven optimization.   Major   optimization objectives in combinational logic design are *area* and *delay* reduction. Optimization of multilevel logic networks is related to *testability* properties. *Two-level logic optimization*, both exact and heuristic, have received much attention, since they are critical in logic synthesis. *ESPRESSO* is the most popular heuristic two-level logic optimizer. *Multilevel logic optimization* techniques aim at the synthesis logic networks with more than two levels. Multilevel networks often require fewer logic gates and fewer connections than two-level logic networks. Multilevel logic optimization techniques also provide flexibility in optimizing area and delay, and satisfying specific design constraints. The drawback to multilevel logic networks is the difficulty of modeling and optimizing. Exact optimization is not considered to be practical in contemporary multilevel synthesis. Techniques of multilevel logic synthesis have no established algorithms compared with two-level logic techniques.

Topic 2: Multivalued logic function optimization. Multivalued logic optimization is the generalization of a binary logic network design for multivalued logic. The overall level of understanding of multilevel logic network design based on multivalued logic gates is at a much less mature stage than binary multilevel networks. Details can be found in the *Proceedings of the Annual IEEE Symposium on Multiple-Valued Logic.*

Topic 3: The advent of nano scale technologies. For earlier technologies, the relevant problems were primarily concerned with component minimization. With the development of VLSI and ULSI logic network technologies and the advent of nano scale technologies, the digital design problems concerned with the minimization of components have become less relevant. These types of problems have been replaced by less well-defined and much more difficult problems such as physical design including partitioning, layout and routing, structural simplicity, and uniformity of modules. The last problem, uniformity, and the related data structures is of particular interest to logic design in nano scale. Many of these problems cannot be solved based on traditional approaches. Investigation of better solutions in this area is ongoing work, of which results there are reported in many journals and conference publications.

Topic 4: Decision diagram-based decomposition. Details on using decision diagrams on minimization can be found, in particular, in "Multi-Level Logic Optimization" by M. Fujita, Y. Matsunaga, and M. Ciesielski, in *Logic Synthesis and Verification* edited by S. Hassoun and T. Sasao. consulting editor R. K. Brayton, Kluwer, 2002.

Topic 5: Decision diagrams for multivalued logic functions are a generalization of binary decision diagrams. Details can be found in the *Proceedings of the Annual IEEE Symposium on Multiple-Valued Logic.*

Topic 6: Lattice decision diagrams are constructed by embedding decision diagrams, representing of symmetric functions, into a *lattice array*. For this, property of symmetric Boolean functions are generalized for an arbitrary Boolean function using an approach called pseudo-symmetry.

An arbitrary Boolean function can be represented using the elementary symmetric functions (see, for example, "Transformation of Switching Functions to Completely Symmetric Switching Functions" by R. C. Born and A. K. Scidmore, *IEEE Transactions on Computers*, volume C-17, number 6,

1968, pages 596–599). This implies that a decision diagram, representing an arbitrary Boolean function, can be transformed into a lattice decision diagram.

Lattice decision diagrams are based on the lattice array, which is defined as a rectangular array of identical cells, each of them being a multiplexer, where every cell obtains signals from two neighbor inputs and gives them to two neighbor outputs.

Topic 7: Computing using Pascal and transeunt triangles. Table 10.3 contains the following graphical data structures for the representation of five elementary Boolean functions: both Pascal and the transeunt triangles (for symmetric functions), decision trees, and decision diagrams.

TABLE 10.3
Graphical data structures: Pascal and transeunt triangles, decision tree, and decision diagram for the representation of elementary Boolean functions of two variables

Further reading

[1] Butler JT, Dueck G, Shmerko VP, and Yanushkevich SN. Comments on SYMPATHY: fast exact minimization of fixed polarity Reed-Muller expansion for symmetric functions. *IEEE Trans. Computer-Aided Design of Integrated Circuits and Systems*, 19(11):1386–1388, 2000.

[2] Coudert O and Sasao T. Two-level logic minimization. In Hassoun S and Sasao T, Eds., Brayton RK., Consulting Ed., *Logic Synthesis and Verification*, Kluwer, Dordrecht, 2002.

[3] Drechsler R and Becker B. *Binary Decision Diagrams. Theory and Implementation*. Kluwer, Dordrecht, 1998.

[4] Dueck GW, Maslov D, Butler JT, Shmerko VP, and Yanuskevich SN. A method to find the best mixed polarity Reed-Muller expression using transeunt triangle. In *Proc. of the 5th Int. Workshop on Applications of the Reed-Muller Expansion in Circuit Design.* Mississippi State University, pp. 82–92, 2001.

[5] Fujita M, Matsunaga Y, and Ciesielski M. Multi-level logic optimization. In Hassoun S and Sasao T, Eds., Brayton RK., Consulting Ed., *Logic Synthesis and Verification*, Kluwer, Dordrecht, 2002.

[6] Hachtel GD and Somenzi F. *Logic Synthesis and Verification Algorithms*, Kluwer, Dordrecht, 1996.

[7] Meinel C and Theobald T. *Algorithms and Data Structures in VLSI Design.* Springer, Heidelberg, 1998.

[8] De Micheli G. *Synthesis and Optimization of Digital Circuits.* McGraw-Hill, New York, 1994.

[9] Minato S. *Binary Decision Diagrams and Applications for VLSI CAD.* Kluwer, Dordrecht, 1996.

[10] Mukhopadhyay A and Schmitz G. Minimization of exclusive-OR and logical-equivalence switching circuits, *IEEE Trans. Comput.*, 19(2):132–140, 1970.

[11] Posthoff Ch and Steinbach B. *Logic Functions and Equations.* Springer, Heidelberg, 2004.

[12] Sasao T and Butler J. A design method for look-up table type FPGA by pseudo-Kronecker expansion. In *Proc. of the 23rd Int. Symp. on Multiple-Valued Logic.* pp. 97–106, 1994.

[13] Sasao T and Butler JT. Planar decision diagrams for multiple-valued functions, *Int. J. Multiple-Valued Logic,* 1:39–64, 1996.

[14] Sasao T. *Switching Theory for Logic Synthesis*, Kluwer, Dordrecht, 1999.

[15] Stanković RS and Astola JT. *Spectral Interpretation of Decision Diagrams.* Springer, Heidelberg, 2003.

[16] Stanion T and Sechen C. Boolean division and factorization using binary decision diagrams. *IEEE Trans. Computer-Aided Design of Integrated Circuits and Systems*, 13:1179–1184, September 1994.

[17] Wu H, Perkowski MA, Zeng X, and Zhuang N. Generalized partially-mixed-polarity Reed-Muller expansion and its fast computation. *IEEE Trans. Comput.*, 45:1084–1088, Sep., 1996.

[18] Yang C and Ciesielski M. BDS: A BDD-based logic optimization system. *IEEE Trans. Computer-Aided Design of Integrated Circuits and Systems*, 21(7):866–876, 2002.

[19] Yanushkevich SN, Shmerko VP, and Steinbach B. Spatial interconnect analysis for predictable nanotechnologies. *J. Computational and Theoretical Nanoscience*, American Scientific Publishers, 4(8):56–69, 2007.

[20] Yanushkevich SN, Butler JT, Dueck GW, and Shmerko VP. Experiments on FPRM expressions for partially symmetric logic functions. In *Proc. of the IEEE 30th Int. Symp. on Multiple-Valued Logic*, pp. 141–146, 2000.

[21] Yanushkevich SN, Miller DM, Shmerko VP, and Stanković RS. *Decision Diagram Techniques for Micro- and Nanoelectronic Design*. Taylor & Francis/CRC Press, Boca Raton, FL, 2006.

[22] Yanushkevich SN and Shmerko VP. *Introduction to Logic Design*, Taylor & Francis/CRC Press, Boca Raton, FL, 2008.

11

Multivalued Data Structures

11.1 Introduction

Logic network design has traditionally been associated with Boolean algebra where the two logic levels are represented by two discrete values. Multivalued logic networks allow more than two levels of a signal and can improve various computational and fabrication characteristics such as number of operations, performance, and power dissipation.

There are many examples of natural processing that utilize the multilevel signals. For example, the advanced study of neuron networks of the brain addresses the hypotheses of multilevel signal representation in biomolecular processing. Another motivation to consider a multilevel encoding of information is that it provides new possibilities in modeling, such as compact representation and efficient transmission. A multilevel signal representation offers efficient methods and techniques for the implementation of algorithms for the manipulation of these signals.

Multivalued algebra is a generalization of Boolean algebra, based upon a set of m elements $M = \{0, 1, 2, \ldots, m\}$, corresponding to multivalued signals and the corresponding operations. This means that multivalued logic circuits operate with multivalued logic signals (Figure 11.1). The primary advantage of multivalued signals is the ability to encode more information per variable than a binary system is capable of doing. Hence, less area is used for interconnections since each interconnection carries more information. Furthermore, reliability of systems is also relevant to the number of connections because there are are sources of wiring error and weak connections. Thus, the adoption of a m-valued signals enables n pins to pass q^n combinations of values rather than just the 2^n limited by the binary representation.

In particular, a 32×32 bit multiplier based on quaternary signed-digit number system consists of only three-stage signed-digit full adders using a binary-tree addition scheme, as shown by Kawahito et. al. [23].

Another application of multivalued logic is residue arithmetic. Each residue digit can be represented by a multivalued code. For example, by this coding, *mod m* multiplication can be performed by a shift operator, and *mod m*

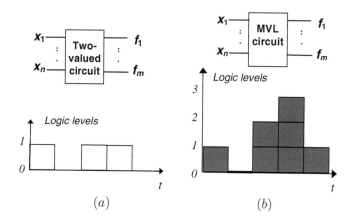

FIGURE 11.1

A Boolean network operates with two-level logic signals (a), and a quaternary network operates with four-valued logic signals (b).

addition can be executed using radix-arithmetic operators. Kameyama et al. [22] proposed implementation of this approach on multivalued bidirectional current-node MOS technology.

The operations of Boolean algebra have their analogs in multivalued algebra. The multivalued counterparts of binary AND, OR, and NOT are the multivalued *conjunction, disjunction,* and *cycling* operators introduced by Post in 1920. The other basis, *modulo 2,* was extended by Bernstein in 1928 by introducing *modulo m* addition and multiplication of the integers *modulo m.*

To synthesize logic networks, a functionally complete, or universal, set of basic operations is needed.

Example 11.1 *There are four **1-place operations** in binary logic circuits design: constants 0 and 1, identity, and complement, and 16 **2-place operations**. There are two operations, NOR and NAND, each one of which constitutes a functionally complete set. In quaternary logic, there are 256 1-place functions and $4^{4^2} = 4^{16} \approx 4.3 \times 10^9$ 2-place functions.*

There are various universal (functionally complete) sets of operations for multivalued algebra: *Post algebra, Webb algebra, Bernstein algebra,* and *Allen and Givone algebra* There are a lot of other algebras oriented mostly toward circuit implementations.

11.2 Representation of multivalued functions

An n-variable m-valued function f is defined as a mapping of a finite set $\{0, \ldots, m-1\}^n$ into a finite set $\{0, \ldots, m-1\}$, that is,

$$f : \{0, \ldots, m-1\}^n \to \{0, \ldots, m-1\}.$$

Example 11.2 *Function $f : \{0,1,2\}^n \to \{0,1,2\}$ is called a* **ternary** *logic function; $f : \{0,1,2,3\}^n \to \{0,1,2,3\}$ is called a* **quaternary** *logic function; $f : \{0,1,2\}^n \to \{0,1\}$ is called* **multivalued** *input* **binary-valued** *output logic function.*

Specifically, m is considered to be a primary number. There are m^{m^n} different possible functions.

Example 11.3 *If $m = 3$ and the number of variables is equal to $n = 2$, then there are $3^{3^3} = 19.683$ possible ternary functions of two variables.*

We will use the following techniques and associated data structures for representing logic functions:

(a) Symbolic (algebraic) notations

(b) Vector notations, that is, truth vector and vector of coefficients

(c) Matrix (two dimensional) notations for word-level representation

(d) Graph-based representations such as direct acyclic graphs (DAG) and decision diagram techniques,

(e) Embedded graph-based 3D data structures

Algebraic data structures include multivalued sum-of-products, polynomial expressions, and word-level representations.

Truth table and truth vector. The simplest way to represent a multivalued logic function is the truth table. The truth table of a logic function is the representation that tabulates all possible input combinations with their associated output values.

A truth vector of a multivalued logic function f of n m-valued variables $x_1, x_2, \ldots, x_n$ is defined as

$$\mathbf{F} = [f(0), f(1), \ldots, f(m^n - 1)]^T.$$

The index i of the element $x^{(i)}$ corresponds to the assignments $i_1 i_2 \ldots i_n$ of variables $x_1, x_2, \ldots, x_n$ ($i_1, i_2, \ldots, i_n$ is binary representation of i, $i = 0, \ldots, m^n - 1$).

Example 11.4 *The* **truth vector** $\mathbf{F}$ *of a ternary MIN function of two variables is* $\mathbf{F} = [000011012]^T$.

Symbolic, or algebraic notations include sum-of-products and polynomial forms. The sum-of-products form is represented as follows:

$$f = s_0 \varphi_0(x_1, \ldots, x_n) + \ldots + s_{m^n-1} \varphi_{m^n-1}(x_1, \ldots, x_n),$$

where $s_i \varphi_i(x_1, \ldots, x_n)$ is a literal function such that

$$\varphi_i(x_1, \ldots, x_n) = \begin{cases} 1, \text{ if } x_1, \ldots, x_n = 1; \\ 0, \text{ otherwise.} \end{cases}$$

Appendix D contains truth tables of ternary and quaternary functions of one and two variables.

Graphical data structures, adopted for multivalued functions, include multivalued networks, cube-based representation, decision trees and diagrams, and spatial topological structures resulting from embedding decision trees and diagrams into spatial configurations. Graph-based representations of networks such as DAGs are used at gate-level design phases. Figure 11.2 shows the gate libraries for constructing both Boolean (a) and multivalued (b) circuits.

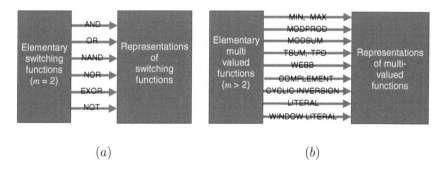

(a) $\qquad\qquad\qquad\qquad\qquad (b)$

FIGURE 11.2
Representation of Boolean (a) and multivalued logic (b) functions.

Graph-based representations of networks such as directed grapphs are used at gate level design phases. Multivalued trees and decision diagrams can be considered as an extension of binary decision diagram techniques for multivalued functions. Shannon and Davio decomposition exist for multivalued logic as well. Any m-valued function can be given by a multivalued decision tree and decision diagram.

Example 11.5 *A function of two ternary variables is represented by its* **truth table** *$\mathbf{F}= [111210221]^{T}$. In algebraic form the function is expressed by*

$$f = 0 \cdot x_1^1 x_2^2 + 1 \cdot (x_1^0 x_2^0 + x_1^0 x_2^1 + x_1^0 x_2^2 + x_1^1 x_2^1 + x_1^2 x_2^2)$$
$$+ 2 \cdot (x_1^1 x_2^0 + x_1^2 x_2^0 + x_1^2 x_2^1).$$

The map and the decision diagram of the function are given in Figure 11.3 (the nodes of the diagram implement the Shannon expansion for the ternary logic function).

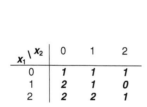

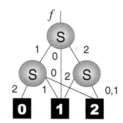

FIGURE 11.3
The map of the ternary function and its graphical representation (Example 11.5).

Compared to Boolean logic, multivalued logic has some advantages when designing high-speed arithmetic components in avoiding the time-consuming carry propagation inherent in switching gates. However, there are several disadvantages, since multivalued logic circuits often have to be embedded in a conventional digital system and therefore additional circuitry is needed to transform multivalued logic into digital signals and vice versa.

Multivalued logic is considered as an algebraic basis for threshold gates design. Threshold gates with various interpretations of weights and threshold functions are the components of artificial neural networks. The prototypes of these neural networks include circuits on CMOS neurons, and also recent nanoelectronic and molecular implementations.

One of the motivations to develop design techniques such as decision diagrams for multivalued functions is that nanotechnologies provide an opportunity to utilize the concept of multivalued signals in nanodevices. For instance, multivalued logic has been used for storing synaptic weights in neural circuits. This preserves robust information processing and reduces the number of circuit components per artificial synaptic circuit. In particular, multistate resonant tunnelling device (RTD) memory cells are a promising way to implement area-efficient multivalued logic circuits as components for monolithically integrated neural systems. The hope is that the fault tolerance of neural circuits will compensate for the errors caused by smaller noise margins.

11.3   Multivalued logic

Boolean algebra is the mathematical foundation of binary systems. Boolean algebra is defined on a set of two elements, $M = \{0, 1\}$. The operations of Boolean algebra must adhere to certain properties, called laws or axioms, used to prove more general laws about Boolean expressions to simplify expressions or factorize them, for example. Multivalued algebra is a generalization of Boolean algebra towards a set of m elements $M = \{0, 1, 2, \ldots, m\}$ and corresponding operations.

The focus of this section is operators on m-valued logic; algebras that are specified on a universal set of operations, and data structures for the representation and manipulation of multivalued logic functions.

11.3.1   Operations of multivalued logic

A multivalued logic function f of n variables $x_1, x_2, \ldots, x_n$ is a logic function defined on the set $M = \{0, 1, \ldots, m - 1\}$. A multivalued logic circuit operates with multivalued logic signals. Each of the logic operations has a corresponding logic gate. Multivalued logic gates are closely linked to hardware in their implementations, given in Tables 11.1 and Table 11.2.

Below, we list some of the implementation-oriented m-valued logic operations.

The MAX operation is defined as

$$\text{MAX}(x_1, x_2) = \begin{cases} x_1 & \text{if } x_1 \geq x_2 \\ x_2 & \text{otherwise.} \end{cases}$$

When $m = 2$, this operation turns into an OR operation. The properties of MAX operations in ternary logic resemble those of Boolean algebra, that is, $\text{MAX}(x, x) = x$, that is, $x \vee x = x$ in a binary circuit, $x \vee 0 = x$, and $x \vee 2 = 2$,

TABLE 11.1
Library of ternary $(m = 3)$ two-variable elementary functions

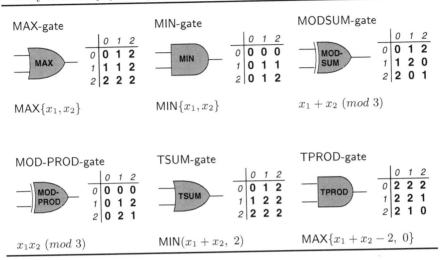

MAX-gate				MIN-gate				MODSUM-gate			

MAX-gate — MAX

	0	1	2
0	0	1	2
1	1	1	2
2	2	2	2

MIN-gate — MIN

	0	1	2
0	0	0	0
1	0	1	1
2	0	1	2

MODSUM-gate — MOD-SUM

	0	1	2
0	0	1	2
1	1	2	0
2	2	0	1

$\text{MAX}\{x_1, x_2\}$ $\text{MIN}\{x_1, x_2\}$ $x_1 + x_2 \ (mod\ 3)$

MOD-PROD-gate — MOD-PROD

	0	1	2
0	0	0	0
1	0	1	2
2	0	2	1

TSUM-gate — TSUM

	0	1	2
0	0	1	2
1	1	2	2
2	2	2	2

TPROD-gate — TPROD

	0	1	2
0	2	2	2
1	2	2	1
2	2	1	0

$x_1 x_2 \ (mod\ 3)$ $\text{MIN}(x_1 + x_2,\ 2)$ $\text{MAX}\{x_1 + x_2 - 2,\ 0\}$

that is, $x \vee 0 = x$ and $x \vee 1 = 1$ in a binary case. A MAX function of n variables is written

$$\text{MAX}(x_1, x_2, \dots, x_n) = x_1 \vee x_2 \vee \dots \vee x_n.$$

The MIN operation of x_1 and x_2 is defined as

$$\text{MIN}(x_1, x_2) = \begin{cases} x_2 & \text{if } x_1 \geq x_2 \\ x_1 & \text{otherwise.} \end{cases}$$

and for n variables is written, $\text{MIN}(x_1, x_2, \dots, x_n) = x_1 \wedge x_2 \wedge \dots \wedge x_n$.

The modulo m product operation is defined by

$$\text{MOD-PROD}(x_1, x_2, \dots, x_n) = x_1 x_2 \dots x_n \quad mod\ (m).$$

The modulo m sum operation is defined below as

$$\text{MODSUM}(x_1, x_2, \dots, x_n) = x_1 + x_2 + \dots + x_n \quad mod\ (m).$$

It can be shown that the modulo m sum operation MODSUM, modulo m product operation MOD-PROD, and the constant 1 constitute a universal set of operations, defined as the Galois algebra $GF(m)$.

The truncated sum operation of n variables is specified by

$$\text{TSUM}(x_1, x_2, \dots, x_n) = \text{MIN}(x_1 \vee x_2 \vee \dots \vee x_n,\ m - 1).$$

The truncated product operation is defined by

$$TPROD(x_1, x_2, \ldots, x_n) = MIN(x_1 \wedge x_2 \wedge \ldots \wedge x_n, \ (m-1)).$$

Example 11.6 (a) *Let* $m = 2$, *then*

$$MOD\text{-}PROD(x_1, x_2, \ldots, x_n) = AND(x_1, x_2, \ldots, x_n),$$
$$MODSUM(x_1, x_2, \ldots, x_n) = x_1 \oplus x_2 \oplus \ldots \oplus x_n.$$

(b) *Let* $m = 2$, $n = 2$, *then*

$$TSUM(x_1, x_2) = MIN(x_1 \vee x_2, 1) = x_1 \vee x_2$$
$$TPROD(x_1, x_2) = MIN(x_1 x_2, \ 1) = x_1 x_2.$$

The Webb function is defined below as

$$x_1 \uparrow x_2 = MAX(x_1, x_2) + 1 \quad (mod \ m).$$

The Pierce operation is a binary analog of the Webb operation.

The complement operation is specified by

$$\overline{x} = (m-1) - x,$$

where $x \in M$ is a unary operation. For example, in ternary logic, $\overline{x} = 2 - x$. Notice that the property $\overline{\overline{x}} = x$ can be used in multivalued logic. This is because $(m-1) - \overline{x} = (m-1) - ((m-1) - x) = x$.

The clockwise cycle operation, or r-order cyclic complement

$$\hat{x}^r = x + r \quad (mod \ m).$$

This implies that

$$\hat{x}^0 = x \quad (mod \ m),$$
$$\hat{x}^m = x + m = x \quad (mod \ m).$$

The operation MIN and the clockwise cycle operation form a complete system as well. Given m-valued variable, $m!$ different complements of the variable can be distinguished.

Example 11.7 (a) *Let* $m = 2$, *then* $\overline{x} = (2-1) - x = 1 - x$.
(b) *Let* $m = 2$, *then the system is* $\{AND, NOT\}$, *that is, NAND that is known to be* **complete**.

TABLE 11.2
Library of ternary $(m = 3)$ logic functions

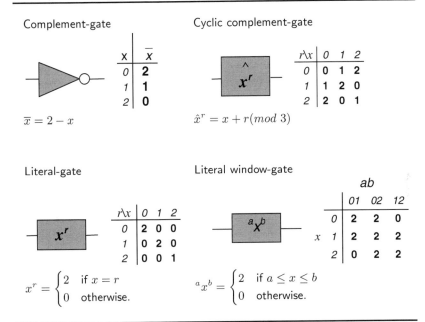

Complement-gate		Cyclic complement-gate

Complement-gate

x	$\bar{x}$
0	2
1	1
2	0

$\bar{x} = 2 - x$

Cyclic complement-gate

r\x	0	1	2
0	0	1	2
1	1	2	0
2	2	0	1

$\hat{x}^r = x + r (mod\ 3)$

Literal-gate

r\x	0	1	2
0	2	0	0
1	0	2	0
2	0	0	1

$x^r = \begin{cases} 2 & \text{if } x = r \\ 0 & \text{otherwise.} \end{cases}$

Literal window-gate

	ab		
	01	02	12
0	2	2	0
x 1	2	2	2
2	0	2	2

$^a x^b = \begin{cases} 2 & \text{if } a \le x \le b \\ 0 & \text{otherwise.} \end{cases}$

The literal operation is specified below

$$x^y = y^x = \begin{cases} m - 1 & \text{if } x = y, \\ 0 & \text{otherwise.} \end{cases}$$

A particular case of a literal multivalued-input binary-output function

$$x^y = \begin{cases} 1 & \text{if } x = y, \\ 0 & \text{otherwise.} \end{cases}$$

The window literal operation is defined as

$$^a x^b = \begin{cases} m - 1 & \text{if } a \le x \le b, \\ 0 & \text{otherwise.} \end{cases}$$

Any m-valued single-output system can be described by a functionally complete set of primitive operations. Various algebras exist to provide functional completeness for $m > 2$.

Example 11.8   *In Table 11.3, the truth tables for elementary ternary logic functions are given. The basic operations for $m = 3$ (ternary) and $m = 4$ (quaternary) are given in Tables 11.4 and 11.5, respectively.*

TABLE 11.3

Truth tables for elementary ternary logic functions (Example 11.8)

x_1	x_2	MAX	MIN	MODSUM	MODPROD	TSUM	TPROD
0	0	0	0	0	0	0	0
0	1	1	0	1	0	1	0
0	2	2	0	2	0	2	0
1	0	1	0	1	0	1	0
1	1	1	1	2	1	2	1
1	2	2	1	0	2	2	2
2	0	2	0	2	0	2	0
2	1	2	1	0	2	2	2
2	2	2	2	1	1	2	2

Symbolic, or algebraic notations include sum-of-products and polynomial forms. The sum-of-products form is represented as follows:

$$f = s_0 \varphi_0(x_1, \ldots, x_n) + \ldots + s_{m^n-1} \varphi_{m^n-1}(x_1, \ldots, x_n),$$

where $s_i \varphi_i(x_1, \ldots, x_n)$ is a literal function such that

$$\varphi_i(x_1, \ldots, x_n) = \begin{cases} 1, & \text{if } x_1, \ldots, x_n = 1; \\ 0, & \text{otherwise.} \end{cases}$$

11.3.2   Multivalued algebras

There are various universal (functionally complete) sets of operations for multivalued algebra. Some are given below.

Post algebra is based on two operations: 1-cycle inversion $(x+1)_{mod\ m}$, and MAX operation $x \vee y = \text{MAX}(x, y)$. Using these operations, one can describe any m-valued logic function. The analogs of Post operations in Boolean algebra are NOT and OR operations, which also constitute a universal system.

Webb algebra is based on one operation, the Sheffer–Stroke operation that is specified as $x|y = \text{MAX}(x, y) + 1 (mod\ m)$. In Webb algebra, a functionally complete set consists of one function.

TABLE 11.4

Basic operations of 3-valued logic functions (Example 11.8)

COMPLEMENT

x	0	1	2
$\bar{x}$	2	1	0

r-CYCLIC COMPLEMENT

	x	0	1	2
r	0	0	1	2
	1	1	2	0
	2	2	0	1

LITERAL

	x	0	1	2
x^0		2	0	0
x^1		0	2	0
x^2		0	0	2

WINDOW LITERAL

	x	0	1	2
$^0x^0$		2	0	0
$^0x^1$		2	2	0
$^0x^2$		2	2	2
$^1x^1$		0	2	0
$^1x^2$		0	2	2
$^2x^2$		0	0	2

MAX

	0	1	2
0	0	1	2
1	1	1	2
2	2	2	2

$\overline{MAX}$

	0	1	2
0	2	1	0
1	1	1	0
2	0	0	0

TSUM

	0	1	2
0	0	1	2
1	1	2	2
2	2	2	2

$\overline{TSUM}$

	0	1	2
0	2	1	0
1	1	0	0
2	0	0	0

TPROD

	0	1	2
0	0	0	0
1	0	0	1
2	0	1	2

$\overline{TPROD}$

	0	1	2
0	2	2	2
1	2	2	1
2	2	1	0

MIN

	0	1	2
0	0	0	0
1	0	1	1
2	0	1	2

$\overline{MIN}$

	0	1	2
0	2	2	2
1	2	1	1
2	2	1	0

MODSUM

	0	1	2
0	0	1	2
1	1	2	0
2	2	0	1

$\overline{MODSUM}$

	0	1	2
0	2	1	0
1	1	0	2
2	0	2	1

MODPROD

	0	1	2
0	0	0	0
1	0	1	2
2	0	2	1

$\overline{MODPROD}$

	0	1	2
0	2	2	2
1	2	1	0
2	2	0	1

Bernstein algebra or modulo-sum and modulo-product algebra includes the modulo m sum: $x_1 + x_2 \pmod m$, and the modulo m product: $xy \pmod m$.

Allen and Givone algebra's universal set consists of $\text{MIN}(x_1, x_2)$, $\text{MAX}(x_1, x_2)$, and the *window literal operation*

$$^a x^b = \begin{cases} m - 1 & \text{if } a \leq x \leq b \\ 0 & \text{otherwise.} \end{cases}$$

Allen and Givone algebra is often used in multivalued logic circuit design in many variations, for example, MIN, MAX, truncated sum, and appropriate subset of 1-place functions.

There are a lot of other algebras oriented mostly toward circuit implementations. For example, there are algebras based on MIN, MAX, and CYCLE operations. Other examples are MIN, TSUM, and WINDOW LITERAL operations.

TABLE 11.5

Basic operations of 4-valued logic functions (Example 11.8)

COMPLEMENT

x	0	1	2	3
$\bar{x}$	3	2	1	0

r-CYCLIC COMPLEMENT

	x 0	1	2	3
0	0	1	2	3
1	1	2	3	0
r 2	2	3	0	1
3	3	0	1	2

LITERAL

x	0	1	2	3
x^0	3	0	0	0
x^1	0	3	0	0
x^2	0	0	3	0
x^3	0	0	0	3

WINDOW LITERAL

x	0	1	2	3
$^0x^0$	3	0	0	0
$^0x^1$	3	3	0	0
$^0x^2$	3	3	3	0
$^0x^3$	3	3	3	3
$^1x^1$	0	3	0	0
$^1x^2$	0	3	3	0
$^1x^3$	0	3	3	3
$^2x^2$	0	0	3	0
$^2x^3$	0	0	3	3
$^2x^3$	0	0	3	3
$^3x^3$	0	0	0	3

MAX

	0	1	2	3
0	0	1	2	3
1	1	1	2	3
2	2	2	2	3
3	3	3	3	3

$\overline{MAX}$

	0	1	2	3
0	3	2	1	0
1	2	2	1	0
2	1	1	1	0
3	0	0	0	0

TSUM

	0	1	2	3
0	0	1	2	3
1	1	2	3	3
2	2	3	3	3
3	3	3	3	3

$\overline{TSUM}$

	0	1	2	3
0	3	2	1	0
1	2	1	0	0
2	1	0	0	0
3	0	0	0	0

TPROD

	0	1	2	3
0	0	0	0	0
1	0	0	0	1
2	0	0	1	2
3	0	1	2	3

$\overline{TPROD}$

	0	1	2	3
0	3	3	3	3
1	3	3	3	2
2	3	3	2	1
3	3	2	1	0

MIN

	0	1	2	3
0	0	0	0	0
1	0	1	1	1
2	0	1	2	2
3	0	1	2	3

$\overline{MIN}$

	0	1	2	3
0	3	3	3	3
1	3	2	2	2
2	3	2	1	1
3	3	2	1	0

MODSUM

	0	1	2	3
0	0	1	2	3
1	1	2	3	0
2	2	3	0	1
3	3	0	1	2

$\overline{MODSUM}$

	0	1	2	3
0	3	2	1	0
1	2	1	0	3
2	1	0	3	2
3	0	3	2	1

MODPROD

	0	1	2	3
0	0	0	0	0
1	0	1	2	3
2	0	2	0	2
3	0	3	2	1

$\overline{MODPROD}$

	0	1	2	3
0	3	3	3	3
1	3	2	1	0
2	3	1	3	1
3	3	0	1	2

11.4   Galois fields $GF(m)$

A Boolean algebraic system $\langle \mathbf{B}, \cdot, 0, 1 \rangle$ is called a *Boolean ring* if it satisfies the conditions that there exist elements x, $(x \neq 0)$ and y such that $xy = yx = 1$. This is the condition of the field called *Galois field*, denoted by $GF(2)$. The space of the functions

$$f : C_2^n \rightarrow GF(2),$$

where $C_2 = (\{0, 1\}, \oplus)$ includes polynomial expressions.

The generalization   of the Galois field to multivalued functions

$$f : \{0, \ldots, m-1\}^n \rightarrow \{0, \ldots, m-1\},$$

where m is a prime, leads to $GF(m)$. Every finite set

$$Z_m = \{0, 1, 2, \ldots, m-1\}$$

with modulo m addition and multiplication is a field if and only if m is a *prime* number. Such field is called a *Galois field modulo m* denoted by $GF(m)$.

Example 11.9   *The set $\{0, 1, 2\}$ with addition and multiplication modulo 3 is a field. There is an identity 0 with respect to modulo 3 addition, and identity 1 with respect to modulo 3 multiplication. Every element has a unique additive inverse, and every element other than 0 has a multiplicative inverse.*

Various forms for m-valued functions can be generated using transforms over $GF(m)$. The multivalued polynomial expansion for a one-variable function over $GF(m)$ is expressed as

$$f = r_0 + r_1 x + r_2 x^2 + \ldots + r_{m-1} x^{m-1} \ \ over \ \ GF(m)$$

Given $GF(3)$ this resolves to

$$f = r_0 + r_1 x + r_2 x^2 \ \ over \ \ GF(3)$$

For two variables, the ternary Reed–Muller expansion is

$$\begin{aligned} f = {} & r_0 + r_1 x_2 + r_2 x_2^2 + r_3 x_1 + r_4 x_1 x_2 \\ & + r_5 x_1 x_2^2 + r_6 x_1^2 + r_7 x_1^2 x_2 + r_8 x_1^2 x_2^2 \ \ over \ \ GF(3) \end{aligned}$$

Galois field $GF(3)$. A ternary logic function of n variables can be represented as a polynomial in Galois field $GF(3)$.

Example 11.10 *A ternary function $MAX(x_1, x_2)$, truth table given in Table 11.3, can be represented in **algebraic** (sum-of-products) form as follows:*

$$MAX(x_1, x_2) = 0 \cdot x_1^0 x_2^0 + 1 \cdot x_1^0 x_2^1 + 2 \cdot x_1^0 x_2^2 + 1 \cdot x_1^1 x_2^0$$
$$+ 1 \cdot x_1^1 x_2^1 + 2 \cdot x_1^1 x_2^2 + 2 \cdot x_1^2 x_2^0 + 2 \cdot x_1^2 x_2^1$$
$$+ 2 \cdot x_1^2 x_2^2$$

Galois field $GF(4)$. It should be noted that *modulo* 4 addition and multiplication do not form a field, because the element 2 has no multiplicative inverse (there is no element a such that $2 \cdot a = 1$ so that $a = 2^{-1}$). However, a field $GF(2^k)$ can be introduced, where the element of the field are k-bit binary numbers. For example, an algebra for the quaternary field $GF(4)$ with elements {0,1,A,B} can be constructed (see the Further Reading section for references), and $GF(4)$ addition and multiplication tables can be specified for that algebra.

A 4-valued logic function of n variables can be represented as a Galois field polynomial

$$f = a_0 + \sum_{i=1}^{4^n - 1} a_i g(i), \tag{11.1}$$

where coefficients $a_i \in \{0, 1, 2, 3\}$, and $g(i)$ are the product terms defined as elements of the vector

$$\mathbf{X}(n) = \bigotimes_{i=1}^{n} [1 \ x_i \ x_i^2 \ x_i^3],$$

where $\bigotimes$ denotes the Kronecker product, and addition and multiplication are carried out in $GF(4)$.

Given a truth vector $\mathbf{F}[f(0) \ldots f(4^n - 1)]^T$ of a 4-valued logic function of n variables, the vector of coefficients $\mathbf{A} = [a_0 \ldots a_{4^n - 1}]^T$ can be computed by the matrix equation

$$\mathbf{A} = \mathbf{GF},$$

where the $(4^n - 1) \times (4^n - 1)$ matrix $\mathbf{G}$ is constructed as

$$\mathbf{G}(n) = \bigotimes_{i=1}^{n} \mathbf{G}(1, \ \mathbf{G}(1=) \begin{bmatrix} 1 & 0 & 0 & 0 \\ 0 & 1 & 3 & 2 \\ 0 & 1 & 2 & 3 \\ 1 & 1 & 1 & 1 \end{bmatrix}.$$

Equation 11.1 can be written in the form

$$f = \mathbf{XA}. \tag{11.2}$$

There are four permutations of the values of four-valued variables. These permutations are described by $x_i + k$ for $k = 0, 1, 2, 3$, where the addition is in $GF(4)$, and can be identified as the four different complements of a four-valued variable.* By using these different complements of variables, k^n different polynomial expressions for a given 4-valued logic function can be derived. The Reed–Muller expression with the smallest number of nonzero coefficients is called the *optimal* polynomial expression for a given logic function.

Example 11.11 *Given a truth vector* $\mathbf{F} = [0000013203210213]^T$ *of a 4-valued logic function of two variables* $n = 2$. *Figure 11.4 illustrates the* **polynomial expression** *construction.*

$$\mathbf{G}(2) = \mathbf{G}(1) \otimes \mathbf{G}(1) = \begin{bmatrix} 1&0&0&0\\ 0&1&3&2\\ 0&1&2&3\\ 1&1&1&1 \end{bmatrix} \otimes \begin{bmatrix} 1&0&0&0\\ 0&1&3&2\\ 0&1&2&3\\ 1&1&1&1 \end{bmatrix}$$

$$\mathbf{A} = \mathbf{GF} = \underbrace{\left[\begin{array}{cccc|cccc|cccc|cccc}
1&0&0&0&0&0&0&0&0&0&0&0&0&0&0&0\\
0&1&3&2&0&0&0&0&0&0&0&0&0&0&0&0\\
0&1&2&3&0&0&0&0&0&0&0&0&0&0&0&0\\
1&1&1&1&0&0&0&0&0&0&0&0&0&0&0&0\\ \hline
0&0&0&0&1&0&0&0&3&0&0&0&2&0&0&0\\
0&0&0&0&0&1&3&2&0&3&1&2&0&2&2&0\\
0&0&0&0&0&1&2&3&0&3&2&1&0&2&0&2\\
0&0&0&0&1&1&1&1&3&3&3&3&2&2&2&2\\ \hline
0&0&0&0&1&0&0&0&2&0&0&0&3&0&0&0\\
0&0&0&0&0&1&3&2&0&2&2&0&0&3&1&2\\
0&0&0&0&0&1&2&3&0&2&0&2&3&3&2&1\\
0&0&0&0&1&1&1&1&2&2&2&2&3&3&3&3\\ \hline
1&0&0&0&1&0&0&0&1&0&0&0&1&0&0&0\\
0&1&3&2&0&1&3&2&0&1&3&2&0&1&3&2\\
0&1&2&3&0&1&2&3&0&1&2&3&0&1&2&3\\
1&1&1&1&1&1&1&1&1&1&1&1&1&1&1&1
\end{array}\right]}_{\mathbf{G}(2) = \mathbf{G}(1) \otimes \mathbf{G}(1)}
\begin{bmatrix} 0\\0\\0\\0\\0\\1\\3\\2\\0\\3\\2\\1\\0\\2\\1\\3 \end{bmatrix} =
\begin{bmatrix} 0\\0\\0\\0\\0\\0\\0\\0\\0\\0\\1\\0\\0\\0\\0\\0 \end{bmatrix}$$

$$\begin{aligned} f &= \mathbf{XA} = [1\ x_i\ x_i^2\ x_i^3]\mathbf{A} \\ &= x_1^2 x_2^2 \end{aligned}$$

FIGURE 11.4
Deriving the polynomial expression for a 4-valued logic function of two variables (Example 11.11).

*The other possible permutations up to the total number of 4! do not affect the number of coefficients.

11.4.1   Algebraic structure for Galois field representations

For a set whose elements can be identified with first four nonnegative integers $\{0, 1, 2, 3\}$, the Galois field $GF(4)$ structure is provided by addition and multiplication defined in Table 11.6.

TABLE 11.6
Addition and multiplication in $GF(4)$

$+$	0	1	2	3		$\cdot$	0	1	2	3
0	0	1	2	3		0	0	0	0	0
1	1	0	3	2		1	0	1	2	3
2	2	3	0	1		2	0	2	3	1
3	3	2	1	0		3	0	3	1	2

11.4.2   Galois field expansions

The set of elementary functions $1, x, x^2, x^3$ is a basis in the space of one-variable functions over $GF(4)$. Therefore, each quaternary function f given by the truth vector $\mathbf{F} = [f(0), \ldots, f(3)]^T$ can be represented by Fourier-like Galois field expansion given in the matrix form by

$$f = \begin{bmatrix} 1 & x & x^2 & x^3 \end{bmatrix} \mathbf{G}_4(1)\mathbf{F},$$

where $\mathbf{G}_4(1) = \begin{bmatrix} 1 & 0 & 0 & 0 \\ 0 & 1 & 3 & 2 \\ 0 & 1 & 2 & 3 \\ 1 & 1 & 1 & 1 \end{bmatrix}$, and the all calculations are carried out in $GF(4)$. In

this notation, the basis functions $1, x, x^2, x^3$ can be considered as columns of the matrix $\mathbf{G}_4^{-1}(1)$, inverse to $\mathbf{G}_4(1)$, $\mathbf{G}_4^{-1}(1) = \begin{bmatrix} 1 & 0 & 0 & 0 \\ 1 & 1 & 1 & 1 \\ 1 & 2 & 3 & 1 \\ 1 & 3 & 2 & 1 \end{bmatrix}$.

Extension of Galois field representations to n-variable functions is straightforward with the Kronecker product. GF-representation of an n-variable quaternary function f given by its truth vector $\mathbf{F} = [f(0), \ldots, f(4^n - 1)]^T$ is given by

$$f = \left(\bigotimes_{i=1}^{n} \begin{bmatrix} 1 & x_i & x_i^2 & x_i^3 \end{bmatrix} \right) \left(\bigotimes_{i=1}^{n} \mathbf{G}_{4i} \right) \mathbf{F},$$

where $\otimes$ denotes the Kronecker product and $\mathbf{G}_{4i} = \mathbf{G}_4(1)$. The calculations are carried out in $GF(4)$.

11.5 Fault models based on the concept of change

Fault models for MVL circuits depend on the style and the technology with which the circuit is implemented. There are several approaches to designing logic models for physical defects or failures. In this section, a fault is described by a set of changes of a multivalued signal. For a formal description of these changes, logic differences can be used. Note that a set of changes can be minimized. This approach can be especially useful for local testing of multivalued devices.

In this chapter, direct changes to a signal are considered for the representation of various faults.

> **Example 11.12** *In a ternary system, signal changes from logic level 0 to logic level 1 or 2 are denoted by $D_{0 \to 1}$, $D_{0 \to 2}$, and $D_{0 \to 2}$ for direct changes. Grouping these changes, it is possible to describe various logic* **models of faults** *in multivalued networks.*

Below, several common types of faults in multivalued system are listed:

▶ A *stuck-at-k* fault occurs in a line x if x generates the output signal k for all input signals.

▶ A $\beta_1 - \beta_2$ *window* fault occurs in a line x if x operates correctly for input signals $\beta_1 \le t \le \beta_2$, and x is *stuck-at-β_1* for input signal $t < b1$ and *stuck-at-β_2* $t > \beta_2$, where $\beta_1, \beta_2 \in \{0, 1, ..., m - 1\}$.

▶ An *r-order input signal variation* fault occurs in a line x, $r \in \{0, 1, ..., m - 1\}$, if x generates the output signal $t' = t + r$ for input signal t.

▶ A *shift* fault

$$p^+, \quad p \in \{0, 1, ..., m - 1\} \text{ or}$$
$$p^-, \quad p \in \{1, ..., m - 1\}$$

occurs in a line x if x generates the output signal $t' = (p + 1)$ or $t' = (p - 1)$ if $t = p$, and x operates correctly for input signals $t \ne p$.

An elementary event in a MVL gate is a direct change of the logical value on line x from α to β $\alpha, \beta \in \{0, ..., m - 1\}$, and $\alpha \ne \beta$.

> **Example 11.13** *A fault* **shift** r^+ *in a m-valued gate can be described by the direct change $D_{r \to r+1}$.*

Figure 11.5 illustrates the above faults by a sample of 4-valued signals.

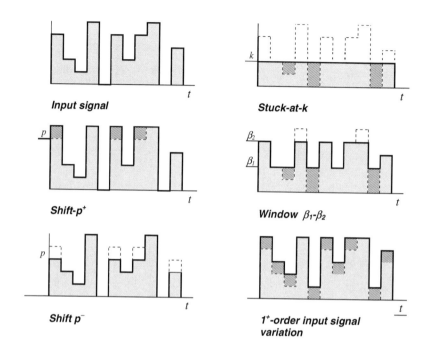

FIGURE 11.5

Changing of the 4-valued signal by various types of faults.

Logic faults in a binary network are often described as *stuck-at-0* and *stuck-at-1* faults. These faults can be considered as elementary changes and described, for example, by Boolean differences. The analog of these faults in a k-valued network is *stuck-at-k* faults. They can also be described in terms of change. To describe *stuck-at-k* faults, it is necessary to analyze and take into account all possible types of changes.

> **Example 11.14** *In the case of a 4-valued gate the minimal number of elementary changes is 12. Hence, a set of* **stuck-at-k** *faults in a 4-valued combinational network is described by 12 types of various changes $D_{r \to r+1}$.*

A relationship between different types of faults can be described in terms of change. Figure 11.6 illustrates the relationships between *stuck-at-k*, r-order signal variation, shift r^{+} and window $\beta_1 - \beta_2$ faults for a 4-valued circuit, and rules for detection of these faults.

Example 11.15 *Particular cases of fault detection in a 4-valued circuit are given below. These use the rules given in Figure 11.6:*

(a) *A test to detect the fault **window** 1-2 in a quaternary circuit can be derived from the following changes: $D_{0\to1}$ and $D_{3\to1}$. The same events are used to detect a **stuck-at-1** fault.*

(b) *Tests to detect 1^+ input **signal variation** are derived from the following changes: $D_{0\to1}$, $D_{1\to2}$, $D_{2\to3}$, and $D_{3\to0}$.*

(c) *The changes $D_{0\to2}$, $D_{1\to2}$, and $D_{3\to2}$ are used in the logic equation for detection of **stuck-at-2** faults.*

11.6 Polynomial representations of multivalued logic functions

Polynomial representations of multivalued logic functions possess the following virtues:

▶ Polynomial expressions are associated with the analysis of multivalued logic functions in terms of change through logic Taylor expansion,

▶ The corresponding decision tree and diagram provide a useful opportunity for detailed analysis of multivalued logic functions, including switching activity,

▶ The decision tree embedded into a hypercube-like structure allows word-wise computation and manipulation of polynomial expressions of various polarities,

▶ The cost of implementation using polynomial expression is often less then that of sum-of-products expressions, and

▶ Polynomial expressions can be efficiently computed using matrix transforms and, thus, the calculations are mapped onto massive parallel tools.

An m-valued logic function f of n variables is described by a sum-of-products (SOP)

$$f = \bigvee_{j=0}^{m^n-1} \varphi_{j_1}(x_1) \times \cdots \times \varphi_{j_n}(x_n), \qquad (11.3)$$

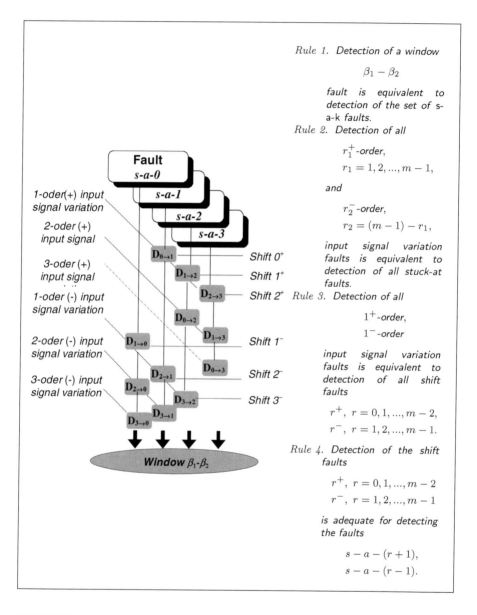

FIGURE 11.6
The relationship between s-a-k, r-order input signal variation, and *window* $\beta_1 - \beta_2$ faults for a 4-valued circuit and the rules for detecting several faults (Example 11.15).

where

$$\varphi_{j_i}(x_i) = \begin{cases} 0, & \text{if } j_i \neq x_i; \\ m-1, & \text{if } j_i = x_i. \end{cases}$$

Direct and inverse transforms over $GF(m)$ are defined by the matrix equations (Figure 11.7)

$$\begin{aligned} \mathbf{R} &= \mathbf{R}_{m^n}^{(c)} \mathbf{F} && over \ GF(m) \\ \mathbf{F} &= \mathbf{R}_{m^n}^{-1(c)} \mathbf{R} && over \ GF(m) \end{aligned} \tag{11.4}$$

where $c = c_1 c_2 \ldots c_n$ is an m-valued representation of $c = 1, 2, \ldots, m^n$. The pair of matrices $\mathbf{R}_{m^n}^{(c)}$ and $\mathbf{R}_{m^n}^{-1(c)}$ in Equation 11.4 are calculated as

$$\begin{aligned} \mathbf{R}_{m^n}^{(c)} &= \mathbf{R}_m^{(c_1)} \otimes \mathbf{R}_m^{(c_2)} \otimes \cdots \otimes \mathbf{R}_m^{(c_n)} \\ \mathbf{R}_{m^n}^{-1(c)} &= \mathbf{R}_m^{-1(c_1)} \otimes \mathbf{R}_m^{-1(c_2)} \otimes \cdots \otimes \mathbf{R}_m^{-1(c_n)} \end{aligned} \tag{11.5}$$

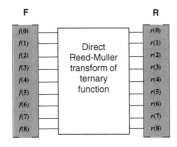

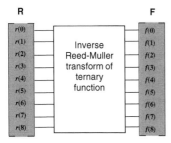

The pair of direct and inverse transforms

$$\begin{aligned} \mathbf{R} &= \mathbf{R}_{32}^{(c)} \mathbf{F} && over \ GF(3) \\ \mathbf{F} &= \mathbf{R}_{32}^{-1(c)} \mathbf{R} && over \ GF(3) \end{aligned}$$

where

$$\begin{aligned} \mathbf{R}_{32}^{(c)} &= \mathbf{R}_3^{(c_1)} \otimes \mathbf{R}_3^{(c_2)} \\ \mathbf{R}_{32}^{-1(c)} &= \mathbf{R}_3^{-1(c_1)} \otimes \mathbf{R}_3^{-1(c_2)} \end{aligned}$$

Basic transform matrices $\mathbf{R}_{32}^{(c)}$ and $\mathbf{R}_{32}^{-1(c)}$ for polarity $c = 0, 1, 2$ are given in Table 11.7

FIGURE 11.7
Direct and inverse transforms for a ternary function ($m = 3$) of two variables ($n = 2$).

The components of matrix $\mathbf{R}_m^{(c_j)}$ and $\mathbf{R}_m^{-1(c_j)}$, $j = 1, 2, \ldots, n$, are obtained as the solution to the logic equation

$$\mathbf{R}_m^{-1(c_j)} \mathbf{R}_m^{(c_j)} = \mathbf{I}_m \quad over \ GF(m) \tag{11.6}$$

where $\mathbf{I}_m$ is $m \times m$ identity matrix.

In Table 11.7, the basic arithmetic transform matrices for ternary $(m = 3)$ logic functions are given.

Example 11.16 *Given $m = 3$ and $c = 2$, matrices $\mathbf{R}_3^{-1(2)}$ and $\mathbf{R}_3^{(2)}$ meets the condition from Equation 11.6*

$$\mathbf{R}_3^{-1(2)} \mathbf{R}_3^{(2)} = \begin{bmatrix} 1 & 2 & 1 \\ 1 & 0 & 0 \\ 1 & 1 & 1 \end{bmatrix} \begin{bmatrix} 0 & 1 & 0 \\ 1 & 0 & 2 \\ 2 & 2 & 2 \end{bmatrix} = \mathbf{I}_3 \quad over \ GF(3).$$

TABLE 11.7
Basic transform matrices for polarity $c = 0, 1, 2$, of a ternary logic function

Polarity c	Direct	Inverse
0	$\mathbf{R}_3^{(0)} = \begin{bmatrix} 1 & 0 & 0 \\ 0 & 2 & 1 \\ 2 & 2 & 2 \end{bmatrix}$	$\mathbf{R}_3^{-1(0)} = \begin{bmatrix} 1 & 0 & 0 \\ 1 & 1 & 1 \\ 1 & 2 & 1 \end{bmatrix}$
1	$\mathbf{R}_3^{(1)} = \begin{bmatrix} 0 & 0 & 1 \\ 2 & 1 & 0 \\ 2 & 2 & 2 \end{bmatrix}$	$\mathbf{R}_3^{-1(1)} = \begin{bmatrix} 1 & 1 & 1 \\ 1 & 2 & 1 \\ 1 & 0 & 0 \end{bmatrix}$
2	$\mathbf{R}_3^{(2)} = \begin{bmatrix} 0 & 1 & 0 \\ 1 & 0 & 2 \\ 2 & 2 & 2 \end{bmatrix}$	$\mathbf{R}_3^{-1(2)} = \begin{bmatrix} 1 & 2 & 1 \\ 1 & 0 & 0 \\ 1 & 1 & 1 \end{bmatrix}$

Polarity

Polarity plays an important role in decision diagram techniques. Equations 11.5 are a formal justification of the statement that an arbitrary logic function can be represented by m^n different polynomial expressions, or *polarities*. Also, Equations 11.5 are a formal notation of the problem of optimal representation of multivalued functions by polynomial expressions (see Example 11.18). This is because it is possible to find an optimal (in terms of minimal number of literals) representation among the m^n different polynomial expressions. Equations 11.5 provide a formal description of the behavior of a multivalued function in terms of change.

The polynomial expression of polarity c is described as follows:

$$R^{(c)} = \sum_{j=0}^{m^n-1} r_j(x_1+c_1)^{j_1}(x_2+c_2)^{j_2}\cdots(x_n+c_n)^{j_n} \quad GF(m) \quad (11.7)$$

$$(x_i+c_i)^{j_i} = \begin{cases} x_i+c_i = \hat{x}_i^{c_i}, & j_i \neq 0; \\ 1, & j_i = 0, \end{cases} \quad (11.8)$$

where $\hat{x}_i^{c_i}$ is a c_i-order cyclic complement of the variable x_i, $c_i \in (0,1,\ldots, m-1)$. The coefficient r_j is the j-th component of the vector $\mathbf{R}$ calculated by a matrix-vector transform (Equation 11.4). That is,

$$r_j = \sum_{k=0}^{m^n-1} f_k r_{k,j} \quad over \ GF(m), \quad (11.9)$$

where $f(k)$ is the k-th component of the truth vector $\mathbf{F}$, and $r_{j,k}$ is the $(k_{j,k})$-th element in the matrix $\mathbf{R}_{m^n}^{(c)}$.

> **Example 11.17** *The truth vector* $\mathbf{F} = [201000102]^T$ *represents a ternary logic function of two variables. There are nine **polynomial** expressions generated for this function, corresponding to polarities* $c_1 c_2 = \{00, 01, 02, 10, 11, 12, 20, 21, 22\}$. *Figure 11.8 shows the computing of polynomial expressions given* $c_1 c_2 = 01$. *The matrix transform implies that the coefficients* r_j, $j = 0, 1, 2, \ldots, 8$, *are calculated by Equation 11.9. Networks to implement this function in SOP are* $f = 2\varphi_0(x_1)\varphi_0(x_2) \vee 1\varphi_0(x_1)\varphi_2(x_2) \vee 1\varphi_2(x_1)\varphi_0(x_2) \vee 2\varphi_2(x_1)\varphi_2(x_2)$ *and polynomial forms are shown in Figure 11.9.*

Techniques for computing the coefficients of polynomial expression of a ternary logic function	

Given:
A ternary function (m=3) of two variables (n=2):
$\mathbf{F} = [201000102]^T$
Polarity of a polynomial expression:
$c = 1$, $c = c_1 c_2 = 01$,
$c_1 = 0$, $c_2 = 1$

x_1	x_2	F
0	0	2
0	1	0
0	2	1
1	0	0
1	1	0
1	2	0
2	0	1
2	1	0
2	2	2

Elementary transform matrices for $c_1 = 0$ and $c_2 = 1$ are given in Table 11.7. Using Equation 11.5, form the transform matrix

$$\mathbf{R}^{(2)} = \mathbf{R}^{(2)}_{32} \, \mathbf{F} = \left(\mathbf{R}^{(0)}_3 \otimes \mathbf{R}^{(1)}_3 \right) \mathbf{F}$$

$$= \left(\begin{bmatrix} 1 & 0 & 0 \\ 0 & 2 & 1 \\ 2 & 2 & 2 \end{bmatrix} \otimes \begin{bmatrix} 0 & 0 & 1 \\ 2 & 1 & 0 \\ 2 & 2 & 2 \end{bmatrix} \right) \mathbf{F}$$

$$= \begin{bmatrix} 0\,0\,1 & 0\,0\,0 & 0\,0\,0 \\ 2\,1\,0 & 0\,0\,0 & 0\,0\,0 \\ 2\,2\,2 & 0\,0\,0 & 0\,0\,0 \\ 0\,0\,0 & 0\,0\,2 & 0\,0\,1 \\ 0\,0\,0 & 1\,2\,0 & 2\,1\,0 \\ 0\,0\,0 & 1\,1\,1 & 2\,2\,2 \\ 0\,0\,2 & 0\,0\,2 & 0\,0\,2 \\ 1\,2\,0 & 1\,2\,0 & 1\,2\,0 \\ 1\,1\,1 & 1\,1\,1 & 1\,1\,1 \end{bmatrix} \begin{bmatrix} 2 \\ 0 \\ 1 \\ 0 \\ 0 \\ 0 \\ 1 \\ 0 \\ 2 \end{bmatrix} = \begin{bmatrix} 1 \\ 1 \\ 0 \\ 2 \\ 2 \\ 0 \\ 0 \\ 0 \\ 0 \end{bmatrix} \quad over \ GF(3)$$

Note that by Equation 11.9:

$r_j = \sum_{k=0}^{3^2-1} f_k r_{j,k}$ over $GF(3)$
$r_0 = f(0)r_{0,0} + \ldots + f(8)r_{0,8} = 2 \times 0 + \cdots + 2 \times 0 = 1$
$r_1 = f(0)r_{1,0} + \ldots + f(8)r_{1,8} = 2 \times 2 + \cdots + 2 \times 0 = 1$
$r_2 = f(0)r_{2,0} + \ldots + f(8)r_{2,8} = 2 \times 2 + \cdots + 2 \times 0 = 0$
$r_3 = f(0)r_{3,0} + \ldots + f(8)r_{3,8} = 2 \times 0 + \cdots + 2 \times 1 = 2$
$r_4 = f(0)r_{4,0} + \ldots + f(8)r_{4,8} = 2 \times 0 + \cdots + 2 \times 0 = 2$
$r_5 = f(0)r_{5,0} + \ldots + f(8)r_{5,8} = 2 \times 0 + \cdots + 2 \times 2 = 0$
$r_6 = f(0)r_{6,0} + \ldots + f(8)r_{6,8} = 2 \times 0 + \cdots + 2 \times 2 = 0$
$r_7 = f(0)r_{7,0} + \ldots + f(8)r_{7,8} = 2 \times 1 + \cdots + 2 \times 0 = 0$
$r_8 = f(0)r_{8,0} + \ldots + f(8)r_{8,8} = 2 \times 1 + \cdots + 2 \times 1 = 0$

Equation 11.9 for $m = 3$, $n = 2$, $c = 1$:

$$R^{(1)} = \sum_{j=0}^{3^2-1} r_j (x_1 + 0)^{j_1} (x_2 + 1)^{j_2} = \sum_{j=0}^{3^2-1} r_j x_1^{j_1} \hat{x}_2^{j_2}$$

$$= r_0 x_1^0 \hat{x}_2^0 + r_1 x_1^0 \hat{x}_2^1 + r_3 x_1^1 \hat{x}_2^0 + r_4 x_1^1 \hat{x}_2^1$$

$$= 1 + \hat{x}_2 + 2x_1 + 2x_1 \hat{x}_2 \quad over \ GF(3)$$

FIGURE 11.8
Computing coefficients of the polynomial expression (Example 11.17).

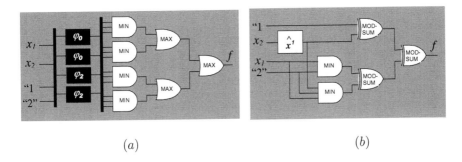

(a)   (b)

FIGURE 11.9
A ternary network that implements an SOP expression (a) and polynomial expression (b) (Example 11.17).

Example 11.18 *Multivalued expressions can be manipulated in different polarities; for example, the optimal form can be found using optimization techniques. Table 11.8 contains all 9 forms for the function given by truth vector $[020120000]^T$. The optimal (in terms of number of literals) polarity is $c = 8$.*

Example 11.19 *Table 11.9 represents a typical library of elementary multivalued gates in polynomial form.*

In signal processing, the matrix representation in the form of Equation 11.5 is known as a *factorized* representation of a transform matrix. These equations play the central role in synthesis of so-called *fast algorithms*.

A ternary variable x can be represented in polynomial form in one of the 28 bases which correspond to 28^n mixed polarity forms of a ternary function of n variables [10, 11] (Table 11.10, Figure 11.10).

Example 11.20 *Let*

$$\mathbf{A}_1 = [1 \ x_1 \ x_1^2] \quad and \quad \mathbf{A}_2 = [1 \ \hat{x}_2 \ \hat{x}_2^2]$$

The terms of a ternary function of two variables are defined as follows:

$$\mathbf{A}_1 \otimes \mathbf{A}_2 = [1 \ x_1 \ x_1^2] \otimes [1 \ \hat{x}_2 \ \hat{x}_1^2]$$
$$= [1 \ \hat{x}_2 \ \hat{x}_2^2 \ x_1 \ x_1\hat{x}_2 \ x_1\hat{x}_2^2 \ x_1^2 \ x_1^2\hat{x}_2 \ x_1^2\hat{x}_2^2 \ ]$$

TABLE 11.8

Polynomial expression in polarities $c = 0, 1, \ldots, 8$ of ternary logic functions of two variables

Polarity	Polynomial expression over $GF(3)$
$c = 0 \; (c_1 c_2 = 00)$	$R^{(0)} = 2x_1 + x_1^2 + x_1 x_2 + 2x_1^2 x_2$
$c = 1 \; (c_1 c_2 = 01)$	$R^{(1)} = 2 + 2x_1 + 2x_1^2 + 2\hat{x}_2 + 2x_1 \hat{x}_2 + 2x_1^2 \hat{x}_2$
$c = 2 \; (c_1 c_2 = 02)$	$R^{(2)} = 2x_1 \hat{x}_2 + 2x_1^2 \hat{x}_2 + 2\hat{x}_2$
$c = 3 \; (c_1 c_2 = 10)$	$R^{(3)} = 1 + 2\hat{x}_1 + \hat{x}_1^2 + 2x_2 + \hat{x}_1 x_2 + 2\hat{x}_1^2 x_2$
$c = 4 \; (c_1 c_2 = 11)$	$R^{(4)} = 2 + \hat{x}_1 + 2\hat{x}_1^2 + 2\hat{x}_2 + \hat{x}_1 \hat{x}_2 + 2\hat{x}_1^2 \hat{x}_2$
$c = 5 \; (c_1 c_2 = 12)$	$R^{(5)} = 2\hat{x}_1 \hat{x}_2 + \hat{x}_1^2 \hat{x}_2 + 2\hat{x}_2$
$c = 6 \; (c_1 c_2 = 20)$	$R^{(6)} = 2\hat{x}_1^2 x_2 + \hat{x}_1$
$c = 7 \; (c_1 c_2 = 21)$	$R^{(7)} = 2\hat{x}_1^2 \hat{x}_2 + 2\hat{x}_1$
$c = 8 \; (c_1 c_2 = 22)$	$R^{(8)} = 2\hat{x}_1^2 \hat{x}_1$

Example 11.21 *Given the* **vector of polarity** *zero coefficients*

$$\mathbf{R} = [211010010]^T$$

of a ternary function of two $(n = 2)$ variables. We use the result of Example 11.20 to derive the polynomial expression

$$\mathbf{R} = 2 + \hat{x}_2 + \hat{x}_2^2 + x_1 \hat{x}_2 + x_1^2 \hat{x}_2 \quad GF(3)$$

11.7 Polynomial representations using arithmetic operations

Arithmetic expressions of multivalued functions are an alternative approach to the description of logic circuits. They share properties of polynomial expressions and, at the same time, simplify representations of multioutput functions. In many applications arithmetic expressions provide a better insight into the analysis of multivalued functions. Examples of such applications are satisfiability, tautology, and equivalence checking.

TABLE 11.9
Polynomial representation of elementary ternary $(m = 3)$ functions
of two variables

Ternary gate		Polynomial representation

Ternary gate	Table	Polynomial representation
	$\begin{array}{c\|c} x & \overline{x} \\ \hline 0 & 2 \\ 1 & 1 \\ 2 & 0 \end{array}$	$2 + 2x + x^2 \ (mod \ 3)$
x^r	$\begin{array}{c\|ccc} r/x & 0 & 1 & 2 \\ \hline 0 & 0 & 1 & 2 \\ 1 & 1 & 2 & 0 \\ 2 & 2 & 0 & 1 \end{array}$	$\left\{\begin{array}{l} 1+x \ (mod \ 3), \ r=1; \\ 2+x \ (mod \ 3), \ r=2. \end{array}\right.$
MAX	$\begin{array}{c\|ccc} & 0 & 1 & 2 \\ \hline 0 & 0 & 1 & 2 \\ 1 & 1 & 1 & 2 \\ 2 & 2 & 2 & 2 \end{array}$	$x_1 + x_2 + 2x_1x_2 + x_1x_2^2 + x_1^2x_2 + x_1^2x_2^2 \ (mod \ 3)$
MIN	$\begin{array}{c\|ccc} & 0 & 1 & 2 \\ \hline 0 & 0 & 0 & 0 \\ 1 & 0 & 1 & 1 \\ 2 & 0 & 1 & 2 \end{array}$	$x_1x_2 + 2x_1x_2^2 + 2x_1^2x_2 + 2x_1^2x_2^2 \ (mod \ 3)$
MOD-SUM	$\begin{array}{c\|ccc} & 0 & 1 & 2 \\ \hline 0 & 0 & 1 & 2 \\ 1 & 1 & 2 & 0 \\ 2 & 2 & 0 & 1 \end{array}$	$x_1 + x_2 \ (mod \ 3)$
MOD-PROD	$\begin{array}{c\|ccc} & 0 & 1 & 2 \\ \hline 0 & 0 & 0 & 0 \\ 1 & 0 & 1 & 2 \\ 2 & 0 & 2 & 1 \end{array}$	$x_1x_2 \ (mod \ 3)$
TSUM	$\begin{array}{c\|ccc} & 0 & 1 & 2 \\ \hline 0 & 0 & 1 & 2 \\ 1 & 1 & 2 & 2 \\ 2 & 2 & 2 & 2 \end{array}$	$x_1 + x_2 + 2x_1x_2^2 + 2x_1^2x_2 + 2x_1^2x_2^2 \ (mod \ 3)$
TPROD	$\begin{array}{c\|ccc} & 0 & 1 & 2 \\ \hline 0 & 0 & 0 & 0 \\ 1 & 0 & 1 & 2 \\ 2 & 0 & 2 & 2 \end{array}$	$2x_1x_2 + x_1x_2^2 + 2x_1^2x_2 + 2x_1^2x_2^2 \ (mod \ 3)$

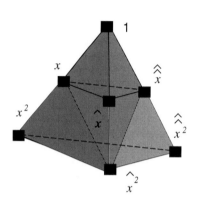

There are 28 paths between nodes pictured on this pyramid, which correspond to 28 polarity vectors for ternary function of single variable:

$$\overbrace{[1\ x\ x^2], [1\ \hat{x}\ \hat{x}^2], \ldots, [1\ \hat{\hat{x}}\ \hat{x}^2]}^{12\ paths},$$

$$\overbrace{[x\ \hat{x}\ x^2], [x\ \hat{x}\ \hat{\hat{x}}^2], \ldots, [x\ x^2\ \hat{\hat{x}}^2]}^{5\ paths},$$

$$\overbrace{[\hat{x}\ \hat{\hat{x}}\ \hat{x}^2], [\hat{x}\ \hat{\hat{x}}\ x^2], \ldots, [\hat{x}\ \hat{x}^2\ x^2]}^{5\ paths},$$

$$\overbrace{[\hat{\hat{x}}\ x\ \hat{x}^2], [\hat{\hat{x}}\ x\ x^2], \ldots, [\hat{\hat{x}}\ \hat{x}^2\ \hat{x}^2]}^{5\ paths}$$

$$\overbrace{[x^2\ \hat{x}^2\ \hat{\hat{x}}^2]}^{1\ path}$$

FIGURE 11.10

Graphical representation of polarities of a ternary variable x.

TABLE 11.10

Possible vectors of polarities of a ternary variable x

1	x	x^2	x	$\hat{x}$	x^2									
1	$\hat{x}$	$\hat{x}^2$	$\hat{x}$	$\hat{\hat{x}}$	$\hat{x}^2$									
1	$\hat{\hat{x}}$	$\hat{\hat{x}}^2$	$\hat{\hat{x}}$	x	$\hat{\hat{x}}^2$									

1	x	$\hat{x}^2$	x	$\hat{x}$	$\hat{x}^2$	1	x^2	$\hat{x}^2$	x	x^2	$\hat{x}^2$	x^2	$\hat{x}^2$	$\hat{\hat{x}}^2$
1	$\hat{x}$	$\hat{\hat{x}}^2$	$\hat{x}$	$\hat{\hat{x}}$	$\hat{\hat{x}}^2$	1	$\hat{x}^2$	$\hat{x}^2$	$\hat{x}$	$\hat{x}^2$	$\hat{\hat{x}}^2$			
1	$\hat{\hat{x}}$	x^2	$\hat{\hat{x}}$	x	x^2	1	$\hat{\hat{x}}^2$	x^2	$\hat{\hat{x}}$	$\hat{\hat{x}}^2$	x^2			

1	x	$\hat{\hat{x}}^2$	x	$\hat{x}$	$\hat{\hat{x}}^2$	x	x^2	$\hat{\hat{x}}^2$		
1	$\hat{x}$	x^2	$\hat{x}$	$\hat{\hat{x}}$	x^2	$\hat{x}$	$\hat{x}^2$	x^2		
1	$\hat{\hat{x}}$	$\hat{x}^2$	$\hat{\hat{x}}$	x	$\hat{x}^2$	$\hat{\hat{x}}$	$\hat{x}^2$	$\hat{x}^2$		

11.7.1 Direct and inverse arithmetic transforms

Direct and inverse arithmetic transforms are defined as follows (Figure 11.11):

$$\mathbf{P} = \frac{1}{(m-1)^n} \, \mathbf{P}_{m^n}^{(c)} \mathbf{F},$$

$$\mathbf{F} = \mathbf{P}_{m^n}^{-1(c)} \mathbf{P}.$$

(11.10)

The pair of matrices $\mathbf{P}_{m^n}^{(c)}$ and $\mathbf{P}_{m^n}^{-1(c)}$ of arithmetic transforms in Equation 11.10 are calculated by the Kronecker product

$$\begin{aligned}
\mathbf{P}_{m^n}^{(c)} &= \mathbf{P}_m^{(c_1)} \otimes \mathbf{P}_m^{(c_2)} \otimes \cdots \otimes \mathbf{P}_m^{(c_n)}, \\
\mathbf{P}_{m^n}^{-1(c)} &= \mathbf{R}_m^{-1(c_1)} \otimes \mathbf{P}_m^{-1(c_2)} \otimes \cdots \otimes \mathbf{P}_m^{-1(c_n)}.
\end{aligned}$$

(11.11)

Elements of the matrices $\mathbf{P}_m^{(c_j)}$ and $\mathbf{P}_m^{-1(c_j)}$, $j = 1, 2, \ldots n$, are obtained as solutions of the equation

$$\mathbf{P}_m^{-1(c_j)} \mathbf{P}_m^{(c_j)} = \mathbf{I}_m.$$

(11.12)

In Table 11.11, the basic arithmetic transform matrices for ternary $(m = 3)$ logic functions are given.

Example 11.22 *Given $m = 3$ and $c = 2$, the basic matrices of the* **direct** *and* **inverse** *transform satisfy Equation 11.12:*

$$\mathbf{P}_3^{-1(2)} \mathbf{P}_3^{(2)} = \begin{bmatrix} 1 & 2 & 1 \\ 1 & 0 & 0 \\ 1 & 1 & 1 \end{bmatrix} \begin{bmatrix} 0 & 2 & 0 \\ -1 & -3 & 4 \\ 1 & 1 & -2 \end{bmatrix} = \begin{bmatrix} 1 & 0 & 0 \\ 0 & 1 & 0 \\ 0 & 0 & 1 \end{bmatrix} = \mathbf{I}_3.$$

11.7.2 Polarity

Equation 11.11 is a formal description of forming different polarities for the manipulation of logic functions in arithmetic form:

▶ An arbitrary m-valued logic function can be represented by m^n different generalized arithmetic expressions, or *polarities.*

▶ A formal description of the behavior of a multivalued function in terms of change can be derived from this equation, since rows of the transform matrices describe arithmetic analogs of logic differences, as will be shown in the following section.

TABLE 11.11
Basic arithmetic transform matrices for a ternary logic
function of polarity $c = 0, 1, 2$

Polarity c	Direct	Inverse
0	$\mathbf{P}_3^{(0)} = \begin{bmatrix} 2 & 0 & 0 \\ -3 & 4 & -1 \\ 1 & -2 & 1 \end{bmatrix}$	$\mathbf{P}_3^{-1(0)} = \begin{bmatrix} 1 & 0 & 0 \\ 1 & 1 & 1 \\ 1 & 2 & 1 \end{bmatrix}$
1	$\mathbf{P}_3^{(1)} = \begin{bmatrix} 0 & 0 & 2 \\ 4 & -1 & -3 \\ -2 & 1 & 1 \end{bmatrix}$	$\mathbf{P}_3^{-1(1)} = \begin{bmatrix} 1 & 1 & 1 \\ 1 & 2 & 1 \\ 1 & 0 & 0 \end{bmatrix}$
2	$\mathbf{P}_3^{(2)} = \begin{bmatrix} 0 & 2 & 0 \\ -1 & -3 & 4 \\ 1 & 1 & -2 \end{bmatrix}$	$\mathbf{P}_3^{-1(2)} = \begin{bmatrix} 1 & 2 & 1 \\ 1 & 0 & 0 \\ 1 & 1 & 1 \end{bmatrix}$

$$P^{(c)} = \frac{1}{(m-1)^n} \sum_{j=0}^{m^n-1} p_j (x_1 + c_1)^{i_1} (x_2 + c_2)^{i_2} \cdots (x_n + c_n)^{i_n} \quad (11.13)$$

where

$$(x_i + c_i)^{j_i} = \begin{cases} x_i + c_i, & j_i \neq 0 \ (mod \ m); \\ 1, & j_i = 0. \end{cases} \quad (11.14)$$

Coefficient p_j is the j-th component of the vector $\mathbf{P}$ calculated by a matrix-vector transform (Equation 11.10). That is,

$$p_j = \sum_{k=0}^{m^n-1} f_k p_{k,j}, \quad (11.15)$$

where $f(k)$ is the k-th component of the truth vector $\mathbf{P}$, and $p_{j,k}$ is the $(k_{j,k})$-th element in the matrix $\mathbf{P}_{m^n}^{(c)}$.

Note that the coefficients p_j are also cofactors in the Taylor expansion.

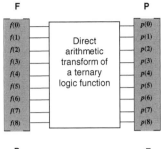

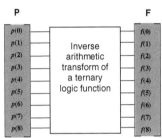

The pair of direct and inverse arithmetic transforms

$$\mathbf{P} = \frac{1}{(3-1)^2} \, \mathbf{P}_{3^2}^{(c)} \mathbf{F},$$

$$\mathbf{F} = \mathbf{P}_{3^2}^{-1(c)} \mathbf{P},$$

where

$$\mathbf{P}_{3^2}^{(c)} = \mathbf{P}_3^{(c_1)} \otimes \mathbf{P}_3^{(c_2)},$$
$$\mathbf{P}_{3^2}^{-1(c)} = \mathbf{R}_3^{-1(c_1)} \otimes \mathbf{P}_3^{-1(c_2)}.$$

Basic transform matrices $\mathbf{P}_{3^2}^{(c)}$ and $\mathbf{P}_{3^2}^{-1(c)}$ for polarity $c = 0, 1, 2$ are given in Table 11.11

FIGURE 11.11
Direct and inverse arithmetic transforms for a ternary function of two variables.

Example 11.23 *In Figure 11.12, the calculation of an* **arithmetic expression** *of polarity* $c_1 c_2 = 02$ *for a ternary function of two variables given truth vector* $\mathbf{F} = [010211202]^T$ *is given. There are nine arithmetic expressions to represent this function that correspond to the polarities* $c_1 c_2 = \{00, 01, 02, 10, 11, 12, 20, 21, 22\}$.

Example 11.24 *Table 11.12 represents a typical library of elementary multivalued gates in arithmetic form.*

11.7.3 Word-level representation

Similar to the word-level representation of Boolean functions, the properties of linearity and superposition are utilized in computing word-level arithmetic expressions of multioutput logic functions.

Techniques for computing the coefficients of polynomial using arithmetic operations

Elementary transform matrices for $c_1 = 0$ and $c_2 = 2$ are given in Table 11.7. Using Equation 11.5, form the transform matrix.

$$\mathbf{P}^{(2)} = {}^1/_4 \times \mathbf{P}_{3^2}^{(2)} \mathbf{F} = {}^1/_4 \times \left(\mathbf{P}_3^{(0)} \otimes \mathbf{P}_3^{(2)} \right) \mathbf{F}$$

$$= {}^1/_4 \times \left(\begin{bmatrix} 2 & 0 & 0 \\ -3 & 4 & -1 \\ 1 & -2 & 1 \end{bmatrix} \otimes \begin{bmatrix} 0 & 2 & 0 \\ -1 & -3 & 4 \\ 1 & 1 & -2 \end{bmatrix} \right) \mathbf{F}$$

Given:
A ternary function ($m=3$) of two variables ($n=2$):
$\mathbf{F} = [010211202]^T$
Polarity of an arithmetic expression:
$c = 2$
$c = c_1 c_2 = 02,$
$c_1 = 0,\ c_2 = 2$

x_1

f

x_2

x_1	x_2	$\mathbf{F}$
0	0	0
0	1	1
0	2	0
1	0	2
1	1	1
1	2	1
2	0	2
2	1	0
2	2	2

$$= {}^1/_4 \times \begin{bmatrix} 0 & 4 & 0 & & & & & & \\ -2 & -6 & 8 & & & & & & \\ 2 & 2 & -4 & & & & & & \\ 0 & -6 & 0 & 0 & 8 & 0 & 0 & -2 & 0 \\ 3 & 9 & -12 & -4 & -12 & 16 & 1 & 3 & -4 \\ -3 & -3 & 6 & 4 & 4 & -8 & -1 & -1 & 2 \\ 0 & 2 & 0 & 0 & -4 & 0 & 0 & 2 & 0 \\ -1 & -3 & 4 & 2 & 6 & -8 & -1 & -3 & 4 \\ 1 & 1 & -2 & -2 & -2 & 4 & 1 & 1 & -2 \end{bmatrix} \begin{bmatrix} 0 \\ 1 \\ 0 \\ 2 \\ 1 \\ 1 \\ 2 \\ 0 \\ 2 \end{bmatrix} = \begin{bmatrix} 4 \\ -6 \\ 2 \\ 2 \\ -1 \\ 3 \\ -2 \\ 5 \\ -3 \end{bmatrix}$$

Note that by Equation 11.15:

$p_j = \sum_{k=0}^{3^2-1} f_k p_{j,k}$

$p_0 = f(0)p_{0,0} + \ldots + f(8)p_{0,8} = 0 \times 0 + \cdots + 2 \times 0 \quad = \boxed{4}$

$p_1 = f(0)p_{1,0} + \ldots + f(8)p_{1,8} = 1 \times (-2) + \cdots + 2 \times 0 = \boxed{-6}$

$p_2 = f(0)p_{2,0} + \ldots + f(8)p_{2,8} = 0 \times 2 + \cdots + 2 \times 0 \quad = \boxed{2}$

$p_3 = f(0)p_{3,0} + \ldots + f(8)p_{3,8} = 0 \times 0 + \cdots + 2 \times 0 \quad = \boxed{2}$

$p_4 = f(0)p_{4,0} + \ldots + f(8)p_{4,8} = 0 \times 3 + \cdots + 2 \times -4 \quad = \boxed{-1}$

$p_5 = f(0)p_{5,0} + \ldots + f(8)p_{5,8} = 0 \times (-3) + \cdots + 2 \times 2 = \boxed{3}$

$p_6 = f(0)p_{6,0} + \ldots + f(8)p_{6,8} = 0 \times 0 + \cdots + 2 \times 0 \quad = \boxed{-2}$

$p_7 = f(0)p_{7,0} + \ldots + f(8)p_{7,8} = 0 \times (-1) + \cdots + 2 \times 4 = \boxed{5}$

$p_8 = f(0)p_{8,0} + \ldots + f(8)p_{8,8} = 0 \times 1 + \cdots + 2 \times (-2) = \boxed{-3}$

Equation 11.9 for $m = 3$, $n = 2$, $c = 2$:

$$P^{(2)} = {}^1/_4 \times \sum_{j=0}^{8} p_j (x_1 + 0)^{j_1} (x_2 + 2)^{j_2}$$

$$= {}^1/_4 \times (4 - 6\hat{x}_2 + 2\hat{x}_2^2 + 2x_1 - x_1\hat{x}_2 + 3x_1\hat{x}_2^2$$
$$ - 2\hat{x}_1^2 + 5x_1^2\hat{x}_2 - 3x_1^2\hat{x}_2^2)$$

A SOP expression can be derived from the truth table:

$$f = 1\varphi_0(x_1)\varphi_1(x_2) + 2\varphi_1(x_1)\varphi_0(x_2) + 1\varphi_1(x_1)\varphi_1(x_2)$$
$$+ 1\varphi_1(x_1)\varphi_2(x_2) + 2\varphi_2(x_1)\varphi_0(x_2) + 2\varphi_2(x_1)\varphi_2(x_2)$$

FIGURE 11.12
Computing coefficients of the polynomial representation of a ternary logic function using arithmetic operations (Example 11.23).

TABLE 11.12

Arithmetic representation of elementary ternary $(m = 3)$
two-variable functions

Ternary gate	Arithmetic representation

$$\begin{array}{c|c} x & \overline{x} \\ \hline 0 & 2 \\ 1 & 1 \\ 2 & 0 \end{array} \qquad 2 - x$$

$$\begin{array}{c|ccc} r/x & 0 & 1 & 2 \\ \hline 0 & 0 & 1 & 2 \\ 1 & 1 & 2 & 0 \\ 2 & 2 & 0 & 1 \end{array} \qquad \begin{cases} x, & r=0; \\ 1 + \frac{5}{2}x - \frac{3}{2}x^2, & r=1; \\ 2 - \frac{7}{2}x + \frac{3}{2}x^2, & r=2. \end{cases}$$

MAX
$$\begin{array}{c|ccc} & 0 & 1 & 2 \\ \hline 0 & 0 & 1 & 2 \\ 1 & 1 & 1 & 2 \\ 2 & 2 & 2 & 2 \end{array} \qquad x_1 + x_2 - \frac{10}{4}x_1 x_2 + x_1 x_2^2 + x_1^2 x_2 - \frac{1}{2}x_1^2 x_2^2$$

MIN
$$\begin{array}{c|ccc} & 0 & 1 & 2 \\ \hline 0 & 0 & 0 & 0 \\ 1 & 0 & 1 & 1 \\ 2 & 0 & 1 & 2 \end{array} \qquad \frac{10}{4}x_1 x_2 - x_1 x_2^2 - x_1^2 x_2 + \frac{1}{2}x_1^2 x_2^2$$

MOD-SUM
$$\begin{array}{c|ccc} & 0 & 1 & 2 \\ \hline 0 & 0 & 1 & 2 \\ 1 & 1 & 2 & 0 \\ 2 & 2 & 0 & 1 \end{array} \qquad x_1 + x_2 + \frac{21}{4}x_1 x_2 - \frac{15}{4}x_1 x_2^2 - \frac{15}{4}x_1^2 x_2 + \frac{9}{4}x_1^2 x_2^2$$

MOD-PROD
$$\begin{array}{c|ccc} & 0 & 1 & 2 \\ \hline 0 & 0 & 0 & 0 \\ 1 & 0 & 1 & 2 \\ 2 & 0 & 2 & 1 \end{array} \qquad \frac{1}{4}x_1 x_2 + \frac{3}{4}x_1 x_2^2 + \frac{3}{4}x_1^2 x_2 - \frac{3}{4}x_1^2 x_2^2$$

TSUM
$$\begin{array}{c|ccc} & 0 & 1 & 2 \\ \hline 0 & 0 & 1 & 2 \\ 1 & 1 & 2 & 2 \\ 2 & 2 & 2 & 2 \end{array} \qquad x_1 + x_2 + \frac{6}{4}x_1 x_2 - x_1 x_2^2 - x_1^2 x_2 + \frac{1}{2}x_1^2 x_2^2$$

TPROD
$$\begin{array}{c|ccc} & 0 & 1 & 2 \\ \hline 0 & 0 & 0 & 0 \\ 1 & 0 & 1 & 2 \\ 2 & 0 & 2 & 2 \end{array} \qquad \frac{6}{4}x_1 x_2 - \frac{2}{4}x_1^2 x_2^2$$

Example 11.25 *A three-output ternary logic function of two variables given by truth vectors can be represented by a word-level arithmetic expression (Figure 11.13). The first method is based on the direct arithmetic transform (Equation 11.10) of truth vectors $\mathbf{F}_0, \mathbf{F}_1$ and $\mathbf{F}_2$. The resulting vectors of coefficients $\mathbf{P}_0, \mathbf{P}_1$ and $\mathbf{P}_2$ form the vector $\mathbf{D}$ calculated as a weighted sum (the first method). Alternatively, the direct arithmetic transform (Equation 11.10) is applied to the vector $\mathbf{F}$ calculated as a weighted sum of $\mathbf{F}_0$, $\mathbf{F}_1$, and $\mathbf{F}_2$ (the second method).*

A nonlinear word-level expression can be represented by a linear word-level expression.

11.8   Fundamental expansions

Consider a three-valued signal with three logic values 0, 2, and 3 (Figure 11.14). There are four possible situations (for simplification, the direction of change is not considered): change $0 \leftrightarrow 1$, change $0 \leftrightarrow 2$, change $1 \leftrightarrow 2$, and no change ($0 \leftrightarrow 0$, $1 \leftrightarrow 1$, $2 \leftrightarrow 2$). The problem is formulated as detection of changes in a ternary function f if the ternary variable x_i is changed.

In contrast to formal notation of Boolean difference, where the complement of binary variable x_i is defined as $\overline{x}_i$, in multivalued logic *the cyclic complement* of a multivalued variable x_i is used.

11.8.1   Logic difference

Let f be an m-valued (m is prime) logic function of n variables. The t_i-th order cyclic complement to a variable x_i, $i = 1, 2 \ldots, n$, is

$$\overset{t_i}{\hat{x}_i} = x_i + t_i \quad \text{mod } (m),$$

where $t_i \in \{0, 1, 2, \ldots, m-1\}$. The logic difference of a function f with respect to the t_i-order cyclic complement of the variable x_i is defined as

$$\partial f / \partial \overset{t_i}{\hat{x}_i} = \sum_{p=0}^{m-1} r_{m-t_i, p} \, f(x_1, \ldots, \overset{p}{\hat{x}_i}, \ldots, x_n) \quad \text{over GF}(m), \qquad (11.16)$$

Techniques for computing a set of ternary functions using polynomial representation

Method 1

$$\mathbf{P}_0 = {}^1/_4 \times \mathbf{P}^{(0)}_{32}\mathbf{F}_0 = {}^1/_4 \times [\,8 - 6\ 2 - 6\ 3\ 3 - 17\ 2 - 17\ 9\,]^T,$$
$$\mathbf{P}_1 = {}^1/_4 \times \mathbf{P}^{(0)}_{32}\mathbf{F}_1 = {}^1/_4 \times [\,4\ 0\ 0\ 6\ 11 - 9\ -2 - 7\ 5\,]^T,$$
$$\mathbf{P}_2 = {}^1/_4 \times \mathbf{P}^{(0)}_{32}\mathbf{F}_2 = {}^1/_4 \times [\,0\ 8 - 4 - 4\ 19 - 9\ 4 - 11\ 5\,]^T.$$
$$\mathbf{D} = 3^0\mathbf{P}_0 + 3^1\mathbf{P}_1 + 3^2\mathbf{P}_2$$
$$= {}^1/_4 \times [20\ 66 - 34\ - 24\ 237\ - 125\ 32\ - 137\ 69]^T$$

Method 2

$$\mathbf{F}_D = [\mathbf{F}_2|\mathbf{F}_1|\mathbf{F}_0] = 3^2\mathbf{F}_2 + 3^1\mathbf{F}_1 + 3^0\mathbf{F}_0$$

$$= 3^2 \begin{bmatrix} 0 \\ 1 \\ 0 \\ 0 \\ 2 \\ 0 \\ 2 \\ 2 \\ 1 \end{bmatrix} + 3^1 \begin{bmatrix} 1 \\ 1 \\ 2 \\ 2 \\ 2 \\ 0 \\ 2 \\ 1 \\ 1 \end{bmatrix} + 3^0 \begin{bmatrix} 2 \\ 1 \\ 1 \\ 1 \\ 2 \\ 0 \\ 1 \\ 0 \\ 1 \end{bmatrix} = \begin{bmatrix} 5 \\ 13 \\ 4 \\ 7 \\ 26 \\ 0 \\ 25 \\ 21 \\ 13 \end{bmatrix}$$

$$\mathbf{D} = {}^1/_4 \times \mathbf{P}^{(0)}_{32}\mathbf{F}_D$$
$$= [20\ 66\ - 34\ - 24\ 237\ - 125\ 32\ - 137\ 69]^T$$

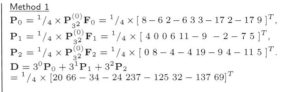

$$= \frac{1}{4} \begin{bmatrix} 4 & 0 & 0 & & & & & & \\ -6 & 8 & -2 & & & & & & \\ 2 & -4 & 2 & & & & & & \\ -6 & 0 & 0 & 8 & 0 & 0 & -2 & 0 & 0 \\ 9 & -12 & 3 & -12 & 16 & -4 & 3 & -4 & 1 \\ -3 & 6 & -3 & 4 & -8 & 4 & -1 & 2 & -1 \\ 2 & 0 & 0 & -4 & 0 & 0 & 2 & 0 & 0 \\ -3 & 4 & -1 & 6 & -8 & 2 & -3 & 4 & -1 \\ 1 & -2 & 1 & -2 & 4 & -2 & 1 & -2 & 1 \end{bmatrix}$$

$$\mathbf{D} = {}^1/_4 \times (20 + 66x_2 - 34x_2^2 - 24x_1 + 237x_1x_2$$
$$- 125x_1x_2^2 + 32x_1^2 - 137x_1^2x_2 + 69x_1^2x_2^2)$$

A SOP expression can be derived from the truth table:

$$f_0 = 2\varphi_0(x_1)\varphi_0(x_2) + 1\varphi_0(x_1)\varphi_1(x_2) + 1\varphi_0(x_1)\varphi_2(x_2)$$
$$+ 1\varphi_1(x_1)\varphi_0(x_2) + 2\varphi_1(x_1)\varphi_1(x_2) + 1\varphi_2(x_1)\varphi_0(x_2)$$
$$+ 1\varphi_2(x_1)\varphi_2(x_2)$$
$$f_1 = 1\varphi_0(x_1)\varphi_0(x_2) + 1\varphi_0(x_1)\varphi_1(x_2) + 2\varphi_0(x_1)\varphi_2(x_2)$$
$$+ 2\varphi_1(x_1)\varphi_0(x_2) + 2\varphi_1(x_1)\varphi_1(x_2) + 2\varphi_2(x_1)\varphi_0(x_2)$$
$$+ 1\varphi_2(x_1)\varphi_1(x_2) + 1\varphi_2(x_1)\varphi_2(x_2)$$
$$f_3 = 1\varphi_0(x_1)\varphi_1(x_2) + 2\varphi_1(x_1)\varphi_1(x_2) + 2\varphi_2(x_1)\varphi_0(x_2)$$
$$+ 2\varphi_2(x_1)\varphi_1(x_2) + 1\varphi_2(x_1)\varphi_2(x_2)$$

x_1	x_2	F_2	F_1	F_0
0	0	0	1	2
0	1	1	1	1
0	2	0	2	1
1	0	0	2	1
1	1	2	2	2
1	2	0	0	0
2	0	2	2	1
2	1	2	1	0
2	2	1	1	1

$$\mathbf{P}^{(0)}_{32} = \frac{1}{4}\left(\begin{bmatrix} 2 & 0 & 0 \\ -3 & 4 & -1 \\ 1 & -2 & 1 \end{bmatrix} \otimes \begin{bmatrix} 2 & 0 & 0 \\ -3 & 4 & -1 \\ 1 & -2 & 1 \end{bmatrix} \right)$$

FIGURE 11.13

Representation of a three-output ternary function of two variables by a word-level arithmetic expression of polarity $c = 0$ (Example 11.25).

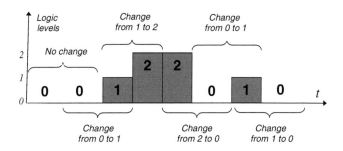

FIGURE 11.14
Change of three-valued signals.

where $r_{m-t_i,p}$ is the $(m - t_i, p)$-th element of the transform matrix $R_m^{(0)}$. It follows from Equation 11.16 that logic difference reflects the change of the value of the multivalued function f with respect to t_i-th cyclic complement of the multivalued variable x_i. There exist $m - 1$ different logic differences with respect to a given variable x_i for an m-valued logic function because there exist $m - 1$ complements to x_i. In contrast to Boolean difference, the Equation 11.16 involves m cofactors in the sum over $GF(m)$.

   Given a Boolean function $(m = 2)$, Equation 11.16 turns into a Boolean difference $\partial f / \partial \hat{x}_i = 1 \cdot \overset{0}{f}(x_1, ..., \hat{x}_i, ..., x_n) + 1 \cdot \overset{1}{f}(x_1, ..., \hat{x}_i, ..., x_n) =$

$$\underbrace{f(x_1, ..., x_i, ..., x_n)}_{Initial\ function} \oplus \underbrace{f(x_1, ..., \overline{x}_i, ..., x_n)}_{x_i\ replaced\ by\ \overline{x}_i} = \frac{\partial f}{\partial x_i},$$

since $\overset{1}{\hat{x}}_i = \overline{x}_i = x_i \oplus 1$, and the

coefficients $r_{2-1,0} = r_{2-1,1} = 1$ are taken from the matrix $R_2^{(0)} = \begin{bmatrix} r_{00} & r_{01} \\ r_{10} & r_{11} \end{bmatrix} = \begin{bmatrix} 1 & 0 \\ 1 & 1 \end{bmatrix}.$

> **Example 11.26** *Figure 11.15 illustrates changes in Boolean and ternary functions described by Equation 11.16. The logic differences $\partial f / \partial \hat{x}_i$, $\partial f / \partial \hat{\hat{x}}_i$, $\partial f / \partial \hat{\hat{\hat{x}}}_i$ correspond to the behavior of a quaternary function $f(\hat{x}_i)$, $f(\hat{\hat{x}}_i)$, $f(\hat{\hat{\hat{x}}}_i)$ for $x_i \rightarrow \{\hat{x}_i, \hat{\hat{x}}_i, \hat{\hat{\hat{x}}}_i\}$.*

   Change of a Boolean function f (a change in the value of f) caused by a change of the variable x_i to $\overline{x}_i$ is detected by the Boolean difference. In the ternary logic function f, a combination difference $\partial f / \partial \hat{x}_i = 2$ and $\partial f / \partial \hat{\hat{x}}_i$ recognizes the type of change.

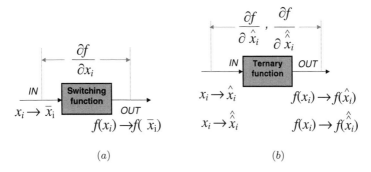

FIGURE 11.15
Logic differences of Boolean (a) and ternary (b) functions (Example 11.26).

Example 11.27 *Two* **logic differences** *with respect to a variable x_i for a ternary system are calculated by the Equation 11.16:*

$$\partial f/\partial \hat{x}_i = \sum_{p=0}^{2-1} r_{3-1,p}\ f(x_1, ..., \overset{p}{\hat{x}_i}, ..., x_n)\ over\ GF(3),$$

$$\partial f/\partial \hat{\hat{x}}_i = \sum_{p=0}^{2-1} r_{3-2,p}\ f(x_1, ..., \overset{p}{\hat{x}_i}, ..., x_n)\ over\ GF(3).$$

Since

$$R_3^{(0)} = \begin{bmatrix} r_{00}\ r_{01}\ r_{02} \\ r_{10}\ r_{10}\ r_{10} \\ r_{20}\ r_{21}\ r_{22} \end{bmatrix} = \begin{bmatrix} 1\ 0\ 0 \\ 0\ 2\ 1 \\ 2\ 2\ 2 \end{bmatrix},$$

the coefficients $r_{m-t_i,p}$ are derived as follows:

    <u>*1-order cyclic complement*</u> *of a variable x_i: $t_i = 1$, and we take coefficients from the last row of $R_3^{(0)}$*
        $r_{3-1,0} = r_{3-1,1} = r_{3-1,2} = 2;$
    <u>*2-order cyclic complement*</u> *of a variable x_i: $t_i = 2$, and we take coefficients from the middle row of $R_3^{(0)}$ $r_{3-2,0} = 0$, $r_{3-2,1} = 2$, $r_{3-2,2} = 1$.*

Figure 11.16 illustrates the changes of x_i and f that are involved in calculation of the logic differences. Note that $\hat{x}_i = x_i + 1$ (mod 3) and $\hat{\hat{x}}_i = x_i + 2$ (mod 3).

Techniques for the change computing

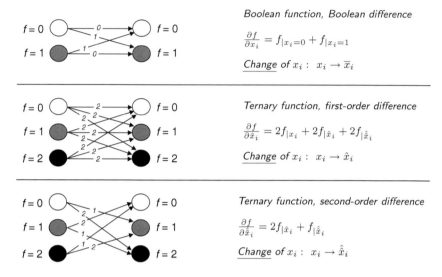

Boolean function, Boolean difference

$$\frac{\partial f}{\partial x_i} = f|_{x_i=0} + f|_{x_i=1}$$

Change of x_i : $x_i \to \overline{x}_i$

Ternary function, first-order difference

$$\frac{\partial f}{\partial \hat{x}_i} = 2f|_{x_i} + 2f|_{\hat{x}_i} + 2f|_{\hat{\hat{x}}_i}$$

Change of x_i : $x_i \to \hat{x}_i$

Ternary function, second-order difference

$$\frac{\partial f}{\partial \hat{\hat{x}}_i} = 2f|_{\hat{x}_i} + f|_{\hat{\hat{x}}_i}$$

Change of x_i : $x_i \to \hat{\hat{x}}_i$

FIGURE 11.16
Change of Boolean and ternary functions with respect to a variable (Example 11.27) and logic differences.

Computing logic differences. The matrix interpretation of the logic difference of an m-valued function f of n-variables with respect to a variable x_i with the t_i-order cyclic complement, $i = 1, 2, \ldots, n$, is given below:

$$\frac{\partial \mathbf{F}}{\partial \hat{x}_i^{t_i}} = \hat{D}_{m^n}^{(i)}{}^{t_i} \mathbf{F}, \tag{11.17}$$

where the matrix $\hat{D}_{m^n}^{(i)}{}^{t_i}$ is formed by the Kronecker product

$$\hat{D}_{m^n}^{(i)}{}^{t_i} = (m-1)I_{m^{i-1}} \otimes \left(\sum_{p=0}^{m-1} r_{m-t_i,p}\, I_m^{(p\to)} \right) \otimes I_{m^{n-i}}, \tag{11.18}$$

and $I_m^{(p\to)}$ is the $m \times m$ matrix generated by p-th right cyclic shift of elements of the identity matrix I_m. Note that the denotation of matrix $\hat{D}_{m^n}^{(i)}{}^{t_i}$ carries information about the size of the matrix (m^n), the number of variables (n), the order of the cyclic complement (t_i), and the variable with respect to which the difference is calculated (x_i).

Example 11.28 *Let* $m = 3$, $t_i = 2$, *then* $\sum_{p=0}^{m-1} r_{m-t_i,p} I_m^{(p\rightarrow)} = \sum_{p=0}^{2} r_{1,p} I_3^{(p\rightarrow)} = 0 \cdot I_3^{(0\rightarrow)} + 2 \cdot I_3^{(1\rightarrow)} + 1 \cdot I_3^{(2\rightarrow)}$

$$= 2 \cdot \begin{bmatrix} 0 & 1 & 0 \\ 0 & 0 & 1 \\ 1 & 0 & 0 \end{bmatrix} + 1 \cdot \begin{bmatrix} 0 & 0 & 1 \\ 1 & 0 & 0 \\ 0 & 1 & 0 \end{bmatrix} = \begin{bmatrix} 0 & 2 & 1 \\ 1 & 0 & 2 \\ 2 & 1 & 0 \end{bmatrix}.$$

Given a Boolean function ($m = 2$), Equation 11.17 is the Boolean differences in matrix form

$$\frac{\partial \mathbf{F}}{\partial x_i} = D_{2^n}^{(i)} \mathbf{F}, \tag{11.19}$$

where matrix $D_{2^n}^{(i)}$ is formed by Equation 11.18 $D_{2^n}^{(i)} = I_{2^{i-1}} \otimes D_2 \otimes I_{2^{n-i}}$, $D_2 = \begin{bmatrix} 1 & 1 \\ 1 & 1 \end{bmatrix}$.

Example 11.29 *The structure of matrix* $\hat{D}_{m^n}^{(i)\,t_i}$ *for the parameters below is illustrated in Figure 11.17.*

Example 11.30 *Given the truth vector* $\mathbf{F} = [0123112322233333]^T$ *of a quaternary ($m = 4$) logic function of two variables ($n = 2$), the logic difference* $\partial \mathbf{F}/\partial \hat{x}_1$ *is calculated by Equation 11.17 and Equation 11.18 as follows:*

$$\frac{\partial \mathbf{F}}{\partial \hat{x}_1} = \hat{D}_{4^2}^{(1)} \mathbf{F} = \left(\begin{bmatrix} 0 & 1 & 2 & 3 \\ 1 & 0 & 3 & 2 \\ 2 & 3 & 0 & 1 \\ 3 & 2 & 1 & 0 \end{bmatrix} \otimes \begin{bmatrix} 1 & & & \\ & 1 & & \\ & & 1 & \\ & & & 1 \end{bmatrix} \right) \mathbf{F}$$

$$= \begin{bmatrix}
 & & & & 1 & & & & 2 & & & & 3 & & & \\
 & & & & & 1 & & & & 2 & & & & 3 & & \\
 & & & & & & 1 & & & & 2 & & & & 3 & \\
 & & & & & & & 1 & & & & 2 & & & & 3 \\
1 & & & & & & & & 3 & & & & 2 & & & \\
 & 1 & & & & & & & & 3 & & & & 2 & & \\
 & & 1 & & & & & & & & 3 & & & & 2 & \\
 & & & 1 & & & & & & & & 3 & & & & 2 \\
2 & & & & 3 & & & & & & & & 1 & & & \\
 & 2 & & & & 3 & & & & & & & & 1 & & \\
 & & 2 & & & & 3 & & & & & & & & 1 & \\
 & & & 2 & & & & 3 & & & & & & & & 1 \\
3 & & & & 2 & & & & 1 & & & & & & & \\
 & 3 & & & & 2 & & & & 1 & & & & & & \\
 & & 3 & & & & 2 & & & & 1 & & & & & \\
 & & & 3 & & & & 2 & & & & 1 & & & &
\end{bmatrix} \begin{bmatrix} 0 \\ 1 \\ 2 \\ 3 \\ 1 \\ 1 \\ 2 \\ 3 \\ 2 \\ 2 \\ 2 \\ 3 \\ 3 \\ 3 \\ 3 \\ 3 \end{bmatrix} = \begin{bmatrix} 0 \\ 0 \\ 3 \\ 0 \\ 0 \\ 1 \\ 2 \\ 0 \\ 0 \\ 2 \\ 1 \\ 0 \\ 0 \\ 3 \\ 0 \\ 0 \end{bmatrix}$$

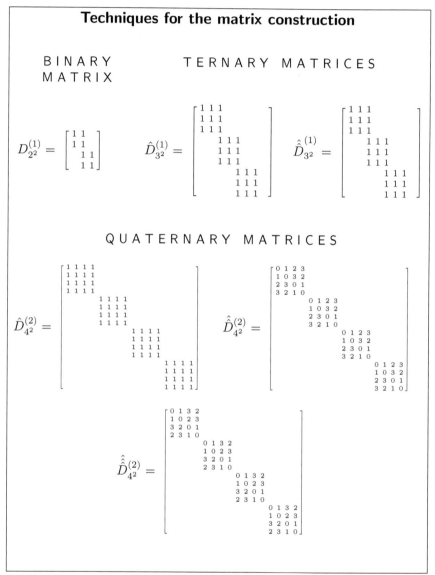

FIGURE 11.17
Logic difference matrices with respect to variable x_2 for Boolean, ternary, and quaternary functions of two variables (Example 11.29).

11.8.2 Logic Taylor expansion of a multivalued function

The logic analog of the Taylor series for an m-valued function f of n variables at the point $c \in 0, 1, \ldots, m^n - 1$, is defined as

$$f = \sum_{i=0}^{m^n - 1} r_i^{(c)} \underbrace{(x_1 \oplus c_1)^{i_1} \ldots (x_n \oplus c_n)^{i_n}}_{i\text{-}th\ term} \quad \text{mod } (m). \quad (11.20)$$

In this expression

▶ m is a prime number.

▶ $c_1 c_2 \ldots c_n$ (polarity) and $i_1 i_2 \ldots i_n$ is the m-valued representation of c and i correspondingly.

▶ $r_i^{(c)}$ is the i-th coefficient, the value of the multiple (n-ordered) logic difference at the point $d = m - c$

$$r_i^{(c)} = \left. \frac{\partial^n f(d)}{\overset{m-i_1}{\partial} \hat{x}_1 \overset{m-i_2}{\partial} \hat{x}_2 \ldots \overset{m-i_n}{\partial} \hat{x}_n} \right|_{d=m-c} \quad (11.21)$$

▶ $\overset{m-i_j}{\partial} \hat{x}_i$ indicates with respect to which variables the multiple logic difference is calculated, and is defined by

$$\overset{m-i_j}{\partial} \hat{x}_i = \begin{cases} 1, & m = i_j, \\ \overset{m-i_j}{\partial} \hat{x}_j, & m \neq i_j. \end{cases} \quad (11.22)$$

11.8.3 Computing polynomial expressions

It follows from Equation 11.20 that (a) logic Taylor expansion produces m^n polynomial expressions that correspond to m^n polarities, (b) a variable x_j is 0-polarized if it enters into the expansion uncomplemented, and c_j-polarized otherwise, and (c) the coefficients in the logic Taylor series are logic differences.

While the i-th coefficient r_i is described by a logical expression, it can be calculated in different ways, for example, by matrix transformations, cube-based technique, decision diagram technique, and probabilistic methods. It is possible to calculate separate coefficients or their arbitrary sets using logic differences.

> **Example 11.31** *By Equation 11.20, the polynomial expression of an arbitrary ternary ($m = 3$) function of two variables ($n = 2$) and the 7-th polarity $c = 7$, $c_1, c_2 = 2, 1$, is defined as a **logic Taylor expansion** of this function (Figure 11.18).*

Techniques for logic differences computing

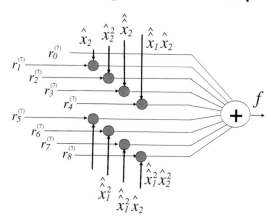

Step 1. *Apply Equation 11.20 for $m = 2$, $n = 2$:*

$$f = \sum_{i=0}^{3^2-1} r_1^{(7)} (x_1 \oplus 2)^{i_1} (x_2 \oplus 1)^{i_2} = \sum_{i=0}^{8} r_i^{(7)} \hat{\hat{x}}_1^{i_1} \hat{x}_1^{i_2}$$

Step 2. *Polynomial expression:*

$$f = r_0^{(7)} + r_1^{(7)} \hat{x}_2 + r_2^{(7)} \hat{x}_2^2 + r3^{(7)} \hat{\hat{x}}_2 + r_4^{(7)} \hat{x}_1 \hat{x}_2 + r_5^{(7)} \hat{x}_1 \hat{x}_2^2 + r_6^{(7)} \hat{x}_1{}^2$$
$$+ r_7^{(7)} \hat{x}_1{}^2 \hat{x}_2 + r_8^{(7)} \hat{x}_1{}^2 \hat{x}_2^2$$

Step 3. *Logic derivatives*

$$r_i^{(c)} = \frac{\partial^2 f(7)}{\partial \hat{x}_1^{3-i_1} \partial \hat{x}_2^{3-i_2}}$$

$$\frac{\partial^{3-i_j}}{\partial \hat{x}_i} = \begin{cases} 1, & 3 = i_j \\ \frac{\partial^{3-i_j}}{\partial \hat{x}_j}, & 3 \neq i_j \end{cases}$$

$r_1 = \partial f(7)/\partial \hat{x}_2$	$r_5 = \partial^2 f(7)/\partial \hat{x}_1 \partial \hat{x}_2$
$r_2 = \partial f(7)/\partial \hat{x}_2$	$r_6 = \partial f(7)/\partial \hat{x}_2$
$r_3 = \partial^2 f(7)/\partial \hat{x}_1 \partial \hat{x}_2$	$r_7 = \partial^2 r(7)/\partial \hat{x}_1 \partial \hat{x}_2$
$r_4 = \partial^2 f(7)/\partial \hat{x}_1 \partial \hat{x}_2$	$r_8 = \partial^2 r(7)/\partial \hat{x}_1 \partial \hat{x}_2$

FIGURE 11.18

Constructing the logic difference for a logic Taylor expansion of an arbitrary ternary ($m = 3$) function of two ($n = 2$) variables for polarity $c = 7$ (Example 11.31).

Example 11.32 *(Continuation of Example 11.31) Consider the function $f = MAX(x_1, x_2)$. The values of the first three coefficients at the point $c = 7$ are given in Figure 11.19. The other differences can be calculated in a similar way. Finally, the vector of coefficient is* $\mathbf{R} = [200012201]^T$, *which yields*

$$f = 2 + \hat{x}_1 \hat{x}_2 + 2\hat{x}_1 \hat{x}_2^2 + 2\hat{x}_2^2 + \hat{x}_2^2 \hat{x}_2^2.$$

Techniques for computing coefficients of the polynomial given polarity

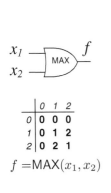

$f = \text{MAX}(x_1, x_2)$

Polynomial coefficients (logic differences)

$$r_0 = f(7) = f(2,1) = 2$$

$$r_1 = \frac{\partial f(7)}{\partial \hat{x}_2}$$

$$= 2f(x_1, \hat{x}_2) + f(x_1, \hat{\hat{x}}_2)$$

$$= 2f(2, \hat{1}) + f(2, \hat{\hat{1}})$$

$$= 2f(2,2) + f(2,0) = 2 \cdot 2 + 2 = 0 \ (\text{mod } 3)$$

$$r_2 = \frac{\partial f(7)}{\partial \hat{x}_2}$$

$$= 2f(x_1, x_2) + 2f(x_1, \hat{x}_2) + 2f(x_1, \hat{\hat{x}}_2)$$

$$= 2f(2,1) + 2f(2, \hat{1}) + 2f(2, \hat{\hat{1}})$$

$$= 2f(2,1) + 2f(2,2) + 2f(2,0)$$

$$= 2 \cdot 2 + 2 \cdot 2 + 2 \cdot 2 = 0 \ (mod \ 3)$$

FIGURE 11.19

Taylor expansion of the ternary ($m = 3$) function MAX of two ($n = 2$) variables for polarity $c = 7$ (Example 11.32).

11.8.4   Computing polynomial expressions in matrix form

The logic Taylor expansion consists of n logic differences with respect to each variable and $m^n - n - 1$ multiple logic differences.

11.8.5 $\mathcal{N}$-hypercube representation

Let us utilize Davio tree and hypercube-like structure, which implements positive Davio expansion in the nodes (Table 11.13), to compute Boolean differences. The positive Davio expansion is given in the form

$$f = f_{x=0}$$
$$+ x \cdot (f_{x=1} + 3f_{x=2} + 2f_{|x=3})$$
$$+ x^2 \cdot (f_{x=1} + 2f_{x=2} + 3f_{x=3})$$
$$+ x^3 \cdot (f_{x=0} + f_{x=1} + f_{x=2} + f_{x=3}).$$

Figure 11.20 illustrates the computing of logic differences by different data structures: decision tree and hypercube-like structure. It follows from this form that:

▶ Branches of the Davio tree carry information about logic differences.

▶ Terminal nodes are the values of logic differences for corresponding variable assignments.

▶ Computing of polynomial coefficients can be implemented on the Davio tree as a data structure.

▶ The Davio tree includes values of all single and multiple logic differences given a variable assignment $x_1 x_2 ... x_n = 00 ... 0$. This assignment corresponds to calculation of polynomial expression of polarity 0, so in the Davio tree, positive Davio expansion is implemented at each node.

▶ Representation of a logic function in terms of change is a unique representation; it means that the corresponding decision diagram is canonical.

▶ The values of terminal nodes correspond to coefficients of logic Taylor expansion.

The Davio tree can be embedded into a hypercube-like structure, and the above mentioned properties are valid for that data structure as well.

11.9   Further study

A variety of implementations have been proposed during the last decades. Multivalued logic has been successfully used for many practical problems, in particular, multivalued flash memory, multi-input binary-output PLAs. Multivalued logic provides new possibilities for design and implementation of adders, multiplies, and dividers. In multivalued logic, m-level signals are used to carry information instead of two levels, 0 and 1, in conventional computers.

TABLE 11.13

Analogues of Shannon and Davio expansions in $GF(4)$

Type	Rule of expansion

f

S

1 2 3 4

$$f = \overbrace{J_0(x) \cdot f_{x=0}}^{Leaf\ 1} + \overbrace{J_1(x) \cdot f_{x=1}}^{Leaf\ 2}$$
$$+ \underbrace{J_2(x) \cdot f_{x=2}}_{Leaf\ 3} + \underbrace{J_3(x) \cdot f_{x=3}}_{Leaf\ 4}$$

f

pD

1 2 3 4

$$f = f_{x=0} + x \cdot (f_{x=1} + 3f_{x=2} + 2f_{x=3})$$
$$+ x^2 \cdot (f_{x=1} + 2f_{x=2} + 3f_{x=3})$$
$$+ x^3 \cdot (f_{x=0} + f_{x=1} + f_{x=2} + f_{x=3})$$

f

nD′

1 2 3 4

$$f = f_{x=1} + {}^{1-}x \cdot (f_{x=0} + 2f_{x=2} + 3f_{x=3})$$
$$+ {}^{1-}x^2 \cdot (f_{x=0} + 3f_{x=2} + 2f_{x=3})$$
$$+ {}^{1-}x^3 \cdot (f_{x=0} + f_{x=1} + f_{x=2} + f_{x=3})$$

f

nD″

1 2 3 4

$$f = f_{x=2} + {}^{2-}x \cdot (3f_{x=0} + 2f_{x=1} + f_{x=3})$$
$$+ {}^{2-}x^2 \cdot (2f_{x=0} + 3f_{x=1} + f_{x=3})$$
$$+ {}^{2-}x^3 \cdot (f_{x=0} + f_{x=1} + f_{x=2} + f_{x=3})$$

f

nD‴

1 2 3 4

$$f = f_{x=3} + {}^{3-}x \cdot (2f_{x=0} + 3f_{x=1} + f_{x=2})$$
$$+ {}^{3-}x^2 \cdot (3f_{x=0} + 2f_{x=1} + f_{x=2})$$
$$+ {}^{3-}x^3 \cdot (f_{x=0} + f_{x=1} + f_{x=2} + f_{x=3})$$

Design example: a 3D computing of logic differences

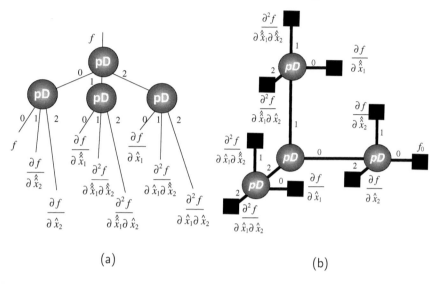

(a)   (b)

FIGURE 11.20
Computing logic differences by Davio tree (a) and hypercube-like structure for ternary logic function of two variables (b).

Plenty of useful information on the theory and application of multivalued logic can be found in the *Proceedings of the Annual IEEE International Symposium on Multiple-Valued Logic* that has been held since 1970. In addition, a good source of information on the above mentioned topics is the *International Journal on Multiple-Valued Logic and Soft Computing*. The fundamentals of multivalued logic can be found in the books by Davio [5], Epstein [8], Hurst [17], Muzio [28], and Rine [30] on the applied problems of multivalued logic. An overview of multivalued logic and its application can be found in papers by Smith [34], Hurst [18], Butler [3], Hanyu et al. [14, 15], and Brayton [2].

Further reading

[1] Aoki T, Kameyama M, and Higuchi T. Design of interconnection-free biomolecular computing systems. In *Proc. 21st IEEE Int. Symp. on Multiple-Valued Logic*, pp. 173–180, 1991.

[2] Brayton RK and Kharti SR. Multi-valued logic synthesis. In *Proc. 12th Int. Conf. on VLSI Design*, pp. 196–206, 1999.

[3] Butler JT. Multiple-valued logic. *IEEE Potentials*, 14(2):11–14, 1995.

[4] Butler JT. On the number of locations required in the content-addressable memory implementation of multiple-valued functions. In *Proc. 13th Int. Symp. on Multiple-Valued Logic*, pp. 94–102, 1983.

[5] Davio M, Deschamps J-P, and Thayse A. *Discrete and Switching Functions.*

McGraw–Hill, Maidenhead, 1978.

[6] Deng X, Hanyu T, and Kameyama M. Quantum-device-oriented multiple-valued logic system based on a super pass gate. *IEICE Trans. Information and Systems,* E78-D(8):951–958, 1995.

[7] Dziurzanski P, Malyugin VD, Shmerko VP, and Yanushkevich SN. Linear models of circuits based on the multivalued components. *Automation and Remote Control,* 63(6):960–980, 2002.

[8] Epstein G. *Multi-Valued Logic Design.* Institute of Physics Publishing, London, UK, 1993.

[9] Green DH and Taylor IS. Multiple-valued switching circuits by means of generalized Reed–Muller expansion. *Digital Processes,* 3:63–81, 1976.

[10] Green DH. Ternary Reed–Muller switching functions with fixed and mixed polarity. *Int. J. of Electronics,* 67:761–775, 1989.

[11] Green DH. Families of Reed–Muller canonical forms. *Int. J. of Electronics,* 70(2):259–280, 1991.

[12] Guima TA and Katbab A. Multivalued logic integral calculus. *Int. J. of Electronics,* 65:1051–1066, 1988.

[13] Guima TA and Tapia MA. Differential calculus for fault detection in multivalued logic networks. In *Proc. 17th IEEE Int. Symp. on Multiple-Valued Logic,* pp. 99–108, 1987.

[14] Hanyu T, Kameyma M, and Higuchi T. Prospects of multiple-valued VLSI processors. *IEIE Trans. Electronics,* E76-C(3):383–392, 1993.

[15] Hanyu T. Challenge of a multiple-valued technology in recent deep-submicron VLSI. In *Proc. 31st IEEE Int. Symp. on Multiple-Valued Logic,* pp. 241–247, 2001.

[16] Higuchi T and Kameyama M. Synthesis of multiple-valued logic networks based on tree-type universal logic modules. In *Proc. 5th Int. Symp. on Multiple-Valued Logic,* pp. 121–130, 1975.

[17] Hurst S. *The Logical Processing of Digital Signals.* Arnold, London, 1978.

[18] Hurst SL. Multiple-valued logic – its status and its future. *IEEE Trans. Comput.,* 33(12):1160–1179, 1984.

[19] Hurst S, Miller D, and Muzio J. *Spectral Technique in Digital Logic.* Academic Press, New York, 1985.

[20] Kam T, Villa T, Brayton RK, and Sagiovanni-Vincentelli AL. Multi-valued decision diagrams: theory and applications. *Int. J. on Multiple-Valued Logic,* 4(1-2):9–62, 1998.

[21] Karpovsky MG, Stanković RS, and Astola JT. *Spectral Logic and Its Applications for the Design of Digital Devices.* John Wiley & Sons, Hoboken, NJ, 2008.

[22] Kameyama M, Hanyu T, and Higuchi T. Design and implementation of quaternary NMOS integrated circuits for pipelined image processing. *IEEE J. of Solid-State Circuits,* 22(1):20–27, 1987.

[23] Kawahito S, Kameyama M, Higuchi T, and Yamada H. A 32×32-bit multiplier using multiple-valued MOS current-mode circuits. *IEEE J. of Solid-State Circuits,* 23(1):124–132, 1988.

[24] Miller DM. Spectral signature testing for multiple-valued combinational networks. In *Proc. 12th IEEE Int. Symp. on Multiple-Valued Logic,* pp. 152–158, 1982.

[25] Miller DM. Multiple-valued logic design tools. In *Proc. of the 23rd IEEE Int. Symp. on Multiple-Valued Logic,* pp. 2–11, 1993.

[26] Miller DM and Muranaka N. Multiple-valued decision diagrams with symmetric variable nodes. In *Proc. 26th Int. Symp. on Multiple-Valued Logic,* pp. 242–247, 1996.

[27] Moraga C. Systolic systems and multiple-valued logic. In *Proc. 14th IEEE Int. Symp. on Multiple-Valued Logic,* pp. 98–108, 1984.

[28] Muzio JC and Wesselkamper TS. *Multiple-Valued Switching Theory.* Adam Higler Ltd., Bristol and Boston, 1986.

[29] Perkowski M, Sarabi A, and Beyl F. Fundamental theorems and families of forms for binary and multiple-valued linearly independent logic. In *Proc. IFIP WG 10.5 Int. Workshop on Applications of the Reed–Muller Expansions in Circuit Design,* pp. 288–299, Japan, 1995.

[30] Rine DC, Ed., *Computer Science and Multiple-Valued Logic. Theory and Applications.* 2nd ed., Amsterdam, North-Holland, 1984.

[31] Rudell R and Sangiovanni-Vincentelli A. ESPRESSO-MV: algorithm for multiple-valued logic minimization. In *Proc. IEEE Custom Integrated Circuits Conf.,* pp. 230–234, 1985.

[32] Sasao T. Multiple-valued decomposition of generalized Boolean functions and the complexity of programmable logic arrays. *IEEE Trans. Comput.,* 30(9):635–643, 1981.

[33] Sasao T. Ternary decision diagram – survey. In *Proc. 27th IEEE Int. Symp. on Multiple-Valued Logic,* pp. 241–250, 1997.

[34] Smith KC. The prospects for multivalued logic: a technology and applications view. *IEEE Trans. Comput.,* 30(9):619–634, 1981.

[35] Stanković R, Stankovic M, Moraga C, and Sasao T. Calculation of Reed–Muller–Fourier coefficients of multiple-valued functions through multiple-place decision diagrams. In *Proc. 24th IEEE Int. Symp. on Multiple-Valued Logic,* pp. 82–87, 1994.

[36] Stanković RS and Astola JT. *Spectral Interpretation of Decision Diagrams,* Springer, Heidelberg, 2003.

[37] Shmerko VP, Yanushkevich SN, and Levashenko VG. Test pattern generation for combinational MVL networks based on generalized *D*-algorithm. In *Proc. 22nd IEEE Int. Symp. on Multiple-Valued Logic,* pp. 139–144, 1997.

[38] Shmerko VP, Yanushkevich SN, and Levashenko VG. Techniques of computing logic derivatives for MVL functions. In *Proc. IEEE 26th Int. Symp. on Multiple-Valued Logic,* pp. 267–272, 1996.

[39] Shmerko VP, Yanushkevich SN, Levashenko VG, and Bondar I. Test pattern generation for combinational MVL networks based on generalized D-algorithm. In *Proc. IEEE 27th Int. Symp. on Multiple-Valued Logic*, pp. 139–144, 1997.

[40] Spillman RJ and Su SYH. Detection of single, stuck-type failures in multivalued combinational networks. *IEEE Trans. Comput.*, 26(12):1242–1251, 1977.

[41] Tabakow IG. Using *D*-algebra to generate tests for *m*-logic combinational circuits. *Int. J. of Electronics*, 75(5):897–906, 1993.

[42] Tapia MA, Guima TA, and Katbab A. Calculus for a multi-valued logic algebraic system. *Applied Mathematics and Computation*, pp. 225–285, 1991.

[43] Yanushkevich SN. Systolic algorithms to synthesize arithmetical polynomial forms for *k*-valued logic functions. *Automation and Remote Control*, 55(12):812–1823, 1994.

[44] Yanushkevich SN, Shmerko VP, and Dziurzanski P. Linearity of word-level models: new understanding. In *Proc. IEEE/ACM 11th Int. Workshop on Logic and Synthesis*, pp. 67–72, New Orleans, LA, 2002.

[45] Yanushkevich SN, Dziurzanski P, and Shmerko VP. Word-level models for efficient computation of multiple-valued functions. Part 1: LAR based models. In *Proc. IEEE 32nd Int. Symp. on Multiple-Valued Logic*, pp. 202–208, 2002.

[46] Yanushkevich SN, Shmerko VP, and Lyshevski SE. *Logic Design of NanoICs*. CRC Press, Boca Raton, FL, 2005.

[47] Zilic Z and Vranesic Z. Multiple valued Reed–Muller transform for incomplete functions. *IEEE Trans. Comput.*, 44(8):1012–1020, 1995.

12

Computational Networks

12.1 Introduction

This chapter introduces the techniques for designing computing networks over the library of the elementary computing elements. These techniques are based on various *design styles* and *paradigms*. For example, different design paradigms are utilized in the multilevel logic networks and decision diagrams. This is because of different data structures: a logic network utilizes logic gates from the library of available gates, and the decision diagram corresponds to the multiplexer tree. Moreover, decision diagram can be considered as the particular case of the homogeneous network because the nodes implement the same function.

Complex digital system design is based on a modular principle, that is, systems are built using *standard* modules. These modules (basic blocks, components) correspond to subfunctions (subtasks) of the system's functionality. The set of subfunctions has been identified as useful for a large variety of logic networks, and its elements are available as *library* components. The design of a system using standard modules consists of decomposition of the overall function of the system into subfunctions of standard components and the interconnection of these components.

The concept of the *assembly* of complex computing logic networks from basic elements is well developed in contemporary logic design and widely implemented based on advanced technology. This concept is also intensively developed for predictable technologies. In this chapter, the following standard modules are introduced: multiplexers and demultiplexers, encoders and decoders, code converters, comparators, and shifters.

An n-input, m-output *combinational* network is a network of logic elements that implements a set of $m \geq 1$ Boolean functions of n variables $x_1, x_2, \ldots, x_n$, $f = \{f_1, f_2, \ldots, f_n\}$. The simplest example of an n-input, 1-output combinational network is a logic gate. The key property of a combinational logic network is that the output signals of a combinational network at any time are completely determined by the combination of values assigned to the input signals $x_1, x_2, \ldots, x_n$ at that time, and hence the term "combinational." Any changes in these inputs are assumed to change the network output signals.

An idealized logic network is built from elementary logic elements (gates) and standard logic blocks (multiplexers, demultiplexers, etc.), and is idealized in the sense that all signal changes are assumed to take place instantaneously. In physical networks, signals cannot be transmitted instantaneously, so there is a delay between a network input change and the output change it produces. This *delay*, or *response time*, can be measured. Typically, this time is on the order of nanoseconds (10^{-9} s) per logic element. When addressing the functional or logical behavior of a logic network, we assume in this chapter that all lines and logic elements have zero delay. That is, combinational logic networks are equated with delay-free networks in which all signal changes occur instantaneously.

In this chapter, the following computational networks are introduced: binary and decimal adders, comparators, and error detection and error correction networks.

12.2   Data transfer logic

Data transfer logic aims at controlling the transfer of binary data between several points in a logic network. This is a *moving* or *copying* operation, in which no changes are made to the logic values of signals; that is, no data processing takes place. Data transfer logic is useful for proving datapaths between logic networks and for implementing designs for Boolean functions.

12.2.1   Shared data path

In data transfer logic, inputs are associated with *data sources*, and outputs are associated with *data destinations*. There are two typical tasks of data transfer (Figure 12.1): (a) several data sources have a common destination and (b) one source must send data to several destinations. Because a particular destination can take signals from only one source at a time, the source-to-destination data transfer paths must be shared to some degree. Shared data paths are also employed primarily to reduce the number of lines needed for communication purposes. The devices that enable the communication of data are the *multiplexers* and the *demultiplexers*.

The multiplexer allows only one of many sources to be logically connected to a given common destination at any time (Figure 12.1a). Multiplexers are used to implement designs for Boolean functions and to provide data-flow paths between logic networks. An arbitrary Boolean function of n variables can be implemented by using at most $2^n - 1$ 2-to-1 multiplexers. The inverse of a multiplexer is a 1-to-2^n *demultiplexer* (Figure 12.1b), which serves to connect a common source I to one of 2^n destinations:

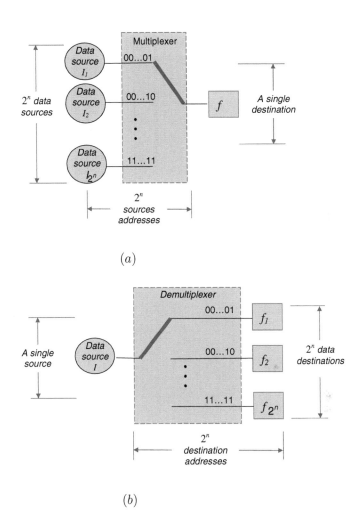

(a)

(b)

FIGURE 12.1

Data transfer connections: a switch model of the multiplexer for transferring data from one source to many simultaneous destinations (a), and a model of the demultiplexer that transfers data from one source to one of many destinations (b).

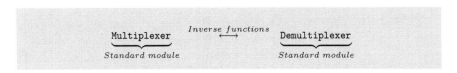

A set of selected bits (addresses) specifies which destination is to be selected at any time. The simplest demultiplexer, which connects the common source to one of two destinations, is called a *single-bit* or *1-to-2* demultiplexer. Also, a *decoder* with an enabled input is called a demultiplexer. The decoder is a logic network that converts a binary code applied to n input lines to one of 2^n different outputs lines.

Logic symbols for the multiplexer (MUX) and demultiplexer (DMUX) are shown in Figure 12.2a,b. In Figure 12.2c, the multiplexer's output is fed into the demultiplexer input, and it creates a router network. Depending on the settings of the control signals, any one input can be routed to any one output. This router network is not a full crossbar switch because at any time only one output is active. This is because a 1-bit bus is used for the transmission of 2^n logic signals.

12.2.2   Multiplexer

A logic network called a 2^n-to-1 *multiplexer* or *data selector* allows one and only one of 2^n sources to be logically connected to a common destination f at a time. The source I_i, $i = 1, 2, ..., ... 2^n$, is transferred to the output if the address input is i. The address signal is a binary integer i, which controls the selection of the desired source I_i. By changing the address from i to j, $i, j \in \{1, 2, ... 2^n\}$, the input source for the multiplexer is changed from I_i to I_j.

Let the number of sources be $k = 2^n, k \in \{1, 2, ..., K\}$. Then, n address bits allow 2^n distinct addresses to be specified. The output of the multiplexer is defined by the following relationship of its data input and the address

$$f = \bigvee_{i=0}^{2^n-1} \underset{\underset{In}{\uparrow}}{I_{i+1}} \cdot \underbrace{S_1^{i_1} S_2^{i_2} \cdots S_n^{i_n}}_{Address} \tag{12.1}$$

where

$$S_j^{i_j} = \begin{cases} \overline{S_j}, \text{ if } i_j = 0 \\ S_j, \text{ if } i_j = 1 \end{cases}$$

The address $S_1^{i_1} S_2^{i_2} \cdots S_n^{i_n}$ in Equation 12.1 is a minterm or product of n select variables S_j, $j = 1, 2, ..., n$, and the corresponding data input I_{i+1}, $i = 0, 1, ..., 2^n - 1$.

Example 12.1   *It follows from Equation 12.1 that the 2^n-to-1* **multiplexer** *has 2^n sources of data and $n = \log_2 2^n$ selected lines. 2-to-1, 4-to-1, and 8-to-1 multiplexers transmit 2, 4, and 8 inputs to one output, respectively (Table 12.1).*

TABLE 12.1
Multiplexers and their formal descriptions (Example 12.1)

Design techniques for multiplexers

Graphical representation   **Formal description**

2ⁿ-to-1 m u l t i p l e x e r

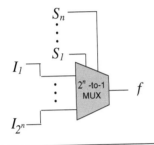

$$f = \bigvee_{i=0}^{2^n-1} I_{i+1} \cdot \overbrace{S_1^{i_1} S_2^{i_2} \cdots S_n^{i_n}}^{\text{n-bit address}}$$

2-to-1 m u l t i p l e x e r

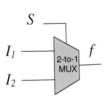

$$f = \bigvee_{i=0}^{2^1-1} I_{i+1} S$$

$$= I_1 \overline{S} \vee I_2 S = \begin{cases} I_1, \text{ if } S = 0; \\ I_2, \text{ if } S = 1. \end{cases}$$

4-to-1 m u l t i p l e x e r

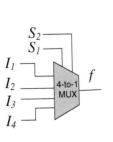

$$f = \bigvee_{i=0}^{2^2-1} I_{i+1} \cdot S_1^{i_1} S_2^{i_2}$$

$$= I_1 \overline{S}_1 \overline{S}_2 \vee I_2 \overline{S}_1 S_2 \vee I_3 S_1 \overline{S}_2 \vee I_4 S_1 S_2$$

$$= \begin{cases} I_1, \text{ if } S_1 S_2 = 00 \\ I_2, \text{ if } S_1 S_2 = 01 \\ I_3, \text{ if } S_1 S_2 = 10 \\ I_4, \text{ if } S_1 S_2 = 11 \end{cases}$$

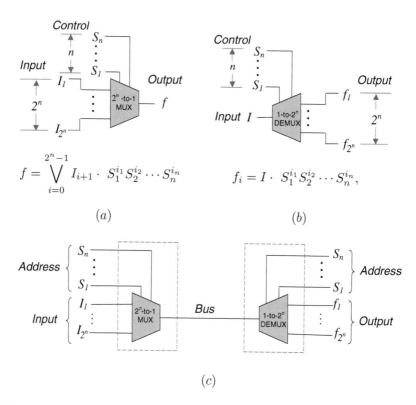

FIGURE 12.2

Data transfer logic and its formal description: 2^n-to-1 multiplexer (a), 1-to-2^n demultiplexer (b), and a router network (c).

12.2.3 Multiplexers and the Shannon expansion theorem

Consider the Shannon expansion $f = \overline{x}_i f_0 \vee x_i f_1$. This expansion means that if $x_i = 0$, then f_0 is selected; and if $x_i = 1$, then f_1 is selected. Thus, the variable x_i in a Shannon expansion can be interpreted as the select input to a 2-to-1 multiplexer (Figure 12.3) that has f_0 and f_1 as the data inputs:

$$f_0 = f(x_1, \ldots, x_{i-1} \;\boxed{0}\; x_{i+1}, \ldots, x_n)$$

$$f_1 = f(x_1, \ldots, x_{i-1} \;\boxed{1}\; x_{i+1}, \ldots, x_n)$$

Consider the Shannon expansion with respect to two variables:

$$f = \overline{x}_1 \overline{x}_2 f_{00} \vee \overline{x}_1 x_2 f_{01} \vee x_1 \overline{x}_2 f_{10} \vee x_1 x_2 f_{11}$$

If $x_1, x_2 = 0,0$, then f_{00} is chosen, and for the remaining selections of x_1, x_2 (01,10 and 11), the values f_{01}, f_{10}, and f_{11} are chosen, respectively. This corresponds to the function of the 4-to-1 multiplexer (Figure 12.3).

12.2.4 Single-bit (2-to-1) multiplexer

For a 2-to-1 multiplexer, only one address bit S is needed to identify the selected source: $S = 0$ selects I_0 to connect to output, and $S = 1$ selects I_1. The single-bit multiplexer is defined by the equation

$$f = I_0 \overline{S} \vee I_1 S \tag{12.2}$$

In this multiplexer, the output is equal I_0 when S is 1, and it is equal to I_1 when S is 0.

Example 12.2 *The single-bit multiplexer is shown in Figure 12.4. In this network, two AND gates determine whether to pass their respective data inputs to the OR gate. The upper AND gate passes signal I_0 when S is 0 (since the other input to the gate is $\overline{S}$). The lower AND gate passes signal I_1 when S is 1.*

12.2.5 Word-level multiplexer

A single word-level multiplexer generates a word **F** equal to one of the two input words, **X** and **Y**, depending on the control input bit S. There are multiplexers that allow one the selection of a word from a number of sources depending on the control signals:

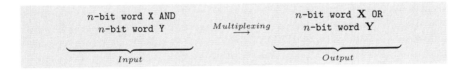

n-bit word X AND		n-bit word **X** OR
n-bit word Y	*Multiplexing* $\longrightarrow$	n-bit word **Y**
Input		*Output*

Example 12.3 *Given the 4-bit words* $\mathbf{X} = \{x_1 x_2 x_3 x_4\}$ *and* $\mathbf{Y} = \{y_1 y_2 y_3 y_4\}$, *design a logic network for the selection of one of the words, **X** or **Y**, using the control signal $S \in \{0, 1\}$ (0 for selecting **X** and 1 selecting **Y**) and 2-to-1 multiplexers. The solution is shown in Figure 12.5a. The four logic signals that perform a common function can be grouped together to form a bus (Figure 12.5b).*

Design techniques for multiplexers

Shannon   expansion
with respect to the variable x_i

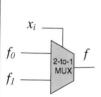

$$f = \overline{x}_i f \overbrace{(x_1, \ldots, x_{i-1}0\ x_{i+1}, \ldots, x_n)}^{n-1\ variables}$$
$$\lor\ x_i f \underbrace{(x_1, \ldots, x_{i-1}1\ x_{i+1}, \ldots, x_n)}_{n-1\ variables}$$
$$= \overline{x}_i f_0 \lor x_i f_1$$

where

$$f_0 = f(x_1, \ldots, x_{i-1}0\ x_{i+1}, \ldots, x_n)$$
$$f_1 = f(x_1, \ldots, x_{i-1}1\ x_{i+1}, \ldots, x_n)$$

Shannon   expansion
with respect to a group of variables

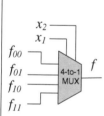

$$f = \bigvee_{j=0}^{m} f_j\ x_1 x_2 \ldots \underbrace{x_{i_1}^{c_{i_1}} x_{i_2}^{c_{i_2}} \ldots x_{i_m}^{c_{i_m}}}_{Control\ variables} \cdots\ x_{n-1}x_n$$

where $j = c_{i_1} c_{i_2} \ldots c_{i_m}$,

$$x_j^{c_j} = \begin{cases} x_j, & c_j = 1 \\ \overline{x}_j, & c_j = 0 \end{cases}$$

Let $m = 2^2 - 1 = 3$, and let the group of variables x_1 and x_2 be chosen for expansion. Then

$$f = f_{00} \lor f_{01} \lor f_{10} \lor f_{11},$$
$$f_{00} = \overline{x}_1 \overline{x}_2 f(0, 0, x_3, \ldots, x_n)$$
$$f_{01} = \overline{x}_1 x_2 f(0, 1, x_3, \ldots, x_n)$$
$$f_{10} = x_1 \overline{x}_2 f(1, 0, x_3, \ldots, x_n)$$
$$f_{11} = x_1 x_2 \underbrace{f(1, 1, x_3, \ldots, x_n)}_{n-2\ inputs}$$

FIGURE 12.3

Multiplexers implement Shannon expansion with respect to a variable or a group of variables.

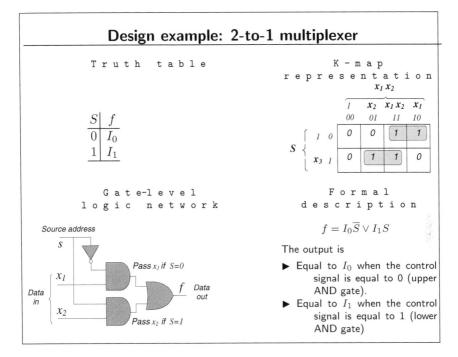

Design example: 2-to-1 multiplexer

Truth table

S	f
0	I_0
1	I_1

K-map representation

S \ $x_1 x_2$	1 (00)	x_2 (01)	$x_1 x_2$ (11)	x_1 (10)
1 (0)	0	0	1	1
x_3 (1)	0	1	1	0

Gate-level logic network

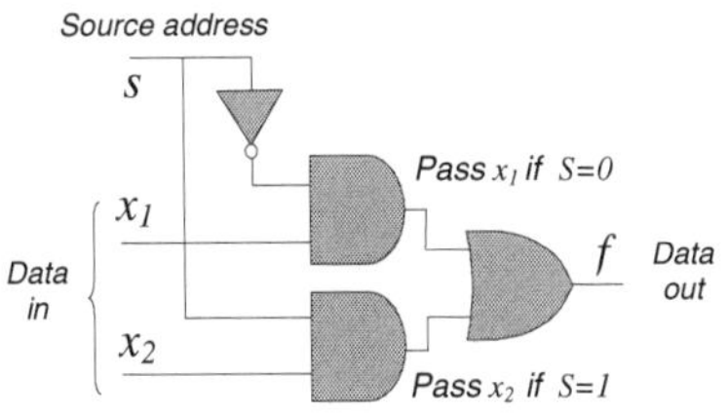

Source address
S
Pass x1 if S=0
x1
Data in
f
Data out
x2
Pass x2 if S=1

Formal description

$$f = I_0\overline{S} \vee I_1 S$$

The output is

- Equal to I_0 when the control signal is equal to 0 (upper AND gate).
- Equal to I_1 when the control signal is equal to 1 (lower AND gate)

FIGURE 12.4
Single-bit multiplexer: description using the truth table, K-map, and gate-level network (Example 12.2).

In Equation 12.1, data inputs and control variables can be functions themselves. That is, they can be connected to other networks.

12.3 Implementation of Boolean functions using multiplexers

In the design of a large multiplexer-based network, the following resources can be used:

- Connections of multiplexers in a multiplexer tree
- Extension of the number of selection (address) lines
- Additional logic gates

There are two approaches to the design of networks for computing Boolean functions using multiplexers:

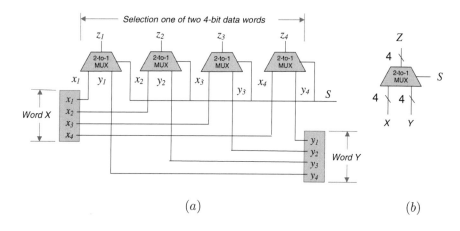

FIGURE 12.5
A logic network of 2-to-1 multiplexers for selecting one of two 4-bit words (a),
and one of the multibit multiplexers (b) (Example 12.3).

(*a*) Decomposition of the truth table

(*b*) Shannon expansion of the implemented Boolean function

If a Boolean function is given by its truth table or a truth vector, then the
first approach is appropriate. However, this approach is limited by the size of
the truth table. The second approach utilizes the flexibility of the Shannon
expansion and is acceptable for algebraic representations of Boolean functions
of many variables.

Algorithm for computing Boolean functions using multiplexers

    Given: A Boolean function of n variables given by its truth
           vector.
    Step 1: Choose k variables for selected lines.
    Step 2: Divide the truth vector into 2^k groups of $2^{n-k} \times 1$
           subvectors.
    Step 3: These subvectors are the truth vectors of the functions
           that are the inputs of the multiplexer.
    Step 4: Repeat steps 1-3 for each function obtained in step 3
           until the new function (cofactors) contains less than $k+1$
           variables.
    Result: A logic network of multiplexers.

It should be noted that if the cofactors are functions of more than one
variable but less than k (where $k = 2, 3, \ldots, K$), then AND, OR, and others

additional logic gates can be used to form the inputs of the first level of multiplexers.

12.3.1 Multiplexer tree

The 2^n-to-n multiplexer can implement an arbitrary Boolean function of n variables. This is because any element of the input truth vector of a Boolean function can be transferred to the output using the mechanism of addressing. The disadvantage of this implementation of a Boolean function is that the number of inputs is equal to 2^n. To resolve this problem, a 2^n-to-1 multiplexer can be represented as a multilevel network of 2^m-to-1 multiplexers, $m < n$. This network is called a *multiplexer tree*.

Example 12.4 *In Figures 12.6a,b, two 2-to-1 multiplexers are combined to provide the desired functionalities. Figure 12.6c shows a multiplexer tree.*

Example 12.5 *In Figure 12.7, a two-level **multiplexer tree** using 4-to-1 multiplexers is shown. This network realizes the function of a 16-to-1 multiplexer.*

The number of data inputs to the multiplexer can be reduced if the inputs are from the set of constants and literals, $\{0, 1, x_i^{i_j}\}$. In this way, a 2^n-input multiplexer can implement a Boolean function of $n + 1$ variables.

Example 12.6 *Given a Boolean function of three variables $f = \bigvee m(1, 2, 4, 6, 7)$, let us implement it using an 8-to-1 multiplexer (Figure 12.8a). The standard SOP expression for f is factored as follows:*

$$f = \bigvee m(1, 2, 4, 6, 7)$$
$$= \overline{x}_3(x_1\overline{x}_2) \vee \overline{x}_3(\overline{x}_1 x_2) \vee x_3(\overline{x}_1\overline{x}_2) \vee x_3(\overline{x}_1 x_2) \vee x_3(x_1 x_2)$$
$$= \overline{x}_3(x_1\overline{x}_2) \vee \underbrace{(\overline{x}_3 \vee x_3)}_{Equal\ to\ 1}(\overline{x}_1 x_2) \vee x_3(\overline{x}_1 x_2) \vee x_3(\overline{x}_1\overline{x}_2) \vee x_3(x_1 x_2)$$
$$= \overline{x}_3(x_1\overline{x}_2) \vee 1 \cdot (\overline{x}_1 x_2) \vee x_3(\overline{x}_1 x_2) \vee x_3(\overline{x}_1\overline{x}_2) \vee x_3(x_1 x_2)$$

A K-map interpretation of this transform and its implementation using a 4-to-1 multiplexer are given in Figures 12.8b,c.

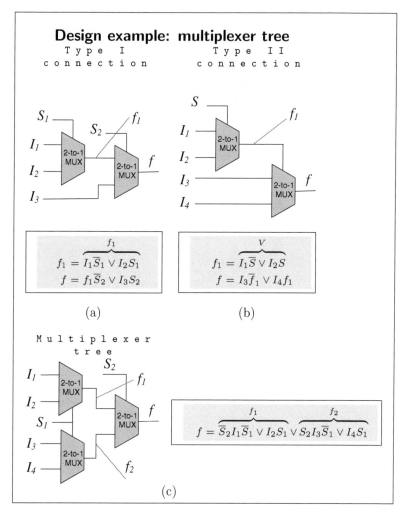

FIGURE 12.6
Connection of 2-to-1 multiplexers (Example 12.4).

12.3.2 Combination of design approaches using multiplexers

Two design approaches for computing Boolean functions using multiplexers are given in Tables 12.2 and 12.3. These approaches are introduced by several design examples.

Design example 1

Implementation of the EXOR function of two variables can be done via pairwising its truth table rows. This yields two cofactors, which are mapped

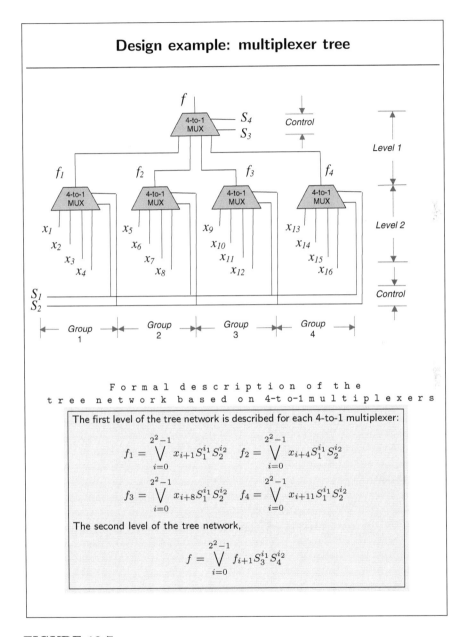

FIGURE 12.7
Multiplexer tree for implementing a 16-to-1 multiplexer using 4-to-1 multiplexers (Example 12.5).

Design example: reduction of the number of inputs

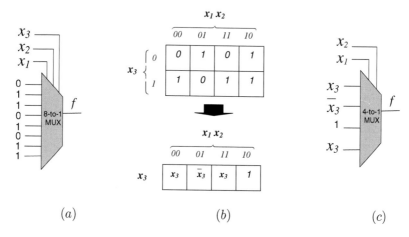

(a)   (b)   (c)

FIGURE 12.8
Interpretation of the Shannon expansion theorem for the reduction of the
number of inputs in the multiplexer for implementing a Boolean function
(Example 12.6).

into the 2-to-1 multiplexer directly. If a Boolean function is given in algebraic
form, a solution is provided by the application of the Shannon expansion with
respect to the variable x_1, which plays the role of the selected line of the
multiplexer (Table 12.2).

> **Example 12.7**   *To implement a two-input EXOR function using a 2-to-1 multiplexer, let us use the Shannon expansion $f = x_1 \oplus x_2 = \overline{x}_1 f_0 \vee x_1 f_1$, where $f_0 = x_2$ and $f_1 = \overline{x}_1$. The network solution is given in Table 12.2.*

Design example 2

Using the first approach, a Boolean function of three variables $f = x_1 x_2 \vee x_2 x_3 \vee x_1 x_3$ is implemented using the 4-to-1 multiplexer via decomposition
of the truth table into four rows. The same result can be obtained using the
Shannon expansion with respect to the variables x_1 and x_2, which become the
selected lines in the 4-to-1 multiplexer. For this, the Boolean function must
be represented in algebraic form, while the K-map is useful for decreasing the
number of terms (Table 12.2).

TABLE 12.2
Logic network design techniques for computing Boolean functions using multiplexers

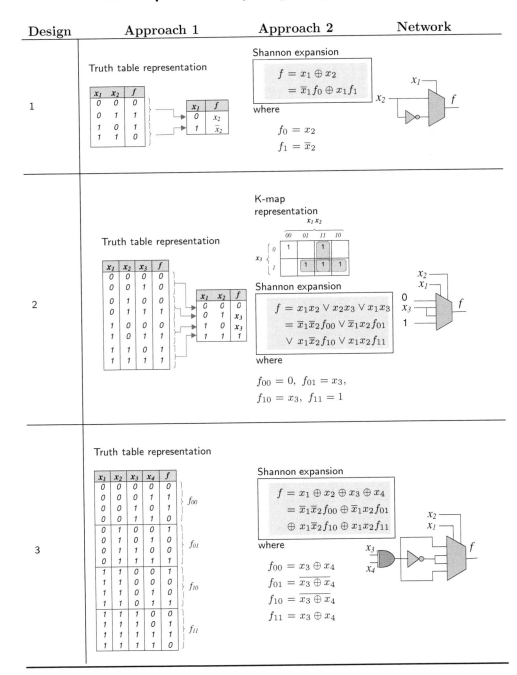

Techniques for computing using multiplexers

TABLE 12.3

Logic network design techniques for computing Boolean functions using multiplexers (continuation of Table 12.2)

Techniques for computing using multiplexers (continuation)

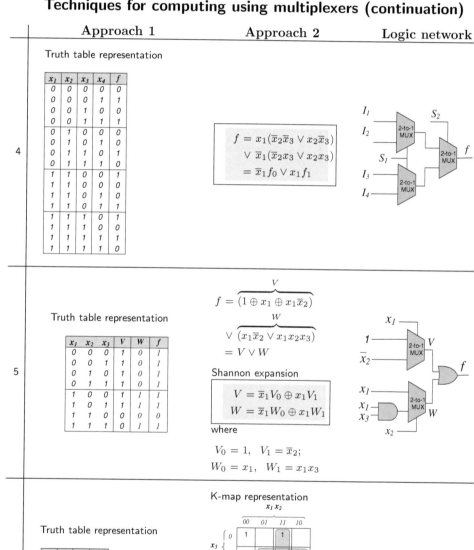

	Approach 1	Approach 2	Logic network

4

Approach 1 — Truth table representation

x_1	x_2	x_3	x_4	f
0	0	0	0	0
0	0	0	1	1
0	0	1	0	0
0	0	1	1	1
0	1	0	0	0
0	1	0	1	0
0	1	1	0	1
0	1	1	1	0
1	1	0	0	1
1	1	0	0	0
1	1	0	1	0
1	1	0	1	1
1	1	1	0	1
1	1	1	0	0
1	1	1	1	1
1	1	1	1	0

Approach 2:

$$f = x_1(\overline{x}_2\overline{x}_3 \vee x_2\overline{x}_3)$$
$$\vee\ \overline{x}_1(\overline{x}_2 x_3 \vee x_2 x_3)$$
$$= \overline{x}_1 f_0 \vee x_1 f_1$$

5

Approach 1 — Truth table representation

x_1	x_2	x_3	V	W	f
0	0	0	1	0	1
0	0	1	1	0	1
0	1	0	1	0	1
0	1	1	1	0	1
1	0	0	1	1	1
1	0	1	1	1	1
1	1	0	0	0	0
1	1	1	0	1	1

Approach 2:

$$f = \overbrace{(1 \oplus x_1 \oplus x_1\overline{x}_2)}^{V}$$
$$\vee\ \overbrace{(x_1\overline{x}_2 \vee x_1 x_2 x_3)}^{W}$$
$$= V \vee W$$

Shannon expansion

$$V = \overline{x}_1 V_0 \oplus x_1 V_1$$
$$W = \overline{x}_1 W_0 \oplus x_1 W_1$$

where

$$V_0 = 1,\quad V_1 = \overline{x}_2;$$
$$W_0 = x_1,\quad W_1 = x_1 x_3$$

6

Approach 1 — Truth table representation

x_1	x_2	x_3	f
0	0	0	0
0	0	1	0
0	1	0	0
0	1	1	1
1	0	0	0
1	0	1	1
1	1	0	1
1	1	1	1

x_1	f
0	$x_2 x_3$
1	$x_2 \vee x_3$

Approach 2 — K-map representation

$$x_1 x_2$$

	00	01	11	10
x_3 0	1		1	
1		1	1	1

Shannon expansion

$$f = x_1 x_2 \vee x_2 x_3 \vee x_1 x_3$$
$$= \overline{x}_1 f_0 \oplus x_1 f_1$$

where

$$f_0 = x_2 x_3$$
$$f_1 = x_2 \vee x_3$$

Design example 3

The Boolean function of five variables $f = x_1 \oplus x_2 \oplus x_3 \oplus x_4$ is given in algebraic form, namely, in polynomial form. Shannon expansion can be applied to this form, for example, with respect to variables x_1 and x_2. Note that in addition to the 4-to-1 multiplexer, AND and NOT gates are required (Table 12.2).

Design example 4

Given the Boolean function $f = x_1 \overline{x}_2 \overline{x}_3 \vee x_1 x_2 \overline{x}_3 \vee \overline{x}_1 \overline{x}_2 x_3 \vee \overline{x}_1 x_2 x_3$, let us factor the function $f = x_1 (\overline{x}_2 \overline{x}_3 \vee x_2 \overline{x}_3) \vee \overline{x}_1 (\overline{x}_2 x_3 \vee x_2 x_3) = \overline{x}_1 f_0 \vee x_1 f_1$. A multiplexer tree of 2-to-1 multiplexers for implementing it is shown in Table 12.3. Other solutions are also possible.

Design example 5

Let the Boolean function f contain the polynomial expression and an SOP expression $V = 1 \oplus x_1 \oplus x_1 \overline{x}_2$ and $W = x_1 \overline{x}_2 \vee x_1 x_2 x_3$, respectively, such that $f = V \vee W$. Shannon expansion is applied to the subfunctions V and W, and then traditional design using the 2-to-1 multiplexers can be used. Finally, the function f is obtained by using the OR gate (Table 12.3).

Design example 6

Given the Boolean function $f = 1 \oplus x_1 x_2 \oplus x_1 x_3 x_4 \oplus x_1 x_2 x_3 x_4$, implement it using 2-to-1 multiplexers and arbitrary additional gates. For this, the truth table is decomposed into two rows (Table 12.3). The same result is obtained using Shannon expansion with respect to variable x_1. The K-map is used for decreasing the number of terms (minimization is not required in this approach).

Design example 7

An arbitrary logic gate can be modeled using a multiplexer tree. Figure 12.9 illustrates the design techniques for the AND, OR, and EXOR functions.

12.4   Demultiplexers

A 2^n *demultiplexer* is a combinational logic network with n selected (address) inputs, one data input, and 2^n data outputs:

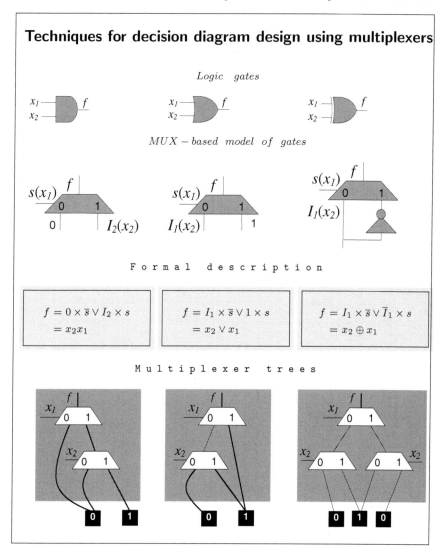

FIGURE 12.9
Multiplexer trees of the AND, OR, and EXOR functions.

A logic signal	$\xrightarrow{Distribution}$	2^n copies of logic signal
Input		*Output*

A demultiplexer is also called a *distributor* or selector. It performs the function inverse to that of the multiplexer. That is, a demultiplexer routes the input data to one of the outputs selected using the select lines; all other outputs are zero.

Example 12.8 *Examples of **demultiplexers** denoted DMUX are given in Table 12.4. The two-output demultiplexer requires one select input. The four-output demultiplexer has two select inputs, and 8-output demultiplexer has three select inputs.*

TABLE 12.4
Demultiplexers and their formal descriptions (Example 12.8)

Design example: demultiplexers

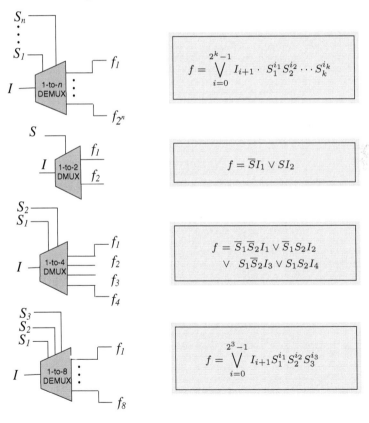

$$f = \bigvee_{i=0}^{2^k-1} I_{i+1} \cdot S_1^{i_1} S_2^{i_2} \cdots S_k^{i_k}$$

$$f = \overline{S}I_1 \vee SI_2$$

$$f = \overline{S}_1\overline{S}_2I_1 \vee \overline{S}_1S_2I_2$$
$$\vee \ S_1\overline{S}_2I_3 \vee S_1S_2I_4$$

$$f = \bigvee_{i=0}^{2^3-1} I_{i+1}S_1^{i_1}S_2^{i_2}S_3^{i_3}$$

The formal model of a demultiplexer is defined by the following equation:

$$\text{Demultiplexer output } f_i = \underset{\underset{In}{\uparrow}}{I} \cdot \overbrace{S_1^{i_1} S_2^{i_2} \cdots S_n^{i_n}}^{Address\ i}, \qquad (12.3)$$

where

$$S_j^{i_j} = \begin{cases} \overline{S}_j, & \text{if } i_j = 0 \\ S_j, & \text{if } i_j = 1 \end{cases}$$

Address $S_1^{i_1} S_2^{i_2} \cdots S_n^{i_n}$ in Equation 12.3 is a minterm of k selected variables S_j, $j = 1, 2, \ldots, n$.

Example 12.9 *It follows from Equation 12.3 that 1-to-2 and 1-to-4 demultiplexers transmit their inputs to two or four outputs, respectively:*

$$n = 1 : \ f = I \cdot S_1^{i_1} = I \cdot \overline{S}_1 \vee I S_2$$
$$n = 2 : \ f = I \cdot S_1^{i_1} S_2^{i_2}$$
$$= I \cdot \overline{S}_1 \overline{S}_2 \vee I \cdot \overline{S}_1 S_2 \vee I \cdot S_1 \overline{S}_2 \vee I \cdot S_1 S_2$$

Implementation of decision diagrams on multiplexers

A binary decision diagram can be directly mapped into a multiplexer network. The multiplexer network operates in the bottom-up fashion. The bottom-up design of the network corresponds to the so-called "timed Shannon circuits with multiplexers." A decision diagram using pD and nD nodes can be mapped into a multiplexer network with additional EXOR gates (Table 12.5).

Example 12.10 *Examples of multiplexer trees for implementing a binary decision diagram for Boolean functions of three variables ($n = 3$) are given in Figure 12.10.*

12.5 Decoders

A decoder is a data path device that implements functions similarly to multiplexers and demultiplexers. The n-to-2^n *decoder* is a logic network that decodes n inputs into 2^n outputs with the property that only one of the 2^n-output lines is asserted at a time, and each output corresponds to one valuation of the inputs (Figure 12.11a). An n-to-2^n decoder is a *minterm generation*.

TABLE 12.5
The nodes of decision trees: denotation, implementation, and formal notation

Techniques for decision diagram design using multiplexers

Node	Realization	Algebraic form	Matrix form

S h a n n o n   e x p a n s i o n

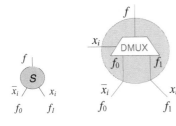

$$f = \overline{x}_i f_0 \vee x_i f_1$$
$$f_0 = f_{x_i=0}$$
$$f_1 = f_{x_i=1}$$

$$f = [\,\overline{x}_i \; x_i\,] \begin{bmatrix} 1 & 0 \\ 0 & 1 \end{bmatrix} \begin{bmatrix} f_0 \\ f_1 \end{bmatrix}$$
$$= [\overline{x}_i \; x_i] \begin{bmatrix} f_0 \\ f_1 \end{bmatrix}$$

P o s i t i v e   D a v i o   e x p a n s i o n

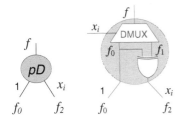

$$f = f_0 \oplus x_i f_2$$
$$f_0 = f_{x_i=0}$$
$$f_2 = f_{x_i=0} \oplus f_{x_i=1}$$

$$f = [\,1 \; x_i\,] \begin{bmatrix} 1 & 0 \\ 1 & 1 \end{bmatrix} \begin{bmatrix} f_0 \\ f_1 \end{bmatrix}$$
$$= [1 \; x_i] \begin{bmatrix} f_0 \\ f_0 \oplus f_1 \end{bmatrix}$$

N e g a t i v e   D a v i o   e x p a n s i o n

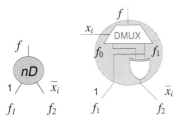

$$f = f_1 \oplus \overline{x}_i f_2$$
$$f_1 = f_{x_i=1}$$
$$f_2 = f_{x_i=0} \oplus f_{x_i=1}$$

$$f = [\,1 \; \overline{x}_i\,] \begin{bmatrix} 0 & 1 \\ 1 & 1 \end{bmatrix} \begin{bmatrix} f_0 \\ f_1 \end{bmatrix}$$
$$= [1 \; \overline{x}_i] \begin{bmatrix} f_1 \\ f_0 \oplus f_1 \end{bmatrix}$$

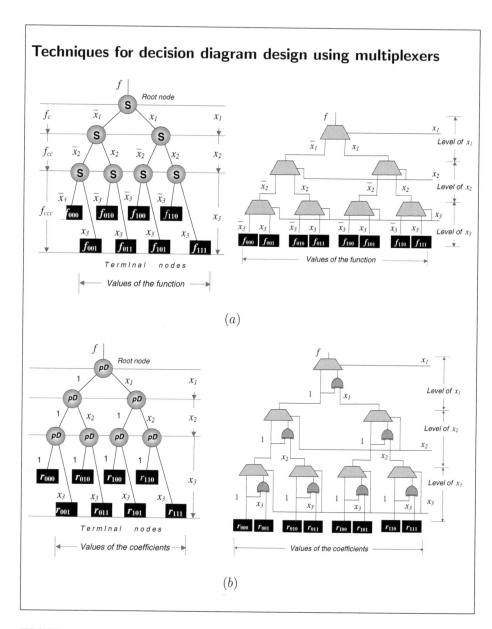

FIGURE 12.10

Implementation of binary decision trees on multiplexers (a) and multiplexers and EXOR gates (b) (Example 12.10).

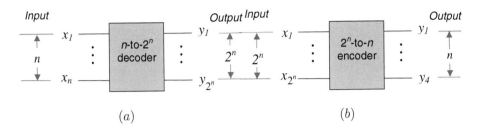

FIGURE 12.11
The n-to-2^n decoder (a) and 2^n-to-n encoder (b).

Example 12.11 *The gate level implementation of a 3-to-8 decoder, its symbol, and truth table are shown in Figure 12.12.*

The cascading decoders aims at decoding larger code words.

Example 12.12 *Figure 12.13 shows how 3-to-8 decoders can be combined to make a 4-to-16 decoder. The x_4 input drives the enable (En) inputs of the two decoders. The first decoder is enabled if $x_4 = 0$, and the second decoder is enabled if $x_4 = 1$.*

The concept of cascading that is introduced in Example 12.12 can be applied for the design of decoders of any size. Because of the treelike structure, the resulting networks are referred to as *decoder tree*.

Example 12.13 *Figure 12.14 shows how 3-to-8 decoders can be combined to make a 4-to-16 decoder.*

Example 12.14 *A block diagram and logic network of a 4-to-2 decoder is shown in Figure 12.15. The truth table shows that for $E_n = 0$, the outputs are set to 0 regardless of the input values x_1 and x_2. In this decoder,*

$$y_1 = \overline{x}_1\overline{x}_2E, \quad y_2 = \overline{x}_1x_2E,$$
$$y_3 = x_1\overline{x}_2E, \quad y_4 = x_1x_2E$$

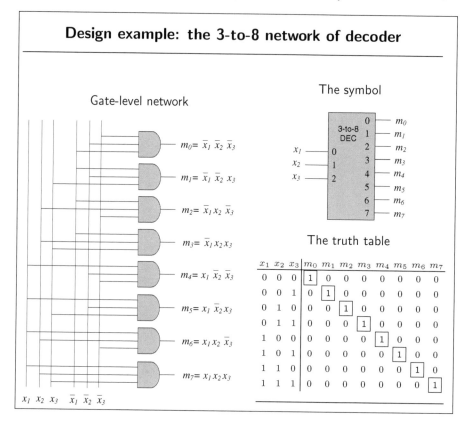

Design example: the 3-to-8 network of decoder

Gate-level network

$m_0 = \bar{x}_1 \, \bar{x}_2 \, \bar{x}_3$

$m_1 = \bar{x}_1 \, \bar{x}_2 \, x_3$

$m_2 = \bar{x}_1 \, x_2 \, \bar{x}_3$

$m_3 = \bar{x}_1 \, x_2 \, x_3$

$m_4 = x_1 \, \bar{x}_2 \, \bar{x}_3$

$m_5 = x_1 \, \bar{x}_2 \, x_3$

$m_6 = x_1 \, x_2 \, \bar{x}_3$

$m_7 = x_1 \, x_2 \, x_3$

$x_1 \; x_2 \; x_3 \quad \bar{x}_1 \; \bar{x}_2 \; \bar{x}_3$

The symbol

The truth table

x_1	x_2	x_3	m_0	m_1	m_2	m_3	m_4	m_5	m_6	m_7
0	0	0	1	0	0	0	0	0	0	0
0	0	1	0	1	0	0	0	0	0	0
0	1	0	0	0	1	0	0	0	0	0
0	1	1	0	0	0	1	0	0	0	0
1	0	0	0	0	0	0	1	0	0	0
1	0	1	0	0	0	0	0	1	0	0
1	1	0	0	0	0	0	0	0	1	0
1	1	1	0	0	0	0	0	0	0	1

FIGURE 12.12

A 3-to-8 decoder: gate-level network, symbol, and truth table (Example 12.11).

12.6   Implementation of Boolean functions using decoders

The n-to-2^n decoder can implement an arbitrary Boolean function of n variables. This is because the decoder is a minterm generator. The minterms generated by the decoder can then be combined using additional OR gates.

Given an m-to-2^n decoder, where $m < n$ and n divisible to m, the $\frac{n}{m}$-level tree of decoders can be used.

Design example: 4-to-16 decoder

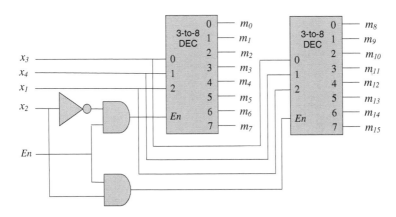

FIGURE 12.13

A 4-to-16 decoder design using 3-to-8 decoders (Example 12.12).

Algorithm for computing Boolean functions using decoders

Given: A Boolean function of n variables given by its truth
vector, and the size of $m - to - 2^m$ decoder.
Step 1: Divide the n variables into groups of m variables
Step 2: Design the logic network level by level.
Result: A logic network of $\frac{n}{m}$-levels of decoders (decoder
tree).

In the above design technique, one decoder is in the first level, and 2^1 decoders
in the next level, etc. The last level includes $2^{\frac{n}{m}}$ decoders. The output of the
last-level decoder that corresponds to the k minterms are OR-ed using k-input
OR (or network of OR gates).

Note that the enable input of the first level is 1, and the enable input of
the other levels are the outputs of decoders of the previous levels. It should
be noted that if n is not divisible by m, then a specific design can be used.

Design example 1

Given a 2-to-4 decoder and the SOP expression for the Boolean function of
two variables $(n = 2)$, $f = \bar{x}_1 x_2 \vee x_1 \bar{x}_2 = \bigvee m(1, 2)$, only one 2-to-4 decoder
and a two-input OR gate are required. This is because $n = m = 2$ (Table
12.6).

Design example: 4-to-16 decoder tree

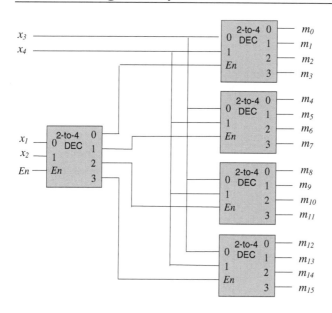

FIGURE 12.14
A 4-to-16 decoder tree (Example 12.13).

Design example: 2-to-4 decoder

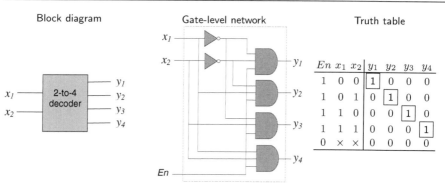

FIGURE 12.15
2-to-4 decoder (Example 12.14).

Design example 2

Consider the Boolean function of four variables ($n = 4$), given the SOP expression (Table 12.6)

$$f = \overline{x}_1\overline{x}_2\overline{x}_3 x_4 \vee \overline{x}_1\overline{x}_2 x_3\overline{x}_4 \vee \overline{x}_1 x_2 x_3\overline{x}_4 \vee \overline{x}_1 x_2 x_3 x_4$$
$$\vee\; x_1 x_2\overline{x}_3\overline{x}_4 \vee x_1 x_2\overline{x}_3 x_4 \vee x_1 x_2 x_3\overline{x}_4 \vee x_1 x_2\overline{x}_3 x_4$$
$$= \bigvee m(1, 2, 4, 7, 8, 11, 13, 14)$$

It can be implemented using a decoder tree so that the outputs of the decoder at the first level are used as enable inputs for the second-level decoders.

Design example 3

Consider a Boolean function of three variables ($n = 3$) (Table 12.6) given the SOP expression

$$f = \overline{x}_1 x_2 x_3 \vee x_1\overline{x}_2 x_3 \vee x_1 x_2\overline{x}_3 \vee x_1 x_2 x_3 = \bigvee m(3, 5, 6, 7)$$

We can use two decoders and the four-input OR gate. Variable x_1 is used as enable input, so that minterm $m_3 = \overline{x}_1 x_2 x_3$ is generated by the first decoder, and the minterms $m_5 = x_1\overline{x}_2 x_3, m_6 = x_1 x_2\overline{x}_3$, and $m_7 = x_1 x_2 x_3$ are generated by the second decoder.

12.7   Encoders

An encoder is a data path device that implements functions similarly to multiplexers and demultiplexers. A *binary encoder* encodes data from 2^n inputs into an n-bit number (Figure 12.11b):

$$\underbrace{2^n \text{ bits}}_{Input} \xrightarrow{Encoding} \underbrace{n \text{ bits}}_{Output}$$

If the input signal y_i has a logic value 1 and the other inputs are 0, then the outputs represent a binary number i.

> **Example 12.15** *A 4-to-2 encoder is shown in Figure 12.16. The input patterns that have multiple inputs set to 1 are not shown in the truth table; that is, these patterns are treated as don't-care conditions.*

TABLE 12.6
Logic network design techniques for computing Boolean functions
using decoders

Techniques for computing using decoders

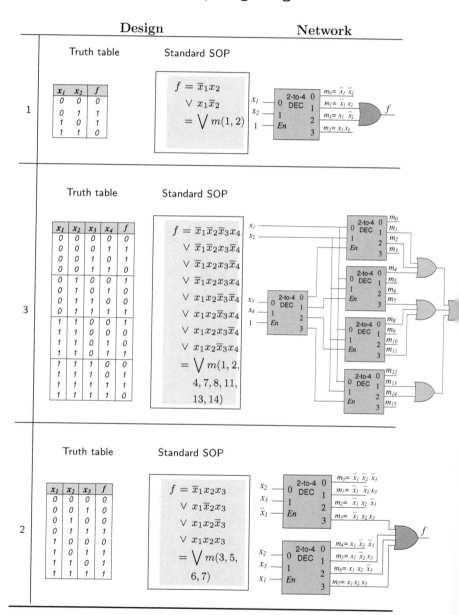

Design example: encoder

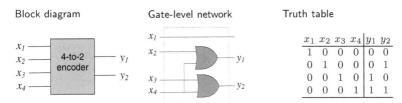

| Block diagram | Gate-level network | Truth table |

| x_1 x_2 x_3 x_4 | 4-to-2 encoder | y_1 y_2 |

Truth table:

x_1	x_2	x_3	x_4	y_1	y_2
1	0	0	0	0	0
0	1	0	0	0	1
0	0	1	0	1	0
0	0	0	1	1	1

FIGURE 12.16
4-to-2 encoder (Example 12.15).

12.7.1 Comparators

One-bit comparator

Given two binary signals, bits x_i and x_j, the detector of their equality is defined by the equation

$$x_i^{\sigma_i} = x_j^{\sigma_j}, \quad i \neq j, \tag{12.4}$$

where $x^1 = x$ and $x^0 = \overline{x}$. Equation 12.4 can be generalized for n signals. For example, given the three signals x_i, x_j, and x_t: $x_i^{\sigma_i} = x_j^{\sigma_j} = x_t^{\sigma_t}, \quad i \neq j \neq t$.

> **Example 12.16** *A detector of bit equality, or a one-bit comparator, for $\sigma_i = \sigma_j$, $i = j = 1$ (Equation 12.4) is shown in Figure 12.17. It has two inputs, x_1 and x_2, and generates a single output f, such that the output is equal to 1 if either x_1 and x_2 are both 1 or are both 0.*

Detector of word equality

Combinational networks that perform word-level computations are constructed using logic gates to compute the individual bits of the output word, based on the individual bits of the input word.

> **Example 12.17** *The detector of an 8-bit word equality is shown in Figure 12.18b. This network tests whether the two 4-bit words **X** and **Y** are equal. That is, the output is equal to 1 if and only if each bit of **X** is equal to the corresponding bit of **Y**.*

Design example: equality detection

x_1 x_2	f	
0 0	**1**	Equality $x_1 = x_2$
0 1	0	
1 0	0	
1 1	**1**	Equality $\overline{x}_1 = \overline{x}_2$

Network to test for bit equality for $\sigma_1 = \sigma_2$:

$$x_1^1 = x_2^1 \equiv x_1 = x_2$$
$$x_1^0 = x_2^0 \equiv \overline{x}_1 = \overline{x}_2$$

The output is equal to 1 when

▶ *Both inputs x_1 and x_2 are 1 (detected by the upper AND gate), or*
▶ *Both are equal to 0 (detected by lower AND gate)*

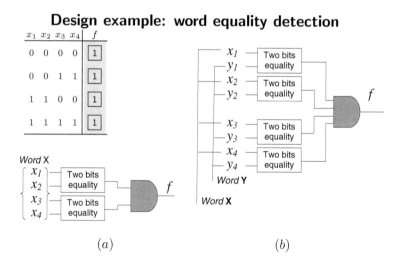

FIGURE 12.17
A combinational network for detecting bit-level equality (Example 12.16).

Design example: word equality detection

x_1 x_2 x_3 x_4	f
0 0 0 0	1
0 0 1 1	1
1 1 0 0	1
1 1 1 1	1

(a)

(b)

FIGURE 12.18
Combinational network for detecting 4-bit word equality (a), and 8-bit equality (b) of two words (Example 12.17).

12.7.2 Code detectors

Detector of word code

A bit-level equality network can be used for the detection of some specific codes, or strings, in a word.

Example 12.18 *The detector of the codes 0000, 0011, 1100, and 1111 in a 4-bit word is shown in Figure 12.18a.*

12.8   Design examples: adders and multipliers

Binary adder

An arithmetic network, or circuit, is a combinational logic network that performs arithmetic operations such as addition, subtraction, multiplication, and division on binary numbers or on decimal numbers coded as binary ones.

Example 12.19 *The addition suggests four possible elementary operations: $0 + 0 = 0$, $0 + 1 = 1$, $1 + 0 = 0$, and $1 + 1 = 2_{10} = 10_2$. The first three operations produce a sum that requires one bit for its representation; when both the augend and addend are equal to 1, the binary sum requires two bits. Because of this, the result is generally represented by two bits, the* **carry** *and the* **sum***. The carry obtained from the addition of two bits is added to the next higher-order pair of significant bits.*

Half adder

A half adder is an arithmetic network that generates the sum of two binary digits. The network has two inputs and two outputs. The input variables are the augend x_1 and the addend x_2 bits to be added, and the output variables produce the sum S and the carry C:

$$\text{SUM } S = x_1 \oplus x_2 \quad \text{CARRY } C = x_1 x_2$$

Example 12.20 *In Figure 12.19, the design of a* **half adder** *is introduced. Also, its implementation using decision diagrams is given. These diagrams implement the SOP representation; that is, the sum S is represented as*

$$\text{SUM } S = x_1 \oplus x_2 = x_1 \overline{x}_2 \vee \overline{x}_1 x_2$$

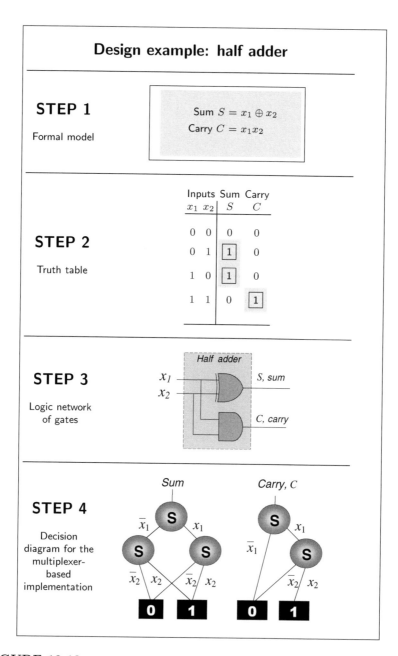

FIGURE 12.19

A half adder design (Example 12.20).

Full adder

A *full adder* is a combinational logic network that forms the arithmetic sum of the following input bits (Figure 12.20): augend and addend, x_1 and x_2, and the carry from the previous (right) significant bit, C_{in}. The full adder forms the two-digit binary number $\text{SUM} = C_{out} \times 2^1 + S \times 2^0$, which, when expressed in bitwise form, yields

$$S = x_1 \oplus x_2 \oplus C_{in}$$
$$C_{out} = x_1 x_2 \vee C_{in}(x_1 \oplus x_2)$$

Example 12.21 *In Figure 12.20, the design of a* **full adder** *is introduced. Figure 12.21 illustrates the implementation of decision diagrams for Boolean functions s_i and c_i using multiplexers MUX.*

Binary cascade adder

A *binary adder* is a logic network that produces the arithmetic sum of two binary numbers using only combinational logic. To add two n-bit numbers, one-bit full adder can be connected in a chain, so that the carry-out of the previous full adder is a carry-in to the next full adder. This forms a *cascade* logic network, also called ripple-carry adder.

Example 12.22 *In Figure 12.22, a 4-bit* **adder** *is shown, where $A = a_0 a_1 a_2 a_3$ and $B = b_0 b_1 b_2 b_3$. The input carry to the adder is C_0, and the output carry is C_4.*

An n-input cascade adder requires n full adders. Such a design is called a computational *array of cells*, where the full adder is referred to as a *cell*.

Carry-look-ahead adder

A nonoptimized four-bit adder can be made by the use of the generic one-bit adder cell connected one to the other (ripple-carry adder). One of the drawbacks of the cascade adder is a long delay due to the many gates in the carry path from the least significant bit to the most significant bit. The delay Δ can be evaluated as $\Delta = 2n + 2$ gate delay units.

The carry propagation can be speeded-up in two ways. The first (most obvious) way is to use a faster logic circuit technology. The second way is to generate carries by means of forecasting logic that does not rely on the carry signal being rippled from stage to stage of the adder. This faster alternative to the carry-ripple adder can be obtained at the cost of more gates with a

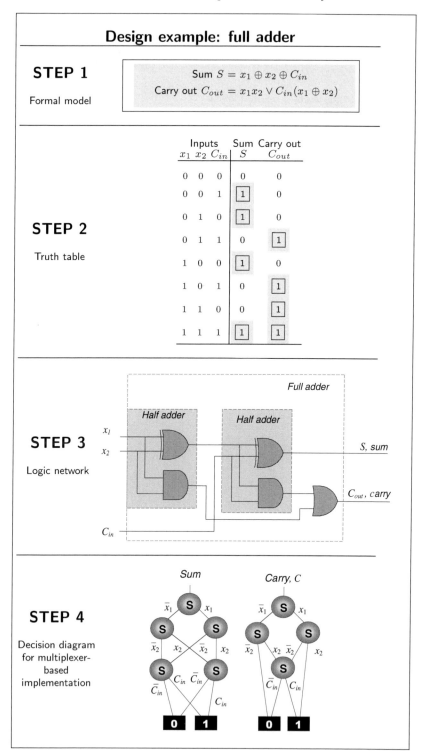

FIGURE 12.20

A full-adder design (Example 12.21).

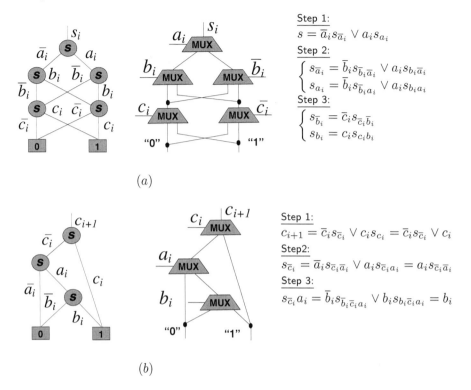

(a)

(b)

FIGURE 12.21

Multiplexer-based synthesis of full adder using decision diagrams of functions sum s_i (a) and carry c_i (b)

larger number of inputs. To reduce the delay, the carry-propagation path is broken so that a compromise between speed and cost is obtained by performing addition as a two-step process. First, the values of all carries into the full-adder modules are determined, and then, simultaneously, all result bits are computed. This adder is called *carry-look-ahead adder* (Figure 12.23).

Ripple-carry adder is a multilevel network designed by the connection of full-adders. For the multilevel adder, the total time required is calculated as the delay from carry-in C_i to carry-out C_{i+1}. Depending on the position at which a carry signal has been generated, the propagation time can be variable. In the best case, when there is no carry generation, the addition time will only take into account the time to propagate the carry signal. With a ripple-carry adder, if the input bits a_i and b_i are different for all position i, then the carry signal is propagated at all positions (thus never generated), and the addition is completed when the carry signal has propagated through the whole adder.

Formally, it is possible to express a carry as a function of all the preceding low-order carry by using the recursivity of the carry function.

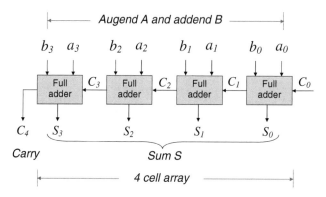

FIGURE 12.22
The 4-bit cascade binary adder (Example 12.22).

Actually, ripple-carry adders are fast only for some configurations of the input words where carry signals are generated at some positions. They can be divided into blocks, where a special circuit detects quickly if all the bits to be added are different.

Carry generator determines the values of all intermediate carries before the corresponding sum bits are computed. *Sum generator* uses precomputed carriers to determine the value of the sum bits. The cost of addition of two numbers of m and n digits is $O(min\{m, n\})$.

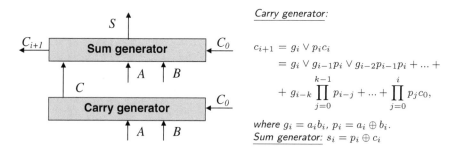

Carry generator:

$$c_{i+1} = g_i \vee p_i c_i$$
$$= g_i \vee g_{i-1}p_i \vee g_{i-2}p_{i-1}p_i + \dots +$$
$$+ g_{i-k} \prod_{j=0}^{k-1} p_{i-j} + \dots + \prod_{j=0}^{i} p_j c_0,$$

where $g_i = a_i b_i$, $p_i = a_i \oplus b_i$.
Sum generator: $s_i = p_i \oplus c_i$

FIGURE 12.23
Carry-look-ahead adder.

One approach to designing a full adder is formulated as follows: separate the parts of the full adder not involving the carry propagation path from those containing the path. This separation is based on the following: assuming $g_i = a_i b_i$ and $p_i = a_i \oplus b_i$, the equation for the i-th cell of the full adder can be written as follows:

$$C_{i+1} = \underbrace{a_i b_i}_{g_i} \vee C_i \underbrace{(a_i \oplus b_i)}_{p_i} = g_i \vee C_i p_i \tag{12.5}$$

It follows from Equation 12.5 that

(a) For the inputs $a_i = b_i = 1$, the i-th full adder *generates* a carry g_i.

(b) For the inputs a_i and b_i, when one of them is equal to 1, the i-th full adder *propagates* a carry p_i.

The resulting network implements the following equations:

$$g_i = a_i b_i$$
$$p_i = a_i \oplus b_i$$
$$S_i = a_i \oplus b_i \oplus C_i$$

Example 12.23 *The design of the 4-bit* **carry-look-ahead** *adder is illustrated in Figure 12.24.*

Multipliers

A combinational multiplier for positive integers is used in floating-point processors and signal processing applications. An $n \times m$ bits combinational multiplier is a combinational circuit that produces the multiplication $0 \leq A \times B \leq (2^n - 1)(2^m - 1)$ (product) of two integer numbers: $0 \leq A \leq 2^n - 1$ (multiplicand) and $0 \leq B \leq 2^m - 1$ (multiplier):

$$A = \sum_{i=0}^{m-1} a_i 2^i, \quad B = \sum_{j=0}^{n-1} b_j 2^j, \quad m \geq n$$

$$A \times B = \sum_{h=0}^{m+n-2} \left(\sum_{i+j=h} a_i b_j \right) 2^h$$

When the operands are interpreted as integers, the product is generally twice the length of the operands in order to preserve the information content.

Multiplication can be considered as a series of repeated additions. This repeated addition method that is suggested by the arithmetic definition is so slow that it is almost always replaced by an algorithm that makes use of positional number representation.

It is possible to decompose multipliers in two parts. The first part is dedicated to the generation of partial products, and the second one collects

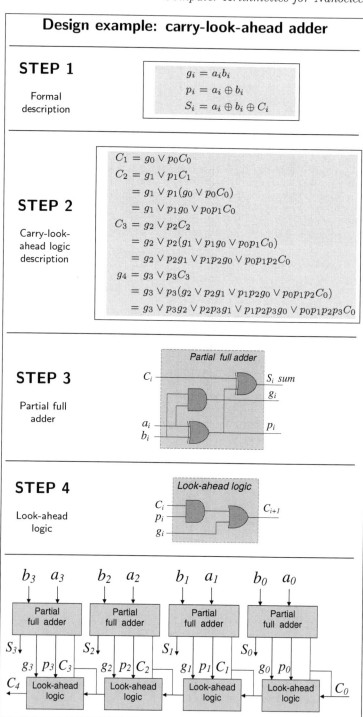

FIGURE 12.24
The 4-bit carry-look-ahead adder design (Example 12.23).

and adds them. As for adders, it is possible to enhance the intrinsic performance of multipliers. Acting in the generation part, the Booth (or modified Booth) algorithm is often used because it reduces the number of partial products. The collection of the partial products can then be made using a regular array, a Wallace tree or a binary tree.

The simplest multiplication can be viewed as repeated shifts and adds (one adder, a shift register, and a small amount of control logic). The disadvantage is that it is slow. One fairly simple improvement to this is to form the matrix of partial products in parallel and then use a 2D array of full adders to sum the rows of partial products.

Example 12.24 *The 8×6 computing structure shown in Figure 12.25 is known as* **array multiplier**. *The multiplier consists of $m - 1 = 5$ $n = 8$-bit carry-ripple adders and $m = 6$ arrays of n AND gates. The delay of the multiplier is defined as the critical path equal to sum of delay of the buffer circuit connecting input signal and AND gates, the delay of AND gate, and the delay of the adders.*

The advantages of the array multiplier are that it is a regular structure and a local interconnect: each cell is connected only to its neighbors. The disadvantage is that the worst-case delay path goes from the upper left corner diagonally down to the lower right corner and then across the ripple carry adder, that is, the delay is linearly proportional to the operand size. The method that can be employed to decrease the delay of the array multiplier is to replace the ripple carry adder with a carry-look-ahead adder. Another approach to collection of the partial products is based on so-called *Wallance tree multiplier*. The number of operations occurring in multiplication is at most $O(mn)$

Arithmetic-logic units

The arithmetic-logic unit (ALU) is a module capable of realizing a set of arithmetic and logic functions. ALU performs the specific operations selected dynamically by the control unit of the processor.

Example 12.25 *In Figure 12.26, a 4-bit* **ALU** *has two 4-bit data inputs A and B, a carry-in input C_0, 4-bit data output S, and also P and G outputs that can be used for computing carry-out signal $C_4 = G \vee P \cdot C_0$. This 4-bit ALU module can be used to construct larger ALUs.*

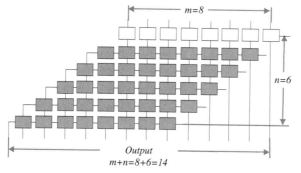

a_7b_0 a_6b_0 a_5b_0 a_4b_0 a_3b_0 a_2b_0 a_1b_0 a_0b_0
a_7b_1 a_6b_1 a_5b_1 a_4b_1 a_3b_1 a_2b_1 a_1b_1 a_0b_1
a_7b_2 a_6b_2 a_5b_2 a_4b_2 a_3b_2 a_2b_2 a_1b_2 a_0b_2
a_7b_3 a_6b_3 a_5b_3 a_4b_3 a_3b_3 a_2b_3 a_1b_3 a_0b_3
a_7b_4 a_6b_4 a_5b_4 a_4b_4 a_3b_4 a_2b_4 a_1b_4 a_0b_4
a_7b_5 a_6b_5 a_5b_5 a_4b_5 a_3b_5 a_2b_5 a_1b_5 a_0b_5

Formal description

$$A \times B = A\left(\sum_{i=0}^{m-1} b_i 2^i\right) = \sum_{i=0}^{m-1} Ab_i 2^i$$

The multiplication is performed by adding the integers $Ab_i 2^i$. Because b_i is either 0 or 1, we get

$$Ab_i = \begin{cases} 0, & if\ b_i = 0; \\ A, & if\ b_i = 1. \end{cases}$$

FIGURE 12.25
A 8×6 multiplier: topology of architecture and multiplication scheme (Example 12.24).

12.9 Design example: magnitude comparator

A logic network that compares two binary numbers and indicates whether they are equal, and/or indicates an arithmetic relationship (greater or less than) between the numbers, is called a *magnitude comparator*. The EXOR and XNOR gates may be viewed as 1-bit comparators. In particular, if the input bits are equal, then the output of the XNOR gate is 1, while otherwise it is 0. Therefore, a straightforward implementation of an n-bit comparator can be accomplished using XNOR gates.

Example 12.26 *A logic network for a 4-bit-comparator using XNOR gates is given in Figure 12.27.*

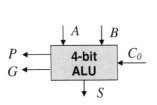

Control	Function, S
Transfer	A
Complement	$\overline{A}$
AND	AB
OR	$A \vee B$
EXOR	$A \oplus B$
Increment	$A + 1$
ADD	$A + B + 1$
	$A + B$
1C subtraction	$A + \overline{B}$
1C subtraction	$A + \overline{B} + 1$
Decrement	$A - 1$

FIGURE 12.26

A 4-bit arithmetic-logic units (Example 12.25).

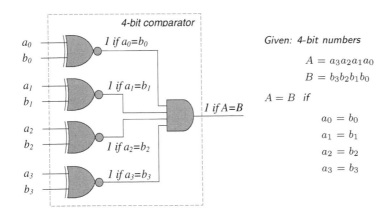

FIGURE 12.27

A 4-bit comparator (Example 12.26).

Example 12.27 *Given the 2-bit numbers $A = a_1a_0$ and $B = b_1b_0$, design a 2-bit* **magnitude comparator** *with three outputs: $A = B$, $A > B$, and $A < B$. The design of a 2-bit comparator is given in Figures 12.28, 12.29, and 12.30.*

Design example: 2-bit magnitude comparator

STEP 1
Problem formalization

Number A		Number B		A=B	A>B	A<B
a_1	a_0	b_1	b_0			
0	0	0	0	1	0	0
0	0	0	1	0	0	1
0	0	1	0	0	0	1
0	0	1	1	0	0	1
0	1	0	0	0	1	0
0	1	0	1	1	0	0
0	1	1	0	0	0	1
0	1	1	1	0	0	1
1	0	0	0	0	1	0
1	0	0	1	0	1	0
1	0	1	0	1	0	0
1	0	1	1	0	0	1
1	1	0	0	0	1	0
1	1	0	1	0	1	0
1	1	1	0	0	1	0
1	1	1	1	1	0	0

STEP 2
Minimization using K-maps

$f(A = B)$ $f(A > B)$ $f(A < B)$

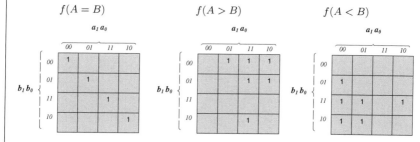

STEP 3
Minimal SOP description

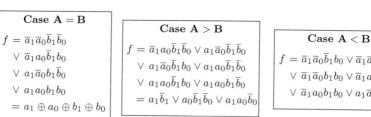

Case A = B

$f = \bar{a}_1 \bar{a}_0 \bar{b}_1 \bar{b}_0$
$\vee\ \bar{a}_1 a_0 \bar{b}_1 b_0$
$\vee\ a_1 \bar{a}_0 b_1 \bar{b}_0$
$\vee\ a_1 a_0 b_1 b_0$
$= a_1 \oplus a_0 \oplus b_1 \oplus b_0$

Case A > B

$f = \bar{a}_1 a_0 \bar{b}_1 \bar{b}_0 \vee a_1 \bar{a}_0 \bar{b}_1 \bar{b}_0$
$\vee\ a_1 \bar{a}_0 \bar{b}_1 b_0 \vee a_1 a_0 \bar{b}_1 \bar{b}_0$
$\vee\ a_1 a_0 \bar{b}_1 b_0 \vee a_1 a_0 b_1 \bar{b}_0$
$= a_1 \bar{b}_1 \vee a_0 \bar{b}_1 \bar{b}_0 \vee a_1 a_0 \bar{b}_0$

Case A < B

$f = \bar{a}_1 \bar{a}_0 \bar{b}_1 b_0 \vee \bar{a}_1 a_0 b_1 \bar{b}_0$
$\vee\ \bar{a}_1 \bar{a}_0 b_1 b_0 \vee \bar{a}_1 a_0 b_1 \bar{b}_0$
$\vee\ \bar{a}_1 a_0 b_1 b_0 \vee a_1 \bar{a}_0 b_1 b_0$

FIGURE 12.28

A 2-bit magnitude comparator (Example 12.27).

Design example: 2-bit magnitude comparator (continuation)

STEP 4: IMPLEMENTATION
Multiplexer-based logic network

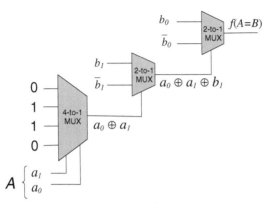

Decoder-based implementation

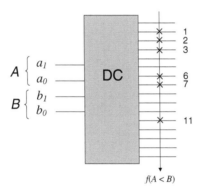

PLA-based implementation

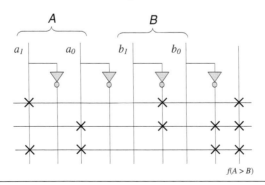

FIGURE 12.29

A 2-bit magnitude comparator, continuation (Example 12.27).

Design example: 2-bit magnitude comparator (continuation)

D e c i s i o n   d i a g r a m   i m p l e m e n t a t i o n

Using Shannon expansion

Using Davio expansion

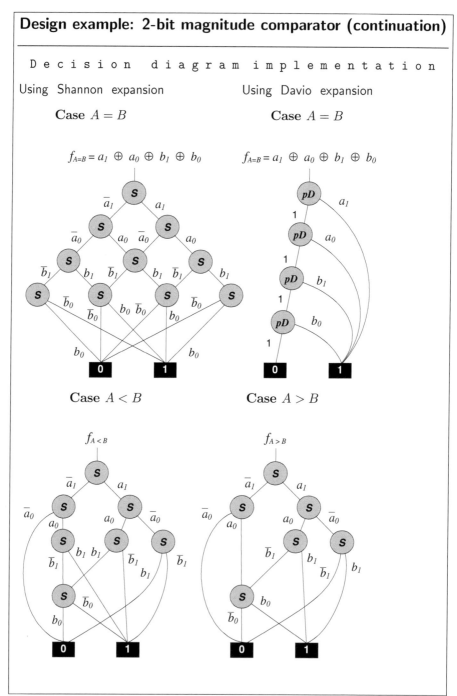

FIGURE 12.30

A 2-bit magnitude comparator, continuation (Example 12.27).

12.10 Design example: BCD adder

The code known as *binary-coded decimal* (BCD) encodes the digits 0 through 9. Because there are 10 digits to encode, it is necessary to use four bits per digit. Each digit is encoded by the binary pattern that represents its unsigned value (Table 12.7). Only 10 of the 16 available patterns are used in a BCD. This means that the remaining six patterns should not occur in logic networks that operate on BCD numbers; these patterns are usually treated as don't-care conditions in the design process.

TABLE 12.7
Binary coded decimal digits

Decimal	BCD code	Decimal	BCD code
0	0000	5	0101
1	0001	6	0110
2	0010	7	0111
3	0011	8	1000
4	0100	9	1001

Conversion between BCD and decimal representations is trivial, that is, a direct substitution of four bits for each decimal digit is all that is necessary. The addition of two BCD digits is complicated by the fact that the sum may exceed 9, in which case a correction will have to be made. Let $X = x_3x_2x_1x_0$ and $Y = y_3y_2y_1y_0$ be the BCD digits. Denote as $S = s_3s_2s_1s_0$ the desired sum of X and Y, $S = X + Y$. The algorithm for addition of two BCD digits is as follows:

BCD addition

Given: BCD X and Y
Calculate $Z = X + Y$
Step 1: If $Z \le 9$, then the desired sum digit is equal to the result $S = Z$ and CARRY-OUT $= 0$
Step 2: (*a*) If $Z > 9$, then correct the desired sum digit $S = Z + 6$
    (*b*) If $Z > 15$, and CARRY-OUT $= 1$, then correct the desired sum digit $S = Z + 6$ and keep CARRY-OUT $= 1$
Result: BCD representation of Z

The technique for adding the BCD digits is illustrated in Table 12.8.

Consider a block diagram of a BCD adder (Figure 12.31). This includes two 4-bit binary adders and a correction circuit. The first 4-bit adder produces the

TABLE 12.8
Addition of BCD digits

Techniques for computing in BCD

Example	Comments

(4) 0 1 0 0 (3) 0 0 1 1 (5) + 0 1 0 1 (2) + 0 0 1 0 (9) 1 0 0 1 (5) 0 1 0 1	These results are correct.

(5)   0 1 0 1
 (9) + 1 0 0 1
 (14) 1 1 1 0
    + 0 1 1 0 Correction
 (10 + 4) 1 0 1 0 0
      ↑
    Carry
	The 4-bit addition yields
	
	 $Z = 5 + 9 = 14$
	
	 which is wrong in BCD format (numbers greater than 9 are not allowed). Adding 6 to the intermediate sum Z provides a necessary correction.

(8)   1 0 0 0
 (9) + 1 0 0 1
 (17) 1 0 0 0 1
    + 0 1 1 0 Correction
 (10 + 7) 1 0 1 1 1
      ↑
    Carry
	The 4-bit addition yields
	
	 $Z = 8 + 9 = 17$
	
	 which is wrong in BCD format. Adding 6 to the intermediate sum Z provides a necessary correction.

(9)   1 0 0 1
 (9) + 1 0 0 1
 (18) 1 0 0 1 0
    + 0 1 1 0 Correction
 (10 + 8) 1 1 0 0 0
      ↑
    Carry
	The 4-bit addition yields
	
	 $Z = 9 + 9 = 18$
	
	 which is wrong in BCD format. Adding 6 to the intermediate sum Z provides a necessary correction.

sum of two 4-bit numbers. Its result, the sum $Z = X + Y$, can be correct or incorrect. The block that detects whether $Z > 9$ generates the output signal, ADJUST. This signal controls the multiplexer that provides the correction of the intermediate sum when needed:

(*a*) If ADJUST $= 0$, then the output is $S = Z + 0 = Z$.

(*b*) If ADJUST $= 1$, then the output is $S = Z + 6$ and the carry-out is $C_{out} = 1$.

BCD adder design includes the following steps:

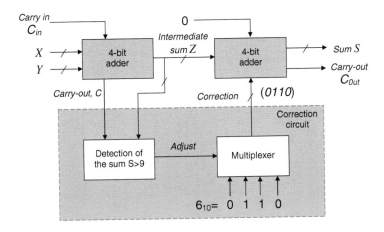

FIGURE 12.31

BCD adder.

Step 1: Formalize the problem in the form of a truth table for the ADJUST function of the detection circuit using don't cares.

Step 2: Minimize the ADJUST function using a K-map.

Step 3: Design the correction circuit.

Example 12.28 *Design the* **correction circuit** *for the 4-bit BCD adder.*

Consider the truth table of the ADJUST *output for the detector of the sum $S > 9$. The input variables are the bits $z_3 z_2 z_1 z_0$ of sum Z, and the carry-out C of the 4-bit adder. The K-map of the* ADJUST *function is given in Figure 12.32a. The minimized function is* ADJUST $= C \vee z_1 z_3 \vee z_2 z_3$. *The design network is given in Figures 12.32b and 12.33.*

12.11 The verification problem

Verification is the comparison of two models for consistency. There are various classes of verification techniques in logic design. In particular, they can be classified as follows:

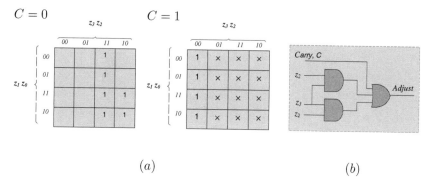

(a) (b)

FIGURE 12.32
BCD adder design: the K-maps of the ADJUST function (Example 12.28).

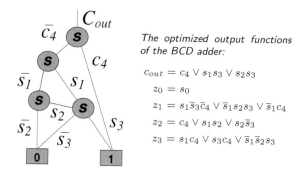

The optimized output functions of the BCD adder:

$$c_{out} = c_4 \lor s_1 s_3 \lor s_2 s_3$$
$$z_0 = s_0$$
$$z_1 = s_1 \overline{s}_3 \overline{c}_4 \lor \overline{s}_1 s_2 s_3 \lor \overline{s}_1 c_4$$
$$z_2 = c_4 \lor s_1 s_2 \lor s_2 \overline{s}_3$$
$$z_3 = s_1 c_4 \lor s_3 c_4 \lor \overline{s}_1 \overline{s}_2 s_3$$

FIGURE 12.33
Decision diagram of the correction function of the BCD adder (Example 12.28).

(a) *Implementation verification techniques*: these techniques aim at verifying the correctness of the synthesis task

(b) *Design verification techniques*: these techniques are related to the highest (architectural) level of design.

In the logic synthesis of combinational networks, the functionality of the networks to be designed is given in terms of a network of logical gates. An important task in logic synthesis is to optimize the representation of this network on the gate level with respect to several factors. These optimization criteria include the number of gates, chip area, energy consumption, delay, clock period, etc.

12.11.1 Formal verification

Formal verification of logical networks demands that a mathematical proof of correctness implicitly covers all possible input patterns (stimuli). It has to be proven formally that the *specification* and *implementation* are functionally equivalent, that is, that both compute exactly the same Boolean function. The verification problem is formulated as follows:

Given: The specifications of a network to be implemented in terms of a function f, a verified network realizing f, and a new realization that is claimed to realize f.

Problem: Verify that the realization and specification are equivalent.

Solution: Prove mathematically that the input–output behavior of the network and the new realization are equal.

The most basic requirement of a new design is that it be *functionally correct*. That is, the design must meet the problem specification. To verify functional correctness, logic simulation software packages are used to create and test a computational model of the proposed design.

Practical design verification means validating that an implemented gate-level design matches its desired behavior as specified at the register-transfer level. This is accomplished in practical design by means of *equivalence checking*. This is not the same as formal verification, which is proving that a design has a desirable property, namely, in equivalence checking:

► Correctness is defined as the equivalence of two designs.

► Equivalence is usually localized by finding the correspondence between latches, that is, checking if they have the same next-state function.

12.11.2 Equivalence-checking problem

Equivalence checking for combinational logic networks is formulated as follows:

Given: Two logic networks.

Problem: Check if their corresponding outputs are equal for all possible input patterns.

Solution: Since two logic networks are equivalent if and only if the canonical representations of their output functions are the same,

(a) Derive the canonical representation (standard SOP expressions).

(b) Construct the canonical graphical form, such as a full tree or decision diagram.

Example 12.29 *Given the specified and optimized networks (Figure 12.34), the simplest approach to verifying that these networks are equivalent is to derive truth tables independently for each network and compare the values of the outputs for corresponding inputs. From the truth table comparison, it follows that specified and optimized networks are not equivalent. The error in optimization occurs because the input x_4 must be complemented. Correction of this error results in a network that is equivalent to the specified network (implemented network).*

A complete binary tree and a (reduced and ordered) binary decision diagram are canonical forms. Thus, two logic networks are equivalent if their decision diagrams are isomorphic. In practice, subfunctions of two logic networks (called specification and implementation) are transformed into a decision diagram by simulating the networks gate by gate, normally in a depth-first manner. The decision diagrams to be checked for equivalence must be reduced and ordered.

Example 12.30 *Figure 12.35 shows the two simplest logic networks: an AND gate and a network consisting of three NOT gates and one AND gate. Let us check their equivalence using (a) a comparison of the standard SOPs of both networks, and (b) a comparison of their decision diagrams for isomorphism. The first network implements the function $f = x_1 x_2$, which is the standard SOP expression. The second network implements the function $f = \overline{\overline{x}_1 \vee \overline{x}_2}$. Converting it into the standard SOP form yields $g = \overline{\overline{x}_1 \vee \overline{x}_2} = xy$.*

12.11.3   Design example 1: Functionally equivalent networks

A decision diagram is shown in Figure 12.35a. Derivation of a decision diagram requires two steps: generation of the decision diagram of the AND gate with inverted inputs, and finding the complement of this decision diagram, that is, inverting the terminal node values (Figure 12.35b).

Since SOP expressions for f and g are equal, these networks are equivalent. By inspection, it can be seen that the decision diagram of the AND gate and the last-derived decision diagram are isomorphic and, therefore, that both networks are functionally equivalent.

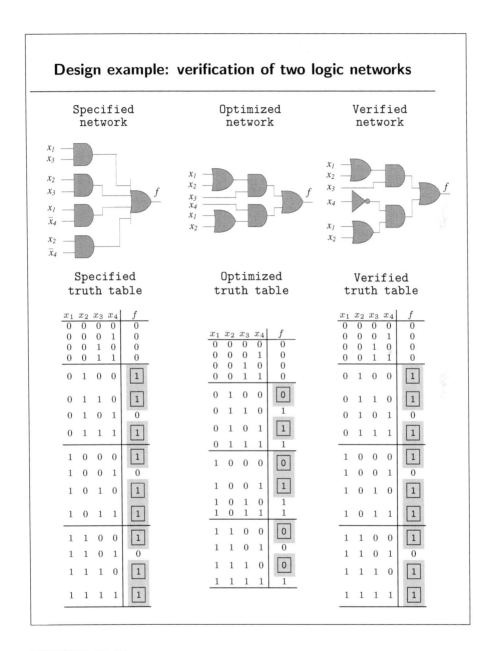

FIGURE 12.34

Verification of two networks using truth tables (Example 12.29).

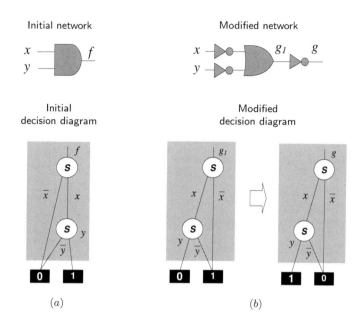

FIGURE 12.35
Two logic networks and the derivation of their decision diagrams.

12.11.4   Design example 2: Verification of logic networks using decision diagrams

Consider the two networks shown in Figure 12.36a,b:

The first network: A decision diagram of the inverter is used to construct the decision diagram of the AND gate $(x_1 \overline{x}_3)$; and in the same manner the decision diagram of the other AND gate is created $(x_2 x_3)$ (Figure 12.36b). Next, their two AND functions are considered to be the inputs of the OR gate, which forms the function $f_1 = x_1 \overline{x}_3 \lor x_2 x_3$. This manipulation results in the final decision diagram for f_1.

The second network: The decision diagrams for both AND gates in the first level of the network are derived first. Next, decision diagrams for an OR gate $(x_1 \lor x_2 x_3)$ and for an EXOR gate are constructed $(x_1 \lor x_2 x_3) \oplus x_1 x_3$. The variable ordering is fixed to x_1, x_2, x_3. A check for the isomorphism of the decision diagrams presented proves their equivalence, and thus, the functional equivalence of the given networks.

Another approach is to manipulate decision diagrams without isomorphism comparison. To verify that two combinational networks with outputs F and G are equivalent, the decision diagram for $\overline{f \oplus g}$ is constructed, where f

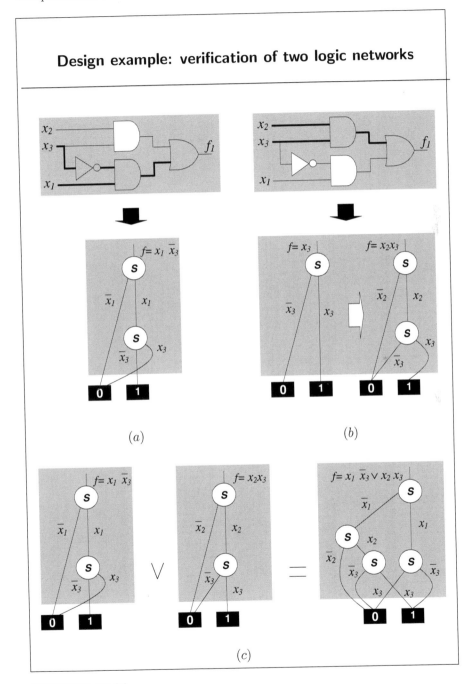

FIGURE 12.36

Construction of the decision diagram for the Boolean function $f_1 = x_1\overline{x}_3 \vee x_2x_3$.

**Design example:
verification of two logic networks (continuation)**

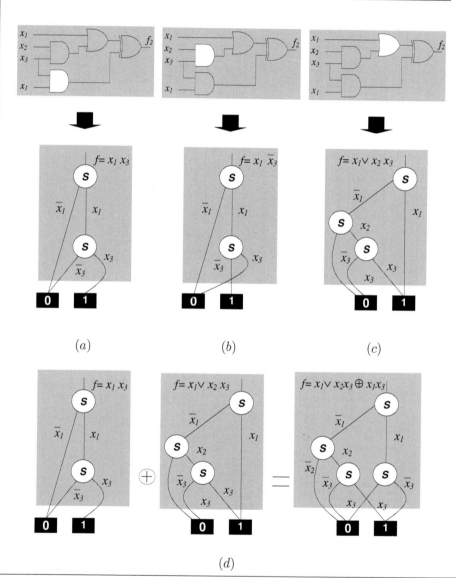

(a)   (b)   (c)

(d)

FIGURE 12.37
The logic network and the decision diagram of its two AND gates (a) and
(b), and OR gate (the function $x_1 \vee x_2 x_3$) (c); construction of the decision
diagram for the function $(x_1 \vee x_2 x_3) \oplus x_1 x_3$.

and g represent the Boolean functions for F and G, respectively. Due to the canonicity of the decision diagram, the two circuits implement the same Boolean function if and only if the resulting decision diagram is identical to the terminal 1.

Example 12.31 *Consider the functions f and g from Example 12.30. To check their equivalence, let us derive the decision diagram for the EXOR function $\overline{f \oplus g}$. Figure 12.38 illustrates the calculation using decision diagrams for f and $\overline{g}$ using the equation $\overline{f \oplus g} = f \oplus \overline{g}$. Since the resulting diagram is a constant terminal "1", both networks are considered to be equivalent.*

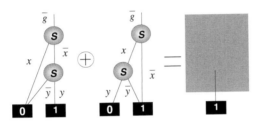

FIGURE 12.38

Calculation using decision diagrams for f and $\overline{g}$ (Example 12.31).

12.12   Decomposition

Simplification of logic networks by factoring is a satisfactory technique in many cases, but it does not have systematic methods for all cases. It is evident that a more systematic method for optimization would be very useful. This process is called *decomposition*. Decomposition is the central problem of the representation, manipulation, optimization, and implementation of Boolean functions. For example, a two-level logic network can be decomposed into a multilevel network. In some cases, multilevel networks may reduce the cost of implementation.

Decomposability is a property of a Boolean function $f(\mathbf{X})$, which can be described in the form

$$f(\mathbf{X}) = h(\mathbf{X}_1, g(\mathbf{X}_2))$$

where $\mathbf{X}_1$ and $\mathbf{X}_2$ are sets of variables,

$$\mathbf{X} = \mathbf{X}_1 \cup \mathbf{X}_2,$$

and g is called a *subfunction*. If such a decomposition exists, it is called a *functional* decomposition of f. Generally, the functions h and g are less complex than f.

12.12.1   Disjoint and nondisjoint decomposition

A decomposition is called *disjoint* if the sets of n_1 and n_2 variables, $\mathbf{X}_1$ and $\mathbf{X}_2$, respectively, are not overlapped; that is,

$$\mathbf{X}_1 \cap \mathbf{X}_2 = \emptyset.$$

A decomposition is called *nondisjoint* if the sets of variables $\mathbf{X}_1$ and $\mathbf{X}_2$ are overlapped; that is,

$$\mathbf{X}_1 \cap \mathbf{X}_2 \neq \emptyset.$$

Example 12.32 *Let f be a Boolean function of six variables $x_1, x_2, \ldots, x_6$.*

    (a) The **partition** *$(\mathbf{X}_1, \mathbf{X}_2)$, where $\mathbf{X}_1 = (x_1, x_2, x_6)$ and $\mathbf{X}_2 = (x_3, x_4, x_5)$, results in a* **disjoint** *decomposition because the sets $\mathbf{X}_1$ and $\mathbf{X}_2$ are not overlapped: $\mathbf{X}_1 \cap \mathbf{X}_2 = (x_1, x_2, x_6) \cap (x_3, x_4, x_5) = \emptyset$.*

    (b) The partition $(\mathbf{X}_1, \mathbf{X}_2)$, where $\mathbf{X}_1 = (x_1, x_2, x_3, x_6)$ and $\mathbf{X}_2 = (x_2, x_3, x_4, x_5)$, results in a **nondisjoint** *decomposition because the sets $\mathbf{X}_1$ and $\mathbf{X}_2$ are overlapped: $\mathbf{X}_1 \cap \mathbf{X}_2 = (x_1, x_2, x_3, x_6) \cap (x_2, x_3, x_4, x_5) \neq \emptyset$.*

Decomposition can be applied to various data structures: decomposition of Boolean function (in algebraic, matrix, or cube-based forms), decomposition of logic networks, and decomposition using decision diagrams.

12.12.2   Decomposition chart

The *decomposition chart* of a Boolean function f is a table with 2^{n_1} columns and 2^{n_2} rows. A binary number corresponding to a variable assignment is assigned to each row and column, One element of the table corresponds to one value of a Boolean function f. In other words, the decomposition chart is a rearranged truth table of the function.

Example 12.33 *A decomposition chart for the Boolean function f of five variables is given in Figure 12.39. A set of variables X is partitioned into two sets,* $\mathbf{X}_1 = \underbrace{(x_1, x_2, x_3)}_{n_1=3}$ *and* $\mathbf{X}_2 = \underbrace{(x_4, x_5)}_{n_2=2}$.

Partition
$\mathbf{X}_1 = (x_1, x_2, x_3)$

		000	*001*	*010*	*011*	*100*	*101*	*110*	*111*
Partition $\mathbf{X}_2 = (x_4, x_5)$	*00*		1			1	1	1	
	01								
	10	1		1	1				1
	11	1				1	1	1	

FIGURE 12.39
A decomposition chart for a Boolean function of five variables (Examples 12.33 and 12.34).

The number of distinct columns (rows) in the decomposition chart is called the *column (row) multiplicity*.

Example 12.34 *The column* **multiplicity** *in the decomposition chart in Figure 12.39 is two, and the row multiplicity is three.*

12.12.3 Disjoint bi-decomposition

Consider the simplest case of decomposition, disjoint *bi-decomposition*, which means the decomposition of a Boolean function into exactly two subfunctions of variables from two partitions, $\mathbf{X}_1$ and $\mathbf{X}_2$, with respect to one of three operations:

▶ OR type, $f = g_1(\mathbf{X}_1) \vee g_2(\mathbf{X}_2)$
▶ AND type $f = g_1(\mathbf{X}_1) \wedge g_2(\mathbf{X}_2)$,
▶ EXOR type $f = g_1(\mathbf{X}_1) \oplus g_2(\mathbf{X}_2)$

A simple check for disjoint bi-decomposition of a Boolean function can be accomplished using the decomposition chart. A Boolean function f has a disjoint bi-decomposition if and only if the row and column multiplicities of its decomposition chart are less than or equal to two. This is a *necessary*, but *not a sufficient* condition.

> **Example 12.35** *The Boolean function, whose decomposition chart is given in Figure 12.39, does not have a disjoint* **bi-decomposition** *with respect to the partitions* (X_1, X_2) *since its row multiplicity is three.*

A variety of approaches to verifying this and other types of decompositions have been developed, according to various criteria of design, for single-output and multioutput functions. One of the approaches for verification of existence of OR and AND type decompositions given a single-output Boolean function is considered below.

A Boolean function has OR-type disjoint bi-decomposition if every product in the minimal SOP for this function consists of literals from $\mathbf{X}_1$ only or $\mathbf{X}_2$ only. The minimal SOP expression is defined in this case as an SOP form consisting of the prime implicants only (and some of them can be essential prime implicants), so that no product can be deleted without changing the function represented by this expression.

Algorithm for finding OR type disjoint decomposition

Given: a minimal SOP expression consisting of t products.
Step 1 Start with a trivial partition, so that each partition includes one variable only.
Step 2 Form another partition by merging two blocks of the previous partition if at least one literal from each block occurs in the first product. Repeat for all t products of the SOP expression.
Step 3 If the t-th partition has at least two blocks, $\mathbf{X}_1$ and $\mathbf{X}_2$, then the function has a disjoint bi-decomposition of the form

$$f(\mathbf{X}_1, \mathbf{X}_2) = g_1(\mathbf{X}_1) \vee g_2(\mathbf{X}_2)$$

12.12.4   Design example: the OR type bi-decomposition

Consider the Boolean function $f = \overline{x}_1 \overline{x}_2 x_3 \vee \overline{x}_1 x_2 \overline{x}_3 \vee x_1 x_2 \overline{x}_3 \vee \overline{x}_1 x_4 x_5 \vee x_4 x_5$. The process of finding out if the function has an OR type disjoint bi-decomposition is illustrated in Figure 12.40.

12.12.5   Design example: AND type bi-decomposition

OR type decomposition is relevant to AND type decomposition as follows from the following statement: a Boolean function f has AND type disjoint bi-decomposition with respect to $\mathbf{X}_1, \mathbf{X}_2$, if and only if $\overline{f}$ has an OR-type

Design example: OR type bi-decomposition

INPUT DATA $\qquad f = \overline{x}_1 \overline{x}_2 x_3 \vee \overline{x}_1 x_2 \overline{x}_3 \vee x_1 x_2 \overline{x}_3 \vee \overline{x}_1 x_4 x_5 \vee x_4 x_5$

PRELIMINARY STEP

Minimization

$f = \overline{x}_1 \overline{x}_2 x_3 \vee (\overline{x}_1 \vee x_1) x_2 \overline{x}_3$
$\quad \vee x_4 x_5 (\overline{x}_1 \vee 1)$
$= \overline{x}_1 \overline{x}_2 x_3 \vee x_2 \overline{x}_3 \vee x_4 x_5$

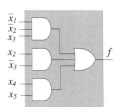

STEP 1

(a) Assign the partitions

$$f = \{(x_1), (x_2), (x_3), (x_4), (x_5)\}$$

(b) Consider the first product of the minimal SOP expression, $\overline{x}_1 \overline{x}_2 x_3$ Form the new partition

$$f = \{(x_1, x_2), (x_3), (x_4), (x_5)\}$$

(c) Consider the second product of the minimal SOP expression, $x_2 \overline{x}_3$ and form another partition

$$f = \{(x_1, x_2, x_3), (x_4), (x_5)\}$$

(d) Consider the third product of the minimal SOP expression, $x_4 x_5$ and form the last partition

$$f = \{(x_1, x_2, x_3), (x_4, x_5)\}$$

STEP 2

Necessity condition
check

Partition $\mathbf{X}_1 = (x_1, x_2, x_3)$

Partition $\mathbf{X}_2 = (x_4, x_5)$		000	001	010	011	100	101	110	111
	00	1	1					1	
	01	1	1					1	
	10	1	1					1	
	11	1	1	1	1	1	1	1	1

STEP 3
OR bi-decomposition

$f = (\overline{x}_1 \overline{x}_2 x_3 \vee x_2 \overline{x}_3) \vee x_4 x_5$

$\qquad \underbrace{\qquad}_{\mathbf{X}_1} \qquad \underbrace{\qquad}_{\mathbf{X}_2}$

$= g_1 \underbrace{(x_1, x_2, x_3)}_{} \vee g_2 \underbrace{(x_4, x_5)}_{}$

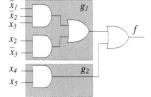

FIGURE 12.40

OR bi-decomposition for a Boolean function of five variables.

disjoint bi-decomposition with respect to this partition. Let $f = x_1 x_2 \overline{x}_4 \vee x_1 x_2 x_5 \vee \overline{x}_3 \overline{x}_4 \vee \overline{x}_3 x_5$. The decomposition chart has a column and row multiplicity of two. However, this function does not have an OR type disjoint bi-decomposition. This can be proved using the algorithm given above: the products $x_1 x_2 \overline{x}_4$ and $\overline{x}_3 \overline{x}_4$ shows that x_1, x_2, x_3, x_4 must be included in the same set, and the products $x_1 x_2 x_5$ and $\overline{x}_3 x_5$ shows that x_1, x_2, x_3, x_5 must be included in the same block, so all variables must be included in one block.

However, the complement of the function f has OR-type disjoint decomposition with respect to the partition $f = (\{x_1, x_2, x_3\}, \{x_4, x_5\})$ (Figure 12.41). Therefore, the function f has AND-type disjoint bi-decomposition. To find it, let us convert the SOP form of f into the POS form:

$$f = x_1 x_2 (\overline{x}_4 \vee x_5) \vee \overline{x}_3 (\overline{x}_4 \vee x_5) = (x_1 x_2 \vee \overline{x}_3)(\overline{x}_4 \vee x_5)$$

This form clearly corresponds to the AND-type decomposition $f = \underbrace{g_1 (x_1, x_2, x_3)}_{\mathbf{X}_1} \wedge \underbrace{g_2 (x_4, x_5)}_{\mathbf{X}_2}$.

12.12.6   Functional decomposition using decision diagrams

Decision diagrams are used in decomposition techniques as follows:

▶ The results of functional decomposition are converted into decision diagrams; decision diagrams are used as a data structure for implementation; they are derived from the decomposition chart; and/or

▶ An initial decision diagram is decomposed using the specific rules of partitioning decision diagrams; in this approach, the decision diagram structure is used in all steps of the decomposition.

Table 12.9 shows two types of decomposition (out of many). Shannon decomposition is used to implement a network based on multiplexers. Davio decomposition is used in the form of the *positive Davio* (*pD*) expansion $f = f_0 \oplus x_i f_2$ and *negative Davio* (*nD*) expansion $f = f_1 \oplus \overline{x}_i f_2$, where $f_0 = f_{x_i=0}$, $f_1 = f_{x_i=1}$, and $f_2 = f_{x_i=1} \oplus f_{x_i=0}$; it is useful in the case of gate level implementation using polynomial representations of Boolean functions.

12.12.7   Design example: Shannon decomposition of Boolean function with respect to a subfunction

The Shannon decomposition of the Boolean function $f = x_1 \overline{x}_2 x_3 \vee \overline{x}_1 x_2 x_3 \vee \overline{x}_1 \overline{x}_2 x_4 \vee x_1 x_2 x_4$ with respect to the subfunction $g = x_1 \oplus x_2$ can be found as shown in Figure 12.42. It also shows the reduced ordered decision diagram, and decomposed decision diagram using multiplexers. The inputs of the multiplexer are the functions x_3 and x_4, and the control signal is generated by the function g, which is implemented as a decision diagram itself.

Design example: **AND type bi-decomposition**

INPUT DATA

$$f = x_1 x_2 \overline{x}_4 \vee x_1 x_2 x_5 \vee \overline{x}_3 \overline{x}_4 \vee \overline{x}_3 x_5$$

Find the complement of the function f:

PRELIMINARY STEP

$$\overline{f} = \overline{x_1 x_2 \overline{x}_4 \vee x_1 x_2 x_5 \vee \overline{x}_3 \overline{x}_4 \vee \overline{x}_3 x_5}$$
$$= \overline{x_1 x_2 (\overline{x}_4 \vee x_5) \vee \overline{x}_3 (\overline{x}_4 \vee x_5)}$$
$$= \overline{(x_1 x_2 \vee \overline{x}_3)(\overline{x}_4 \vee x_5)}$$
$$= \overline{x_1 x_2 \vee \overline{x}_3} \vee \overline{\overline{x}_4 \vee x_5}$$
$$= \overline{x_1 x_2} x_3 \vee x_4 \overline{x}_5$$
$$= (\overline{x}_1 \vee \overline{x}_2) x_3 \vee x_4 \overline{x}_5$$
$$= \overline{x}_1 x_3 \vee \overline{x}_2 x_3 \vee x_4 \overline{x}_5$$

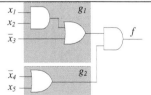

STEP 1

(a) Assign the partition $f = \{(x_1),(x_2),(x_3),(x_4),(x_5)\}$

(b) Consider the first product of the SOP expression of $\overline{f}$, $\overline{x}_1 x_3$, and form the new partition $f = \{(x_1, x_3),(x_2),(x_4),(x_5)\}$

(c) Consider the second product of the SOP expression, $\overline{x}_2 x_3$, and form another partition $f = \{(x_1, x_2, x_3),(x_4),(x_5)\}$

(d) Consider the third product of the SOP expression, $x_4 \overline{x}_5$, and form the last partition $f = \{(x_1, x_2, x_3),(x_4, x_5)\}$

Decision: Since $\overline{f}$ has OR-type decomposition, then f has AND-type decomposition

STEP 2

Necessity condition check: decomposition chart

Partition $X_2 = (x_4, x_5)$

Partition $X_1 = (x_1, x_2, x_3)$

	000	001	010	011	100	101	110	111
00	1		1		1		1	1
01	1		1		1		1	1
10								
11	1		1		1		1	1

Row and column multiplicity are equal to two

STEP 3
Form the POS form

$$f = x_1 x_2 (\overline{x}_4 \vee x_5) \vee \overline{x}_3 (\overline{x}_4 \vee x_5)$$
$$= (x_1 x_2 \vee \overline{x}_3)(\overline{x}_4 \vee x_5)$$
$$= (x_1 \vee \overline{x}_3)(x_2 \vee \overline{x}_3)(\overline{x}_4 \vee x_5)$$

STEP 4
AND-type bi-decomposition

$$\overline{f} = g_1 \overbrace{(x_1, x_2, x_3)}^{\mathbf{X}_1} \wedge g_2 \overbrace{(x_4, x_5)}^{\mathbf{X}_2}$$

FIGURE 12.41
AND bi-decomposition for a Boolean function of five variables.

Design example: Shannon decomposition

INPUT DATA

$$f = x_1 \overline{x}_2 x_3 \vee \overline{x}_1 x_2 x_3$$
$$\vee \ \overline{x}_1 \overline{x}_2 x_4 \vee x_1 x_2 x_4$$

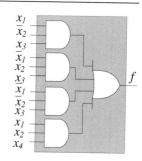

STEP 1

The Shannon decomposition of the Boolean function f with respect to the subfunction $g = x_1 \oplus x_2$ can be found via factoring:

$$f = x_3(x_1 \overline{x}_2 \vee \overline{x}_1 x_2) \vee x_4(\overline{x}_1 \overline{x}_2 \vee x_1 x_2)$$

Since

$$x_1 \overline{x}_2 \vee \overline{x}_1 x_2 = \overline{\overline{x}_1 \overline{x}_2 \vee x_1 x_2} x_1 \oplus x_2 = g,$$

then $f = x_3 \overline{g} \vee x_4 g$.

STEP 2
Logic network
implementation

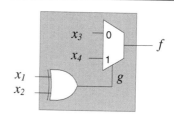

STEP 3
Decision diagram
implementation

Initial decision diagram

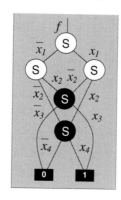

Decomposed decision diagram

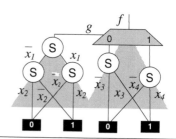

FIGURE 12.42

Shannon decomposition for a Boolean function of four variables.

TABLE 12.9
Analysis of decomposition techniques

Type	Definition	Implementation
Shannon decomposition	$f_1(X_0) = f(X_0, x_i = 0)$ $f_2(X_0) = f(X_0, x_i = 1)$ $f(f_1, f_2, x_i) = \overline{x}_i f_1 \vee x_i f_2$	
Positive Davio pD and negative Davio nD decomposition	$f_1(X_0) = f(X_0, x_i = 0)$ $f_2(X_0) = f_{x_i=0} \oplus f_{x_i=1}$ pD node: $f(f_1, f_2, x_i) = f_1 \oplus x_i f_2$ nD node: $f(f_1, f_2, x_i) = f_1 \oplus \overline{x}_i f_2$	

12.13 Further study

Advanced topics of computational networks

Topic 1: Threshold elements. A threshold function f of n variables $x_1, \ldots, x_n$ is a Boolean function that can be represented in the form: $f = 1$ iff $\sum_{i=1}^{n} w_i x_i - \theta \geq 0$, where the summation is arithmetic rather than Boolean; w_i are *weights* (which may be assumed to be positive or negative integers without loss of generality); and the integer θ is the *threshold*. While not all Boolean functions are threshold functions, any Boolean function can be composed of threshold functions. This composition corresponds to network interconnections of two or more threshold elements, each of which realizes a particular threshold function.

Topic 2: Cellular logic is a direction of logic design that deals with the analysis and synthesis of logic networks in the form of cellular arrays. A *cellular array* consists of a 1-, 2-, or 3-dimensional identical cells with uniform interconnection. Study of cellular arrays took on a new importance with the advent of nanotechnology. In this technology, a cell may contain a large number of logic elements. From manufacturing points of view, it is advantageous to have the "nanochip" produced in the form of identical cells. In addition, packing density is higher not only because of the reduction of size of the cells but also because of elimination of much of the interconnection wiring between the nanochips. The reliability of devices based on cellular arrays is higher compared with logic networks using other approaches. The most distinctive feature of cellular arrays is their flexibility in performance.

There are two main approaches to synthesis in logic design: (*a*) those based on formal *exact* algorithms, and (*b*) those based on informal methods, also called *heuristic* algorithms; most of them provide quasi-optimal solutions.

Topic 3: Multilevel logic network design. Two-level logic networks based on SOP and POS expressions are sometimes impractical. Logic networks that have more than two levels often have fewer gates and meet lower fan-in and fan-out limits. The minimization of two-level networks is based on the minimization of SOP expressions. No practical and exact minimization techniques are known for general multilevel networks. The design of multilevel networks is more complex than that of two-level ones. However, many *heuristic* techniques are used in practice for multilevel logic network optimization. Heuristic minimization techniques can provide minimal or at least near-minimal designs. Techniques for multilevel network design are based on the following methods: (a) decomposition, (b) factoring, and (c) local transformations.

An extension of these techniques is based on the notation functional flexibility. *Functional flexibility* is defined as the condition in which an alternative Boolean function can replace a function at a certain point in a logic network. There are many optimization methods in advanced logic design that use functional flexibility. One of them is called *set of pairs of functions to be distinguished (SPFD)* (see the paper: S. Yamashita, H. Sawada, and A. Nagoya, "SPFD: A Method to Express Functional Flexibility" *IEEE Transactions on Computer-Aided Design of Integrated Circuits and Systems*, volume 19, number 8, pages 840–849, 2000). The SPFD approach can be well understood in terms of Shannon information, that is, as the information content of the Boolean function, since it indicates what information contributes to the network performance.

Topic 4: Heuristic optimization techniques. An optimization problem is defined as a problem whose solution can be measured in terms of a cost (or objective) function. It is usually multi-iterative procedures based on heuristic approach.

Optimization methods are distinguished with respect to the type of data structure:

▶ Minimization algorithms for Boolean functions based on algebraic manipulations.

▶ Optimization algorithms for multilevel logic networks based on graphical-based network models.

▶ Optimization techniques for decision diagrams. The size of a decision diagram depends on the order of the variables. Optimization techniques are based on various strategies, in particular, (a) optimization of the order using the specific properties of Boolean function, and (b) dynamic reordering.

That is, the algorithms for optimization of Boolean functions in cube form and in the form of decision diagrams are different. Also, different techniques are required for the optimization of logic networks of logic gates, and networks of multiplexers and threshold elements.

Exact minimization of Boolean functions is considered a classic problem of logic design; it was addressed first by W. Quine (1952) and E. McCluskey (1956). The Quine-McCluskey algorithm is the exact minimization algorithm. The major problem with this algorithm is that all prime implicants for a

Boolean function have to be computed. This becomes computationally very expensive for a function with a large number of inputs. The Quine–McCluskey algorithm often fails to simplify medium-size SOP expressions.

The exact algorithms are often unsuitable for the minimization of Boolean functions and optimizations of logic networks. Heuristic algorithms give acceptable solutions and sometimes even optimal ones in a short time for large-size problems, but there is no guarantee that they will do so in any particular case.

Heuristic minimization of Boolean functions is motivated by the need to reduce the size of two-level forms (SOP and POS expressions). For example, the ESPRESSO package developed at IBM and the University of California at Berkeley utilizes various heuristics for avoiding the cost of generating all prime implicants. ESPRESSO can be viewed as applying a set of the following operators to minimize a Boolean function: EXPAND, REDUCE, and IRREDUNDANT. The EXPAND operator aims to maximize the size of cubes in a *cover*. A cover of a Boolean function is a set of cubes, none of which is contained by any cube in the OFF-set of the function. The ON-set of a Boolean function is defined as a set of cubes, and each cube in the set produces a logic 1 for the function; cubes that are not in the ON-set belong to the OFF-set or the DC-set (don't-care set). The bigger a cube, the more minterms it covers, thereby making them redundant. The REDUCE operator decreases the size of each cube in the ON-set of a Boolean function. The IRREDUNDANT operator removes redundant implicants from the cover of a Boolean function.

ESPRESSO performs the minimization of a Boolean function specified in terms of its ON-set, OFF-set, and DC-set. The implicants in the ON-set represent the initial (nonminimum) cover of the Boolean function. By applying EXPAND, REDUCE, and IRREDUNDANT operators, ESPRESSO finds the near-minimum cover of the function.

Heuristic optimization methods can provide optimal or at least near-optimal designs for transferring two-level logic networks into multilevel networks. Two-level logic networks are sometimes impractical, usually because they require too many gates even after minimization. This is because they often fail to meet the fan-in and fan-out constraints of the implementation technology. Many-level logic networks that are obtained using heuristic algorithms often contain fewer gates than any minimal two-level design for the same function. No practical and exact optimization techniques are known for general multilevel logic networks. Heuristic techniques are usually used to solve the problem of multilevel logic network design.

Topic 5: 3D computing structures. An embedding of a *guest* graph G into a *host* graph H is a one-to-one mapping $\varphi\colon V(G) \to V(H)$, along with a mapping α that maps an edge $(u,v) \in E(G)$ to a path between $\varphi(u)$ and $\varphi(v)$ in H. A guest structure is a *technology-independent* data structure with ability for computing. A host topology is *technology-dependent* data structure. Hypercube-like topology and particular $\mathcal{N}$-hypercube topology is of particular interest in embedding techniques because of a number of useful properties. $\mathcal{N}$-hypercube is shown in Figure 12.43.

Decomposition based on splitting variable is the base of contemporary decision diagram techniques. Shannon decomposition is a restricted type

TABLE 12.10
Analysis of decomposition techniques

Type	Implementation
Decomposition with respect to a single variable	

Shannon decomposition

$$f_1(X_0) = f(X_0, x_i = 0)$$
$$f_2(X_0) = f(X_0, x_i = 1)$$
$$f(f_1, f_2, x_i) = \overline{x}_i f_1 \vee x_i f_2$$

Positive Davio pD **and negative Davio** nD **decomposition**

$$f_1(X_0) = f(X_0, x_i = 0)$$
$$f_2(X_0) = f_{x_i=0} \oplus f_{x_i=1}$$
$$pD: \quad f(f_1, f_2, x_i) = f_1 \oplus x_i f_2$$
$$pD: \quad f(f_1, f_2, x_i) = f_1 \oplus \overline{x}_i f_2$$

Positive arithmetic pD_A **and negative arithmetic** nD_A **decomposition**

$$f_1(X_0) = f(X_0, x_i = 0)$$
$$f_2(X_0) = f_{x_i=0} - f_{x_i=1}$$
$$pD_A: \quad f(f_1, f_2, x_i) = f_1 + x_i f_2$$
$$pD_A: \quad f(f_1, f_2, x_i) = f_1 + \overline{x}_i f_2$$

| Decomposition with respect to a set variables | |

Ashenhurst decomposition

$$f(X) = f(X_1, X_0)$$
$$= f(f_1(X_1), X_0)$$

Curtis decomposition

$$f(X) = f(X_1, X_0)$$
$$= f(f_1(X_1), ..., f_k(X_1), X_0)$$

AND-bi-
OR-bi-
EXOR-bi- decomposition

$$f(X) = f(X_a, X_b, X_c)$$
$$= f(g(X_a, X_c), h(X_b, X_c))$$

Weak-bi-decomposition

$$f(X_a, X_c) = f(g(X_a, X_c), h(X_c))$$

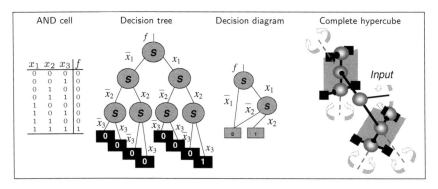

FIGURE 12.43
An AND function of three variables given by the truth table, corresponding decision tree , and hypercube representation.

of decomposition $f(X) = f(f_1(X_0), f_2(X_0), x_i)$ (Table 12.10). It consists of three decomposition components: two of them depend on the variables $X_0 = X \backslash x_i$, and the third one is equal to x_i. Davio decomposition is used in the form of the *positive Davio (pD)* expansion $f = f_0 \oplus x_i f_2$ and *negative Davio (nD)* expansion $f = f_1 \oplus \overline{x}_i f_2$, where $f_0 = f_{x_i=0}$, $f_1 = f_{x_i=1}$, and $f_2 = f_{x_i=1} \oplus f_{x_i=0}$.

Further reading

A survey of advanced logic design can be found in the special issue "Electronic Design Automation at the Turn of Century," *IEEE Transactions on Computer-Aided Design of Integrated Circuits and Systems*, volume 19, number 12, 2000.

[1] Ashenden PJ. *Digital Design. An Embedded Systems Approach Using VHDL.* Elsevier, Burlington, MA, 2008.

[2] Brayton RK. The future of logic synthesis and verification. In Hassoun S and Sasao T, Eds., Brayton RK., Consulting Ed., *Logic Synthesis and Verification*, Kluwer, Dordrecht, 2002.

[3] DeMicheli G. *Synthesis and Optimization of Digital Circuits.* McGraw Hill, New York, 1994.

[4] Fabricius ED. Modern Digital Design and Switching Theory. CRC Press, Boca Raton, FL, 1992.

[5] Fujita M, Matsunaga Y, and Ciesielski M. Multi-level logic optimization. In Hassoun S and Sasao T, Eds., Brayton RK, Consulting Ed., *Logic Synthesis and Verification*, Kluwer, Dordrecht, 2002.

[6] Givone DD. *Digital Principles and Design.* McGraw-Hill, New York, 2003.

[7] Hachtel GD and Somenzi F. *Logic Synthesis and Verification Algorithms*, Kluwer, Dordrecht, 1996.

[8] Hayes JP. *Introduction to Digital Logic Design.* Addison-Wesley, Reading, MA, 1993.

[9] Mano MM and Kime C. *Logic and Computer Design Fundamentals.* 3rd edition. Prentice Hall, Upper Saddle River, NJ, 2005.

[10] Meinel C and Theobald T. *Algorithms and Data Structures in VLSI Design.* Springer, Heidelberg, 1998.

[11] Sasao T and Butler JT. On bi-decomposition of logic functions. In *Proc. ACM/IEEE Int. Workshop on Logic Synthesis*, Tahoe City, California, pp. 85–102, May, 1997.

[12] Sandige RS. *Digital Design Essentials.* Prentice Hall, Upper Saddle River, NJ, 2002.

[13] Sasao T. *Switching Theory for Logic Synthesis*, Kluwer, Dordrecht, 1999.

[14] Siu KY, Roychowdhury VP, and Kailath T. Depth-size tradeoffs for neural computation. *IEEE Trans. Comput.*, 40(12):1402–1411, 1991.

[15] Unger SH. *The Essence of Logic Circuits.* Prentice Hall, Englewood Cliffs, NJ, 1989.

[16] Yanushkevich SN, Shmerko VP, and Steinbach B. Spatial interconnect analysis for predictable nanotecnologies. *J. Computational and Theoretical Nanoscience*, 5(1):56–69, 2008.

13

Sequential Logic Networks

13.1 Introduction

The logic networks can be classified as combinational or sequential. Sequential logic networks consist of two different types of logic elements: the elements that operate on data values, and the elements that contain a *state*. Given the same input, a combinational element always produces the same output. An element contains a state if it has some internal storage. These elements are called *state elements*. In a state element, the required inputs are the data values to be written into the element, and the clock, which determines when the data value is written. The output from a state element provides the value that was written in an earlier clock cycle.

Sequential networks have the property that the output depends not only on the present input but also on the past sequence of inputs. A sequential network contains combinational networks and cannot be described completely by a truth table. In sequential networks, two different types of description are used: *state* description and *time behavior*.

Sequential networks are designed using libraries of sequential and combinational standard modules. In this chapter, the standard modules of sequential networks are introduced. Standard modules that contain state are also called *sequential modules* because their outputs depend on both their inputs and the contents of the internal state. The storage properties of these modules are provided by physical phenomena. These phenomena are well studied, and widely used in modern logic design, although various new chemical and physical phenomena are being actively investigated as technology is progresses (details are given in the "Further reading" section).

13.2 Physical phenomena and data storage

Data storage can be implemented using various physical principles. The operating principle of storage elements may be based upon various physical effects and phenomena. Mechanical, electrical, magnetic, optic, acoustic, molecular, and atomic effects are utilized in storage elements.

Physical memories often store information in the form of two energy states, allowing storage and retrieval to be accomplished by a transfer of energy. In predictable technology of the future, these storage properties are studied in nanospace, that is, at the molecular and atomic levels. These phenomena can be abstracted to obtain a technology-independent model of a primitive storage element. This element can be defined as a storage device that has two configurations, or states: storing 0 and storing 1.

A binary storage element requires two stable states to represent 0 and 1, and a mechanism for writing and rewriting this information (Figure 13.1). Their storage capacity is one binary digit, or one bit. Devices with more than two states can be treated as a combination of two-state devices. In this way the "bit" may be used as a general measure for storage capacity and information content.

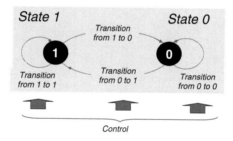

FIGURE 13.1

A binary storage element requires two stable states to represent 0 and 1, and mechanism for writing and rewriting this information.

13.3   Basic principles

Memory elements have the capability of storing the value of a binary quantity Q: state $Q = 0$, the memory element stores the value 0, and state $Q = 1$, the memory element stores the value 1. The variable Q is called the *state variable*. State $Q = 0$ and state $Q = 1$ are required to be *stable*. A change in Q requires some control in the form of additional input signals. Changing from the stable state $Q = 0$ to the state $Q = 1$ and vice versa may not be direct; that is, a memory element may pass through one or more unstable (transient) states. An element that has precisely two states is called a *bistable* element, or a bistable network.

13.3.1 Feedback

Combinational logic networks have no *feedback*; that is, no copy of an output signal goes back to the input part of the network. In simple cases, networks with feedback can be analyzed by tracing signals through the network.

Example 13.1 *Consider the simplest case of an inverter with feedback (Figure 13.2). **Step 1:** If, at some instance in time, the inverter input is 0, this value is propagated through the inverter and causes the output to become 1, after some delay at the inverter. **Step 2:** The logic value 1 is fed back into the input, and after the propagation delay, the inverter output will become 0. **Step 3:** When logic value 0 feeds back into the input, the output will again switch to 1. The inverter output will continue to oscillate back and forth between 0 and 1; that is, it never reaches a stable state.*

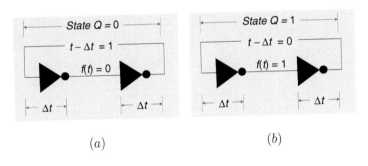

(a)   (b)

FIGURE 13.2
An ideal model of a bistable memory element with state 0 (a) and state 1 (b) (Example 13.1).

Cascading of two inverters in a row forms a network with two stable states called the *bistable* network. An ideal model of a bistable logic network is illustrated in Figure 13.2. It consists of two cascaded inverters with a closed loop. The states $Q = 0$ and $Q = 1$ are associated with $f(t) = 0$ and $f(t) = 1$, respectively. In this model, there is no control mechanism for changing the state from $Q = 0$ to $Q = 1$, and vice versa.

Example 13.2   *Consider a feedback loop with two inverters in it (Figure 13.2).* **Stable state A:** *If the input to the first inverter is 0, its output will be logical 1. The input to the second inverter will be 1, and its output will be 0. This 0 will feed back into the first inverter. No changes will occur, because this input is already 0. This is a stable state.* **Stable state B:** *The second stable state of the network occurs when the input to the first inverter is 1, and the input to the second inverter is 0.*

As follows from Example 13.2, in the cascade of two inverters there is a mechanism to get it to change from whichever initial condition it started from. No finite delay can cause it to oscillate. This feedback system is locked up at one of its two states. Extra logic is needed to set this network to a specific value; that is, the feedback path must be broken while a new value is connected to the input.

Cascaded inverters with an odd number of inverters are called *ring oscillators*. They provide useful functions, such as the generation of sequences of logical 0s and 1s. This is because of oscillating behavior of this network. The signals that a ring oscillator generates are repeated every period. The odd number of inverters leads to the oscillatory behavior that repeats every $t_P = \sum_i \Delta t_i$ time units (an inverter delay is of 1 time unit). The time t_P is called the *period*. The duration of the period depends on the number of inverters in the chain.

Example 13.3   *In Figure 13.3, five cascaded inverters generate sequences of logical 0s and 1s with a period of $5\Delta t$ time units.*

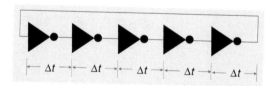

FIGURE 13.3
Cascaded inverters for generating sequences of logical 0s and 1s (Example 13.3).

13.3.2 Clocking techniques

Clocking techniques defines when signals can be read and when they can be written. It is important to specify the timing of reads and writes because, if a signal is written at the same time it is read, the value of the read could correspond to the old value, the newly written value, or even some mix of the two.

In this chapter, we assume an *edge-triggered* clocking technique. Ideally, the sides of clock pulses rise and fall in zero time. In practice, the sides rise and fall in nonzero time, and the slopes of the sides do not change from zero to nonzero (or vice versa) in zero time. One period of the clock waveform includes an interval of time when the clock pulse is 1 ("high" in positive logic) and another interval of time when it is 0 ("low" in positive logic).

An edge-triggered clocking technique means that any values stored in the network are updated only on a clock edge. Thus, the state elements all update their internal storage on the clock edge. Because only state elements can store a data value, any collection of combinational logic must have its inputs coming from a set of state elements and its outputs written into a set of state elements. The inputs are values that were written in a previous clock cycle, while the outputs are values that can be used in a following clock cycle.

13.4 Data structures for sequential logic networks

The operation of a sequential network is characterized by a sequential and continuous changing of its states, which are specified by contents of the network's registers, or memory storage. This behavior is described by the formal model called *finite state machine*. This model is represented as follows:

▶ A sequence of *state transitions* controlled by the network's inputs.

▶ *State table*, which defines the network's next-state functions.

▶ *State diagram*, which represents all internal states and relationships between the states.

▶ *Characteristic equations* that describe the behavior of the flip-flops in the network.

13.4.1 Characteristic equations

The functional behavior of a state machine can be described by *characteristic equations*, that specify their next states as functions of their current states and inputs.

Example 13.4   *Given a network with the input D and output Q, characteristic equation $Q^+ = D$ means that the next state of the output Q, denoted by Q^+, will be equal to the value of the input D in the present state.*

13.4.2   State tables and diagrams

The number of input/output sequences in a sequential network is infinite. However, the number of combinations of the primary input and internal state values is finite, hence, "finite state machine." This fact is utilized by the state table, which contains all input/output sequences in implicit form. The *state* of a logic network at time t is defined as the current logic values of some set of signals of interest. Two types of states are distinguished: the *internal* state, and the *total* state. The internal state represents what the network remembers from its behavior prior to time t. The total state completely determines the next action to be taken by the network.

A *state table* specifies the next state and the outputs of a sequential network in terms of its present state and inputs.

A *state diagram* is a graphical representation of the states (assigned to the nodes of the graph) and the transition between the states (denoted by the directed edges of the graph). It is a directed graph that represents all internal states and possible state transitions.

State tables and diagrams defined at the same level of abstraction contain exactly the same information about the sequential network's behavior. Given a state table, one can construct acorresponding state diagram, and vice versa.

Example 13.5   *Figure 13.4 shows the **state table** and the **state diagram** for the simplest sequential module.*

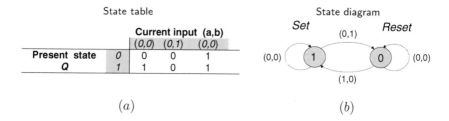

(a)   *(b)*

FIGURE 13.4

Excerpts from the state table (a) and the state diagram (b) (Example 13.5).

13.5 Latches

Latches and flip-flops are the basic building blocks of sequential logic networks. Latches provide a mechanism for the simplest control of the storage of one bit:

$$\underbrace{\text{1 bit memory}}_{Physical\ phenomenon} \quad \overset{Design}{\longrightarrow} \quad \underbrace{\text{1 bit memory control}}_{Latch}$$

There are several types of latches, distinguished by their method of controlling their functions. The latch is specified by a characteristic table and equation, a state table and state diagram, and a timing diagram.

13.5.1 SR latch

The *SR latch* is the basic memory element, defined as follows:

SR latch

▶ Is a bistable memory network.
▶ Has two inputs labeled *set* S and *reset* R.
▶ Has two outputs labeled Q and $\overline{Q}$.

In an SR latch based on NAND gates, the NAND gates are considered to be "cross-coupled," with each NAND feeding back its output to the other NOR gate.

Example 13.6 *An **SR latch** with two cross-coupled NAND gates is shown in Figure 13.5. The condition that is undefined for this latch is when both inputs are equal to 0 at the same time. This input combination must be avoided.*

The SR latch operates as follows:

(a) It has two stable states defined as $Q = 0$, which is called the *reset state*, and $Q = 1$, which is called the *set state*.

(b) Latch operation is defined by the following input combinations:

$(S, R) = (0, 0)$ leaves the latch in either of its stable states indefinitely. This state may be either $(Q, \overline{Q}) = (0, 1)$ or $(Q, \overline{Q}) = (1, 0)$ which is indicated in the truth table by stating that the $Q, \overline{Q}$ outputs have values 0/1 and 1/0, respectively.

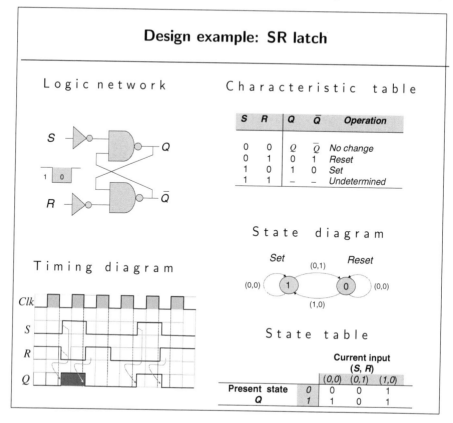

FIGURE 13.5

The SR latch using NAND gates, timing diagram (failing-edge clock-controlled), characteristic table, state diagram, and state table (Example 13.6).

$(S, R) = (1, 0)$ sets the latch by changing its state from $Q = 0$ to $Q = 1$. If the initial state is already $Q = 1$, setting the latch has no effect.

$(S, R) = (0, 1)$ resets the latch by changing its state from $Q = 1$ to $Q = 0$. If the initial state is already $Q = 0$, setting the latch has no effect.

$(S, R) = (1, 1)$ must be avoided by making sure that 1s are not applied to both inputs simultaneously. This is because an SR latch has unpredictable behavior under this operating condition.

The outputs Q and $\overline{Q}$ of a latch are complementary only after the state has stabilized, because unstable states can appear only temporarily during state transition.

13.5.2 Gated SR latch

The operations of the SR latch can be modified by providing an additional control input. This input determines when the state of the latch can be changed. This latch is called a *gated SR latch*. A gated SR latch is defined as follows:

Gated SR latch

▶ Is a bistable memory network.
▶ Includes two NAND gates and, in addition, has two other gates controlled by S, R, and clock Clk signals.
▶ The control input Clk acts as an enable signal for the S and R inputs.
▶ The output of the NAND gates stays at logical 1 as long as the control input remains at 0.
▶ When the control input goes to 1, data from the S or R input is allowed to affect the SR latch.

Note that an indeterminate condition occurs when all three inputs are equal to logical 1.

Example 13.7 *The **gated SR latch** with NAND gates and its data structures is given in Figure 13.6.*

13.5.3 D latch

Assume that the inputs S and R of the SR latch are fed into one data input D (Figure 13.7). A D latch is defined as follows:

D latch

▶ Is a bistable memory network with a single input.
▶ Is based on the gated SR latch with connected inputs.
▶ Has the data input, D, and control signal, Clk.
▶ Has two outputs, Q and $\overline{Q}$.

In the D latch, it is impossible to have a troublesome situation such as in the SR latch when $S = R = 1$. This is because the output Q merely tracks the value of the input D while $Clk = 1$. As soon as Clk goes to 0, the state of the latch is "frozen" until the next time the clock signal goes to 1. Therefore, the D latch scores the value of the D input seen at the time the clock changes from 1 to 0.

Design example: gated SR latch

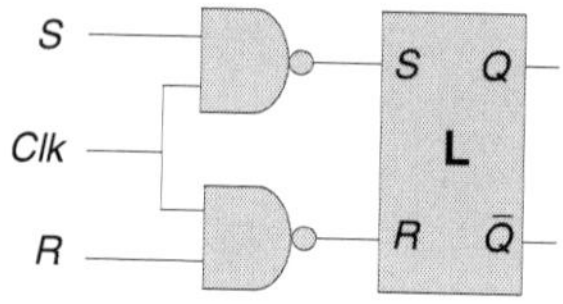

Characteristic table

Clk	S	R	Q	$\bar{Q}$	Operation
0	×	×	Q	$\bar{Q}$	No change
1	0	0	Q	$\bar{Q}$	No change
1	0	1	0	1	Reset
1	1	0	1	0	Set
1	1	1	–	–	Undetermined

State diagram

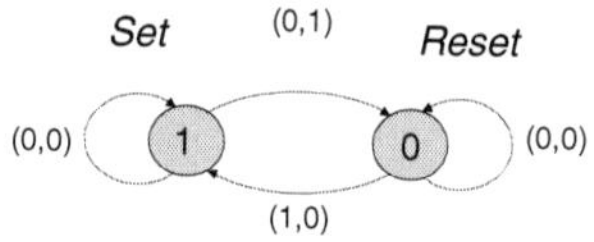

Logic network

State table

Present state	Current input (S, R)		
	(0,0)	(0,1)	(1,0)
Q 0	0	0	1
1	1	0	1

Timing diagram

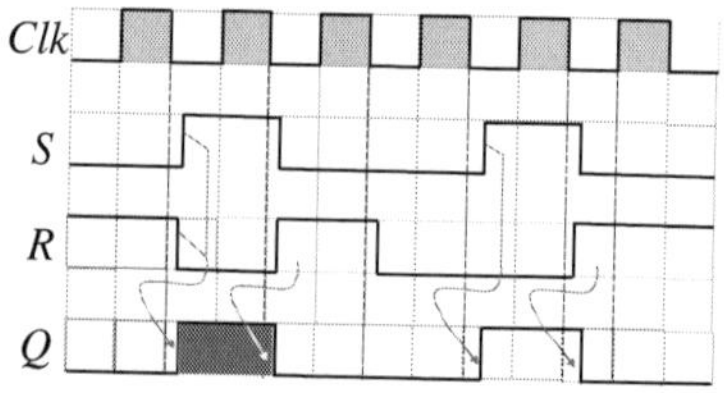

FIGURE 13.6

The SR gated latch is based on latch with additional control logic (Example 13.7).

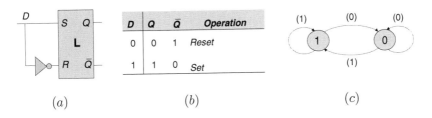

FIGURE 13.7
The D latch based on SR latch.

Example 13.8 *Figure 13.8 shows an implementation of* **D latches** *using multiplexers. For a positive-edge-triggered latch, the D input is selected when the clock signal Clk is high, and the output is held using feedback when the clock signal is low. For a negative-edge-triggered latch, input 0 of the multiplexer is selected when the clock is low, and the D input is passed to the output. When the clock signal is high, input 1 of the multiplexer is selected.*

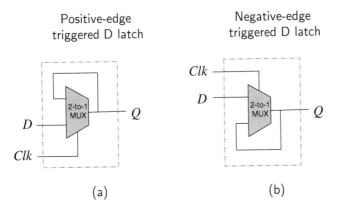

FIGURE 13.8
Multiplexer-based D latches: negative-edge-triggered (a) and positive-edge-triggered D latch (Example 13.8).

13.6   Flip-flops

The most important property of sequential logic networks is that they have global feedback. The implementation of the global feedback requires a flexible control functions; that is, the memory elements must efficiently elaborate timing logic. Latches do not satisfy this requirement. They have a common property of immediate output responses (to within the propagation delay times). This property is an undesirable property in certain applications: it is necessary that output changes occur only coincident with changes on a control input line. However, a combination of two latches called *flip-flops*, can produce the desired behavior. Flip-flops employ the clock signal to precisely control the times at which data are transferred.

A flip-flop is defined as follows:

Flip-flop

▶ Is a bistable memory device, with inputs, that remains in a given state as long as power is applied and until input signals are applied to cause its output to change.

▶ Consists of a latch, a basic bistable element, to which appropriate logic is added in order to control its state.

▶ Employs special control signals in order to specify

    (*a*) The times at which the network responds to changes to its input data signals,

    (*b*) The times at which the network changes its output data signal.

The process of storing a logic 1 into a flip-flop is called *setting* the flip-flop. The process of storing a logic 0 into flip-flop is called *resetting* or *clearing* the flip-flop.

13.6.1   The master–slave principle in flip-flop design

The property of having the timing of a flip-flop response being related to a control input signal is achieved with *master–slave* and *edge-triggered* flip-flops.

A master-slave flip-flop consists of two cascaded sections, each capable of storing either a 0 or a 1. The first section is referred to as the *master* and the second section as the *slave*. Information is entered into the master on one edge of a control signal and is transferred to the slave on the next edge of the control signal.

1 bit memory	$\xrightarrow{Design}$	Master	$\xrightarrow{Control}$	Slave
Physical phenomenon		*Latch*		*Latch*

The momentary change is called a *trigger*. The transition a trigger causes is refereed to *trigger* the flip-flop. The key to the proper operation of a flip-flop is to trigger it only during a signal transition. Two types of edge triggered flip-flops are distinguished:

► *Positive-edge-triggered* (change takes place when the clock goes from 0 to 1, that is, rising edge) flip-flops, and

► *Negative-edge-triggered* (change takes place when the clock goes from 1 to 0, that is, falling edge) flip-flops.

The positive transition is defined as the positive-edge and the negative transition as the negative-edge. In positive-edge-triggered flip-flop, the positive or rising edge of the clock to initiate the entire state transition process is used. A negative-edge-triggered flip-flop behaves like the positive analog, except that the negative or falling edge of the clock initiates the state transition process. The stored value after the transition depends on the inputs and what was stored prior to the transition.

Example 13.9 *Figure 13.9 shows a* **positive-edge-triggered** *flip-flop based on a master–slave configuration. The flip-flop is built of the cascaded negative latch (master) with a positive latch (slave). A* **negative-edge-triggered** *flip-flop can be constructed by using the positive latch first.*

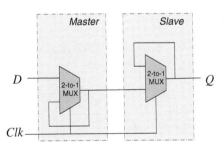

► The master is transparant on the low phase of the clock, and the D input is passed to the master output
► The slave is in the hold mode, keeping its previous value by using feedback
► The master stops sampling the input, and the slave starts sampling
► The slave samples the output of the master during the high phase of the clock while the master remains in a hold mode
► The value Q is the value of D right before the rising edge of the clock, achieving the positive-edge-triggered effect.

FIGURE 13.9

Multiplexer-based master–slave positive-edge-triggered flip-flop (Example 13.9).

13.6.2   D flip-flop

The D flip-flop is the counterpart of the D latch. The D flip-flop can be constructed from a pair of D latches connected as shown in Figure 13.10. This particular configuration of a network is known as a *master–slave* network. The D flip-flop is defined as follows:

D flip-flop

▶ Is a bistable memory network with data input and control clock input signals.
▶ Is based on D latches, one of which is called a *master* and is controlled by the clock, *Clk*, while the other is called *slave*.
▶ Is negatively or positively clocked; the first case corresponds to the inverted clock *Clk* for the slave latch, and the second case corresponds to the inverted clock $\overline{Clk}$ for the master latch.

The D flip-flop operates as follows:

(a) When the clock signal *Clk* is 1, the master latch is disabled and its output remains stable; the data output of the slave latch is stable, too.

(b) When the clock signal *Clk* changes to 0, the slave latch is disabled and its data outputs remain unchanged. The master latch is enabled and begins to respond to the input data *D*.

Example 13.10   *The behavior of both negative- and positive-edge-triggered* **D flip-flops** *is shown in Figure 13.10. The initial value Q is unknown. In particular, for the positive-edge triggered D flip-flop, when the first raising edge of the clock occurs, the state of the D flip-flop is established. Since D = 0 at this time, Q goes to 0. There is a slight delay in the output. Usually the input D changes shortly after transition. The output $\overline{Q}$ is the opposite of the output Q. At the second raising edge, D = 1, and Q = 1 for the next clock period. At the third raising edge, D = 1, and Q = 1 for another clock period. If the signal D were to go back and forth between clock transitions, the output Q would not be affected. The characteristic equations, state tables, and the state diagram are given in Figure 13.10 as well.*

13.6.3   JK flip-flop

The functioning of the JK flip-flop is identical to that of the SR flip-flop in the SET, RESET, and no change conditions of operation. The difference is

Design example: D flip-flops

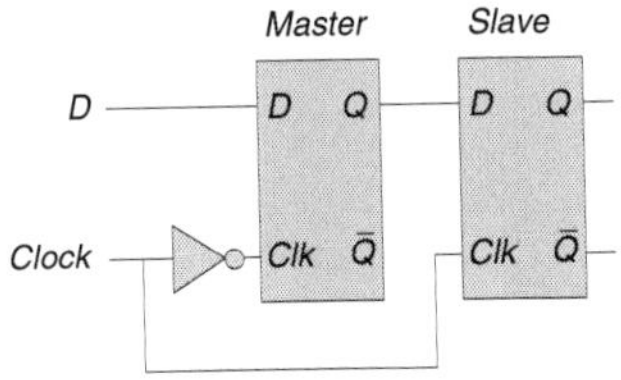

Characteristic table

Clk	D	Q	$\bar{Q}$	Operation
0	×	Q	$\bar{Q}$	No change
1	0	0	1	Reset
1	1	1	0	Set

Logic network

Negative-edge clocked D flip-flop

State diagram

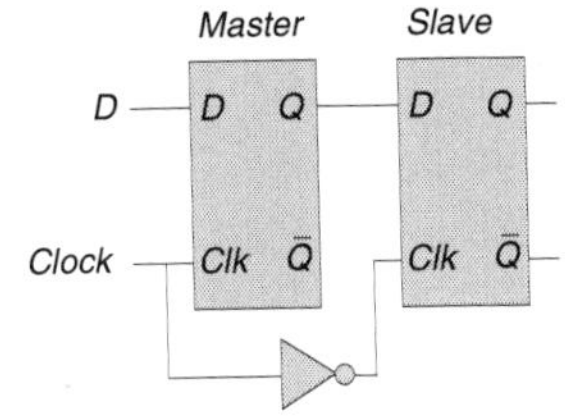

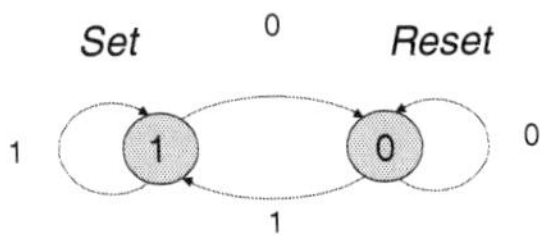

Positive-edge-clocked D flip-flop

State table

Present state		Current input D	
		(0)	(1)
Present state Q	0	0	1
	1	0	1

Timing diagram

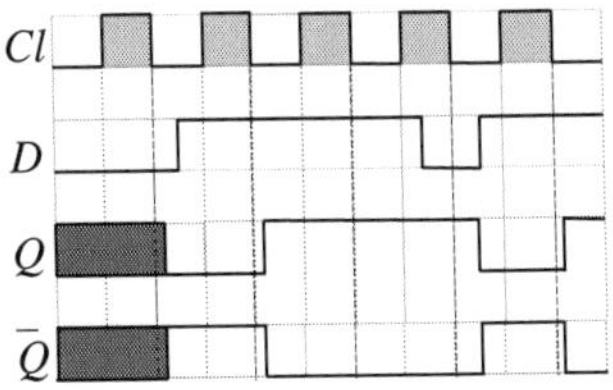

FIGURE 13.10

A D flip-flop is based on the master–slave network configuration (Example 13.10).

that the JK flip-flop has no invalid state as does the SR flip-flop. The JK flip-flop is defined as follows:

JK flip-flop

▶ Is a bistable memory network with data input and control clock input signals.
▶ Can be constructed from both of SR or D latches in a master–slave configuration, with additional logic.
▶ Is negative- or positive-edge-triggered.

The next state of each JK flip-flop is evaluated from the corresponding J and K inputs and characteristic equation $Q = J\overline{Q} \vee \overline{K}Q$. There are four cases:

$JK = 00$: no change, the next state is same as the present state, Next state = Present state.

$JK = 11$: the next state is the complement of the present state, Next state = $\overline{\text{Present state}}$.

$JK = 01$: the next state is 0.

$JK = 10$: the next state is 1.

> **Example 13.11** *Figure 13.11 shows the waveforms applied to the J, K, and clock Clk inputs of the negative-edge-triggered **JK flip-flop**. The Q output is determined, assuming that the flip-flop is initially in RESET. The Q output is determined, assuming that the flip-flop starts out RESET and the clock is active LOW.*

> **Example 13.12** *A negative-edge-triggered JK flip-flop based on SR latches is shown in Figure 13.12.*

13.6.4 T flip-flop

The T flip-flop is a complementing flip-flop which is defined as follows:

T flip-flop

▶ Is a bistable memory network with data input and control clock input signals.
▶ Toggles the output signal values on each active clock edge (negative for negative-edge-triggered flip-flop, and positive for positive-edge-triggered flip-flop) when the input $T = 1$.
▶ Can be obtained from a JK flip-flop when the inputs J and K are tied together, or using a D flip-flop and an EXOR gate.

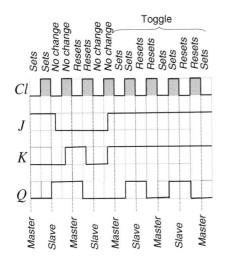

The master latch is assumed to be in the state determined by the J and K inputs beginning at the leading (positive going) edge of the clock pulse.

► The state of the master latch is transferred to the slave latch on the trailing edge of the clock pulse

► The state of the slave latch appears on the Q and $\overline{Q}$ outputs at the trailing edge.

► The feedback specifies the characteristic toggle operation when $J = K = 1$.

► The J and K inputs cannot change while the clock pulse is active because the state of the master latch can change during this time.

FIGURE 13.11

Timing diagram of the master–slave JK flip-flop (Example 13.12).

The T flip-flop is operated as follows:

► Its characteristic equation is $Q^+ = Q \oplus T$.
► When $T = 0$ ($J = K = 0$), a clock edge does not change its output.
► When $T = 1$ ($J = K = 1$), a clock edge complements its output.

Example 13.13 *A negative-edge-triggered* **T flip-flop** *is shown in Figure 13.13.*

13.7 Registers

A flip-flop stores one bit of information. A set of n flip-flops stores n bits of information; for instance, an n-bit number. Registers are classified as

► Storing registers and
► Shift registers.

13.7.1 Storing register

An n-bit storing *register* with parallel load is defined as follows (Figure 13.14):

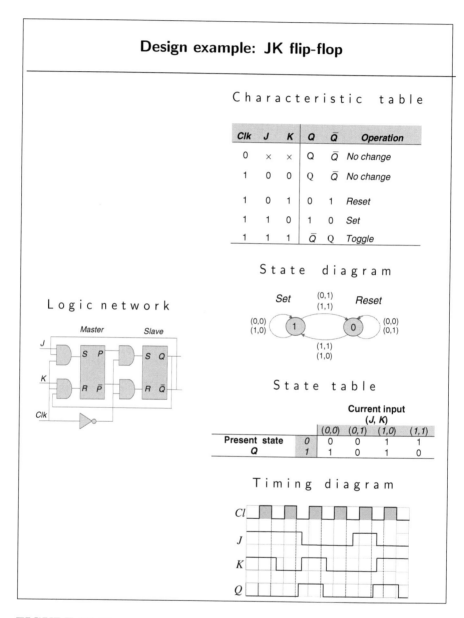

FIGURE 13.12

A negative-edge-triggered JK flip-flop based on SR latches in master–slave configuration.

Design example: **T flip-flop**

Characteristic table

Clk	T	Q	$\bar{Q}$	Operation
0	×	Q	$\bar{Q}$	No change
1	0	Q	$\bar{Q}$	No change
1	1	$\bar{Q}$	Q	Toggle

State diagram

Logic network

T flip-flop based on JK flip-flop

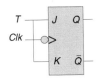

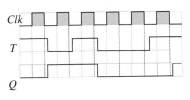

State table

		Current input T	
		(0)	(1)
Present state	0	0	1
Q	1	1	0

Timing diagram

FIGURE 13.13

A negative-edge-clocked T flip-flop based on the JK flip-flop (Example 13.13).

Storing n-bit register

▶ Is a network of flip-flops for storing an n-bit number or n-bit vector.
▶ Has

    ▶ n inputs $I = I_1, I_2, \ldots, I_n$
    ▶ n outputs $I = Z_1, Z_2, \ldots, Z_n$
    ▶ Control inputs *Load*, *Clear*, and *Clock* (*Clk*)

The control input *Clear* is asynchronous; that is, it affects the output immediately rather than when the clock signal is received. The control signal *Clear* forces the value $00 \ldots 0$ into the register. This operation is useful for initialization purposes to guarantee that the register contains a predefined value.

The cell design principle provides the possibility for design and implementation using copies.

Example 13.14 *In Figure 13.15, a 4-bit* **register** *using D flip-flops and 2-to-1 multiplexers is given. The register can be constructed using four copies of 1-bit registers.*

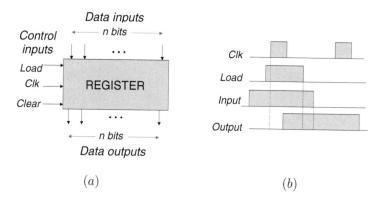

(a) (b)

FIGURE 13.14
An n-bit register module (a) and time-behavior diagram (b).

13.7.2 Shift register

A register that provides the ability to shift its contents is called a *shift register*. A n-bit shift register is capable of transferring data among adjacent flip-flops. These transfers can be *bidirectional* or *unidirectional* (either to the left or to

Design example: cell   design   principle

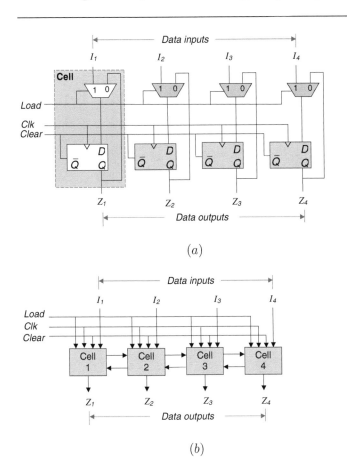

(a)

(b)

FIGURE 13.15
A 4-bit register (a) and its use for storing data in the terminal nodes of decision tree (b) (Example 13.14).

the right). An n-bit shift register is defined as follows (Figure 13.16):

Shift register

▶ n inputs $I = I_1, I_2, \ldots, I_n$
▶ n outputs $I = Z_1, Z_2, \ldots, Z_n$
▶ Control inputs *Load*, *Left*, *Right*, and *Clock* (*Clk*)

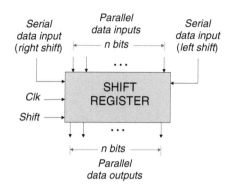

FIGURE 13.16

An n-bit parallel-in, parallel-out bidirectional shift-register module.

> **Example 13.15** *In Figure 13.15, a four-bit parallel-in, parallel-out bidirectional* **shift register** *using D flip-flops and 4-to-1 multiplexers is given.*

The functionality of shift registers can vary. For example, an n-bit sequential-in, parallel-out shift-register has one input to shift data in, and n outputs.

13.7.3   Other shift registers: FIFO and LIFO

Two special but very often used types of register with shift function are first-in-first-out (FIFO) and last-in-first-out (LIFO) registers. The FIFO register acts as a pipeline, or a queue, as the data enter the register sequentially, bit by bit, and exits sequentially, starting from the first bit entered. The LIFO register acts as a stack, since the bit pushed first into the LIFO is fetched last. FIFO and LIFO registers can form wordwide arrays. The control of such arrays is not trivial, and requires additional control circuitry.

13.8   Counters

A *counter* is defined as a sequential network with n binary outputs and $p \leq 2^n$ states. Counters are classified using *functional* (direction of count, coding of binary sequences, etc.) and *implementation* (ripple, or asynchronous counter, synchronous counter, etc.) criteria.

Design example: cell design principle

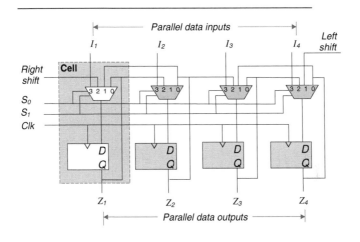

(a)

(b)

FIGURE 13.17
A 4-bit bidirectional shift-register (Example 13.15).

The following types of counters are distinguished with respect to the direction of the count:

▶ *Upward* counter; state i is followed by state $(i + 1)$ *mod* p.
▶ *Downward* counter; state i is followed by state $(i - 1)$ *mod* p.
▶ *Up/down* counters count both ways; control input determines the direction.

Counters are also classified with respect to (a) number of states and (b) type of output encoding. Using these criteria, the following counters are distinguished:

▶ *Binary* counter (2^n states, binary representation of integers $0, 1, \ldots, 2^n - 1$)

▶ *Decimal* counter (10 states, decimal code)

▶ *Gray-code* counter (2^n states, Gray code)

▶ *Ring* counter

13.8.1 Binary counters

A counter that follows the binary number sequence is called a *binary* counter. An n-bit binary counter consists of n flip-flops and can count in binary from 0 through $2^n - 1$. Binary counters can be implemented using *asynchronous* or *synchronous* design principles.

Asynchronous binary counters

An asynchronous, or *ripple* counter, is a cascade of T flip-flops such that the input of the i-th flip-flop is connected to the output of the $(i-1)$ the flip-flop, $i = 1, 2, \ldots, n$, and the input for the first flip-flop is connected to the clock signal.

> **Example 13.16** *In Figure 13.18, a 4-bit binary ripple **counter** is shown. In this counter, T flip-flops are used because a T flip-flop changes state (toggles) on every rising edge of its clock input. Thus, each bit of the counter toggles if and only if the immediately preceding bit changes from 1 to 0.*

The disadvantage of ripple counters is that the propagation delay between the input and the output of a flip-flop is summed up to maximum n times since toggle signals must propagate through the entire counter. The worst case occurs when the counter outputs change from 11...1 to 00...0.

Synchronous binary counter

In a synchronous binary counter, all of the flip-flops' clock inputs are connected to the same common clock signal and all of the flip-flops' outputs change at the same time. This allows the avoidance of the delays present in asynchronous counters. All the flip-flops in a synchronous counter change simultaneously after a single flip-flop propagation delay.

Design example: ripple counter

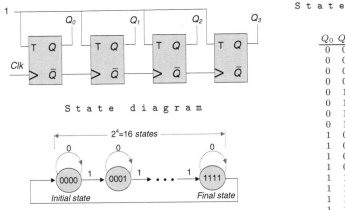

State table

Q_0	Q_1	Q_2	Q_3
0	0	0	0
0	0	0	1
0	0	1	0
0	0	1	1
0	1	0	0
0	1	0	1
0	1	1	0
0	1	1	1
1	0	0	0
1	0	0	1
1	0	1	0
1	0	1	1
1	1	0	0
1	1	0	1
1	1	1	0
1	1	1	1

FIGURE 13.18
A 4-bit binary ripple counter (Example 13.16).

Example 13.17 *Figure 13.19 shows a 4-bit synchronous binary counter. The design of the counter starts with the derivation of the state diagram, and then the excitation equations are specified for each T flip-flop.*

The binary counters are also called *modulo-2^n counters*. For example, a 4-bit binary counter is a modulo-16 counter.

Modulo-m counters

Both asynchronous and synchronous counters can be modified to count up to a certain number, $m \leq 2^n$. Such a counter is built on the n-bit binary counter, which can be redesigned to accommodate counting from 0 to $m-1$ and then wrap around.

Example 13.18 *Figure 13.20 shows a 4-bit synchronous modulo-10 counter. The counter is built on the 4-bit binary counter, with a special connection of its data load and count inputs. This connection ensures that when the counter output reaches the value $1001_2 = 9_{10}$, the counter should be reset to 0000.*

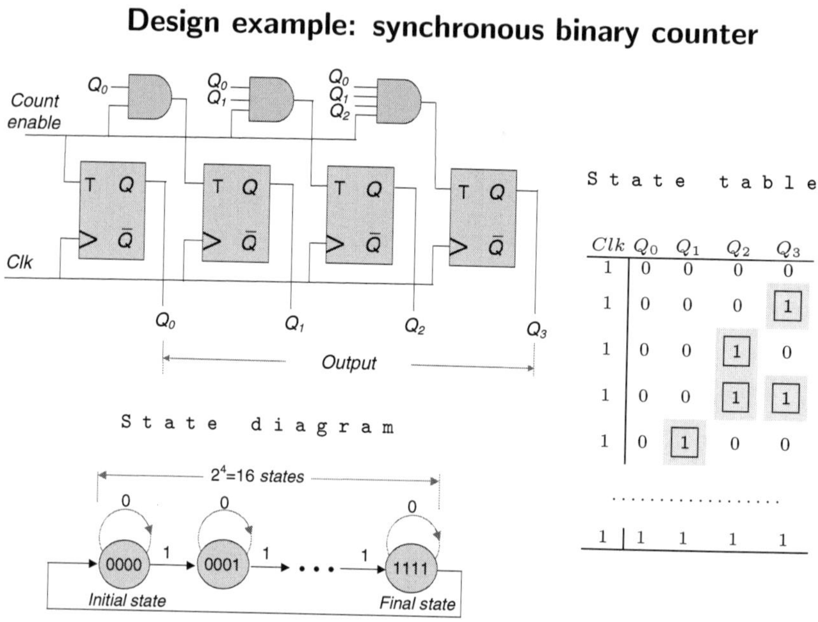
Design example: synchronous binary counter
Count enable
Clk
T Q
Q0 Q1 Q2 Q3
Output
State table
Clk Q0 Q1 Q2 Q3
1 0 0 0 0
1 0 0 0 1
1 0 0 1 0
1 0 0 1 1
1 0 1 0 0
1 1 1 1 1
State diagram
2^4=16 states
0000 0001 1111
Initial state
Final state

FIGURE 13.19
A 4-bit synchronous binary counter (Example 13.17).

Multibit counters can be assembled from smaller counters, connected as a cascade. For example, an 8-bit binary synchronous counter can be built based on two 4-bit synchronous counters.

Counters based on shift registers

Shift registers can be used for the design of special types of counters called *ring counters* and *Johnson counter*. The ring counter includes a shift register that is initialized so that only one of its flip-flops is set to 1, while the others are set to 0. Upon each count pulse, the single 1 is shifted right. An n-bit ring counter has n states in its counting sequence; this counter circulates a single bit among the flip-flops to provide n distinguishable states.

Example 13.19 *Figure 13.21 shows a 4-bit synchronous ring counter.*

13.8.2 Countdown chains

Cascaded counters are often used to divide a high-frequency clock signal to obtain highly accurate pulse frequencies. Cascaded counters used for such purposes are called *countdown chains*.

Design example: modulo-10 counter

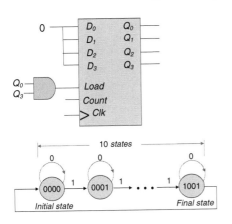

Q_0	Q_1	Q_2	Q_3
0	0	0	0
0	0	0	1
0	0	1	0
0	0	1	1
0	1	0	0
0	1	0	1
0	1	1	0
0	1	1	1
1	0	0	0
1	0	0	1

FIGURE 13.20

A modulo-10 counter (Example 13.18).

Example 13.20 *Let a basic clock frequency be of 1MHz. To generate signals of 100kHz, 10kHz, and 1kHz, three cascaded decade counters can be used (forming a divide-by-1000 with intermediate divide-by-10 and divide-by-100 outputs).*

13.9   Sequential logic network design

In combinational network designs, the design process typically starts with the problem statement, which is often a verbal description of the intended behavior of the system. The goal is to develop a logic network utilizing the available components and meeting the design objectives and constraints.

Sequential network design is based on the concept of assembling a complete sequential network from basic simple elements; both elements with storage properties and combinational elements. Therefore, circuits are at the top level in the design hierarchy, and their design supersedes the design of combinational components.

The operation of a sequential network is described as a sequence of *state transitions* controlled by the inputs. *State tables* define the network's next-state functions. The *state diagram* represents all internal states and

Design example: modulo 4 ring counter

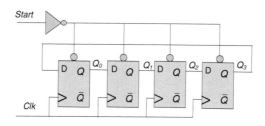

FIGURE 13.21
A 4-bit ring counter (Example 13.19).

relationships between the states. The state behavior of the sequential circuits is described by the state table, state diagram, as well as excitation and output equations.

Comparison of data structures that are used in analysis and synthesis of combinational and sequential logic networks is given in Table 13.1.

TABLE 13.1
Data structures in logic network design

Combinational networks based on SOP forms	Combinational network based on polynomial forms	Sequential networks
Truth table	Functional table	State table
SOP expressions	Polynomial expressions	State equation, excitation equation, and output equation
Decision trees and diagrams	Functional decision trees and diagrams	State diagrams
Gates	Gates	Gates and flip-flops
Cubes	Functional cubes	

Analysis and synthesis of sequential networks

The two facets of sequential logic network design are analysis and synthesis. The analysis problem is formulated as follows. Given a sequential network, provide a tabular description of this network. The sequential networks can be synchronous (globally clocked) or asynchronous (locally clocked). A sequential network at any given time is described by its inputs, its outputs, and the state of its flip-flops at this time.

The flip-flop outputs changed at every clock pulse, and the outputs and the next state of the flip-flops are both a function of the inputs and the present state. The analysis of a sequential network consists of obtaining a table or a diagram for the time sequence of inputs, outputs, and internal states. It is also possible to write Boolean expressions that describe the behavior of the sequential network. These expressions must include the necessary time sequence.

The steps involved in the synthesis of sequential networks are basically the reverse of those involved in analysis. Synthesis involves the establishment of a sequential network realization that satisfies a set of input/output specifications. Given a word specification of a network, design a sequential network.

13.10   Mealy and Moore models of sequential networks

Finite state machines are the formal models of sequential logic networks, and they are divided into two classes: the *Mealy* model, and the *Moore* model. The Mealy model is characterized as follows (Figure 13.22):

▶ The output is a function of both the present state and the inputs.
▶ The outputs of a sequential network may change if the inputs change during the clock cycle; the new outputs are available by the next clock edge.

The specific properties of the Moore model are listed below (Figure 13.22):

▶ The output is a function of the present state only.
▶ The outputs are synchronized with the clock; that is, the new outputs are not available by the next clock edge.

Both models are used for the analysis and synthesis of sequential networks, and are represented using various data structures.

13.11   Data structures for analysis of sequential networks

A sequential network consists of

▶ A network with memory properties using flip-flops
▶ A combinational network using available logic gates

Models of sequential logic networks
Mealy model

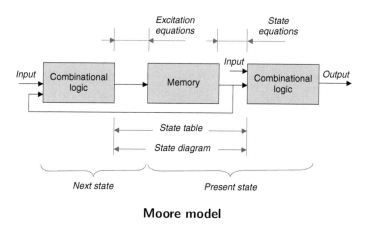

Moore model

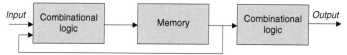

FIGURE 13.22
Mealy and Moore models of a synchronous sequential network.

Data structures for analyzing combinational networks were considered in the previous chapters. For analyzing a network consisting of a combinational network and a memory-based network, specific data structures are used, such as

▶ *State equations,*

▶ *State tables,* or *transition tables*

▶ *State diagrams*

▶ *Excitation equations*

▶ *Output equations*

13.11.1　State equations

The behavior of a clocked sequential network can be described in algebraic form by means of state equations. The state equations, also called *transition equations*, specify the next state as a function of the present state and the inputs:

Past state	*Transition* $\longrightarrow$	Present state	*Transition* $\longrightarrow$	Next state

13.11.2   Excitation and output equations

Knowledge of the type of flip-flop present provides necessary information for deriving a description of the memory part of a sequential network. Combinational networks in sequential networks are distinguished with respect to their functions. They are described as follows:

(*a*) The part of the combinational network that generates external outputs; this part is described by a set of Boolean equations called *output equations*.

(*b*) The part of the combinational network that generates the input signals to flip-flops; this part is described by a set of Boolean equations called *excitation equations*.

Example 13.21 *The sequential network shown in Figure 13.23 consists of two D flip-flops, input x, output y, and the clock signal Clk. This network is a **Mealy model** because the output y is a function of both input x and the present state of D flip-flops. Let us derive the state and excitation equations for the sequential network:*

Step 1: *Derive the excitation which are the input functions for both flip-flops, $D_1 = Q_1 x \vee Q_2 x$ and $D_2 = \overline{Q}_1 x$.*

Step 2: *Derive the output equation: $y = (Q_1 \vee Q_2)\overline{x}$.*

Step 3: *Since the characteristic equations for D flip-flops are $D_1 = Q_1^+$ and $D_2 = Q_2^+$, then the state equations are as follows: $Q_1^+ = Q_1 x \vee Q_2 x$ and $Q_2^+ = \overline{Q}_1 x$.*

13.11.3   State table

Derivation of the state table requires listing all possible binary combinations of the present states and inputs. This creates the truth table for the functions of the next-states and outputs, whose values are calculated for the table from the state and output equations.

Design example: analysis of D flip-flop-based logic network

STEP 1

Initial sequential network

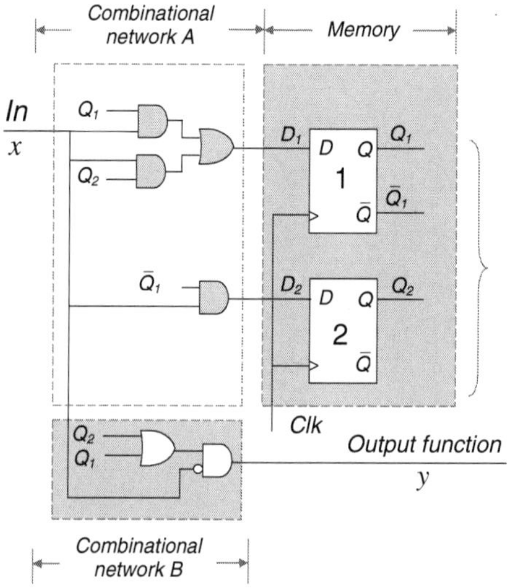

STEP 2

State equations

$$Q_1^+ = Q_1 x \vee Q_2 x$$
$$Q_2^+ = \overline{Q_1} x$$

Output function

$$y = (Q_1 \vee Q_2)\overline{x}$$

Excitation equations

$$D_1 = Q_1 x \vee Q_2 x$$
$$D_2 = \overline{Q_1} x$$

STEP 3

State table

Present state Q_1 Q_2		Input x	Next state Q_1^+ Q_2^+		Output y
0	0	0	0	0	0
0	0	1	0	1	0
0	1	0	0	0	1
0	1	1	1	1	0
1	0	0	0	0	1
1	0	1	1	0	0
1	1	0	0	0	1
1	1	1	1	0	0

STEP 4

State diagram

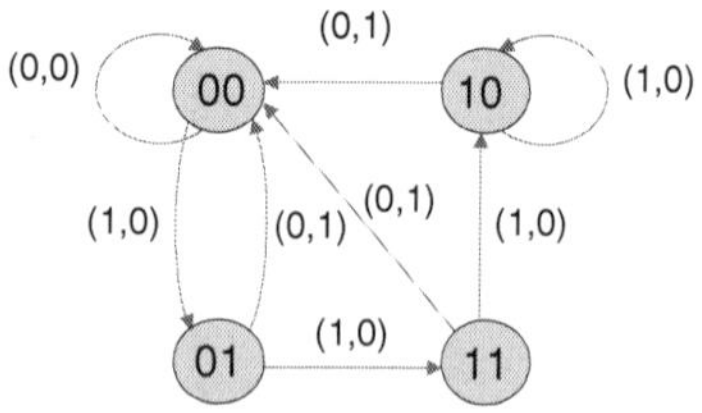

FIGURE 13.23
Sequential network analysis (Examples 13.21, 13.22, and 13.23).

Example 13.22 *(Continuation of Example 13.21). Since there are three variables on the right side of the state equations (two present states, Q_1 and Q_2 and one input; Figure 13.23), there are 8 binary combinations of them. The values of the next-states of the first and second flip-flops and the output can be calculated from the state and output equations: $Q_1^+ = Q_1 x \vee Q_2 x$, $Q_2^+ = \overline{Q_1} x$ and $y = (Q_1 \vee Q_2)\overline{x}$, respectively, by substitution of all 8 assignments into them.*

13.11.4 State diagram

The state diagram is a graphical data structure (directed graphs) for the representation of sequential logic networks. It provides the same information as a state table. The binary number inside each node identifies the state of the flip-flops. The direct links are labeled with two binary numbers:

(a) The input value during the present state

(b) The output during the present state with the given input

Example 13.23 *(Continuation of Example 13.21). Given: the state table and the assignments of the flip-flop outputs, Q_1 and Q_2. The number of positive codes specifies the number of states. This example deals with the codes 00, 01, 10, and 11, and therefore four states are distinguished. This implies a state diagram, consisting of four nodes, and links between them labeled by pairs (input, output). For example, the directed link from state 00 to 01 is labeled (1,0), meaning that when the network is in the present state 00 and the input is 1, the output is 0. The network makes transition to the next state 01 after the next clock cycle: if the input changes to 0, then the output becomes 1; if the input remains at 1, the output stays at 0.*

The foregoing examples and practice problems have dealt, so far, with networks with D flip-flops. In general, the derivation of state equations depends on the type of flip-flops in the network. This analysis is considered in the next section.

13.12   Analysis of sequential networks with various types of flip-flops

A logic network with n flip-flops represent a finite state machine with 2^n different states. Analysis of this state machine can be performed based on knowledge of the types of flip-flops and a list of the Boolean expressions describing the combinational parts of the logic network. It provides the necessary information for deriving the state, output, and excitation equations, as well as the state diagrams. Summarizing, the algorithm for the analysis of a sequential network includes four major steps, as shown below:

Step 1: Determine the flip-flops' input equations by inspection of the network, in terms of the present state and input variables. These form the excitation equations.

Step 2: Determine the next-state equation using the corresponding flip-flop characteristic equations.

Step 3: Create the truth table of the values of the input equations for each combination of the present state and inputs (state table).

Step 4: Build the state diagram using the state table.

13.12.1   Analysis of a sequential network with D flip-flops

A logic network with D flip-flops can be analyzed using the general steps described above. The specific feature of the network with D flip-flops is that the characteristic equation for D flip-flop is $Q^+ = D$.

> **Example 13.24** *Consider a sequential network with D flip-flop (Figure 13.24). Its analysis involves the following steps:* **Step 1:** *Inspection of the network and derivation of the input equation for the D-flip-flops.* **Step 2:** *Determination of the next-state values using the D flip-flop characteristic equation $Q^+ = D$.* **Step 3:** *Creation of the state table using the next-state equation determined in Step 2.* **Step 4:** *Drawing the state diagram.*

13.12.2   Analysis of a sequential network with JK flip-flops

The analysis of a network with JK flip-flops is performed in a slightly different way from the previous techniques for networks with D flip-flops. For a D flip-flop, the state equation is the same as the input equation.

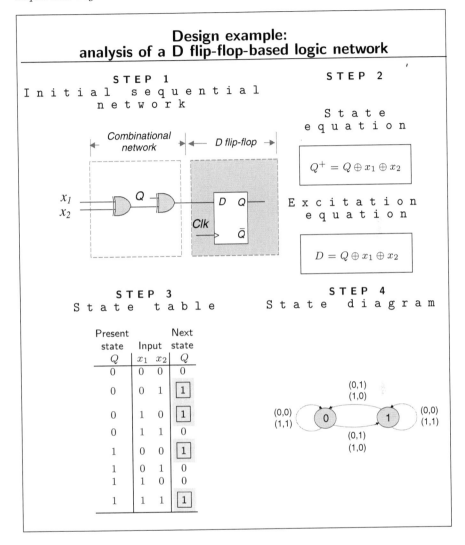

FIGURE 13.24

Analysis of a sequential network with D flip-flops (Example 13.24).

Deriving the state equations for JK flip-flops involves the following steps:

Step 1: Determine the excitation equations.

Step 2: Substitute the excitation equation into the characteristic equation for JK flip-flop, or, instead, list all combinations for the values of each excitation equation and determine the next-state values using the JK flip-flop characteristics table.

Example 13.25 *Consider a sequential network with JK flip-flops (Figure 13.25). This is a **Moore model**, since the output is a function of the present state only. By inspection of the network, the inputs of the JK flip-flop are derived as follows: $J_1 = Q_2$, $K_1 = Q_2\overline{x}$, and $J_2 = \overline{x}$, $K_2 = Q_1 \oplus x$. The state equations are derived by substituting the above expressions into the characteristic equation for JK flip-flop ($Q^+ = J\overline{Q} \vee \overline{K}Q$):*

$$Q_1^+ = J_1\overline{Q_1} \vee \overline{K_1}Q_1 = \overline{Q_1}Q_2 \vee Q_1\overline{Q_2} \vee Q_1x$$
$$Q_2^+ = J_2\overline{Q_2} \vee \overline{K_2}Q_2 = \overline{Q_2}\overline{x} \vee Q_1Q_2x \vee \overline{Q_1}Q_2\overline{x}$$

*The state diagram is derived from the state equations. Note that at the links of the state diagram, only input values are indicated as (**Input**).*

13.12.3 Analysis of a sequential network with T flip-flops

Analysis of a network with T flip-flops is similar to the analysis of the networks with JK flip-flops:

Step 1: Determine the excitation equations.

Step 2: Substitute the excitation equation into the characteristic
   equation of T flip-flops, or, instead, list all combinations
   for the values of each excitation equation and determine the
   next-state values using the T flip-flop characteristic table.

Example 13.26 *Analysis of a sequential network with T flip-flops requires knowledge of the characteristic equations of the T flip-flops and includes the following main steps (Figure 13.26). This is a Moore model. The output depends only on the T flip-flop values, that is, the output is a function of the present state only. The state equations and the state diagram, whose links are labeled by input values (**Input**), are shown in Figure 13.26. Note that if the input x is equal to 1, the network behaves as a binary counter with states 00, 01, 10, 11, and back to 00. When $x = 0$, the network remains in the same state.*

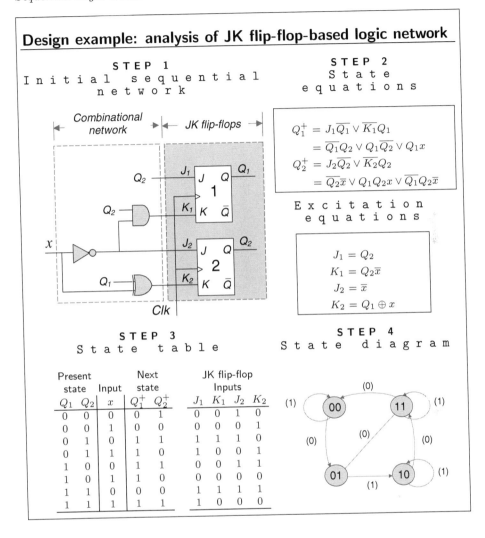

Design example: analysis of JK flip-flop-based logic network

STEP 1
Initial sequential network

STEP 2
State equations

$$Q_1^+ = J_1\overline{Q_1} \vee \overline{K_1}Q_1$$
$$= \overline{Q_1}Q_2 \vee Q_1\overline{Q_2} \vee Q_1 x$$
$$Q_2^+ = J_2\overline{Q_2} \vee \overline{K_2}Q_2$$
$$= \overline{Q_2}\overline{x} \vee Q_1 Q_2 x \vee \overline{Q_1}Q_2\overline{x}$$

Excitation equations

$$J_1 = Q_2$$
$$K_1 = Q_2\overline{x}$$
$$J_2 = \overline{x}$$
$$K_2 = Q_1 \oplus x$$

STEP 3
State table

Present state		Input	Next state		JK flip-flop Inputs			
Q_1	Q_2	x	Q_1^+	Q_2^+	J_1	K_1	J_2	K_2
0	0	0	0	1	0	0	1	0
0	0	1	0	0	0	0	0	1
0	1	0	1	1	1	1	1	0
0	1	1	1	0	1	0	0	1
1	0	0	1	1	0	0	1	1
1	0	1	1	0	0	0	0	0
1	1	0	0	0	1	1	1	1
1	1	1	1	1	1	0	0	0

STEP 4
State diagram

FIGURE 13.25
Analysis of a sequential network with JK flip-flops (Example 13.25).

13.13 Techniques for the synthesis of sequential networks

The main steps to design a sequential network are as follows: (a) state behavior specification, (b) state assignment, and (c) combinational function specification and combinational network design:

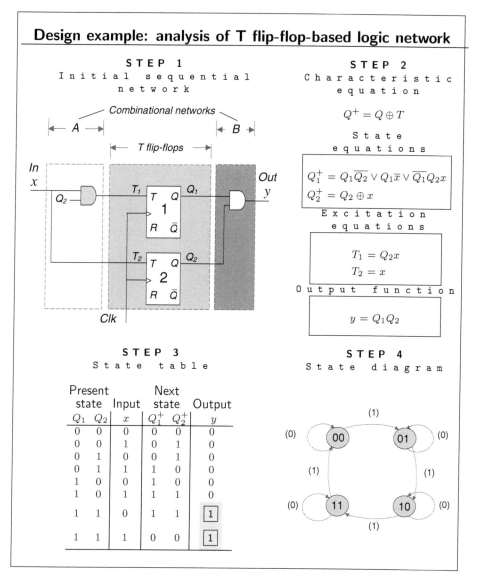

FIGURE 13.26
Analysis of a sequential network with T flip-flops (Example 13.26).

State behavior	*Specification* $\longrightarrow$	State assignment	*Design* $\longrightarrow$	Logic network

The design procedure for synchronous sequential networks consists of the following main steps:

Step 1: Derive a state diagram using the word description of the problem.

Step 2: Reduce the number of states, if necessary.

Step 3: Assign binary values to the states and obtain the binary-coded state table.

Step 4: Choose the type of flip-flops and derive excitation equations and the output function.

Step 5: Draw the sequential logic network.

13.13.1 Synthesis of a sequential network using D flip-flops

Given a state diagram or the state table, a network can be synthesized using the above algorithm, while taking into account the specified type of the flip-flop.

Example 13.27 *Design a sequential network with D flip-flops given the state diagram (Figure 13.27). The given diagram represents the Moore model. Using this diagram, we derive a state table with four states S_0, S_1, S_2, and S_3. We choose to encode the states as follows: $S_0 = 00$, $S_1 = 01$, $S_2 = 10$, $S_3 = 11$. Note that the equations are derived for the functions Q_1^+, Q_2^+, and y from the table using K-maps for minimization. Note that $Q_1^+ = D_1$ and $Q_2^+ = D_2$.*

13.13.2 Synthesis of sequential networks using JK flip-flops

In design of the networks with JK flip-flops, the state table and equations are derived using the JK flip-flop characteristic equations.

Example 13.28 *Design a sequential network using JK flip-flops given the state diagram shown in Figure 13.28. The states S_0, S_1, S_2, and S_3 are encoded as 00,01,10, and 11, respectively. The state table is derived using the characteristic table for JK flip-flops. The table for the next states, Q_1^+ and Q_2^+, is derived first Using the pairs $Q_1 Q_2^+$ and $Q_1^+ Q_2^+$, the corresponding J_1, K_1, J_2, K_2 are placed in the last four columns of the state table J_1, K_1, J_2, K_2 are determined using the characteristic table.*

Design example: synthesis using D flip-flops

STEP 1
State diagram

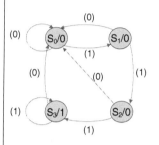

STEP 2
State table

Present state		Input	Next state		Output
Q_1	Q_2	x	Q_1^+	Q_2^+	y
0	0	0	0	0	0
0	0	1	0	1	0
0	1	0	0	0	0
0	1	1	1	0	0
1	0	0	0	0	0
1	0	1	1	1	0
1	1	0	0	0	1
1	1	1	1	1	1

STEP 3
Excitation equations

The first D flip-flop

$$D_1 = Q_1^+$$
$$= \overline{Q}_1 Q_2 x \vee Q_1 \overline{Q}_2 x$$
$$\vee \; Q_1 Q_2 x$$

The second D flip-flop

$$D_2 = Q_2^+$$
$$= \overline{Q}_1 \overline{Q}_2 x \vee Q_1 \overline{Q}_2 x$$
$$\vee \; Q_1 Q_2 x$$

Output function

$$y = Q_1 Q_2 \overline{x} \vee Q_1 Q_2 x$$

STEP 4
Minimization of excitation equations

$$D_1 = Q_1 x \vee Q_2 x$$

$$D_2 = Q_1 x \vee \overline{Q}_2 x$$

$$y = Q_1 Q_2$$

STEP 5
Synthesized sequential network

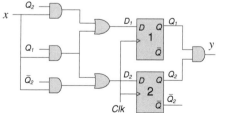

State equation

$$Q_1^+ = \overline{Q}_1 Q_2 x \vee Q_1 \overline{Q}_2 x \vee Q_1 Q_2 x$$
$$Q_2^+ = \overline{Q}_1 \overline{Q}_2 x \vee Q_1 \overline{Q}_2 x \vee Q_1 Q_2 x$$

FIGURE 13.27
Sequential network design using D flip-flops (Example 13.27).

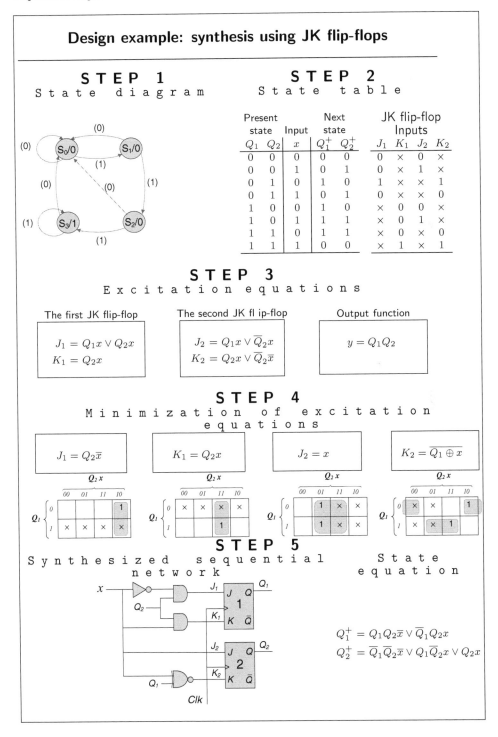

FIGURE 13.28

Sequential network using JK flip-flops (Example 13.28).

13.13.3   Synthesis of sequential networks using T flip-flops

Designing a network with T flip-flops is similar to the design of a network with JK flip-flops.

Example 13.29 *Design a sequential network using T flip-flops given the state diagram (Figure 13.29). This is the state diagram of a 3-bit binary counter. State transitions occur during a clock edge, so the counter remains in its present state if no clock is applied. The number of states is 8, so $\log_2 8 = 3$ T flip-flops are needed. The excitation equations are derived from the excitation table of the T flip-flop and from inspection of the state transition of the present state to the next state. Next, the excitation equations are simplified using K-maps, and, finally, the sequential network is drawn.*

13.14   Redesign

Sometimes a design need a particular type of flip-flop for a specific application, but all you have available is another type. This often happens with an application needing T flip-flops, since these are not generally available in commercial packages.

This can be accomplished through redesigning the flip-flops. This means to re-wire an available type to perform as a targeted flip-flop. We have already seen that a JK flip-flop with its J and K inputs connected to a logic 1 will operate as a T flip-flop. Converting a D flip-flop to T operation is quite similar; the $\overline{Q}$ output is connected back to the D input. To convert a D flip-flop into JK operation, some gates must be added to implements the logical truth that $D = J\overline{Q} \vee \overline{K}Q$. CMOS flip-flops are typically constructed as D types because of the nature of their internal operation. Commercial CMOS JK flip-flops then add this circuit to the input in order to get JK operation. This approach eliminates the internal latching effect, that occurs with the general JK master-slave flip-flop: The J and K input signals must be present at the time the clock signal falls to logic 0, in order to affect the new output state.

The other approach is the total circuit redesign, that may reduce the general number of logic gates in the network.

The redesign problem is formulated as follows. Given a sequential logic network based on a particular type of flip-flops, redesign it using another type of flip-flops.

Design example : synthesis using T flip-flops

STEP 1
State diagram

STEP 2
State table

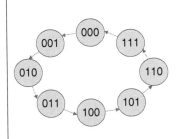

Present state			Next state			T flip-flop Inputs		
Q_3	Q_2	Q_1	Q_3	Q_2	Q_1	T_3	T_2	T_1
0	0	0	0	0	1	0	0	1
0	0	1	0	1	0	0	1	1
0	1	0	0	1	1	0	0	1
0	1	1	1	0	0	1	1	1
1	0	0	1	0	1	0	0	1
1	0	1	1	1	0	0	1	1
1	1	0	1	1	1	0	1	1
1	1	1	0	0	0	1	1	1

STEP 3
Excitation equations

The third
T flip-flop

$$T_3 = \overline{Q}_1 Q_2 Q_3 \vee Q_1 Q_1 Q_1$$

The second
T flip-flop

$$T_2 = \overline{Q}_1 \overline{Q}_2 Q_3 \vee Q_1 \overline{Q}_1 \overline{Q}_1$$
$$\vee \, Q_1 Q_2 \overline{Q}_3 \vee Q_1 Q_2 Q_3$$

The first
T flip-flop

$$T_1 = 1$$

STEP 4
Minimization of excitation equations

$$T_3 = Q_2 Q_1$$

$$T_2 = Q_1$$

$$T_1 = 1$$

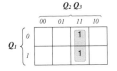

STEP 5
Synthesized sequential network

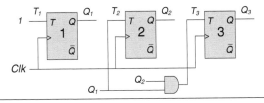

FIGURE 13.29
Sequential network design using T flip-flops (Example 13.29).

Example 13.30 *Consider a network with two D flip-flops (Figure 13.30, Step 1). Redesigning of this network using JK flip-flops is shown at the fifth step in Figure 13.30.*

13.15 Further study

Advanced topics of sequential logic networks

Topic 1: Neural networks such as the Hopfield networks are capable of storing the state of the system. The advantages of Hopfield networks for the computing and implementation of an elementary logic function include the small number of neuron cells and the iterations to achieve minimum energy, simple interconnect topology.

Topic 2: Multilevel memory. In 1992, Intel began a research effort to reduce the amount of silicon required to store a bit of data to a fraction of transistor by storing more than 1 bits of information per cell. The research resulted in manufacturing in 1997 a four-valued 64 Mbit memory device, StrataFlash, storing two bits of information per cell. Intels innovation was followed by a series of announcement of development of multilevel flash memories, including Samsung Electrons 128Mbit 3.3V four-level NAND flash memory, and Hitachi and Mitsubishis 256Mbit four-level NAND flash memory. In 1997 NEC announced development of a 4Gbit four-level DRAM with four-level sensing and restoring operations. An overview of CMOS-related multiple-valued memory technologies has been done in "A review of multiple-valued memory technology" by Glenn Gulak, *28th International Symposium on Multiple-Valued Logic*, pages 222–233, 1998. Nonvolatile multiple-valued memory technologies have been considered in [11]. Recent advances in nanotechnology led to a series of prospective proposals on multiple-valued memory, in particular, single-electron multiple-valued memory, as well as analogue information storage using self-assembled nanoparticle films.

Topic 3: Performance and lower-power sequential logic network design remains an open problem. Much of the previous work in sequential logic synthesis have targeted minimization of the implementation area. The state encoding was mostly fixed, and the logic optimization attempted to improve the logic implementation under the given state assignment. Further exploration of the design space have proved that a proper state encoding can improve design quality and performance.

Topic 4: Verification of sequential circuits is much more complex that that of combinational networks. It is aggravated by the increasing design complexity, which means that the number of states is too large to be represented by the finite state machine. Decomposition of the design to the smaller state machines is the approach used in practice.

Design example: redesign

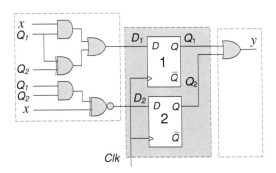

STEP 1
Initial sequential network

STEP 2
State equations

$$Q_1^+ = D_1 = (Q_1 \oplus Q_2) \vee Q_1 X$$
$$Q_2^+ = D_2 = \overline{x \oplus Q_1 Q_2}$$

Excitation equations

$$D_1 = Q_1$$
$$D_2 = Q_2$$

Output equations

$$y = Q_1 \vee Q_2$$

STEP 3
State table

Present state		Input	Next state		JK flip-flop Inputs			
Q_1	Q_2	x	Q_1^+	Q_2^+	J_1	K_1	J_2	K_2
0	0	0	0	1	0	0	1	0
0	0	1	0	0	0	0	0	1
0	1	0	1	1	1	1	1	0
0	1	1	1	0	1	0	0	1
1	0	0	1	1	0	0	1	1
1	0	1	1	0	0	0	0	0
1	1	0	0	0	1	1	1	1
1	1	1	1	1	1	0	0	0

STEP 4
Excitation equations

$$J_1 = Q_2$$
$$K_1 = Q_2 \overline{x}$$
$$J_2 = \overline{x}$$
$$K_2 = Q_1 \oplus x$$

Output equations

$$y = Q_1 \vee Q_2$$

STEP 5
Redesigned network

FIGURE 13.30
Redesign of a sequential network with D flip-flops to a network with JK flip-flops (Example 13.30).

Further reading

[1] Akers SB. Binary decision diagrams. *IEEE Trans. Comput.*, 27(6):509–516, 1978.

[2] Ashenden PJ. *Digital Design. An Embedded Systems Approach Using VHDL.* Elsevier, Burlington, MA, 2008.

[3] Carroll J and Long DT. *Theory of Finite Automata with an Introduction to Formal Languages.* Prentice Hall, Englewood Cliffs, 1989.

[4] Clarke EM, Emerson EA, and Sistla AP. Automatic verification of finite-state concurrent systems using temporal logic specifications. *ACM Trans. Prog. Lang. Syst.*, 8(2):244–263, 1986.

[5] Gill A. *Introduction to the Theory of Finite-State Machines.* McGraw-Hill, New York, 1962.

[6] Givone DD. *Digital Principles and Design.* McGraw-Hill, New York, 2003.

[7] Katz RH and Borriello G. *Contemporary Logic Design.* Pearson Prentice Hall, Upper Saddle River, NJ, 2005.

[8] Kohavi Z. *Switching and Finite Automata Theory.* McGraw-Hill, New York, 1978.

[9] Mano MM and Kime CR. *Logic and Computer Design Fundamentals.* 3rd edition. Pearson Prentice Hall, Upper Saddle River, NJ, 2004.

[10] Marcovitz AB. *Introduction to Logic Design.* McGraw Hill, New York, 2007.

[11] Ricco B, et al. Non-volatile multilevel memories for digital applications. *Proceedings of the IEEE*, 86(12):2399–2421, 1998.

[12] Villa T, Kam T, Brayton RK, and Sangiovanni-Vincentelli A. *Synthesis of Finite State Machines: Logic Optimization.* Kluwer, Dordrecht, 1997.

14

Memory Devices for Binary Data

14.1 Introduction

Memory and programmable devices can extend the functionality of combinational logic networks. These devices are of particular interest in nanotechnology because they can be implemented using the massively copies of memory cells. The *crossbar* computing configuration is attractive for the design and fabrication of memory devices in nano scale. The basic crossbar structure consists of the combination of planes of parallel wires laid out in orthogonal directions. Storage elements are implemented at the molecular scale junction switches formed at the crosspoints of the wires. Note that in microelectronics, the operation or memory cell and the wires are clear-cut distinguished. This distinction does not exist in the crossbar computing structures because nanowires form the memory cells and also connect these cells to one another.

A memory unit is a collection of cells capable of storing a large quantity of binary data. Binary data is transferred to memory for storage. When data is needed for processing, they are transferred from memory to selected registers in the processing unit. Similarly, binary data, received from an input device, are stored in memory, and data, transferred to an output device, are taken from memory. In computer system, a *memory hierarchy* consists of multiple levels of memory with different speeds and sizes.

There are two common operations on data once memory is accessed:

▶ The process of storing new information into memory, referred to as a memory *write* operation.

▶ The process of transferring the stored information out of memory, referred to as a memory *read* operation.

Medium and large scale integration electronic devices are implemented using the concept of arrays of logic elements, or memory cells. In term of organization, memory arrays are classified as follows:

▶ Read-only memory (ROM)

▶ Random-access memory (RAM)

Random-access memory can perform both the write and read operations. Read-only memory can perform only the read operation. This means that suitable binary information is already stored inside the memory, which can be retrieved or read at any time. However, the existing information cannot be altered by writing because read-only memory can only read; it cannot write.

The read-only memory is implemented as an array of homogeneous cells connected in a regular manner and configured to implement some logic. It is, therefore, a *programmable* logic device. The binary data that is stored within a programmable logic device is embedded within the hardware, and this process is referred to as *programming* the device. The word "programming" here refers to a hardware procedure that specifies the bits that are inserted into the hardware configuration of the device.

14.2 Programmable devices

Combinational logic arrays, extended sometimes using sequential elements, are organized based on the following concepts:

▶ Fine-granularity devices, or the "sea of gates" concept, which enables the implementation of any Boolean function using logic cells such as NAND gates only

▶ Medium-granularity devices, which are based on logic blocks, look-up tables, and programmable input-output blocks with flip-flops (medium-grain field-programmable gate arrays (FPGAs))

▶ Large-granularity devices such as complex programmable logic devices (CPLDs) and sequential programmable logic devices (SPLDs), which include AND/OR arrays and universal input/output logic blocks

Nowadays, mostly medium- and large-granularity programmable devices are employed in electronic designs, and they will be considered in detail in this chapter. It should be noted that ROM is sometimes classified as programmable array logic; it is used to implement the look-up-table parts of FPGAs.

A device based on the principle of *programmable logic* is defined as a general-purpose chip containing logic gates and programmable switches for implementing logic network. It contains a collection of logic network elements. Programmable switches allow the logic gates inside the chip to be connected together to implement desired functions.

Simple programmable logic devices include:

▶ Programmable logic array (PLA)
▶ Programmable array logic (PAL)

▶ Gate array logic (GAL)
▶ Programmable logic devices (PLD)
▶ Sequential programmable logic devices (SPLD).

As the system design evolves in response to the system requirements, some functions may be identified for implementation in complex electronic devices such as:

▶ Field programmable gate arrays (FPGA)
▶ Complex programmable logic devices (CPLD)
▶ Application specific integrated circuits (ASIC)
▶ System-on-chip (SoC)
▶ Field programmable system chip (FPSC)

Example 14.1 *Figure 14.1 shows how conventional symbols of logic AND gates can be replaced by array logic symbols. This simplification of a graphical representation can be used for other logic gates.*

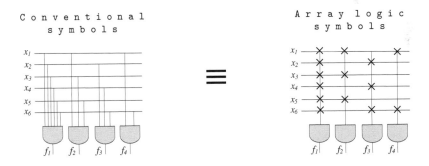

FIGURE 14.1
Excerpt of a network in which conventional symbols of logic gates are replaced by array logic symbols (Example 14.1).

A field-programmable gate array is a semiconductor device containing programmable logic components called "logic blocks," and programmable interconnects. Logic blocks can be programmed to perform the function of basic logic gates such as AND and EXOR, or more complex combinational functions such as decoders or simple mathematical functions. In most FPGAs, the logic blocks also include memory elements, which may be simple flip-flops or more complete blocks of memories.

A hierarchy of programmable interconnects allows logic blocks to be interconnected as needed by the designer. Logic blocks and interconnects can be programmed by the customer/designer, after the FPGA is manufactured, to implement any logical function (hence the name "field-programmable").

FPGAs are usually slower than their ASIC counterparts, as they are not optimized for particular applications, and draw more power. But their advantages include a shorter time to market, ability to be reprogrammed, and lower costs.

The ASIC methodology is the *gate array*. Gate arrays are characterized by their topologies, some of which are shown in Figure 14.2.

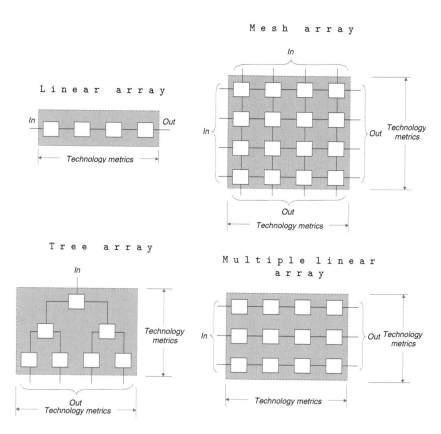

FIGURE 14.2

Array topologies: linear, mesh, tree, and multiple linear.

14.3 Random-access memory

RAM can be thought of as a collection of storage cells, or words of cells (registers), with associated logic networks needed to transfer information in and out of the device. The term *random-access memory* remains to distinguish the sequential type of access in early memories (tape and drum drives) from the immediate access to any address in contemporary transistor-based memories.

14.3.1 Memory array

Each storage element in a memory retains either a 1 or a 0 and is called a *binary storage cell.*

Example 14.2   *An example of a binary storage cell is shown in Figure 14.3. The storage part of the cell is modeled by an SR latch with associated gates. The cell stores one bit in its internal latch.*

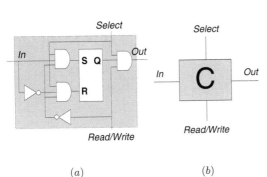

The selected input enables the cell for reading or writing, and the read/write input determines the cell operation when it is selected:
▶ A 1 in the read/write input provides the read operation by forming a path from the latch to the output terminal
▶ A 0 in the read/write input provides the write operation by forming a path from the input terminal to the latch.

(a)   (b)

FIGURE 14.3
A memory cell (a) and its symbol in the memory array (b) (Example 14.2).

A memory consists of $k \times m$ binary storage cells, where k is the number of words and m is the number of bits per word.

Example 14.3   *A* 4 × 4 **RAM** *is shown in Figure 14.4a. A memory with four words needs two address lines. The two address inputs are applied to a* 2 × 4 *decoder to select one of the four words. During the read operation, the four bits of the selected word are transmitted through OR gates to the output terminals. During the write operation, the data available in the input lines are transferred into the four binary cells of the selected word. The cells that are not selected, are disabled, and their previous binary values remain unchanged. When the memory select input, which goes into the decoder, is equal to 0, none of the words are selected, and the contents of all cells remain unchanged regardless of the value of the read/write input.*

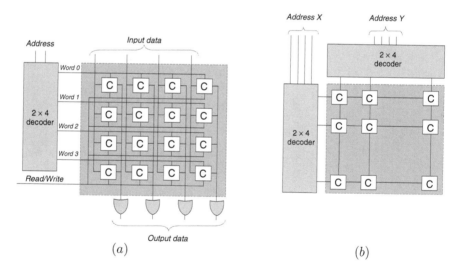

(a) (b)

FIGURE 14.4

A 4 × 4 RAM (a) (Example 14.3) and coincident decoding (b) (Example 14.7).

14.3.2   Words

Memory stores binary data in groups of bits called *words*. A word in memory is an entity of bits that move in and out of storage as a unit. A memory word is a group of 1s and 0s and may represent a number, an instruction, one or more alphanumeric characters, or any other binary-coded information. A group of eight bits is called a *byte*. Most computer memories use words that are multiples of eight bits in length.

Example 14.4 *A 16-bit* **word** *contains two bytes, a 32-bit word is made up of four bytes, and a 64-bit word contains eight bites.*

14.3.3 Address

Communication between a memory and the connected units is achieved through (Figure 14.5):

▶ Data *input* and *output* lines
▶ *Addresses*, or selection lines
▶ *Control* lines that specify the direction of transfer, and other functions

The memory unit is specified by the number of words it contains and the number of bits in each word. The address lines select one particular word. The location of a unit of data in a memory array is called its *address*. Each word in memory is assigned an identification number, that is, its address, starting from 0 up to $2^n - 1$, where n is the number of address lines. The selection of a specific word inside memory is done by applying the n-bit address to the address lines. A decoder accepts this address and opens the paths needed to select the specified word.

It is customary to refer to the number of words (or bytes) in a memory with one of the letters K (kilo), meaning 2^{10}; M (mega), meaning 2^{20}; G (giga), meaning 2^{30}.

The k data input lines provide the information to be stored in memory, and the m data output lines supply the information coming out of memory. The n address lines specify the particular word chosen among the many available. The two control inputs specify the direction of transfer:

▶ The *write* input causes binary data to be transferred into memory
▶ The *read* input causes binary data to be transferred out of memory

14.3.4 Memory capacity

A memory which contains M m-bit words is referred to as memory of size $M \times m$. The number of address lines required to address M words is $log_2 M$.

Example 14.5 *A memory of size $1M \times 16$ includes $2^{10} = 1,024$ words and requires an address of 10 bits.*

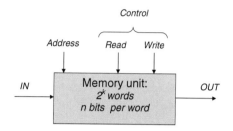

FIGURE 14.5

A generic memory unit consists of n data input and n data output lines, k address lines, and control lines to specify the direction of transfer.

The *capacity* of a memory is the total number of data units that can be stored. It is usually stated as the total number of bits, and sometimes bytes that it can store.

Example 14.6 *A* $1M \times 16$ *memory accommodates* $2^{10} \times 16 = 2^{14}$ *bits, or* 2^{11} *bytes.*

14.3.5 Write and read operations

The two operations that memory performs are the write and read operations. The write signal specifies a *transfer-in* operation. The read signal specifies a *transfer-out* operation. Writing, or transferring a new word to be stored in memory, is performed as follows:

> **Step 1:** Apply the binary address of the desired word to the address lines.
> **Step 2:** Apply the data bits that must be stored in memory to the data input lines.
> **Step 3:** Activate the ''write'' input.
> **Step 4:** The bits are taken from the input data lines and are stored in the word, specified by the address lines.

Reading, or transferring a stored word out of memory, is implemented as follows:

> **Step 1:** Apply the binary address of the desired word to the address lines.
> **Step 2:** Activate the ''read'' input.
> **Step 3:** The bits are taken from the word that has been selected by the address, and are applied to the output data lines.

Normally, the content of the selected word does not change after reading.

14.3.6 Address management

A decoder with n inputs requires 2^n AND gates with k inputs per gate. The total number of gates and the number of inputs per gate can be reduced by employing two decoders in a 2D selection scheme. The basic idea in 2D decoding is to arrange the memory cells in an array that is as close as possible to square. In this configuration, two $n/2$-input decoders are used instead of one n-input decoder. One decoder performs the row selection and the column selection in a 2D matrix configuration.

Example 14.7 *Figure 14.4a shows regular word address decoding (the whole word is read from such a memory), and Figure 14.4b illustrates 2D, or* **coincident decoding** *of the address (a single bit can be accessed in such a memory).*

14.4 Read-only memory

Read-only memory (ROM) is a combinational logic array in which permanent or semipermanent binary data are stored. The data are then embedded in the array using an interconnection pattern. This pattern is established by programming the array, and remains unchanged ("stored"), even when power is turned off and on again. A ROM has:

(a) n address lines

(b) m outputs, from which the data bits of the stored word are read when selected by the address

The number of words in a ROM that can be addressed by n address lines is 2^n. Therefore, the ROM size is $2^n \times m$. A $2^n \times m$ ROM has:

(a) An $n \times 2^n$ decoder

(b) A $2^n \times m$ OR array

Note that ROM does not have data inputs because it does not have a write operation. However, some types of ROM allow the reconfiguration of the array pattern and, thus, the "re-writing" memory. These include:

▶ Programmable ROM (PROM)

▶ Erasable programmable ROM (EPROM)

▶ Electrically erasable programmable ROM (EEPROM)

The internal binary storage of a ROM is specified by a truth table, or a *connection matrix*. They determine the word content at each address.

Example 14.8   *A 32 × 8 ROM consists of 32 words of 8 bits each (Figure 14.6a). There are five address lines that are decoded into 32 outputs by means of a 5 × 32 decoder. The 32 lines from the decoder's output form memory addresses, called **word lines**, which are binary numbers from 0 through 31. The horizontal word lines together with the vertical lines form the OR array. To reflect this fact, the OR gates are shown at the bottom end of each vertical line, so that each OR gate must be considered as having 32 inputs (the gate itself does not physically exist).*

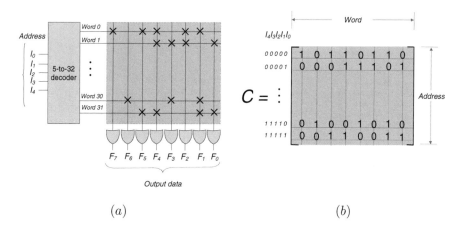

(a)   (b)

FIGURE 14.6
Internal logic and programming of the 32 × 8 ROM (a) (Example 14.8) and connection matrix C (b) (Example 14.9).

14.4.1   Programming ROM

ROM interconnections are programmable. A programmable connection between two lines is logically equivalent to a switch that can be altered to either *closed* (meaning that the two lines are connected), or *open* (meaning that the two lines are disconnected).

The programmable intersection between two lines is also called a *crosspoint*. To implement crosspoint switches, several technologies can be used. In particular, a fuse can be employed, which normally connects the two points, but is open or "blown" by applying a high-voltage pulse to the fuse.

Example 14.9 *Figure 14.6b shows the* **connection** **matrix** *C corresponding to the ROM in Figure 14.6a. This matrix shows that at each address, an 8-bit word is stored. In the connection matrix C, every 0 specifies no connection, and every 1 specifies a path that is obtained by connection. The complete matrix includes 8 columns and 32 rows, thus storing 32 words (only the first and last two words in the ROM are shown in Figure 14.6b).*

14.4.2 Programming the decoder

The decoder in a ROM can be programmed as well. The programming of the decoder corresponds to the implementation of an AND type array, which forms all possible products (minterms) of the decoder's inputs.

Example 14.10 *Figure 14.7a shows the programmed 3-to-8 decoder. This array is comprised of three input lines and three complemented input lines, to form an input variable and its complement for each of the inputs, as well as $2^3 = 8$ output lines that must be seen as 8 AND gates (which do not physically exist) with six inputs each. The AND gates form 8 three-variable products, which are all possible minterms, $\overline{I}_2, \overline{I}_1, \overline{I}_0$ through I_2, I_1, I_0.*

14.4.3 Combinational logic network implementation

The $n \times 2^n$ decoder of the ROM generates all possible 2^n products of its n binary inputs. These products can be interpreted as the minterms of n input variables. By programming a line in the OR array of the ROM, some of the minterms can be logically added to form a canonical SOP. In this way m arbitrary logic functions can be implemented using the decoder and $2^n \times m$ OR array. The algorithm for mapping a system of m Boolean functions of n variables into a ROM is given below:

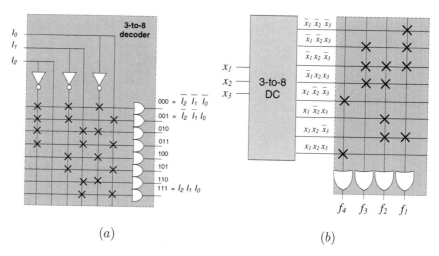

(a)   (b)

FIGURE 14.7
Programming the 3-to-8 decoder for a ROM (a) (Example 14.10) and implementation of four Boolean functions on a ROM with three inputs and four outputs (b) (Example 14.11).

Step 1: Represent each of the m Boolean functions in a canonical SOP form.
Step 2: Specify the $2^n \times m$ decoder.
Step 3: Determine the OR connection matrix in accordance with SOP expressions of the given m functions.

Example 14.11 *A ROM with three inputs and four outputs is given in Figure 14.7b. The decoder with three inputs has $2^3 = 8$ output lines, which together with the OR gates form an 8×4 OR matrix that implements the four canonical SOPs: $f_1 = \bigvee m(0, 1, 2, 6)$, $f_2 = \bigvee m(2, 3, 5, 6)$, $f_3 = \bigvee m(1, 2, 3)$, $f_4 = \bigvee m(4, 7)$.*

14.5 Memory expansion

Available ROM or RAM memory can be expanded to increase word length (the number of bits in a word at each address), or word capacity (number of different addresses) or both word length and capacity. Memory expansion is accomplished by adding an appropriate number of memory arrays while arranging additional address and control lines.

Example 14.12 *Consider an $8K \times 8$ **RAM**. To design $8K \times 16$ RAM, two $8K \times 8$ chips can be connected as shown in Figure 14.8.*

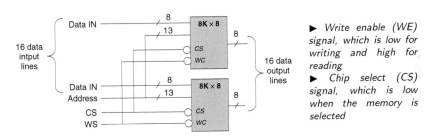

FIGURE 14.8
Implementation of $8K \times 16$ RAM using $8K \times 8$ RAM (Example 14.12).

Example 14.13 *Design a $32K \times 8$ RAM given an $8K \times 8$ RAM with Write Enable (WE) and Chip Select (CS) control inputs. The desired word capacity can be achieved by using four $8K \times 8$ chips (Figure 14.9).*

Expansion of ROM capacity can be achieved by using 2D decoding of the address and multiplexing the data output.

Example 14.14 *Design a $16K \times 8$ ROM. The maximum allowed size of the decoder in a ROM is 8-to-256, and the maximum allowed multiplexer is 64-to-1. Choose the appropriate size of the OR array. A solution with 2D addressing is shown in Figure 14.10.*

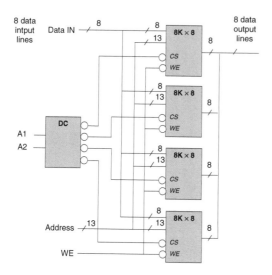

► To address 32K words, an address length of 2^{15} is required

► Using 13 bits to address the words in each chip, the remaining 3 bits of the address can be used to select between four chips

► To read from the chip that is selected, a low WE signal activates the data output for one of four chips only, and those data are transferred to the 8-bit bus using three-state buffers.

FIGURE 14.9
Implementation of $32K \times 8$ RAM with $8K \times 8$ RAM (Example 14.13).

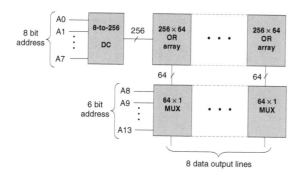

► 14 address lines are required since $16K = 2^{14}$

► The upper part of the address is decoded using an 8-to-256 decoder

► The lower part is used to select one of the 64 outputs of each of 8 256×64 OR arrays

► The desired memory capacity of $16K \times 8 = 256 \times 64 \times 8$ is achieved

FIGURE 14.10
Implementation of $16K \times 8$ ROM with 2D addressing (Example 14.14).

14.6 Programmable logic

Programmable logic devices, PLAs and PALs, are multiinput, multioutput devices, typically organized into an AND gate array, and an OR gate array. Both devices are suited to the SOP representation of Boolean functions. The AND array is used to compute particular product terms of the inputs, depending on the programmed connections. The OR array logically adds these terms together to produce the final SOP expression. The difference between PLA and PAL is the following:

▶ In PLA, both the AND and OR arrays are programmed.
▶ In PAL, the AND array is programmed and OR array is fixed.

It should be noted that, according to the above classification, ROM has a fixed AND array and a programmed OR array.

14.6.1 Programmable logic array (PLA)

A PLA implements Boolean functions in SOP form. It comprises a collection of (Figure 14.11):

(a) AND gates, called an *AND array*
(b) OR gates, called an *OR array*

The AND array consists of literal lines and product lines. It generates a set of k products $P_1, P_2, \ldots P_k$ on its product lines using n inputs $x_1, x_2, \ldots, x_n$ and their complements $\overline{x}_1, \overline{x}_2, \ldots, \overline{x}_n$ available on the literal lines. The product terms serve as the inputs to an OR array, which produces m outputs $f_1, f_2, \ldots, f_m$ on its output lines. Each output can be configured to realize any sum of products, and hence, any SOP representation of a Boolean function. This SOP, in general, is not canonical but rather optimized to get a PLA of minimal size.

The size of the PLA is defined as follows:

$$\mathrm{P\,L\,A\quad s\,i\,z\,e} = (2 \times n + m) \times k$$

where n is the number of data inputs (input variables), m is the number of data outputs, and k is the number of product lines. The upper bound for k is 2^n, but since implementation is intended to minimize the size of the PLA, k is equal to the sum of the number of products (minus the number of shared products) in the SOPs of the m implemented functions.

The PLA is characterized by the following features: (a) regular structure (this feature is intended to simplify the manufacturing process); (b) minimal area of the array needed for implementation, compared to ROM implementation; and (c) efficiency in implementation of systems of functions (not single functions) with shared products.

14.6.2 The PLA's connection matrices

A convenient way to describe Boolean functions in the form acceptable for PLA implementation is by a *connection matrix*. There are two types of connection matrices:

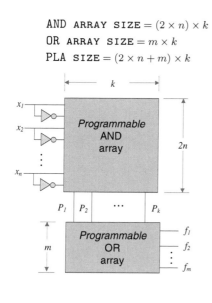

Programmable AND array:

▶ n inputs $x_1, x_2, \ldots, x_n$
▶ n complemented inputs $\overline{x}_1, \overline{x}_2, \ldots, \overline{x}_n$
▶ k output products $P_1, P_2, \ldots, P_k$

Programmable OR array:

▶ k inputs $P_1, P_2, \ldots, P_k$
▶ m outputs $f_1, f_2, \ldots, f_m$

Binary connection matrix for AND array:

$$C_{AND} = \begin{array}{c} \\ P_1 \\ P_2 \\ \vdots \\ P_k \end{array} \begin{bmatrix} x_1 & \overline{x}_1 & \cdots & x_n & \overline{x}_n \\ & & & & \\ & & & & \\ & & & & \\ & & & & \end{bmatrix}$$

Binary connection matrix for OR array:

$$C_{OR} = \begin{array}{c} \\ P_1 \\ P_2 \\ \vdots \\ P_k \end{array} \begin{bmatrix} f_1 & f_2 & \cdots & f_m \\ & & & \\ & & & \\ & & & \\ & & & \end{bmatrix}$$

AND ARRAY SIZE $= (2 \times n) \times k$
OR ARRAY SIZE $= m \times k$
PLA SIZE $= (2 \times n + m) \times k$

FIGURE 14.11
Structure of the n-input m-output PLA and connection matrices for AND gate and OR gate arrays.

▶ The binary $(2 \times n) \times k$ matrix C_{AND} for specifying AND gate array connections; the matrix C_{AND} describes which literal should be included in each product term (the rows determine product terms and the columns represent literals (inputs)).

▶ The binary matrix C_{OR} $m \times k$ for specifying OR gate array connections; matrix C_{OR} describes which product should be included in each output to form the desired functions (the rows determine product terms and the columns represent outputs).

If there are multiple 1s in a row in the connection matrix, it means that the corresponding product term participates in more than one function. Sharing products is a resource for minimizing the PLA's area.

Example 14.15 *Let us derive the* **connection matrix** *for the 4-output Boolean function of three variables: $f_1 = x_1 \vee \overline{x}_2\overline{x}_3$, $f_2 = x_1\overline{x}_3 \vee x_1x_2$, $f_3 = \overline{x}_2\overline{x}_3 \vee x_1x_2$, and $f_4 = \overline{x}_2x_3 \vee x_1$. These functions contain the following product terms: $P_1 = x_1$, $P_2 = \overline{x}_2\overline{x}_3$, $P_3 = x_1\overline{x}_3$, $P_4 = x_1x_2$, $P_5 = \overline{x}_2\overline{x}_3$, $P_6 = x_1x_2$, $P_7 = \overline{x}_2x_3$, $P_8 = x_1$. The set of product terms after deleting the shared products is $P_1 = x_1$, $P_2 = \overline{x}_2\overline{x}_3$, $P_3 = x_1\overline{x}_3$, $P_4 = x_1x_2$, $P_5 = \overline{x}_2x_3$. The connection matrices for AND, C_{AND}, and OR, C_{OR}, gate arrays:*

$$C_{AND} = \begin{array}{c} \\ P_1 \\ P_2 \\ P_3 \\ P_4 \\ P_5 \end{array}
\begin{array}{cccccccc} x_1 & \overline{x}_1 & x_2 & \overline{x}_2 & x_3 & \overline{x}_3 & x_4 & \overline{x}_4 \\ \left[\begin{array}{cccccccc} 1 & & & & & & 1 & \\ & & & 1 & & 1 & & \\ 1 & & & & & 1 & & \\ 1 & & 1 & & & & & \\ & & & & 1 & 1 & & \end{array} \right] \end{array}$$

$$C_{OR} = \begin{array}{c} \\ P_1 \\ P_2 \\ P_3 \\ P_4 \\ P_5 \end{array}
\begin{array}{cccc} f_1 & f_2 & f_3 & f_4 \\ \left[\begin{array}{cccc} 1 & & & 1 \\ 1 & & 1 & \\ & 1 & & \\ & 1 & 1 & \\ & & & 1 \end{array} \right] \end{array}$$

14.6.3   Implementation of Boolean functions using PLAs

Given an m-output Boolean function of n variables or a system of m functions, it can be implemented on a PLA according to the following algorithm.

Step 1: Specify a Boolean function in SOP form (minimize the SOP
   representation if necessary).
Step 2: Determine the PLA connection matrix.
Step 3: Map the matrix into the topology of AND and OR arrays.

Example 14.16 *The* **PLA** *with three inputs, four product terms, and two outputs is given in Figure 14.12. Each AND gate (which physically does not exist) in the* **AND plane** *has six inputs, corresponding to the true and complemented versions of the three input signals. Programming the connections is shown as follows: A signal that is connected to an AND gate is indicated with a wavy line, and a signal that is not connected to the gate is shown with a broken line.*

It follows from Example 14.16 that the circuitry is designed such that any unconnected AND-gate inputs do not affect the output of the AND gate.

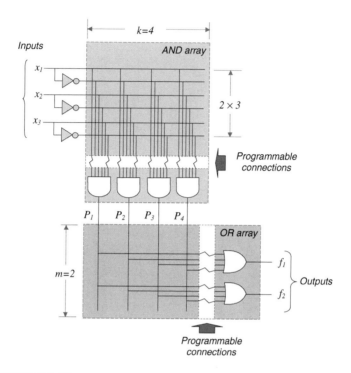

FIGURE 14.12
Gate-level PLA with three inputs, four product terms, and two outputs (Example 14.16).

14.6.4 Programmable array logic

The programmable switches in large PLA arrays may reduce the speed performance of the implemented circuits. This drawback led to the development of a similar device known as a *programmable array logic (PAL)*, in which (a) the AND arrays are programmable, and (b) the OR arrays are fixed (Figure 14.13a).

> **Example 14.17** *Figure 14.13b shows the* **PAL** *with three inputs, four product terms, and two outputs that implements a system of two functions:* $f_1 = x_1 x_2 \vee x_1 \overline{x}_3$ *and* $f_2 = \overline{x}_1 \overline{x}_2 x_3 \vee x_1 x_3$, *where* $P_1 = x_1 x_2$, $P_2 = x_1 \overline{x}_3$, $P_3 = \overline{x}_1 \overline{x}_2 x_3$, *and* $P_4 = x_1 x_3$. *The product terms* P_1 *and* P_2 *are hardwired to one OR gate, and* P_3 *and* P_4 *are hardwired to the other OR gate.*

PALs, in general, offer less flexibility in the implementation of Boolean functions than PLAs.

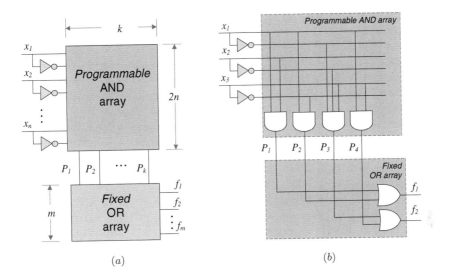

FIGURE 14.13
Structure of an n-input m-output PAL (a), and example of the implementation of a Boolean function (b) (Example 14.17).

14.6.5 Using PLAs and PALs for EXOR polynomial computing

Using PLAs or PALs with internal EXOR gates, it is possible to compute EXOR expressions of Boolean functions using PLAs and PALs.

Example 14.18 *Figure 14.14 shows how to compute the* **linear** *Boolean function* $x_1 \oplus x_2 \oplus x_3 \oplus x_4$ *using a PAL containing an internal EXOR gate.*

Example 14.19 *Figure 14.15 shows how to compute the* **Boolean difference** *of the function* $f = x_1 x_2 \vee \overline{x}_3 x_4$ *using a PAL containing an internal EXOR gate.*

14.7 Field programmable gate arrays

A *field programmable gate array (FPGA)* is a programmable logic device that supports the implementation of large logic networks. The primary

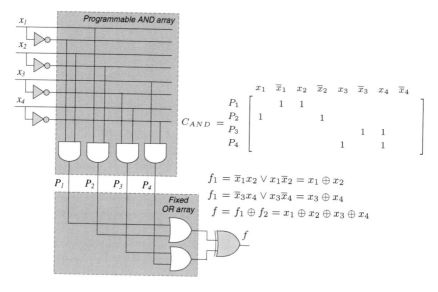

FIGURE 14.14

Computing the linear Boolean function $x_1 \oplus x_2 \oplus x_3 \oplus x_4$ (Example 14.18).

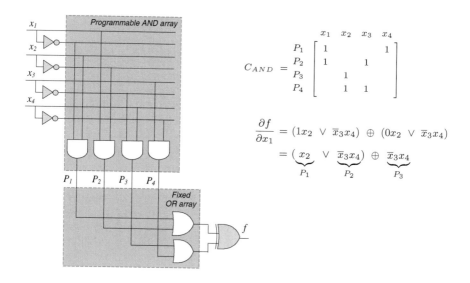

FIGURE 14.15

Computing Boolean difference using a PAL with an EXOR gate extension (Example 14.19).

advantage of FPGAs is their ability to implement any circuit by appropriate programming. The use of a standard FPGA, rather than custom technologies such as application-specific integrated circuits (ASICs), has two key benefits: lower cost and faster time-to-market.

The topological structure of an FPGA is defined as a mesh array (Figure 14.16a). An FPGA includes

▶ *Logic blocks*, denoted as □
▶ *Interconnection switches*, denoted as ■
▶ Input/Output (I/O) blocks

The logic blocks are arranged in a 2D array. Input/Output blocks are used for connecting the pins of the package. The interconnection wires are organized as horizontal and vertical *routing channels* between rows and columns of logic blocks. In Figure 14.16a, two locations for programmable switches are shown:

▶ Switches (black boxes) adjacent to logic blocks (white boxes) hold switches that connect the logic block input and output terminals to the interconnection wires.
▶ Switches (black boxes) that are diagonally between logic blocks connect one interconnection wire to another (such as a vertical wire to a horizontal wire).

Programmable connections also exist between the input/output blocks and the interconnection wires.

Logic blocks

Each logic block has a small number of inputs and one output. The most commonly used logic block is a *lookup table (LUT)*. FPGAs connect circuitry via programmable switches. These switches add significant capacitance to connections, reducing circuit speed. A circuit implemented in an FPGA is typically 10 times larger and roughly 3 times slower than the same circuit implemented on programmed arrays such as PLAs or PALs. The larger size of FPGA circuitry makes FPGA implementations more expensive than other programmable designs.

FPGA architecture

All FPGAs consist of a large number of programmable logic blocks, each of which implements a small amount of digital logic and programmable routing that allow the logic block inputs and outputs to be connected to form larger circuits. The global routing architecture of an FPGA specifies the relative width of the various wiring channels within the chip.

Logic blocks are groups of LUTs and flip-flops along with local routing to interconnect the LUTs within a group. The logic block used in an FPGA strongly influences the FPGA speed and area efficiency. Most FPGAs use logic blocks based on LUTs.

Example 14.20 *Figure 14.16b shows how a 2-input* **LUT** *can be implemented in a RAM-based FPGA.*

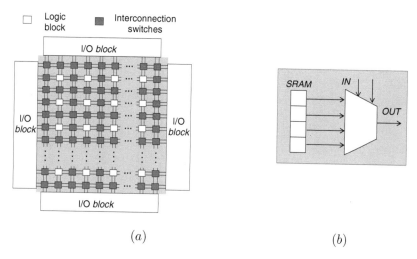

(a) (b)

FIGURE 14.16
Topological structure of an FPGA (a) and a two-input LUT implemented in a RAM-based FPGA (b) (Example 14.20).

A k-input LUT requires 2^k RAM cells and a 2^k-input multiplexer. A k-input LUT can implement any function of k-inputs; one simply programs the 2^k RAM cells to be the truth table of the desired function.

14.8 Further study

Advanced topics of design techniques using memory and programmable devices

Topic 1: Hierarchical programmable devices. Advanced topics involve the design of specific topologies, including 3D structures. A hierarchical, or binary-tree-like FPGA is based on single-input two-output switches. A 2×2 cluster of processing elements can be connected using switches, and four copies of the cluster organized into a "macro" cluster. The structure of this FPGA is represented by a binary decision tree of depth 4, in which the root and levels correspond to switching blocks, and 16 terminal nodes correspond to processing elements.

Another architecture of the hierarchical FPGA is based on a cluster of logic blocks connected by single-input 4-output switch blocks. Possible configurations can be described by a complete binary tree or a multirooted k-ary tree. This tree can be embedded into hypercube-like structures.

In Figure 14.17, two topologies of FPGA are illustrated, where ■ denotes a processing element and ○ denotes a switch block. Four copies of the cluster are organized into a "macro cluster" of different configurations. The first topology (Figure 14.17a) is based on H-tree construction and described by a complete binary tree.

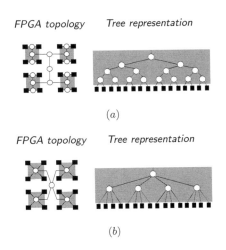

FIGURE 14.17
Topological representations of hierarchical FPGAs by 2D structure, tree, and hypercube-like structure.

Topic 2: Alternative memory devices. Currently, the new technologies for non-volatile memories are being developed to achieve a real RAM-like role, or to erase the border between the both types, ROM and RAM.

One example is Ferroelectric RAM (FeRAM). FeRAM uses a ferroelectric layer in a cell that is otherwise similar to conventional DRAM, this layer holding the charge in a 1 or 0 even with the power removed. To date, FeRAM are still larger than Flash devices.

Another example is Magnetoresistive Random Access Memory, or MRAM, which uses magnetic elements and generally operates in a fashion similar to core. Another technique, known as STT-MRAM, appears to allow for much higher densities.

One more new development is phase-change RAM, or PRAM. PRAM is based on the same storage mechanism as writable CDs and DVDs, but reads them based on their changes in electrical resistance rather than changes in their optical properties.

A number of more esoteric devices have been proposed, including Nano-RAM based on quantum dots, carbon nanotubes, and nanowires, as well as molecular-scale memory devices have been developed recently, but these are currently far from commercialization.

Further reading

[1] Ashenden PJ. *Digital Design. An Embedded Systems Approach Using VHDL.* Elsevier, Burlington, MA, 2008.

[2] Givone DD. *Digital Principles and Design.* McGraw-Hill, New York, 2003.

[3] Brown S and Vranesic Z. *Fundamentals of Digital Logic with VHDL Design.* McGraw-Hill, 2000.

[4] Lai YT and Wang PT. Hierarchical interconnection structures for field programmable gate arrays. *IEEE Trans. VLSI, Systems,* 5(2):186–196, 1997.

[5] Sasao T. *Switching Theory for Logic Synthesis.* Kluwer, Dordrecht, 1999.

[6] Yanushkevich SN and Shmerko VP. *Introduction to Logic Design,* Taylor & Francis/CRC Press, Boca Raton, FL, 2008.

15

Spatial Computing Structures

15.1 Introduction

Logic design of nanodevices in spatial dimensions is based on selected methods of advanced logic design and appropriate spatial topologies. This chapter focuses on the relationship of decision diagrams and 3D graph models such as hypercube data structures, and the hypercube-like topology, $\mathcal{N}$-hypercube.

In computing nanostructure design, the topology, and chemical or/and physical phenomena that correspond to the computational properties, are the key points of design procedure. In biomolecular technology, biomolecular phenomena are considered in a 3D space. Computing biomolecular structure design can be formulated as the problem of delegation of computational properties into 3D biomolecular structures. In this generic formulation of the problem, the guest and host computing structures must be specified, and the procedure of delegation of computing abilities can be reformulated as the problem of embedding of a guest computing structure into the host computing structure. In this formulation, graphical data structures can be used. Decision diagrams can be considered as a guest computing structure that can be embedded into 3D topological structures. The efficiency of embedding depends on the particular properties of a 3D configuration. A hypercube-like topology can be considered as a reasonable template of a 3D configuration because properties of this topology are well studied. In particular:

▶ An arbitrary decision tree is isomorphic to the hypercube-like topology, that is, there is isomorphism between intermediate (operational) nodes and terminal nodes in decision tree and hypercube-like structure.

▶ A decision diagram is isomorphic to the incomplete hypercube-like structure; this is because it is a reduced (optimized) decision tree.

▶ A hypercube-like topology can be easily extended to achieve the required size in 3D.

▶ A hypercube-like topology can be reduced, including the nodes and links, which results in an incomplete hypercube-like configuration.

Figure 15.1 illustrates a 3D computing structure design for the AND

Boolean function of three variables $f = x_1x_2x_3$. At the first step, Boolean function f is represented by a decision tree with $2^3 = 8$ terminal nodes. The optimized decision tree is a decision diagram that is embedded into hypercube. The result of the embedding is a 3D computing structure with topological properties of an incomplete hypercube. Note that an arbitrary Boolean function can be represented by a decision diagram without an intermediate decision tree representation; here we use the decision tree only for the sake of clear explanation.

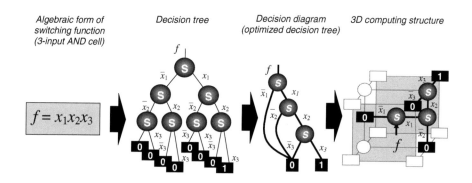

FIGURE 15.1

A 3D representation of a Boolean function AND of three variables.

Three-dimensional design has been explored at the macrolevel for a long time, for example, in the design of distributed systems. It was inspired by nature; for example, the brain, with its "distributed computing and memory", was the prototype in the case of the "connection machine" [20]. The 3D computing architecture concept has been employed by the creators of the supercomputers Cray T3D, NEC's Ncube, and cosmic cube. The components of these supercomputers, as systems, are designed based on classical paradigms of 2D architecture that become 3D because of the *3D topology* (of interconnects), or *3D data structures* and corresponding algorithms, or *3D communication flow*.

Example 15.1 *The topology of today's silicon integrated circuits at a system level varies:*

▶ 1D arrays, e.g., *pipelines, linear systolic processors*

▶ 2D arrays, e.g., *matrix processors systolic arrays*

▶ 3D arrays, e.g., *of hypercube architecture*

At the physical level, very large-scale integration circuits, for instance, are 3D devices because they have a layered structure, that is, interconnection between layers while each layer has a 2D layout. On the way to the top VLSI hierarchy (the most complex VLSI systems), linear and 2D arrays eventually evolved to multiunit architectures such as 3D arrays. Their properties can be summarized as follows:

(a) As stated in the theory of parallel and distributed computing, processing units are packed together and communicate best only with their nearest neighbors.

(b) In the optimal organization, each processing unit will have a diameter comparable to the maximum distance a signal can travel within the time required for some reasonable minimal amount of processing of a signal, for example to determine how it should be routed or whether it should be processed further.

(c) 3D architectures need to contain a fair number of cells before the advantages of the multicell organization become significant compared to competing topologies.

In current multiprocessor designs, 3D structures are not favored since they suffer from gate and wire delay. Internally to each processing unit, 2D architecture is preferable since, at that smaller scale, communication times will be short compared with the cost of computation.

Nanostructures are associated with a molecular/atomic physical platform. This has a truly 3D structure instead of the 3D layout of silicon integrated circuits composed of 2D layers, with interconnections forming the third dimension. At the nano scale level, the advantages of the 3D architecture will begin to become apparent for the following reasons (see, for example, study by Frank [13, 14]):

(a) The distance light can travel in 1 ns in a vacuum is only around 30 cm and two times less in a solid (for example, 1 ns is the cycle of a computer with clock speed of 1GHz), which means that components of such a computer must be packed in a single chip of several centimeter size; thus, a reasonable number of 3D array elements of nanometer size must be integrated on a single tiny chip.

(b) They are desirable for their ideal nature for large-scale computing.

(c) There are many 3D algorithms and designs for existing microscale components that are arranged in 3D space, which computer designers already have experience with.

(d) There are limits to information density as well which imply a direct limit on the size of, in particular, a memory cell.

Thus, the speed-of-light limit (that is information transfer speed limit) and information density pose the following implications for computer architecture:

(a) Traditional architecture will be highly inefficient since most of the processor and memory will not be accessed at any given time due to the size limit.

(b) Interconnection topology must be scalable; most of the existing multiprocessor topologies (hypercubes, binary trees) do not scale since communication times start to dominate as the machine sizes are scaled up.

The solutions to this is use of a parallel, multiprocessing architecture, where each processor is associated with some local memory, in contrast to the traditional von Neumann architecture. The number of processing nodes reachable in n hops from any node cannot grow faster than order n^3 and still embed the network in 3D space with a constant time per hop [47].

This leads us to the conclusion that there is only one class of network topologies that is asymptotically optimal as the machine size is scaled up: namely, a 3D structure, where each processing element is connected to a constant number of physical neighbors. In fact, the processing elements of this 3D mesh can simulate both the processors and wires of the alternative architecture, such as, in particular, a randomly connected network of logic gates. The processing elements must be spread through the structure at a constant density.

Therefore, as processor speeds increase, the speed-of-light limit will cause communication distances to shrink, and the idea of mesh-connected processing elements and memory is, perhaps, the most reasonable and feasible solution.

15.2   The fundamental principles of a 3D computing

The fundamental principles of 3D computing are as follows:

Principle 1: Functionality of the third dimension. The computational models (arithmetics, interconnections of processing elements, data transmission) must be 3D, and computing (processing) must be performed in 3D. The third dimension, Z, carries the functional information, that is, $f = f(X, Y, Z)$. Here, X, Y, and Z are the functionalities (decomposition of f with respect to X, Y, and Z) of Boolean function f corresponding to dimensions of the computing structure in 3D. The third dimension Z leads to ability to accomplish massive parallelism and an efficient 3D processing. The functionality Z typifies natural data processing.

Principle 2: Topological plastisity. The 3D models and processing structures must be flexible to various spatial configurations and organizations. This will ensure flexibility and adaptability. In particular, topological plastisity soundness of enabling configurations and organizations in 3D typify the plastisity of natural neurons and their aggregates.

Principle 3: Fault tolerance and robustness. The 3D models must be fault tolerant and robust. That is, a 3D computational model must perform the desired functions under stochastic behavior of basic components.

Principle 4: Compatibility. Computing in 3D must be supported by computer aided design tools. These tools must be compatible with state-of-the-art modeling techniques and tools.

These computational principles imply that a 3D computing paradigm must satisfy the following properties:

(*a*) Efficient delegation of the computational properties (functionalities) into an arbitrary 3D configuration.

(*b*) Computational characteristics, such as topological redundancy, critical paths, performance, as well as computational capabilities, must be optimized with respect to the spatial configuration.

(*c*) Efficient control, testing, and verification of 3D computing structures.

(*d*) Efficient utilization of the state-of-the-art modeling techniques and tools (traditionally applicable to the 2D computing and designs).

15.3   Spatial structures

Several network topologies have been developed to fit different styles of computation, including massive parallel computation in spatial dimensions. A spatial network topology intended for implementing Boolean functions should possess several characteristics, in particular, minimal degrees of nodes, ability to extend the size of structure with minimal changes to the existing configuration, ability to increase reliability and fault tolerance with minimal changes to the existing configuration, good embedding capabilities, flexibility of design methods, and flexibility of technology.

Based on these criteria, a number of topologies can be considered as potentially useful for solving the problems of spatial logic design, namely; hypercube topology, cycle-connected cube, known as CCC-topology (hypercube), X-hypercube topology, and specific topologies (hyper-Peterson, hyper-star, Fibonacci cube, etc.). Some of these are shown in Figure 15.2.

Hypercube topology (Figure 15.2a) has received considerable attention in classical logic design due mainly to its ability to interpret logic formulas (small diameter, high regularity, high connectivity, and good symmetries). Hypercube-based structures are at the forefront of massive parallel computation because of the unique characteristics of hypercubes (fault tolerance, ability to efficiently permit the embedding of various topologies, such as lattices and trees).

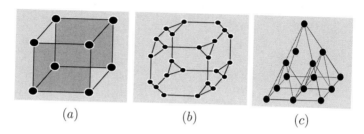

(a)   (b)   (c)

FIGURE 15.2
Spatial configurations: hypercube (a), CCC-hypercube (b), and pyramid (c).

The binary n-hypercube is a special case of the family of k-ary n-hypercubes, which are hypercubes with n dimensions and k nodes in each dimension. The total number of nodes in such a hypercube is $N = k^n$. Parallel computers with direct-connect topologies are often based on k-ary n-cubes or isomorphic structures such as rings, meshes, tori, and direct binary n-cubes.

The CCC-hypercube is created from a hypercube by replacing each node with a cycle of s nodes (Figure 15.2b). Hence it increases the total number of nodes from 2^n to $s \cdot 2^n$ and preserves all features of the hypercube. The CCC-hypercube is closely related to the butterfly network. As has been shown in the previous sections, "butterfly" data flowgraphs are the "nature" of most transforms of Boolean functions in matrix form.

Pyramid topology (Figure 15.2c) is suitable for computations that are based on the principle of hierarchical control, for example, decision trees and decision diagrams. An arbitrarily large pyramid can be embedded into the hypercube with a minimal load factor. Pyramid P_n has nodes on levels $0, 1, \ldots, n$. The number of nodes on level i is 4^i. So, the number of nodes of P_n is $(4^{n+1}/3)$. The unique node on level 0 is called the *root* of the pyramid. The subgraph of P_n induced by the nodes on level i is isomorphic to a mesh $2^i \times 2^i$. The subgraph of P_n induced by the edges connecting different levels is isomorphic to a 4^n-leaf quad-tree. This structure is very flexible for extension. Note that pyramid topology is relevant also to fractal-based computation, which is effective for symmetric functions and is used in digital signal processing, image processing, and pattern recognition.

15.4   Hypercube data structure

Traditionally, hypercube topology is used in logic design for Boolean function manipulation (minimization, representation), communication problems for traffic representation and optimization, and multiprocessor system design for optimization of parallel computing.

In the first approach, each variable of a Boolean function is associated with one dimension in hyperspace. The manipulation of the function is based on a special encoding of the vertices and edges in the hypercube. In the second approach used in communication problems and multiprocessor systems design, the hypercube is the underlying computational model. To design this model, a decision tree or a decision diagram must be constructed and embedded into a hypercube. In this approach the hypercube is utilized as a topological structure for computing in 3D space.

Hypercubes of different dimensions are shown in Figure 15.3. Hypercube properties are given in Table 15.1.

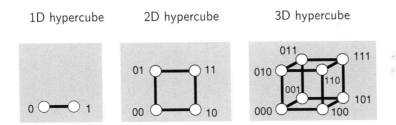

FIGURE 15.3
Hypercubes of different dimensions.

Gray code is used for encoding the indexes of the nodes. There are several reasons to encode the indexes this way. The most important of them is to simplify analysis, synthesis, and embedding of topological structures. Let $b_n...b_1b_0$ be a binary representation of an integer positive number B and $g_n...g_1g_0$ be its Gray code.

$$\text{Binary representation } b_n...b_1b_0 \iff \text{Gray code } g_n...g_1g_0$$

The Hamming distance is a useful measure in hypercube topology. The Hamming sum is defined as the bitwise operation

$$(g_{d-1}\cdots g_0) \oplus (g'_{d-1}\cdots g'_0) = (g_{d-1} \oplus g'_{d-1}),\ldots,(g_1 \oplus g'_1),(g_0 \oplus g'_0) \quad (15.1)$$

TABLE 15.1

Hypercube properties

Property	Definition
Extension	A hypercube is an extension of a graph. The dimensions are specified by the set $\{0, 1, \ldots, n-1\}$. An n-dimensional binary hypercube is a network with $N = 2^n$ nodes and diameter n. There are $d \times 2^{d-1}$ edges in a hypercube of d dimensions.
Specification	Each node of an n-dimensional hypercube can be specified by the binary address $(g_{n-1}, g_{n-2}, \ldots, g_0)$ of length n, where the bit g_i corresponds to the i-th dimension in a Boolean space. Two nodes with addresses $(g_{n-1}, g_{n-2}, \ldots, g_0)$ and $(g'_{n-1}, g'_{n-2}, \ldots, g'_0)$ are connected by an edge (or link) if and only if their addresses differ by exactly one bit.
Hamming distance	There are $\binom{n}{x}$ nodes at Hamming distance of x from a given node, and n node-disjoint paths between any pair of nodes of the hypercube.
Recursion	Hypercube Q_n can be defined recursively.
Fan-out	The *fan-out* (i.e., degree) of every node is n, and the total number of communication links is $\frac{1}{2}N \log N$.
Decomposition	A k-dimensional subcube (k-subcube) of hypercube Q_n, $k \leq n$, is a subgraph of Q_n, that is, a k-dimensional hypercube. A k-subcube of Q_n is represented by a ternary vector $A = a_1 a_2 \ldots, a_n$, where $a_i \in \{0, 1, *\}$, and *denotes an element that can be either 0 or 1.
Intercube distance	Given two subcubes $A = a_1 a_2 \ldots, a_n$ and $B = b_1 b_2 \ldots, b_n$, the *intercube distance* $D_i(A, B)$ between A and B along the i-th dimension is 1 if $\{a_i, b_i = \{0, 1\}\}$; otherwise, it is 0. The distance between two subcubes A, B is given by $D(A, B) = \sum_{i=1}^{n} D_i(A, B)$.
Path	A *path* P of length l is an ordered sequence of nodes $x_{i_0}, x_{i_1}, x_{i_2}, \ldots, x_{i_l}$, where the nodes are labeled with x_{i_j}, $0 \leq j \leq l$, and $x_{i_k} \neq x_{i_{k+1}}$, for $0 \leq k \leq l-1$.

where $\oplus$ is an exclusive or operation. In the hypercube, two nodes are connected by a link (edge) if and only if they have labels that differ by exactly one bit. The number of bits by which labels g_i and g_j differ is denoted by $h(g_i, g_j)$; this is the Hamming distance between the nodes.

Example 15.2 *Hamming distance on two hypercubes is illustrated in Figure 15.4.*

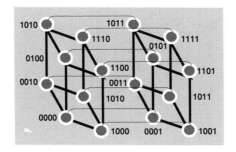

FIGURE 15.4
Hamming sum operation (Example 15.2).

15.5 Assembling of hypercubes

Assembling is the basic topological operation that we apply to synthesize hypercube and hypercube-like data structures. Assembling is the first phase of the development of self-assembly, that is, the process of constructing a unity from components acting under forces/motives internal or local to the components themselves.

To apply an assembly procedure, the following items must be defined: (a) the structural topological components, (b) formal interpretation of the structural topological components in terms of the problem, and (c) the rules of assembly. Assembling is a key philosophy of building complex systems. In this section, the assembling of classical hypercubes is considered. Assembling a hypercube of Boolean functions is accomplished by (a) generating the products as enumerated points (nodes) in the plane, (b) encoding the nodes by Gray code, (c) generating links using Hamming distance, (d) assembling the nodes and links, and (e) Joining a topology of hypercube in n dimensions.

Example 15.3   *The assembly of hypercubes is as follows: 2
points $\Longleftrightarrow$ 1D hypercube ($n = 1$); 4 points $\Longleftrightarrow$ 2D
hypercube ($n = 2$); 8 points $\Longleftrightarrow$ 3D hypercube ($n = 3$);
16 points $\Longleftrightarrow$ 4D hypercube ($n = 4$); 32 points $\Longleftrightarrow$ 5D
hypercube ($n = 5$) (Figure 15.5a).*

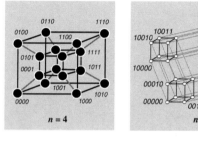
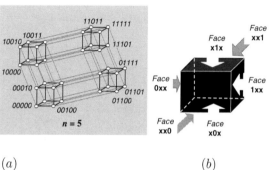

(a)   (b)

FIGURE 15.5
Assembling 3D hypercubes for $n = 4$ and $n = 5$ (a) (Example 15.3) and faces
of the hypercube (b) (Example 15.4).

Notice that the 0-dimensional hypercube ($n = 0$) represents the constant 0.
The line segment connects vertices 0 and 1, and these vertices are called the
face of 1D hypercube and denoted by **x**. A 2D hypercube has four faces, **0x**,
1x, **x0**, and **x1**. The total 2D hypercube can be denoted by **xx**.

Example 15.4   *Six faces of the 3D hypercube, **xx0**, **xx1**, **0xx**, **1xx**,
x1x, and **x0x** (Figure 15.5b).*

15.6 $\mathcal{N}$-hypercube

In this section the extension of the traditional hypercube is considered. This
extension is called the $\mathcal{N}$-hypercube. The classic hypercube structure is the
basis for $\mathcal{N}$-hypercube design and inherits most of the properties of classic
hypercubes.

15.6.1 Extension of a hypercube to $\mathcal{N}$-hypercube

Based on the foregoing, the new, hypercube-like topology called the $\mathcal{N}$-hypercube is introduced. The following reasons advocate developing $\mathcal{N}$-hypercube-based topologies:

▶ $\mathcal{N}$-hypercubes ideally reflect all properties of decision trees and decision diagrams, popular in advanced logic design data structure, enhancing them to more than two dimensions.

▶ $\mathcal{N}$-hypercubes inherit the classic hypercube's properties.

▶ $\mathcal{N}$-hypercubes satisfy a number of nanotechnology requirements.

Several features distinguish the $\mathcal{N}$-hypercube from a hypercube, in particular, the existence of additional nodes, including a unique node called the root. Thanks to that, an arrangement of information flows that is suitable from the point of view of certain technologies can be achieved.

The extension of a hypercube is made by

▶ Embedding additional nodes

▶ Distinguishing the types of nodes

▶ Special space coordinate distribution of the additional nodes

▶ New link assignments

Additional embedded node and link assignments correspond to embedding decision trees in a hypercube and thus convert a hypercube from the passive representation of a function to a connection-based structure, that is, a structure in which calculations can be accomplished. In other words, information connectivity is introduced into the hypercube. Distinguishing the type of nodes satisfies the requirements of graphical data structures of Boolean functions.

An $\mathcal{N}$-hypercube's components are shown in Figure 15.6.

15.6.2 Degree of freedom and rotation

The degree of freedom of each intermediate node can be used for variable order manipulation, as the order of variables is a parameter to adjust in decision trees and diagrams. Additional intermediate nodes, the root node, and corresponding links in the $\mathcal{N}$-hypercube are associated with

▶ The polarity of variables in a Boolean function representation

▶ The structure of the decision tree and decision diagram, and the variable order

▶ The degree of freedom and rotation of the $\mathcal{N}$-hypercube-based topology

Consider the Boolean function of a single variable x. Corresponding 1D $\mathcal{N}$-hypercubes are shown in Figure 15.7.

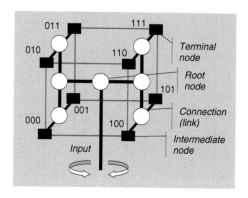

▶ The node is a demultiplexor element, i.e., the node performs Shannon expansion
▶ Each intermediate node has a degree of freedom
▶ Terminal nodes carry the results of computing
▶ The root node resembles the root node of a decision tree
▶ The nodes (their functions and coordinates) and links carry information about the function implemented by the $\mathcal{N}$-hypercube.

FIGURE 15.6
An $\mathcal{N}$-hypercube.

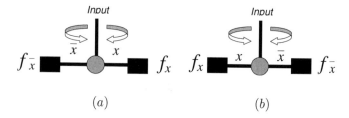

(a)　　　　　　　　　　　　(b)

FIGURE 15.7
The Boolean function of an uncomplemented and complemented variable x, and corresponding 1D $\mathcal{N}$-hypercubes (a) and (b).

Only the intermediate and root nodes in an $\mathcal{N}$-hypercube have freedom. An intermediate node in the 1D $\mathcal{N}$-hypercube has two degrees of freedom (Figure 15.8). The 2D $\mathcal{N}$-hypercube is assembled from two 1D $\mathcal{N}$-hypercubes, and includes three intermediate nodes. The $\mathcal{N}$-hypercube in 2D has $2 \times 2 \times 2 = 8$ degrees of freedom. There are four decision trees with different orders of variables.

Consider an $\mathcal{N}$-hypercube in 3D. This $\mathcal{N}$-hypercube is assembled from two 2D $\mathcal{N}$-hypercubes and includes seven intermediate nodes having $8 \times 8 \times 2 = 128$ degrees of freedom. The degree of freedom of an intermediate node in i-th dimension, $i = 2, 3, \ldots, n$, is equal to $DF_i = 2^{n-i} + 1$.

15.6.3　Coordinate description

Consider an $\mathcal{N}$-hypercube. There are two possible configurations of the intermediate nodes (Figure 15.9):

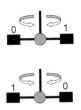

An intermediate node in the 1D $\mathcal{N}$-hypercube has two degrees of freedom.

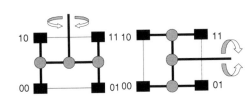

The 2D $\mathcal{N}$-hypercube is assembled from two 1D $\mathcal{N}$-hypercubes, and includes three intermediate nodes The $\mathcal{N}$-hypercube in 2D has

$$2 \times 2 \times 2 = 8 \text{ degrees of freedom.}$$

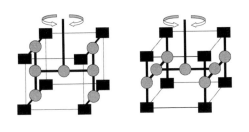

This $\mathcal{N}$-hypercube is assembled from two 2D $\mathcal{N}$-hypercubes, includes 7 intermediate nodes, and has

$$8 \times 8 \times 2 = 128 \text{ degrees of freedom.}$$

The degree of freedom of an intermediate node at i-th dimension, $i = 2, 3, \ldots n$, is equal to

$$DF_i = 2^{n-i} + 1.$$

In general, the degree of freedom DF of the n-dimensional $\mathcal{N}$-hypercube is defined as

$$DF = \sum_i DF_i = \sum_i (2^{n-i} + 1).$$

FIGURE 15.8
Degree of freedom and rotation of the $\mathcal{N}$-hypercube.

$$\text{x00} \iff \text{x01} \iff \boxed{\text{xx1}} \iff \text{x11} \iff \text{x10} \iff \boxed{\text{xx0}} \iff \text{x00}.$$
$$\text{00x} \iff \boxed{\text{0xx}} \iff \text{01x} \iff \text{11x} \iff \boxed{\text{1xx}} \iff \text{10x} \iff \text{00x}.$$

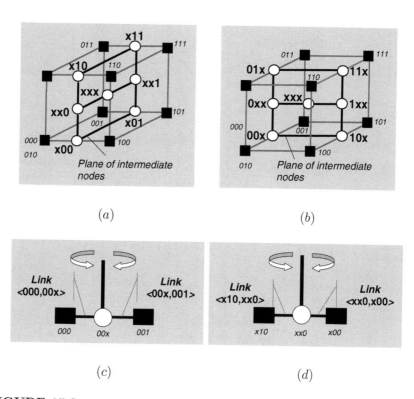

FIGURE 15.9

Coordinate description of the $\mathcal{N}$-hypercube: (a) the first plane, (b) the second plane, (c) the links of an intermediate node, and (d) links of the root (Example 15.5).

An $\mathcal{N}$-hypercube includes two types of links with respect to the root node:

$$\text{Link 1: } \boxed{\text{xx0}} \iff \text{xxx} \iff \boxed{\text{xx1}},$$
$$\text{Link 2: } \boxed{\text{0xx}} \iff \text{xxx} \iff \boxed{\text{1xx}}.$$

The root node coordinate is xxx. There are two types of links in an $\mathcal{N}$-hypercube: links between terminal nodes and intermediate nodes, and links between intermediate nodes, including the root node.

Example 15.5 *In Figure 15.9, link <000,00x> indicates the connection of the terminal node 000 and intermediate node 00x. By analogy, if two intermediate nodes x10 and xx0 are connected, we indicate this fact by <x10,xx0> (Figure 15.9d).*

The number of terminal nodes in the $\mathcal{N}$-hypercube is always equal to the number of nodes in the hypercube. Therefore, the classic hypercube can be considered the basic data structure for representing Boolean functions in which the $\mathcal{N}$-hypercube can be embedded.

There are direct relationships between the hypercube and $\mathcal{N}$-hypercube.

Example 15.6 *The coordinate of a link (face) in the hypercube corresponds to the coordinate of an intermediate node located in the middle of this link (face in the $\mathcal{N}$-hypercube) (Table 15.2).*

TABLE 15.2
Relationship of the components of the hypercube and $\mathcal{N}$-hypercube
(Example 15.6)

Hypercube	$\mathcal{N}$-hypercube
Links x00, 0x0, x10, 10x	Intermediate nodes x00, 0x0, x10,10x
Faces xx0, xx1, 0xx, 1xx, x1x	Intermediate nodes 0xx, 1xx, x1x,x0x

15.6.4 $\mathcal{N}$-hypercube design for $n > 3$ dimensions

Consider the two 3D $\mathcal{N}$-hypercubes shown in Figure 15.9b. To design a 4D $\mathcal{N}$-hypercube, two $\mathcal{N}$-hypercubes must be joined by links. There are seven possibilities for connecting two $\mathcal{N}$-hypercubes since links are allowed between intermediate nodes, intermediate nodes and root nodes, and between the root nodes. The new root node is embedded in the link <xxx0, xxx1>.

Therefore, the number of bits in the coordinate description of both $\mathcal{N}$-hypercubes must be increased by one bit. Suppose that $\mathcal{N}$-hypercubes are connected via link <xxx0, xxx1> between the root nodes xxx0 and xxx1. The resulting topological structure is called a 4D $\mathcal{N}$-hypercube.

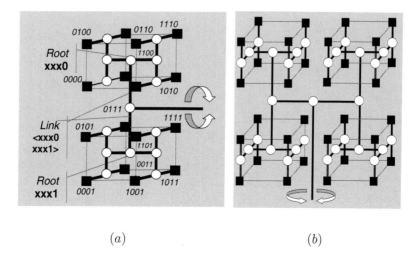

(a)   (b)

FIGURE 15.10
Connections between $\mathcal{N}$-hypercubes in n-dimensional space (Example 15.7).

Example 15.7   *Figure 15.10 shows the possibility of connecting a given $\mathcal{N}$-hypercube to another $\mathcal{N}$-hypercube. This connection property follows from the properties of intermediate nodes.*

15.7 Embedding a binary decision tree into an $\mathcal{N}$-hypercube

A binary tree that represents a Boolean function can be embedded into a hypercube that also represents this function. The $\mathcal{N}$-hypercube can be specified as a hypercube with the following properties:

Property 1: An n-dimensional $\mathcal{N}$-hypercube is derived by embedding an n-level 2^n-leaf complete binary tree into an n-dimensional hypercube.

Property 2: An $\mathcal{N}$-hypercube includes $k = 2^n$ terminal nodes labelled from 0 to $2^n - 1$ so that there is an edge and an intermediate node between any two terminal nodes if the binary representations of their labels differ by only one bit.

Property 3: Each intermediate node is assigned a label that corresponds to a binary representation of the adjacent nodes with the don't-care value for the only different bit.

The leaf vertex, or a terminal node of the complete binary decision tree with n levels, can be embedded into a hypercube with 2^n vertices and $n \times 2^{n-1}$ edges. This is because the complete binary decision tree with n levels has 2^n leaves. This is exactly the number of nodes in the hypercube structure, where each node is connected to $n-1$ neighbors and assigned the n-bit binary code that satisfies the Hamming encoding rule, and, thus has $n \times 2^{n-1}$ edges.

The number of possible embeddings of a complete n-level binary decision tree representing a Boolean function of n variables into an $\mathcal{N}$-hypercube is $n!$. The number of possible variable orders given n variables is equal to $n!$, so the possible number of embeddings is $n!$

Recurrence is a general strategy to generate spatial and homogeneous structures. This strategy can be used for embedding a binary decision tree into a $\mathcal{N}$-hypercube:

Given: A binary decision tree.
Step 1. Embed 2^n leaves of the binary tree (nodes of the level $n + 1$) into the 2^n-node of the n-dimensional hypercube; assign a code to the node so that each node is connected to q Hamming-compatible nodes.
Step 2. Embed 2^{n-1} nodes of the binary tree (nodes of the level n) into edges connecting the existing nodes of the hypercube, taking into account the polarity of the variable.
Step 3. Repeat recursively till we embed the root of the tree into the center of the hypercube.
Result: The structure obtained is a $\mathcal{N}$-hypercube.

Example 15.8   *Let $n = 1$; then a binary decision tree represents a Boolean function of one variable (Figure 15.11). The function takes a value of 1 while $x = 0$ and value 0, while $x = 1$. These values are assigned to two leaves of the binary decision diagram.*

FIGURE 15.11
Embedding a binary decision tree of one variable into the $\mathcal{N}$-hypercube (Example 15.8).

Example 15.9   *Let* $n = 2$; *then the four leaf nodes of the complete binary decision tree of a two-variable function can be embedded into a* $\mathcal{N}$-*hypercube with one root and four intermediate nodes and four leaves (Figure 15.12, second hypercube). To implement a binary decision tree, the root and two intermediate nodes are used, so that there are* $n! = 2! = 2$ *possible embeddings (last two hypercubes in Figure 15.12).*

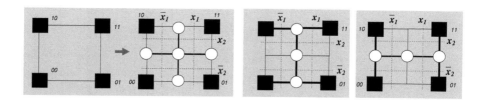

FIGURE 15.12
Embedding a binary decision tree into a 2D $\mathcal{N}$-hypercube (Example 15.9).

Example 15.10   *Let* $n = 3$, *then 8 leaves of the complete binary decision tree of the three-variable function are embedded into the 3D* $\mathcal{N}$-*hypercube (Figure 15.13).*

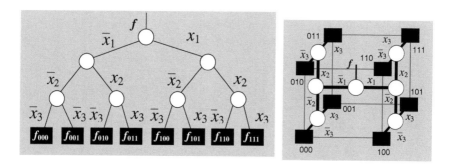

FIGURE 15.13
Embedding the complete binary tree into the 3D $\mathcal{N}$-hypercube (Example 15.10).

15.8 Assembling $\mathcal{N}$-hypercubes

Assembling is one possible approach to $\mathcal{N}$-hypercube design. There are two assembly procedures for the $\mathcal{N}$-hypercube design:

▶ Assembling an $\mathcal{N}$-hypercube from $\mathcal{N}$-hypercubes of smaller dimensions; this is a recursive procedure based on several restrictions (rules).

▶ Assembling a shared set of $\mathcal{N}$-hypercube-based structures; in this approach, some extensions of the above-mentioned rules are used.

The following rules are the basis of the assembly procedure:

Rule 1 (Connections of leaves). A terminal node is connected to one intermediate node only. In Figure 15.14, 32 paired terminal nodes are connected to 16 intermediate nodes.

Rule 2 (Connections). Each intermediate node is connected to a node in the upper dimension and to the lower dimension node, and the root node is connected to two intermediate nodes located symmetrically in opposite faces. Figure 15.14 explains this for a 5D structure.

Rule 3 (Symmetry). Configurations of the terminal and intermediate nodes on the opposite faces are symmetric. In the assembly, Figure 15.14, two pairs of 3D $\mathcal{N}$-hypercubes are connected via their root nodes, forming two new root nodes, and then two pairs are symmetrically connected via a new root nodes, etc.

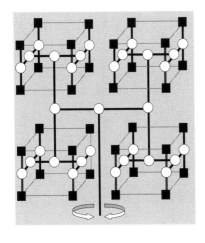

Rule 1 A terminal node is connected to one intermediate node only.

Rule 2 The root node is connected to two intermediate nodes located symmetrically in opposite faces.

Rule 3 Configurations of the terminal and intermediate nodes on the opposite face are symmetric. Two symmetrical planes include terminal nodes only.

FIGURE 15.14
Assembly rules for $\mathcal{N}$-hypercube design.

If the values of some terminal nodes in two $\mathcal{N}$-hypercubes assigned with the same codes are equal, then these $\mathcal{N}$-hypercubes can share some nodes. The $\mathcal{N}$-hypercubes are called *shared* $\mathcal{N}$-hypercubes

15.8.1 Incomplete $\mathcal{N}$-hypercubes

The embedding technique uses the properties of *host* and *guest* graphs. The host graph describes the interaction between the computing nodes; this is a decision tree or diagram. The topology of the computer system is captured by the host graph. The embedding function maps each node and edge in the guest graph (decision tree or diagram) into a unique node and edge, respectively, in the host graph (computer system configuration).

> **Example 15.11** *Incomplete $\mathcal{N}$-hypercubes are constructed by deleting given nodes (Figure 15.15).*

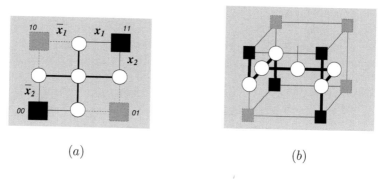

(a) (b)

FIGURE 15.15
Incomplete 2D (a) and 3D (b) $\mathcal{N}$-hypercubes (Example 15.11).

15.8.2 Embedding technique

Embedding decision diagrams into incomplete $\mathcal{N}$-hypercubes consists of the following steps: (a) representation of original decision diagram in the form of multiterminal decision diagrams, (b) embedding multiterminal decision diagrams into $\mathcal{N}$-hypercubes, and (c) deleting nonused intermediate and terminal nodes in an $\mathcal{N}$-hypercube.

Example 15.12 *Figure 15.16 shows the technology of embedding of decision diagrams with different topologies into incomplete 3D $\mathcal{N}$-hypercubes.*

(a) *In Figure 15.16a, a decision tree is transformed into a decision diagram, and then to a linear decision diagram, which is embedded into an incomplete 3D $\mathcal{N}$-hypercube.*

(b) *In Figure 15.16b,c a decision tree is transformed into a decision diagram, then it is restructured to a multiterminal diagram, and embedded into an incomplete 3D $\mathcal{N}$-hypercube.*

15.9 Representation of $\mathcal{N}$-hypercubes using H-tree

2D H-tree is topologically isomorphic to a binary tree. The 2D H-tree topology is constructed recursively from 2D elementary H-clusters.

Example 15.13 *2D H-tree-based topological structures are shown in Fig. 15.17, where Figure 15.17(a) corresponds to 2-input Boolean function, and Figure 15.17(b) is a 4-input Boolean function. The number of input variables in the function equals that of the number of levels in the structure. In Figure 15.17, n equals the number of input variables or levels.*

Example 15.14 *In Figure 15.18, an H-tree-based topological structures for the benchmark C17 (a logic network from standard data base) are shown. Solution is obtained using the NeuroNet Package.*

There are several configurations related to the H-tree. In particular, an H-fractal is a Peano curve and a Mandelbrot tree. The H-tree structure is used in phased-array radar antennae since they have many radiating elements of different sizes together with long wire packed into a small volume. The flat, thick-stemmed H-tree and the ratio between the stem and the branches, and width of the stem and the branches are important geometrical parameters in

Techniques for embedding decision trees and diagrams into hypercube-like topology

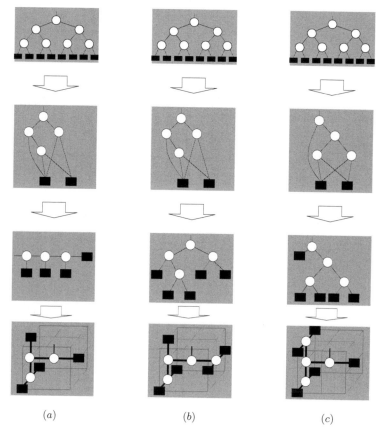

(a)   (b)   (c)

FIGURE 15.16

Embedding decision trees into incomplete $\mathcal{N}$-hypercubes (Example 15.12).

radiation. A complete binary tree of $n = 2^k - 1$ vertices requires (Fig. 15.17a)

$$\texttt{Area} = \texttt{Width} \times \texttt{Height} = O(n) \times O(\lg n) = O(n \lg n)$$

The number of wires is halved from one level to the next, but the length of the wires is doubled. Hence, the amount of wire devoted to each level of the tree is the same. The recurrence that describes the area required by by H-tree is

$$A(n) = 4A(\lfloor \tfrac{n}{4} \rfloor) + 4\sqrt{\tfrac{n}{4}} + 1, \ A(n) = 1 \text{ for } n = 1, \ n = 2 \times 4^k - 1 \text{ and } k \geq 1$$

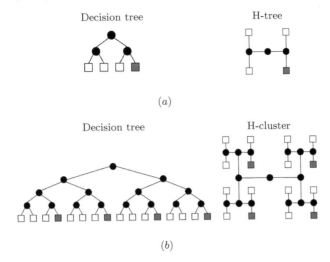

FIGURE 15.17

Representation of decision tree using *H*-trees for $n = 2$ (a) and $n = 4$ (c) (Example 15.13).

The solution of this recurrence is $O(\sqrt{n})$, that is, less than the area of a complete binary tree.

Example 15.15 *(Continuation of Example 15.14) 2D H-trees shown in Figure 15.18 are converted into $\mathcal{N}$-hypercube (Figure 15.19).*

15.10   Spatial topological measurements

There exist a number of basic measures for describing the $\mathcal{N}$-hypercube-based structures (Table 15.3). The *diameter* of a network is defined as the maximum distance between any two nodes in the network. *Link complexity* or *node degree* is defined as the number of physical links per node. For a regular network, where all nodes have the same number of links, the node degree of the network is that of a node. In an $\mathcal{N}$-hypercube, the node degree is 3, except for terminal nodes, whose degree is one.

Given two $\mathcal{N}$-hypercubes $A = a_1 a_2 \ldots, a_n$ and $B = b_1 b_2 \ldots, b_n$, the *intercube distance* $D_i(A, B)$ between A and B along the i-th dimension is 1 if $\{a_i, b_i = \{0, 1\}\}$; otherwise, it is 0. The *distance* between two hypercubes A, B is given by $D(A, B) = \sum_{i=1}^{n} D_i(A, B)$.

Given graphs G and H, an *embedding* of a graph G into a graph H is a

Non-optimized H-tree for the Boolean function

Output 1   Output 2

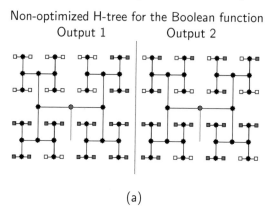

(a)

Optimized H-tree for the Boolean function

Output 1   Output 2

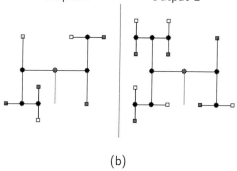

(b)

FIGURE 15.18

Nonoptimized (a) and optimized (b) H-tree for the 2-output Boolean function
of five variables (Example 15.14).

one-to-one mapping α: $V(G) \rightarrow V(H)$, along with a mapping of β, an edge
$(u, v) \in E(G)$, to a path between $\alpha(u)$ and $\alpha(v)$ in H.

> **Example 15.16** *Embedding multiterminal decision diagrams of
> different shape into $\mathcal{N}$-hypercubes is shown in Figure
> 15.20:*
>
> *(a) 2-level S_2–$2S_4$ tree into 2D $\mathcal{N}$-hypercube*
> *(b) 3-level S_2–$2S_2$–$4S_4$ tree into 3D $\mathcal{N}$-hypercube*
> *(c) 2-level S_2–$(S_2 + S_4)$ tree into 2D $\mathcal{N}$-hypercube*
> *(d) 2-level S_4–$4S_2$ tree into 2D $\mathcal{N}$-hypercube*

Note that the bisection width indicates the volume of communication

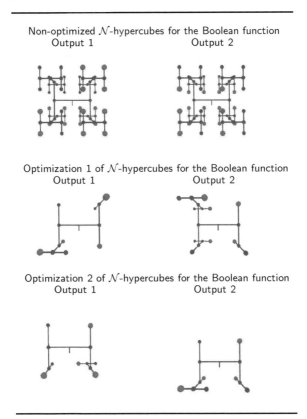

Non-optimized $\mathcal{N}$-hypercubes for the Boolean function
Output 1   Output 2

Optimization 1 of $\mathcal{N}$-hypercubes for the Boolean function
Output 1   Output 2

Optimization 2 of $\mathcal{N}$-hypercubes for the Boolean function
Output 1   Output 2

FIGURE 15.19
Conversion H-trees into $\mathcal{N}$-hypercubes and their optimization (Example 15.15).

allowed between any two halves of the network with an equal number of nodes. The diameter of a network represents the communication delay in the network. For network with the same diameter and degree, the greater number of nodes it connects, the more powerful it is. A high degree of the node represents a high level of communication channels and physical connections required between the computing elements, which would increase both the complexity in constructing the network and the costs in physically implementing the connections between these elements. The size of a hypercube can only be increased by doubling the number of nodes; that is, the granularity of size scaling in an n-dimensional hypercube is 2^n. Other important measures for interconnection networks are scalability, routing simplicity, and fault tolerance. A network is said to be high-performance if it has good properties against all the foregoing measures.

TABLE 15.3
Metrics on $\mathcal{N}$-hypercube based structures

Metric	Characteristic
Diameter	The maximum distance between any two nodes in the network
Link complexity	The number of physical links per node
Dilation of an edge	The length of the path $\alpha(e)$ in H. The dilation of an embedding is the maximum dilation with respect to all edges in G
Average message distance	The average number of links that a carrier should travel between any two nodes
Total number of primitives	The number of $\mathcal{N}$-hypercube primitives in the network
Effectiveness	The average number of variables that represent an $\mathcal{N}$-hypercube
Active nodes	The nodes connected to nonzero terminals through a path
Connectivity	The number of paths from the root
Average path length	The number of links connecting the root node to a nonzero terminal
Bisection width	The minimum number of links to be removed in order to partition the network into two equal halves
The granularity of size scaling	The ability of the system to increase in size with minor or no change to the existing configuration

15.11   Further study

Advanced topics of spatial computational structures

The 3D computing architecture has been explored for the design of macro-scale distributed systems [3], [19], [20], [32]. The 3D computing architecture concept has been employed by the creators of the supercomputers Cray T3D, NEC's Ncube, and cosmic cube [39]. The components of these supercomputers, as systems, are designed based on classical paradigms of 2D architecture that becomes 3D because of the *3D topology* (of interconnects), or *3D data structures* and corresponding algorithms, or *3D communication flow*. An example is Q3M (Quantum 3D Mesh) model of a parallel quantum computer in which each cell has a finite number of quantum bits (qubits) and interacts

Techniques for embedding multiterminal decision trees into hypercubes

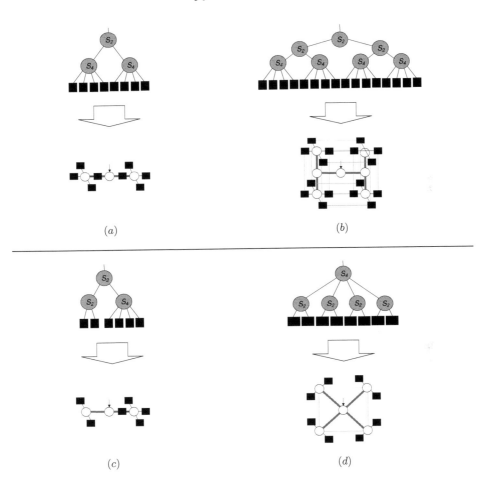

FIGURE 15.20
Embedding multiterminal decision diagrams into $\mathcal{N}$-hypercubes (Example 15.16).

with neighbors, exchanging particles by means of Hamiltonian derivable from Schrödinger equation [13].

The topology of 1D and 2D parallel-pipelining, that is, systolic processors, has been studied since the 80s, starting from the work by Kung [27]. *3D arrays*, e.g., hypercube architecture, was considered in a few papers, such as by Hayes [19].

3D VLSI hierarchy was considered by Lai et.al. [29], including the most complex ones with multiunit architectures such as 3D arrays [11]. 3D cellular automata and VLSI applications were considered by Tsalides et al. [43]. Also, various aspects of a 3D VLSI design have been studied by Kurino et. al. [28], Jain et al. [24, 25], and Horiguchi and Ooki [22].

The fundamental aspects of hypercube graph theory are discussed by Cormen et al. and Saad et al. in [6] and [38], respectively. The properties of the different topologies are studied in a number of papers, in particular, by Becker et al. [1], Ohring et al. [35], Shen et al. [40], and Wagner [48].

Fibonacci topology was studied by Hsu [23], Jiang [26], and Wu [50]. A Fibonacci cubes consists of two smaller Fibonacci cubes of unequal sizes and are related to incomplete $\mathcal{N}$-hypercubes topology.

Lattice structures. Perkowski et al. [36] developed an approach for embedding decision diagrams into a lattice structure. Butler et al. [2], Popel and Dani [37] studied the representation of Boolean and multivalued functions by a transeunt triangle topology. Requirements of the topology of nanostructures have been studied by Collier et al. [5], Ellenbogen et al. [10], Lyshevski [33], Yanushkevich et al. [52], and in [8]. Margolis et al. [34], Smith [42] and Frank [14, 15] advocated for 3D memory array as the way to accommodate the requirements for speed of a nanometer-size computing structure to be integrated on a single tiny chip. In addition, there are many 3D algorithms and designs for existing microscale components that are arranged in 3D space, which computer designers already have experience with (see, for example, works by Greenberg et al. [18] and Leiserson et al. [31].

Further reading

[1] Becker B and Simon HU. How Robust is the *n*-cube? *Information and Computation*, 77:162–178, 1988.

[2] Butler JT, Dueck GW, Yanushkevich SN, and Shmerko VP. On the number of generators for transeunt triangles. *Discrete Applied Mathematics*, 108:309–316, 2001.

[3] Campbell ML, Toporg ST, and Taylor SL. 3D wafer stack neurocomputing. In *Proceedings of the IEEE International Conference on Wafer Scale Integration*, pp. 67–74, 1993.

[4] Chaudhuri PP, Chowdhury DP, Nandi S, and Chattopadhyay S. *Additive Cellular Automata: Theory and Applications*, IEEE Computer Society – Wiley, 1997.

[5] Collier CP, Wong EW, Belohradsk M, Raymo FM, Stoddart JF, Kuekes PJ, Williams RS, and Heath JR. Electronically configurable molecular-based logic gates. *Science*, 285:391–394, July 1999.

[6] Cormen TH, Leiserson CE, Riverst RL, and Stein C. *Introduction to Algorithms*. MIT Press, 2001.

[7] Crawley D, NikolicK, Forshaw M, Ackermann J, Videlot C, Nguyen TN, Wang L, and Sarro PM. 3D molecular interconnection technology. *J. Micromechanics and Microengineering*, 13(5):655–662, 2003.

[8] Crawley D, Nikolić K, and Forshaw M., Eds., *3D Nanoelectronic Computer Architecture and Implementation.* Institute of Physics Publishing, UK, 2005.

[9] DeHon A. Array-based architecture for FET-based nanoscale electronics *IEEE Trans. Nanotechnology*, 2(23):23–32, 2003.

[10] Ellenbogen JC and Love JC. Architectures for molecular electronic computers: logic structures and an adder designed from molecular electronic diodes. In *Proceedings of the IEEE*, 88:386–426, 2000.

[11] Endoh T, Sakuraba H, Shinmei K, and Masuoka F. New three dimensional (3D) memory array architecture for future ultra high density DRAM. In *Proc. of the 22nd Int. Conf. on Microelectronics*, 442:447–450, 2000.

[12] Feringa BL, Ed., *Molecular Switches*, Wiley-VCH, Weinheim, Germany, 2001.

[13] Frank MP and Knight TF Jr. Ultimate theoretical models of nanocomputers. *Nanotechnology*, 9:162–176, 1998.

[14] Frank MP. Physical Limits of Computing. *Computing in Science and Engineering*, 4(3):16–25, 2002.

[15] Frank MP. Approaching the physical limits of computing. In *Proc. of the IEEE 35th Int. Symp. on Multiple-Valued Logic*, pp. 168–185, 2005.

[16] Gerousis C, Goodnick SM, and Porod W. Toward nanoelectronic cellular neural networks. *Int. J. of Circuits Theory and Applications* 28(6):523-535, 2000.

[17] Goser K, Pacha C, Kanstein A, and Rossmann ML. Aspects of systems and circuits for nanoelectronics. *Proceedings of the IEEE*, 85:558–573, April, 1997.

[18] Greenberg RI. The fat-pyramid and universal parallel computation independent of wire delay. *IEEE Trans. Comput.*, 43(12):1358–1364, 1994.

[19] Hayes JP, Mudge T, Stout QF, Colley S, and Palmer J. A microprocessor-based hypercube supercomputer. *IEEE Micro*, 6(5):6–17, 1986.

[20] Hillis WD. *The Connection Machine.* MIT Press, Cambridge, MA, 1985.

[21] Ho C-T and Johnsson SL. Embedding hyperpyramids into hypercubes. *IRM J. Res. Develop.*, 38(1):31–45 1994.

[22] Horiguchi S and Ooki T. Hierarchical 3D-torus interconnection network. In *Proc.of the Int. Simp. on Parallel Architectures, Algorithms, and Networks*, Dallas, TX, pp. 50–56, 2000.

[23] Hsu WJ. Fibonacci cubes – a new interconnection topology. *IEEE Trans. on Parallel and Distributed Systems*, 4(1):3–12, 1993.

[24] Jain VK, Ghirmai T, and Horiguchi S. TESH: a new hierarchical interconnection network for massively parallel computing. *IEICE Trans. on Information and Systems*, E80D(9):837–846,1997.

[25] Jain VK and Horiguchi S. VLSI considerations for TESH: a new hierarchical interconnection network for 3D integration. *IEICE Trans. on VLSI Systems*, 6(3):346–353,1998.

[26] Jiang FS. Embedding of generalized Fibonacci cubes in hybercubes with faulty nodes. *IEEE Trans. on Parallel and Distributed Systems*, 8(7):727–737, 1997.

[27] Kung HT and Leiserson CE. Systolic arrays for VLSI. In Mead C and Conway L, Eds., *Introduction to VLSI Systems*, Addison-Wesley, pp. 260–292, 1980.

[28] Kurino H, Matsumoto T, Yu KH, Miyakawa N, Itani H, Tsukamoto H, and Koyanagi M. Three-dimensional integration technology for real time microvision systems. In *Proc. of the IEEE Int. Conf. on Innovative Systems in Silicon*, pp. 203–212, 1997.

[29] Lai YT and Wang PT. Hierarchical interconnection structures for field programmable gate arrays. *IEEE Trans. VLSI Systems*, 5(2):186–196, 1997.

[30] Leighton FT. *Introduction to Parallel Algorithms and Architectures: Arrays, trees and Hypercubes.* Kaufmann, San Mateo, 1992.

[31] Leiserson CH. Fat-trees: universal networks for hardware-efficient supercomputing. *IEEE Trans. Comput.*, 34(10):892–901, 1985.

[32] Little MJ, Grinberg J, Laub SP, Nash JG, and Jung MW. The 3D computer. In *Proc. of the IEEE Int. Conf. on Wafer Scale Integration*, pp. 55–64, 1989.

[33] Lyshevski SE. 3D multi-valued design in nanoscale integrated circuits. In *Proc. of the 35th IEEE Int. Symp. on Multiple-Valued Logic*, pp. 82–87, 2005.

[34] Margolus N and Levitin L. The maximum speed of dynamical evolution. *Physica D*, 120:188–195, 1998.

[35] Ohring S and Das SK. Incomplete hypercubes: embeddings of tree-related networks. *J. Parallel and Distributed Computing*, 26:36–47, 1995.

[36] Perkowski MA, Chrzanowska-Jeske M, and Xu Y. Lattice diagrams using Reed–Muller logic. In *Proc. of the IFIP WG 10.5 Int. Workshop on Applications of the Reed–Muller Expansion in Circuit Design*. Japan, pp. 85–102, 1997.

[37] Popel DV and Dani A. Sierpinski gaskets for logic function representation. In *Proc. of the IEEE 32nd Int. Symp. on Multiple-Valued Logic*, pp. 39–45, 2002.

[38] Saad Y and Schultz MH. Topological properties of hypercubes. *IEEE Trans. Comput.*, 37(7):867–872, 1988.

[39] Seitz CL. The cosmic cube. *Communications of the ACM*, 28(1):22–33, 1985.

[40] Shen X, Hu Q, and Liang W. Embedding k-ary complete trees into hypercubes. *J. Parallel and Distributed Computing*. 24:100–106, 1995.

[41] Shmerko VP and Yanushkevich SN. Three-dimensional feedforward neural networks and their realization by nano-devices. *Artificial Intelligence Review Int. J.*, Special Issue on Artificial Intelligence in Logic Design, pp. 473–494, 1994.

[42] Smith W. Fundamental physical limits on computation. *Technical report, NECI*, May 1995. http://www.neci.nj.nec.com/ - homepages/wds/fundphys.ps.

[43] Tsalides Ph, Hicks PJ, and York TA. Three-dimensional cellular automata and VLSI applications. *IEE Proceedings*, Pt.E, 136(6):490–495, 1989.

[44] Tzeng N-F and Chen H-L. Structural and tree embedding aspects of incomplete hypercubes. *IEEE Trans. Comput.*, 43(312):1434–1439, 1994.

[45] Sasao T and Butler JT. Planar decision diagrams for multiple-valued functions, *Int. J. Multiple-Valued Logic*, vol. 1, pp. 39–64, 1996.

[46] Vichniac G. Simulating physics with cellular automata. *Physica D*, 10:96–115, 1984.

[47] Vitanyi PMB. Multiprocessor architectures and physical law. In *Proc. of the 2nd IEEE Workshop on Physics and Computation*, Dallas, TX, pp. 24–29, 1994.

[48] Wagner AS. Embedding the complete tree in hypercube. *J. Parallel and Distributed Computing*, 26:241–247, 1994.

[49] Wu AY. Embedding tree networks into hypercubes. *J. Parallel and Distributed Computing*, 2:238–249, 1985.

[50] Wu J. Extended Fibonacci cubes. *IEEE Trans. Parallel and Distributed Systems*, 8(12):1203–1210, 1997.

[51] Yanushkevich SN, Shmerko VP, and Lyshevski SE. *Logic Design of NanoICs*, CRC Press, Boca Raton, FL, 2004.

[52] Yanushkevich SN, Shmerko VP, Guy L, and Lu DC. Three dimensional multiple-valued circuit design based on single-electron logic. In *Proc. 34th IEEE Int. Symp. on Multiple-Valued Logic*, pp. 275–280, 2004.

16

Linear Cellular Arrays

Linear arrays is a particular case of computational networks. There are various types of linear arrays that can be classified using criteria of the computing algorithm, function of the node, memory requirements, and control functions. Linear arrays are potential candidates for nanotechnology because they can be designed using multiple copies of the operation node. However, linear arrays are different from decision diagrams (they are also constructed on the copy principle) because they have a regular interconnection. A particular case of decision diagrams has linear topology and, therefore, can be embedded into a linear network.

In this chapter, we introduce a linear array design based on mapping a linear decision diagram into the linear multiplexer tree, which is a linear cellular array. The following properties characterize a linear array: (a) an arbitrary logic function or a multilevel logic network can be represented by a set of linear arrays, (b) an arbitrary path in logic network can be represented by a linear array, and (c) the set of linear arrays can be represented as a 3D array.

This chapter introduces design techniques for linear cellular arrays based on various computing paradigms. The designed linear arrays are characterized by the common properties, such as identical cells, local transitions, and scalability, and are distinguished by a complexity of the function of the cell, data transmission between cells, type of the parallelism, local and global control functions, flexibility for the 3D array extension, memory size, and memory organization.

16.1 Introduction

Cellular arrays show promise in the area of predictable technologies. They can be one, two, or three-dimensional, as shown in Figure 16.1. Cellular arrays are characterized by the following common features: a large number of simple *identical* cells, a small number of *local* transition instructions between cells, high *scaling* properties, *massively parallel* computing, and *universality*; that is, an arbitrary Boolean function can be implemented using cellular arrays.

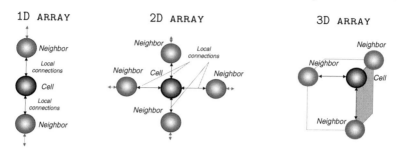

FIGURE 16.1
Cellular arrays of various dimensions.

Cellular arrays are distinguished using detailed criteria as follows:

Distinguished criteria of cellular arrays

Criterion 1: Complexity of the function of the cell; this function must be very simple for the implementation using molecular switches.

Criterion 2: Data transmission between cells and arrays (underlying physical or chemical phenomena, speed, direction, etc.).

Criterion 3: Scalability techniques; changes of the data size do not cause changes at the architectural level.

Criterion 4: Type of the parallelism with respect to the level of abstraction.

Criterion 5: Local and global control functions, the complexity of local transactions, testing, and verification.

Criterion 6: Flexibility for the 3D array extension; molecular devices can be fabricated as 2D or 3D computing structures; design techniques must provide flexibility in 2D to 3D mapping and vice versa.

Criterion 7: Memory size and memory organization.

Hence, cellular arrays are intrinsically parallel computing structures consisting of a regular latticework of identical processors that compute in lockstep while exchanging information with nearby processors. The cellular arrays overcome interconnect limitations and mitigate other problems. This allows the scaling properties of the devices to be fully exploited. However, there are several disadvantages to cellular arrays, in particular: (a) the efficiency of implementation of different Boolean functions varies over a wide range, from unacceptable due to complexity to nearly optimal computation; and (b) universality can be achieved by expensive technical and technological solutions that are often not acceptable in practice. These disadvantages are the motivation for the development of *specific-purpose* cellular arrays, which are efficient for only one or several classes of Boolean functions and/or technological constraints. An example of specific-purpose cellular arrays are the 2D *systolic arrays*, a subclass of cellular arrays. Systolic arrays are defined as arrays of synchronized, sequentially connected processors that process data in parallel by transmitting them, in rhythmic fashion, from one processor to its close neighbor. In the *parallel-pipelined* paradigm, the input/output organization is bit-serial, whereas bit-level processing is implemented in parallel. The main disadvantage of systolic arrays is that there parallel input/output of systolic arrays must be compatible with the other devices in a system. *Linear* (1D) systolic arrays avoid this disadvantage. The implementation is suited also for hybrid technology, such as a CMOS/nanowire/molecular technology.

The key reason for using 3D networks in fabrication of discrete devices is its compactness. In the contemporary technology, this effect is achieved by *multilayered* techniques.

Example 16.1 *In chips, logic networks are typically assembled from layers, where logic cells occupy the bottom layers and their interconnections are routed in upper metal layers.*

This approach can be useful for nano scale design. However, the third dimension for some nanotechnologies can be considered beyond the multilayered concept, but as a dimension that relates to physical phenomena.

Example 16.2 *In molecular technologies, the third dimension plays the role of operation factor (spatial angles correspond to various operations or functions of molecular structures). Hence, 3D network design techniques are in demand for molecular electronics.*

The problem of a 3D computing structure design can be formulated as embedding the *guest* (through network design) into the *host* (required by technology) topology. Computing abilities of the guest structure are delegated to the host structure under some constraints of the host topology.

Similar to cellular array, a linear systolic array consists of an array of processing cells, each cell connected to its nearest neighbors only. Linear systolic arrays are different from linear systolic arrays by information flow processing that results in more efficient utilization of computing resources (systolic arrays implement matrix-based algorithms). The specific requirement for the implementation of linear systolic arrays is that a local memory is needed in each computing element. The role of this memory is trivial and can be implemented using appropriate technology because of progress in memory devices. Linear systolic arrays are a suitable candidate for the hybrid technologies, such as conventional silicon electronics and molecular electronics.

Each design methodology addresses specific properties of cellular arrays. In this chapter, we introduce a computing array design method that provides a number of useful features for the molecular electronics, such as, flexibility in 2D and 3D topological manipulations, switch-based implementation of cells, and memory structure.

16.2   Linear   arrays   based   on   systolic   computing paradigm

The main disadvantage of systolic arrays is that the massively parallel input/output of systolic arrays must be compatible with the other devices in a system: if all components of the system are based on massively-parallel principles, the compatibility of the input/outputs is ideal; otherwise, the advantages of systolic arrays can be lost in practice because of the imbalance of input/output interfaces in the system. A particular class of systolic arrays is called *linear* systolic arrays. Linear systolic arrays avoid this disadvantage. The attractive features of linear systolic arrays, which are 1D arrays, include the following:

Properties of linear systolic arrays

(*a*) Linear systolic arrays operate in a massively-parallel, pipelined fashion with data and results streaming through in regular, rhythmic patterns.

(*b*) Linear systolic arrays do not require specific compatibility on input/output interfaces when implemented in a system.

(*c*) The disadvantages of sequential input/output in systolic arrays are alleviated by parallel-pipeline processing.

(*d*) Systolic arrays can be *embedded* into an arbitrary 3D or 3D topology.

16.2.1   Terminology

Input data structure. The input and output of a systolic array are streams of binary values. The input of a systolic array is a binary vector $\mathbf{F}$ that is the *truth vector* of a Boolean function f. The truth vector $\mathbf{F}$ is the truth table of f. The length of $\mathbf{F}$ is 2^n, where n is the number of variables $x_1 \ldots x_n$ in f. The j-th element of $\mathbf{F}$ corresponds to the j-th *assignment* of variables $x_1 \ldots x_n$. The vector $\mathbf{F}$ is processed in a systolic array in a parallel-pipeline fashion.

The output data structure. The output of a systolic array is a binary vector of length 2^n. The values of the output vector are generated sequentially. There is a *delay* between the i-th input element of $\mathbf{F}$ and the j-th output element. The structure of the output vector depends on the implemented transform.

Example 16.3 *If the systolic array implements a Reed–Muller transform, the j-th element of $\mathbf{R}$ corresponds to the j-th coefficient of the polynomial expression. The polynomial expressions are different in parameters called **polarities** and type of operation (arithmetic).*

Data structure compatibility. A Boolean function can be represented in many forms, called *data structures*. In logic network design, various data structures can be used. The efficiency of collaboration between networks of different designs is defined as the *compatibility* of data structures.

Processing. Input data are manipulated by processing cells, PEs. The i-th PE_i includes a computing cell CC_i and a storage cell SC_i. Each PE_i is connected with a local memory SC_i and consists of logic elements, such as AND, OR, EXOR, and switches. The linear network of PEs implements a *Taylor-like expansion*, the fundamental expansion of logic design. *Differences* of Boolean functions are the coefficients in *polynomial-like* expansions (using Taylor expansion and various arithmetics).

Flexibility. We distinguish *algorithmic, topological,* and *functional* flexibility. The kind of flexibility depends on the criteria with respect to which flexibility is considered.

Example 16.4 *Cell arrays are flexible with respect to the implemented algorithm and the functions of computing cells.*

16.2.2 Design principles of parallel-pipeline computing structures

The basic principles of systolic array design and their correspondence to implementation strategies are given in Table 16.1. For example, the principle of identical processing elements means that the process of fabrication is simplified by copying this element. For simplification, the synchronization of cells is not considered. This and other details can be found in the literature related to this topic.

TABLE 16.1

Basic principles of systolic array design under the constraints of predictable nanotechnologies

Design principle	Implementation
Identical processing elements	Copies of processing element
Parallel-pipeline processing	Sequential connections of identical processing elements
Sequential input/output	Standard interfaces
Simple logic operations	Design under library of elements based on switches
Storage cell	FIFO register
Local interconnections	Simple instructions and synchronization

16.2.3 Design phases

Systolic array design is accomplished by three phases as follows:

Systolic array design phases

Phase 1: Representation of the algorithm in a recursive form, referred to as a factorized form of matrix transforms.

Phase 2: Mapping the matrix equations into a linear array or a 2D array.

Phase 3: Designing the processing elements (PEs), or cells, based on their formal model (size of memory for input and intermediate data, their operation and synchronization, etc.).

A linear systolic array consists of n PEs (Figure 16.2a). Each PE includes a computing cell (CC) and a storage cell (SC) organized as a first-in-first-out (FIFO) register (Figure 16.2b).

Example 16.5 *An example of the connections between three PEs is given in Figure 16.2c.*

16.2.4 Formal description of a linear systolic array

Let $\mathbf{F}$ and $\mathbf{Y}$ be the input and output binary vectors of size 2^n. The input data for the systolic array is a set of the $2^n \times 1$ binary vector $\mathbf{F}$, its l-th component $f(l)$ is loaded at l-th time ($l = 1, ..., 2^n - 1$). The resulting vector $\mathbf{Y}$ appears as an output of the n-th CC.

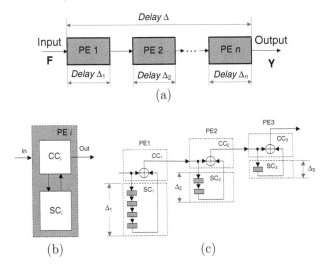

FIGURE 16.2
Topology and structure of a linear systolic array: it consists of n identical cells, PE_i (a), a processing element; PE_i, consisting of a computing cell CC_i and a local storage cell SC_i (b); and an example of a linear systolic array with three PEs (c) (Example 16.5).

Generic description a linear systolic array

The formal model of computation performed by the i-th PE is the matrix equation based on the Kronecker product.

$$\mathbf{Y} = U_{2^n} \odot \mathbf{F} \tag{16.1}$$

$$= \overbrace{(U_{2^n}^{(1)} \otimes U_{2^n}^{(2)} \otimes \cdots \otimes U_{2^n}^{(n)})}^{n \ processing \ elements} \odot \mathbf{F} \tag{16.2}$$

where $\otimes$ denotes the Kronecker product used in the factorization of matrix U_{2^n}, and $\odot$ is a vector operation specified for each particular case. Factorization provides the mapping of Equation 16.1 into a computing structure.

Description of processing element

The i-th PE processes the input vector $\mathbf{F}_i$ as follows:

$$\mathbf{Y}_i = U_{2^n}^{(i)} \odot \mathbf{F}_i$$

where $U_{2^n}^{(i)}$ is defined by the equation

$$U_{2^n}^{(i)} = I_{2^{n-i}} \otimes U_2 \otimes I_{2^{i-1}}$$

U_2 is an elementary 2×2 transform matrix, and $I_{2^{n-i}}$ and $I_{2^{i-1}}$ are identity matrices. The formal model of computation in the array of n PEs corresponds to the product of n matrices.

Functional flexibility of linear systolic array

The *functional* flexibility of systolic arrays is defined their ability to compute Boolean functions in various forms without a change in the structural and architectural properties of the systolic array. Functional flexibility is achieved by the parameters of the model. It follows from Equation 16.1, that a change in the parameters below provides the functional flexibility:

▶ The operation $\odot$, which can be EXOR, AND, OR, or other.
▶ The elementary matrix U_2 of a size 2×2 (see the last column in Table 16.2).

Processing phases

A systolic array operates in a parallel and pipelined fashion with data and results streaming through in regular, rhythmic patterns. The systolic array processes data in a similar fashion to an assembly line:

Acceleration: The first phase is *speeding-up* (acceleration), a process whereby input data is processed by the first PE, and the result is passed to the second PE, etc. (Figure 16.3).

Stationary processing: The second phase corresponds to a *stationary* processing of data. At this phase, all PEs are involved.

Deceleration: The third phase is a *slowdown* (deceleration) process. At this phase, there is no input data, but the rest of the PEs continue the processing.

16.2.5   Implementation

In this section, we utilize the functional flexibility of systolic arrays under the constraints of predictable nanotechnologies. The most feasible elementary logic function in most nanotechnologies is a *switch*. Based on these results provided by technology, we have developed a library of nanoelements that includes (a) MUX implementation of computing cell, CC_i, that is, Boolean functions AND, OR, EXOR, and NOT, and (b) A decision diagram-based implementation of a computing cell, CC_i.

TABLE 16.2
Formal models of systolic arrays and processing cells for the polynomial forms of Boolean functions and differences of two Boolean functions

Function	Model of systolic array	Model of cell
Polynomial form*	$\mathbf{R}_{2^n}^{(c)} = \mathbf{R}_{2^n}^{(c_1)} \times \cdots \times \mathbf{R}_{2^n}^{(c_n)}$	$\mathbf{R}_{2^n}^{(c_i)} = \mathbf{I}_{2^{n-i}} \otimes \mathbf{R}_2^{(c_i)} \otimes \mathbf{I}_{2^{i-1}}$
		$\mathbf{R}_2^{(0)} = \begin{bmatrix} 1 & 0 \\ 1 & 1 \end{bmatrix} \quad \mathbf{R}_2^{(1)} = \begin{bmatrix} 0 & 1 \\ 1 & 1 \end{bmatrix}$
Polynomial form**	$\mathbf{P}_{2^n}^{(c)} = \mathbf{P}_{2^n}^{(c_1)} \times \cdots \times \mathbf{P}_{2^n}^{(c_n)}$	$\mathbf{P}_{2^n}^{(c_i)} = \mathbf{I}_{2^{n-i}} \otimes \mathbf{P}_2^{(c_i)} \otimes \mathbf{I}_{2^{i-1}}$
		$\mathbf{P}_2^{(0)} = \begin{bmatrix} 1 & 0 \\ -1 & 1 \end{bmatrix} \quad \mathbf{P}_2^{(1)} = \begin{bmatrix} 0 & 1 \\ 1 & -1 \end{bmatrix}$
Polynomial form**	$\mathbf{W}_{2^n}^{(c)} = \mathbf{W}_{2^n}^{(c_1)} \times \cdots \times \mathbf{W}_{2^n}^{(c_n)}$	$\mathbf{W}_{2^n}^{(c_i)} = \mathbf{I}_{2^{n-i}} \otimes \mathbf{W}_2^{(c_i)} \otimes \mathbf{I}_{2^{i-1}}$
		$\mathbf{W}_2^{(0)} = \begin{bmatrix} 1 & 1 \\ -1 & 1 \end{bmatrix} \quad \mathbf{W}_2^{(1)} = \begin{bmatrix} 1 & 1 \\ -1 & 1 \end{bmatrix}$
Difference*	$\dfrac{\partial \mathbf{F}}{\partial x_i} = \mathbf{D}_{2^n}^{(i)} \mathbf{F} \quad (mod\ 2)$	$\mathbf{D}_{2^n}^{(i)} = \mathbf{I}_{2^{i-1}} \otimes \mathbf{D}_2 \otimes \mathbf{I}_{2^{n-i}}$
		$\mathbf{D}_2 = \begin{bmatrix} 1 & 1 \\ 1 & 1 \end{bmatrix}$
Difference**	$\dfrac{\tilde{\partial} \mathbf{F}}{\tilde{\partial} x_i} = \tilde{\mathbf{D}}_{2^n}^{(i)} \mathbf{F}$	$\tilde{\mathbf{D}}_{2^n}^{(i)} = \mathbf{I}_{2^{i-1}} \otimes \tilde{\mathbf{D}}_2 \otimes \mathbf{I}_{2^{n-i}}$
		$\tilde{\mathbf{D}}_2 = \begin{bmatrix} -1 & 1 \\ 1 & -1 \end{bmatrix}$
Difference**	$\dfrac{\tilde{\tilde{\partial}} \mathbf{F}}{\tilde{\tilde{\partial}} x_i} = \tilde{\tilde{\mathbf{D}}}_{2^n}^{(i)} \mathbf{F}$	$\tilde{\tilde{\mathbf{D}}}_{2^n}^{(i)} = \mathbf{I}_{2^{i-1}} \otimes \tilde{\tilde{\mathbf{D}}}_2^{(c_i)} \otimes \mathbf{I}_{2^{n-i}}$
		$\tilde{\tilde{\mathbf{D}}}_2^{(0)} = \begin{bmatrix} 1 & 1 \\ 1 & 1 \end{bmatrix} \quad \tilde{\tilde{\mathbf{D}}}_2^{(1)} = \begin{bmatrix} 1 & -1 \\ -1 & 1 \end{bmatrix}$

Note: * – logic operations; ** – arithmetic operations

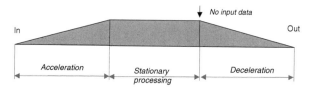

FIGURE 16.3

Three phases of the parallel-pipeline processing in a systolic array: acceleration, stationary processing, and slowdown.

To satisfy the requirements of nanotechnology, operation performed by the cells of a systolic array must be simple and must also be implemented by switches. In our approach, we used MUX-based implementation (Figure 16.4). Let I_1 and I_2 be binary values at the inputs of MUX and s be a variable, then

$$f = I_1 \times \bar{s} \vee I_2 \times s \qquad (16.3)$$

Example 16.6   *An AND gate can be represented by the MUX $f = 0 \times \bar{s} \vee I_2 \times s = x_2 x_1$. This model is a good candidate for implementation using molecular switches, SET devices, composed of nanowires and wrap-gates controlled by a voltage signal.*

The library given in Figure 16.4 can be extended to implement three and/or more input gates. These models correspond to the decision diagrams of the gates, and can be implemented on a systolic array.

16.2.6   Computing polynomial forms using logical operations

Systolic arrays for computing Boolean function in the polynomial form using logical operations (Reed–Muller expressions) are used for the solution of the compatibility of data in the form of a truth vector and coefficients of the polynomial expressions:

<Truth vector> $\Longleftrightarrow$ <EXOR expressions>

The basic operation of a polynomial expression of Boolean functions is EXOR. The logic EXOR can be implemented in some nanotechnologies in a similar manner to the AND or OR. Representation and manipulation of the EXOR expression can provide less interconnects in some cases compared with other forms.

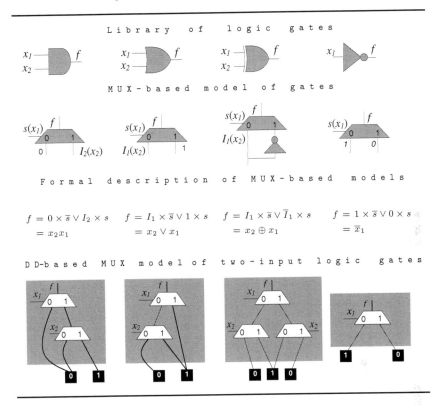

FIGURE 16.4
Nanotechnology-oriented library of elementary Boolean functions AND, OR, EXOR, and NOT, and corresponding decision diagrams.

The formal model for computing the **R** vector of polynomial coefficients given the truth vector **F** of a Boolean function of n variables is as follows:

$$\mathbf{R} = \mathbf{R}_{2^n}^{(c)}\mathbf{F} \ (mod\ 2)$$

To design a systolic array, factorization of the transform matrix $\mathbf{R}_{2^n}^{(c)}$ is needed. Using Equation 16.1, we obtain

$$\mathbf{R}_{2^n}^{(c)} = \mathbf{R}_{2^n}^{(c_1)} \times \mathbf{R}_{2^n}^{(c_2)} \times \cdots \times \mathbf{R}_{2^n}^{(c_n)} \ (mod\ 2).$$

where $i = 1, 2, \ldots, n$ and $(c_1 c_2 \ldots c_n)$ is the binary representation of c, $c \in 0, 1, \ldots, 2^n - 1$ is the *polarity* (Table 16.2). Hence, polynomial coefficients are computed in n iterations.

Example 16.7   *A formal description of a systolic array for computing the Reed–Muller transform of a Boolean function of three variables ($n = 3$) is given in Figure 16.5. Its flowgraph and the PE for the first iteration are shown in Figure 16.6.*

Note that there are 2^n polynomial expressions given Boolean functions. An example of computing various polynomial expressions for a given elementary Boolean function is given in Table 16.3.

TABLE 16.3

Representation of elementary Boolean functions by polynomial (Reed–Muller) expressions and Boolean differences

Function	Boolean differences				Polynomial
	f	$\frac{\partial f}{\partial x_2}$	$\frac{\partial f}{\partial x_1}$	$\frac{\partial^2 f}{\partial x_1 \partial x_2}$	expression
$f = x_1 \wedge x_2$	0	0	0	1	$x_1 x_2$
	0	0	1	1	$x_1 \oplus x_1 \bar{x}_2$
	0	1	0	1	$x_2 \oplus \bar{x}_1 x_2$
	1	1	1	1	$1 \oplus \bar{x}_2 \oplus \bar{x}_1 \oplus \bar{x}_1 \bar{x}_2$
$f = x_1 \vee x_2$	0	1	1	1	$x_2 \oplus x_1 \oplus x_1 x_2$
	1	1	0	1	$1 \oplus \bar{x}_2 \oplus x_1 \bar{x}_2$
	1	0	1	1	$1 \oplus \bar{x}_1 \oplus \bar{x}_1 x_2$
	1	0	0	1	$1 \oplus \bar{x}_1 \bar{x}_2$
$f = x_1 \oplus x_2$	0	1	1	0	$x_2 \oplus x_1$
	1	1	1	0	$1 \oplus \bar{x}_2 \oplus x_1$
	1	1	1	0	$1 \oplus x_2 \oplus \bar{x}_1$
	0	1	1	0	$\bar{x}_2 \oplus \bar{x}_1$

16.2.7   Computing differences using logical operations

Systolic arrays for computing Boolean differences are based on the representation of a truth vector by a set of Boolean differences:

<Truth vector> ⟺ <Logic Taylor expansion>

Boolean differences are defined as the coefficients of the logic Taylor series for a Boolean function. They detect the change in a Boolean function if a variable is changed (Table 16.4).

The matrix form of the Boolean difference with respect to the i-th variable x_i is defined by the equation given in Table 16.2. This equation describes one iteration of the m-iteration procedure for calculation of the m-order Boolean

Techniques for linear systolic array computing

$$\mathbf{R}_{2^3} = \mathbf{R}_{2^3}^{(c_1)} \mathbf{R}_{2^3}^{(c_2)} \mathbf{R}_{2^3}^{(c_3)}$$

$$= \underbrace{\left(\mathbf{I}_{2^2} \otimes \mathbf{R}_2^{(c_1)} \otimes 1\right)}_{PE1} \underbrace{\left(\mathbf{I}_2 \otimes \mathbf{R}_2^{(c_2)} \otimes \mathbf{I}_2\right)}_{PE2} \underbrace{\left(1 \otimes \mathbf{R}_2^{(c_3)} \otimes \mathbf{I}_{2^2}\right)}_{PE3}$$

$$= \begin{bmatrix} \mathbf{R}_2^{(c_1)} & & & \\ & \mathbf{R}_2^{(c_1)} & & \\ & & \mathbf{R}_2^{(c_1)} & \\ & & & \mathbf{R}_2^{(c_1)} \end{bmatrix} \begin{bmatrix} \mathbf{I}_2 & & \\ \mathbf{I}_2 & \mathbf{I}_2 & \\ & \mathbf{I}_2 & \\ & \mathbf{I}_2 & \mathbf{I}_2 \end{bmatrix} \begin{bmatrix} \mathbf{I}_4 & & \\ & & \\ \mathbf{I}_4 & \mathbf{I}_4 & \\ & & \end{bmatrix}$$

$$= \underbrace{\begin{bmatrix} 1 & & & & & & \\ 1\,1 & & & & & & \\ & 1 & & & & & \\ & 1\,1 & & & & & \\ & & 1 & & & & \\ & & 1\,1 & & & & \\ & & & 1 & & & \\ & & & 1\,1 & & & \end{bmatrix}}_{FIFO_{SC_1}} \underbrace{\begin{bmatrix} 1 & & & & \\ & 1 & & & \\ 1 & 1 & & & \\ & 1 & 1 & & \\ & & 1 & & \\ & & & 1 & \\ & & 1 & 1 & \\ & & & 1 & 1 \end{bmatrix}}_{FIFO_{SC_2}} \underbrace{\begin{bmatrix} 1 & & & & \\ & 1 & & & \\ & & 1 & & \\ & & & 1 & \\ 1 & & & & 1 \\ & 1 & & & 1 \\ & & 1 & & 1 \\ & & & 1 & 1 \end{bmatrix}}_{FIFO_{SC_3}}$$

FIGURE 16.5
Systolic array computation for the Reed–Muller transform and Boolean differences, $n = 3$ (Example 16.7).

$$\mathbf{R}_{2^3}^{(0)} = \begin{bmatrix} 1 & & & & & \\ 1 & & & & & \\ & 1 & & & & \\ & 1 & & & & \\ 1 & & 1 & & & \\ & 1 & & 1 & & \\ & & & 1 & & \end{bmatrix}$$

(a)

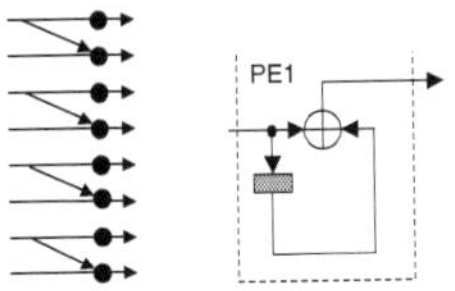

(b)

$$\mathbf{I}_{2^2} \otimes \mathbf{R}_{2^1} \otimes 1$$
$$i = 0: \quad f_0 = f(0)$$
$$i = 1:$$
$$f_1 = f(0) \oplus f(1)$$
$$\cdots\cdots\cdots\cdots\cdots$$
$$i = 7:$$
$$f_7 = f(6) \oplus f(7)$$

(c)

FIGURE 16.6
Design of PEs for computing polynomial expressions: transform matrix (a), truncated "butterfly" flowgraph and the structure of a PE (b), and formal model (c).

difference. It is implemented by a systolic array consisting of m PEs. Each PE implements the matrix-vector $\mathbf{Y}_i = D_2^{(i)} \mathbf{F}$, and the structure of the PE is similar to the one considered for the Reed–Muller transform, except that two switches are required instead of one.

TABLE 16.4

Change detection by Boolean
differences

Change of f	$\frac{\partial f}{\partial x_i}$
if $f_0 = 0$ and $f_1 = 0$ then	0
if $f_0 = 0$ and $f_1 = 1$ then	1
if $f_0 = 1$ and $f_1 = 0$ then	1
if $f_0 = 1$ and $f_1 = 1$ then	0

Example 16.8   *Figure 16.7 illustrates the design of a PE given $i = 2$ of systolic array for computing a **multiple Boolean difference** given a Boolean function of three variables. Note, there are 2^n Boolean differences given a Boolean function of n variables.*

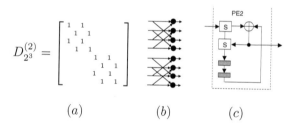

$$D_{2^3}^{(2)} = \begin{bmatrix} 1 & 1 & & & & & & \\ & 1 & 1 & & & & & \\ 1 & & 1 & & & & & \\ & & 1 & 1 & & & & \\ & & & & 1 & 1 & & \\ & & & & & 1 & 1 & \\ & & & & 1 & & 1 & \\ & & & & & & 1 & 1 \end{bmatrix}$$

(a) (b) (c)

FIGURE 16.7

Computing the Boolean differences of a Boolean function of three variables with respect to the variable x_2: transform matrix (a), "butterfly" flowgraph (b), and PE (c) (Example 16.8).

Example 16.9   *An example of the computing of Boolean differences in algebraic form for some elementary Boolean functions is given in Table 16.3.*

The Reed–Muller expression of an arbitrary Boolean function of two variables ($n = 2$) and the polarity $c = 1(c_1 c_2 = 01$, x_1 is uncomplemented and x_2 is complemented) is the logic Taylor expansion of this function $f = f(1) \oplus \frac{\partial f(1)}{\partial x_2}\overline{x}_2 \oplus \frac{\partial f(1)}{\partial x_1}x_1 \oplus \frac{\partial^2 f(1)}{\partial x_1 \partial x_2}x_1\overline{x}_2.$

16.2.8 Computing polynomial forms using arithmetic operations

Systolic arrays can be used for computing Boolean functions in a polynomial form using arithmetic operations. This form is known as an *arithmetic polynomials*. The computing of the arithmetic expressions of Boolean functions is useful in a word-level representation, for example, in compact description of a multioutput network and in representation of networks by linear decision diagrams.

Consider mapping of the arithmetic transform algorithm

$$\mathbf{P} = \mathbf{P}_{2^n}^{(c)} \mathbf{F}$$

to systolic array. To design a systolic array, factorization of the transform matrix $\mathbf{P}_{2^n}^{(c)}$ is needed. Using Equation 16.1, we obtain

$$\mathbf{P}_{2^n}^{(c)} = \mathbf{P}_{2^n}^{(c_1)} \times \mathbf{P}_{2^n}^{(c_2)} \times \cdots \times \mathbf{P}_{2^n}^{(c_n)}$$

where $i = 1, 2, \ldots, n$ and $(c_1 c_2 \ldots c_n)$ is the binary representation of c, $c \in 0, 1, \ldots, 2^n - 1$ is a *polarity* (Table 16.2).

Example 16.10 *Given $n = 3$ the flowgraph includes three iterations expressed by the factorization relations as shown in Figure 16.8. Note that there are 2^n arithmetic expressions given a Boolean function of n variables. Details are given in Table 16.2.*

Example 16.11 *An example of computing arithmetic expressions in algebraic form is given in Table 16.5.*

16.2.9 Computing differences using arithmetic operations

Computing the arithmetic differences of Boolean functions is useful in analysis of sensitivity of a network when direction of a change is important and in event-driven techniques.

Arithmetic differences are defined as the coefficients of the arithmetic Taylor series for a Boolean function. They detect the change and direction of change of a Boolean function if the variable is changed (Table 16.6). The arithmetic difference with respect to the variable x_i is denoted by $\tilde{\partial} f / \tilde{\partial} x_i$ and describes the behavior of the Boolean function in terms of change of its variable x_i, as illustrated by Table 16.6.

A matrix form of the arithmetic difference with respect to the i-th variable x_i is defined by the equation given in Table 16.2. Note that there are 2^n arithmetic differences given a Boolean function of n variables.

TABLE 16.5

Representation of elementary Boolean functions by arithmetic expressions and arithmetic differences

Function	Arithmetic difference				Arithmetic expression
	f	$\frac{\tilde{\partial} f}{\partial x_2}$	$\frac{\tilde{\partial} f}{\partial x_1}$	$\frac{\tilde{\partial}^2 f}{\partial x_1 \partial x_2}$	
x_1 —[f] x_2 — $y = x_1 \wedge x_2$	0 0 0 1	0 0 1 -1	0 1 0 -1	1 -1 -1 1	$x_1 x_2$ $x_1 - x_1 \bar{x}_2$ $x_2 - \bar{x}_1 x_2$ $1 - \bar{x}_2 - \bar{x}_1 + \bar{x}_1 \bar{x}_2$
x_1 —[f] x_2 — $y = x_1 \vee x_2$	0 1 1 1	1 -1 0 0	1 0 -1 0	-1 1 1 -1	$x_2 + x_1 - x_1 x_2$ $1 - \bar{x}_2 + x_1 \bar{x}_2$ $1 - \bar{x}_1 + \bar{x}_1 x_2$ $1 - \bar{x}_1 \bar{x}_2$
x_1 —[f] x_2 — $y = x_1 \oplus x_2$	0 1 1 0	1 -1 -1 1	1 -1 -1 1	-2 2 2 -2	$x_2 + x_1 - 2x_1 x_2$ $1 - \bar{x}_2 - x_1 + 2x_1 \bar{x}_2$ $1 - x_2 - \bar{x}_1 + 2\bar{x}_1 x_2$ $\bar{x}_2 + \bar{x}_1 - 2\bar{x}_1 \bar{x}_2$

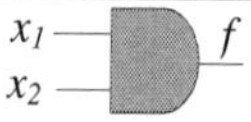

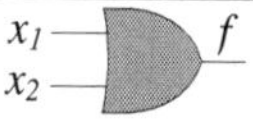

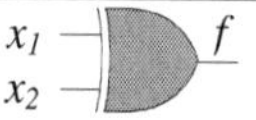

Techniques for linear systolic array computing

FIGURE 16.8

Example of a systolic array for computing the arithmetic transform and arithmetic differences, $n = 3$.

Example 16.12 *For example, the arithmetic expression of an arbitrary Boolean function of two variables ($n = 2$) and the polarity $c = 1(c_1 c_2 = 01)$ is the arithmetic Taylor expansion of this function: $f = f(1) + \frac{\tilde{\partial} f(1)}{\partial x_2} \bar{x}_2 + \frac{\tilde{\partial} f(1)}{\partial x_1} x_1 + \frac{\tilde{\partial}^2 f(1)}{\partial x_1 \partial x_2} x_1 \bar{x}_2$. It is illustrated in Table 16.5 for some logic gates.*

TABLE 16.6
Change detection by arithmetic differences

Change of f	$\frac{\tilde{\partial} f_0}{\partial x_i}$	$\frac{\tilde{\partial} f_1}{\partial x_i}$
if $f_0 = 0$ and $f_1 = 0$ then	0	0
if $f_0 = 0$ and $f_1 = 1$ then	1	-1
if $f_0 = 1$ and $f_1 = 0$ then	-1	1
if $f_0 = 1$ and $f_1 = 1$ then	0	0

Note that $\frac{\tilde{\partial} f}{\partial x_i}$ depends on x_i, since $-\overline{x}_i + x_i = (-1)^{\overline{x}_i}$ and $-x_i + \overline{x}_i = (-1)^{x_i}$, and $\frac{\tilde{\partial} f}{\partial x_i} = 0$ if f does not depend on x_i.

16.2.10 Computing Walsh expressions

Computing of Walsh expressions are useful in word-level representation of Boolean functions. One approach to mapping the Walsh transform $\mathbf{W} = \mathbf{W}_{2^n}^{(c)} \mathbf{F}$ to a systolic array is similar to the approach to mapping Reed–Muller and arithmetic transforms in factorized form. The factorization of the matrix $\mathbf{W}_{2^n}^{(c)}$ is given in Table 16.2. The flowgraph of the algorithm corresponds to the n-iteration processing of the input data (components of the truth table in Figure 16.9). The structure of the PE is similar to one the that is used for the calculation of the Boolean difference, except that control is organized in a different way. The Walsh differences are defined as coefficients of a Taylor-like series for a Boolean function. The matrix form of the Walsh difference with respect to the i-th variable x_i is defined by the equation given in Table 16.2.

Example 16.13 *The Walsh expression of an arbitrary Boolean function of two variables ($n = 2$), given the polarity $c = 1(c_1 c_2 = 01)$, is the arithmetic Taylor expansion:*
$$f = f(1) + \frac{\tilde{\partial} f(1)}{\tilde{\partial} x_2}\overline{x}_2 + \frac{\tilde{\partial} f(1)}{\tilde{\partial} x_1}x_1 + \frac{\tilde{\partial}^2 f(1)}{\tilde{\partial} x_1 \tilde{\partial} x_2}x_1 \overline{x}_2.$$
Table 16.7 contains the Walsh representation in terms of Walsh differences of several logic gates ($\sum f_i$ means the sum of values in the truth table of f).

16.2.11 Compatibility of polynomial data structures

Each of the systolic arrays considered previously provides the compatibility of a truth vector and one of its polynomial forms. The compatibility of polynomial forms is as follows:

Techniques for linear systolic array computing

$$\mathbf{W}_{2^3} = \mathbf{W}_{2^3}^{(c_1)}\mathbf{W}_{2^3}^{(c_2)}\mathbf{W}_{2^3}^{(c_3)}$$

$$= \underbrace{(\mathbf{I}_{2^2} \otimes \mathbf{W}_2^{(c_1)} \otimes 1)}_{PE1} \underbrace{(\mathbf{I}_2 \otimes \mathbf{W}_2^{(c_2)} \otimes \mathbf{I}_2)}_{PE2} \underbrace{(1 \otimes \mathbf{W}_2^{(c_3)} \otimes \mathbf{I}_{2^2})}_{PE3}$$

$$= \begin{bmatrix} \mathbf{W}_2^{(c_1)} & & & \\ & \mathbf{W}_2^{(c_1)} & & \\ & & \mathbf{W}_2^{(c_1)} & \\ & & & \mathbf{W}_2^{(c_1)} \end{bmatrix} \begin{bmatrix} \mathbf{I}_2 & & \mathbf{I}_2 & \\ \mathbf{I}_2 & \mathbf{I}_2 & & \\ & & \mathbf{I}_2 & \\ & \mathbf{I}_2 & \mathbf{I}_2 & \end{bmatrix} \begin{bmatrix} \mathbf{I}_4 & & \\ & \mathbf{I}_4 & \mathbf{I}_4 \\ & & \end{bmatrix}$$

$$= \underbrace{\begin{bmatrix} 1 & 1 & & & & & \\ -1 & 1 & & & & & \\ & & 1 & 1 & & & \\ & & -1 & 1 & & & \\ & & & & 1 & 1 & \\ & & & & -1 & 1 & \\ & & & & & & 1 & 1 \\ & & & & & & -1 & 1 \end{bmatrix}}_{FIFO_{SC_1}} \underbrace{\begin{bmatrix} 1 & & 1 & & & \\ & 1 & & 1 & & \\ -1 & & 1 & & & \\ & -1 & & 1 & & \\ & & & & 1 & & 1 \\ & & & & & 1 & & 1 \\ & & & & -1 & & 1 \\ & & & & & -1 & & 1 \end{bmatrix}}_{FIFO_{SC_2}} \underbrace{\begin{bmatrix} 1 & & & & 1 & & & \\ & 1 & & & & 1 & & \\ & & 1 & & & & 1 & \\ & & & 1 & & & & 1 \\ -1 & & & & 1 & & & \\ & -1 & & & & 1 & & \\ & & -1 & & & & 1 & \\ & & & -1 & & & & 1 \end{bmatrix}}_{FIFO_{SC_3}}$$

FIGURE 16.9

Example of a systolic array for computing Walsh transform and Walsh differences, $n = 3$.

TABLE 16.7

Representation of elementary Boolean functions by Walsh expressions and Walsh differences

Function	$\sum f_i$	Walsh differences $\frac{\tilde{\partial} f}{\tilde{\partial} x_2}$	$\frac{\tilde{\partial} f}{\tilde{\partial} x_1}$	$\frac{\tilde{\partial}^2 f}{\tilde{\partial} x_1 \tilde{\partial} x_2}$	Walsh expression
$y = x_1 \wedge x_2$	1	-1	-1	1	$\frac{1}{4}[1 - (-1)^{x_2} - (-1)^{x_1} + (-1)^{x_1 x_2}]$
	1	1	-1	-1	$\frac{1}{4}[1 + (-1)^{\overline{x}_2} - (-1)^{x_1} - (-1)^{x_1 \overline{x}_2}]$
	1	-1	1	-1	$\frac{1}{4}[1 - (-1)^{x_2} + (-1)^{\overline{x}_1} - (-1)^{\overline{x}_1 x_2}]$
	1	1	1	1	$\frac{1}{4}[1 + (-1)^{\overline{x}_2} + (-1)^{\overline{x}_1} + (-1)^{\overline{x}_1 \overline{x}_2}]$
$y = x_1 \vee x_2$	3	-1	-1	-1	$\frac{1}{4}[3 - (-1)^{x_2} - (-1)^{x_1} - (-1)^{x_1 x_2}]$
	3	1	-1	1	$\frac{1}{4}[3 + (-1)^{\overline{x}_2} - (-1)^{x_1} + (-1)^{x_1 \overline{x}_2}]$
	3	-1	1	1	$\frac{1}{4}[3 - (-1)^{x_2} + (-1)^{\overline{x}_1} + (-1)^{\overline{x}_1 x_2}]$
	3	1	1	-1	$\frac{1}{4}[3 + (-1)^{\overline{x}_2} + (-1)^{\overline{x}_1} + (-1)^{\overline{x}_1 \overline{x}_2}]$
$y = x_1 \oplus x_2$	2	0	0	-2	$\frac{1}{2}[2 - 2(-1)^{x_1 x_2}]$
	2	0	0	2	$\frac{1}{4}[2 + 2(-1)^{x_1 \overline{x}_2}]$
	2	0	0	2	$\frac{1}{4}[2 + 2(-1)^{\overline{x}_1 x_2}]$
	2	0	0	-2	$\frac{1}{4}[2 - 2(-1)^{\overline{x}_1 \overline{x}_2}]$

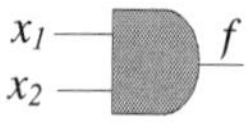

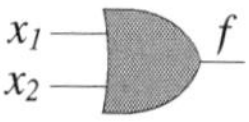

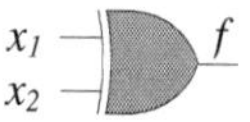

$$\langle \text{Reed-Muller expressions} \rangle \Longleftrightarrow \langle \text{Truth vector} \rangle \tag{16.4}$$
$$\Longleftrightarrow \langle \text{Arithmetic expressions} \rangle$$

$$\langle \text{Walsh expressions} \rangle \Longleftrightarrow \langle \text{Truth vector} \rangle \tag{16.5}$$
$$\Longleftrightarrow \langle \text{Reed-Muller expressions} \rangle$$

$$\langle \text{Walsh expressions} \rangle \Longleftrightarrow \langle \text{Truth vector} \rangle \tag{16.6}$$
$$\Longleftrightarrow \langle \text{Arithmetic expressions} \rangle$$

It follows from schemes 16.5 – 16.7 that there is no direct compatibility between polynomial forms, and the truth vector is used as an intermediate form.

16.3   Spatial systolic arrays

16.3.1   3D cellular array design

A 3D cellular array consists of a regular lattice of identical cells in a cubic space using 3D interconnects. Interconnections in a 3D cellular array can be implemented through via-holds between the different layers. 3D cellular arrays are characterized by 100% parallel processing, local interconnections (near/nearest neighbors only), and the processors in the 3D network being identical.

16.3.2   3D systolic array design using embedding techniques

We consider a 3D systolic array design at the *local* level, at which the cells can be implemented in 3D, and the *global* level, at which the cells can be implemented in 2D or 3D; the systolic topology is represented in 3D.

The optimal 2D array for computing is a tree. The tree-based structure is characterized as follows: the tree, embedded into a hypercube, will form a multidimensional architecture for parallel-pipelined computing (Table 16.8). The 2D tree topology used in FPGA design, in particular, is called an *H-tree*.

TABLE 16.8
Tree-like and spatial systolic arrays, decision trees

Characteristic	Tree-like systolic array	Spatial systolic array	2D decision tree	Spatial decision tree
Sum-of-products	✓	Embedded	✓	Embedded
Reed–Muller	✓	Embedded	✓	Embedded
Arithmetic	✓	Embedded	✓	Embedded
Walsh	✓	Embedded	✓	Embedded

Example 16.14 *An example of H-tree structure is given in Figure 16.10a, and Figure 16.10b shows the accuracy of the array during computation.*

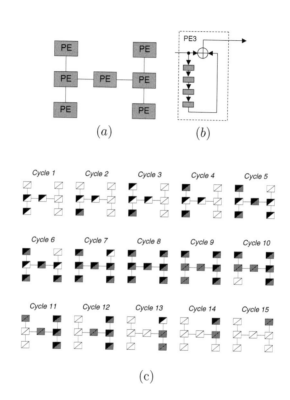

FIGURE 16.10
Computing based on H-tree systolic array: The Reed–Muller transform of functions of three variables (a), a processing element, PE, for a Davio expansion with respect to the first variable (b), 15 cycles of an H-tree network systolic array (black, gray, and white states correspond to loading, computing, and noncomputing processes, respectively).

$\mathcal{N}$-hypercubes-based models can be mapped into the so-called *hypercube arrays* since their structure is homogeneous, the function of nodes is identical, information flow is organized in a parallel-pipelined fashion, and they have sequential and parallel inputs and outputs, due to constraints on locality in 3D space.

16.3.3 3D hypercube systolic arrays

The design of a hypercube systolic array for computing a two-variable Boolean function is given in Figure 16.11. To implement a 0-polarity Reed–Muller transform for a two-variable Boolean function, we (a) design a Davio decision tree, and (b) embed the tree into a 2D hypercube. The two PEs implement the first and the second iteration of the matrix transform; these PEs are the central and intermediate nodes of the 2D hypercube, respectively (Figure 16.11c).

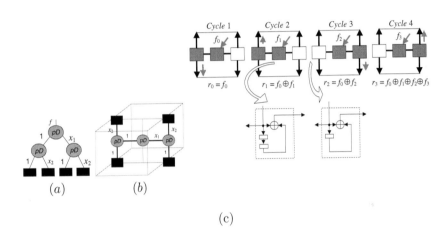

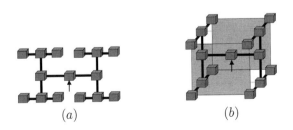

FIGURE 16.11
Graphical representation of a two-input Boolean function in sum-of-products form: decision positive Davio pD tree (a), decision pD tree embedded in a hypercube (b), and cellular topology (c) for calculation of the 0-polarity Reed–Muller expression of the function.

FIGURE 16.12
Four-iteration systolic computing using a 2D H-tree (a) and a 3D $\mathcal{N}$-hypercube (b).

16.4   Linear arrays based on linear decision diagrams

In this section, we introduce linear decision diagrams, that is, graphical data structure for computing Boolean functions. Our goal is twofold: (a) to create a library of elementary functions (gates) in the form acceptable for aggregation of these gates into a large logic network, and (b) to satisfy the restrictions of cellular structures. This goal was formulated to satisfy the requirements of molecular electronics (computing cells must be very simple and identical, mapping a 2D into a 3D computing structure and vice versa, flexibility at various levels of abstraction under constraints of molecular fabrication, etc.). Linear diagrams are considered as a special class of decision diagrams, widely used in logic design for optimization and verifications.

Various techniques of logic design of discrete devices are utilized and developed in our approach, namely, partitioning (decomposition), composition (aggregation), decision diagram techniques focusing on topological properties, and aggregation in spatial dimensions. These techniques provide the representation, manipulation, and transformations of logic networks under constraints of molecular technology.

We solve the problem of partitioning under constraints of linearity of algebraic expressions represented cofactors (subfunctions). In this approach, an arbitrary multioutput Boolean function f of n variables $x_1, \ldots, x_n$ can be decomposed into a set of *linear models*. A linear model y of n variables is an algebraic polynomial with $n + 1$ integer coefficients $d_0, d_1, \ldots, d_n$ and arithmetic operations

$$y = d_0 + d_1 x_1 + \ldots + d_n x_n \qquad (16.7)$$

A linear model (Equation 16.9) can be mapped into a linear decision diagram, in which the coefficients are assigned to terminal nodes. A restricted class of Boolean functions can be represented by a linear model (Equation 16.9), and therefore, by decision diagrams of linear topology. The approach that guarantees that an arbitrary Boolean function can be represented by a set of linear models (Equation 16.9) is introduced in this chapter. In terms of molecular devices design, this statement means that computational properties can be delegated to the structures of linear topology. The flexibility of the proposed partitioning is provided by the fact that any part of a logic network, or any subfunction of a Boolean function, can be represented by a linear model.

In this approach, the fundamental properties of an arbitrary linear data structure are utilized. The most important fact is that any elementary n-input Boolean function such as AND, OR, and NOT can be represented by linear model (16.9) and the corresponding linear decision diagram . Since AND, OR, and NOT functions form the universal set, then an arbitrary Boolean function can be described by a set of linear models. The above is the *composition* (aggregation) problem, which forms the background of the proposed approach. The following fundamental property of linear data structure is the basis of our approach: Given a set of linear expressions and the corresponding linear decision diagrams, it is possible to derive a new linear expression or the respective decision diagram using the principle of composition.

Hence, it is desirable to define the simplest linear decision diagram and the corresponding linear model. We define the sets of linear models and the corresponding linear decision diagrams of elementary Boolean functions, and then use a composition of those to form linear models and linear decision diagrams of arbitrary Boolean function. However, there are two main obstacles that seem to prevent a wide application of these techniques in practice.

The first problem addresses the type of data structure used for linearization. There were many attempts to obtain linear decision diagrams by manipulation of decision diagrams attributes. However, linear topology was observed only for particular Boolean functions, for example, adders and multipliers can be represented by binary moment diagrams of linear topology. Our linearization paradigm states that particular linear solutions (coefficients in Equation 16.9) given a Boolean function, exist and, hence, it is possible to derive linear decision diagrams for these solutions. Another approach utilizes the properties of composition of the arithmetic representations of Boolean functions, resulting in a linear model. The techniques for algebraic linearization, i.e., linear model design combined with decision diagram techniques are the basic of our approach.

The second obstacle for application of linearization paradigm in practice was exponential growth of coefficient values of Equation 16.9 for the Boolean functions of the large number of variables. A methods for encoding the coefficients in Equation 16.9 to prevent this effect has been developed. This method is very effective in terms of time of calculation for functions of hundreds of variables.

The main assumption of the introduced in this chapter approach is that elementary functions (AND, OR, EXOR, etc.) can be described by linear decision diagrams. Therefore, an arbitrary Boolean function can be decomposed to the simplest linear models and their linear combinations. To implement this idea, *arithmetic* representation of Boolean operations is used. Arithmetic representation of Boolean functions is known from George Boole, the founder of Boolean algebra.

Example 16.15 *The arithmetic expression of OR function $x_1 \vee x_2$ is* $x_1 + x_2 - x_1 x_2$, *and* $x_1 \oplus x_2 = x_1 + x_2 - 2x_1 x_2$.

Arithmetic expressions were used for designing the first computers [20]. The purpose of linearization is to find the sum of unary literals, for example,

$x_1 + x_2 + 1$. This sum is an arithmetic form of a multioutput, or word-level function. Not every arrangement of bits (functions) at the word-level results in a linear arithmetic form. Thus, some manipulation is required. Word-level manipulation of Boolean functions $f_1, f_2, \ldots, f_r$ are based on Shannon, Davio, and the arithmetic analogs of Davio expansion.

A linear model is based on linear decision diagrams. Linear decision diagrams is the result of the direct mapping of linear expressions into a word-level decision diagram. A linear decision diagram can be considered as a boundary case of word-level diagrams and has a number of useful features.

The most important and promising property for space representation is the simple embedding procedure of linear decision diagrams into 3D structures (hypercubes, hypercube-like topology, pyramids, etc.). For arithmetic expressions, the main goal is to minimize the effects of large value coefficients in linear models. The crucial idea is to replace the computation by the manipulation of codes of coefficients. This is possible in some cases because of the regular structure of coefficients.

An arbitrary Boolean function can be represented by a unique arithmetic expression, that is, polynomial expression using arithmetic operations.

Example 16.16 *Boolean function $x_1 \vee x_2$ can be represented in arithmetic form as $x_1 + x_2 - x_1 x_2$.*

The remarkable property of arithmetic expressions is that they can be applied to an r-output (word-level) function f with outputs $f_1, f_2, \ldots, f_r$. We focus on the effects of grouping the functions with the goal of representing a word-level expression in linear form.

16.4.1 Grouping

Consider the problem of grouping several Boolean functions in a word-level format. Let an r-output function f with outputs $f_1, f_2, \ldots, f_r$ be given. This function is described by the word-level arithmetic expression

$$f = 2^{r-1} f_r + \ldots + 2^1 f_2 + 2^0 f_1 \qquad (16.8)$$

and the outputs can be restored in a unique way. Therefore, the outputs of a circuit can be grouped together by using a weighted sum of the outputs. Given the simplest commutator function, the direct transmission of input data to outputs is described by $f = 2^{r-1} x_1 + 2^{r-2} x_2 + \ldots + x_r$.

Example 16.17 *Assume $n = 2$, then $f = 2x_1 + x_2$ (Figure 16.13). This expression does not include product terms of variables, therefore, it is **linear**.*

The linear model of a Boolean function f of n variables $x_1, \ldots, x_n$ is an

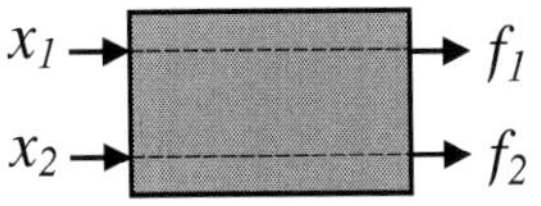

x_1 x_2	f_1	f_2
0 0	0	0
0 1	0	1
1 0	1	0
1 1	1	1

$x_1 \rightarrow$ [■■■■] $\rightarrow f_1$

$x_2 \rightarrow$ [■■■■] $\rightarrow f_2$

Word-level expression

$$f = 2^{r-1}x_1 + 2^{r-2}x_2 + \ldots + x_r$$
$$n = 2:$$
$$f = 2x_1 + x_2$$

FIGURE 16.13

The direct transmission of input data to outputs, truth table, and word-level representation (Example 16.17).

expression with $(n+1)$ integer coefficients $d_0^*, d_1^*, \ldots, d_n^*$

$$f = d_0^* + \sum_{i=1}^{n} d_i^* x_i = d_0^* + d_1^* x_1 + \ldots + d_n^* x_n \qquad (16.9)$$

Note that the word-level arithmetic expression

$$f = \sum_{i=0}^{2^n-1} d_i \cdot (x_1^{i_1} \cdots x_n^{i_n})$$

can be linear (Equation 16.9) in two cases: either the arithmetic expression of each f_j is linear, or no f_j generates linear expressions separately, but their combination produces a linear model.

Linearization generally means the transformation of a nonlinear expression to a linear model (Equation 16.9), with no more than $(n+1)$ nonzero coefficients. The idea of linearization can be explained by the following example.

Example 16.18 *The function $f = x_1 \vee x_2 = x_1 + x_2 - x_1 x_2$ is extended to the two-output Boolean function $f_1 = 1 \oplus x_1 \oplus x_2$, and $f_2 = x_1 \vee x_2$, that derives from the linear word-level representation $f = 2^1 f_2 + 2^0 f_1 = x_1 + x_2 + 1$. The position of f_2 (the most significant bit) in this linear expression is indicated by the masking operator*

$$\Xi^2\{f\} = \Xi^2\{x_1 + x_2 + 1\}$$

*In other words, to obtain a linear model given the Boolean function $f_2 = x_1 \vee x_2$, a **garbage function** $f_1 = 1 \oplus x_1 \oplus x_2$ has to be added. Then, f_2 can be extracted using the masking operator. The problem is how to find this additional function.*

In the absence of such a technique, a small amount of multioutput functions can generate linear models using the naive approach. We use such functions to form a fixed library of primitive cells.

The outputs of a Boolean function in arithmetic form:

$$f_1 = x_1 \oplus x_2 = x_1 + x_2 - 2x_1x_2$$
$$f_2 = x_1x_2$$

Word-level expression

$$f = 2^1 f_2 + 2^0 f_1$$
$$= 2^1 x_1x_2 + 2^0(x_1 + x_2 - 2x_1x_2)$$
$$= x_1 + x_2$$

x_1	x_2	f_2	f_1	f
0	0	0	0	0
0	1	0	1	1
1	0	0	1	1
1	1	1	0	2

FIGURE 16.14
Half-adder circuit, its truth table, and the word-level representation (Examples 16.19 and 16.22).

Example 16.19 *The **half-adder** (Figure 16.14) can be represented by the linear model $f = x_1 + x_2$. Permutation of the outputs f_1 and f_2 generates the **nonlinear** expression: $f = 2x_1 + 2x_2 - 3x_1x_2$.*

The above example demonstrates the high sensitivity of a linear model (Equation 16.9) to any permutation of outputs in the word-level description. On the other hand, it is a unique representation given the order of the Boolean function f_1 or f_2.

16.4.2 Computing the coefficients

As an example, Equation 16.9 describes a set of Boolean functions. If $n = 1$, then $f = d_0^* + d_1^* x_1$, and the function f is single-output. The coefficients d_0^* and d_1^* can be calculated by the equation

$$\mathbf{D} = \mathbf{P}_{2^1} \cdot \mathbf{F} = \begin{bmatrix} 1 & 0 \\ -1 & 1 \end{bmatrix} \begin{bmatrix} 0 \\ 1 \end{bmatrix} = \begin{bmatrix} 0 \\ 1 \end{bmatrix}$$

i.e., $d_0^* = 0$, $d_1 = 1$. In general, f is the r-output Boolean function $f_1, \ldots, f_r$. Let f be a three-output Boolean function: $f_1 = x_1$, $f_2 = \overline{x}_1$, $f_3 = x_1$, with the truth vector $\mathbf{F} = [2\ 5]^T$. Calculation of the coefficients implies

$$\mathbf{D} = \mathbf{P}_{2^1} \cdot \mathbf{F} = \begin{bmatrix} 1 & 0 \\ -1 & 1 \end{bmatrix} \begin{bmatrix} 2 \\ 5 \end{bmatrix} = \begin{bmatrix} 2 \\ 3 \end{bmatrix}$$

and $f = 2 + 3x_1$. Assuming $n = 2$, Equation 16.9 yields $f = d_0^* + d_1^* x_1 + d_2^* x_2$.

Example 16.20 *The linear word-level arithmetic expression for a half adder function is defined as shown in Figure 16.15.*

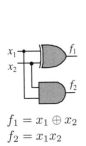

$f_1 = x_1 \oplus x_2$
$f_2 = x_1 x_2$

Truth vector

$$\mathbf{F} = [\, \mathbf{F_2} | \mathbf{F_1} \,] = \begin{bmatrix} 0 & 0 \\ 0 & 1 \\ 0 & 1 \\ 1 & 0 \end{bmatrix} = \begin{bmatrix} 0 \\ 1 \\ 1 \\ 2 \end{bmatrix}$$

Vector of coefficients

$$\mathbf{D} = \mathbf{P}_{2^2} \cdot \mathbf{F} = \begin{bmatrix} 1 & 0 & 0 & 0 \\ -1 & 1 & 0 & 0 \\ -1 & 0 & 1 & 0 \\ 1 & -1 & -1 & 1 \end{bmatrix} \begin{bmatrix} 0 \\ 1 \\ 1 \\ 2 \end{bmatrix} = \begin{bmatrix} 0 \\ 1 \\ 1 \\ 0 \end{bmatrix}$$

Word-level linear arithmetic expression $f = x_1 + x_2$

FIGURE 16.15
Constructing the linear word-level expression for a half adder by the matrix method (Example 16.20).

16.4.3 Weight assignment

There are two kinds of weight assignments in a linear model in the following formulation:

▶ The weight assignment to each function in a set of Boolean functions, and
▶ The weight assignment to each linear model in a set of expressions.

The weight assignment to the set of Boolean functions is defined by Equation 16.8. The weight assignment to the set of linear models is defined as follows. Let f_i be the i-th, $i = 1, 2, \ldots, r$, linear word-level arithmetic expression of n_i variables. A linear expression of an elementary Boolean function is represented by

$$t_i = \lceil \log_2 n_i \rceil + 1 \ \ bits \tag{16.10}$$

where $\lceil x \rceil$ is a ceiling function (the least integer greater than or equal to x). Suppose that f_i is the description of some primitive. In this formulation, the problem is to find a representation of an arbitrary level of a combinational circuit from a linear model as an n-input r-output Boolean function. To construct a word-level expression of r linear models, the weight assignment must be made appropriately. This means that applying any pattern to the inputs of f, an output of each expression f_i cannot affect the outputs of others.

Figure 16.16a illustrates the problem. Formally, the weight assignments such that functions do not overlap are determined by the equations

$$f = \sum_{i=0}^{r-1} 2^{T_i} f_{i+1}$$

$$T_i = \begin{cases} 0 & \text{for } i = 0 \\ T_{i-1} + t_i & \text{for } i > 0 \end{cases}$$

and t_i is calculated by Equation 16.10.

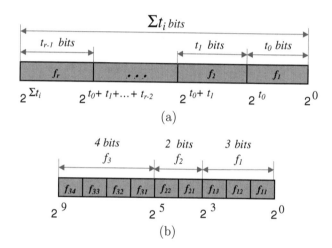

FIGURE 16.16

The word-level format for the set of r linear models f_i (a) and an example for f_1, f_2, and f_3 that are respectively 3rd, 2nd, and 4th subfunctions (b).

Example 16.21 *Let the word-level arithmetic expression f consist of three linear models f_1, f_2, and f_3 (Figure 16.16b). The expressions are constructed as follows: $f_1 = 2^0 f_{11} + 2^1 f_{12} + 2^2 f_{13}$, $f_2 = 2^3 f_{21} + 2^4 f_{22}$, $f_3 = 2^5 f_{31} + 2^6 f_{32} + 2^7 f_{33} + 2^8 f_{34}$ by Equation 16.4.3, the weight assignment of linear expressions f_1, f_2 and f_3 results in*

$$f = 2^0 f_1 + 2^3 f_2 + 2^5 f_3$$

16.4.4 Masking

The masking operator $\Xi^t\{f\}$ indicates the position $t \in (1, \dots, r)$ of a given function f in the word-level expression that represents a set of r Boolean functions.

> **Example 16.22** *The arithmetical expression of a half adder $f = x_1 + x_2$ is a two-output Boolean function: $f_1 = x_1 \oplus x_2$, $f_2 = x_1 x_2$ (Figure 16.14). The most significant bit (f_2) can be extracted by the masking operator $\Xi^2\{x_1 + x_2\}$, whereas the function f_1 is encoded by the least significant bit that can be recovered by the masking operator $\Xi^1\{x_1 + x_2\}$.*

16.5 Linear models of elementary functions

As was shown earlier, the majority of functions cannot be converted to a linear model, since their arithmetic equivalent includes nonlinear products. In this section, we focus on linearizing elementary Boolean functions.

16.5.1 Boolean functions of two and three variables

An arbitrary Boolean function of two variables can be described, in a unique way, by a linear model. In Table 16.9 the linear expressions for the two-input primitive Boolean functions are shown. The linear expressions for the three-input primitives are given in Table 16.9.

> **Example 16.23** *A Boolean function $x_1 \oplus x_2$ is represented by the nonlinear arithmetic expression $x_1 + x_2 - 2x_1x_2$. The nonlinear part is the product term $2x_1x_2$. To linearize, this function is expanded to the two-input function $f_1 = x_1 \oplus x_2$ and $f_2 = x_1x_2$. The Boolean function f_1 is extracted from the linear expression by the masking operator $f = \Xi^1\{x_1 + x_2\}$.*

16.5.2 Fundamental theorems of linearization

The linear combination of linear expressions for some elementary Boolean functions produces linear expressions. An approach to designing linear expressions for two-input and three-input elementary Boolean functions is

TABLE 16.9

Linear models for two-input and three-input gates

Function	2-input	3-input
x_1 — f x_2 — $f = x_1 x_2$	$\Xi^2 \{x_1 + x_2\}$	$\Xi^3 \{x_1 + x_2 + x_3\}$
x_1 — f x_2 — $f = x_1 \vee x_2$	$\Xi^2 \{1 + x_1 + x_2\}$	$\Xi^3 \{3 + x_1 + x_2 + x_3\}$
x_1 — f x_2 — $f = x_1 \oplus x_2$	$\Xi^1 \{x_1 + x_2\}$	$\Xi^1 \{x_1 + x_2 + x_3\}$
x_1 — f x_2 — $f = \overline{x_1 x_2}$	$\Xi^2 \{3 - x_1 - x_2\}$	$\Xi^3 \{6 - x_1 - x_2 - x_3\}$
x_1 — f x_2 — $f = \overline{x_1 \vee x_2}$	$\Xi^2 \{2 - x_1 - x_2\}$	$\Xi^3 \{4 - x_1 - x_2 - x_3\}$

discussed in the previous section. An elegant method for designing linear arithmetic expressions for many-input elementary Boolean functions has been developed by *Malyugin* [9]. We introduce *Malyugin's theorem* without proof.

Let the input variable of a primitive gate be x_j or $\overline{x}_j$. Denote the j-th input, $j = 1, 2, \ldots, n$, as

$$x_j^{i_j} = \begin{cases} x_j \text{ if } j_i = 0 \\ \overline{x}_j \text{ if } i_j = 1 \end{cases}$$

Theorem 16.1 *The n-variable AND function* $x_1^{i_1} \ldots x_n^{i_n}$ *can be represented by the linear model*

$$f = 2^{t-1} - n + \sum_{j=1}^{n} (i_j + (-1)^{i_j} x_j) \tag{16.11}$$

generated by an r-output function, in which the function AND is the most significant bit, as indicated by the masking operator $\Xi^r \{f\}$.

Theorem 16.2 *The n-variable OR function $x_1^{i_1} \vee \ldots \vee x_n^{i_n}$ can be represented by the linear model*

$$f = 2^{t-1} - 1 + \sum_{j=1}^{n} (i_j + (-1)^{i_j} x_j) \qquad (16.12)$$

of an r-output function, so that the function OR is the most significant bit $f = \Xi^r\{f\}$.

Theorem 16.3 *The n-variable EXOR function $x_1^{i_1} \oplus \ldots \oplus x_n^{i_n}$ can be represented by the linear model*

$$f = \sum_{j=1}^{n} (i_j + (-1)^{i_j} x_j) \qquad (16.13)$$

of an r-output function, in which the function EXOR is in the least significant bit, $\Xi^1\{f\}$.

In the above statements, the parameter t (the number of bits in a linear word-level representation of a given Boolean function) is defined by Equation 16.10: $t = \lceil \log_2 n \rceil + 1$. Note that the expression $1 \oplus x_j^{i_j}$ must be avoided in Equation 16.13. Before applying Equation 16.13, we have to replace $\overline{x}_j$ with $x_j \oplus 1$, or replace $x_j \oplus \overline{x}_j$ with 1 in order to cancel 1's.

Table 16.10 contains three n-input primitives and corresponding linear expressions. For a NOT function, the corresponding linear model equals $\Xi^1\{1-x\}$. Based on the expressions from Table 16.10, it is possible to describe modified gates.

Example 16.24 *Boolean functions $\overline{x}_1 \vee x_2$ and $\overline{x}_1 \oplus x_2$ can be represented by linear arithmetic expressions as follows:*

$$\overline{x}_1 \vee x_2 = \Xi^2\{1 + (1 - x_1) + x_2\} = \Xi^2\{2 - x_1 + x_2\}$$
$$\overline{x}_1 \oplus x_2 = \Xi^1\{(1 - x_1) + x_2\} = \Xi^1\{1 - x_1 + x_2\}$$

16.5.3 "Garbage" functions

Linear models are word-level arithmetic expressions that possess specific properties. First, the linear expression involves extra functions called *garbage functions*. The number of garbage functions G increases with the number of variables in a function that have been linearized:

$$G = t - 1 = \lceil \log_2 n \rceil \qquad (16.14)$$

TABLE 16.10

Linear models for the n-input AND, OR, and EXOR functions

Function	Linear model
$f = x_1^{i_1} \dots x_n^{i_n}$	$\Xi^r\{2^{t-1} - n + \sum_{j=1}^{n}(i_j + (-1)^{i_j}x_j)\}$
$f = x_1^{i_1} \vee \dots \vee x_n^{i_n}$	$\Xi^r\{2^{t-1} - 1 + \sum_{j=1}^{n}(i_j + (-1)^{i_j}x_j)\}$
$f = x_1^{i_1} \oplus \dots \oplus x_n^{i_n}$	$\Xi^1\{\sum_{j=1}^{n}(i_j + (-1)^{i_j}x_j)\}$

Example 16.25 *The linear expression for a two-input AND function (Table 16.9) is $f = x_1 + x_2$. To derive this linear form, the garbage function f_1 has been added, so that the given function AND is the most significant bit in the word-level description, f_2 (Figure 16.17a). To derive a linear representation of a three-input AND function, two garbage functions have been added through the two least significant bits of the 3-bit word (Figure 16.17b).*

16.5.4   Graphical representation of linear models

A *linear* decision diagram is used to represent a multioutput (word-level) Boolean function. A set of linear diagrams is a formal model used to represent a multilevel circuit. In linear word-level diagrams, a node realizes the *arithmetic analog* of a positive Davio expansion

$$pD_A : f = \underbrace{f_{x_i=0}}_{left\ term} + \underbrace{x_i(-f_{x_i=0} + f_{x_i=1})}_{right\ term}$$

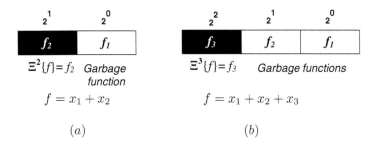

FIGURE 16.17
Garbage functions in linear model of a two-input (a) and three-input AND function (b) (Example 16.25).

and terminal nodes correspond to integer-valued coefficients of the Boolean function f. Thus, linear decision diagrams consist of pD_A nodes. A linear decision diagram is used to represent an arbitrary network described by a linear word-level arithmetic expression; the nodes correspond to a pD_A expansion and the terminal nodes are assigned to the coefficients of the linear expression. A linear decision diagram for n-input linear model includes n nonterminal and $n + 1$ terminal nodes.

> **Example 16.26** *Figure 16.18a,b shows a linear decision diagram for an AND function. This diagram includes three nodes. Lexigraphic order of variables is used: x_1, x_2.*

The linear decision diagram embedded in a 2D $\mathcal{N}$-hypercube is given in Figure 16.18c. In Table 16.11, the linear decision diagrams for two-input gates from the typical gate library are given.
Summarizing,

▶ Elementary Boolean functions can be represented by linear models as shown in Tables 16.9 and 16.10.

▶ Linear compositions of linear expressions (gates in the level of a circuit) produce linear models.

▶ Linear expressions directly map into linear decision diagrams (Table 16.11).

16.6 Logic networks and linear decision diagrams

Suppose a multilevel logic network with respect to a typical library of gates is given. The problem is formulated as follows: represent an arbitrary level

TABLE 16.11
Linear decision diagrams derived from linear word-level arithmetic
expressions for two-variable functions

Function	Linear decision diagram	Mask
x_1 — f x_2 $f = x_1 x_2$ $y = x_1 + x_2$	y — pD_A — 0 — pD_A — 0 — 0 x_1 : 1 x_2 : 1	$f = \Xi^2 \{x_1 + x_2\}$
x_1 — f x_2 $f = x_1 \vee x_2$ $y = 1 + x_1 + x_2$	y — pD_A — 0 — pD_A — 0 — 1 x_1 : 1 x_2 : 1	$f = \Xi^2 \{1 + x_1 + x_2\}$
x_1 — f x_2 $f = x_1 \oplus x_2$ $y = x_1 + x_2$	y — pD_A — 0 — pD_A — 0 — 0 x_1 : 1 x_2 : 1	$f = \Xi^1 \{x_1 + x_2\}$
x_1 — f x_2 $f = \overline{x_1 x_2}$ $y = 3 - x_1 - x_2$	y — pD_A — 0 — pD_A — 0 — 3 x_1 : -1 x_2 : -1	$f = \Xi^2 \{3 - x_1 - x_2\}$
x_1 — f x_2 $f = \overline{x_1 \vee x_2}$ $y = 2 - x_1 - x_2$	y — pD_A — 0 — pD_A — 0 — 2 x_1 : -1 x_2 : -1	$f = \Xi^2 \{2 - x_1 - x_2\}$

Techniques for linear decision diagram construction

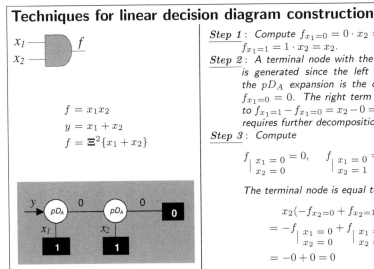

x_1 ——⊐ f
x_2 ——

$f = x_1 x_2$

$y = x_1 + x_2$

$f = \boxminus^2 \{x_1 + x_2\}$

Step 1: Compute $f_{x_1=0} = 0 \cdot x_2 = 0$ and
$f_{x_1=1} = 1 \cdot x_2 = x_2$.

Step 2: A terminal node with the value 1
is generated since the left term of
the pD_A expansion is the constant
$f_{x_1=0} = 0$. The right term is equal
to $f_{x_1=1} - f_{x_1=0} = x_2 - 0 = x_2$ and
requires further decomposition.

Step 3: Compute

$$f\Big|_{\substack{x_1=0\\x_2=0}} = 0, \qquad f\Big|_{\substack{x_1=0\\x_2=1}} = 0$$

The terminal node is equal to

$$x_2(-f_{x_2=0} + f_{x_2=1})$$
$$= -f\Big|_{\substack{x_1=0\\x_2=0}} + f\Big|_{\substack{x_1=0\\x_2=1}}$$
$$= -0 + 0 = 0$$

FIGURE 16.18
Design of a linear decision diagram for the AND function (Example 16.26).

of this network by linear word-level equation and linear wordlevel decision
diagram. Let the level of multilevel logic network is described by n inputs
$x_1, \ldots, x_n$ and r gates. The solution is based on the following theorem.

Theorem 16.4 *An arbitrary level of logic network with n inputs $x_1, \ldots, x_n$
and r gates (r outputs) is modeled by a linear diagram with n nodes assigned
input variables, and $n + 1$ terminal nodes, assigned coefficients of the linear
expression.*

The proof follows immediately from the fact that an arbitrary n-input r-
output function can be represented by a weighted arithmetic expression.

Example 16.27 *A level of a network is shown in Figure 16.19. The
linear models describing the first, second, and third
gates are given in Table 16.9. Combining these
expressions, we compile f where parameters T_0, T_1,
and T_2 are calculated by (Equation 16.4.3). The final
result is*

$$f = 2^0(x_1 + x_2) + 2^2(\overline{x}_1 + x_2) + 2^4(\overline{x}_2 + x_3)$$
$$= 2^0(x_1 + x_2) + 2^2(1 - x_1 + x_2) + 2^4(1 - x_2 + x_3)$$
$$= -3x_1 - 12x_2 + 17x_3 + 20$$

Techniques for linear decision diagram construction

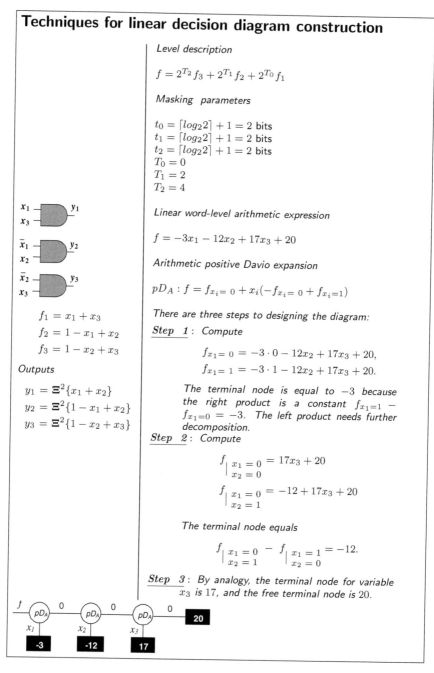

Level description

$$f = 2^{T_2} f_3 + 2^{T_1} f_2 + 2^{T_0} f_1$$

Masking parameters

$$t_0 = \lceil log_2 2 \rceil + 1 = 2 \text{ bits}$$
$$t_1 = \lceil log_2 2 \rceil + 1 = 2 \text{ bits}$$
$$t_2 = \lceil log_2 2 \rceil + 1 = 2 \text{ bits}$$
$$T_0 = 0$$
$$T_1 = 2$$
$$T_2 = 4$$

Linear word-level arithmetic expression

$$f = -3x_1 - 12x_2 + 17x_3 + 20$$

Arithmetic positive Davio expansion

$$pD_A : f = f_{x_i = 0} + x_i(-f_{x_i = 0} + f_{x_i = 1})$$

There are three steps to designing the diagram:

Step 1: Compute

$$f_{x_1 = 0} = -3 \cdot 0 - 12x_2 + 17x_3 + 20,$$
$$f_{x_1 = 1} = -3 \cdot 1 - 12x_2 + 17x_3 + 20.$$

The terminal node is equal to -3 because the right product is a constant $f_{x_1=1} - f_{x_1=0} = -3$. The left product needs further decomposition.

Step 2: Compute

$$f \Big|_{\substack{x_1 = 0 \\ x_2 = 0}} = 17x_3 + 20$$

$$f \Big|_{\substack{x_1 = 0 \\ x_2 = 1}} = -12 + 17x_3 + 20$$

The terminal node equals

$$f \Big|_{\substack{x_1 = 0 \\ x_2 = 1}} - f \Big|_{\substack{x_1 = 1 \\ x_2 = 0}} = -12.$$

Step 3: *By analogy, the terminal node for variable x_3 is 17, and the free terminal node is 20.*

Left-side diagram labels:

x_1 ―▷ y_1
x_3

$\bar{x}_1$ ―▷ y_2
x_2

$\bar{x}_2$ ―▷ y_3
x_3

$$f_1 = x_1 + x_3$$
$$f_2 = 1 - x_1 + x_2$$
$$f_3 = 1 - x_2 + x_3$$

Outputs

$$y_1 = \Xi^2\{x_1 + x_2\}$$
$$y_2 = \Xi^2\{1 - x_1 + x_2\}$$
$$y_3 = \Xi^2\{1 - x_2 + x_3\}$$

f ―(pDA)―0―(pDA)―0―(pDA)―0―[20]
   x_1 | x_2 | x_3 |
[-3]   [-12]   [17]

FIGURE 16.19

Technique for representing a level of a logic network by linear word-level arithmetic expression and decision diagrams (Examples 16.27 and 16.28).

Let us design a set of linear decision diagrams for the circuit from Example 16.27, i.e., $f_1 = x_1 + x_3$, $f_2 = 1 - x_1 + x_2$, and $f_3 = 1 - x_2 + x_3$. Note that the order of variables in the diagram can be arbitrary. Let us choose lexigraphical order: x_1, x_2.

Example 16.28 *The linear decision diagram for the expression* $-3x_1 - 12x_2 + 17x_3 + 20$ *consists of three nodes (Figure 16.19).*

From Examples 16.27 and 16.28, one can observe that the coefficients in linear expressions are quite large even for small logic networks. Therefore, a special technique is needed to alleviate this effect.

16.7 Linear models for logic networks

In this section, an arbitrary r-level combinational circuit is represented by r linear decision diagrams, that is, for each level of a circuit, a linear diagram is designed. The complexity of this representation is $O(G)$, where G is the number of gates in the circuit. The outputs of this model are calculated by transmitting data through this set of diagrams. This approach is the basis for representing circuits in spatial dimensions:

<2D circuit> $\Rightarrow$ < A set of linear diagrams> $\Rightarrow$ <Hypercube-like topology>.

An arbitrary m-level logic network can be uniquely described by a set of m linear decision diagrams, and, vice versa, this set of linear decision diagrams corresponds to a unique network. To prove this statement, let one of m levels with r n-input gates from a fixed library be described by one linear model. Fixing the order of the gates in the level, that is, keeping unambiguity in the structure, we can derive the unique linear decision diagram for this level, as well as for other $m - 1$ levels of the network.

From this statement, it follows that

▶ The order of gates in a level of circuit must be fixed.
▶ The complexity of linear decision diagram does not depend on the order of variables.
▶ Data transmission through linear decision diagrams must be provided.

Example 16.29 *Figure 16.20 shows the three-level Boolean network and its representation by a set of three linear decision diagrams.*

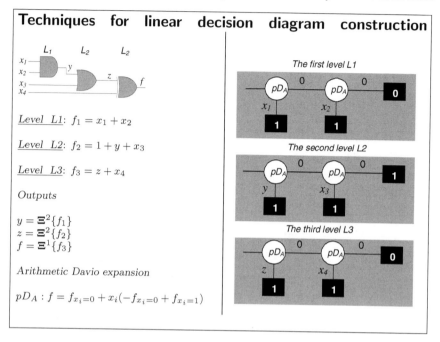

FIGURE 16.20

Representation of a three-level circuit by a set of linear decision diagrams (Example 16.29).

16.8   Linear models for multivalued logic networks

In this section, the generalization of linear word-level data structures toward multivalued functions is considered. In a similar manner to binary functions, linear word-level expressions and decision diagrams are distinguished by their type of decomposition (expansion). The focus of this section is an approach to the representation of m-valued functions of n variables by the linear word-level expression

$$f = d_0 + d_1 x_1^{\circ} + d_2 x_2^{\circ} + \cdots + d_n x_n^{\circ} \qquad (16.15)$$

An arbitrary multivalued function can be represented in linear form (Equation 16.15). However, in this section, the library of linear models (linear expressions, decision diagrams, and hypercube-like structures) includes elementary functions only. Thus, different techniques can be used to design an arbitrary multivalued circuit with respect to this library of gates.

16.8.1 Approach to linearization

The approach to linearization includes the following steps:

Given: The truth vector of multivalued logic function.
Step 1: Partitioning of the truth vector F of the m-valued
function f of n m-valued variables to a set of
subvectors F_j°.
Step 2: Encoding the multivalued variables x_i. The new, binary
variables x_i° are called *pseudo-variables*.
Step 3: Representation of the multivalued function f by a
linear word-level arithmetic expression that depends on
binary pseudo-variables x_i°.
Result: Linear arithmetic expression.

Specifically, these algorithm includes three main steps: partitioning, encoding, and representation of logic function in an algebraic form.

Step 1: Partition. Given the truth vector $\mathbf{F}$ of an m-valued n-variable logic function f. Let us partition this vector into τ subvectors,

$$\tau = \left\lceil \frac{m^n}{n+1} \right\rceil \tag{16.16}$$

where $\lceil a \rceil$ denotes the least integer greater than or equal to a. The order of the partition is fixed (with respect to assignments of variables). The index μ of subvector $\mathbf{F}_\mu$ that contains the i-th element of the initial truth table is equal to

$$\mu = \left\lfloor \frac{i}{n+1} \right\rfloor \tag{16.17}$$

where $\lfloor a \rfloor$ is the greatest integer less than or equal to a.

> **Example 16.30** *Partitioning the truth vector* $\mathbf{F}$ *of length* $3^3 = 27$ *of a ternary* $(m = 3)$ *function of three variables* $(n = 3)$ *is illustrated in Figure 16.21. The location of the 20-th element of the truth vector* $\mathbf{F}$ *is determined by the index* $\mu = \left\lfloor \frac{20}{3+1} \right\rfloor = 5$ *of subvector* $\mathbf{F}_\mu$*. This element belongs to subvector* $\mathbf{F}_5$*.*

Step 2: Encoding. Consider the μ-th subvector $\mathbf{F}_\mu$, $\mu = 0, 1, \ldots, \tau - 1$. The length of the subvector $\mathbf{F}_\mu$ is $n + 1$. Hence, the i-th element is allocated in the subvector $\mathbf{F}_\mu$. Its position inside μ is specified by the index $j = Res\left(\frac{i}{n+1}\right) = 5$. Assignments of n variables $x_1^\circ, x_2^\circ, \ldots, x_n^\circ$ in $\mathbf{F}_\mu$ are called *pseudo-variables*. The pseudo-variables are the *binary* variables valid for assignments with at most one 1.

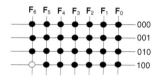

The vector $\mathbf{F}$ is partitioned to

$$\tau = \left\lceil \frac{3^3}{3+1} \right\rceil = \left\lceil \frac{27}{4} \right\rceil = \lceil 6.7 \rceil = 7$$

subvectors $\mathbf{F}_0, \mathbf{F}_1, \ldots, \mathbf{F}_6$
The 21-th element is located in the truth
vector $\mathbf{F}_\mu$,

$$\mu = \left\lfloor \frac{21}{3+1} \right\rfloor = 5$$

FIGURE 16.21

Representation of the truth vector of a multivalued function by 2D data structure (Example 16.30).

Example 16.31 *Assignments of* **pseudo-variables** *of a three-valued function of two, three, and four variables are as follows:*

(a) $x_1^\circ x_2^\circ = \{00, 01, 10\}$; *given* $i = 1$, $\mu = \left\lfloor \frac{1}{2+1} \right\rfloor = 0$;

(b) $x_1^\circ x_2^\circ x_3^\circ = \{000, 001, 010, 100\}$; *given* $i = 20$, $\mu = \left\lfloor \frac{20}{3+1} \right\rfloor = 5$;

(c) $x_1^\circ x_2^\circ x_3^\circ x_4^\circ = \{0000, 0001, 0010, 0100, 1000\}$; *given* $i = 10$, $\mu = \left\lfloor \frac{10}{4+1} \right\rfloor = 2$.

Step 3: Representation of a function by linear word-level arithmetic expression. This phase consists of

(a) Forming a word-level vector $\mathbf{F}^\circ$ from subvectors $\mathbf{F}_1^\circ$, $\mathbf{F}_2^\circ$, $\ldots$ $\mathbf{F}_\tau^\circ$
(b) A truncated arithmetic transform of vector $\mathbf{F}^\circ$

Let $\mathbf{W} = \begin{bmatrix} m^{\tau-1} & m^{\tau-2} & \cdots & m^1 & m^0 \end{bmatrix}^T$ be the weight vector. A truth vector $\mathbf{F}^\circ$ of a function f of n pseudo-variables $x_1^\circ, \ldots, x_n^\circ$ includes $n+1$ elements and is calculated by

$$\mathbf{F}^\circ = [\mathbf{F}_{\tau-1}| \ldots |\mathbf{F}_1|\mathbf{F}_0]\mathbf{W} \tag{16.18}$$

The truncated transform of $\mathbf{F}^\circ$ yields the vector of arithmetic coefficients $\mathbf{D}$. The relationship between the $\mathbf{F}^\circ$ and vector of coefficients $\mathbf{D} = [d_0 d_1 \ldots d_n]$ is defined by the pair of transforms

$$\mathbf{D} = \mathbf{T}_{n+1} \cdot \mathbf{F}^\circ \tag{16.19}$$

$$\mathbf{F}^\circ = \mathbf{T}_{n+1}^{-1} \cdot \mathbf{D} \tag{16.20}$$

where $(n + 1) \times (n + 1)$ direct $\mathbf{T}_{n+1}$ and inverse $\mathbf{T}_{n+1}^{-1}$ truncated arithmetic transform matrices are formed by truncation of $2^n \times 2^n$ arithmetic transform matrices P_{2^n} and $P_{2^n}^{-1}$, respectively. The truncated rule is as follows: (a) remove all rows that contain more than one 1 and (b) Remove the remaining columns that consist of all 0s. The vector of coefficients $\mathbf{D}$ yields the linear word-level arithmetic expression

$$D = d_0 + d_1 x_1^\circ + d_2 x_2^\circ + \cdots + d_n x_n^\circ$$

Example 16.32 *Given a two-variable three-valued function, the 3×3 direct and inverse arithmetic* **truncated matrices** *are equal to*

$$\mathbf{T}_3 = \begin{bmatrix} 1 & 0 & 0 \\ -1 & 1 & 0 \\ -1 & 0 & 1 \end{bmatrix} \qquad \mathbf{T}_3^{-1} = \begin{bmatrix} 1 & 0 & 0 \\ 1 & 1 & 0 \\ 1 & 0 & 1 \end{bmatrix}$$

Example 16.33 *(Continuation of Example 16.32.) Arithmetic expressions for subvectors* $\mathbf{D}_0$, $\mathbf{D}_1$, *and* $\mathbf{D}_2$ *in Table 16.12 are calculated by the direct truncated transform (Equation 16.19). For example,* $\mathbf{D}_1$ *is calculated as follows:*

$$\mathbf{D}_1 = \mathbf{T}_3 \mathbf{F}_1 = \begin{bmatrix} 1 & 0 & 0 \\ -1 & 1 & 0 \\ -1 & 1 & 1 \end{bmatrix} \begin{bmatrix} 1 \\ 1 \\ 2 \end{bmatrix} = \begin{bmatrix} 1 \\ 0 \\ 1 \end{bmatrix}$$

which yields the algebraic form $d_1 = 1 + x_1^\circ$.

Example 16.34 *(Continuation of Example 16.33.) The truth vector* $\mathbf{F}^\circ$ *of the two-input* $MAX(x_1, x_2)$ *function is calculated as shown in Figure 16.22. The direct truncated transform (Equation 16.19) is used for transformation. The final result is the linear expression*

$$D = 3^2 D_2 + 3^1 D_1 + 3^1 D_0 = 21 + 5x_1^\circ + x_2^\circ$$

16.8.2 Manipulation of the linear model

A linear word-level expression of elementary multivalued functions is a form of representation and computation due to the following properties:

Property 1: It is convertible to the initial function by way of an operator (a control parameter of the linear model);

Property 2: It is an intrinsically parallel model because it is at wordlevel;

TABLE 16.12
Partitioning of the truth vector $\mathbf{F}$ and deriving the
linear word-level arithmetic expression for a ternary
$MAX(x_1, x_2)$ function

	Function MAX				Linear model	
$x_1 x_2$	$\mathbf{F}$	$\mathbf{F}_\mu$		$x_1^\circ x_2^\circ$	$\mathbf{D}_\mu$	D_μ
00	0		$\begin{bmatrix} 0 \\ 1 \\ 2 \end{bmatrix}$	00		
01	1	$\mathbf{F}_0 =$		01	$\mathbf{D}_0 = \begin{bmatrix} 0 \\ 1 \\ 2 \end{bmatrix}$;	$D_0 = x_2^\circ + 2x_1^\circ$
02	2			10		
10	1		$\begin{bmatrix} 1 \\ 1 \\ 2 \end{bmatrix}$	00		
11	1	$\mathbf{F}_1 =$		01	$\mathbf{D}_1 = \begin{bmatrix} 1 \\ 0 \\ 1 \end{bmatrix}$;	$D_1 = 1 + x_1^\circ$
12	2			10		
20	2		$\begin{bmatrix} 2 \\ 2 \\ 2 \end{bmatrix}$	00		
21	2	$\mathbf{F}_2 =$		01	$\mathbf{D}_2 = \begin{bmatrix} 2 \\ 0 \\ 0 \end{bmatrix}$;	$D_2 = 2$
22	2			10		

Property 3: It is extendable to arbitrary logic functions.

The following example illustrates some of these properties by calculation
of the function using the linear model given the input assignments. Let a
three-valued ($m = 3$) two-input ($n = 2$) elementary logic function be given
by the linear expression D. The masking operation

$$f = \Xi^\mu \{D\} \qquad\qquad (16.21)$$

is used to recover the value of the logic function.

Example 16.35 *(Continuation of Example 16.34.)   Calculation of
values of $MAX(x_1, x_2)$ given the linear model and
$x_1 = 2$, $x_2 = 1$ involves several steps (Figure 16.22):*

   (a) *Find the index μ of subvector $\mathbf{F}_\mu$ in a word-
       level representation. Here, the parameter μ is
       determined as follows: assignment $x_1 x_2 = 21$
       corresponds to the 7-th element of the truth
       vector $\mathbf{F}$; hence $\mu = \lfloor 7/3 \rfloor = 2$.*
   (b) *Use the encoding rule given in Table 16.12: $x_1 x_2 =
       21 \to x_1^\circ, x_2^\circ = 01$.*
   (c) *Calculate the value of $MAX(2,1)$ for the
       assignment of pseudo-variables $x_1^\circ = 0, x_2^\circ = 1$:
       $MAX(2,1) = D_2(0,1) = 2$.*

Techniques for linearization of multivalued logic functions

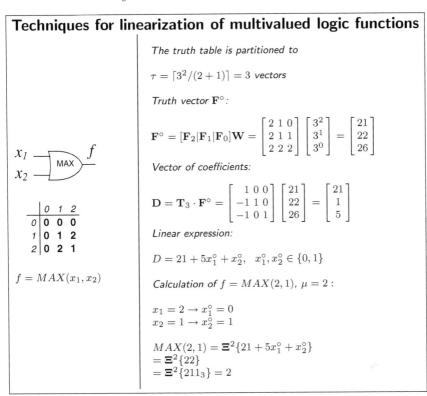

The truth table is partitioned to

$$\tau = \lceil 3^2/(2+1)\rceil = 3 \text{ vectors}$$

Truth vector $\mathbf{F}^\circ$:

$$\mathbf{F}^\circ = [\mathbf{F}_2|\mathbf{F}_1|\mathbf{F}_0]\mathbf{W} = \begin{bmatrix} 2 & 1 & 0 \\ 2 & 1 & 1 \\ 2 & 2 & 2 \end{bmatrix} \begin{bmatrix} 3^2 \\ 3^1 \\ 3^0 \end{bmatrix} = \begin{bmatrix} 21 \\ 22 \\ 26 \end{bmatrix}$$

Vector of coefficients:

$$\mathbf{D} = \mathbf{T}_3 \cdot \mathbf{F}^\circ = \begin{bmatrix} 1 & 0 & 0 \\ -1 & 1 & 0 \\ -1 & 0 & 1 \end{bmatrix} \begin{bmatrix} 21 \\ 22 \\ 26 \end{bmatrix} = \begin{bmatrix} 21 \\ 1 \\ 5 \end{bmatrix}$$

Linear expression:

$$D = 21 + 5x_1^\circ + x_2^\circ, \quad x_1^\circ, x_2^\circ \in \{0,1\}$$

Calculation of $f = MAX(2,1)$, $\mu = 2$:

$$x_1 = 2 \to x_1^\circ = 0$$
$$x_2 = 1 \to x_2^\circ = 1$$

$$MAX(2,1) = \Xi^2\{21 + 5x_1^\circ + x_2^\circ\}$$
$$= \Xi^2\{22\}$$
$$= \Xi^2\{211_3\} = 2$$

x_1 —, x_2 —, MAX, f

	0	1	2
0	0	0	0
1	0	1	2
2	0	2	1

$f = MAX(x_1, x_2)$

FIGURE 16.22
Representation of the three-valued two-variable logic function $f = MAX(x_1, x_2)$ by a linear word-level arithmetic expression (Examples 16.34 and 16.35).

16.8.3 Library of linear models of multivalued gates

Table 16.13 contains the linear models of various ternary gates from a library of gates. The linear models from Table 16.13 can be extended to an arbitrary logic function.

Example 16.36 *The ternary function $f = \overline{x_1 + x_2}$ can be represented by a linear expression as follows:*

$$f = \overline{x_1 + x_2} \pmod 3 = \Xi^\mu\{3^0(2 - x_2^\circ - 2x_1^\circ) + 3^1(1 - x_2^\circ + x_1^\circ) + 3^2(2x_2^\circ + x_1^\circ)\}$$
$$= \Xi^\mu\{5 + 14x_2^\circ + 10x_1^\circ\}$$

TABLE 16.13
Library of linear word-level arithmetic models of three-valued gates

Function			Vector of coefficients
$\overline{x}$	$=$	$2 - x$	$\mathbf{D} = [2 \ -1]^{\mathrm{T}}$
$x_1 \cdot x_2 \pmod 3$	$=$	$15x_1^{\circ} + 21x_2^{\circ}$	$\mathbf{D} = [0 \ 21 \ 15]^{\mathrm{T}}$
$MIN(x_1, x_2)$	$=$	$21x_1^{\circ} + 12x_2^{\circ}$	$\mathbf{D} = [0 \ 12 \ 21]^{\mathrm{T}}$
$TSUM(x_1, x_2)$	$=$	$21 + 5x_1^{\circ} + 4x_2^{\circ}$	$\mathbf{D} = [21 \ 4 \ 5]^{\mathrm{T}}$
$MAX(x_1, x_2)$	$=$	$21 + 5x_1^{\circ} + x_2^{\circ}$	$\mathbf{D} = [21 \ 1 \ 5]^{\mathrm{T}}$
$TPROD(x_1, x_2)$	$=$	$21x_1^{\circ} + 9x_2^{\circ}$	$\mathbf{D} = [0 \ 9 \ 21]^{\mathrm{T}}$
$(x_1 + x_2) \pmod 3$	$=$	$21 - 10x_1^{\circ} - 14x_2^{\circ}$	$\mathbf{D} = [21 \ -14 \ -10]^{\mathrm{T}}$
$x_1 \vert x_2$	$=$	$1 - x_1^{\circ} + 5x_2^{\circ}$	$\mathbf{D} = [1 \ 5 \ -1]^{\mathrm{T}}$

16.8.4   Representation of multivalued logic networks

Let D be a level of a multivalued, multilevel network and consist of r two-input multivalued gates. The level implements an n-input r-output logic function, or subnetwork over the library of gates. Since each gate is described by a linear model, this subnetwork can be described by a linear expression too. The strategy for representation of a multivalued logic network by a set of linear expressions is as follows:

$$\text{Gate model } D_j \Longleftrightarrow f = \Xi^{\mu}\{D\}$$

$$\text{Network level model } \mathbf{D} \Longleftrightarrow f_j = \Xi^{3(j-1)+\mu}\{L\}$$

$$\text{Logic network model (set of D)} \Longleftrightarrow \text{Set of level outputs}$$

To simplify the formal notation, let us consider the library of ternary gates given in Table 16.13.

Let D_j, $j = 1, 2, \ldots, r$, be a linear arithmetic representation of the j-th gate and its output correspond to the j-output of a subnetwork. Assume that the order of gates in the subcircuit is fixed. A linear word-level arithmetic of an n-input r-output of a ternary subnetwork (level) is defined as

$$D = \sum_{j=1}^{r} 3^{3(j-1)} D_j \tag{16.22}$$

Example 16.37 *Let a level of a ternary network be given as shown in Figure 16.23. This figure explains the calculation of the linear expression using Equation 16.22.*

To calculate the value of the j-th output f_j, $j \in \{1, \ldots, r\}$, a masking operator is utilized:

$$f_j = \Xi^\xi\{D\} \tag{16.23}$$

where $\xi = 3(j-1) + \mu$. This recovers the ξ-th digit in a word-level value D.

Example 16.38 *(Continuation of Example 16.37.) For the assignment*

$$x_1 x_2 x_3 x_4 x_5 x_6 = 201112 \implies x_1^\circ x_2^\circ x_3^\circ x_4^\circ x_5^\circ x_6^\circ = 000110,$$

the outputs f_j, $j \in \{1, 2, 3\}$, are calculated as follows:
$f_1 = 2$, $f_2 = 1$, and $f_3 = 2$.

Techniques for linearization of multivalued logic functions

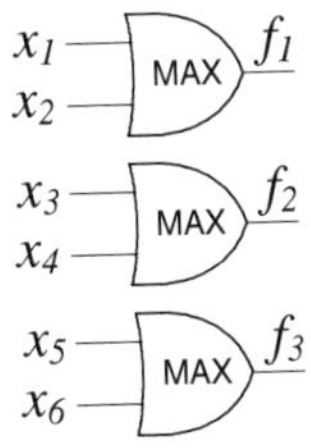

$$D = \sum_{j=1}^{3} 3^{3(j-1)} D_j = 3^0 D_1 + 3^1 D_2 + 3^2 D_3$$

$$= 3^0 (21 + 5x_1^\circ + x_2^\circ)$$
$$+ 3^3 (21 + 5x_3^\circ + x_4^\circ)$$
$$+ 3^6 (21 + 5x_5^\circ + x_6^\circ)$$
$$= 15897 + 5x_1^\circ + x_2^\circ + 135x_3^\circ + 27x_4^\circ + 3645x_5^\circ + 729x_6^\circ.$$

x_1 — MAX — f_1
x_2

x_3 — MAX — f_2
x_4

x_5 — MAX — f_3
x_6

The relationship of the assignments of variables and pseudo-variables:

$$x_1 x_2 x_3 x_4 x_5 x_6 = 201112 \implies x_1^\circ x_2^\circ x_3^\circ x_4^\circ x_5^\circ x_6^\circ = 000110$$

$f_1 \to D_1$
$f_2 \to D_2$
$f_3 \to D_3$
$D_1 = 21 + 5x_1^\circ + x_2^\circ$
$D_2 = 21 + 5x_3^\circ + x_4^\circ$
$D_3 = 21 + 5x_5^\circ + x_6^\circ$

Given the assignments $\mu_1 = 2$, $\mu_2 = 1$, $\mu_3 = 1$,
$D = 15897 + 5 \cdot 0 + 1 \cdot 0 + 135 \cdot 0 + 27 \cdot 1 + 3645 \cdot 1 + 729 \cdot 0 = 19569$

The outputs f_j, $j \in \{1, 2, 3\}$, are recovered by

$$f_1 = \Xi^{3 \cdot 0 + 2}\{19569\} = \left\lfloor \frac{19569}{3^2} \right\rfloor \ (mod\,3) = 2$$

$$f_2 = \Xi^{3 \cdot 1 + 1}\{19569\} = \left\lfloor \frac{19569}{3^4} \right\rfloor \ (mod\,3) = 1$$

$$f_3 = \Xi^{3 \cdot 2 + 1}\{19569\} = \left\lfloor \frac{19569}{3^7} \right\rfloor \ (mod\,3) = 2$$

FIGURE 16.23
Recovery of the MAX function from a word-level linear model (Examples 16.37 and 16.38).

16.8.5   Linear decision diagrams

There are three hierarchical levels in the representation of multivalued functions. The first level corresponds to the description of a gate:

Gate $\Longleftrightarrow$
> Linear expression $\Longleftrightarrow$
>> Linear decision diagram

The second level corresponds to the description of a level in a multilevel circuit:

Network level $\Longleftrightarrow$
> Linear expression $\Longleftrightarrow$
>> Linear decision diagram

The third level corresponds to the description of the logic network:

Network $\Longleftrightarrow$
> Set of linear expressions $\Longleftrightarrow$
>> Set of linear decision diagrams

Based on the aforementioned statements, an arbitrary multivalued network can be modeled by a set of linear word-level decision diagrams.

> **Example 16.39** *The linear decision diagram and its embedding in a $\mathcal{N}$-hypercube for the ternary MAX gate are represented in Figure 16.24.*

16.9   Linear word-level representation of multivalued functions using logic operations

It has been shown in Chapter 7 that an arbitrary Boolean function can be represented by a linear nonarithmetic word-level expression. In this section, an extension of this technique to multivalued functions is presented.

16.9.1   Linear word-level for MAX expressions

Let us denote

Variable x_i by $x_{i,0}$ $(q = 0)$,

The complement of variable $\overline{x}_i = (m-1) - x_i$ by $x_{i,1}(q = 1)$,

The cyclic complement of variable $\widehat{x_i} = x_i + 1$ by $x_{i,2}$ $(q = 2)$,

The integer positive values that correspond to the i-th variable x_i by $w_{i,q}$,

MAX function by $\vee$.

Techniques for linearization of multivalued logic functions

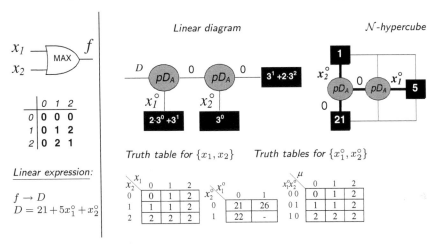

Linear diagram | $\mathcal{N}$-hypercube

Truth table for $\{x_1, x_2\}$ | Truth tables for $\{x_1^\circ, x_2^\circ\}$

Linear expression:

$f \rightarrow D$

$D = 21 + 5x_1^\circ + x_2^\circ$

FIGURE 16.24

Representation of the ternary function MAX by a linear decision diagram and $\mathcal{N}$-hypercube (Example 16.39).

A linear word-level expression for the MAX operation of an n-variable multivalued function is defined by

$$f = \overset{n}{\underset{i=1}{\widehat{\bigvee}}} w_{i,q} x_{i,q} \tag{16.24}$$

Example 16.40 *Examples of word-level representation are given as follows: a single-output ternary MAX function of two variables $f = x_1 \widehat{\vee} x_2 = x_1 \vee x_2$ is a linear expression because it does not contain any product of variables. A two-output ternary function $f_1 = \overline{x}_1 \vee \overline{x}_2$, $f_2 = x_1 \vee x_2$ of two variables can be represented by the linear expression $f = 3\overline{x}_1 \widehat{\vee} 3\overline{x}_2 \widehat{\vee} x_1 \widehat{\vee} x_2$. Details are given in Figure 16.25.*

To recover the initial data from the linear expression, we apply a masking operator. A value f_j of a j-th multivalued function, $j \in \{1, \ldots, r\}$, can be recovered from a linear expression (Equation 16.24) by the masking operator $f_j = \Xi^j\{f\}$.

Techniques for linearization of Boolean functions

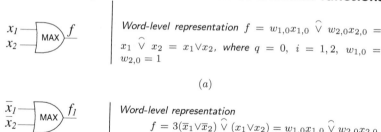

(a)

Word-level representation $f = 3(\overline{x}_1 \vee \overline{x}_2) \overset{\frown}{\vee} (x_1 \vee x_2) = w_{1,0} x_{1,0} \overset{\frown}{\vee} w_{2,0} x_{2,0}$

$$= 3\overline{x}_1 \overset{\frown}{\vee} 3\overline{x}_2 \overset{\frown}{\vee} x_1 \overset{\frown}{\vee} x_2$$

where $q = \{0, 1\}$, $i = 1, 2$, $w_{1,0} = w_{2,0} = 1$, $w_{1,1} = w_{2,1} = 3$

(b)

FIGURE 16.25

Linear word-level nonarithmetic representation of a ternary MAX function
(a) and two ternary MAX functions (Example 16.40).

Example 16.41 *(Continuation of Example 16.40.) (a) A single-output ternary MAX function of two variables is recovered:*
$f = \Xi^1\{x_1 \overset{\frown}{\vee} x_2\} = x_1 \vee x_2$. *(b) A two-output ternary function of two variables is recovered:* $f_1 = \Xi^1\{3\overline{x}_1 \overset{\frown}{\vee} 3\overline{x}_2 \overset{\frown}{\vee} x_1 \overset{\frown}{\vee} x_2\} = \overline{x}_1 \vee \overline{x}_2$ *and* $f_2 = \Xi^2\{3\overline{x}_1 \overset{\frown}{\vee} 3\overline{x}_2 \overset{\frown}{\vee} x_1 \overset{\frown}{\vee} x_2\} = x_1 \vee x_2$.

A multilevel multivalued logic network can be described by a linear word-level logic, once each level consists of gates of the same type.

Example 16.42 *The two-input, three-output level of a ternary network given in Figure 16.26 is described by the expression* $0x_1 \overset{\frown}{\vee} 4x_2 \overset{\frown}{\vee} 3\overline{x}_1 \overset{\frown}{\vee} 9\overline{x}_2$. *The linear diagram that corresponds to this expression consists of four nodes implementing the Shannon expansion. The values of the outputs given truth vectors* $\mathbf{F_1}$, $\mathbf{F_2}$, *and* $\mathbf{F_3}$ *are calculated in Figure 16.26. Given the assignment* $x_1 x_2 \overline{x}_1 \overline{x}_2 = \{0022\}$, *the outputs are equal to*

$$f_1(0022) = 0 \vee 0 \vee 0 \vee 0 = 0(x_1 \vee x_2 = 0 \vee 0 = 0)$$
$$f_2(0022) = 0 \vee 0 \vee 2 \vee 0 = 2(\overline{x}_1 \vee x_2 = 2 \vee 0 = 2)$$
$$f_3(0022) = 0 \vee 0 \vee 0 \vee 2 = 2(x_1 \vee \overline{x}_2 = 0 \vee 2 = 2)$$

Techniques for linearization of Boolean functions

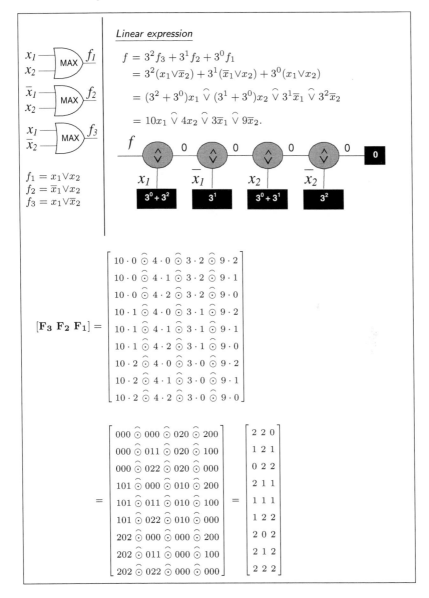

FIGURE 16.26

A two-level ternary logic network and its linear diagram (Example 16.42).

16.10   3D computing arrays design

In this section, we show how to transform 2D arrays into 3D arrays based on linear decision diagrams. The generated 3D computing arrays have several particular properties that distinguish 3D linear arrays from known devices targeting massive parallelism. A 3D computing array consists of operational and memory arrays of elements (Figure 16.27a). The arrays are demanded by the need for functional and efficient design and optimization at various implementation levels. For example, switches, storage, and data transmission are characterized by different requirements and implementation technology. An advantage of the array architecture is its ability to perform aggregation of 2D and 3D arrays. This concept suits hybrid molecular, biomolecular, and micro technologies.

Hence, there are some similarities between the proposed 3D linear array and a 3D classic cellular arrays. In particular, both are composed of identical cells connected in a regular fashion. However, each cell in a classic 3D cellular array has six neighbors. In a 3D linear array the number of neighbors can be less. That is, linear array is a particular case of a cellular array and can be viewed as a fine-grained array.

Operational arrays can be implemented using switches (each cell performs the arithmetic analog of Davio expansion), and each cell of the memory array is a small-size storing element (several bits).

Our goal is to increase the computational power of 3D arrays. We typify the cell by its multivalued analog, which processes signals on 3, 4, ,16 levels.

16.11   Further study

Advanced topics of linear arrays

Topic 1: Systolic arrays and reversible computing. A new horizon for the systolic array can be predicted, which will be based on reversible elements. Recall that an m-input, m-output totally-specified Boolean function $f(x_1, x_1, \ldots, x_m)$ is *reversible* if it maps each input assignment to a unique output assignment, that is, for $B = \{0, 1\}$ $f{:}B^m \rightarrow B^m$, $f^{-1}{:}B^m \rightarrow B^m$. The notion of reversibility arises from the fact that, since each input pattern is associated with a unique output pattern, a reversible function always has an inverse. A reversible function can be realized by a cascade of reversible gates in which there is no fan-out or feedback. Frequently, the function is its own inverse, particularly for primitive reversible gates. If one replaces each gate in the cascade with a gate realizing the inverse of the function realized by the original gate (it may be itself), then the cascade applied in reverse order

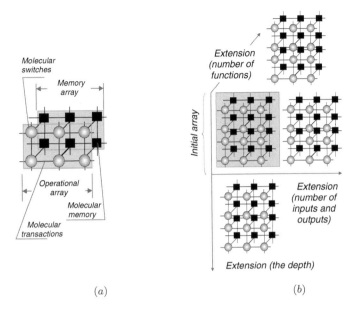

Molecular
switches

Memory
array

Operational
array

Molecular
memory

Molecular
transactions

Initial array

Extension
(number of
functions)

Extension
(number of
inputs and
outputs)

Extension (the depth)

(a)   (b)

FIGURE 16.27
Universal linear array (ULA) design: an arbitrary Boolean function can be
represented by operational and memory arrays (a); extension of the ULA in
3D (b)

realizes the inverse of the function realized by the original cascade. The most
commonly used set of reversible gates is the *Toffoli* family. A second family of
interest is the set of *Fredkin gates*. Such a gate with no control lines is a swap
gate. Toffoli and Fredkin gates are self-inverse. To the best of our knowledge,
the design of systolic arrays using reversible elements has not being developed
as yet.

Topic 2: Linear transformation of variables. is a method for optimizing the
representation of a Boolean function. In terms of a spectral technique, the
linear transformation of variables is a method for reducing the number of
nonzero coefficients in the spectrum of a Boolean function [3, 19].

Topic 3: Linearization of Boolean and multivalued functions.
Representation of a given system of logic functions in the form of the
superposition of a system of linear functions and a residual nonlinear part of
minimal complexity is considered in [3].

Further reading

[1] Aizenberg IN, Aizenberg NN, and Vandewalle J. *Multi-Valued and Universal
Binary Neurons – Theory, Learning, Applications.* Kluwer, Dordrecht, 2000.

[2] Ancona MG. Systolic processor design using single-electron digital circuits.

Superlattices and Microstructures, 20(4):461–472, 1996.

[3] Karpovsky MG, Stanković RS, and Astola JT. *Spectral Logic and Its Applications for the Design of Digital Devices*. John Wiley & Sons, Hoboken, NJ, 2008.

[4] Kukharev GA, Tropchenko AY, and Shmerko VP. *Systolic Signal Processors*. Publishing House "Belarus," Minsk, Belarus, 1988.

[5] Kumar VKP and Tsai Y-C. Designing linear systolic arrays. *J. Parallel and Distributed Computing*, 7:441–463, 1989.

[6] Kung SY. *VLSI Array Processors*, Prentice Hall, Englewood Cliffs, 1988.

[7] Lai Y-T, Pedram M, and Vrudhula SBK. EVBDD-based algorithms for integer linear programming, spectral transformation, and function decomposition. *IEEE Trans. CAD of Integrated Circuits and Systems*, 13(8):959–975, 1994.

[8] Levashenko VG, Shmerko VP, and Yanushkevich SN. Solution of Boolean differential equations on systolic arrays. *Cybernetics and Systems Analysis*, 32(1):26–40, 1996.

[9] Malyugin VD. Realization of Boolean function's corteges by means of linear arithmetical polynomial. *Automation and Remote Control*, 45(2):239–245, 1984.

[10] Minato S. *Binary Decision Diagrams and Applications for VLSI CAD*, Kluwer, Dordrecht, 1996.

[11] Moraga C. Systolic systems and multiple-valued logic. *Proc. 14th IEEE Int. Symp. Multiple-Valued Logic*, pp. 98–108, 1984.

[12] Negrini R and Sami MG. *Fault Tolerance Trough Reconfiguration in VLSI and WSI Arrays*. The MIT Press, Cambridge, MA, 1989.

[13] Papaioannou SG. Optimal test generation in combinational network by pseudo-Boolean programming. *IEEE Trans. Comput.*, 26(6):553–560, 1977.

[14] Popel DV and Yanushkevich SN. Modeling combinational circuits using linear word-level structures. *Automation and Remote Control*, 65(6):1018–1032, 2004.

[15] Sasao T. Design methods for multi-rail cascades. In *Proc. Int. Workshop on Boolean Problems*, Germany, 2002.

[16] Shmerko VP. Synthesis of arithmetic forms of Boolean functions using the Fourier transform. *Automation and Remote Control*, 50(5):684–691, Pt2, 1989.

[17] Shmerko VP. Malyugin's theorems: a new concept in logical control, VLSI design, and data structures for new technologies. *Automation and Remote Control*, 65(6):893–912, 2004.

[18] Sinha BP and Srimani PK. Fast parallel algorithms for binary multiplication and their implementation on systolic architectures. *IEEE Trans. Comput.*, 38(3):424–431, 1989.

[19] Stanković RS and Astola JT. Some remarks on linear transform of variables in representation of adders by word-level expressions and spectral transform decision diagrams. In *Proc. IEEE 32nd Int. Symp. on Multiple-Valued Logic*, pp. 116–122, 2002.

[20] Synthesis of Electronic Computing and Control Circuits. *The Annals of the Computation Laboraratory of Harvard University*, vol. XXVII, Cambridge, MA, 1951.

[21] Yanushkevich SN. Systolic algorithms to synthesize arithmetical polynomial forms for k-valued logic functions. *Automation and Remote Control*, 55(12):1812–1823, 1994.

[22] Yanushkevich SN, Shmerko VP, and Dziurzanski P. Linearity of word-level models: new understanding. In *Proc. IEEE/ACM 11th Int. Workshop on Logic and Synthesis*, pp. 67–72, New Orleans, LA, 2002.

[23] Yanushkevich SN. Spatial systolic arrays design for predictable nanotechnologies. *J. Computational and Theoretical Nanoscience*, 4(3):467–481, 2007.

[24] Yanushkevich SN, Miller DM, Shmerko VP, and Stanković RS. *Decision Diagram Techniques for Micro- and Nanoelectronic Design*, Taylor & Francis/CRC Press, Boca Raton, FL, 2006.

17

Information and Data Structures

17.1 Introduction

Information theory provide the fundamentals for data transmission. Because of data transmission between gates in logic network, techniques of information theory can be used in various design tasks, in particular, in optimization problems. The keystone of this theory is the Shannon entropy and information. Shannon information describes the uncertainty eliminated by the measurement. In digital systems, information has the unit bit (binary digit). Information or entropy measures the average number of binary digits that are needed to code a given message with a set of symbols with a given probability. This information measures the average length of the binary sequences.

In nanotechnology, the thermodynamic notation of entropy is used, that is, physical and chemical phenomena and their interpretation in terms of computing can be expressed in terms of thermodynamic entropy. There is the relationship between Shannon communication entropy and thermodynamic entropy. Note that information can be interpreted as *organization* and entropy as *disorganization*. One of the key concepts of nanotechnology the concept of *self-organization* is closely related to entropy and information.

This chapter focuses on information measures. In physical systems, information, in a certain sense, is a measurable quantity that is independent of the physical medium by which it is conveyed. The most appropriate measure of information is mathematically similar to the measure of entropy, that is, it is measured in bits of digits. The technique of information theory is applied to the problems of the extraction of information from systems containing an element of randomness.

17.2 Information-theoretic measures

A computing system can be seen as a process of communication between computer components. The classical concept of information advocated by

Shannon is the basis for this. However, some adjustment must be done to it in order to capture a number of features of the design and processing of a computing system.

The information-theoretic standpoint on computing is based on the following notations. *Source of information* is a stochastic process where an event occurs at time point i with probability p_i. That is, the source of information is defined in terms of the probability distribution for signals from this source. The problem is usually formulated in terms of sender and receiver of information and used by analogy with communication problems. *Information engine* is the machine that deals with information. *Quantity of information* is a value of a function that occurs with the probability p; this quantity is equal to $(-\log_2 p)$. *Entropy* $H(f)$ is the measure of the information content of X. The greater the uncertainty in the source output, the higher is its information content. A source with zero uncertainty would have zero information content and, therefore, its entropy would be equal to zero.

The information and entropy can be calculated with respect to the given sources:

Information carried by the value of a variable or function,
Conditional entropy of function f values given function g,
Relative information of the value of a function given the value of a variable,
Mutual information between the variable and function,
Joint entropy over a distribution of jointly specified functions f and g.

Many useful characteristics are derived from the entropy, namely, the conditional entropy, mutual information, joint information, and relative information. Figure 17.1 illustrates the basic principles of input and output information measures in a logic circuit, where the shared arrows mean that the value of $x_i(f)$ carries the information, the shared arrow therefore indicating the direction of the information stream. Obviously, we can compare the results of the input and output measures and calculate the loss of information. Information and entropy can be measured on decision trees and diagrams.

Let $A = \{a_1, a_2, \ldots, a_n\}$ be a complete set of events with the probability distribution

$$\{p(a_1), p(a_2), \ldots, p(a_n)\}$$

The *entropy* of the finite field A is given by

$$H(A) = -\sum_{i=1}^{n} p(a_i) \cdot \log p(a_i) \tag{17.1}$$

where the logarithm is base 2. The entropy can never be negative, i.e., $\log p(a_i) \leq 0$, and thus $H(A) \geq 0$. The entropy is zero if and only if A contains one event only.

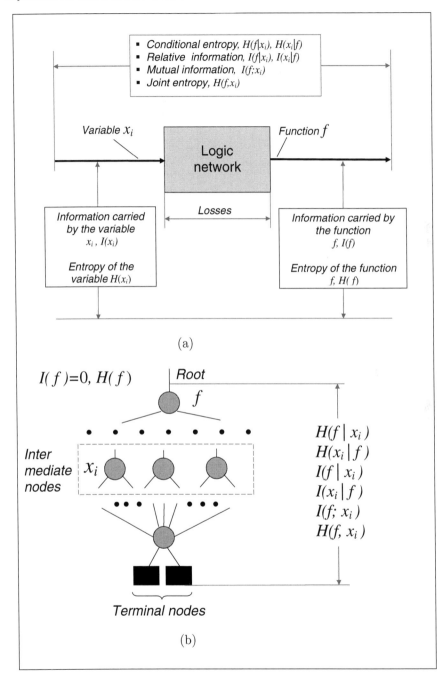

FIGURE 17.1
Information measures and computing input/output relationships of information in logic network (a) and decision diagram (b).

Example 17.1 *For a Boolean function that takes the value 1 with the probability p_1 and the value 0 with the probability p_0, the entropy is minimally $H(A) = 0$ bit/pattern, when $p_0 = 0$ or $p_0 = 1$, and the entropy reaches its maximum $H(A) = 1$ bit/pattern, when $p_0 = p_1 = 0.5$.*

Example 17.2 *The entropy for the completely specified Boolean function f of three variables given by the truth vector $\mathbf{F} = [10111110]^T$ is calculated as follows:*

$$H(f) = -0.25 \cdot \log 0.25 - 0.75 \cdot \log 0.75$$
$$= 0.8113 \ bit/pattern$$

The entropies of variables

$$H(x_1) = H(x_2) = H(x_3)$$
$$= -0.5 \cdot \log 0.5 - 0.5 \cdot \log 0.5$$
$$= 1 \ bit/pattern$$

that is, uncertainty in the variables of this function is the same.

Let A and B be finite fields of events with the probability distributions $\{p(a_i)\}$, $i = 1, 2, \ldots, n$ and $\{p(b_j)\}$, $j = 1, 2, \ldots, m$, respectively. The *conditional entropy* of the finite field A with respect to B is defined by

$$H(A|B) = -\sum_{i=1}^{n} \sum_{j=1}^{m} p(a_i, b_j) \cdot \log \frac{p(a_i, b_j)}{p(b_j)} \qquad (17.2)$$

Example 17.3 *(Continuation of Example 17.2.) The conditional entropy of Boolean function f with respect to the variable x_1, x_2, and x_3 is (Figure 17.2)*

$$H(f|x_1) = 0.8113 \ bit/pattern$$
$$H(f|x_2) = 0.8113 \ bit/pattern$$
$$H(f|x_3) = 0.5 \ bit/pattern$$

The variable x_3 conveys more information than the variables x_1 and x_2.

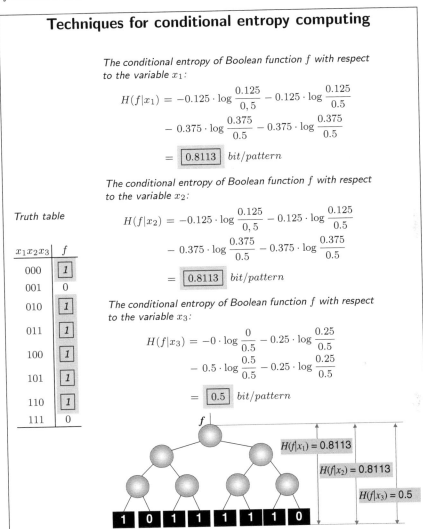

The conditional entropy of Boolean function f with respect to the variable x_1:

$$H(f|x_1) = -0.125 \cdot \log \frac{0.125}{0,5} - 0.125 \cdot \log \frac{0.125}{0.5}$$
$$- 0.375 \cdot \log \frac{0.375}{0.5} - 0.375 \cdot \log \frac{0.375}{0.5}$$
$$= \boxed{0.8113} \; bit/pattern$$

The conditional entropy of Boolean function f with respect to the variable x_2:

$$H(f|x_2) = -0.125 \cdot \log \frac{0.125}{0,5} - 0.125 \cdot \log \frac{0.125}{0.5}$$
$$- 0.375 \cdot \log \frac{0.375}{0.5} - 0.375 \cdot \log \frac{0.375}{0.5}$$
$$= \boxed{0.8113} \; bit/pattern$$

The conditional entropy of Boolean function f with respect to the variable x_3:

$$H(f|x_3) = -0 \cdot \log \frac{0}{0.5} - 0.25 \cdot \log \frac{0.25}{0.5}$$
$$- 0.5 \cdot \log \frac{0.5}{0.5} - 0.25 \cdot \log \frac{0.25}{0.5}$$
$$= \boxed{0.5} \; bit/pattern$$

Truth table

$x_1 x_2 x_3$	f
000	1
001	0
010	1
011	1
100	1
101	1
110	1
111	0

$H(f|x_1) = 0.8113$

$H(f|x_2) = 0.8113$

$H(f|x_3) = 0.5$

FIGURE 17.2

The conditional entropy of a completely specified Boolean function f with respect to the variables x_1, x_2, and x_3 (Example 17.3).

17.3   Information-theoretic measures

Information-theoretic measures include information, or information quantity, and entropy.

17.3.1   Quantity of information

Let us assume that all combinations of values of variables occur with equal probability. A value of a function that occurs with the probability p carries a quantity of information equal to

$$< \texttt{Quantity of information} > \ = \ -\log_2 p \ \ bit$$

where p is the probability of that value occurring. Note that information is measured in bits.

The information carried by the value of a of x_i is equal to

$$I(x_i)_{|x_i=a} = -\log_2 p \ \ bit$$

where p is the quotient between the number of tuples whose i-th components equal a and the total number of tuples. Similarly, the information carried by a value b of f is

$$I(f)_{|f=b} = -\log_2 q \ \ bit$$

where q is the quotient between the number of tuples in the domain of f and the number of tuples for which f takes the value b.

> **Example 17.4**   *The information carried by the values of variable x_i for a Boolean function f given by a truth table is calculated in Figure 17.3.*

17.3.2   Conditional entropy and relative information

Conditional entropy. This is a measure of a random variable f given a random variable x. To compute the conditional entropy, the conditional probability of f must be calculated. The conditional probability of value b of the function f, the input value a of x_i being known, is

$$p(f = b | x_i = a) = \frac{p_{|f=b \atop x_i=a}}{p_{|x_i=a}},$$

where $p_{|f=b \atop x_i=a}$ is the probability of f being equal to b and x_i being equal to a. Similarly, the conditional probability of value a of x_i given value b of the function f is

$$p(x_i = a | f = b) = \frac{p_{|f=b \atop x_i=a}}{p_{|f=b}}$$

Conditional entropy $H(f|g)$ of the function f given the function g is

$$H(f|g) = H(f,g) - H(g) \qquad\qquad (17.3)$$

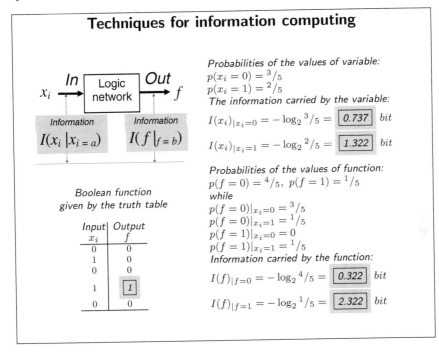

Techniques for information computing

FIGURE 17.3
The information carried by the variable x_i and the Boolean function f (Example 17.4).

Joint entropy. In Equation 17.3, $H(f, g)$ is the *joint entropy*:

$$H(f, g) = - \sum_{a=0}^{m-1} \sum_{b=0}^{m-1} p_{\substack{f=a \\ g=b}} \times \log p_{\substack{f=a \\ g=b}} \qquad (17.4)$$

where $p_{\substack{f=a \\ g=b}}$ denotes the probability that f takes value a and g takes value b, simultaneously. It follows from Equation 17.3 that the joint entropy is the sum of the source entropy, $H(g)$, and conditional entropy given g, that is, $H(f, g) = H(g) + H(f|g)$. The conditional entropy $H(f|g)$ represents the information loss in the circuit in going from input to output. It is how much must be added to the source entropy to get the joint entropy. Since f and g play symmetric roles, Equation 17.3 can be rewritten in the form $H(f, g) = H(f) + H(g|f)$.

Chain rule. In circuit analysis and decision tree design, the so-called *chain rule* is useful

$$H(f_1, \ldots, f_n | g) = \sum_{i=1}^{n} H(f_i | f_1, \ldots, f_{i-1}, g) \qquad (17.5)$$

Additivity of entropy, or the chain rule for entropies, is defined as $H(f,g) = H(f) + H(g|f)$. The conditional entropy $H(f|g)$ is non-negative and equal to zero if and only if a function u such that $f = u(g)$ exists with probability one.

Relative information. For value b of function f given value a_i of the input variable x_i this is

$$I(f = b|x_i = a) = -\log_2 p \ (f = b|x_i = a)$$

The relative information of value a_i of the input variable x_i given value b of the function f is $I(x_i = a|f = b) = -\log_2 p \ (x_i = a|f = b)$. Once the probability is equal to 0, we suppose that the relative information is equal to 0.

> **Example 17.5**   *Figure 17.4 illustrates the calculation of the conditional and relative information given the truth table of a Boolean function.*

17.3.3   Entropy of a variable and a function

Let the input variable x_i be the outcome of a probabilistic experiment, and the random function f represent the output of some step of computation. Each experimental outcome results in different conditional probability distributions on the random f. Shannon's entropy of the variable x_i is defined as

$$H(x_i) = \sum_{l=0}^{m-1} p_{|x_i = a_l} \log_2 p_{|x_i = a_l} \tag{17.6}$$

where m is the number of distinct values assumed by x_i. Shannon's entropy of the function f is

$$H(f) = \sum_{k=0}^{n-1} p_{|f = b_k} \log_2 p_{|f = b_k} \tag{17.7}$$

where n is the number of distinct values assumed by f.

This definition of the measure of information implies that the greater the uncertainty in the source output, the smaller is its information content. In a similar fashion, a source with zero uncertainty would have zero information content and, therefore, its entropy would likewise be equal to zero.

> **Example 17.6**   *Figure 17.5 illustrates the calculation of entropy of the variable and function. The entropy of the variable x_i and function f is 0.971 bits and 0.722 bits.*

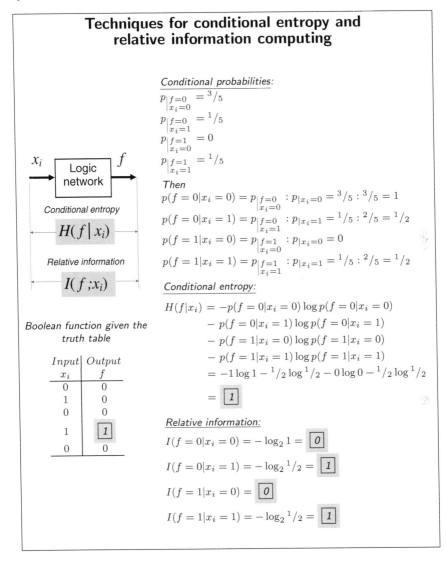

Techniques for conditional entropy and relative information computing

Conditional probabilities:

$$p_{\substack{f=0 \\ x_i=0}} = {}^3/_5$$

$$p_{\substack{f=0 \\ x_i=1}} = {}^1/_5$$

$$p_{\substack{f=1 \\ x_i=0}} = 0$$

$$p_{\substack{f=1 \\ x_i=1}} = {}^1/_5$$

Then

$$p(f = 0 | x_i = 0) = p_{\substack{f=0 \\ x_i=0}} : p_{|x_i=0} = {}^3/_5 : {}^3/_5 = 1$$

$$p(f = 0 | x_i = 1) = p_{\substack{f=0 \\ x_i=1}} : p_{|x_i=1} = {}^1/_5 : {}^2/_5 = {}^1/_2$$

$$p(f = 1 | x_i = 0) = p_{\substack{f=1 \\ x_i=0}} : p_{|x_i=0} = 0$$

$$p(f = 1 | x_i = 1) = p_{\substack{f=1 \\ x_i=1}} : p_{|x_i=1} = {}^1/_5 : {}^2/_5 = {}^1/_2$$

Conditional entropy:

$$
\begin{aligned}
H(f|x_i) = {} & -p(f = 0 | x_i = 0) \log p(f = 0 | x_i = 0) \\
& - p(f = 0 | x_i = 1) \log p(f = 0 | x_i = 1) \\
& - p(f = 1 | x_i = 0) \log p(f = 1 | x_i = 0) \\
& - p(f = 1 | x_i = 1) \log p(f = 1 | x_i = 1) \\
= {} & -1 \log 1 - {}^1/_2 \log {}^1/_2 - 0 \log 0 - {}^1/_2 \log {}^1/_2 \\
= {} & \boxed{1}
\end{aligned}
$$

Relative information:

$$I(f = 0 | x_i = 0) = -\log_2 1 = \boxed{0}$$

$$I(f = 0 | x_i = 1) = -\log_2 {}^1/_2 = \boxed{1}$$

$$I(f = 1 | x_i = 0) = \boxed{0}$$

$$I(f = 1 | x_i = 1) = -\log_2 {}^1/_2 = \boxed{1}$$

Left side of figure:

x_i → **Logic network** → f

Conditional entropy

$$H(f \mid x_i)$$

Relative information

$$I(f ; x_i)$$

Boolean function given the truth table

Input	Output
x_i	f
0	0
1	0
0	0
1	1
0	0

FIGURE 17.4

Computing conditional entropy and relative information (Example 17.5).

In general,

► For any variable x_i it holds that $0 \le H(f) \le 1$; similarly, for any function f, $0 \le H(x_i) \le 1$. Note that for patterns of variables it may not hold.

► The entropy of any variable in a completely specified function is 1.

► The entropy of a constant is 0.

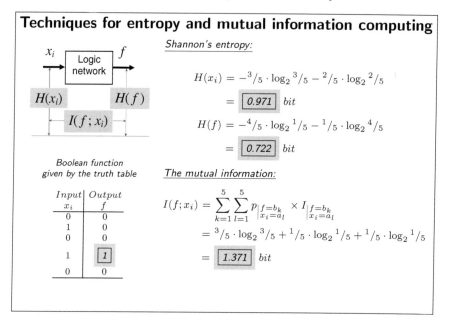

Techniques for entropy and mutual information computing

Shannon's entropy:

$$H(x_i) = -{}^3/_5 \cdot \log_2 {}^3/_5 - {}^2/_5 \cdot \log_2 {}^2/_5$$

$$= \boxed{0.971} \; bit$$

$$H(f) = -{}^4/_5 \cdot \log_2 {}^1/_5 - {}^1/_5 \cdot \log_2 {}^4/_5$$

$$= \boxed{0.722} \; bit$$

The mutual information:

$$I(f;x_i) = \sum_{k=1}^{5}\sum_{l=1}^{5} p_{\substack{f=b_k \\ x_i=a_l}} \times I_{\substack{f=b_k \\ x_i=a_l}}$$

$$= {}^3/_5 \cdot \log_2 {}^3/_5 + {}^1/_5 \cdot \log_2 {}^1/_5 + {}^1/_5 \cdot \log_2 {}^1/_5$$

$$= \boxed{1.371} \; bit$$

Boolean function given by the truth table

Input x_i	Output f
0	0
1	0
0	0
1	1
0	0

FIGURE 17.5

Shannon's entropy and mutual information (Examples 17.6 and 17.8).

17.3.4   Mutual information

Mutual information is used to measure the dependence of the function f on the values of the variable x_i and vice-versa, i.e., how statically distinguishable distributions of f and x_i are. If the distributions are different, then the amount of information f carries about x_i is large. If f is independent of x_i, then f carries zero information about x_i. Figure 17.6 illustrates the mutual information between two variables f and g.

The mutual information between the value b of the function and the value a of the input variable x_i is:

$$I(f;x_i) = I(f;x_i)_{|f=b} - I(f = b|x_i = a)$$

$$= -\log_2 p_{|f=b} + \log_2 \frac{p_{|\substack{f=b \\ x_i=a}}}{p_{|x_i=a}}$$

Useful relationships are

$$I(g;f) = I(f;g) = H(f) - H(f|g)$$
$$= H(g) - H(g|f)$$
$$= H(f) + H(g) - H(f,g)$$

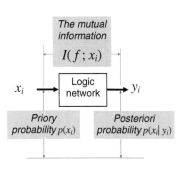

Mutual information is the difference between the entropy and conditional entropy

$$I(f;g) = H(f) - H(f|g)$$
$$= H(f) + H(g) - H(f,g)$$

that is, the difference of the uncertainty of f and the remaining uncertainty of f after knowing g. This quantity is the information of f obtained by knowing g.

The mutual information is a degree of the dependency between f and g and always takes positive values.

The additivity for the mutual information of three random variables is

$$I(f;g,z) = I(f;g) + I(f;z|g)$$

FIGURE 17.6

Mutual information.

17.3.5 Interpretation of mutual information

Mutual information can also be interpreted as follows. A logic network can be represented as a transmission system. The input and output signals are x_i and y_j, $i, j \in 1, 2, \ldots, q$, respectively. Prior to reception the probability of the signal x_i was $p(x_i)$. This is the *a priori* probability of x_i (Figure 17.6). After transmission of x_i and reception of y_j, the probability that the input signal was x_i becomes $p(x_i|y_j)$. This is the *a posteriori* probability of x_i.

The change in the probabilities $p(x_i)$ and $p(x_i|y_j)$ can be understood as how much the receiver (output) learned from the reception of y_j.

> **Example 17.7** *If the a posteriori probability $p(x_i|y_j) = 1$, the output signal is exactly the same the input signal.*

The difference between the a priori probabilities and the a posteriori probabilities is the difference of information uncertainty before and after the reception of a signal y_j. This difference measures the gain in information and can be described using the notation of a mutual information

$$I(x_i; y_j) \quad = \quad \underbrace{\log_2 \frac{1}{p(x_i)}}_{A\ priori\ information} \quad - \quad \underbrace{\log_2 \frac{1}{p(x_i|y_j)}}_{A\ posteriori\ information} \quad = \quad \log_2 \frac{p(x_i|y_j)}{p(x_i)}$$

If $p(x_i) = p(x_i|y_j)$, then the mutual information is zero and no information has been transmitted. If some additional knowledge is obtained about x_i from the output signal y_j, the mutual information is a positive value.

17.3.6   Conditional mutual information

The *conditional mutual* information between g and f given z, $I(g;f|z)$, is defined as follows

$$I(g;\underbrace{f_1,\ldots,f_n}_{f}|z) = \sum_{i=1}^{n} I(g;f_i|f_1,\ldots,f_{i-1},z) \tag{17.8}$$

If g and f in Equation 17.8 are independent, then $I(g;f) \geq 0$. The mutual information is a measure of the correlation between g and f. For example, if g and f are equal with a high probability, then $I(g;f)$ is large. If f_1 and f_2 carry information about g and are independent given g, then $I(z(f_1,f_2);g) \leq I(f_1;g) + I(f_2;g)$ for any Boolean function z.

> **Example 17.8**   *Figure 17.5 illustrates the calculation of the mutual information. The variable x_i carries 0.322 bits of information about the function f.*

17.4   Information measures of elementary Boolean function of two variables

There are two approaches to information measures of elementary functions of two variables:

▶ The values of input variables are considered as random patterns; for a two-input elementary function, there are four random patterns $x_1x_2 \in \{00.01, 10, 11\}$.
▶ The values of input variables are considered as noncorrelated random signals; for a two-input elementary function, there are random signals $x_1 \in \{0,1\}$ and $x_2 \in \{0,1\}$.

Information measures based on pattern. Consider a two-input AND function with four random combinations of input signals: 00 with probability p_{00}, 01 with probability p_{01}, 10 with probability p_{10}, and 11 with probability p_{11} (Figure 17.7a).

Using Shannon's formula (17.6), we can calculate the entropy of the input signals, denoted by H_{in}, as follows:

$$\begin{aligned}
H_{in} = & - p_{00} \times \log_2 p_{00} - p_{01} \times \log_2 p_{01} \\
& - p_{10} \times \log_2 p_{10} - p_3 \times \log_2 p_{11} \ \ bit/pattern
\end{aligned}$$

Approaches to probabilities computing for two binary signals

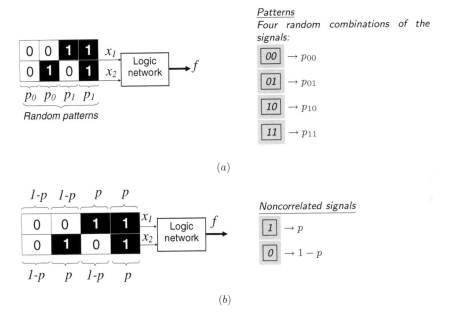

(a)

(b)

FIGURE 17.7
Measurement of probabilities: random patterns (a) and noncorrelated signals (b).

The maximum entropy of the input signals can be calculated by inserting into the above equation $p_i = 0.25$, $i = 0, 1, 2, 3$ (Figure 17.8).

The output of the AND function is equal to 0 with probability 0.25, and equal to 1 with probability 0.75. The entropy of the output signal, H_{out}, is calculated by Equation 17.7

$$H_{out} = -0.25 \times \log_2 0.25 - 0.75 \times \log_2 0.75 = 0.81 \; bit/pattern$$

The following example demonstrates a technique of computing information measures for input signals that are not correlated. Information measures of the two-variable functions AND, OR, EXOR, and NOT are given in Table 17.1 for $p(x_1) = p(x_2) = 0.5$.

Information measures based on noncorrelated signals. Let the input signal be equal to 1 with probability p, and 0 with probability $1 - p$ (Figure 17.7b). The entropy of the input signal is

$$H_{in} = -(1 - p)^2 \times \log_2(1 - p)^2 - 2(1 - p) \times \log_2(1 - p)p - p^2 \times \log_2 p^2$$
$$= -2(1 - p) \times \log_2(1 - p) - 2p \times \log_2 p \; bit$$

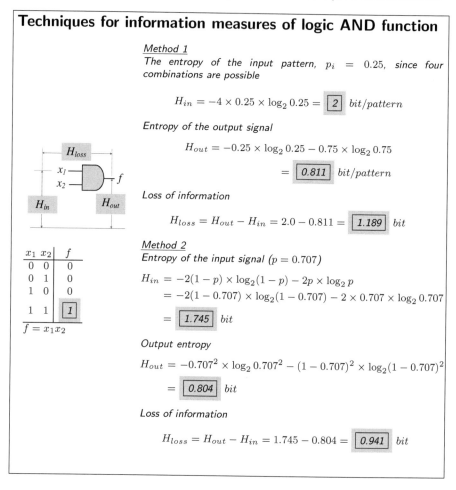

Techniques for information measures of logic AND function

Method 1
The entropy of the input pattern, $p_i = 0.25$, since four combinations are possible

$$H_{in} = -4 \times 0.25 \times \log_2 0.25 = \boxed{2} \; bit/pattern$$

Entropy of the output signal

$$H_{out} = -0.25 \times \log_2 0.25 - 0.75 \times \log_2 0.75$$

$$= \boxed{0.811} \; bit/pattern$$

Loss of information

$$H_{loss} = H_{out} - H_{in} = 2.0 - 0.811 = \boxed{1.189} \; bit$$

Method 2
Entropy of the input signal $(p = 0.707)$

$$H_{in} = -2(1 - p) \times \log_2(1 - p) - 2p \times \log_2 p$$
$$= -2(1 - 0.707) \times \log_2(1 - 0.707) - 2 \times 0.707 \times \log_2 0.707$$

$$= \boxed{1.745} \; bit$$

Output entropy

$$H_{out} = -0.707^2 \times \log_2 0.707^2 - (1 - 0.707)^2 \times \log_2(1 - 0.707)^2$$

$$= \boxed{0.804} \; bit$$

Loss of information

$$H_{loss} = H_{out} - H_{in} = 1.745 - 0.804 = \boxed{0.941} \; bit$$

x_1 x_2	f
0 0	0
0 1	0
1 0	0
1 1	$\boxed{1}$

$f = x_1 x_2$

FIGURE 17.8
Information measures of AND functions of two variables.

The output of the AND function is equal to 1 with probability p^2, and equal to 0 with probability $1 - p^2$. Hence, the entropy of the output signal is

$$H_{out} = -p^2 \times \log_2 p^2 - (1 - p)^2 \times \log_2(1 - p)^2 \quad bit$$

The maximum value of the output entropy is equal to 1 when $p = 0.707$. Hence, the input entropy of the AND function is 0.745 *bits* (Figure 17.8). We observe that in the case of noncorrelated signals, information losses are less.

TABLE 17.1
Information measures of elementary Boolean functions of two variables

Function	Information estimations
x_1 — f x_2 —	$f = x_1 x_2$ $H(f) = -p_{\|f=0} \cdot \log_2 p_{\|f=0} - p_{\|f=1} \cdot \log_2 p_{\|f=1}$ $p(f) = 0.5 \cdot 0.5 = 0.25$ $H(f) = -0.25 \cdot \log_2 0.25 - 0.75 \cdot \log_2 0.75 = \boxed{0.8113}$ *bit*
x_1 — f x_2 —	$f = x_1 \vee x_2$ $H(f) = -p_{\|f=0} \cdot \log_2 p_{\|f=0} - p_{\|f=1} \cdot \log_2 p_{\|f=1}$ $p(f) = 1 - (1 - 0.5) \cdot (1 - 0.5) = 0.75$ $H(f) = -0.75 \cdot \log_2 0.75 - 0.25 \cdot \log_2 0.25 = \boxed{0.8113}$ *bit*
x_1 — f x_2 —	$f = x_1 \oplus x_2$ $H(f) = -p_{\|f=0} \cdot \log_2 \|f = 0 - p_{\|f=1} \cdot \log_2 p_{\|f=1}$ $p(f) = 0.5 \cdot 0.5 + 0.5 \cdot 0.5 = 0.5$ $H(f) = -0.5 \cdot \log_2 0.5 - 0.5 \cdot \log_2 0.5 = \boxed{1}$ *bit*
x — f	$f = \bar{x}$ $H(f) = p_{\|f=0} \cdot \log_2 p_{\|f=0} - p_{\|f=1} \cdot \log_2 p_{\|f=1}$ $p(f) = 1 - 0.5 = 0.5$ $H(f) = -0.5 \cdot \log_2 0.5 - 0.5 \cdot \log_2 0.25 = \boxed{1}$ *bit*

17.5 Information-theoretic measures in decision trees and diagrams

In this section, we address the design of decision trees with nodes of three types: Shannon (S), positive Davio (pD), and negative Davio (nD) based on information-theoretic approaches. An approach revolves around choosing the "best" variable and the "best" expansion type with respect to this variable for any node of the decision tree in terms of information measures. This means that in any step of the decision making on the tree, we have an opportunity to choose both the variable and the type of expansion based on the criterion of minimum entropy.

An entropy-based optimization strategy can be thought of as generation of *optimal paths* in a decision tree, with respect to the minimum entropy criterion.

17.5.1   Decision tree induction

The best known use of entropy and information measures on decision trees is the induction of decision trees (ID3) algorithm for optimization.

> **Example 17.9**   *Figure 17.9 illustrates the calculation of entropy on a decision tree.*

Free binary decision trees are derived by permitting the permutation of the order of variables in a subtree independently of the order of variables in the other subtrees related to the same nonterminal node.

Another way of generalizing decision trees is to use different expansions at the nodes in the decision tree. This decision tree is designed by arbitrarily choosing any variable and any of the S, pD, or nD expansions for each node.

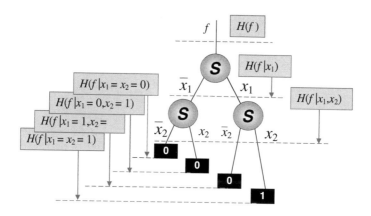

FIGURE 17.9

Measurement of entropy on a decision tree (Example 17.9).

17.5.2   Information-theoretic notation of Shannon and Davio expansion

In the top-down decision tree design, two information measures are used: conditional entropy, $H(f|Tree$, and mutual information, $I(f;Tree)$. The *initial state* of this process is characterized by the maximum value for the conditional entropy $H(f|Tree) = H(f)$. Nodes are recursively attached to the decision tree by using the top-down strategy. In this strategy the entropy $H(f|Tree)$ of the function is reduced, and the information $I(f;Tree)$

increases, since the variables convey the information about the function. Each *intermediate state* can be described in terms of entropy by the equation

$$I(f; Tree) = H(f) - H(f|Tree) \qquad (17.9)$$

The goal of such an optimization is to maximize the information $I(f; Tree)$ that corresponds to the minimization of entropy $H(f|Tree)$, in each step of the decision tree design. The final state of the decision tree is characterized by $H(f|Tree) = 0$ and $I(f; Tree) = H(f)$, that is, *Tree* represents the Boolean function f (Figure 17.10).

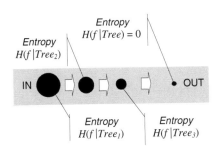

Maximizing the information $I(f; Tree_i)$ corresponds to the minimization of entropy $H(f|Tree_i)$.

The final state of the decision tree design corresponds to $H(f|Tree_3) = 0$ and $I(f; Tree) = H(f)$, that is, < Tree > represents the Boolean function f

FIGURE 17.10
Four steps to minimization of the entropy $H(f|Tree)$ in designing a decision tree for reduction of uncertainty.

The decision tree design process is a recursive decomposition of a Boolean function. A step of this recursive decomposition corresponds to the expansion of a Boolean function f with respect to the variable x. Assume that the variable x in f conveys information that is, in some sense, the rate of influence of the input variable on the output value for f.

The initial state of the expansion $\omega \in \{S, pD, nD\}$ can be characterized by the entropy $H(f)$ of f, and the final state by the conditional entropy $H^\omega(f|x)$. The ω expansion of the Boolean function f with respect to the variable x is described in terms of entropy as follows:

$$I^\omega(f; x) = H(f) - H^\omega(f|x) \qquad (17.10)$$

A formal criterion for completing the sub-tree design is $H^\omega(f|x) = 0$.

Information notation of S expansion

The designed decision tree based on the S expansion is mapped into a sum-of-products expression as follows: a leaf with the logic value 0 is mapped into $f = 0$, and with the logic value 1 into $f = 1$; a nonterminal node is mapped into $f = \overline{x} \cdot f_{x=0} \vee x \cdot f_{x=1}$ (Figure 17.11). The information measure of S expansion for a Boolean function f with respect to the variable x is represented by the equation

$$H^S(f|x) = p_{|x=0} \cdot H(f_{x=0}) + p_{|x=1} \cdot H(f_{x=1}) \qquad (17.11)$$

The information measure of S expansion is equal to the conditional entropy $H(f|x)$:

$$H^S(f|x) = H(f|x) \qquad (17.12)$$

Shannon expansion

$$f = \overline{x} \cdot f_{x=0} \oplus x \cdot f_{x=1}$$

Information-theoretic notation

$$H^S(f|x) = \underbrace{p_{|x=0}H(f_{x=0})}_{Left\ leaf} + \underbrace{p_{|x=1}H(f_{x=1})}_{Right\ leaf}$$

Positive Davio expansion

$$f = f_{x=0} \oplus x \cdot (f_{x=0} \oplus f_{x=1})$$

Information-theoretic notation

$$H^{pD}(f|x) = \underbrace{p_{|x=0}H(f_{x=0})}_{Left\ leaf} + \underbrace{p_{|x=1}H(f_{x=0} \oplus f_{x=1})}_{Right\ leaf}$$

Negative Davio expansion

$$f = f_{x=1} \oplus \overline{x} \cdot (f_{x=0} \oplus f_{x=1})$$

Information-theoretic notation

$$H^{nD}(f|x) = \underbrace{p_{|x=1}H(f_{x=1})}_{Left\ leaf} + \underbrace{p_{|x=0}H(f_{x=0} \oplus f_{x=1})}_{Right\ leaf}$$

FIGURE 17.11
Shannon and Davio expansions and their information measures for a Boolean function.

Information notation of pD and nD expansion

The information measure of a pD expansion of a Boolean function f with respect to the variable x is represented by

$$H^{pD}(f|x) = p_{|x=0} \cdot H(f_{x=0}) + p_{|x=1} \cdot H(f_{x=0} \oplus f_{x=1}) \qquad (17.13)$$

The information measure of the nD expansion of a Boolean function f with respect to the variable x is

$$H^{nD}(f|x) = p_{|x=1} \cdot H(f_{x=1}) + p_{|x=0} \cdot H(f_{x=0} \oplus f_{x=1}) \qquad (17.14)$$

Note that, since $H^S(f|x) = H(f|x)$ and $I^S(f;x) = I(f;x)$, the mutual information of f and x is $I(f;x) = H(f) - H(f|x)$. This is because

$$H(f|x) = p_{|x=0} \cdot H(f_{x=0}) + p_{|x=1} \cdot H(f_{x=1}) + p_{|x=1} \cdot H(f_{x=0} \oplus f_{x=1})$$
$$- p_{|x=1} \cdot H(f_{x=0} \oplus f_{x=1})$$
$$= H^{pD}(f|x) + p_{|x=1} \cdot (H(f_{x=1}) - H(f_{x=0} \oplus f_{x=1}))$$

Thus

$$I(f;x) = H(f) - H^{pD}(f|x) - p_{|x=1} \cdot (H(f_{x=1}) - H(f_{x=0} \oplus f_{x=1}))$$
$$I(f;x) = H(f) - H^{nD}(f|x) - p_{|x=0} \cdot (H(f_{x=0}) - H(f_{x=0} \oplus f_{x=1}))$$

Example 17.10 *Given a Boolean function of three variables, calculate the entropy of Shannon and Davio expansions with respect to all variables. The results are summarized in Table 17.2. We observe that the minimal value of the information-theoretic measure corresponds to Shannon expansion with respect to the variable x_2.*

17.5.3 Optimization of variable ordering in a decision tree

The entropy-based design of an optimal decision tree can be described as the optimal (with respect to the information criterion) node selection process. A path in the decision tree starts from a node and finishes in a terminal node. Each path corresponds to a term in the final expression for f. The criterion for choosing the decomposition variable x and the expansion type $w \in \{S, pD, nD\}$ is that the conditional entropy of the function with respect to this variable has to be minimum.

The entropy-based algorithm for the minimization of AND/EXOR expressions is introduced in the following example. In this algorithm, the ordering restriction is relaxed. This means that (a) each variable appears

TABLE 17.2

Choosing the type of expansion for a Boolean function of
three variables (Example 17.10)

	Shannon, S	Positive Davio, pD	Negative Davio, nD
	$H^S(f_1\|x)$	$H^{pD}(f_1\|x)$	$H^{nD}(f_1\|x)$
x_1	0.88	0.95	0.88
x_2	0.67	0.88	0.75
x_3	0.98	0.98	0.95

once in each path, and (*b*) the orderings of variables along the paths may be
different.

Example 17.11 *The design of an AND/EXOR decision tree for the
hidden weighted bit function is given in Figure 17.12.
The order of variables in the tree is evaluated
based on a measure of entropy of the Boolean
function f with respect to variables x_1 x_2, and x_3.
According to the criterion of minimum entropy, x_1
is assigned to the root. The other assignments
are shown in Figure 17.12. The quasi-optimal
polynomial expression corresponding to this tree is
$f = x_2 x_3 \oplus x_1 \overline{x}_3$.*

17.6   Information-theoretic measures in multivalued functions

Information-theoretic measures can be applied to multivalued logic functions.
The focus is S, pD, and nD expansions and decision tree design in terms of
entropy and information.

Information-theoretic measures can be applied to m-valued functions.
For calculation, the logarithm base m is applied, for example, $\log_3$
for ternary function, $\log_4$ for quaternary function, etc. The following
example demonstrates the technique for computing information-theoretic
characteristics for the function given by a truth table.

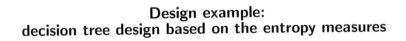

**Design example:
decision tree design based on the entropy measures**

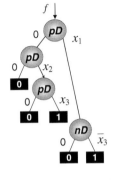

Step 1. *Choose the variable x_1 and pD expansion for the root node*

$$H^{pD}(f|x_1) = \boxed{0.91} \quad \text{bit}$$

Decision: select the $f_0 = f_{x_1=0}$

Step 2. *Choose the variable x_2 and pD expansion for the next node*

$$H^{pD}(f|x_2) = \boxed{0.5} \quad \text{bit}$$

$f_0 = f_{x_2=0} = 0$
Decision: select
$f_1 = f_{x_2=0} \oplus f_{x_2=1}$

x_1 x_2 x_3	f
0 0 0	0
0 0 1	0
0 1 0	0
0 1 1	1
1 0 0	1
1 0 1	0
1 1 0	1
1 1 1	1

Step 3. *Select pD expansion for the variable x_3*

$f_0 = f_{x_3=0} = 0$ and
$f_1 = f_{x_3=0} \oplus f_{x_3=1} = 1$
Decision: select
$f_0 = f_{x_1=0} \oplus f_{x_1=1}$

Step 4. *Choose the variable x_3 and select nD expansion*

$f_0 = f_{x_3=1} = 0$ and
$f_1 = f_{x_3=0} \oplus f_{x_3=1} = 1$

FIGURE 17.12
AND/EXOR decision tree design (Example 17.11).

Example 17.12 *Computation of entropy, conditional entropy, and mutual information for a four-valued function f given its truth vector $[0000\ 0231\ 0213\ 0321]^T$ is shown in Figure 17.13.*

17.6.1 Information notation of S expansion

The information measures of Shannon expansion in the field $GF(4)$

$$f = J_0(x) + xJ_1(x) + x^2 J_2(x) + x^3 J_3(x)$$

Techniques for entropy measures of a four-valued logic function

Truth table

x_1 x_2	f
0 0	0
0 1	0
0 2	0
0 3	0
1 0	0
1 1	3
1 2	1
1 3	0
2 0	2
2 1	0
2 2	1
2 3	3
3 0	0
3 1	3
3 2	2
3 3	1

The probabilities of the logic function values are

$$p_{|f=0} = {}^7/_{16}, \quad p_{|f=1} = p_{|f=2} = p_{|f=3} = {}^3/_{16}$$

The entropy of the logic function f is

$$H(f) = -{}^7/_{16} \cdot \log_2 {}^7/_{16} - 3 \cdot {}^3/_{16} \cdot \log_2 {}^3/_{16}$$

$$= \boxed{1.88} \; bit$$

The conditional entropy of the logic function f with respect to the variable x_1 is

$$H(f|x_1) = -{}^4/_{16} \cdot \log_2 1 - 12 \cdot {}^1/_{16} \cdot \log_2 {}^1/_4$$

$$= \boxed{1.5} \; bit$$

The conditional entropy with respect to variable x_2 is

$$H(f|x_2) = \boxed{1.25} \; bit$$

The mutual information for the logic function f and the variables x_1 and x_2 is

$$I(f; x_1) = \boxed{0.38} \; bit$$

$$I(f; x_2) = \boxed{0.63} \; bit$$

FIGURE 17.13

Information-theoretic measures of a four-valued logic function (Example 17.12).

are given in Table 17.3 (first row), where $J_i(x)$, $i = 0, \ldots, k - 1$, are the characteristic functions, denoted by $J_i(x) = 1$, if $x = i$ and $J_i(x) = 0$, otherwise. The average Shannon entropy is equal to conditional entropy $H(f|x)$ of function f with respect to x:

$$H^S(f|x) = H(f|x) \tag{17.15}$$

This is because the terminal nodes of Shannon expansion represent the values of the function. The information notation of Shannon expansion is specified as follows:

$$I^S(f; x) = I(f; x) \tag{17.16}$$

Since $H^S(f|x) = H(f|x)$ and $I^S(f; x) = I(f; x)$, then for S expansion, we can write $I(f; x) = H(f) - H(f|x)$.

TABLE 17.3
Information measures of Shannon and Davio expansions in GF(4)

Type	Information-theoretic measures

$$H^S(f|x) = \overbrace{p_{|x=0} \cdot H(f_{x=0})}^{Leaf\ 1} + \overbrace{p_{|x=1} \cdot H(f_{x=1})}^{Leaf\ 2}$$
$$+ \underbrace{p_{|x=2} \cdot H(f_{x=2})}_{Leaf\ 3} + \underbrace{p_{|x=3} \cdot H(f_{x=3})}_{Leaf\ 4}$$

$$H^{pD}(f|x) = p_{|x \neq 0} \cdot \frac{H(f_1) + H(f_2) + H(f_3)}{3}$$
$$+ p_{|x=0} \cdot H(f_{x=0})$$

$$H^{nD'}(f|x) = p_{|x \neq 1} \cdot \frac{H(f_0) + H(f_2) + H(f_3)}{3}$$
$$+ p_{|x=1} \cdot H(f_{x=1})$$

$$H^{nD''}(f|x) = p_{|x \neq 2} \cdot \frac{H(f_0) + H(f_1) + H(f_3)}{3}$$
$$+ p_{|x=2} \cdot H(f_{x=2})$$

$$H^{nD'''}(f|x) = p_{|x \neq 3} \cdot \frac{H(f_0) + H(f_1) + H(f_2)}{3}$$
$$+ p_{|x=3} \cdot H(f_{x=3})$$

17.6.2   Information notations of pD and nD expansion

Consider $m = 4$. Then the positive and negative Davio expansions of a four-valued logic function are defined as follows:
Positive Davio expansion pD

$$
\begin{aligned}
f &= f_{x=0} + x(f_{x=1} + 3f_{x=2} + 2f_{x=3}) \\
&\quad + x^2(f_{x=1} + 2f_{x=2} + 3f_{x=3}) \\
&\quad + x^3(f_{x=0} + f_{x=1} + f_{x=2} + f_{x=3}) \\
&= f_{x=0} + xf_1 + x^2 f_2 + x^3 f_3
\end{aligned}
$$

Negative Davio expansion nD'

$$
f = f_{x=1} + \widehat{x} f_0 + \widehat{x}^2 f_2 + \widehat{x}^3 f_3
$$

Negative Davio expansion nD''

$$
f = f_{x=2} + \widehat{\widehat{x}} f_0 + \widehat{\widehat{x}}^2 f_1 + \widehat{\widehat{x}}^3 f_3
$$

Negative Davio expansion nD'''

$$
f = f_{x=3} + \widehat{\widehat{\widehat{x}}} f_0 + \widehat{\widehat{\widehat{x}}}^2 f_1 + \widehat{\widehat{\widehat{x}}}^3 f_2
$$

The entropy associated with the preceding expansions is shown in Table 17.3.

17.6.3   Information criterion for decision tree design

The main properties of the information measure are (a) the recursive character of S, pD, and nD expansions and their generalization for the four-valued case, and (b) the possibility of choosing a decomposition variable and expansion type based on the information measure. Decision tree design can be interpreted as an optimized (with respect to information criterion) node selection process. The criterion for choosing the decomposition variable x and expansion type $\omega \in \{S, pD, nD\}$ is that the conditional entropy of the logic function given a variable x_i has to be minimal

$$
H^\omega(f|x) = MIN(H^{\omega_j}(f|x_i) \mid \forall \; pairs \; (x_i, \omega_j)) \tag{17.17}
$$

In the algorithm, the ordering restriction is relaxed. This means that (i) each variable appears once in each path and (ii) the order of variables along with each path may be different.

> **Example 17.13** *Decision tree design for the logic function f from Example 17.12 is given in Figure 17.14.*

Design example:
entropy-based a free ternary decision tree design

Step 1. *Choose variable x_2 and $4-pD$ expansion for root node, because the minimal entropy is*

$$H^{pD}(f|x_2) = \boxed{0.75} \text{ bit}$$

Functions $f_0 = f_{x_2=0}$ and $f_3 = f_{x_2=0} + f_{x_2=1} + f_{x_2=2} + f_{x_2=3}$ both take logic value 0. Select the function

$$f_1 = f_{x_2=1} + 3f_{x_2=2} + 2f_{x_2=3}.$$

Step 2. *Choose variable x_1 and pD expansion for the next node. The successors are constant: $f_0 = 0$, $f_1 = 0$, $f_2 = 3$, and $f_3 = 1$. Select the function*

$$f_2 = f_{x_2=1} + 2f_{x_2=2} + 3f_{x_2=3}$$

Step 3. *Select $H^{nD'}$ expansion for variable x_1. The successors are constant: $f_0 = 0$, $f_1 = 1$, $f_2 = 0$, and $f_3 = 1$*

The decision tree obtained is

Truth table

x_1 x_2	f
0 0	0
0 1	0
0 2	0
0 3	0
1 0	0
1 1	2
1 2	3
1 3	1
2 0	0
2 1	2
2 2	3
2 3	1
3 0	0
3 1	3
3 2	2
3 3	1

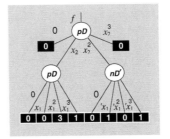

The corresponding polynomial expression is

$$f = 3 \cdot x_2 \cdot x_1^2 + x_2 \cdot x_1^3 + x_2^2 \cdot {}^{1-}x_1 + x_2^2 \cdot {}^{3-}x_1$$

By analogy:

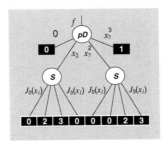

$$f = 2 \cdot x_2 \cdot J_1(x_1) + 3 \cdot x_2 \cdot J_2(x_1)$$
$$+ 2 \cdot x_2^2 \cdot J_2(x_1) + 3 \cdot x_2^2 \cdot J_3(x_1)$$

FIGURE 17.14

Decision tree design (Example 17.13).

17.6.4 Remarks on information-theoretic measures in decision diagrams

Given a completely specified Boolean function

$$f : \ p(x = 0) = p(x = 1) = \frac{1}{2}.$$

Since each node of the decision diagram is an instance of the Shannon expansion, a probability assignment algorithm in a top-down fashion works as follows:

$$p(f) = \frac{1}{2}p(f_{x=0}) + \frac{1}{2}p(f_{x=1}).$$

Another approach is the following. Consider $p(leaf|f = 1) = 1$ and $p(leaf|f = 0) = 0$, and output probability $p(f = 1) = p(root)$ and $p(f = 0) = 1 - p(root)$. Application of the recursive strategy:

$$p(node) = p(x = 0)p(edgel) + p(x = 1)p(edger)$$

provides calculation of conditional and joint probabilities for computing conditional entropy. To find the joint probability $p(f = 1, x = 1)$, it is necessary to set $p(x = 1) = 1$ and $p(x = 0) = 0$ before decision diagram traversal. This allows the calculation of the whole range of probabilities using only one diagram traversal.

Example 17.14 *Given the Boolean function* $f = x_3 \lor x_2 \lor x_1$ *by its truth vector* $\mathbf{F} = [10001111]$*:*

$$H(f) = -5/8 \cdot log_2(5/8) - 3/8 \cdot log_2(3/8) = \boxed{0.96} \ bit$$
$$H(f|x_1) = -1/8 \cdot log_2(1/4) - 3/8 \cdot log_2(3/4)$$
$$- 4/8 \cdot log_2(4/4) - 0 = \boxed{0.41} \ bit$$

Similarly, $H(f|x_2) = \boxed{0.91}$ *bit,* $H(f|x_3) = \boxed{0.91}$ *bit. A top-down approach of assigning* $p(leaf|f = 1) = 1$ *yields* $p(f = 1) = p(root) = 0.625$ *(diagram with three nodes). The result of setting* $p(x_2 = 0) = 1$ *is a conditional probability* $p(f_{x_2=0}) = 0.75$*.*

17.7 Ternary and pseudo-ternary decision trees

The information model of the S-node corresponding to the Shannon expansion with respect to the variable x of a Boolean function f is represented by the relation

$$H^S(f|x) = H(f_{x=0}) + H(f_{x=1}) \qquad (17.18)$$

A criterion to choose a variable x in S-decomposition is that the conditional entropy of the Boolean function f with respect to the given variable x has to be minimum.

Ternary decision trees are a generalization of the binary decision trees derived by permitting nodes with three outgoing edges. In ternary decision trees derived as a generalization of the binary decision trees, it is assumed that

▶ The first two outgoing edges point to the cofactors of f determined as in the Shannon decomposition for f.

▶ The third outgoing edge points to a subfunction $f_0 \# f_1$, where $\#$ denotes a binary operation.

By choosing different operations, different ternary decision trees are defined. Examples are AND-, OR-, EXOR-ternary decision trees, and Kleene-ternary decision trees, defined by using the logic operations AND, OR, EXOR, and the Kleene operation, respectively. In these ternary decision trees, the correspondence between a ternary decision tree and a Boolean function f is defined as follows:

(a) For a terminal node v, $v = \begin{cases} f_v = 1, & \text{if } v=1 \\ f_v = 0, & \text{if } v=0 \end{cases}$

(b) For a nonterminal node $f_v = \overline{x} \cdot f_0 \vee x \cdot f_1 \vee 1 \cdot f_2$, where subfunctions f_0, f_1, and f_2 are computed as follows: $f_0 = f_{x=0}$, $f_1 = f_{x=1}$, and $f_2 = f_0 \# f_1$. The symbol $\#$ denotes any of the operations AND, OR, EXOR, or the Kleene operator.

Example 17.15 *The EXOR ternary decision tree based on an entropy-based technique for the Boolean function f given by truth vector $\mathbf{F} = [10111110]^T$ is constructed as shown in Figure 17.15. Construction is started with variable x_3, because it follows from Example 17.3 that the order of decomposed variables is $\{x_3, x_2, x_1\}$.*

Design example:
entropy-based strategy for a free ternary decision tree design

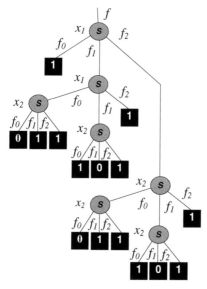

The first edge corresponds to $x_3 = 0$,
    $f_0 = \{000, 010, 100, 110\}$
The second edge corresponds to
    $x_3 = 1$,
    $f_1 = \{001, 011, 101, 111\}$
The third edge is formed as
    $f_2 = f_0 \oplus f_1$

*The result of building the decision tree
yields the following AND/EXOR form:*

$$f = \bar{x}_3 \oplus x_1 x_3 \oplus x_2 x_3$$

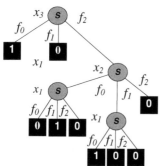

FIGURE 17.15

Example of entropy-based decision tree design for a free ternary decision tree with final result $f = \bar{x}_3 \oplus x_1 x_3 \oplus x_2 x_3$ (a) and free pseudo-ternary decision tree with resulting function $f = \bar{x}_3 \vee \bar{x}_1 x_2 \vee x_1 \bar{x}_2$ (b) given a Boolean function of three-variables (Example 17.15).

17.8   Further study

Advanced topics of information-theoretic measures

Topic 1: Shannon entropy. Shannon suggested a measure to represent the information in numerical values, denoted as the *Shannon entropy* [39]. The Shannon information theory has been developed for many applications in circuit design. The latest characterization of a computing system as a communication system is consistent with the von Neumann concept of a computer. The bit strings of information are understood as messages to be communicated from a messenger to a receiver. Each message i is an event that has a certain probability of occurrence p_i with respect to its inputs. The measure of information produced when one event is chosen from the set of N events is the entropy of message i: $-\sum_{i \in N} p_i \log p_i$.

The amount of randomness in a probability distribution is measured by its entropy (or information). In a fundamental sense, the concept of information proposed by Shannon captures only the case when unlimited computing power is available. However, computational cost may play a central role. This and other aspects of information theory can be found in [4, 13, 17, 24].

Topic 2: Applications in logic design. Useful results can be found in the book *Artificial Intelligence in Logic Design* edited by S.N. Yanushkevich, Kluwer, 2004, which includes nine papers on the fundamentals of logic function manipulation based on the artificial intelligence paradigm, evolutionary circuit design, information measures in circuit design, and logic design of nanodevices.

Testing. The analysis is based a model in which all signals are assumed to have certain statistical properties. The dynamic flavor of entropy has been studied in many papers to express testability (observability and controllability) measures for gate-level circuits. For example, Agraval has shown that the probability of fault detection can be maximized by choosing test patterns that maximize the information at output [1]. The problem of the construction of sequential fault location for permanent faults has been considered by Varshney et al. [50].

State-of-the-art decision diagram techniques. Designing an optimal decision tree based on information-theoretic measures was studied by Goodman and Smyth [12], and by Hartmann et al. [14]. There are some approaches to designing an optimal decision tree from a given decision table [12]. Among them, we note the branch-and-bound technique and dynamic programming methods that are exhaustive search methods. Decomposition allows us to reduce the search space. However, the methods do not guarantee the optimal solution.

Hartmann et al. [14] applied the information theoretic criterion to convert decision tables with don't cares into near-optimal decision trees. The truth table of a logic function is considered as a special case of a decision table with variables replacing the tests in the decision table. The technique to convert the decision table into a decision tree and its optimization has been applied for Boolean and multivalued function minimization by Kabakcioglu et al. [16], Lloris et al. [18] and Lloris-Ruiz et al. [19]. The search for the best implicants to appear in the final solution is realized by using a decision tree, so that Shannon expansion formula $\overline{x} \cdot f_{x=0} \oplus x \cdot f_{x=1}$ is applied at each node in the decision tree. Cheushev et al. [6] gave a justification for the use of the information measure for Boolean and multivalued logic functions. It was shown that in the simplest understanding, the information theory methods give the same results as the probabilistic approach [42].

Yanushkevich et al. [54] and Shmerko et al. [43] studied AND/EXOR and SOP minimization through ternary decision trees [36]. The nodes of such a decision tree have three outputs and the third edge assigns don't care values with respect to Morreale's operator [27] for generating the prime-irredundant cubes.

In state-of-the-art decision diagram techniques, the information theoretical notation of a Shannon expansion of a Boolean function f with respect to a variable x is used in the form $H^S(f|x) = p_{|x=0} \cdot H(f_{x=0}) + p_{|x=1} \cdot H(f_{x=1})$, where $H(f_{x=0})$ and $H(f_{x=1})$ is the entropy of function f given $x = 0$ and $x = 1$, respectively.

Power dissipation. Existing techniques for power estimation at gate and circuit levels can be divided into dynamic and static. These techniques rely on probabilistic information in the input stream. The average switching activity per node (gate) is the main parameter that needs to be correctly determined. These and related problems are the focus of many researchers. For example, in [22, 33], it is demonstrated that the average switching activity in a circuit can be calculated using either entropy or information energy averages.

Finite state machines. Most of the algorithms for minimization of state assignments in finite state machines target reduced average Boolean per transition, that is, average Hamming distance between states. Several papers have used entropy-based models to solve these problem. In particular, Tyagi's paper [49] provides theoretical lower bounds on the average Hamming distance per transition for finite state machines based on information-theoretic methods.

Evolutionary circuit design. There have already been some approaches to evolutionary circuit design. The main idea is that an evolutionary strategy would inevitably explore a much richer set of possibilities in the design spaces that are beyond the scope of traditional methods. In [2, 7, 8, 20, 21] the evolutionary strategy and information-theoretic measures were used in circuit design.

Functional decomposition using information relationship measures was studied by Rawski et al. [34].

Topic 3: Information engine, computational work, and complexity. A deep and comprehensive analysis of computing systems' information engines has been done by Watanabe [51]. The relationship between function complexity and entropy is conjectured by Cook and Flynn [9]. The complexity of a Boolean function is expressed by the cost of implementing the function as a combinational network.

Hellerman has proposed so-called *logic entropy* [15]. Computation is considered as a process that reduces the disorder (or entropy) in the space of solutions while finding a result. The number of decisions required to find one correct answer in the space of solutions has been defined as *entropy of computation* or *logic entropy* calculated as $log\frac{S}{A}$, where s is the number of solutions and A is the number of answers. This definition is consistent with Shannon entropy provided that the space of solutions is all possible messages (bit strings) of a given length. The answer is one of the messages, so the entropy is the numbers of bits required to specify the correct answer. The term *logic entropy* owes its name to the fact that it depends on the number of logic operations required to perform the computation. In the beginning of the computation, the entropy (disorder) is maximum; at the end of computation, the entropy is reduced to zero.

The other form of entropy is *spatial entropy*, and it is relevant to mapping the computation onto a domain where data travels over a physical distance. The data communication process is a process of removal of spatial entropy, whereas performing logical operations is aimed at the removal of logical entropy (disorder). The spatial entropy of a system is a measure of the effort needed to bring data from the input location to the output locations. The removal of spatial entropy corresponds to reduction of the distance between the input and the output.

Topic 4: Information, entropy, and self-organization. The second law of thermodynamics states that the entropy has to increase in closed systems, that is, it indicates that disorder or disorganization in closed systems has to increase. In contrast to the second law of thermodynamics, there exist living organisms that can create organized or self-organized structures in the disorganized environment and thus produce information. This means that they diminish their entropy. They do not contradict the second law of thermodynamics, because the entropy increases only in closed systems; self-organized systems are not closed systems, they are open systems.

Let S be a system with four possible states a_1, a_2, b_1, and b_2 at the lowest level of its behavior (Figure 17.16). At the highest level of system description, the states are aggregated by two, that is, $A = \{a_1, a_2\}$ and $B = \{b_1, b_2\}$. The behavior of system at the highest level is characterized by two states A and B. The system is in the state A if it is either in state a_1 or in state a_2. Let the probabilities of states at the lowest level are $p(a_1) = p(b_1 = 0.1)$ and $p(a_2) = p(b_2 = 0.4)$. Hence, the probabilities of states at the highest level are $p(A) = p(B) = 0.5$. Shannon entropy at the lowest and highest level are $H = 1.72$ and $H = 1$, respectively.

Topic 5: Complexity and information. A. Kolmogorov and A. Turing studied the complex structures from the information standpoint. The basic assumptions are as follows. A *random sequence* can be defined using the notation of some feature that is not specified in terms of certain structure or patterns and that enables us to reduce such a sequence. *Nonrandom sequence* consists of a sufficient number of patterns and it can be reduced. The complexity of a data sequence can be defined using random and nonrandom sequences. In this notation, the complexity is the minimum data that is required for the description of a given sequence.

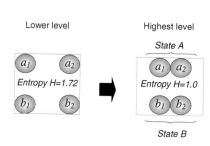

The lowest level: *states* $a_1, a_2, b_1,$ *and* b_2. The probabilities of states:

$$p(a_1) = p(b_1) = 0.1)$$
$$p(a_2) = p(b_2 = 0.4)$$

The highest level: states $A = \{a_1, a_2\}$ *and* $B = \{b_1, b_2\}$.
Probabilities of states:

$$p(A) = p(a_1) + p(a_2) = 0.1 + 0.4 = 0.5$$
$$p(B) = p(b_1) + p(b_2) = 0.1 + 0.4 = 0.5$$

Shannon entropy at the lowest level: $H = 1.72$
Shannon entropy at the highest level: $H = 1$.

FIGURE 17.16

A system with four possible states a_1, a_2, b_1, b_2 at the lowest level is self-organized into system with two states $A = \{a_1, a_2\}$ and $B = \{b_1, b_2\}$ at the highest level.

Topic 6: Other applications. In machine learning, information theory has been recognized as a useful criterion [28]. To classify objects from knowledge of a training set of examples whose classes are previously known, a decision tree rule induction method known as the *ID3 algorithm* was introduced by Quinlan [30]. The method is based on recursive partitioning of the sample space and defines classes structurally by using decision trees.

A number of improved algorithms-exist that use general to specific learning in order to build simple knowledge-based systems by inducing decision trees from a set of examples [31, 32], and the method of quantitative information of logical expressions developed by Zhong and Ohsuga [57].

Further reading

[1] Agraval V. An information theoretic approach to digital fault testing. *IEEE Trans. Comput.*, 30(8):582–587, 1981.

[2] Aguirre AH and Coello CA. Evolutionary synthesis of logic circuits using information theory. In Yanushkevich SN, Ed., *Artificial Intelligence in Logic Design*. Kluwer, Dordrecht, pp. 285–311, 2004.

[3] Arnd C. *Information Measures: Information and Its Description in Science and Engineering*. Springer, Heidelberg, 2001.

[4] Ash RB. *Information Theory*. John Wiley and Sons, New York, 1967.

[5] Brayton R. Understanding SPFDs: a new method for specifying flexibility. In *Notes of the Int. Workshop on Logic Synthesis*, Tahoe City, CA, May, 1997.

[6] Cheushev VA, Shmerko VP, Simovici D, and Yanushkevich SN. Functional

entropy and decision trees. In *Proc. IEEE 28th Int. Symp. on Multiple-Valued Logic*, pp. 357–362, 1998.

[7] Cheushev VA, Yanushkevich SN, Moraga C, and Shmerko VP. Flexibility in logic design. An approach based on information theory methods. *Research Report*, Forschungsbericht 741, University of Dortmund, Germany, 2000.

[8] Cheushev VA, Yanushkevich SN, Shmerko VP, Moraga C, and Kolodziejczyk J. Information theory method for flexible network synthesis. In *Proc. IEEE 31st Int. Symp. on Multiple-Valued Logic*, pp. 201–206, 2001.

[9] Cook RW and Flynn MJ. Logical network cost and entropy. *IEEE Trans. Comput.*, 22(9):823-826, 1973.

[10] Ganapathy S and Rajaraman V. Information theory applied to the conversion of decision tables to computer programs. *Communications of the ACM*, 16:532–539, 1973.

[11] Gershenfeld N. Signal entropy and the thermodynamics of computation. *IBM Systems J.*, 35:577–586, 1996.

[12] Goodman RM and Smyth P. Decision tree design from a communication theory standpoint. *IEEE Trans. Information Theory*, 34(5):979–994, 1988.

[13] Hamming RW. *Coding and Information Theory*. Prentice-Hall, New York, 1980.

[14] Hartmann CRP, Varshney PK, Mehrotra KG, and Gerberich CL. Application of information theory to the construction of efficient decision trees. *IEEE Trans. Information Theory*, 28(5):565–577, 1982.

[15] Hellerman L. A measure of computation work. *IEEE Trans. Comput.*, 21(5):439–446, 1972.

[16] Kabakcioglu AM, Varshney PK, and Hartman CRP. Application of information theory to switching function minimization. *IEE Proceedings*, Pt E, 137(5):389–393, 1990.

[17] Lo H, Spiller T, and Popescu S. *Introduction to Quantum Computation and Information*. World Scientific, Hackensack, New York, 1998.

[18] Lloris L, Gomez JF, and Roman R. Using decision trees for the minimization of multiple-valued functions. *Int. J. on Electronics*, 75(6):1035–1041, 1993.

[19] Lloris-Ruiz A, Gomez-Lopera JF, and Roman-Roldan R. Entropic minimization of multiple-valued functions. In *Proc. of the IEEE 23rd Int. Symp. on Multiple-Valued Logic*, pp. 24–28, 1993.

[20] Luba T, Moraga C, Yanushkevich SN, Shmerko VP, and Kolodziejczyk J. Application of design style in evolutionary multi-level networks synthesis. In *Proc. IEEE Symp. on Digital System Design*, pp. 156–163, Maastricht, Netherlands, 2000.

[21] Luba T, Moraga C, Yanushkevich SN, Opoka M, and Shmerko VP. Evolutionary multi-level network synthesis in given design style. In *Proc. IEEE 30th Int. Symp. on Multiple-Valued Logic*, pp. 253–258, 2000.

[22] Marculescu D, Marculesku R, and Pedram M. Information theoretic measures for power analysis. *IEEE Trans. Computer Aided Design of Integrated Circuits and Systems*, 15(6):599–610, 1996.

[23] Marculescu R, Marculesku D, and Pedram M. Sequence compaction for power estimation: theory and practice. *IEEE Transactions on Computer Aided Design of Integrated Circuits and Systems*, 18(7):973–993, 1999.

[24] Martin NFG and England JW, *Mathematical Theory of Entropy*. Addison-Wesley, Reading, MA, 1981.

[25] Mitaim S and Kosko B. Adaptive stochastic resonance in noisy neurons based on mutual information. *IEEE Trans. Neural Networks*. 15(6):1526–1540, 2004.

[26] Moore EF and Shannon CE. Reliable circuits using less reliable relays. *J. Franklin Institute*, 262:191–208, Sept. 1956. 262:281–297, Oct. 1956.

[27] Morreale E. Recursive operators for prime implicant and irredundant normal form determination. *IEEE Trans. Comput.*, 19(6):504–509, 1970.

[28] Principe JC, Fisher III JW, and Xu D. Information theoretic learning. In Haykin S, Ed., *Unsupervised Adaptive Filtering*, John Wiley and Sons, New York, 2000.

[29] Popel DV. Conquering uncertaity in multiple-valued logic design. Evolutionary synthesis of logic circuits using information theory. In Yanushkevich SN, Ed., *Artificial Intelligence Review, the Int. J.*, 20(3-4):419–433, 2003.

[30] Quinlan JR. Induction of decision trees. In *Machine Learning*, Vol. 1, pp. 81–106, Kluwer, Dordrecht, 1986.

[31] Quinlan JR. Probabilistic decision trees. In Kockatoft Y and Michalshi R, Eds., *Machine Learning, Vol. 3, An AI Approach*, Kluwer, Dordrecht, pp. 140–152, 1990.

[32] Quinlan JR. Improved use of continuos attributes in C4.5. *J. of Artificial Intelligence Research*, 4:77–90, 1996.

[33] Ramprasad S, Shanbhag NR, and Hajj IN. Information-theoretic bounds on average signal transition activity. *IEEE Trans. Very Large Scale Integration (VLSI) Systems*, 7(3):359–368, 1999.

[34] Rawski M, Józwiak L, and Łuba T. Functional decomposition with an efficient input support selection for sub-functions based on information relationship measures. *J. of Systems Architecture*, 47:137–155, 2001.

[35] Sadek AS, Nikolić K, and Forshaw M. Parallel information and compuation with restriction for noise-tolerant nanoscale logic networks. *Nanotechnology*, 15:192–210, 2004.

[36] Sasao T. Ternary decision diagrams and their applications. Chapter 12, pp. 269–292. In Sasao T and Fujita M, Eds., *Representations of Discrete Functions*. Kluwer, Dordrecht, 1995.

[37] Sentovich E and Brand D. Flexibility in logic. In Hassoun S and Sasao T, Eds., Brayton RK, consulting Ed. *Logic Synthesis and Verification*, Kluwer, Dordrecht, pp. 65–88, 2002.

[38] Shannon CE. A Symbolic analysis of relay and switching circuits. *Trans. AIEE*, 57:713–723, 1938.

[39] Shannon CE. A Mathematical theory of communication. *Bell Systems Technical J.*, 27:379–423, 623–656, 1948.

[40] Shannon CE. The synthesis of two-terminal switching circuits. *Bell Sys. Tech. J.*, 28:59–98, 1949.

[41] Shannon CE, Reliable machines from unreliable components. In *Notes by W. W. Peterson of Seminar at MIT*, MIT, March, 1956.

[42] Shen A and Devadas S. Probabilistic manipulation of Boolean functions using free Boolean diagrams. *IEEE Trans. Computer-Aided Design of Integrated Circuits and Systems*, 14(1):87-94, 1995.

[43] Shmerko VP, Popel DV, Stanković RS, Cheushev VA, and Yanushkevich SN. Information theoretical approach to minimization of AND/EXOR expressions of switching functions. In *Proc. IEEE Int. Conf. on Telecommunications*, pp. 444–451, 1999.

[44] Shmerko VP, Popel DV, Stanković RS, Cheushev VA, and Yanushkevich SN. Entropy based algorithm for 4-valued functions minimization. In *Proc. IEEE 30th Int. Symp. on Multiple-Valued Logic*, pp. 265–270, 2000.

[45] Sinha S, Mishchenko A, and Brayton RK. Topologically constrained logic synthesis. In *Proc. Int. Conf. on Computer-Aided Design*, pp. 679–686, 2002.

[46] Sinha S, Khatri S, Brayton RK, and Sangiovanni-Vincentelli AL. Binary and Multi-valued SPFD-based wire removal in PLA network. In *Proc. Int. Conf. on Computer-Aided Design*, pp. 494–503, 2000.

[47] Sinha S and Brayton RK. Impementation and use of SPFD in optimizing Boolean networks. In *Proc. of the Int. Conf. on Computer-Aided Design*, pp. 103–110, 1998.

[48] Tomaszewska A, Dziurzanski P, Yanushkevich SN, and Shmerko VP. Two-phase exact detection of symmetries. In *Proc. IEEE 31st Int. Symp. Multiple-Valued Logic*, pp. 213–219, 2001.

[49] Tyagi A. Entropic bounds of FSM switching. *IEEE Trans. Very Large Scale Integration (VLSI) Systems*, 5(4):456–464, 1997.

[50] Varshney P, Hartmann C, and De Faria J. Application of information theory to sequential fault diagnosis. *IEEE Trans. Comput.*, 31:164–170, 1982.

[51] Watanabe H. A basic theory of information network. *IEICE Trans. Fundamentals*, E76-A(3):265–276, 1993.

[52] Winograd S and Cowan JD. *Reliable Computation in the Presence of Noise*. MIT Press, Cambridge, MA, 1963.

[53] Yamashita S, Sawada H, and Nagoya A. SPFD: a method to express functional flexibility. *IEEE Trans. Computer-Aided Design of Integrated Circuits and Systems*, 19(8):840–849, 2000.

[54] Yanushkevich SN, Shmerko VP, Dziurzanski P, Stanković RS, and Popel DV. Experimental verification of the entropy based method for minimization of switching functions on pseudo-ternary decision trees. In *Proc. IEEE Int Conf. on Telecommunications in Modern Satellite, Cable and Broadcasting Services*, pp. 452–459, 1999.

[55] Yanushkevich SN, Shmerko VP, and Lyshevski SE. *Logic Design of NanoICs*. CRC Press, Boca Raton, FL, 2005.

[56] Yanushkevich SN, Miller DM, Shmerko VP, and Stanković RS. *Decision Diagram Techniques for Micro- and Nanoelectronic Design*. CRC/Taylor & Francis Group, Boca Raton, FL, 2006.

[57] Zhong N and Ohsuga S. On information of logical expression and knowledge refinement. *Trans. of Information Processing Society of Japan*, 38(4):687–697, 1997.

18

Design for Testability

18.1   Introduction

Testability is the property of a discrete device to be efficiently tested. If the testability properties is the focus of all steps of the design cycle, this design style is called *design for testability*. Design for testability provides the necessary conditions for testing a device.

Logic networks can fail to perform their assigned functions, if

▶ They have been designed incorrectly.

▶ Physical defects occur during manufacture or use.

Design errors are addressed by verification of correctness of the network function, and redesign if necessary. The incidence of physical faults can be reduced by quality control in the manufacturing process, by ensuring proper conditions of operation, and by the use of reliable components. Unavoidable failures are addressed by the use of test procedures that determine the presence and location of faults. Testing of the network is performed at the logical and physical levels.

Physical faults have numerous sources: (*a*) faults that occur during manufacturing, such as missing or defective parts and improper assembly; and (*b*) faults occurring after manufacturing, such as network failure due to various wear-and-tear phenomena. *Logic testing* is concerned with *logic models of physical faults*. In this chapter, only logic testing is introduced. At the logical level, we are primarily concerned with testing the logic network's behavior, and for this purpose the test patterns and responses are viewed as sequences of 0s and 1s.

The logic network being tested is called the *network under test*. Sequences of the network input assignments are called *test patterns* or *test vectors*. During testing, a test pattern is used as the stimulus for detecting the presence of a particular fault. A test pattern is used for controlling the logic network so that the presence of a fault in a network can be observed, for example, on at least one of the network's external pins. Therefore, the testability of a logic network is characterized by (*a*) The capability to propagate a test pattern from the primary inputs to a test point of the network, called *controllability*

(*b*) The possibility of observing the responses of a correct and faulty logic network at the primary outputs, called *observability*

The network output is called the *test response*, $R_{network}$ (Figure 18.1). This response is compared to the expected *fault-free response*, $R_{fault\ free}$. If $R_{network} = R_{fault\ free}$ the network passes the test and is assumed to be fault free. If $R_{network} \neq R_{fault\ free}$ the test fails, and the network is known to be faulty. If the test is appropriately designed, the test response $R_{network}$ contains enough information to locate or isolate faults.

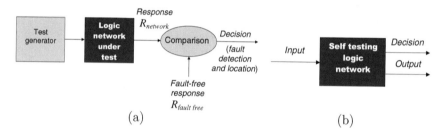

(a)   (b)

FIGURE 18.1

Testing techniques based on external (a) and internal (b) test resources.

Testing techniques can be classified with respect to the way they exploit test resources:

(*a*) Techniques in which *external* test resources are used (Figure 18.1a).

(*b*) Techniques in which *internal* test resources are used; these techniques are called *self-testing* techniques (Figure 18.1b).

The cost of testing grows with increasing density of components and the limited number of input and output pins. For many digital devices, the cost of testing is higher than the cost of design and manufacture.

Consider a combinational logic network. The correctness of the network can be validated by exhaustive application of all possible input patterns and observation of the responses. For an n-input network, this requires the application of 2^n patterns. The use of all 2^n input patterns is called *exhaustive testing*.

Example 18.1   *For a combinational logic network with $n = 20$ inputs, more than 1 million patterns are needed. If the application and observation of a single pattern takes 1 μs, the complete testing of the network requires 1 s.*

Because the number of test patterns of a combinational network with n inputs, 2^n, grows exponentially, exhaustive testing is limited in practice. The testability of a logic network can be improved using various techniques, such as local transformations and decomposition. Efficient testing is accomplished by using automatic test pattern generation (ATPG). ATPG techniques can be divided into *deterministic* and *random* techniques.

18.2   Fault models

Logic testing recognizes gates, latches, and flip-flops as the primitive components, and 0s and 1s as the primitive signals. At this level of complexity, *logic models* of faults are required. These models allow faulty behavior to be defined in terms of the logic components and signals

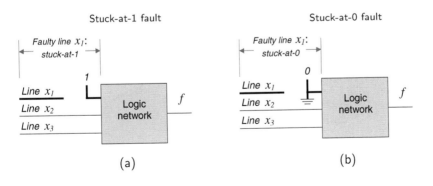

FIGURE 18.2
Single stuck-line models of faults: stuck-at-1 (a) and stuck-at-0 (b) faults.

18.2.1   The single stuck-at model

The *single stuck-at* model is used for gate level testing. This model defines two types of faults for each line x in a logic network (Figure 18.2): the *stuck-at-0* fault, which causes the line x to be permanently stuck at the logic value 0, and the *stuck-at-1* fault, which causes the line x to be permanently stuck at the logic value 1. In this model, only one line is assumed to be faulty at a time (the single-fault assumption), and all gates and other components operate correctly. Presumably, the *stuck-at* model does not cover the complete range of faults that can occur in a complex integrated circuit. However, it has been established that a large number of specific faults, such as *stuck-at-open* and *stuck-at-short*, are covered by the s-a-0 and s-a-1 model. Even though the s-a-0 and s-a-1 models are not perfect, the ease of use and relatively large coverage of the fault space have made it the standard model.

Example 18.2   *Figure 18.3 shows the effect of stuck-at faults on the line x_1, which is an input of the two-input AND gate:*

(a) *The stuck-at-0 fault at the input x_1 causes the output of the AND gate to change from $f = x_1 x_2$ to $f = 0$.*

(b) *The stuck-at-1 fault at the input x_1 causes the output of the AND gate to change from $f = x_1 x_2$ to $f = x_2$.*

Note that the other lines are assumed to be unaffected by faults on the line x_1.

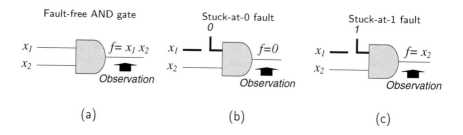

(a)   (b)   (c)

FIGURE 18.3

A fault-free AND gate (a), stuck-at-0 fault (b), and stuck-at-1 fault (c) (Example 18.2).

18.2.2 Fault coverage

The quality of a test is measured by the *fault coverage*:

$$\text{FAULT COVERAGE} = \frac{\text{TOTAL NUMBER OF FAULTS DETECTED}}{2 \times \text{NUMBER OF NODES IN THE NETWORK}}$$

In this equation, the total number of faults detected by the test sequence is divided by two times the number of nodes in the network because each node can give rise to s-a-0 and s-a-1 faults. The coverage number obtained is defined by the fault model employed. In the s-a-0 and s-a-1 model, some of the bridge and short faults are not covered and may not appear in the coverage statistics.

Example 18.3 *Figure 18.4 shows the potential stuck-at faults in the two-input NAND gate. Each row of the table corresponds to a test pattern, and each column corresponds to a potential fault. An × is placed in the table if and only if the test pattern detects a fault; that is, the test pattern covers the (specific) fault. For example, the test pattern 00 detects only a s-a-0 fault in the line f; the test pattern 11 covers s-a-0 faults in the lines x_1 and x_2, as well as the fault s-a-1 in the line f. For test patterns 00 and 10, fault coverage is $\frac{1}{2 \times 1} = \frac{1}{2}$.*

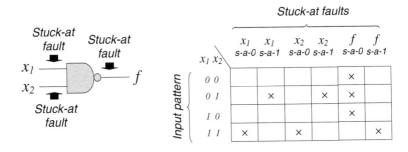

Input pattern $x_1\,x_2$	x_1 s-a-0	x_1 s-a-1	x_2 s-a-0	x_2 s-a-1	f s-a-0	f s-a-1
0 0					×	
0 1		×		×	×	
1 0					×	
1 1	×		×			×

FIGURE 18.4

The two-input NAND gate and its fault coverage table (Example 18.3).

Stuck-at faults can be detected under appropriate controllability and observability conditions.

18.3   Controllability and observability

To generate a test to detect a fault in a combinational network, the following steps are required:

Path sensitization: The inputs must be specified so as to sensitize, or generate the appropriate value (0 for stuck-at-1 and 1 for stuck-at-0 faults) at the site of the fault.

Error propagation: A path must be selected from the faulty line to an output along with remaining signal values, to propagate the fault signal to the output, which must be observable.

Path sensitization is provided by network controllability, and error propagation is ensured by network observability:

Controllability is the ability of a logic network to have a test pattern such that a faulty logic value is produced at the fault line. A node is *easily controllable* if it can be brought to any condition with only a single input vector. A node (or a network) with *low controllability* needs a long sequence of vectors to be brought to a desired state.

Observability is an ability of a logic network to provide the conditions (usually specified by input patterns) for observation of the specified network output. A node with a *high observability* can be monitored directly on the output pins. A node with a *low observability* needs a number of cycles before its state appears on the outputs.

Controllability is provided by setting the values at the node or line under testing to the value opposite to the assumed fault values. For example, if we are testing a line for the stuck-at-0 fault, the line must be set to 1.

Example 18.4   *Consider the fault s-a-1 in line x_1 of the AND gate (Table 18.1). To generate the opposite value (0) at the output of the AND gate, conditions $x_1 = x_2 = 1$ must be held. These conditions mean that the fault s-a-1 in line x_1 is propagated to the output of the AND gate. The corresponding conditions for OR and EXOR gates can be specified by analogy.*

Observability in logic networks refers to the possibility of observing the outputs of the logic network via external outputs (observation points):

TABLE 18.1
Conditions for error propagation through logic gates
(Example 18.4)

Gate	Fault		Condition for propagation
	s-a-0	s-a-1	
x_1 ─── f x_2 ─── $f = x_1 \wedge x_2$	✓	✓	$x_1 = x_2 = 1$ $x_1 = x_2 = 1$
x_1 ─── f x_2 ─── $f = x_1 \vee x_2$	✓	✓	$x_1 = x_2 = 0$ $x_1 = x_2 = 10$
x_1 ─── f x_2 ─── $f = x_1 \oplus x_2$	✓	✓	$x_1 = x_2 = 1$ $x_1 = x_2 = 1$

(In the first gate cell, labels "Fault" and "Propagated fault" appear above the gate symbol.)

Observability problem

Given: The logic network output.
Find: Conditions for observability.
Step 1: Derive a logic equation for the input variables using the
     observable values of the output.
Step 2: Generate inputs from the logic equation.

Example 18.5   *Conditions for observing the output $f = 1$ for the two-input AND gate can be found as shown in Figure 18.5. The solution to the equation $x_1 x_2 = 1$ is $x_1 = x_2 = 1$. That is, in order to observe a logic "1" at the output f, both input signals must be equal to 1. The observed value of the fault-free output must be 1. The alternate value (0) is an indication of a fault at the output line.*

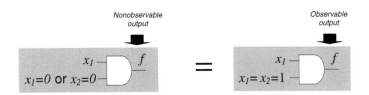

FIGURE 18.5
Observability for the two-input AND gate is defined by the logic equation
$x_1 x_2 = 1$ (Example 18.5).

Example 18.6   *Conditions for observing the output $f = 1$ for the two-input EXOR gate can be found using the following manipulations (Figure 18.6):*

$$f = x_1 \oplus x_2 = 1 \;\Rightarrow\; x_1 = x_2 \oplus 1 \; and \; x_1 = \overline{x}_2$$

That is, to observe a logic "1" at the output, the input signals must satisfy the logic equation $\boxed{x_1 = \overline{x}_2.}$
The solutions to this equation, $x_1 = 0$, $x_2 = 1$, and $x_1 = 1$, $x_2 = 0$, are the desired input signals. The observed value of the output must be 1. The alternate value (0) is an indication of a fault at one of the inputs.

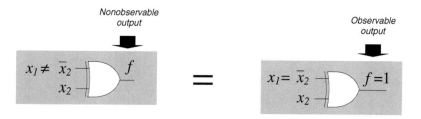

FIGURE 18.6
Observability for the two-input EXOR gate is defined by the logic equation $x_1 = \overline{x}_2$ (Example 18.6).

Example 18.7   *In Table 18.2, the observability conditions for some logic networks are given. For example, for the first network, a logic value of 1 can be observed at the output only if $x_1 = x_2 = 0$. For the second two-output network, the logic equation $x_1 x_2 = 1$ implies that the values 1 and 0 are generated at the outputs f_1 and f_2, respectively, only if $x_1 = x_2 = 1$.*

The controllability and observability of a logic network can be improved using local transformation, decomposition, and additional gates. Combinational logic networks fall into the class of easy observable and controllable networks, since any node can be controlled and observed in a single cycle.

TABLE 18.2
Observability conditions in logic networks (Example 18.7)

Network	Description	Observability
	$f = 1 \oplus x_1 \oplus x_2 \oplus x_1 x_2$	$1 \oplus x_1 \oplus x_2 \oplus x_1 x_2 = 1$ $x_1 \oplus x_2 \oplus x_1 x_2 = 1 \oplus 1 = 0$ $x_1 \oplus x_2 = x_1 x_2$ $x_1 = x_2 = 0$
	$\begin{cases} f_1 = x_1 \oplus x_2 \\ f_2 = x_1 \oplus x_2 \oplus x_1 x_2 \end{cases}$	$\begin{cases} f_1 = x_1 \oplus x_2 = 1 \\ f_2 = x_1 \oplus x_2 \oplus x_1 x_2 = 0 \end{cases}$ $f_1 \oplus x_1 x_2 = 0$ $1 \oplus x_1 x_2 = 0$ $x_1 x_2 = 1$ $x_1 = x_2 = 1$
	$f = x_1(x_2 \oplus 1) \oplus x_1$ After local transformations: $f = x_1 x_2$	$x_1 x_2 = 1$ $x_1 = x_2 = 1$

18.3.1 Observability and Boolean differences

A signal in a binary system is represented by two logical levels, 0 and 1. Let us formulate the task as the detection of a change in this signal. Let f be the output of a fault-free logic network, and let f_{fault} be the output of the faulty logic network. The *difference* function D, which distinguishes the two functions, is defined using the following EXOR operation:

$$D = f \oplus f_{fault} \qquad (18.1)$$

Given a type of fault, the assignment of variables for the function f_{fault} for which $f \oplus f_{fault} = 1$ is called a *test pattern* input for this fault. If not change itself, but direction of change is the matter as issue, then two logical values 0 and 1 can characterize the behavior of the logic signal in terms of change, where 0 means no change in a signal, and 1 indicates that one of two possible changes has occurred $0 \to 1$ or $1 \to 0$.

Let a fault at the i-th input of a Boolean function cause a change from the value x_i to the opposite value, $\overline{x}_i$. This causes the circuit output to be changed from the initial value $f(x_i)$ to $f(\overline{x}_i)$. Note that the values $f(x)$ and $f(\overline{x}_i)$ are not necessarily different. To recognize whether or not they are different requires finding $f(x_i) \oplus f(\overline{x}_i)$.

The latter is the Boolean difference of a Boolean function f of n variables with respect to a variable x_i defined by the equation

$$\frac{\partial f}{\partial x_i} = \underbrace{f(x_1, \ldots, x_i, \ldots, x_n)}_{Initial\ function} \oplus \underbrace{f(x_1, \ldots, \overline{x}_i, \ldots, x_n)}_{Function\ with\ complemented\ x_i}$$

If $\frac{\partial f}{\partial x_i} = 0$, f does not depend on x_i.

Example 18.8   *The Boolean difference specifies the observability of the fault at the input x_1. The Boolean differences with respect to both input variables x_1 and x_2 specify the conditions for observing changes in the output function $f = x_1 \vee x_2$ with respect to changes in the inputs x_1 and x_2 (Figure 18.7).*

s-a-1
fault

x_1
x_2 —— f

Step 1: *Derive the observability option for the line* x_1

$$\frac{\partial f}{\partial x_1} = \frac{\partial (x_1 \vee x_2)}{\partial x_1}$$
$$= (x_1 \vee x_2) \oplus (\overline{x}_1 \vee x_2)$$
$$= (0 \vee x_2) \oplus (1 \vee x_2) = \overline{x}_2$$

x_1
x_2 —— f

s-a-1
fault

Step 2: *Derive the observability option for the line* x_2

$$\frac{\partial f}{\partial x_2} = \frac{\partial (x_1 \vee x_2)}{\partial x_2}$$
$$= (x_1 \vee x_2) \oplus (x_1 \vee \overline{x}_2)$$
$$= (x_1 \vee 0) \oplus (x_1 \vee 1) = \overline{x}_1$$

FIGURE 18.7

Observability options for a two-input OR gate in terms of Boolean differences (Example 18.8).

Example 18.9   *Consider the logic network implementing the Boolean function $x_1 x_2 \vee x_3$. Figure 18.8 shows how to specify controllability and observability conditions for stuck-at faults using Boolean difference with respect to the variable x_3.*

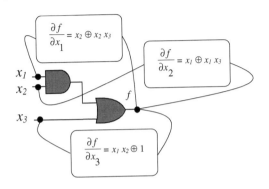

Design example: specification of observability conditions

Step 1: *Observability condition for the line x_1: $\frac{\partial f}{\partial x_1} = x_2 \overline{x}_3$.*

Step 2: *Specify the observability conditions for detecting the changes at x_1 by solving the logic equation $x_2 \overline{x}_3 = 1$: $x_2 x_3 = \{10\}$. When $x_2 x_3 = \{10\}$, a change at x_1 cause a change of either $0 \rightarrow 1$ or $1 \rightarrow 0$ at f.*

Step 3: *Observability condition for the line x_2: $\frac{\partial f}{\partial x_1} = x_1 \overline{x}_3$.*

Step 4: *Specifies the observability conditions for detecting the changes at x_2 by solving the logic equation $x_2 \overline{x}_3 = 1$. When $x_2 x_3 = \{10\}$, a change of x_1 is observable at the output.*

Step 5: *Observability condition for line x_3: $\frac{\partial f}{\partial x_1} = \overline{x_1 x_2}$.*

Step 6: *Specify the observability conditions for detecting changes at x_3 by solving the logic equation $\overline{x_1 x_2} = 1$. When $x_1 x_2 = \{00, 01, 10\}$, any change in x_3 causes changes in the output f.*

FIGURE 18.8
Specification of observability conditions using Boolean differences and logic equations (Example 18.9).

18.3.2 Enhancing observability and controllability

A *test point* is an input or output signal for controlling or observing intermediate signals in a logic network. Suitable locations for the test point in a logic network are defined as places where test points involve: (*a*) signals that are hard to control or observe, or (*b*) signals with values that are particularly useful during test application. In practice, the number of test points is limited. An arbitrary line x in a logic network can be made controllable with respect to either 0 or 1.

Example 18.10 *Given the nonobservable line x (Figure 18.9), this line can be made observable as follows. Suppose two external control signals C_1 and C_2 force a line to 0 and 1, respectively. The task is to replace the line x by a logic network with inputs x, C_0, and C_1 that produces an observable x denoted as* OBSERVABLE x. *The Boolean equation for the observable line x is as follows (Figure 18.9):*

$$\text{OBSERVABLE } x = \overline{C}_0(x \vee C_1) \qquad (18.2)$$

The implementation of Equation 18.2 is called *testability enhancing logic* for the test point.

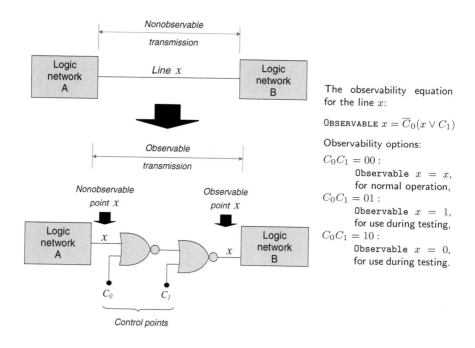

The observability equation for the line x:

OBSERVABLE $x = \overline{C}_0(x \vee C_1)$

Observability options:

$C_0 C_1 = 00$:
Observable $x = x$, for normal operation,

$C_0 C_1 = 01$:
Observable $x = 1$, for use during testing,

$C_0 C_1 = 10$:
Observable $x = 0$, for use during testing.

FIGURE 18.9

Nonobservable and observable transmission line x between two logic networks A and B (Example 18.10).

18.3.3 Detection of stuck-at faults

Consider the output f of a logic network with the stuck-at-0 fault at the input line x_i. This network's output is then described by the Boolean function

$$f_{fault} = f(x_1, \ldots, x_{i-1}, 0, x_{i+1} \ldots, x_n) = f_i(0)$$

The difference (Equation 18.1) that detects a fault in the network caused by the stuck-at-0 fault at the input line x_i is calculated as follows:

$$
\begin{aligned}
f \oplus f_{fault} &= f \oplus f_i(0) \\
&= \overline{x}_i f_i(0) \oplus x_i f_i(1) \oplus f_i(0) = \overline{x}_i f_i(0) \oplus x_i f_i(1) \oplus (\overline{x}_i \oplus 1) f_i(0) \\
&= x_i f_i(1) \oplus x_i f_i(0) = x_i (f_i(0) \oplus f_i(1)) = x_i \frac{\partial f}{\partial x_i}
\end{aligned}
$$

Therefore, all test patterns for the stuck-at-0 fault at the input line x_i can be found as solutions to the followingBoolean equation:

Conditions for detecting the stuck-at-0 fault
at the input line x_i

Controllability
$\downarrow$
$$x_i \quad \times \quad \frac{\partial f}{\partial x_i} \quad = 1 \qquad \qquad (18.3)$$
$\uparrow$
Observability

Equation 18.3 combines the controllability and observability conditions.

Example 18.11 *Assume that we suspect the logic network shown in Figure 18.10 has the stuck-at-0 fault on the line x_2. The test for detecting this fault is derived as shown in Figure 18.10.*

Example 18.12 *Determine the input excitation that exposes an s-a-0 fault, occurring at line A, at the output f of the logic network shown in Figure 18.11. The first requirement of such an excitation is that it should force the fault to occur (Step 1). In this case, we look for a pattern that would set A to 1 under normal circumstances. The only option here is $x_1 = 1$ and $x_2 = 1$. The faulty signal has to propagate to the output f (Step 2). For any change in line A to propagate, it is necessary for line B to be set to 1 and line x_5 to 0. Finally, the test for A_{s-a-0} can now be assembled: $x_1 = x_2 = x_3 = x_4 = 1, x_5 = 0$ (Step 3).*

Design example: detection of the input stuck-at-0 fault

Given:

(*a*) A logic network

(*b*) The line x_2 under potential stuck-at-0 fault

Find:

The test for detection of stuck-at-0 fault on line x_2

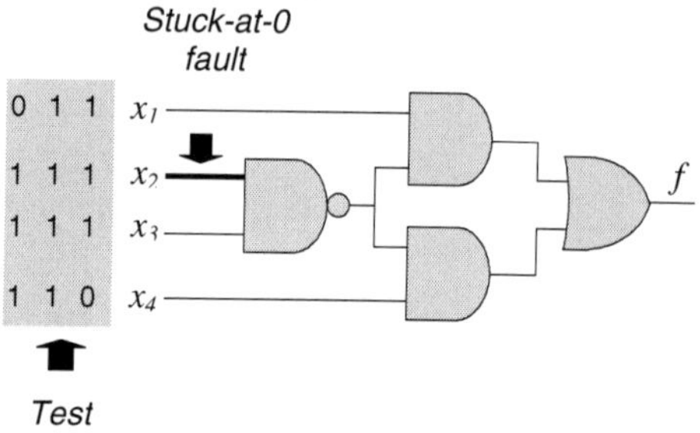

Procedure for the test generation for the stuck-at-0 fault

Step 1: Derive the Boolean expression for the output of the network

$$f = \overline{x_2 x_3}(x_1 \vee x_4)$$

Step 2: Compute the Boolean difference with respect to variable x_2

$$\begin{aligned}
\frac{\partial f}{\partial x_2} &= f_{x_2=0} \oplus f_{x_2=1} \\
&= (x_1 \vee x_4) \oplus \overline{x}_3(x_1 \vee x_4) \\
&= \overline{(x_1 \vee x_4)}\overline{x}_3(x_1 \vee x_4) \vee (x_1 \vee x_4)\overline{\overline{x}_3(x_1 \vee x_4)} \\
&= \overline{x}_1\overline{x}_3\overline{x}_4(x_1 \vee x_2) \vee (x_1 \vee x_4)(x_3 \vee \underbrace{\overline{(x_1 \vee x_4)}}_{\overline{x}_1\overline{x}_4}) \\
&= \underbrace{\overline{x}_1\overline{x}_3\overline{x}_4 x_1}_{0} \vee \underbrace{\overline{x}_1\overline{x}_3\overline{x}_4 x_4}_{0} \vee x_1 x_3 \vee \underbrace{x_1\overline{x}_1\overline{x_4}}_{0} \vee x_3 x_4 \vee \underbrace{x_4\overline{x}_1 x_4}_{0} \\
&= x_3(x_1 \vee x_4)
\end{aligned}$$

Step 3: Derive the logic equation

$$x_2 \frac{\partial f}{\partial x_2} = x_3(x_1 \vee x_4) = 1$$

Step 4: Solve this equation and find the test patterns. The test inputs for the stuck-at-0 fault on line x_2 are

$$\begin{aligned}
\texttt{Test patterns} &= x_2 x_3(x_1 \vee x_2) = 1 \\
(x_1, x_2, x_3, x_4) &= (0,1,1,1), (1,1,1,0), (1,1,1,1)
\end{aligned}$$

FIGURE 18.10

Detection of the input stuck-at-0 fault on line x_2 (Example 18.11).

Design example: controllability and observability

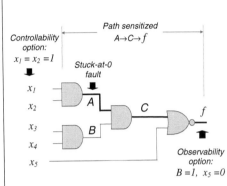

Step 1: Controllability for A. Set A to 1 under normal conditions ($x_1 = 1$ and $x_2 = 1$)..

Step 2: Observability for A. The faulty signal has to propagate to output node f, so that it can be observed (B to be set to 1 and node x_5 to be set to 0):

$$\frac{\partial f}{\partial A} = f_{A=0} \oplus f_{A=1}$$

$$= \overline{0 \cdot x_3 x_4 \vee x_5} \oplus \overline{1 \cdot x_3 x_4 \vee x_5}$$

$$= \overline{x_5} \oplus \overline{x_3 x_4 \vee x_5} = \overline{x_5} \oplus \overline{x_3 x_4}\overline{x_5}$$

$$= \overline{x_5}(1 \oplus \overline{x_3 x_4}) = x_3 x_4 \overline{x_5}$$

Step 3: Combining Steps 1 and 2: $x_1 x_2 \frac{\partial f}{\partial A} = x_1 x_2 x_3 x_4 \overline{x_5}$. Test vector for A: $x_1 = x_2 = x_3 = x_4 = 1, x_5 = 0$

FIGURE 18.11

Controllability and observability (Example 18.12).

Consider the output f of a logic network with the stuck-at-1 fault at the input line x_i. This network is described by the Boolean function

$$f_{fault} = f(x_1, \ldots, x_{i-1}, 1, x_{i+1} \ldots, x_n) = f_i(1)$$

The fault difference function (Equation 18.1) that detects a fault in the network caused by the stuck-at-1 fault at the input line x_i is calculated as follows:

$$f \oplus f_{fault} = f \oplus f_i(1) = \overline{x}_i f_i(0) \oplus x_i f_i(1) \oplus f_i(1)$$
$$= \overline{x}_i f_i(0) \oplus (x_i \oplus 1) f_i(1) = \overline{x}_i f(0) \oplus \overline{x}_i f(1)$$
$$= \overline{x}_i(f_i(0) \oplus f_i(1)) = \overline{x}_i \frac{\partial f}{\partial x_i}$$

Therefore, all tests for the stuck-at-1 fault at the input line x_i are solutions of the following Boolean equation:

Conditions for detecting the stuck-at-1 fault at the input line x_i

Controllability

$$\overline{x}_i \quad \times \quad \frac{\partial f}{\partial x_i} = 1 \qquad (18.4)$$

Observability

Example 18.13 *The logic network in Figure 18.12 implements the Boolean function $f = x_1 x_2 x_3 \lor \overline{x}_1 \overline{x}_2 \overline{x}_3$. Assume a stuck-at-1 fault at the input terminal x_1 of the AND gate occurs. The fault difference is equal to 1 (Equation 18.1) once the test pattern is (011). That is, for the assignments $x_1 = 0, x_2 = 1$, and $x_3 = 1$ the value of f is 0, and the value of f_{fault} is 1.*

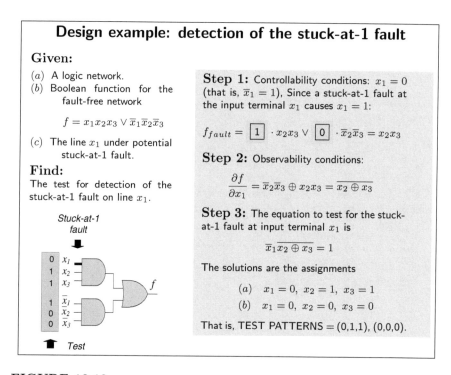

FIGURE 18.12

Logic network with a stuck-at-1 fault at the input terminal x_1 (Example 18.13).

Example 18.14 *Figure 18.13 shows the Boolean function in the form of its truth vector $\mathbf{F}$, and conditions for detecting stuck-at-0 and stuck-at-1 faults. The tests are shown as well.*

Design example: deriving tests to detect stuck-at-0 and stuck-at-1 faults

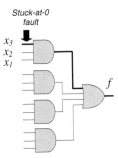

Given: (a) A logic network and the location of potential faults;

(b) The truth vector **F** of this network; and

(c) Conditions for detecting stuck-at-0 and stuck-at-1 faults, $x_3 \frac{\partial F}{\partial x_3}$ and $\overline{x}_3 \frac{\partial F}{\partial x_3}$, respectively.

Step 1: Derive a logic equation for the detection of stuck-at-0 faults. Stuck-at-0 at x_3 causes each value $f_{x_3=1}$ to be changed to $f_{x_3=0}$. Equation to find the test for detecting stuck-at-0 faults is

$$x_3 \frac{\partial F}{\partial x_3} = 1$$

Step 2: Find the test for the detection of stuck-at-0 faults as the solution of the following logic equation: Test= $\{001, 011, 101, 111\}$

Step 3: Derive a logic equation for the detection of stuck-at-1 faults. Stuck-at-1 at x_3 causes each value $f_{x_3=0}$ to be changed to $f_{x_3=1}$. Equation to find the test for detecting stuck-at-1 faults:

$$\overline{x}_3 \frac{\partial F}{\partial x_3} = 1$$

Step 4: Find the test for the detection of stuck-at-0 faults as the solution of the following logic equation: Test= $\{000, 010, 100, 110\}$

$x_1 x_2 x_3$	F	$x_3 \frac{\partial F}{\partial x_3}$	$\overline{x}_3 \frac{\partial F}{\partial x_3}$
0 0 0	0	0	1
0 0 1	1	1	0
0 1 0	1	0	1
0 1 1	0	1	0
1 0 0	1	0	1
1 0 1	0	1	0
1 1 0	0	0	1
1 1 1	1	1	0

FIGURE 18.13

Deriving tests to detect stuck-at-0 and stuck-at-1 faults (Example 18.14).

18.3.4 Testing decision-tree-based logic networks

Decision trees and diagrams are multiplexer networks. Nodes of these trees and diagrams are implemented using multiplexers. The stuck-at model can be applied to the input and output of a node.

Example 18.15 *The fault nodes are shown in Figure 18.14. Under normal conditions, the node performs the Shannon expansion $f = \overline{x}_i f_{x_i=0} \vee x_i f_{x_i=0}$. If the s-a-0 fault causes the f line, the output is $f = 0$. If the s-a-0 fault causes the x_i line, the node produces the output $f_{x_i=0}$.*

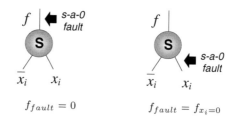

FIGURE 18.14
Fault models of decision diagrams (Example 18.15).

18.4   Functional decision diagrams for computing Boolean differences

Boolean differences are useful for deriving controllability and observability options, as well for detecting faults. However, Boolean differences must be computed. Decision diagrams can be used for: (*a*) computing Boolean differences for modeling logic networks under the fault attack, and (*b*) computing Boolean functions as multiplexer-based networks. In this section, we show that functional decision diagrams can be used not only for computing polynomial forms of Boolean functions but also for computing Boolean differences. It is remarkable that the same decision diagram can perform these two functions: they are are distinguished by different interpretations of the results obtained.

Let us rewrite the positive Davio expansion in the form

$$f = f|_{x_i=0} \oplus x_i \left(f|_{x_i=0} \oplus f|_{x_i=1} \right) = \underbrace{f|_{x_i=0}}_{Left\ branch} \oplus \underbrace{x_i \frac{\partial f}{\partial x_i}}_{Right\ branch}$$

It follows from this form that branches of the functional decision tree carry information about Boolean differences. Terminal nodes are the values of Boolean differences for corresponding variable assignments. The functional decision tree includes the values of all single and multiple Boolean differences given a variable assignment $x_1 x_2 \ldots x_n = 00 \ldots 0$. This assignment corresponds to the calculation of a polynomial expansion of polarity 0.

> **Example 18.16** *Figure 18.15 shows a functional decision tree for an arbitrary Boolean function of two and three variables. The values of the terminal nodes can be interpreted both as the values of Boolean differences and of the polynomial expression.*

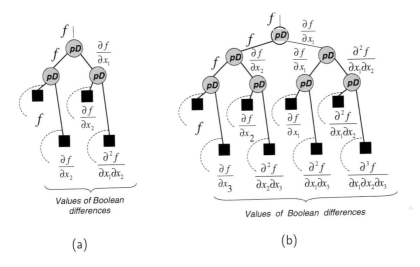

FIGURE 18.15
Computing Boolean differences by functional decision tree for a Boolean function of two (a) and three (b) variables (Example 18.16).

18.5   Random testing

Given a logic network with a large number n of inputs, the exhaustive application of all possible 2^n input patterns and observation of the responses become infeasible in terms of time complexity. An alternative approach is required. The test patterns can be generated using *random* test techniques, which are based on the random selection of tests from the set patterns. Such an approach is based on the following premises:

Premise 1. An exhaustive enumeration of all possible input patterns is *redundant* since a single fault in the logic network is usually covered by a number of input patterns.

Premise 2. A substantial reduction in the number of patterns can be obtained by relaxing the condition that all faults must be detected. The cost of detecting the last single percentage of possible faults might be larger than the eventual replacement cost. For that reason, typical test procedures only attempt a 95–99% fault coverage.

By eliminating redundancy and providing a reduced fault coverage, it is possible to test most combinational logic networks with a limited set of input vectors.

Example 18.17 *The desired function $f = x_1 \oplus x_2$ is implemented by the logic network shown in Figure 18.16. This network is tested under possible stuck-at fault attack on wires $a, b, c, d,$ and e. Each of these single faults causes a new function that is different from the desired EXOR function of two variables. For example, a stuck-at-1 fault on the line d causes the OR function instead of the desired EXOR function.*

Choose an arbitrary assignment of the variables x_1 and x_2 as the first test; say, $x_1 x_2 = (0, 1)$. This test can detect faulty networks, which are caused by the faults $a = 0$ and $b = 1$.

The fragment of a fault detection using random testing is given in the table. The effect of other stuck-at faults, that is, the values of faulty functions is shown in the left column of the table.

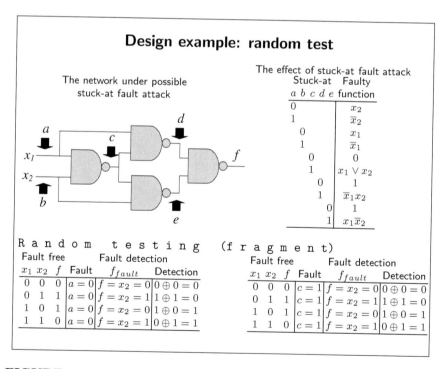

FIGURE 18.16

Random test against stuck-at attack (Example 18.17).

18.6   Design for testability techniques

Design for testability means that the logic network is designed so that it can be easily tested, or has the ability to test itself. These techniques include built-in self-test technique (BIST), scan-path techniques, signature analysis, and boundary-scan techniques.

18.6.1   Self-checking logic networks

An attractive approach to testability is to have the network itself generate the test patterns instead of requiring the application of external patterns. Even more appealing is a technique in which the logic network itself decides if the obtained results are correct. Depending upon the nature of the network, this might require the addition of extra circuitry for generation and analysis of the patterns. Some excising hardware can be used, so that the size overhead of the self-test is not big. A logic network can be designed to make it possible to determine, from the outputs of the network, whether a certain fault exists within the network. In this case, it is unnecessary to explicitly test for this fault. Such a logic network is said to be *self-checking* for this fault.

An additional network, called a *checker*, can be designed to generate a warning signal whenever the outputs of network indicate the presence of a fault within this network. Ideally, logic networks must be self-checking for as many faults as possible. The checker circuit must be designed to be self-checking too.

In previous chapters, various aspects of the application of error detection and error correcting codes were introduced. For instance, the parity check circuit uses Hamming code for error detection. In the parity check code, the Hamming distance is equal to two (the minimum number of bits by which any two code words differ), and the number of check bits is one (independent of the number of outputs in the original network). There are two types of parity checks, even and odd. Any error, affecting a single bit, causes the output to have an even (odd) number of bits, and hence, is automatically detected. Additional hardware, added to the design, can automatically detect and correct errors in the network, though not in the checker itself.

18.6.2   Built-in self-test (BIST)

Built-in self-test technique (BIST) is a testing technique, in which external test resources are not required to apply test patterns and check a logic network's response to those patterns. In BIST techniques, the test patterns are preliminary loaded into the network or generated by the network itself. A BIST design approach is given in Figure 18.17. There are three main components: the logic network under test, stimulus generator, and response analyzer.

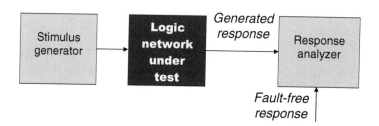

FIGURE 18.17
A built-in self-test (BIST) design approach.

A stimulus generator generates stimuli using *exhaustive* and *random* approaches. In the exhaustive approach, the test length is 2^n, where n is the number of inputs to the network, and is intended to detect all detectable faults, given the space of the available input signals. An example of an exhaustive pattern generator is an n-bit counter. For networks with large n, the time to cycle through the complete input space is unacceptable. An alternative approach is to use *random* testing, which implies the application of a randomly chosen subset of 2^n possible input patterns. This subset should be selected so that a reasonable fault coverage is obtained. An example of a pseudorandom pattern generator is the *linear-feedback shift register*. Some of the outputs are connected to the inputs of EXOR gates, whose output is fed back to the input of the shift register. An n-bit linear-feedback shift register cycles through $2^n - 1$ states before repeating the sequence, which produces a seemingly random pattern.

A response analyzer compares the generated response against the expected response stored in an on-chip memory, though this approach requires impractical area overhead. A cheaper technique called *signature analysis* is to *compress* the responses before comparing them. Storing the compressed response (the signature) of the network requires a minimal amount of memory. Thus, the response analyzer consists of circuitry that compresses the output of a network under test, and a comparator.

18.6.3   Easily testable EXOR logic networks

An *easily testable* logic network is defined as a network that has a small test set. These networks can be designed using the appropriate representations (SOP or polynomial forms), decomposition, and local transformations. An example of an easily testable logic network is the network of AND and EXOR gates that implements a polynomial form of a Boolean function.

Example 18.18 *Figure 18.18 shows the logic network that implements the polynomial expression $f = 1 \oplus x_2 \oplus x_1 x_3 \oplus x_2 x_3 \oplus x_1 x_2 x_3$. Any change at a single line x_1, x_2 or x_3 implies a change at the output f if a single stuck-at-0 or stuck-at-1 fault occurs at the inputs and outputs of any gate. For example, if x_2 has changed from the correct value 0 to 1, then applying the test pattern (0000) or (1111) yields a change of $f = 0$ to $f = 1$. Applying the test patterns (1000) or (1111) causes a change in $f = 1$ to $f = 0$.*

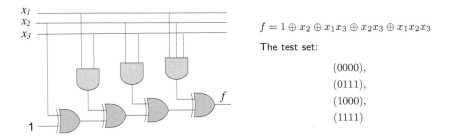

$$f = 1 \oplus x_2 \oplus x_1 x_3 \oplus x_2 x_3 \oplus x_1 x_2 x_3$$

The test set:

$$(0000),$$
$$(0111),$$
$$(1000),$$
$$(1111)$$

FIGURE 18.18

An easily testable logic network that implements the polynomial form of a Boolean function of three variables (Example 18.18).

18.6.4 Improving testability using local transformations

Local transformations can improve the conditions for testability.

Example 18.19 *Figure 18.19 illustrates how the testability of a logic network can be improved using local transformations:*
The area A: The inverter is replaced by an EXOR gate using the identity rule for variables and constants $\overline{x}_2 = x_2 \oplus 1$.
The area B: The inverter is replaced by the EXOR gate using the rule $\overline{x}_1 = x_1 \oplus 1$.
The areas A and B: $x_1(x_2 \oplus 1) \oplus 1 = x_1 x_2 \oplus x_1 \oplus 1$.

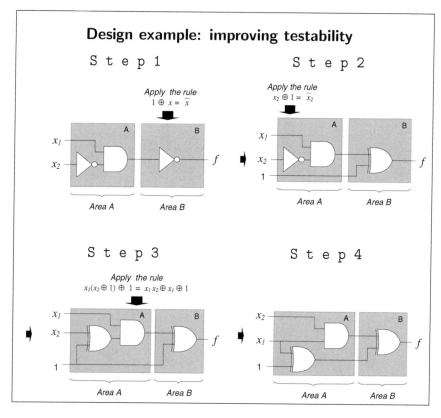

FIGURE 18.19
Improving testability using local transformations in a logic network (Example 18.19).

18.7   Further study

Advanced topics of testability

Topic 1: Fault simulation. It is used to determine the fault coverage to measure the quality of a test program. The most common approach to fault simulation is the parallel fault-simulation technique, in which the correct network is simulated concurrently with a faulty one, each of which has a single fault injected. The results are compared, and a fault is labeled as detected for a given test vector set if the outputs diverge. Most simulators employ a number of techniques, such as selecting the faults with a higher chance of detection first, to expedite the simulation process. Fault simulation can also be performed on hardware devices called *accelerators*, which provide a substantial increase in speed over software-based simulations.

Topic 2: Fault tolerant systems. Assuming reliable computing in the presence of faults is called *fault-tolerance*. In *fault-tolerant* computing paradigms, the effects of faults are mitigated, and correct computations are guaranteed with a certain level of reliability. *Redundancy* is one fault-tolerant techniques that can be used if additional resources are available.

Fault tolerance is introduced to a system through utilizing many design and testing techniques; in particular, error detecting and correcting codes, self checking, and massive redundancy. Fault detection techniques supply warnings of faulty results. However, the use of fault detection techniques does not provide fault tolerance.

Several techniques have been proposed for gate-level fault masking. *Fault masking* employs redundancy, which provides fault tolerance by isolating or correcting fault effects before they are used in computing.

The problem is formulated as follows: given an unreliable logic gate, achieve the correct output values. For example, *Von Neumann's approach* states that it is possible theoretically to achieve an acceptable level of fault tolerance for the simplest imperfect logic gates using massive redundancy under certain constraints (J. von Neumann, "Probabilistic Logics and the Synthesis of Reliable Organisms from Unreliable Components" edited by C. E. Shannon and J. McCarthy, *Automata Studies*, Princeton University Press, Princeton, New York, pages 43–98, 1956).

Topic 3: Testing of quantum cellular automata. Quantum cellular automata (QCA) proposed originally in [25] is a nanotechnology with potential applications for logic network design. QCA is a device that stores logic states not as voltage levels, but rather in the relative position of individual electrons. Since the manufacturing process of nanodevices is still ill-defined, it is difficult to address related manufacturing testing problems. However, in many cases faults in logic networks fabricated in nanotechnology exhibit similar properties to fault models of silicon-based technology [?] For example, a single stuck-at fault model for representing manufacturing defects in QCA can be used.

Further reading

[1] Abramovici M, Breuer MA and Friedman AD. *Digital Systems Testing and Testable Design*. Computer Science Press, Rockville, 1990.

[2] Fujiwara H. *Logic Testing and Design for Testability*. MIT Press, Cambridge, MA, 1985.

[3] Goel P. An implicit enumeration algorithm to generate tests for combinational logic circuits. *IEEE Trans. Comput.*, 30(3):215–222, 1981.

[4] Hamming RW. *Coding and Information Theory*. Prentice-Hall, New York, 1980.

[5] Lent CS, Tougaw PD, Porod W, and Bernstein GH. Quantum cellular automata. *Nanotechnology*, 4:49–57, 1993.

[6] MacWilliams FJ and Sloan NJA. *The Theory of Error-Correcting Codes*. Noth-Holland, 1978.

[7] Negrini R and Sami MG. *Fault Tolerance Trough Reconfiguration in VLSI and WSI Arrays*. The MIT Press, Cambridge, MA, 1989.

[8] Peper F, Lee J, Abo F, Isokawa T, Adachi S, Matsui N, and Mashiko S. Fault-tolerance in nanocomputers: a cellular array approach. *IEEE Trans. Nanotechnology*, 3(1):187–201, 2004.

[9] Reddy SM.   Easily testable realization for logic functions *IEEE Trans. Comput.*, 21:1183–1188, December 1972.

[10] Sasao T. Easily testable realization for generalized Reed-Muller expansions. *IEEE Transa. Comput.*, 46(6):709–716, 1997.

[11] Sellers EF, Hsiao MY, and Bearnson LW. Analyzing errors with the Boolean difference. *IEEE Trans. Comput.*, 17(7):676–683, 1981.

[12] Schulz MH, Trischler E, and Sarfert TM.   SOCRATES: A highly efficient automatic test pattern generation system. *IEEE Trans. Computer-Aided Design*, 7(1):126–136, 1988.

[13] Tahoori MB, Huang J, Momenzadeh M, and Lombardi F. Testing of quantum cellular automata. *IEEE Trans. Nanotechnology*, 3(4):432–442, 2004.

[14] Williams TW and Parker KP. Design for testability – A survey *IEEE Trans. Comput.*, 31:2–15, January 1982.

19

Error Detection and Error Correction

Error rates associated with nanocomputing are extremely high. The use of coding methods for error control is an integral part of nano devices design. This chapter introduces the theory and application of error-control codes, involving both error detection and error correction*.

19.1 Introduction

Coding is the representation of data by *code symbols* or *sequences* of code symbols. Data is said to be placed into the code form by *encoding* and extracted from the code form by *decoding*. The number of code symbols in the encoded data is called *code length*. Codes can have a large average code length. In error-control coding, the code length always exceeds the one required for the unique representation of data. Such codes are said to contain *redundancy*. In many cases, coding redundancy is suitable for nanodevices design.

Error reduction/correction in biocomputing systems becomes extremely important as the complexity of numerous connected biochemical reactions increases. This is needed for scalability and fault tolerance of the computing structures.

The concept of *self-checking* is often used in contemporary logic design. This requires that logic networks be designed such that they have built-in support for on-line fault detection, that is, self-checking. An alternative approach to cope with faults is to incorporate fault tolerance in logic networks and devices. The objective of fault tolerance is either to mask, or recover from, faults once they have been detected.

Reliable computing in the presence of faults is called *fault tolerance*. For example, there are approaches to reliable computing in networks with faulty

---

*This chapter was written by Dr. V. Geurkov and the authors of the book.

nodes and/or interconnects that are based on the application of error control codes (including residual number systems). These approaches state that it is possible to correct a class of faults if a library of *reliable* logic nanocells for implementing such correction is available.

Error control codes can be used at the level of a network of logic nanocells, but they are not acceptable or efficient for improving the reliability of a single logic nanocell. A simple way to correct as well as detect errors is to repeat each bit a certain number of times. The assumption is that the expected bit occurs more often. The scheme can tolerate error rates up to 1 error in every 2 bits transmitted at the expense of the increased bandwidth.

Example 19.1 *Given a stream of data (0s and 1s) that is to be sent. The data are grouped into blocks of bits. Let the stream be 1011. Repeat this block three times each, that is, the blocks 1011 1011 1011 are sent. Suppose that the blocks 1010 1011 1011 are received. One group is not the same as the other two, and the decision can be made that an error has occurred. This scheme is not efficient because it can only detect double errors or correct single errors occured in exactly the same place for each group. For example, errors in 1010 1010 1010 cannot be detected.*

The repetition scheme was studied by von Neumann in his work on the correction of the operation of NAND function [10]. The scheme involves replicating the function to be multiplexed N times. N wires are used to carry the signal of each input and output.

Example 19.2 *Von Neumann demonstrated that this multiplexing technique based on repetition encoding can work in cases where elements have less than $p = 0.0107$ uniform probability of failure. For example, to achieve an overall failure probability of 0.027 for a single multiplexed NAND function, $N = 1000$ for $p = 0.005$.*

Von Neumann's main result is the proof that the complexity of correction and redundancy using repetition encoding at the gate level are so high that it makes redundancy techniques useless for improving the reliability of logic cells that compute elementary logic functions. However, the interest in this simplest encoding is motivated by the fact that self-assembling nanostructures can potentially reach the redundancy levels required for reliable computing.

The origin of error-control coding goes back to 1948 when Claude Shannon presented the fundamental theorem of information theory. The theorem states that, if the rate of data transmission through a noisy channel is less than the capacity of the channel, there exist codes that can make the probability of error at the receiver arbitrarily small. The theorem does not provide for techniques to synthesize these codes; it only shows how to analyze the quality of the code. Since 1950, extensive research has been undertaken to find such codes. The first success was associated with the invention of single error-correcting Hamming codes. However, much closer to the theoretical limit came multiple error-correcting Bose–Chaudhuri–Hocquenghem (BCH) codes for binary channels, and Reed–Solomon (RS) codes for nonbinary channels.

Hamming codes work well in situations where one can reasonably expect errors to be rare events. For example, some memory devices have error ratings in the order of 1 bit per 100 million. The distance 3 Hamming code is capable of correcting this type of errors. However, it is useless in situations when there is a likelihood that multiple random bit errors may occur. In situations such as this, codes with more complicated structures, such as BCH codes, are exercised for error control. A special case of BCH codes oriented to non-binary channels is an RS code. A Reed–Solomon code operates over the entire characters instead of only a few bits. BCH and RS codes remain among the most important classes of codes that possess a powerful algebraic structure.

In some applications, distorted bits may occupy adjacent positions. Such errors are called *burst errors*. Burst errors are common for many computing and memory devices. If errors are expected to occur in physical blocks, an error-correcting code that operates at the block level must be used. The redundancy for a burst error-correcting code is generally lower than that for a random error-correcting code. Note that a single burst error correcting code is opposed to a binary Hamming code which operates at the bit level. As well, it differs from a nonbinary error-correcting Hamming code in that it considers burst of errors, which can "slide" over the entire word. Therefore, the redundancy for a single burst error-correcting code may be greater than that for a nonbinary Hamming code.

Error-control codes were originally intended to efficiently protect the integrity of data communication between digital systems. The notion of efficiency implies maximizing the probability of detecting/correcting errors in the transmitted data while minimizing the required hardware overhead. The similar notion of efficiency has been used for testing/repair of digital systems. Therefore, soon after the introduction of error-control codes, researchers started implementing the basic principles of error-control theory to protect the operational integrity of digital systems themselves. A digital system has been considered as a data transmission channel where the corruption of transmitted data is caused by failures in the system, whereas the "transmitted data" are available at the outputs of the system. This process has had two trends depending on the urgency of error detection. Consequently, two distinct types of logic circuits have been exercised:

The first trend has been related to systems with rigorous requirements to test *latency*, the time between occurrence of an error and its detection in the system. It has exploited combinational logic networks, and the methods developed evolved into various forms of concurrent (or on-line) test solutions. Depending on the type of the code involved (whether it is error-detecting or error-correcting code), these solutions have been able to detect or diagnose/repair faulty units within a system under test during its operation.

The second trend has been characteristic of systems with less severe test latency requirements, such as those that could be turned off and put in an *off-line* test mode. Weakening test latency constraints has allowed test and diagnosis methods to be built upon more economical sequential realization of decoding algorithms for error-control codes. This trend has exploited sequential logic circuits.

High degree of interunit dependency in a digital system significantly raises the multiplicity of errors in the output word(s) of the system (even in the case of a single failure), causing proportional growth of the information redundancy and related test hardware (particularly in the case of error correction).

Special classes of codes (called arithmetic codes) have been developed to protect data processing. These codes provide an additional feature of controlling errors occurring during arithmetic operations. Similarly, arithmetic codes can be used for hardware testing. This is often done when the system is subject to unidirectional faults.

Hence, error-control coding principles are equally important for the enhancement of reliability of communication, as well as reliability of computation.

19.2   Channel models

An instance of an incorrect operation of the system being tested is called an *error*. The causes of observed errors may be design errors, fabrication errors, fabrication defects, and physical failures.

> **Example 19.3** *Errors can be caused by the following factors: (a) wrong specifications, violation of design rules (design errors); (b) wrong components, incorrect wiring (fabrication errors); (c) shorts or opens, improper mask alignments (fabrication defects); or (d) components aging, environmental factors (physical failures).*

Fabrication errors, fabrication defects, and physical failures are referred to as *physical faults*. *Logical faults* represent the effect of the physical faults on the operation of the system and allow a direct mathematical treatment. A fault is *detected* by observing an error caused by it. Errors can also be recovered (or *corrected*), which makes the system fault tolerant.

A communication system connects a data source with a data user through a channel (Figure 19.1). The message produced by the information source is

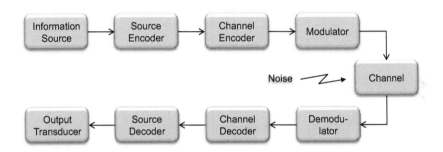

FIGURE 19.1

A simplified model of a communication system.

converted into a nonredundant sequence of binary digits by the source encoder. The sequence from the source encoder called an *information sequence* is passed to the channel decoder. The channel encoder introduces some redundancy into the information sequence that can be used at the receiving end to detect or correct errors introduced by the noise affecting the channel. The redundant sequence of digits is passed to the modulator, which converts it into a sequence of waveforms and passes them through the channel. The channel may transmit a q-ary data symbol at a time, where q is usually the power of two. The demodulator, channel decoder, and source decoder perform the functions that are opposite to the functions of the modulator, channel encoder, and source encoder, respectively. The channel decoder is able to detect/correct the received data. Finally, the signal reconstructed by the source decoder is passed to the data user.

The part of the communication system, that is employed in the design for testability includes the channel encoder and channel decoder only; and the communication channel is represented by the device under test.

The following terminology is used in data communication:

Data transmission

Datastream. The sequence of transmitted symbols generated by the source is called a *datastream*.

Datawords, message words. To encode a stream of data symbols with a *block code*, the stream is broken into blocks of k symbols, called *datawords* (or *message words*).

Codeword. The block code maps each block of k data symbols into a block of n code symbols called a *codeword*.

Message, codestream. A one-to-one mapping of each q-ary k-tuple, which represents the *message*, into a q-ary n-tuple, which represents a codeword, is called an (n, k)-code. Concatenated codewords form a codestream.

Received word. The sequence of n symbols at the output of the channel is called the *received word*. Because of errors, the symbols of the received words may not match those of the transmitted words.

Error correction. The decoder uses the redundancy in the codewords to correct errors in the received words and produces an estimate of the datastream.

Code rate. The ratio of redundant bits to data bits, $(n - k)/k$, is called the *redundancy* of the code; the ratio of data bits to total bits, k/n, is called the *code rate*.

19.3   The simplest error-detecting network

Digital systems use data in the form of a group of bits for their internal operations. Since there is a possibility that, during information processing or storage, data can get corrupted due to faults in the system, there should be some provisions in the system for detecting the erroneous bits. It may also be necessary to correct errors in data in order to restore the system to its normal operating mode. This typically requires additional, or *redundant bits* , to be appended to the data or information bits for *error detection* and/or *error correction* (Figure 19.2). The following terminology is used in error-control coding:

Error detection and error correction problem

▶ The number of bits in the encoded data is called a *codeword length*.
▶ The *length* of a codeword is greater than that of the original data.
▶ The process of appending check bits to the information bits is called *encoding*.
▶ The process of extracting the original information bits from a codeword is called *decoding*.

A single error-detection code is capable of detecting, but not capable of correcting single errors. This code is formed by adding one check symbol to each block of k information symbols. The added symbol is called a *parity*

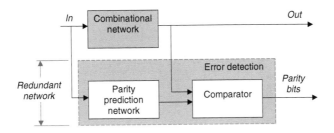

FIGURE 19.2

Principle of error detection using a parity prediction network.

check symbol. The parity check symbol is used as follows:

(a) If the received word contains an *even* number of errors, the decoder will not detect the errors.

(b) If the number of errors is *odd*, the decoder will detect that an odd number of errors, most likely one, has been present.

When the error is detected, the digital system can be designed to request the retransmission of the codestream and/or emit a signal indicating a malfunction. Note that, in the parity bit scheme, double errors are not detected since they do not change the overall parity. However, any odd number of errors (including triple errors) will be detected.

A *linear* Boolean function f of n variables is defined as a function that is represented by a linear positive-polarity polynomial expression:

$$f = r_0 \oplus \bigoplus_{i=1}^{n} r_i x_i = r_0 \oplus r_1 x_1 \oplus \cdots \oplus r_n x_n$$

where $r_i \in \{0, 1\}$ is the i-th coefficient, $i = 1, 2, \ldots, n$, also called *Reed–Muller* polynomial, or linear EXOR expression.

Example 19.4 *Figure 19.3 illustrates the linear and nonlinear components of the polynomial expression of a Boolean function of two variables. Linear polynomials can be implemented using only EXOR gates.*

The equalities

$$\underbrace{x_1 \oplus x_2 \oplus \cdots \oplus x_n = 1}_{Valid\ for\ 1,3,5,...\ variables} \quad \text{and} \quad \underbrace{\overline{x_1 \oplus x_2 \oplus \cdots \oplus x_n} = 1}_{Valid\ for\ 2,4,...\ variables}$$

are valid for odd and even numbers of variables, respectively.

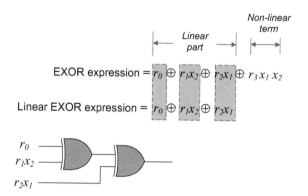

FIGURE 19.3

The linear part of an EXOR expression for a Boolean function of two variables (Example 19.4).

Example 19.5   *The linear EXOR expression of three variables, $x_1 \oplus x_2 \oplus x_3 = 1$, is the three-input parity odd function. This function is equal to 1 when an odd number of variables assume the value 1. It can be implemented by means of two-input EXOR gates (Figure 19.4a). The complement to the odd parity function is the even parity function $\overline{x_1 \oplus x_2 \oplus x_3} = 1$. The even parity function is equal to 1 when an even number of variables is equal to 1. This function can be implemented by EXOR and XNOR gates (Figure 19.4b). Note that the output of this network is equal to 1 when none of the variables are equal to 1.*

Components of error-detection logic network design

▶ The network that generates the parity bit in the transmitter is called a *parity generator.*

▶ The *channel* is a medium through which the transmitter output is sent.

▶ The *receiver's* function is to extract the desired message from the received signal at the channel output. The signal is distorted in the channel. In particular, the signal is contaminated along the path by undesirable signals lumped under the broad term *noise*, which includes random and unpredictable signals.

▶ The network that checks the parity in the receiver is called a *parity checker.*

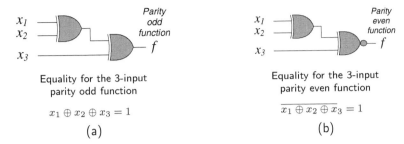

FIGURE 19.4

Parity odd (a) and parity even Boolean functions (b) (Example 19.5).

Example 19.6 *A 3-bit message is to be transmitted together with an even parity bit (Figure 19.5). An error occurs during transmission if the four bits received have an odd number of 1s.*

A simple parity check provides error detection by ensuring that all codewords have the same parity; that is, the same number of 1s mod 2. The parity generator appends an even parity check bit to the transmitted word. If a single error occurs in this word during transmission, the received word will exhibit an odd number of 1s, that is, the parity will change. This can be checked with a similar logic network, a parity checker. If a failure occurs, the system can be prevented from using the corrupted information.

Example 19.7 *The use of a parity generator and parity checker for a 4-bit message is shown in Figure 19.6.*

19.3.1 Error correction

In the single error-correcting Hamming code, information bits along with several parity bits compose a codeword. The values of the parity bits are determined by an even-parity scheme over selected information bits. After a codeword has been transmitted, the parity bits are recalculated at the receiver end to check, whether the correct parity still exists over the selected information bits. By comparing the recalculated parity bits against those in the received word, it is possible to determine

▶ If the received codeword is free of single errors

▶ If a single error has occurred, and exactly which bit has erroneously changed

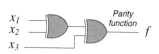

Message Parity function	
$x_1\ x_2\ x_3$	f
0 0 0	0
0 0 1	1
0 1 0	1
0 1 1	0
1 0 0	1
1 0 1	0
1 1 0	0
1 1 1	1

Message Data $x_1\ x_2\ x_3$	Parity f	Parity error function F
0 0 0	0	0
0 0 0	1	1
0 0 1	0	1
0 0 1	1	0
0 1 0	0	1
0 1 0	1	0
0 1 1	0	0
0 1 1	1	1
1 0 0	0	1
1 0 0	1	0
1 0 1	0	0
1 0 1	1	1
1 1 0	0	0
1 1 0	1	1
1 1 1	0	1
1 1 1	1	0

(a)   (b)

FIGURE 19.5

Parity generator (a) and parity checker for a 3-bit message (b) (Example 19.6).

If more than one bit is changed during transmission, then this coding scheme is no longer capable of determining the location of the errors. A single error in a codeword will cause one or more parity checks to fail. The parity check pattern can be used to locate the position of error, provided the checking domains of each parity bit have been chosen correctly. Determining the error position in a binary message is sufficient for correction; the erroneous digit must be readily inverted. Hamming code is capable of correcting all single errors because its *distance*.

Example 19.8   *For the case of 4 information bits, 3 parity bits are included along with the 4 information bits to form a 7 bit codeword. The structure of the codeword in this case is given in Figure 19.7. The following rule is used for the parity bits $p_1, p_2,$ and p_3:*

$$p_1 = 0 \text{ if } b_1 \oplus b_2 \oplus b_4, \text{ and } p_1 = 1 \text{ otherwise.}$$
$$p_2 = 0 \text{ if } b_1 \oplus b_3 \oplus b_4, \text{ and } p_2 = 1 \text{ otherwise.}$$
$$p_3 = 0 \text{ if } b_2 \oplus b_3 \oplus b_4, \text{ and } p_3 = 1 \text{ otherwise.}$$

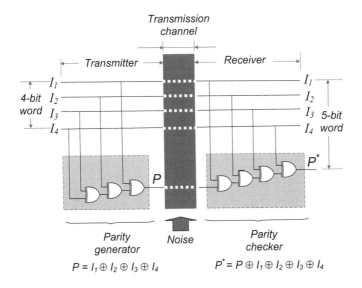

FIGURE 19.6
Using a parity generator and parity checker for a 4-bit message (Example 19.7).

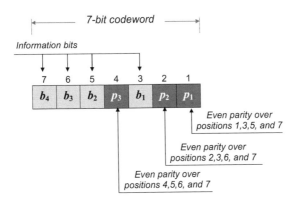

FIGURE 19.7
The structure of the 7-bit Hamming code (Example 19.8).

The Hamming code for the BCD code

The information bits in a Hamming codeword may represent, for example, binary coded decimal (BCD) code bits.

The location of a single error in the Hamming codeword

Upon the receipt of a Hamming codeword, the parity bits are recalculated using the same even parity scheme:

$$\underbrace{\text{Parity bits } p_k \ldots p_1}_{Receiver} \xrightarrow{\textit{Recalculation}} \underbrace{\text{Binary check number } c_k^* \ldots c_1^*}_{Receiver}$$

This recalculation results in a *binary check number* $c_k^* \ldots c_1^*$, which is transmitted along with the codeword.

Example 19.9 *Design example is given in Figures 19.8 and 19.9.*

Example 19.10 *(Continuation of Example 19.9). Design example for the case when the bit p_1 has been erroneously changed from 1 to 0 during transmission is given in Figure 19.10.*

Gray code is used for encoding the indices of the nodes. There are several reasons for encoding the indices. The most important is to simplify the analysis, synthesis, and embedding of topological structures. The conversion of binary code into Gray code and vice versa can be implemented using logic networks.

Example 19.11 *Given a 3-bit binary code, the corresponding Gray code and its implementation are shown in Figure 19.11.*

To build a Gray code for d dimensions, one takes the Gray code for $d-1$ dimensions, reflects it top to bottom across a horizontal line just below the last element, and adds a leading one to each new element below the line of reflection.

19.4 Discrete memoryless channel

A *discrete memoryless channel* is characterized by a q-ary alphabet and conditional probabilities $P(j|i)$, where $P(j|i)$ is the probability of receiving the symbol j given that the symbol i was transmitted. Each output symbol of the channel depends only on the corresponding input. For a given input sequence

Design example: Hamming code for the 4-bit BCD code

Let the following Hamming codeword (Figure 19.9),

7	6	5	4	3	2	1
0	1	1	0	0	1	1
b_4	b_3	b_2	p_3	b_1	p_2	p_1

$(p_3 p_2 p_1 = 011)$ be transmitted (Figure 19.7). Assume that the bit b_1 is erroneously changed from 0 to 1 during transmission:

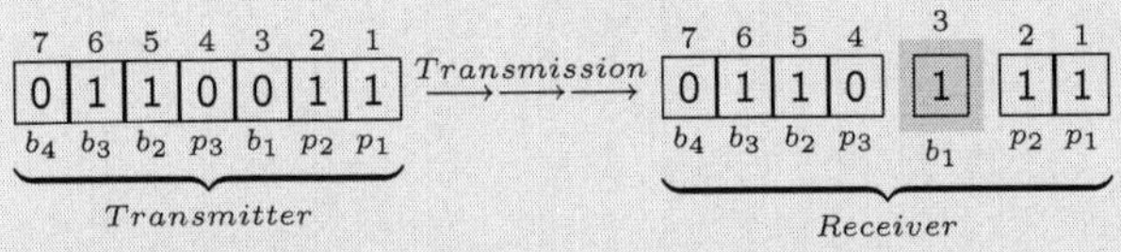

7	6	5	4	3	2	1
0	1	1	0	0	1	1
b_4	b_3	b_2	p_3	b_1	p_2	p_1

$\xrightarrow{Transmission}$

7	6	5	4	3	2	1
0	1	1	0	1	1	1
b_4	b_3	b_2	p_3	b_1	p_2	p_1

Transmitter   *Receiver*

The received word becomes 0110111, which differs from the transmitted codeword 0110011 in the third position. The binary check number (*syndrome*) is formed as follows:

▶ There is an odd number of 1s in the group b_4, b_2, b_1, p_1:

$$p_1 = 1, \quad \text{ODD NUMBER} = \underset{b_4}{\overset{7}{0}} \oplus \underset{b_2}{\overset{5}{1}} \oplus \underset{b_1}{\overset{3}{1}} \oplus \underset{p_1}{\overset{1}{1}} = 1, \quad c_1^* = 1$$

▶ There is an odd number of 1s in the group b_4, b_3, b_1, p_2:

$$p_2 = 1, \quad \text{ODD NUMBER} = \underset{b_4}{\overset{7}{0}} \oplus \underset{b_3}{\overset{6}{1}} \oplus \underset{b_1}{\overset{3}{1}} \oplus \underset{p_2}{\overset{2}{1}} = 1, \quad c_2^* = 1$$

▶ There is an even number of 1s in the group b_4, b_3, b_2, p_3:

$$p_3 = 0, \quad \text{EVEN NUMBER} = \underset{b_4}{\overset{7}{0}} \oplus \underset{b_3}{\overset{6}{1}} \oplus \underset{b_2}{\overset{5}{1}} \oplus \underset{p_3}{\overset{4}{0}} = 0, \quad c_3^* = 0$$

Since $p_3 p_2 p_1 = 011$, the syndrome is $c_3^* c_2^* c_1^* = 011$, thus indicating that the bit 3 is incorrect.

FIGURE 19.8
Hamming code for the 4-bit BCD code (Example 19.9).

$U = u_0, \ldots, u_{n-1}$, the conditional probability of a corresponding output sequence $Z = z_0, \ldots, z_{n-1}$ can be expressed as $P(Z|U) = \prod_{m=0}^{n-1} P(z_m|u_m)$.
The error is defined as

$$\text{Error } e = Z \oplus U = z_o \oplus u_0, \ldots, z_{n-1} \oplus u_{n-1}$$

Design example:
Hamming code for the 4-bit BCD code (continuation)

STEP 1: Hamming code

Decimal digit	b_4	b_3	b_2	p_3	b_1	p_2	p_1
0	0	0	0	0	0	0	0
1	0	0	0	0	1	1	1
2	0	0	1	1	0	0	1
3	0	0	1	1	1	1	0
4	0	1	0	1	0	1	0
5	0	1	0	1	1	0	1
6	0	1	1	0	0	1	1
7	0	1	1	0	1	0	0
8	1	0	0	1	0	1	1
9	1	0	0	1	1	0	0

STEP 2: Minimization

$$p_1 = b_1 \oplus b_2 \oplus b_4 \qquad p_2 = b_1 \oplus b_3 \oplus b_4 \qquad p_3 = b_2 \oplus b_3 \oplus b_4$$

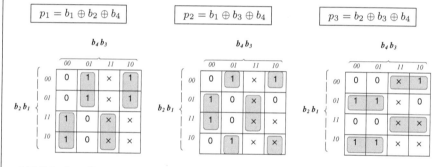

STEP 3: Gate-level logic network

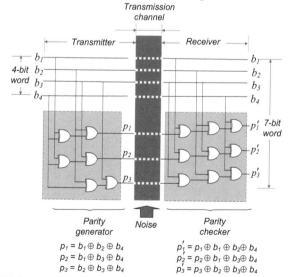

$$p_1 = b_1 \oplus b_2 \oplus b_4$$
$$p_2 = b_1 \oplus b_3 \oplus b_4$$
$$p_3 = b_2 \oplus b_3 \oplus b_4$$

$$p_1' = p_1 \oplus b_1 \oplus b_2 \oplus b_4$$
$$p_2' = p_2 \oplus b_1 \oplus b_3 \oplus b_4$$
$$p_3' = p_3 \oplus b_2 \oplus b_3 \oplus b_4$$

FIGURE 19.9

Hamming code for the 4-bit BCD code (continuation of Example 19.9).

Design example: Hamming code for the 4-bit BCD code

Let the bit p_1 be erroneously changed from 1 to 0 during transmission.

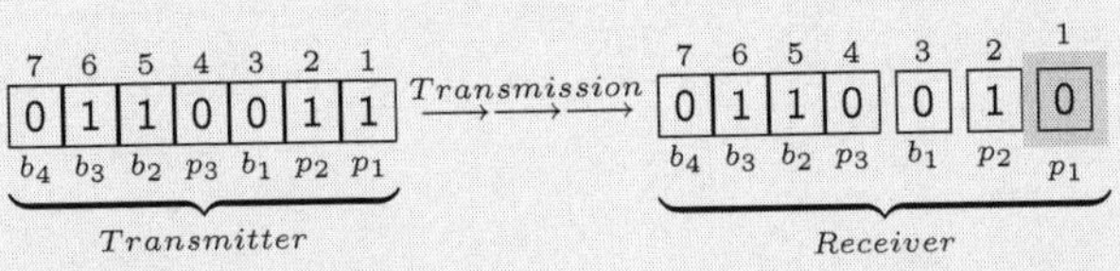

The syndrome is formed as follows:

▶ There is an odd number of 1's in the group b_4, b_2, b_1, p_1:

$$p_1 = 1, \quad \text{ODD NUMBER} = \boxed{0}_{b_4} \oplus \boxed{1}_{b_2} \oplus \boxed{0}_{b_1} \oplus \boxed{0}_{p_1} = 1$$

This requires c_1^* to be set to 1.

▶ There is an odd number of 1's in the group b_4, b_3, b_1, p_2:

$$p_2 = 1, \quad \text{ODD NUMBER} = \boxed{0}_{b_4} \oplus \boxed{1}_{b_3} \oplus \boxed{0}_{b_1} \oplus \boxed{1}_{p_2} = 0$$

This implies $c_2^* = 0$.

▶ There is an even number of 1's in the group b_4, b_3, b_2, p_3:

$$p_3 = 0, \quad \text{EVEN NUMBER} = \boxed{0}_{b_4} \oplus \boxed{1}_{b_3} \oplus \boxed{1}_{b_2} \oplus \boxed{0}_{p_3} = 0$$

This implies $c_3^* = 0$.

Therefore, the syndrome is $c_3^* c_2^* c_1^* = 001$ is indicating that the bit in position 1 is incorrect

FIGURE 19.10

Hamming code for the 4-bit BCD code (Example 19.10).

A *binary symmetric channel* is a discrete memoryless channel where the input and output symbols can be 0 and 1, and the conditional probabilities (or the channel *transition probabilities*) are symmetric: $P(0|1) = P(1|0) = p$, $P(0|0) = P(1|1) = 1 - p$.

A *binary asymmetric channel* is a binary channel where the probability of a 0-error is ϵ, and the probability of a 1-error is p, with $p \gg \epsilon$. In an ideal binary asymmetric channel, only 0-errors or 1-errors can occur in the received word. The error type is known a priori at the receiver. This error category is used for digital hardware testing, where the device is most likely to fail in only one direction. If either 0-errors or 1-errors can occur but not both, the

Design example: Gray code generation

STEP 1
Binary and Gray codes

Binary code	Gray code
000	000
001	001
010	011
011	010
100	110
101	111
110	101
111	100

Binary code	Gray code
000	000
001	001
011	010
010	011
110	100
111	101
101	110
100	111

STEP 2
Equations for conversions

$$g_i = b_i \oplus b_{i+1}$$
$$i = 0 : g_0 = b_0 \oplus b_1$$
$$i = 1 : g_1 = b_1 \oplus b_2$$
$$i = 2 : g_2 = b_2 \oplus 0$$

$$b_i = g_0 \oplus g_1 \oplus \dots g_{3-i} = \bigoplus_{i=0}^{3-i} g_i$$
$$i = 2 : b_2 = g_2 \oplus 0$$
$$i = 1 : b_1 = g_1 \oplus b_2$$
$$= g_1 \oplus g_2$$
$$i = 0 : b_0 = g_0 \oplus b_1$$
$$= g_0 \oplus g_1 \oplus g_2$$

(a)   (b)

STEP 3
Gate-level logic network

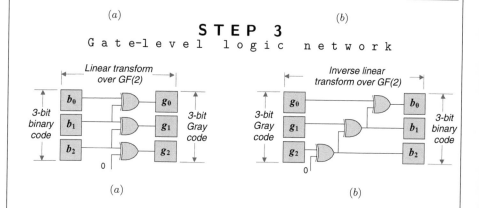

(a)   (b)

FIGURE 19.11

Gray code generation and decoding (Example 19.11).

errors are called *asymmetric errors*. If both 1-errors and 0-errors can occur in the received words but in any particular received word all errors are of one type, these errors are characterized as *unidirectional errors*.

In an *arithmetic channel*, the q-ary symbol is represented as a number N in radix 2: $N = \sum_{i=0}^{n-1} a_i 2^i$. The erroneous number N' is given by $N' = \sum_{i=0}^{n-1} a_i' 2^i$. The error is given by arithmetic difference $N' - N$. The subtraction is carried out as in the set of integers.

19.5 Linear block codes

The encoding procedure uniquely assigns a q^n-tuple to each of the 2^k message k-tuples. This can be done using a look-up table. If V_i and V_j are two codewords in an (n, k)-block code, then the code is said to be *linear* if and only if $V_i \oplus V_j$ is also a codeword.

19.5.1 Vector spaces

The set of all binary n-tuples, V_n, is called a *vector space* over the binary field. The binary field has two operations: addition and multiplication. The rules of these operations coincide with the rules for the modulo 2 addition and multiplication:

Addition	Multiplication
$0 \oplus 0 = 0$	$0 \otimes 0 = 0$
$0 \oplus 1 = 1$	$0 \otimes 1 = 0$
$1 \oplus 0 = 1$	$1 \otimes 0 = 0$
$1 \oplus 1 = 0$	$1 \otimes 1 = 1$

A subset S of the vector space V_n is called a *subspace*, if (i) the all-zero vector exists in S, and (ii) the sum of any two vectors in S also belongs to S.

Example 19.12 *The vector space V_4 includes the following 4-tuples:*

0000 0001 0010 0011 0100 0101 0110 0111
1000 1001 1010 1011 1100 1101 1110 1111

Then a subspace of V_4 is 0000, 0101, 1010, 1111.

A set of 2^k n-tuples is called a *linear block code* if and only if it is a subspace of the vector space V_n of all n-tuples.

Example 19.13 *An example of a (6,3)-code is*

Message	Codeword
0 0 0	0 0 0 0 0 0
1 0 0	1 1 0 1 0 0
0 1 0	0 1 1 0 1 0
1 1 0	1 0 1 1 1 0
0 0 1	1 0 1 0 0 1
1 0 1	0 1 1 1 0 1
0 1 1	1 1 0 0 1 1
1 1 1	0 0 0 1 1 1

19.5.2   Generator matrix, parity check matrix, syndrome

For large k, a look-up table implementation of the decoder requires significant hardware overhead. Instead of storing the codewords, they can be generated as needed. It is possible to find a set of $k < 2^k$ basis vectors (n-tuples) that can generate all 2^k codewords as a linear combination of the basis vectors:

$$U = m_1 V_1 + \cdots + m_k V_k$$

It is expressed in the matrix form as follows:

$$U = mG$$

where $m = m_1, \ldots, m_k$ and $U = u_1, \ldots, u_n$ are the message word and the corresponding codeword, respectively, and the $k \times n$ matrix G is called a *generator matrix*:

$$G = \begin{bmatrix} V_1 \\ \vdots \\ V_k \end{bmatrix} = \begin{bmatrix} v_{11} & \cdots & v_{1n} \\ \vdots & & \vdots \\ v_{k1} & \cdots & v_{kn} \end{bmatrix}$$

Example 19.14   *If* $G = \begin{bmatrix} V_1 \\ V_2 \\ V_3 \end{bmatrix} = \begin{bmatrix} 1 1 0 1 0 0 \\ 0 1 1 0 1 0 \\ 1 0 1 0 0 1 \end{bmatrix}$, *then the codeword* U_4,

corresponding to the fourth message word (dataword, or message vector) $m = 110$, *and the codewords in the above table is*

$$U_4 = \begin{bmatrix} 1 & 1 & 0 \end{bmatrix} \begin{bmatrix} V_1 \\ V_2 \\ V_3 \end{bmatrix} = 1 \cdot V_1 + 1 \cdot V_2 + 1 \cdot V_3 = 101110$$

Therefore, instead of $2^3 = 8$ *vectors, the decoder needs to store only 3 vectors.*

In a *systematic* code, k digits of the codeword coincide with the message digits, and the remaining $r = n - k$ digits are parity digits. Such a structure simplifies the decoding procedure. A generator matrix of a systematic code is represented in the form

$$G = \left[\,P\,|\,I_k\,\right] = \begin{bmatrix} p_{11} & p_{12} & \cdots & p_{1r} & 1 & 0 & \cdots & 0 \\ p_{21} & p_{22} & \cdots & p_{2r} & 0 & 1 & \cdots & 0 \\ & \vdots & & & & & \vdots & \\ p_{k1} & p_{k2} & \cdots & p_{kr} & 0 & 0 & \cdots & 1 \end{bmatrix} \tag{19.1}$$

where P is the parity check portion of G, and I_k is the $k \times k$ identity matrix.

Example 19.15 *Given the (6,3) code considered in Example 19.13, the codewords are described as follows: $U = m_1 + m_3, m_1 + m_2, m_2 + m_3, m_1, m_2, m_3 = u_1, u_2, u_3, u_4, u_5, u_6$.*

For a $k \times n$ generator matrix G, there exists an $r \times n$ matrix H such that $GH^T = 0$, where H^T is the *transpose* of H. This matrix, called a *parity-check matrix*, is represented in the following form:

$$H = \left[\,I_r\,|\,P^T\,\right] = \begin{bmatrix} 1 & 0 & \cdots & 0 & p_{11} & p_{21} & \cdots & p_{k1} \\ 0 & 1 & \cdots & 0 & p_{12} & p_{22} & \cdots & p_{k2} \\ & \vdots & & & & & & \\ 0 & 0 & \cdots & 1 & p_{1r} & p_{2r} & \cdots & p_{kr} \end{bmatrix} \tag{19.2}$$

Generally, U is a codeword if and only if $UH^T = 0$. Let $U = u_1, \ldots, u_n$ and $Z = z_1, \ldots, z_n$ be a transmitted codeword and a received word, respectively. The *syndrome* of r is defined as follows:

$$S = ZH^T = eH^T \tag{19.3}$$

It follows from expression 19.3 that the syndrome depends only on the error itself and not on the received vector. If Z is a codeword, then $S = 0$. If Z contains detectable errors, then $S \neq 0$.

Example 19.16 *Suppose that the codeword $U = 101110$ is transmitted, and the vector $Z = 001110$ is received. Apparently,*

$$S = ZH^T = [001110]\begin{bmatrix} 1 & 0 & 0 \\ 0 & 1 & 0 \\ 0 & 0 & 1 \\ 1 & 1 & 0 \\ 0 & 1 & 1 \\ 1 & 0 & 1 \end{bmatrix} = [1, 1+1, 1+1] = [100]$$

We can verify that the syndrome of the error pattern is the same: $S = eH^T = [100000]H^T = [100]$.

19.5.3   Standard array and error correction

The *standard array* consists of the 2^n n-tuples arranged in a way such that the first row contains all the codewords, and the first column contains all the correctable error patterns. Each row, called a *coset*, consists of an error pattern in the first column, called the *coset leader*, followed by the codewords perturbed by the error pattern. The standard array format for an (n, k) code, where k is the length of the codeword, is as follows:

$$
\begin{bmatrix}
U_1 & U_2 & \cdots & U_i & \cdots & U_{2^k} \\
e_2 & U_2 + e_2 & \cdots & U_i + e_2 & \cdots & U_{2^k} + e_2 \\
e_3 & U_2 + e_3 & \cdots & U_i + e_3 & \cdots & U_{2^k} + e_3 \\
\vdots & \vdots & & \vdots & & \\
e_j & U_2 + e_j & \cdots & U_i + e_j & \cdots & U_{2^k} + e_j \\
\vdots & \vdots & & \vdots & & \\
e_{2^r} & U_2 + e_{2^r} & \cdots & U_i + e_{2^r} & \cdots & U_{2^k} + e_{2^r}
\end{bmatrix}
$$

The decoding algorithm replaces a corrupted vector (any n-tuple excluding those in the first row) with a valid codeword from the top of the column, containing the corrupted vector. Each member of a coset will have the same syndrome, which is equal to the syndrome of the corresponding coset leader.

The error-correction procedure is as follows:

The error-correction procedure

Step 1: Calculate the syndrome $S = ZH^T$.

Step 2: Locate coset leader (error pattern) e_j, whose syndrome satisfies the equation $S = ZH^T$. This error pattern is assumed to be the corruption.

Step 3: The corrected received vector, or codeword, is defined as follows: $U = Z + e_j$.

Example 19.17   *Given the codeword $U = 101110$ of the $(6, 3)$ code, the $2^6 = 64$ 6-tuples are arranged in the following standard array:*

000000	110100	011010	101110	101001	011101	110011	000111
000001	110101	011011	101111	101000	011100	110010	000110
000010	110110	011000	101100	101011	011111	110001	000101
000100	110000	011110	101010	101101	011001	110111	000011
001000	111100	010010	100110	100001	010101	111011	001111
010000	100100	001010	111110	111001	001101	100011	010111
100000	010100	111010	001110	001001	111101	010011	100111
010001	100101	001011	111111	111000	001100	100010	010110

The valid codewords are the eight vectors in the first row, and the *correctable error patterns* are the seven nonzero coset leaders in the first column. All 1-bit errors and one 2-bit error are correctable.

Let us determine all syndromes of the correctable errors: $S = e_j H^T =$

$e_j \begin{bmatrix} 1 & 0 & 0 \\ 0 & 1 & 0 \\ 0 & 0 & 1 \\ 1 & 1 & 0 \\ 0 & 1 & 1 \\ 1 & 0 & 1 \end{bmatrix}$. The results are listed in Table 19.1. Using this look-up table, the

decoder can identify an estimate of the error, $\hat{e}$. For example, the syndrome $S = 110$ will correspond to $\hat{e} = 000100$. In the general case, this mapping can be done by a 2×8 ROM or any other programmable logic device. In Example 19.18, this procedure is realized by a logic combinational network. The decoder then adds $\hat{e}$ to Z to obtain an estimate of the transmitted codeword: $\hat{U} = Z + \hat{e} = U + (e + \hat{e})$. If the estimated error is the same as the actual one $(e = \hat{e})$, then the codeword estimate is equal to the transmitted codeword $(U = \hat{U})$. If this is not true, a decoding error takes place.

Example 19.18 *Let us assume that codeword* $U = 101110$ *is transmitted, and the vector* $Z = 101010$ *is received. The syndrome of* Z *is calculated as follows:* $S = [101010]H^T = [110]$. *According to the look-up table, this syndrome corresponds to the error estimate* $\hat{e} = 000100$, *and the corrected vector is estimated as* $\hat{U} = Z + \hat{e} = 101010 + 000100 = 101110$. *Since* $\hat{U} = U$, *no decoding error occured. Errors in the parity check part can normally be ignored; therefore only message part needs to be corrected; that is,* $010 + 100 = 110$. *The reduced decoder, which implements this operation for the message part only, is shown in Figure 19.12. The upper part of the decoder computes the syndrome* $S = ZH^T$, *that is,*
$s_1 = z_1 + z_4 + z_6$, $s_2 = z_2 + z_4 + z_5$, *and* $s_3 = z_3 + z_5 + z_6$.
The remaining part performs an operation on the look-up table, producing only a message part of an error pattern $\hat{e}$. *This part consists of the three rightmost bits of the error pattern,* e_4, e_5, *and* e_6. *The message bits* z_4, z_5, *and* z_6 *of the received word are being corrected by addition with these error bits.*

19.5.4 Distance and error-control capability

The *Hamming weight* $w(U)$ of a codeword U is the number of nonzero components in U. For a binary vector U, this is the number of ones in it. The *Hamming distance* between two codewords U and V, denoted $d(U, V)$, is defined as the number of elements in which they differ. Obviously, the Hamming weight of a codeword is equal to its Hamming distance from the all-zeros vector.

TABLE 19.1

Look-up table of syndromes of
the correctable errors

Error pattern					Syndrome			
0	0	0	0	0	0	0	0	0
0	0	0	0	0	1	1	0	1
0	0	0	0	1	0	0	1	1
0	0	0	1	0	0	1	1	0
0	0	1	0	0	0	0	0	1
0	1	0	0	0	0	0	1	0
1	0	0	0	0	0	1	0	0
0	1	0	0	0	1	1	1	1

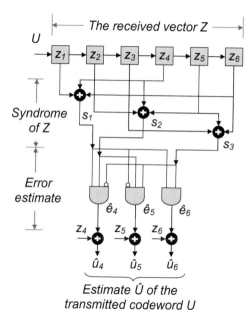

▶ *The codeword $U = 101110$ is transmitted.*

▶ *The vector $Z = 101010$ is received. The syndrome is*

$$S = [101010]H^T = [110]$$

▶ *This syndrome corresponds to $\hat{e} = 000100$, then the corrected vector is obtained as follows:*

$$\hat{U} = Z + \hat{e}$$
$$= 101010 + 000100$$
$$= 101110$$

▶ *Error verification results in $\hat{U} = U$*

Estimate $\hat{U}$ of the
transmitted codeword U
(message part only)

FIGURE 19.12

Implementation of the reduced (6,3) decoder (Example 19.18).

Example 19.19 (*a*) *Given* $U = 101100$, *its Hamming weight is* $w(101100) = 3$; (*b*) *given two words* $U = 110100$ *and* $V = 010101$, *the Hamming distance* d *is* $d(110100, 010101) = 2$; (*c*) $U + V = 100001$ *and* $w(100001) = 2$.

Let a space V_n of codewords of length n be given. Consider all pairs of codewords in the space V_n. The smallest distance between the pairs is called the *minimum distance* of the code and is denoted d_{min}. In a linear code, the sum of two codewords yields another codeword. Hence, the distance between two codewords is equal to the weight of a third codeword. Thus, in order to compute a minimum distance of a code, we only need to examine the weight of each codeword of this code. The minimum weight will correspond to the minimum distance of the code.

19.5.5 Optimal decoder

The task of the decoder is to estimate the transmitted codeword U_i, having received the vector Z. The optimal decoder implements a *maximum likelihood* algorithm. It decides in favor of U_i if the conditional probability $P(Z|U_i) = \max P(Z|U_j)|_{over\ all\ U_j}$. Given the binary symmetric channel, the likelihood of U_i with respect to Z is inversely proportional to the distance between U_i and Z. Then the equation is substituted with $d(Z|U_i) = \max d(Z|U_j)|_{over\ all\ U_j}$. In other words, the decoder determines the distance between r and each of the possible transmitted codewords U_j, and selects U_i, for which the distance is minimal, as the most likely one.

Example 19.20 *Consider two codewords, U and V, of the code with minimum distance 5 (Figure 19.13). Let us assume that codeword U was transmitted.*
(a) If the received word is Z_1, whose distance from U is 1, the decoder will select U and discard Z_1, thus providing error correction of the received word.
(b) The received word Z_2 whose distance from U is 2, is discarded as well.
(c) In case of receiving Z_3 or Z_4, positioned respectively at the distance 3 and 4 from U, these vectors will be mistakenly corrected to the codeword V, and the decoding error will occur. If the distance between the received and transmitted word is 5, then the codeword V will be received, and the error becomes undetectable.

This example demonstrates that the code with minimum distance d_{min} is capable of *detecting*

$$e = (d_{min} - 1)\ \text{errors}$$

or *correcting* up to $t = \lfloor \frac{d_{min}-1}{2} \rfloor$ errors ($\lfloor a \rfloor$ is the largest integer not exceeding a).

Transmitted   Received
codeword   codeword

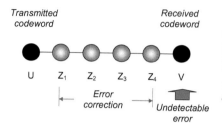

Codewords U (transmitted) and V
(received).
Minimum distance is 5.
The received word: Z_1 with distance
1, Z_2 with distance 2, Z_3 with
distance 3, and Z_4 with distance 4.

FIGURE 19.13
Relationship between distance and error-control capabilities (Example 19.20).

Example 19.21 *Consider the code used in the example 19.18. Let the*
transmitted word be $U = 101110$, and let the received
word be corrupted in the two rightmost positions, that
is, $Z = 101101$. Then, the syndrome $S = 110$.
According to the look-up table, the error estmate will
be $\hat{e} = 000100$, and the corrected vector is estimated
as $\hat{U} = Z + \hat{e} = 101101 + 000100 = 101001$. Since
$\hat{U} \neq U$, a decoding error has occured. However, the
double error has been detected; the syndrome does not
equal 0 anymore. Here, $d_{min} = 3, e = 2$, and $t = 1$.

19.6 Cyclic codes

In cyclic codes, any cyclic shift of a codeword produces another codeword.
Cyclic codes are easily implemented using linear-feedback shift registers, which
are constructed using EXOR gates and memory elements. These codes are
useful in sequential-access computing devices.

An (n, k) linear code is called a *cyclic code* if it satisfies the following
property. If the n-tuple

$$U = (u_0, u_1, \ldots, u_{n-1})$$

is a codeword of a code K, then

$$U^{(1)} = (u_{n-1}, u_0, u_1, \ldots, u_{n-2})$$

obtained by a cyclic shift of U, is also a codeword of the code K. In general,

$$U^{(i)} = (u_{n-i}u_{n-i+1}, \ldots, u_{n-1}, u_0, u_1, \ldots, u_{n-i-1})$$

obtained by i cyclic shifts, is also a codeword of the code K.

An (n, k) cyclic code can detect all single errors in a codeword, all burst errors (multiple adjacent faults) of length $b \leq (n - k)$, and other patterns of errors depending on the particular code. The words of a cyclic code are usually represented by polynomials. For instance, the codeword $U^{(1)} = (u_0, u_1, \ldots, u_{n-1})$ corresponds to the polynomial

$$U(x) = u_0 + u_1 x + \cdots + u_{n-1} x^{n-1}$$

This notation describes a cyclic shift by the following modulo operation:

$$U^{(i)}(x) = x^i U(x) \bmod (x^n + 1)$$

Example 19.22 *Let $U = 1011$ be a 4-bit word that is cyclically shifted twice. The resulting word becomes $U = 1110$. This is described in the polynomial form as follows: $U(X) = 1 + x^2 + x^3$, $x^2 U(x) = x^2 + x^4 + x^5$, $x^2 U(x) \bmod (x^4 + 1) = 1 + x + x^2$, so that the coefficients of the remainder correspond to the bits of the cyclically shifted word.*

It can be proven that a codeword of a cyclic code can be generated using a *generator polynomial* in the same way that a block code with a generator matrix is formed. A cyclic code is uniquely and completely characterized by its generator polynomial of degree $(n - k)$ or greater with the coefficients either 0 or 1.

The necessary requirement for a polynomial to be the generator polynomial of a cyclic code is that it must be divisible by the $x^n + 1$ polynomial. The generator polynomial will have a form $g(x) = g_0 + g_1 x + \cdots + g_r x^r$, where g_0 and g_r must be 1. If the message polynomial is $m(x) = m_0 + m_1 x + \cdots + m_{k-1} x^{k-1}$, then the codeword polynomial of degree $n - 1 = k + r - 1$ is $U(x) = m(x)g(x)$. That is, $r = n - k$. U is a valid codeword if and only if $g(x)$ is divisible by $U(x)$ without a remainder (i.e., $Z(x) \bmod g(x) = 0$). Indeed, if $U(x) = m(x)G(x)$ was transmitted, the corrupted version of it is $Z(x) = U(x) + e(x)$, where $e(x)$ is the *error pattern polynomial*. The generator polynomial is implemented by a linear-feedback shift register.

Example 19.23 *Given the check polynomial $x^{12} \oplus x^{11} \oplus x^3 \oplus x^2 \oplus x \oplus 1$, the linear-feedback shift register for the implementation of this polynomial is shown in Figure 19.14. This shift register will contain the check bits at the end of the encoding process.*

Design example:
the generator polynomial $x^{12} \oplus x^{11} \oplus x^3 \oplus x^2 \oplus x \oplus 1$

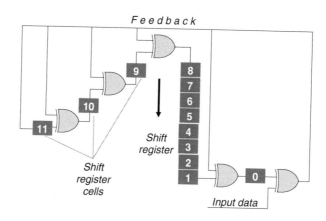

FIGURE 19.14

Implementation of the generator polynomial $x^{12} \oplus x^{11} \oplus x^3 \oplus x^2 \oplus x \oplus 1$ (Example 19.23).

The syndrome calculation for a linear block code is done by performing multiplication (Equation 19.3). For a cyclic code, this operation can be represented as follows:

$$S(x) = Z(x) \bmod g(x) = e(x) \bmod g(x) \qquad (19.4)$$

That is, the syndrome polynomial of the received word coincides with that of the error-pattern polynomial. If there are no errors, the received polynomial is the same as the transmitted one (since $S(x) = 0$).

19.6.1   Systematic encoding

Generally, the cyclic codes are not systematic. This is because the message digits are not present in the codeword explicitly, but are rather calculated using all received digits. In systematic encoding, the message digits are placed in the rightmost position, which simplifies the decoding procedure. To make a cyclic code of a systematic form, the following encoding must be applied:

$$U(x) = x^r m(x) + x^r m(x) \bmod g(x)$$

The multiplication by x^r provides the right-shift of the message polynomial by r positions. Addition of the remainder makes it the codeword. The result is a polynomial of degree $n-1$ or less, and if divided by $g(x)$, there is a zero remainder. The codeword $U(x)$ can be expanded into its polynomial terms as follows:

$$U(x) = p_0 + p_1 x + \cdots + p_{r-1} x^{r-1} + m_0 x^r + m_1 x^{r+1} + \cdots + m_{k-1} x^{n-1}$$

Example 19.24 *Let us generate a systematic codeword from the (7,4) codeword set for the message vector $m = 1011$. Here, $n = 7$ and $r = 3$. It can be seen that $g(x) = 1 + x + x^3$ is divisible by $x^7 + 1$, and so can be used as a generator polynomial. The solution is as follows:*

$$m(X) = 1 + x^2 + x^3,$$
$$x^3 m(x) = x^3(1 + x^2 + x^3) = x^3 + x^5 + x^6,$$
$$x^3 m(x) \bmod (1 + x + x^3) = 1, \quad U(x) = 1 + x^3 + x^5 + x^6.$$

19.6.2 Implementation of modulo $g(x)$ division

Consider the division of polynomials using the following example.

Example 19.25 *Let us divide*

$$V(x) = x^3 + x^5 + x^6 \quad by \quad g(x) = 1 + x + x^3$$

in algebraic and vector forms:

$$
\begin{array}{r|l}
x^6 + x^5 + x^3 & x^3 + x + 1 \\
x^6 + x^4 + x^3 & x^3 + x^2 + x + 1 \\
\hline
x^5 + x^4 \\
x^5 + x^3 + x^2 \\
\hline
x^4 + x^3 + x^2 \\
x^4 + x^2 + x \\
\hline
x^3 + x \\
x^3 + x + 1 \\
\hline
1
\end{array}
$$

$$
\begin{array}{r|l}
1\ 1\ 0\ 1\ 0\ 0\ 0 & 1\ 0\ 1\ 1 \\
1\ 0\ 1\ 1 & 1\ 1\ 1\ 1 \\
\hline
1\ 1\ 0\ 0 \\
1\ 0\ 1\ 1 \\
\hline
1\ 1\ 1\ 0 \\
1\ 0\ 1\ 1 \\
\hline
1\ 0\ 1\ 0 \\
1\ 0\ 1\ 1 \\
\hline
0\ 0\ 1
\end{array}
$$

The logic network that implements this operation is shown in Figure 19.15. The same network can be used to encode the message polynomial, $m(x) = m_0 + m_1 x + m_2 x^2 + m_3 x^3$. Multiplication of $m(x)$ by x^3 is equivalent to padding the message vector by three leftmost zeroes.

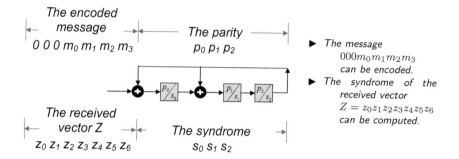

FIGURE 19.15

A scheme for computing $\bmod\,(1+x+x^3)$ division (Examples 19.25 and 19.29).

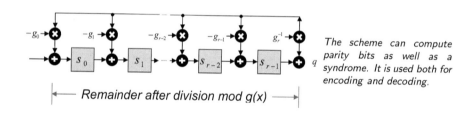

FIGURE 19.16

A scheme for computing $\bmod\,g(x)$ division.

Generally, the network that performs division of the polynomial

$$V(x) = v_0 + v_1 x + v_2 x^2 + \cdots + v_m x^m$$

by the polynomial

$$g(x) = g_0 + g_1 x + g_2 x^2 + \cdots + g_r x^r$$

is presented in Figure 19.16.

After m shifts, it will contain the remainder of $V(x)$ with respect to $g(x)$: $V(x) \bmod g(x)$, and the degree of the remainder polynomial does not exceed $r - 1$. The symbol q in the figure represents a quotient produced during the division.

19.7 Block codes

There exist two different classes of error-control codes:

▶ *Block* codes
▶ *Trellis* codes

In the former a block of k data digits is represented by an n-digit codeword. Codewords are concatenated and sequentially transmitted to the channel to form an infinite stream of code symbols. Trellis codes also encode a stream of data symbols into a stream of code symbols.

The datastream is divided into blocks of length k called *dataframes*, which are relatively small and are encoded into blocks of length n called *codeframes*. Thus, the datastream is broken into

▶ A sequence of *blocks*
▶ A sequence of *frames*

depending on whether it is a block code or a trellis code. However, the most important point is that, for block codes, a single block of the codestream depends only on a single block of the datastream, whereas for trellis codes, a single frame of the codestream depends on multiple frames of the datastream.

The decoder of the trellis code can store m dataframes, where m is called the *frame memory* of the encoder. After each clock time, the encoder computes a single codeframe of length n from the current coming dataframe and the m stored dataframes. Hence, successive codeframes are linked by the encoding procedure, and the concatenated stream of such codeframes forms an infinitely long codeword.

A k-symbol dataframe can affect all succeeding codeword frames, whereas in a block code, a k-symbol datablock determines only the next n-symbol codeblock and no others. Due to this property, trellis codes are said to operate on channels with memory, whereas block codes are considered as operating on memoryless channels. A block code can be defined as a set of M q-ary sequences of length n (codewords). Usually, $M = q^k$.

Block codes are characterized by three parameters:

▶ The blocklength n
▶ The datalength k
▶ The minimum distance d_{min}

19.7.1 Hamming codes

As mentioned earlier, the Hamming code is capable of correcting a single error. The minimal distance of this code is 3. The binary Hamming code is characterized by the following parameters:

$$(n, k) = (2^r - 1, 2^r - 1 - r)$$

where r is the number of parity bits. Each column of a parity check matrix of the Hamming code contains an r-bit binary number. A single error in the message vector produces the syndrome, which coincides with this number. These numbers are arranged in ascending order, so that the syndrome indicates the direct position of the error in the binary form.

Example 19.26 *A parity check matrix for the nonsystematic (7,4) Hamming code can be chosen follows (in terms of computation, it is more convenient to store the check and message bits of the codeword in the positions indicated in the upper row here):*

$$H = \begin{bmatrix} p_1 & p_2 & m_3 & p_4 & m_5 & m_6 & m_7 \\ 0 & 0 & 0 & 1 & 1 & 1 & 1 \\ 0 & 1 & 1 & 0 & 0 & 1 & 1 \\ 1 & 0 & 1 & 0 & 1 & 0 & 1 \end{bmatrix}$$

Let us assume the error vector to be $e = 0000100$. Then

$$S = eH^T = [0000100] \begin{bmatrix} 0 & 0 & 1 \\ 0 & 1 & 0 \\ 0 & 1 & 1 \\ 1 & 0 & 0 \\ 1 & 0 & 1 \\ 1 & 1 & 0 \\ 1 & 1 & 1 \end{bmatrix} = [101]$$

This directly indicates that error has affected the position 5, which happened to be a message bit m_5. The parity check matrix for the systematic cyclic Hamming code, generated by $g(x) = 1 + x + x^3$, is formed as follows:

$$H = \begin{bmatrix} 1 & 0 & 0 & 1 & 0 & 1 & 1 \\ 0 & 1 & 0 & 1 & 1 & 1 & 0 \\ 0 & 0 & 1 & 0 & 1 & 1 & 1 \end{bmatrix}$$

The Hamming code can also be defined for q-ary alphabets, where $q \neq 2$. The parameters of this code are $(n, k) = ((q^r - 1)/(q-1)), (q^r - 1)/(q-1)) - r)$.

Example 19.27 *If $q = r = 3$, then*

$$(n, k) = ((3^3 - 1/3 - 1), (3^3 - 1/3 - 1) - 3) = (13, 10)).$$

19.7.2   BCH codes

The BCH code is a generalization of the Hamming code, that allows multiple error correction. BCH codes are important because at a block length of a few hundred, they outperform all other block codes with the same blocklength and code rate. The most commonly used BCH codes employ a binary alphabet and a codeword length of $n = 2^r - 1$, where $r = 3, 4, \ldots$.

Finite field properties

The following properties of a field of q elements, $GF(q)$, are important in the description of BCH codes.

Finite field properties

Property 1: The $q - 1$ nonzero elements form an Abelian group under multiplication. Let β^i be the multiplication of a string of i $\beta's, \beta \ldots \beta$, and let β^{-i} be the multiplicative inverse. The smallest positive r such that $\beta^r = 1$ is called the *order* of the element β.

Property 2: Every field contains at least one element α, whose order is $q-1$ such that $\alpha, \alpha^2, \ldots, \alpha^{q-1}$ are the $q - 1$ distinct nonzero elements. The element α is called a *primitive element.*

Property 3: The order of a field must be a power of a prime (i.e., $q = p^m$, where p is a prime).

Property 4: Any $GF(p)$, where p is prime, is isomorphic to the field of integers modulo p.

Property 5: Any $GF(p^m)$, where p a prime, is isomorphic to the field of polynomials over $GF(p)$ and irreducible polynomial of degree m.

Property 6: All $q-1$ nonzero elements of the field are the $q-1$ roots of $x^{q-1} - 1$.

Property 7: Let α be a primitive element of $GF(p^m)$, where p is a prime, and let i_j be an element of $GF(p)$. Then any $\beta \in GF(p^m)$ can be represented as follows: $\beta = i_0 + i_1\alpha + \cdots + i_{m-2}\alpha^{m-2} + i_{m-1}\alpha^{m-1}$.

Example 19.28 *Consider the field $GF(2^3)$. According to property 5, it has the structure of the field of polynomials. Seven polynomials are the seven nonzero field elements (polynomials) over $GF(2^3)$:*

$$\alpha = \alpha$$
$$\alpha^2 = \alpha^2$$
$$\alpha^3 = \alpha^3 = 1 + \alpha$$
$$\alpha^4 = \alpha + \alpha^2$$
$$\alpha^5 = \alpha^2 + \alpha^3 = 1 + \alpha + \alpha^2$$
$$\alpha^6 = \alpha + \alpha^2 + \alpha^3 = \alpha + \alpha^2 + 1 + \alpha = 1 + \alpha^2$$
$$\alpha^7 = \alpha + \alpha^3 = \alpha + 1 + \alpha = 1$$

Example 19.29 *The shift-register performs multiplication by $\alpha = y$ modulo $1 + y + y^3$ from left to right. The shift-register contents from left to right are the coefficients (a_0, a_1, a_2) of the field element $a_0 + a_1 y + a_2 y^2$ (Figure 19.15). Multiplication by y is a shift to the right, but in the case of the term y^2, the feedback connections allow replacing it with $1 + y$. This is exactly what is required to perform the modulo $1 + y + y^3$ operation. If one starts with $\alpha = y$ and (010) in the feedback register, the successive shifts produce all seven nonzero field elements. Further shifts would repeat the pattern.*

It can be observed that there are three different ways to represent a nonzero field element of $GF(p^m)$:

(a) By a power of a primitive element

(b) By a polynomial over $GF(p^m)$: $a_0 + a_1 y + \cdots + a_{m-1} y^{m-1}$

(c) By a vector of m components: $a_0, a_1, \ldots, a_{m-1}$

BCH code properties

A general BCH code is specified by the parameters q, r, α, v, and d. Here, r is the number of q-ary check symbols, d is a minimal distance of the code, an element α is in the field $GF(q^r)$, and the code consists of all polynomials over $GF(q)$ of degree $n - 1$, or less, which have the following roots: $\alpha^v, \alpha^{v+1}, \ldots, \alpha^{v+d-2}$. The code length n is the order of α. Usually, α is primitive, and $n = q^r - 1$.

The code can be specified in the matrix form as follows:

$$[c_0 c_1 \ldots c_{n-1}] \begin{bmatrix} \alpha^0 & \alpha^0 & \cdots & \alpha^0 & \alpha^0 \\ \alpha^{v+d-2} & \alpha^{v+d-3} & \cdots & \alpha^{v+1} & \alpha^v \\ \vdots & \vdots & \cdots & \vdots & \vdots \\ \alpha^{(n-1)(v+d-2)} & \alpha^{(n-1)(v+d-3)} & \cdots & \alpha^{(n-1)(v+1)} & \alpha^{(n-1)v} \end{bmatrix} = 0$$

It can be shown that the number of binary check symbols, as well as the degree of the generator polynomial $g(x)$, is less than or equal to $r(d-1)/2$. And also, the minimum distance between any two codewords is at least d. Thus, the code can correct all patterns, containing up to $t = \lfloor \frac{d-1}{2} \rfloor$ errors.

19.7.3   Reed–Solomon codes

A Reed–Solomon (RS) code operates over entire characters instead of only a few bits. In an RS code, the parity bytes are appended to a block of information bytes. The RS (n, k) code is defined using the following parameters: (a) l – the number of bits in a character or symbol, (b) k: the number of l-bit characters comprising the data block, and (c) n: the number of bits in the codeword. The RS(n, k) code can correct errors in the k information bytes. For example, the popular RS(255, 223) code uses 223 8-bit information bytes and 32 check bytes to form a 255-byte codeword. It is able to correct as many as 16 erroneous bytes in the information block.

RS codes are nonbinary codes capable of multiple error corrections. RS codes are constructed using the elements of a finite field with q elements, the *Galois field GF(q)*. The number of elements in the Galois field is p^m, where p is prime number and m is a positive integer. Given $p = 2$, the Galois field elements are derived using a polynomial $p(x)$ of degree m such that the powers of a primitive element α up to $2^m - 2$ are distinct, and $\alpha^{2^m - 1} = 1$. Thus, $0, 1, \alpha, \alpha^2, \ldots, \alpha^{2^m} - 2, \alpha^{2^m} - 1$ is the set of 2^m field elements. As mentioned earlier, each element in the field can be expressed as the sum of the elements $1, \alpha, \alpha^2, \ldots, \alpha^m - 1$.

> **Example 19.30** *Let α be a primitive element of the polynomial $p(x) = x^3 + x + 1$. Then, $\alpha^3 + \alpha + 1 = 0$ or $\alpha^3 = \alpha + 1$. The elements of the field are $0, 1, \alpha, \alpha^2, \alpha^3 = \alpha + 1, \alpha^4 = \alpha(\alpha + 1) = \alpha^2 + \alpha, \ldots \alpha^7 = \alpha^3 + \alpha = \alpha + 1 + \alpha = 1$ (see Example 19.28).*

RS codes are a class of BCH codes where $r = 1$ and $q = p^i$, and p is a prime (in most cases $p = 2$). Then, since $r = 1$, the field of coefficients is the same as the field of roots. To specify a code with the minimum distance d, the generator polynomial is selected as

$$g(x) = \prod_{i=v}^{v+d-2} (x - \alpha^i)$$

The code length is $q - 1$, and the number of check digits is $d - 1$. The minimum distance, d, is the greatest possible for any code having $d - 1$ check digits. Note that the digits here are q-ary, not binary.

> **Example 19.31** *Suppose that $q = 8, d = 4$, and $v = 1$. Let $GF(2^3)$ be polynomials modulo $1 + y + y^3$ over $GF(2)$, with $\alpha = y$. The code length is seven octal digits, with three octal check digits. It can correct all single errors and detect all double errors. It is required that $g(x)$ have the roots $\alpha = y, \alpha^2 = y^2$, and $\alpha^3 = y + 1$. Therefore, $g(x) = (x - y)(x - y^2)(x - y - 1) = 1 + y^2 + yx + (1 + y^2)x^2 + x^3$, or $g(x) = (1 + \alpha^2) + \alpha x + (1 + \alpha^2)x^2 + x^3$.*

The RS (n, k) code is defined as follows: $(n, k) = (2^l - 1, 2^l - 1 - 2t)$, where t is the symbol-error correcting capability of the code, and $n - k = 2t$

is the number of parity symbols. The minimum distance of this code is $d = n - k + 1$. The code is capable of correcting any combination of t or fewer errors, where $t = \lfloor \frac{d-1}{2} \rfloor = \lfloor \frac{n-k}{2} \rfloor$. This equation implies that the RS codes require no more than $2t$ parity symbols to correct t symbols: $g(x) = g_0 + g_1 x + g_2 x^2 + \cdots + g_{2t-1} x^{2t-1} + x^{2t}$.

> **Example 19.32** *For the double-error correcting RS code, $2t = 4$. Let the field-generating polynomial be $1 + y + y^3$. Then, to perform multiplication of field elements, the following RS polynomial is derived: $g(x) = (x - \alpha)(x - \alpha^2)(x - \alpha^3)(x - \alpha^4) = \alpha^3 + \alpha^1 x + \alpha^0 x^2 + \alpha^3 x^3 + x^4$.*

Encoding and decoding

Since RS codes are cyclic codes, their encoding in systematic form is performed similarly to the general encoding procedure for binary codes. That is, a message polynomial $m(x)$ is shifted into rightmost k stages of the encoding register, and then a parity polynomial $p(x)$ is placed to the leftmost $n - k$ stages. Shifting right will be equivalent to multiplication $m(x)$, by x^{n-k}. To obtain a parity polynomial, the shifted message polynomial is divided by $g(x)$. The remainder will constitute $p(x)$. In other words, $p(x) = x^{n-k} m(x) \bmod g(x)$, and the resulting code polynomial will be $U(x) = p(x) + x^{n-k} m(x)$.

> **Example 19.33** *Let us encode a message polynomial using the generator polynomial for RS code obtained in Example 19.32. The input to be encoded is a three-symbol message. The parameters of the code are $(n, k) = (2^3 - 1, 2^3 - 1 - 2 \times 2) = (7, 3)$. Let the message be $\underbrace{010}_{\alpha^1} \underbrace{110}_{\alpha^3} \underbrace{111}_{\alpha^5}$. Therefore, the message polynomial becomes $m(x) = \alpha^1 x^0 + \alpha^3 x^1 + \alpha^5 x^2$. The generator polynomial is $g(x) = \alpha^3 + \alpha^1 x + \alpha^0 x^2 + \alpha^3 x^3 + x^4$. The message polynomial is multiplied by $x^{n-k} = x^4$, and then the product polynomial is divided by the generator polynomial. Using field arithmetic, the remainder after this division can be found as $(\alpha^1 x^4 + \alpha^3 x^5 + \alpha^5 x^6) \bmod g(x) = \alpha^0 + \alpha^2 x + \alpha^4 x^2 + \alpha^6 x^3$. This remainder is added with the right-shifted message polynomial, giving the codeword $U(x) = \alpha^0 + \alpha^2 x + \alpha^4 x^2 + \alpha^6 x^3 + \alpha^1 x^4 + \alpha^3 x^5 + \alpha^5 x^6$. The encoding procedure can be performed using the logic network presented in Figure 19.16.*

Consider the decoding procedure. Suppose there are ν errors in the codeword at locations $x^{j_1}, x^{j_2}, \ldots, x^{j_\nu}$. Then, the error polynomial can be represented as follows:

$$e(x) = e_{j_1} x^{j_1} + e_{j_2} x^{j_2} + \ldots + e_{j_\nu} x^{j_\nu} \tag{19.5}$$

The indices $1, 2, \ldots, \nu$ refer to the error positions, whereas the index j refers to the error location. To correct the corrupted codeword, each error value e_{j_l} and its location x^{j_l}, where $l = 1, 2, \ldots, \nu$, must be determined. An *error locator* number is defined as $\beta_l = \alpha^{j_l}$. Next, the $n - k = 2t$ syndrome symbols are obtained by substituting α^i into the received polynomial for $i = 1, 2, \ldots, 2t$:

$$S_1 = Z(\alpha) = e_{j_1} \beta_1 + \cdots e_{j_\nu} \beta_\nu$$
$$\vdots \tag{19.6}$$
$$S_{2t} = Z(\alpha^{2t}) = e_{j_1} \beta_1^{2t} + \cdots e_{j_\nu} \beta_\nu^{2t}$$

There are $2t$ unknowns and $2t$ equations. However, these equations cannot be solved in the usual way because they are nonlinear. Any technique that solves this system of equations is known as a RS decoding algorithm.

When a nonzero syndrome vector $S = S_1 S_2 \ldots S_{2t}$ has been computed, it indicates that errors have been received. Next, it is necessary to locate errors and correct them. An *error-locator* polynomial can be defined as

$$\sigma(x) = (1 + \beta_1 x)(1 + \beta_2 x) \cdots (1 + \beta_\nu x) = 1 + \sigma_1 x + \sigma_2 x^2 + \cdots + \sigma_\nu x^\nu \tag{19.7}$$

The roots of $\sigma(x)$ are $1/\beta_1, 1/\beta_2, \ldots, 1/\beta_\nu$. The reciprocal of the roots of $\sigma(x)$ are the error-location numbers of the error pattern $e(x)$. If coefficients of $\sigma(x)$ were known, the zeroes of $\sigma(x)$ could have been found to obtain the error locations. Therefore, let us first compute the locator coefficients $\sigma_1, \sigma_2, \ldots, \sigma_\nu$ from syndromes. It can be shown that after applying some transformations to Equation 19.7, we can derive the following set of equations:

$$\sigma_1 S_{j+\nu-1} + \sigma_2 S_{j+\nu-2} + \cdots + \sigma_{nu} S_j = -S_{j+\nu}, j = 1, \ldots, 2t - \nu$$

Writing the first ν of these equations in matrix form will yield

$$\begin{bmatrix} S_1 & S_2 & S_3 & \cdots & S_{\nu-1} & S_\nu \\ S_2 & S_3 & S_4 & \cdots & S_\nu & S_{\nu+1} \\ & & & \vdots & & \\ S_{\nu-1} & S_\nu & S_{\nu+1} & \cdots & S_{2\nu-3} & S_{2\nu-2} \\ S_\nu & S_{\nu+1} & S_{\nu+2} & \cdots & S_{2\nu-2} & S_{2\nu-1} \end{bmatrix} \begin{bmatrix} \sigma_\nu \\ \sigma_{\nu-1} \\ \vdots \\ \sigma_2 \\ \sigma_1 \end{bmatrix} = \begin{bmatrix} -S_{\nu+1} \\ -S_{\nu+2} \\ \vdots \\ -S_{2\nu-1} \\ -S_{2\nu} \end{bmatrix}$$

This equation can be solved with respect to $\sigma_1, \sigma_2, \ldots, \sigma_\nu$ by inverting the left product matrix provided the matrix is nonsingular. It can be proved that if there are t errors, then the matrix is nonsingular. This can be used to determine the number of errors that actually occurred (which may be less than t). Generally, we start from the maximum number of potential errors, t.

If the matrix is nonsingular, we proceed further. If it is singular, we reduce the number of errors to $t - 1$, and check if the reduced matrix is singular. This procedure continues until a nonsingular matrix is obtained. The system of equations is then solved by inverting the matrix.

Next, attempt to find the zeroes of $\sigma(x)$ by trial and error. The error locations are the reciprocals of the zeroes. Once locations are found, the values of errors can be determined using the Equation 19.5. The considered procedure is known as the *Peterson–Gorenstein–Zierler* decoding.

Example 19.34 *Let us assume that the double-symbol error has occurred in the codeword derived in the previous example, such that*

$$e(x) = 0 + 0x + 0x^2 + \alpha^2 x^3 + \alpha^5 x^4 + 0x^5 + 0x^6$$

The received word is

$$Z(x) = \alpha^0 + \alpha^2 x + \alpha^4 x^2 + \alpha^0 x^3 + \alpha^6 x^4 + \alpha^3 x^5 + \alpha^5 x^6$$

The syndrome calculation yields

$$S_1 = Z(\alpha) = \alpha^3, \quad S_2 = Z(\alpha) = \alpha^5$$
$$S_3 = Z(\alpha) = \alpha^6, \quad S_4 = Z(\alpha) = 0$$

Since $S \neq 0$, the next step is initiated. Applying the procedure explained earlier, we will obtain

$$\begin{bmatrix} S_1 & S_2 \\ S_2 & S_3 \end{bmatrix} \begin{bmatrix} \sigma_2 \\ \sigma_1 \end{bmatrix} = \begin{bmatrix} S_3 \\ S_4 \end{bmatrix} \quad or \quad \begin{bmatrix} \alpha^3 & \alpha^5 \\ \alpha^5 & \alpha^6 \end{bmatrix} \begin{bmatrix} \sigma_2 \\ \sigma_1 \end{bmatrix} = \begin{bmatrix} \alpha^6 \\ 0 \end{bmatrix}$$

As it can be seen, the matrix

$$\begin{bmatrix} \alpha^3 & \alpha^5 \\ \alpha^5 & \alpha^6 \end{bmatrix}$$

is nonsingular, which indicates that there were two errors. This system is solved by multiplying both sides of the aforementioned equation by the inverse of the nonsingular matrix. Finally, we will obtain

$$\begin{bmatrix} \sigma_2 \\ \sigma_1 \end{bmatrix} = \begin{bmatrix} \alpha^0 \\ \alpha^6 \end{bmatrix}$$

Having derived the values of σ_1 and σ_2 and using Equation 19.7, we will get $\sigma(x) = \alpha^0 + \sigma_1 x + \sigma_2 x^2 = \alpha^0 + \alpha^6 x + \alpha^0 x^2$.

The roots of $\sigma(x)$ are the reciprocals of the error locations. Once these roots are located, the error locations will be known. By exhaustive search, it can be determined that only $\sigma(\alpha^3) = \sigma(\alpha^4) = 0$. Therefore, the error are located in the inverse of these roots. The equation $\sigma(\alpha^3) = 0$ implies that one root exists at $1/\beta_l = \alpha^3$. Thus, $\beta_l = 1/\alpha^3 = \alpha^4$. Similarly, $\sigma(\alpha^4) = 0$ implies that another root exists at $1/\beta_{l'} = 1/\alpha^4 = \alpha^3$, where l and l' refer to the first and the second errors, respectively. Since there are 2-symbol errors here, the error polynomial is of the form: $e(x) = e_{j_1} x^{j_1} + e_{j_2} x^{j_2}$. After these two errors were found at locations α^3 and α^4, the error values can be calculated. An error has been denoted e_{j_l}, where the index j refers to the error locations and the index l identifies the l-th error. Since each error value is assigned to a particular location, the notation can be simplified by denoting e_{j_l} as e_l. In order to determine the error values e_1 and e_2 associated with the locations $\beta_1 = \alpha^3$ and $\beta_2 = \alpha^4$, any of the four possible syndrome equations can be used. It follows from Equation 19.6 that

$$S_1 = Z(\alpha) = e_1\beta_1 + e_2\beta_2 = e_1\alpha^3 + e_2\alpha^4$$
$$S_2 = Z(\alpha^2) = e_1\beta_1^2 + e_2\beta_2^2 = e_1(\alpha^3)^2 + e_2(\alpha^4)^2$$

Solving this equation yields $\begin{bmatrix} e_1 \\ e_2 \end{bmatrix} = \begin{bmatrix} \alpha^2 \\ \alpha^5 \end{bmatrix}$.

Thus, the estimated error polynomial is $\hat{e}(x) = \alpha^2 x^3 + \alpha^5 x^4$. This is added to the received word, producing the transmitted codeword. It ultimately delivers a decoded message, which is exactly the message that was chosen at the beginning of this example.

Example 19.35 $Z(x) + \hat{e}(x) = \alpha^0 + \alpha^2 x + \alpha^4 x^2 + \alpha^0 x^3 + \alpha^6 x^4 + \alpha^3 x^5 +$
$$\alpha^5 x^6 + \alpha^2 x^3 + \alpha^5 x^4 =$$
$$\alpha^0 + \alpha^2 x + \alpha^4 x^2 + \alpha^6 x^3 + \alpha^1 x^4 + \alpha^3 x^5 + \alpha^5 x^6 = U(x)$$

19.8   Arithmetic codes

Arithmetic codes have been proposed for checking errors that occur during arithmetic operations, as well as for transmission and memory protection. In contrast to algebraic codes considered so far, in these codes, the codewords are integer numbers and ordinary arithmetic operations apply. Arithmetic codes are of special interest in fault-tolerant computing, since they serve to detect or correct errors in the results produced by arithmetic units of computers. Real-time detection of transient and permanent faults can be obtained with some degree of redundancy, but without duplication of the arithmetic units.

The measure of effectiveness is the listing of error values that can be detected or corrected when the code is utilized. These values are determined by the properties of the code and are independent of the network logic structure.

Example 19.36 *An n-digit r-radix code is said to be a single error-detecting or correcting one, if all single errors are detected or corrected in its n-digit radix-r numbers.*

An integer can be expressed in a radix r form as

$$N = a_{n-1}r^{n-1} + a_{n-2}r^{n-2} + \cdots + a_0, \quad 0 \leq a_i < r, \quad i = 0, 1, \ldots, n-1$$

The arithmetic weight of N, denoted by $W(N)$, is the minimum number of nonzero terms when expressed in the form

$$N = b_{n-1}r^{n-1} + b_{n-2}r^{n-2} + \cdots + b_0, \quad 0 \leq b_i \in [0, \pm 1, \pm 2, \ldots, 1 \pm (r-1)]$$

It can be observed that $W(N) = W(-N)$. For binary arithmetic, $r = 2$, and $b \in (0, 1, -1)$.

> **Example 19.37** *The arithmetic weight of 14 in binary is 2, because*
> $$14 = 1 \cdot 2^4 + 0 \cdot 2^3 + 0 \cdot 2^2 - 1 \cdot 2^1 + 0 \cdot 2^0.$$

The arithmetic distance between two numbers N_1 and N_2 is defined as the arithmetic weight of $N_1 - N_2$.

> **Example 19.38** *Given $r = 2$, the distance between 31 and 39 is 1, because $31 - 39 = -8 = -2^3 = -1 \times 2^3$.*

The definition of arithmetic distance previously matches very closely to the types of errors that can occur in an arithmetic operation.

> **Example 19.39** *Consider addition of two numbers 31 and 2 in binary (i.e., $011111 + 000010 = 100001$). Suppose that there is a carry failure at the second digit. Then the result will be $011101 = 29$. This carry failure changes four digits of the result. However, the failure is counted as a single error because the arithmetic weight of $33 - 29 = 4 = 2^2 - 1 \times 2^2$ is defined to be 1.*

Let C be an arithmetic code. It can be shown that

(a) The code C is capable of detecting d or fewer errors if and only if the arithmetic distance of C is at least $d + 1$.

(b) The code C is capable of correcting t or fewer errors if and only if the arithmetic distance of C is at least $2t + 1$.

19.8.1 AN codes

The simplest arithmetic code is the AN code. These codes are formed by multiplying the dataword by a number that is not a power of the radix of the representation (such as two for binary numbers). The redundancy is determined by the chosen multiplier, called the *modulus*.

In an AN code, a given integer N is represented by the product AN and a constant A, also called the *check base*.

Consider the addition of two numbers N_1 and N_2. The sum of the coded numbers is $AN_1 + AN_2 = A(N_1 + N_2)$, which is equal to the coded form of their sum. Therefore, the coded numbers can be added by an ordinary adder. If an error e occurs in the sum, the result R is

$$R = A(N_1 + N_2) + e = AN_3 + e$$

To check for errors, divide R by A and find the remainder. If $e = 0$, there will be no remainder. Designating R mod A as $|R|_A$, we obtain

$$|R|_A = |A(N_1 + N_2) + e|_A = |e|_A$$

Here $|e|_A$ is called the *syndrome*. A nonzero syndrome indicates that errors have occurred.

The syndromes, corresponding to the single errors, are of the type $|Ar^i|_A$, where $1 \le a < r$. In order to detect all single errors, the syndromes corresponding to the single-error patterns must all be nonzero. If A and r are relatively prime, then $|Ar^i|_A \ne 0$. In particular, r and $r+1$ are relatively prime, and, hence, if the check base $A = r + 1$, the AN code will be capable of detecting any single error. The number of extra digits needed to provide this error detection is no larger than 2, since $2 > \log_r(r+1)$. This is true for any number of digits in the codeword.

The class of codes with $A = r^c - 1$ for some positive constant c has certain advantages, because the residue modulo A can be obtained without division. Let $U = (a_{n-1}a_{n-2} \ldots a_0)$ be a codeword of such an AN code. Let this sequence be partitioned into l subsequences of c digits each as follows:

$$U = (a_{n-1} \ldots a_{(i-l)c} | \ldots | a_{2c-1} \ldots a_c | a_{c-1} \ldots a_0) =$$
$$B_{l-1}r^{(l-1)c} + B_{l-2}r^{(l-2)c} + \cdots + B_1 r^c + B_0$$

where $B_i = a_{ic} + a_{ic+1}r + a_{ic+c-1}r^{c-1}$. Since $|r^{jc}|_{r^c-1} = 1$ for all $j \ge 0$, then

$$|U|_{r^c-1} = |B_{l-1} + B_{l-2} + \cdots + B_0|_{r^c-1}$$

It follows from the last expression that the residue can be obtained by adding the bytes modulo $r^c - 1$ by a parallel adder instead of dividing long sequences of bits using an arithmetic dividing circuits. These circuits are similar to shift registers but possess more complicated structure due to arithmetic carries. This scheme is referred to as a low cost implementation. Here, bytes consist of groups of bits that are not necessarily multiples of 8.

Example 19.40 *If $A = 7$ and $N = (100011110)$, then $110 + 011 + 100 = 1101$. Adding the carry out bit to the least significant bit yields $101 + 001 = 110$. Since N is 286 decimal, then $|286|_7 = 6$, which is the decimal equivalent of 6.*

In order to correct all of some set of error patterns, the syndromes corresponding to each pair of error patterns in the set, must be different and nonzero. The syndromes, corresponding to the single errors are of type $|\pm ar^j|_A$ for $0 \le j \le n-1$, so an appropriate check base A must be selected to get the distinct nonzero syndromes.

Assume that A is a prime. Then the integers modulo A form a field. Suppose that 2 is a primitive element of the field. Then its order is $A-1$, and $2^0, 2^1, 2^2, \ldots, 2^{(A-1)/2} = -1, -2, \ldots, -2^{(A-3)/2} = 2^{A-2}$ are all distinct.

> **Example 19.41** *Number 2 is a primitive element of order 10 of GF(11), and $2^0 = 1, 2^1 = 2, 2^2 = 4, 2^3 = 8, 2^4 = 5, 2^5 = 10 = -1, 2^6 = 9 = -2, 2^7 = 7 = -2^2, 2^8 = 3 = -2^3, 2^9 = 6 = -2^4$ are all distinct.*

If the range of integers is restricted so that $AN < 2^{(A-1)/2}$, the syndromes, corresponding to single errors, are all different. Therefore, an AN code given a prime A, a primitive element 2 of $GF(A)$, and the maximum integer range less than $2^{(A-1)/2}$ is capable of correcting all single errors.

> **Example 19.42** *The range of binary coded decimal (BCD) numbers is from 0 to 9. 19 is a prime with a primitive root 2 and the maximum integer range $9 \cdot 19 < 2^{(19-1)/2} = 512$. Thus a 19N code of BCD numbers is capable of correcting all single errors.*

19.8.2 Separate codes

A *separate* code is the code in which the message part and the check part are processed separately. The advantage of a separate code is that the arithmetic and checking operation can be done in parallel. Therefore, the speed of the system is not deteriorated by the addition of redundant bits to the check part.

AN codes, described the previous section, are nonseparate. However, every AN code has a corresponding mod A separate code with the same distance and essentially the same redundancy, as shown here.

Let $s = \lceil \log_r A \rceil$. Then every integer can be represented as an s-digit number in radix-r form. Consider the code in which the check symbol $C(N)$ for N is the residue modulo $-r^s N \bmod A$. Then $-r^s N = Aq + C(N)$ for some q, or $r^s N + C(N) = -Aq$. Since $r^s N + C(N)$ is a multiple of A, this is a codeword in the AN code. But if $r^s N + C(N)$ is represented in radix r, the low s digits correspond to the check $C(N)$ part and the higher-order digits correspond to the message part N. So, this is a separate code. Moreover, the minimum distance of this separate code is at least the minimum distance of the AN code.

19.8.3   Residue codes

The desirable property of residue codes is that for the arithmetic operations such as addition, subtraction, and multiplication, the check bits of the result can be determined from the check bits of the operands. The check bits are defined as follows: $Check = N \bmod m$, where m is the residue of the code and N is the number of information bits. The number of check bits is equal to $\lceil \log_2 m \rceil$.

Example 19.43 *Consider three information bits I_0, I_1, and I_2, and $m = 3$. Since $\lceil \log_2 m \rceil = \lceil \log_2 3 \rceil = 2$, two check bits $Check_0$ and $Check_1$ are needed. They are calculated as follows. For $I_0 = I_1 = I_2 = 0$ $(N = 0)$, $Check = Check_1 Check_0 = 0 \bmod 3 = 0$. For $I_0 = 1$, $I_1 = I_2 = 0$ $(N = 1)$, $Check = Check_1 Check_0 = 1 \bmod 3 = 1$, and so on.*

Information bits			Check bits	
I_2	I_1	I_0	$Check_1$	$Check_0$
0	0	0	0	0
0	0	1	0	1
0	1	0	1	0
0	1	1	0	0
1	0	0	0	1
1	0	1	1	0
1	1	0	0	0
1	1	1	0	1

Figure 19.17 shows the implementation of the adder with error detection and correction using the residue code.

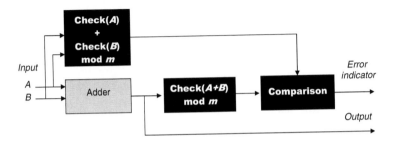

FIGURE 19.17

Application of the residue code for error detection in the adder (Example 19.43).

In a residue code with an odd m, all single bit errors can be detected. As m increases, the number of the required check bits increases. The error-detecting capability remains unchanged for single bit errors.

Figure 19.18 illustrates the principles of implementation of error-detecting and error-correcting codes. The type of code to be used depends on the type of logic network.

Example 19.44 *For data transmission buses, a parity check code can be useful. However, this code is not acceptable for adders and other arithmetic networks. This is because the value of the parity check bits for the sum operation cannot be determined from the check bits of the operands.*

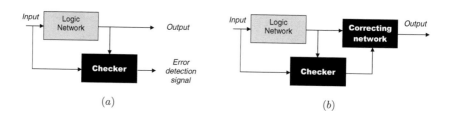

(a)   (b)

FIGURE 19.18
Implementation of error detection (a) and error correction (b) codes.

Example 19.45 *Let $A = 0001$, $B_1 = 0101$, and $B_2 = 0011$. The check bits for the sums $A + B_1 = 0110$ and $A + B_2 = 0100$ are as follows: $Check(A) = 1$, $Check(B_1) = Check(B_2) = 0$, $Check(A+B_1) = Check(A+B_2) = 1$. Thus, addition does not preserve parity, because the parity check must be be performed after each addition.*

19.9　Further study

Advanced topics of error detection and correction

Topic 1: Fault-tolerant design. *Hardware redundancy* can be employed in several forms. *Static* redundancy achieves fault tolerance without actually detecting any faults. *Dynamic* redundancy has a built-in fault detection mechanism and the capability to recover from the effect of a fault.

Hybrid redundancy utilizes the features of static and dynamic redundancy approaches. The degree of reliability improvement using error correction codes depends on the failure modes of a device. For example, a if soft error (random, nonrecurring changes in memory logic states) of a single cell is the dominant failure mode in a memory device, they could be eliminated by using error-checking and error-correction codes. Hard errors (permanent errors due to physical defects in memory) can affect the rows and column of a memory RAM-like hierarchy.

System-level fault detection is aimed at identifying a faulty processing device and or a subsystem. Self-checking devices based on combinational logic networks allow detection of errors not only in the system under test but also in the checking device itself. These devices are normally oriented to the detection of single faults.

Topic 2: Concatenate codes to reduce errors. Block (usually Reed–Solomon) and convolutional codes are frequently combined in concatenated coding schemes. A block code of q-ary symbols of length kK can be broken into K subblocks, each of k symbols, where each subblock is viewed as an element from a q^k-ary alphabet. A sequence of K such q^k-ary symbols can be first encoded with an (N, K) code over $GF(q^k)$ to form a sequence of N such q^k-ary symbols. Then each of the N q^k-ary symbols can be reinterpreted as kq-ary symbols and coded with an (n, k) q-ary code. In this way, a concatenated code has two distinct levels of coding, an inner code and an outer code. The strength of concatenated code is in that they can successfully decode highly broken messages.

Topic 3: Turbo codes. The most recent development in error correction is turbo coding, a scheme that combines two or more relatively simple convolutional codes. Turbo codes are a class of high-performance error correction codes, which allow to achieve maximal information transfer over a limited-bandwidth communication link in the presence of data-corrupting noise.

Turbo codes and low-density parity-check codes come closest to approaching the Shannon limit, the theoretical limit of maximum information transfer rate over a noisy channel. Turbo codes make it possible to increase data rate without increasing the power of a transmission, or they can be used to decrease the amount of power used to transmit at a certain data rate. Their main drawbacks are the relatively high decoding complexity.

Topic 4: Iterative codes. These codes consider multidimensional allocation of transmitted data. Encoding is provided, for example, by rows and by columns. These modifications improve the error-correcting capability of the code.

Topic 5: Error-locating codes. These are codes based on Kronecker product of two parity-check matrices: one of an error-detecting code and the other of an error-correcting code. They possess intermediate properties between error-detecting and error-correcting codes. They allow location of faulty subblocks of a codeword, but does not tell which particular bits are in error. The redundancy of these codes lies in between the redundancies of the corresponding error-detecting and error-correcting codes. These codes find their applications in digital systems diagnosis. They extract information from signatures to locate faulty units in a system under test.

Topic 6: Coding strategy Table 19.2 presents several functional classes of codes.

Depending on logic network characteristics, an appropriate code can be chosen. In particular,

Error detection codes are simple in the implementation; an error is detected, if the received message yields a nonzero syndrome.

Error-correcting codes: In the case of multiple-error correction, decoding complexity grows exponentially with inhering the number of corrected errors.

TABLE 19.2
Coding for error control in logic networks

Error control	Code
Single error detection	Capable of detection of any odd number of random errors. Simple implementation.
Single error detection or/and correction	Capable of detection/correction of any error affecting no more than one bit in a codeblock of n symbols.
Burst error correction	Capable of correcting any span of errors of a fixed length.
Random error correction	Capable of correcting multiple errors.

The most easily implemented are error-detecting codes protecting against random errors as well as burst errors. More complicated are decoding circuits for error-correcting codes. The complexity increases drastically with the growth of the number of errors that may occur simultaneously. Cyclic codes require much less hardware overhead compared with the encoding–decoding circuits for noncyclic codes.

Further reading

Recently, other codes have been developed that come very close to the Shannon's limit, such as turbo codes and low-density parity-check codes. Information about these codes can be found in [1] and [2].

More rigorous discussion on error-control coding can be found in [1]–[4]. Application of error-control codes to testing are studied in [5]. References [5]–[7] consider fundamentals of digital systems testing. References [8] and [9] cover fundamentals of testing, error-control coding, and its influence on testing. Arithmetic codes, circuits implementing decoding procedures for those codes, and self-checking circuits are discussed in [11].

References

[1] Blahut R. *Algebraic Codes for Data Transmission.* Cambridge University Press, Cambridge, UK, 2006.

[2] Lin S and Costello D. *Error Control Coding: Fundamentals and Applications.* Prentice Hall, NY, 2004.

[3] Peterson W and Weldon E. *Error-Correcting Codes.* MIT Press, Cambridge, MA, 1972.

[4] Sklar B. *Digital Communications: Fundamentals and Applications.* Prentice Hall, 2001.

[5] Abramovici M, Breuer M, and Friedman A. *Digital Systems Testing and Testable Design.* Wiley-IEEE, Computer Society Press, 1994.

[6] Bushnell M and Agrawal V. *Essentials of Electronic Testing for Digital, Memory, and Mixed-Signal VLSI Circuits.* Kluwer, Dordrecht, 2000.

[7] Mourad S and Zorian Y. *Principles of Testing Electronic Systems.* John Wiley and Sons, Hoboken, NJ, 2000.

[8] Rao T and Fujiwara E. *Error-Control Coding for Computer Systems.* Prentice Hall, New York, 1989.

[9] Pradhan D. *Fault-Tolerant Computing: Theory and Techniques.* Prentice Hall, Engle-Wood Cliffs, NJ, 1986.

[10] Von Neumann J. Probabilistic logics and the synthesis of reliable organisms from unreliable components. In Shannon CE and McCarthy J, Eds., *Automata Studies,* Princeton University Press, Princeton, NJ, pp. 43–98, 1956.

[11] Wakerly J. *Error Detecting Codes, Self-Checking Circuits and Applications.* Elsevier North-Holland, New York, 1978.

20

Natural Computing

Natural computing is the computational version of natural phenomena. In this chapter, natural computing principles are applied to perform Boolean function. Encoding of natural phenomena is the key aspect of molecular electronics. To apply the encoding, the *intermediate* data structure called the *logic primitives*, is needed. Intermediate structure reflects the property of phenomenon, from one side, and computational properties of Boolean functions, from another side. In this chapter, the techniques for interpretation the following physical and chemical phenomena using logic primitives are introduced

(*a*) *Quantum dot* phenomenon.

(*b*) *Complementary biomolecular* phenomena.

Biomolecular phenomena is often considered with respect to the computational paradigm based on *self-assembly*. Molecular self-assemblies are characterized by the topological properties such as symmetry, similarity, and periodicity. It is reasonable to use these properties in encoding a self-assembling phenomena in terms of Boolean function computing. In this chapter, a particular class of *fractals*, namely, *Sierpinsky* triangles and *transeunt* triangles are introduces as the appropriate models of self-assembly. This is because the relationships of these fractals and data structures for Boolean function representation are well established.

Another examples of techniques of encoding the molecular or physical phenomena used in natural computing, are introduced in this chapter as well:

(*a*) *Evolutionary-based computing*; this paradigm is inspired by the biological phenomena of structural evolving under the pressure of natural and artificial selection; and

(*b*) *Neural-based computing*; this paradigm is inspired by the biological phenomena of data processing in nerve networks.

20.1 Introduction

The area known as *biological inspiration for computing* refers to the computing paradigms based on biological analog of computing, that is, insights from biology that are relevant to aspects of computing. Instead of the traditional silicon-based computer technologies, natural computing utilizes the ideas from nature to develop computational systems or perform computation using natural media, for example, molecular structures. Note that these computational approaches are the highly simplified analogs of natural phenomena.

> **Example 20.1** *Artificial neural networks are based on computing paradigm inspired by the parallel processing capabilities and learning of nerve networks.*

Logic primitives for encoding physical and chemical phenomena.
Logic primitives are defined as an intermediate data structure between chemical or physical phenomena and data structures for Boolean function manipulation and implementation:

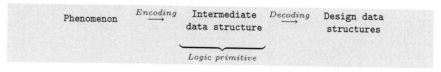

The result of physical or chemical process can be interpreted in terms of logical computing. To get this result, one might read the resulting structure and *decode* in terms of assembled logic primitives.

The *control* of such processes can be implemented at the level of physical or chemical phenomena. That is, the resulting structure must contain the designed properties of logic computing. The control at these levels can be achieved under some constraints. For this, an appropriate logic primitives and assembling rules must be chosen. The problem is formulated as the representation of the universal set of elementary Boolean functions using logic primitives. Using this universal set of functions, an arbitrary Boolean function can be implemented.

Logic primitives are specified with respect to the underlying phenomenon, in particular:

▶ In quantum dot techniques, logic primitives are called **quantum cells**.

▶ In complementary biomolecular computing, logic primitives are called **molecular primitives**.

▶ In DNA computing, the logic primitives are called **tiles**.

▶ In evolutionary logic design, the logic primitives are called **chromosomes**.

▶ In artificial neural network, the logic primitives are called **states**.

Quantum dot phenomenon also known as *quantum cellular automata (QCA)*, is a nanotechnology that has recently been recognized as one of emerging technologies with potential applications in future computers.

Quantum dots have been called *artificial atoms* or boxes for electrons because they have discrete charge states and energy-level structures. They are typically small electrically conducting regions of various topologies. Logic primitives are used for encoding this phenomenon in terms of Boolean function computing.

Example 20.2 *Two logic primitives are associated with the* **quantum dot** *phenomenon. These primitives can be used for computing an arbitrary Boolean network using such logic components as wire, complement, and majority function. Each logic primitive is represented by four quantum dots in a square array. Electrons are able to tunnel between the dots but are unable to leave this square array. The two ground-state polarizations are energeticall equivalent and can be interpreted as logic 0 or 1.*

Complementary biomolecular phenomena. Instead of electrical signals representing streams of binary 1s and 0s, the chemical concentrations of specific inducer molecules act as the input and output signals. These molecules interact with other molecules and bind specific molecular structures. This chemical process can be used to implement Boolean functions. The idea is to *encode* this chemical process in terms of *logic primitives*.

Example 20.3 **Molecular primitive** *is often defined as a switch (two-terminal molecular device). Logic gates (many-terminal molecular devices) can be designed by assembling molecular primitive.*

Evolutionary computation is a family of population-based techniques inspired by natural selection. *Evolutionary algorithms* are search and optimization procedures that find their origin and inspiration in the biological world. Evolutionary algorithms is a general term for related techniques that are based on the natural evolution paradigm, namely, genetic algorithms, evolutionary strategies, and evolutionary programming. In this chapter, we focus on the application of genetic algorithms for logic network design.

The main idea of genetic algorithms is that, in order for a population of individuals to collectively adapt to some environment, it should behave like a natural system; survival, and therefore reproduction, is promoted by the elimination of useless behavior and by rewarding useful behavior. The idea of evolutionary logic network design is very captivating because the designer

needs no special software to implement a given architecture and design style.
These are provided by evolutionary (adaptive) search for the network solution.

Example 20.4   *Evolutionary strategies can be distinguished, in particular, using the criterion of the level of Boolean network description* [36, 37, 47]. *For example, in* **evolutionary logic design**, *gate-level primitive components such as logic gates and flip-flops can be used. Another strategy is based on the evolution at the graph-based abstraction* [4]. *This model operates with data-flow graphs, each of them representing a logic network, and that is why this approach is applicable for complex logic network design.*

Various strategies can be used in evolutionary Boolean network design. In particular, in combinational logic network design, the gate-level functionality and inter-connectivity of the gates between network inputs and outputs are encoded by logic primitives, called *chromosomes*.

Example 20.5   *Encoding of the two-input AND gate results the following netlist of the chromosome:* $\boxed{0}\;\boxed{2}\;\boxed{-1}$, *where the first and the second input are connected to the network inputs 0 and 2, respectively, and the AND gate is encoded by the number -1.*

Neural computations. In the biological neuron, inputs are located in the synapses of the cell. Different inputs can provide different amounts of activation. A membrane potential appears as a result of the integration of the neural inputs. The action potential occurs only when the membrane potential is above a critical threshold level. The computational model of the biological neuron includes inputs and output with corresponding weights and threshold level.

Artificial neural networks are defined as information processing structures designed with inspiration taken from the nervous system. This is a massively parallel-distributed processing structure. Neurons can have *forward* and *feedback* connections to other neurons. These interconnected neurons give rise to forward computing of Boolean functions using neurons or threshold elements.

A small number of interconnected neurons can exhibit complex behaviors and information-processing capabilities that cannot be observed in single neurons. That is, information is *distributed* and processed in a parallel way. No single neuron is responsible for storing a complete data or solution; it has to be distributed over several neurons (connections) in the network. Data is stored in the connection strengths between neurons. The computation (learning) is defined as finding the appropriate connection strengths. Knowledge (data) is distributed over the connections among neurons. In Boolean function computing, the states of neurons (0s or 1s) are associated with the energy function, which can be mapped into the truth table of the implemented Boolean function. The values of Boolean function for given assignments of variables are distributed in neural network.

Example 20.6 *The* **Hopfield** *computing paradigmis is based on the concept of minimization of energy in a stochastic system. The minimization of an energy function is referred to as embedding a "correct solution" into the Hopfield network. For example, the minimization of energy function E_{AND} corresponds to the output of two-input AND function given assignment of input variables. The energy function can be considered as the encoded truth table of the AND function.*

20.2 Intermediate data structures

The structure of the intermediate data is defined by the encoding phenomena and logic design task. For example, logic primitives in the form of quantum cells conveys a specific information about the quantum dot phenomenon. Assembling of these cells are based on various design techniques (two-level or multilevel network, clocking, delay propagation, testing). Logic primitives in the form of chromosomes carry the features of genes assembling and mutation. From the other hand, the logic network can be designed under strong constraints of gene mutation; this is because an uncontrolled mutation can produces unacceptable logic networks.

There are a number of general techniques for constructing of logic primitives, such as logic and spatial specification, modeling, and verification.

Logic primitives

The *intermediate* data structure is defined as the form for representing the particular physical or chemical phenomenon; this form, called *logic primitives* is the intermediate between the phenomena and logic design data structures.

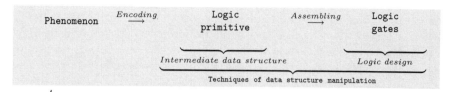

Intermediate data structure is characterized by such properties as the ability to be encoded, recognition by topological properties, rules for assembling, and ability to compose using these rules the universal set of logic operations.

> **Example 20.7** *It is reasonable to represent logic primitives in graphical form. This level of abstraction provides the correponding mechanism for establishing acceptable interpretation in terms of computing.*

Logic primitives as simplest encoding blocks can be interpreted in terms of computing using assembling:

Using the universal set of the elementary Boolean function, an arbitrary logic network can be designed.

Specification, functions, and assembling logic primitives

Logic primitives corresponding to various phenomena are different. To achieve the completeness of operation in terms of Post theorem, assembling rules for these primitive must be applied.

In Table 20.1, various physical and chemical phenomena and corresponding computing paradigms are introduced. These phenomena are encoded using intermediate data structure, – logic primitives. In the first phenomenon, logic primitive is specified as switch. Using this primitives, various universal sets of logic functions can be constructed by assembling.

In the second phenomenon, the library of includes a set of logic primitives, and each of them corresponds to the row a truth table. If the rules of spatial configuration are specified, an arbitrary Boolean function can be represented using these set of logic primitives. For example, this approach is used in the DNA molecular computing.

The third phenomenon in Table 20.1 is encoded by the two types of logic primitives (1-primitive and 0-primitive) under the constraints of spatial configuration. This encoding scheme, which is used for quantum-dot phenomena, results two computing components: the NOT function and the majority function. An arbitrary Boolean function can be implemented using these components.

TABLE 20.1
Phenomena and corresponding intermediate data structures (logic primitives)

| Phenomenon | Logic primitive | | Functionality |
	Specification	Function	of assembling
Phenomenon 1	Switch	Complement	An arbitrary logic network
Phenomenon 2	Two-input one-output primitives	The row of a truth table for two-input one-output Boolean function and the rules of spatial orientation	An arbitrary logic network
Phenomenon 3	1-primitive, 0-primitive	Constants 1 and 0, and rules of spatial orientation	An arbitrary logic network

Computational models

The computing paradigms that utilize an encoding using logic primitives and their assembling can be represented by various computational models. These models are different in the following:

Logical specification. An appropriate labels must be placed into layout of logic primitives to provide an efficient encoding and assembling.

Spatial specification. An appropriate rules for assembling of logic primitives must be defined from the properties of corresponding physical or chemical phenomena.

Verification of the correspondence of physical or chemical phenomena to the logic primitives and the functionality of the assembled structures:

For example, verification of self-assembled DNA structures was studied in [40].

Logical and spatial specification provide the mechanism for modeling and simulation various natural computing paradigms.

Example 20.8   *Assembling rules for logic primitives that utilize the phenomenon of quantum dot computing are complete. That is, these rules satisfy the Post theorem on completeness of logic operations because an arbitrary Boolean function can be implemented using the majority function and the complement (Figure 20.1).*

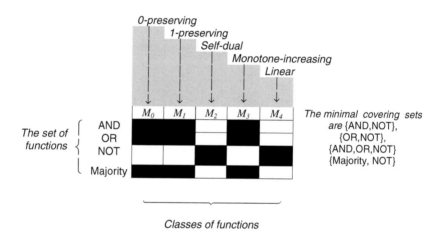

FIGURE 20.1

The universal sets for computing models (Example 20.8).

Figure 20.2 show the logic primitives formed based on quantum dot and complementary molecular phenomena. Each of logic primitives is characterized by particular logical and spatial specification. Logic primitives can be represented in graphical, tabulated, and algebraic forms.

Design for testability

Logic network models based on nanodevices often need to take into account the stochastic nature processing (for example, chemical reactions in cells) and nonpredictable faults caused by fabrication. The following examples can be considered as the motivation for the development stochastic models of Boolean networks.

Design example: logical and spatial specification of logic primitives

Logic primitives of quantum dot phenomenon

Logic primitive of complemenary biomolecular phenomenon

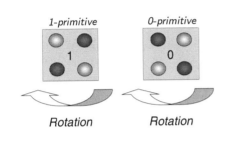

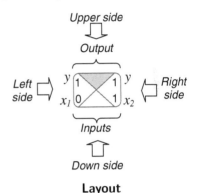

Layout

Square configuration with labeled sides

Layout

Square configuration with labeled sides

Computational model

Encoding. The 1-primitive and 0-primitive, each of specific topology, can be associated with logical constants 1 and 0, respectively.

Assembling model is based on Coulomb interaction. Logic primitives that are positioned diagonally from each other tend to anti-align. The tendency of a particular primitives assembly to move to a ground state ensures that it takes on the polarization of the majority of its neighbors.

Computational model

Encoding. Functionality of a primitive is defined by the assignment of input variables x_1 or x_2 and specified value (0 or 1) of the output $y = f(x_1, x_2)$.

Assembling model is based on a formal model of the molecular assembling, for example, crystal growth. Logic primitives can be assembled using four specifically labeled sides and rules of spatial orientation. An assembly of primitives carries information that are specified by labels.

FIGURE 20.2

Quantum dot and complementary biomolecular phenomena and their interpretation in terms of computational models.

Quantum dot-based Boolean networks. One of the possible technologies for the implementation of quantum dot phenomenon includes the following steps: (a) electron beam lithography, (b) metal deposition, and (c) oxidation. Any of these steps can introduce nonpredictable faults that are caused by defects. Defects during the electron beam lithography stage are most likely to be junctions in the wrong sport, over, or under exposed junctions; this may result in improper switching of cells. A defective cell could be either fluctuating (overexposed), or be stuck to a particular switching position (underexposed). In metal deposition stage, the most likely defect could be dots with wrong size and shape, thereby having an unwanted reduction of energy degeneracy.

Switching networks based on complementary biomolecular devices. There are various logic primitives for biomolecular phenomena. One of possible representation of logic primitives is given in Figure 20.2. This is a square configuration with labeled sides. Down and upper side is interpreted as input variables x_1, x_2 and output y of Boolean function, respectively.

20.3   Quantum dot phenomena encoding

The basic four dot quantum cell represent four possible positions for electrons placed in square configuration. Two extra electrons are loaded in each cell. They can tunnel between dots, but not through cell. Due to electrostatic repulsion, electrons are forced to reside in diagonally opposite dots of the square. Two energetically minimal equivalent arrangements exists for an isolated quantum cell (Figure 20.3).

Logic primitives

There two types of logic primitives in quantum dot phenomena encoding:

▶ *1-primitive*

▶ *0-primitive*

These primitives can be associated with logical constants 1 and 0, respectively. However, 1-primitive and 0-primitive (quantum cells) are distinguished by the configuration of quanted dots. The cells can be rotated by 45° and placed side by side.

Example 20.9   *A 90° cell can be placed between two 45° cells.*

The configurations convey an additional information that are used in assembling of logic primitives.

Interconnections

A useful feature of quantum dot phenomenon for Boolean function implementation is the ability to cross wires in the plane. In silicon-based technology, information is represented by voltages or currents, and wire crossings require a bridge out of the plane. In quantum dot cells different wires carrying different binary values can cross each other at the same level without any interference or crosstalk. Unlike silicon-based technology, where metal layers are used to implement connections without performing of any functions, the extra layers of quantum cells can be used as active components. It is believed that multi-layer quantum dot-based logic networks can potentially require less area than silicon-based planar implementations.

(a)   (b)

FIGURE 20.3
Logic primitives (quantum cells) of quantum-dot phenomenon: 1-cell (a) and 0-cell (b).

Assembling

Logic primitive interconnection in quantum dot phenomena is based on Coulomb interaction, that is, the state of one cell is transferred to its neighbors using physical phenomena of quantum-mechanical tunneling. This phenomena can be defined in terms of encoding as follows (Table 20.2):

Wire: Data is propagated using chain copies of 1-primitives. The binary signal propagates from input to output of the primitives because of the electrostatic interactions between primitives.

Complement: Logic primitives that are positioned diagonally from each other tend to anti-align; this effect is employed in inverter encoding. In Table 20.2, the layout of primitives for the input $x = 0$ is given.

Majority: This is a Boolean function of three variables

$$f = x_1 x_2 \vee x_1 x_3 \vee x_2 x_3$$

A layout of logic primitives for the inputs $x_1 = 0, x_2 = x_3 = 1$ is given in Table 20.2. The tendency of this primitives assembly to move to a ground state ensures that it takes on the polarization of the majority of its neighbors. This configuration of primitives tend to follow the majority polarization because it represents the lowest energy state.

AND function: By fixing the input $x_3 = 0$, the AND function can be implemented using the majority function:

$$f = x_1 x_2 \vee x_1 x_3 \vee x_2 x_3$$
$$= x_1 x_2 \vee x_1 \cdot 0 \vee x_2 \cdot 0 = x_1 x_2$$

A layout of logic primitives for the inputs $x_1 = 1$ and $x_2 = 0$ is given in Table 20.2.

OR function: By fixing the input $x_3 = 1$, the OR function can be implemented using the majority function:

$$f = x_1 x_2 \vee x_1 x_3 \vee x_2 x_3$$
$$= x_1 x_2 \vee x_1 \cdot 1 \vee x_2 \cdot 1$$
$$= x_1 \vee x_2$$

A layout of primitives for the inputs $x_1 = 0$ and $x_2 = 1$ is given in Table 20.2.

Useful property of a majority is a simple extension for more than three variables. For example, a four-input majority function is constructed as follows

$$f = \underbrace{x_1 x_2 \vee x_1 x_3 \vee x_2 x_3}_{3-input\ majority} \overbrace{\vee x_1 x_4 \vee x_2 x_4}^{Extension}$$

Relationship to other data structures

Positioned logic primitives, that is, configurations such as wire, complement, and majority, can be described using various data structures.

Example 20.10 *In Figure 20.4, the relationships between intermediate data structure, logic primitives, and decision trees, decision diagrams, algebraic, and tabulated forms are shown. Note that*

$$f = \overline{x}_1 x_2 x_3 \vee x_1 \overline{x}_2 x_3 \vee x_1 x_2 \overline{x}_3 \vee x_1 x_2 x_3$$
$$= x_1 x_2 \vee x_1 x_3 \vee x_2 x_3$$

Testing

Example 20.11 *Two test vectors suffice to test three-input majority gate:* $\{x_1, x_2, x_3\} = \{110\}$ *(stuck-at-0 faults) and* $\{x_1, x_2, x_3\} = \{001\}$ *(stuck-at-1 faults).*

Clocking

Unlike complementary logic networks, quantum cells have no inherent directionality for information flow. This problem can be eliminated by controlling the flows; that is, to utilize the property of quantum cells that data flows in one direction, from input to output. This can be achieved by a selected sequence of clocking signals, which ensures unidirectional data flow. The logic network using quantum cells is partitioned into clocking zones, such that all cells in a zone are controlled by the same clock signal. Cells in each zone perform a specific calculation [26].

20.4 Complementary biomolecular phenomena encoding

There are various molecular phenomena which can be useful in complementary biomolecular computing based on logic primitives. Logic primitives are characterized by their relationships with other data structures (tabulated, algebraic, and graphical forms), input and output extensions (fan-in and fan out), and acceptability for self-assembly molecular models.

TABLE 20.2
Encoding of logic primitives based on the quantum dot phenomenon.

Design example: library of logic primitives based on the quantum dot phenomenon

Function	Truth table	Assignment	Logic primitive (given assignments)

Wire

$f = f$

x	$f = x$
0	0
1	1

Input assignment:
$x = 1$
Output: $y = 1$
x	f
1	1

Invertor

$f = \overline{x}$

x	$f = \overline{x}$
0	1
1	0

Input assignment:
$x = 1$
Output: $y = 0$
x	f
1	0

Majority

$f = x_1 x_2$
$\lor\ x_1 x_3$
$\lor\ x_2 x_3$

x_1	x_2	x_3	f
0	0	0	0
0	0	1	0
0	1	0	0
0	1	1	1
1	0	0	0
1	0	1	1
1	1	0	1
1	1	1	1

Input assignments:
$x_1 = 0, x_2 = x_3 = 1$
Output: $y = 1$
x_1	x_2	x_3	f
0	1	1	1

Input $x_1{=}0$

Output f=1

Input x$_3$=1

AND

$f = x_1 x_2$

x_1	x_2	f
0	0	0
0	1	0
1	0	0
1	1	1

Input assignments:
$x_1 = 1, x_2 = 0$
Output: $y = 0$
x_1	x_2	f
1	0	0

Input $x_1{=}1$

Output f=0

Input x$_3$=0

OR

$f = x_1 \lor x_2$

x_1	x_2	f
0	0	0
0	1	1
1	0	1
1	1	1

Input assignments:
$x_1 = 0, x_2 = 1$
Output: $y = 1$
x_1	x_2	f
0	1	1

Input $x_1{=}0$

Output f=1

Input x$_3$=1

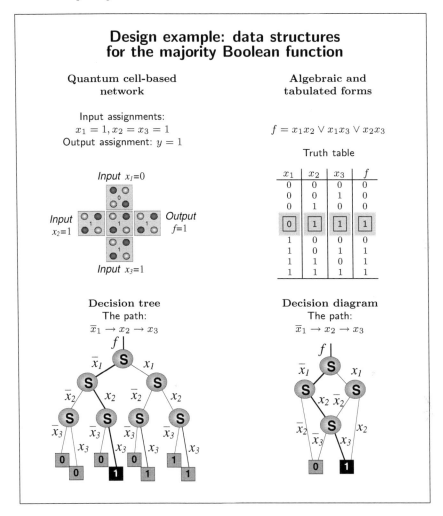

FIGURE 20.4
Quantum dot-based network of the majority Boolean function and related data structures (Example 20.10).

Logic primitives

The properties of logic primitives for the complementary biomolecular phenomena are as follows:

Properties of the logic primitive computing

Property 1: Encoding. A logic primitive conveys data using attributes of logical and spatial specification.

Property 2: Functions. The logic primitive can be considered as an elementary two-input single-output encoding device. Functionality of a molecular primitive is defined by the assignment of input variables x_1 or x_2 and specified value (0 or 1) of the output $y = f(x_1, x_2)$.

Property 3: Assembling. The logic primitives can be assembled using the rules of spatial orientation. These rules are developed using particular molecular specifications and labeling.

Library of logic primitives

Libraries of logic primitives can be developed using various requirements of computing Boolean functions. A universal library of logic primitives is given in Table 20.3. However, four molecular primitives are needed for the representation of Boolean function of two variables.

Example 20.12 *In Table 20.3, the row*

$$\begin{array}{ccc} x_1 & x_2 & y \\ \boxed{0} & \boxed{0} & \boxed{0} \end{array}$$

is implemented by a logic primitive $y = x_1 \vee x_2$ for the assignments $x_1 = x_2 = 0, y = 0$.

An arbitrary Boolean function can be computed over this library using the principle of assembling of the primitives.

Example 20.13 *An arbitrary logic network can be implemented using four logic primitives* $\begin{array}{ccc} x_1 & x_2 & y \\ \boxed{0} & \boxed{0} & \boxed{1} \end{array}$, $\begin{array}{ccc} x_1 & x_2 & y \\ \boxed{0} & \boxed{1} & \boxed{1} \end{array}$, $\begin{array}{ccc} x_1 & x_2 & y \\ \boxed{1} & \boxed{0} & \boxed{1} \end{array}$, and $\begin{array}{ccc} x_1 & x_2 & y \\ \boxed{1} & \boxed{1} & \boxed{0} \end{array}$ *from Table 20.3. This is because the NAND gate can be implemented using these primitives, and the NAND function constitutes a universal set.*

TABLE 20.3

All logic primitives: the tabulated form of a Boolean function y for the given assignments of variables x_1 and x_2 (the row of the truth table), the logic primitive, and the algebraic form

Design example: library of logic primitives for the implementation of Boolean functions

Tabulated form of Boolean function	Logic primitive	Algebraic form of Boolean function
$x_1=0$, $x_2=0$, $y=0$		$y = x_1 \vee x_2$ for $x_1 = x_2 = 0$
$x_1=0$, $x_2=0$, $y=1$		$y = \bar{x}_1 \bar{x}_2$ for $x_1 = x_2 = 0$
$x_1=0$, $x_2=1$, $y=0$		$y = x_1 \vee \bar{x}_2$ for $x_1 = 0, x_2 = 1$
$x_1=0$, $x_2=1$, $y=1$		$y = \bar{x}_1 x_2$ for $x_1 = 0, x_2 = 1$
$x_1=1$, $x_2=0$, $y=0$		$y = \bar{x}_1 \vee x_2$ for $x_1 = 1, x_2 = 0$
$x_1=1$, $x_2=0$, $y=1$		$y = x_1 \bar{x}_2$ for $x_1 = 1, x_2 = 0$
$x_1=1$, $x_2=1$, $y=0$		$y = \bar{x}_1 \vee \bar{x}_2$ for $x_1 = 1, x_2 = 1$
$x_1=1$, $x_2=1$, $y=1$		$y = x_1 x_2$ for $x_1 = x_2 = 1$

Relationship to other data structures

Logic primitives can be converted into various data structures and interpreted using switches, tabulated forms, decision trees and diagrams.

Example 20.14 *Figure 20.5 shows the relationship of the two-input, single-output logic primitive with assignments $x_1 = 0, x_2 = 1$ for which the output is $y = 1$. Logic primitive can be described in tabulated form for given assignment of the variables (the raw 001 of the truth table), algebraic form (a), the switch-based form (b), and by the path in decision tree (c).*

Representation of Boolean functions

A logic primitive functionally corresponds to a particular configuration of two switches given the assignments of the input and output variables (Figure

Design example: relationship between the intermediate and other data structures

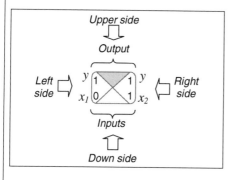

Logic primitive of the molecular computing. Two-input tile

▶ Input assignment:
$$x_1 = 0, x_2 = 1$$
▶ Output $y = 1$

Relationship of data structures

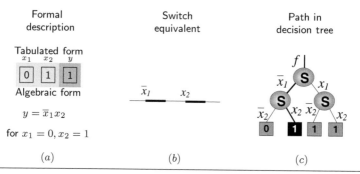

Formal description

Tabulated form

x_1	x_2	y
0	1	1

Algebraic form

$$y = \overline{x}_1 x_2$$

for $x_1 = 0, x_2 = 1$

(a)

Switch equivalent

$\overline{x}_1$ x_2

(b)

Path in decision tree

(c)

FIGURE 20.5

Two-input, single-output logic primitive can be represented in tabulated form, algebraic form, switch-based network, decision tree and diagram (Example 20.14).

20.5). An arbitrary Boolean function of two variables can be represented by two switches.

Example 20.15 *In Table 20.4, technique for the representation of the two-input AND and OR functions by logic primitives is introduced. Each row in the truth table is represented by a logic primitive and interpreted in terms of switches.*

TABLE 20.4
Logic primitives for the representation of elementary Boolean functions of
two variables (Example 20.15)

Design example: logic primitives for the representation of the elementary Boolean functions

Function	Gate symbol	Truth table	Logic primitive
AND $f = x_1 x_2$	x_1 ⟶⊐ f x_2 ⟶	$\begin{array}{cc|c} x_1 & x_2 & f \\ \hline 0 & 0 & 0 \\ 0 & 1 & 0 \\ 1 & 0 & 0 \\ 1 & 1 & 1 \end{array}$	Assignments $x_1 = 0, x_2 = 0$ Assignments $x_1 = 0, x_2 = 0$ Assignments $x_1 = 0, x_2 = 0$ Assignments $x_1 = 0, x_2 = 0$
OR $f = x_1 \vee x_2$	x_1 ⟶⊐ f x_2 ⟶	$\begin{array}{cc|c} x_1 & x_2 & f \\ \hline 0 & 0 & 0 \\ 0 & 1 & 1 \\ 1 & 0 & 1 \\ 1 & 1 & 1 \end{array}$	Assignments $x_1 = 0, x_2 = 0$ Assignments $x_1 = 0, x_2 = 0$ Assignments $x_1 = 0, x_2 = 0$ Assignments $x_1 = 0, x_2 = 0$

Implementation of three-input Boolean functions

Assembling of the logic primitives for three-input Boolean functions AND, OR, NAND, NOR, and EXOR is shown in Table 20.5. The simplest rules for spatial orientation are used, such as matching labels between the sides of the primitives. The non-labeled primitives implement wire-function, that is, direct transmission of the input signal to the output. In the column "Primitive assembly," the primitives are shown only out of three from eight possible assignments of variables from the truth table.

Example 20.16 *Three-input AND gate can be constructed using two two-input AND gates (Table 20.5). This restriction of the inputs is described by the Boolean function $f = x_1 x_2 x_3$. The truth table for the 3-input AND function contains $2^3 = 8$ rows, that is, 8 logic primitive configurations are needed for computing this function.*

It follows from Example 20.16, that assembled logic primitives form various topological structures. Properties of these spatial structures can be utilized in molecular computing based on self-assembly paradigm. For example, under appropriate control, these spatial structures can be described as fractal or fractal-like structures.

Interpretation computing using decision trees

Logic primitives as an intermediate data structure can be interpreted by various data structures which are used in contemporary logic design. Decision trees and lattice decision diagrams [5] are of a particular interest because of their symmetric properties. These properties can be used in effective representation and manipulation of logic primitives in control self-assembly.

Computing using the logic primitives can be associated with the paths in decision trees and diagrams. The path in a decision tree or a decision diagram (from the rooted node to the terminal node) of a Boolean function of two variables corresponds to the logic primitive. There are four different paths in this decision tree or a decision diagram, and each path is terminated at the 0- or 1-terminal node.

Example 20.17 *Table 20.6 shows the correspondence between a logic primitive and a path in the decision trees for a Boolean function of two variables. There are four paths in this tree, and each path can be represented by a logic primitive.*

Self-assembly in terms of logic primitives

Intermediate data structure, such as logic primitives, is useful in computational interpretation of the control molecular self-assembly. Self-assembly is defined as a process where nanostructures spontaneously nucleate or assemble into a suitable medium under the proper conditions. Depending on the medium and the external environment, the nanostructures can sometimes *self order*, that is, they can automatically arrange themselves into a regimented (periodic) array in spatial dimensions.

TABLE 20.5
Implementation of the three-input logic gates using the logic primitives
(Example 20.16)

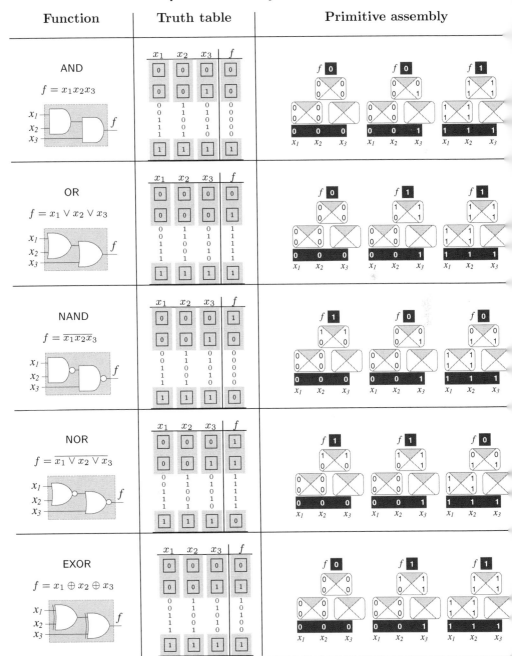

Design example: assemblies of logic primitives for the implementation of 3-input elementary Boolean functions

Function	Truth table	Primitive assembly

TABLE 20.6

Interpretation of computing using decision tree: computing the path in the decision tree using the logic primitives (Example 20.17)

Decision tree	Path description	Logic primitive
	x_1 x_2 y $\boxed{0}$ $\boxed{0}$ $\boxed{0}$ $y = x_1 \vee x_2$ for $x_1 = x_2 = 0$	$y = x_1 \vee x_2$
	x_1 x_2 y $\boxed{0}$ $\boxed{0}$ $\boxed{1}$ $y = \overline{x}_1 x_2$ for $x_1 = 0, x_2 = 1$	$y = \overline{x}_1 x_2$
	x_1 x_2 y $\boxed{1}$ $\boxed{0}$ $\boxed{1}$ $y = x_1 \overline{x}_2$ for $x_1 = 1, x_2 = 0$	$y = x_1 \overline{x}_2$
	x_1 x_2 y $\boxed{1}$ $\boxed{1}$ $\boxed{1}$ $y = x_1 x_2$ for $x_1 = x_2 = 1$	$y = x_1 x_2$

Molecular self-assembly is an approach to spontaneously forming highly ordered monolayers on various substrate surfaces. The key engineering principle for molecular self-assembly is to design molecular building blocks that are able to undergo spontaneous stepwise interactions. In this design, the instructions are incorporated into the structural framework of each molecular component.

Self-assembly is considered with respect to the control techniques: (a) *untemplated* and (b) *templated* structures. Controllable self-assembly and robust binding/paring can be implemented using techniques of *templates*. The templates carry the information about computational properties. Fractals that can be interpreted in terms of Boolean functions are useful as templates for self-assembly.

Example 20.18 *Step-wise asembly model for local parallel biomolecular computing is introduced in* [38]. *In* [40], *properties of* **untemplated** *crystals are studied with respect to the qualitative appeareance of Sierpinski triangles. In particular, cellular automata with EXOR local computing can be considered as a reasonable model of self-assembly. At each time step, each cell is computed as the XOR of its two neighbors.*

Self-assembled structures can be inspected visually using microscopy, and data can be read out.

Example 20.19 *The DNA self-assembly is based on constructing the special forms of DNA in which strands cross over between multiple double helices, creating two-dimensional logic primitives (tiles).*

In contrast to classical computing paradigms, the random phenomena of self-assembly create a randomness in the time required to perform a given computation. Molecular self-assembly may have various types of errors.

20.5   Fractal-based models for self-assembly

Fractal geometry is the geometry of nature. Its core concept is *self-similarity*. Self-similarity can be defined as an invariance with respect to scaling. In this section, we consider *strictly* self-similar fractal structures, such as Sierpinski triangle and transeunt triangle. These fractals are composed of small but exact copies of itself.

In molecular computing, *self-assembly* is defined as the process, by which an organized structure can be formed spontaneously from simple parts. Fractals are useful models for the self-assembly because of a regular and periodic structures. Fractals describe the assembly of natural structures such as crystals, DNA helices, and microtubules. Logic primitives are used for interpretation of self-assembly process in terms of computing the Boolean functions. The relationship between the assembled structures and fractals can be considered as follows:

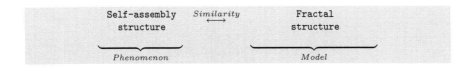

The basic building blocks in this model are logic primitives. For instance, Sierpinski triangles can be assembled from primitive triangles.

In this section, fractals are interpreted in terms of Boolean functions. The basic idea is that there is a relationship between the tabulated form of a Boolean function distributed in the fractal and properties of this function, such as symmetry and polynomial forms.

Regular and periodic templates are needed for fabrication of DNA structures. In this section, two types of templates are considered: *Sierpinski gasket*, or *Sierpinski triangles*, and *transeunt triangles*.

20.5.1 Sierpinski triangle for Boolean functions

The Sierpinski triangle is a type of fractal structure. A Sierpinski triangle can be seen as a *Pascal triangle* in which the EXOR operation is used instead of arithmetic addition while forming the Pascal triangle.

The Sierpinski triangle for a Boolean function $f(x_1, x_2, \ldots, x_n)$ is a triangle formed by modulo 2 addition of 0s and 1s starting from the bottom row, which is the truth vector of f. A Sierpinski triangle for an n-variable function has a width of 2^n and a height of 2^n.

There are $(4^n + 2^n)/2$ elements in total. Besides the defining relation, there are other relations among the elements in the Sierpinski triangle. Since the truth vector for a function of n variables has an even number of entries (2^n), it can be divided evenly into two parts. Each part produces, on its own, two subtriangles.

Let T_0 be a Sierpinski triangle for a Boolean function f of a single variable, that is, it is 3-element triangle $T_0 = (0, x_1)$. To form T for a function of two variables, another two 3-element triangles must be added. Here $T_0 = (0, x_2)$,

$$T_1 = (1, x_2)$$
$$T_2 = T_1 \oplus T_2$$

and T_0 turns to $T_0 = (0, x_2)$. The elements of T_2 are $f(0, x_2) \oplus f(1, x_2)$. This recursive procedure can be applied to a function of any number of variables.

Example 20.20 *Let T be a Sierpinski triangle for a Boolean function $f(x_1, x_2, \ldots, x_n)$. Representation of T by 3 triangles T_0, T_1, and T_2 is shown in Figure 20.6.*

Nonbottom elements of T_2 by the construction of T are corresponding sums. So, T_2 is the Sierpinski triangle that represents the Boolean function

$$f(0, x_2, \ldots, x_n) \oplus f(1, x_2, \ldots, x_n)$$

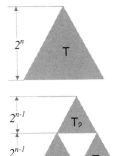

▶ T_0 is the transeunt triangle for the function

$$f(0, x_2, \ldots, x_n)$$

▶ T_1 is the triangle for the function

$$f(1, x_2, \ldots, x_n)$$

▶ T_2 is the triangle for the function

$$f(0, x_2, \ldots, x_n) \oplus f(1, x_2, \ldots, x_n)$$

The bottom of T_2 is the "exclusive or" of the elements of the bottom row of triangles for functions

$$f(0, x_2, \ldots, x_n) \text{ and } f(1, x_2, \ldots, x_n),$$

FIGURE 20.6

Representation of Sierpinski triangle T by combination of three subtriangles T_0, T_1, and T_2 (Example 20.20).

Example 20.21 *Let a Boolean function f of three variables x_1, x_2, x_3 be given by a truth table or truth vector $\mathbf{F} = [00101011]^T$ (Figure 20.7). This function can be represented by a decision tree whose terminal nodes represent the truth vector. The Sierpinski triangle can be constructed from the truth vector, and it results in the coefficient vectors of the polynomial forms of this function.*

Properties of Sierpienski triangle in terms of Boolean data structures

Property 1 (0-polarity polynomial): The bits along triangle's left side are coefficients of zero polarity polynomial $\mathbf{RM(0)}$.

Property 2 (1-polarity polynomial): The bits along the right side of each subtriangle on the left side of the triangle are the coefficients of polynomial of the polarity 1 called the vector of coefficients $\mathbf{RM(1)}$.

Property 3 $(2, 3, \ldots 2^n - 2$-polarity polynomial): The bits along left or right side of each following row of subtriangles such that the bottom element is j-th element of the truth vector $\mathbf{F}$, and the coefficients of polynomial of polarity j are represented by the vector of coefficients $\mathbf{RM(j)}$, $j = 1, 2, \ldots, 2^n - 2$.

Property 4 $((2^n - 1)$-polarity polynomial): The bits along the right side of the triangle are the coefficients of polynomial of polarity $2^n - 1$ called the vector of coefficients $\mathbf{RM(2^n - 1)}$.

Design example: data structure manipulations

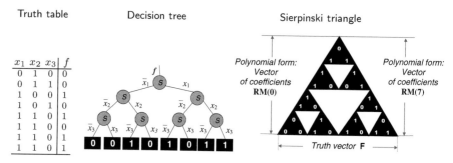

| Truth table | Decision tree | Sierpinski triangle |

x_1	x_2	x_3	f
0	1	0	0
0	1	1	0
1	0	0	1
1	0	1	0
1	1	0	1
1	1	0	0
1	1	0	1
1	1	0	1

FIGURE 20.7

Relationship between the truth table of a Boolean function $f(x_1, x_2, x_3)$ given a truth vector $\mathbf{F} = [00101011]^T$, its decision tree, and Sierpinski triangle (Example 20.21).

Example 20.22 *Let the Boolean function $f = x_1\overline{x}_2$ be given by its truth vector $\mathbf{F} = [0010]^T$ (Figure 20.8). The vectors of coefficients of the polynomials for polarities 0,1,2, and 3, $\mathbf{RM}(0) = [0011]^T$, $\mathbf{RM}(1) = [1111]^T$, $\mathbf{RM}(2) = [0001]^T$, $\mathbf{RM}(3) = [0101]^T$ can be retrived from the Sierpinski triangle.*

20.5.2   Transeunt triangles for Boolean functions

Transeunt triangles are data structures for symmetric Boolean functions. A Boolean function f of n variables $x_1, x_2, \ldots, x_n$ is *totally symmetric* if and only if it is unchanged by any permutation of variables.

A totally symmetric function is completely specified by a *carrier* vector $\mathbf{F}_c = [a_0 a_1, \ldots, a_n]$, such that Boolean function f is a_i for all assignments to $x_1, x_2, \ldots, x_n$ that have i 1s, where $0 \leq i \leq n$.

Example 20.23 *A Boolean function $f = \overline{x}_1\overline{x}_2\overline{x}_3 \vee x_1x_2x_3$ is totally symmetric. A carrier vector of this function is $\mathbf{F}_c = [1\ 0\ 0\ 1]$.*

Representation of a symmetric Boolean function consisting of 1s and 0s, arranged in a triangle, is based on the function's $(n+1)$-bit *carrier* vector $\mathbf{F}_c$.

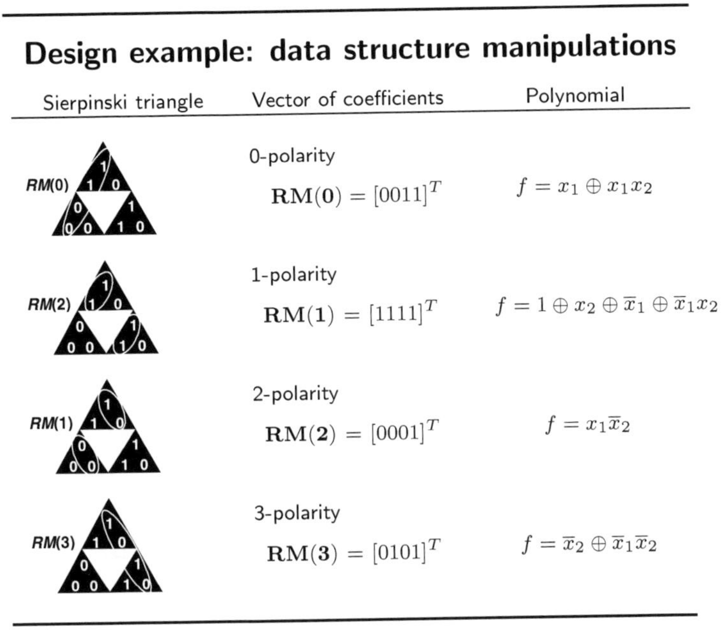

Design example: data structure manipulations

Sierpinski triangle	Vector of coefficients	Polynomial

0-polarity

$$\mathbf{RM(0)} = [0011]^T \qquad f = x_1 \oplus x_1 x_2$$

1-polarity

$$\mathbf{RM(1)} = [1111]^T \qquad f = 1 \oplus x_2 \oplus \bar{x}_1 \oplus \bar{x}_1 x_2$$

2-polarity

$$\mathbf{RM(2)} = [0001]^T \qquad f = x_1 \bar{x}_2$$

3-polarity

$$\mathbf{RM(3)} = [0101]^T \qquad f = \bar{x}_2 \oplus \bar{x}_1 \bar{x}_2$$

FIGURE 20.8
Location of the coefficient vectors of the polynomial forms of the Boolean function $f = x_1 \bar{x}_2$ in the Sierpinski triangle defined from its truth table (bottom side of the triangle) (Example 20.22).

Thus, the triangle is formed as follows:

Transeunt triangle construction using carrier vector

Given: A Boolean function.
Step 1: The carrier vector is located at the base of triangle.
Step 2: A vector of n 1s and 0s is formed by the exclusive OR of adjacent bits in the carrier vector.
Step 3: A vector of $(n-1)$ 1s and 0s is formed by the exclusive OR of adjacent bits in the previous vector, etc.
Step 4: At the apex of the triangle is a single 0 or 1.
Result: Transeunt triangle.

Example 20.24 *In Figure 20.9a,b,c, the Sierpinski and transeunt triangles for the symmetric function $f = \overline{x_1 x_2}$ are compared. Carrier vector $\mathbf{F}_c = [1\ 1\ 0]^T$ is formed from the truth vector $\mathbf{F} = [1\ 1\ 1\ 0]^T$.*

Design example: data structure manipulations

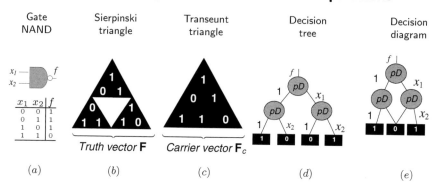

FIGURE 20.9

The AND function computing: two-input AND gate (a), Sierpinski triangle (b), transeunt triangle (c), decision tree (d), and decision diagram (e) (Examples 20.24 and 20.28).

20.5.3   Forward and inverse procedures

Manipulation of transeunt triangles can be accomplished using *forward* and *inverse* procedures.

Forward procedure is defined as follows:

> Carrier vector $\longrightarrow$ Carrier vector of coefficients
>
> *Forward procedure*

The forward procedure can be specified as the converting of a symmetric Boolean function given in the tabulated form into a polynomial form (vector of coefficients).

Inverse procedure   is defined as follows:

> Carrier vector of coefficients $\longrightarrow$ Carrier vector
>
> *Inverse procedure*

The inverse procedure can be specified as computing of a symmetric Boolean function given in the polynomial form (represented by vector of coefficients). The result of this computation is the values of polynomial expressions (carrier vector).

Example 20.25 *The forward transform in vector and algebraic form for the NAND function of two variables is given in Figure 20.9c); for 0-polarity polynomial (left side of the triangle):*

$$\mathbf{F}_c = [110] \quad \rightarrow \quad \mathbf{RM(0)}_c = [101]$$

$$\underbrace{\phantom{\mathbf{F}_c = [110]}}_{Carrier\ vector} \qquad \underbrace{\phantom{\mathbf{RM(0)}_c = [101]}}_{Carrier\ vector\ of\ coefficients}$$

$$f = \quad \overline{x_1 x_2} \quad = \quad 1 \oplus x_1 x_2$$

$$\underbrace{\phantom{f = \overline{x_1 x_2}}}_{SOP\ form} \qquad \underbrace{}_{0-polarity\ polynomial}$$

and for 3-polarity polynomial (right side of the triangle):

$$\mathbf{F}_c = [110] \quad \rightarrow \quad \mathbf{RM(3)}_c = [011]$$

$$\underbrace{\phantom{\mathbf{F}_c = [110]}}_{Carrier\ vector} \qquad \underbrace{\phantom{\mathbf{RM(3)}_c = [011]}}_{Carrier\ vector\ of\ coefficients}$$

$$f = \quad \overline{x_1 x_2} \quad = \quad \overline{x}_2 \oplus \overline{x}_1 \oplus \overline{x}_1 x_2$$

$$\underbrace{\phantom{f = \overline{x_1 x_2}}}_{SOP\ form} \qquad \underbrace{\phantom{\overline{x}_2 \oplus \overline{x}_1 \oplus \overline{x}_1 x_2}}_{3-polarity\ polynomial}$$

This is because $\overline{x_1 x_2} = 1 \oplus x_1 x_2 = 1 \oplus (x_1 \oplus 1 \oplus 1)(x_2 \oplus 1 \oplus 1) = 1 \oplus (\overline{x}_1 \oplus 1)(\overline{x}_2 \oplus 1) = \overline{x}_2 \oplus \overline{x}_1 \oplus \overline{x}_1 x_2.$

Example 20.26 *The forward and inverse procedures for the symmetric Boolean function of three variables* $f = \overline{x}_1 \overline{x}_2 \overline{x}_3 \oplus x_1 x_2 x_3$ *(n = 3) using the transeunt triangle are shown in Figure 20.10.*

The useful computational properties of the transeunt triangles are listed in Table 20.7. Proofs of these properties are given in [9, 10]. The basic property is as follows. The transeunt triangle can be generated from the carrier vector, from the carrier vector of coefficients in $\mathbf{RM(0)}_c$ or in $\mathbf{RM(2^n-1)}_c$ by the exclusive OR of adjacent bits repeatedly until a single bit is obtained (Example 20.26). This is because the exclusive OR is *self-invertible*. That is, given $A \oplus B$ and the value of A, we can find the value of $B = A \oplus (A \oplus B)$.

Among the polynomial expansions, one with a minimal number of non-zero coefficients (called *minimal expansion*) can be found. An algorithm for computing the minimal (fixed) polarity polynomial expansion using the transeunt triangle is as follows:

TABLE 20.7
Properties of transeunt triangles

Property	Content
Forward generation	Symmetric Boolean function is given by the carrier vector $\mathbf{F}_c$ (the base of triangle), carrier vectors of coefficients of 0-polarity $\mathbf{RM(0)}_c$ and $\mathbf{RM(2^n-1)}_c$-polarity polynomials are generated as the left and the right sides of the triangle (read from the bottom to top)
Inverse generation	Symmetric Boolean function is given by the carrier vector of coefficients of 0-polarity $\mathbf{RM(0)}_c$ (the base of triangle), and the carrier vector $\mathbf{F}_c$ is generated as the right side of the triangle (read from the bottom to top)
Complemented form	If T is a transeunt triangle of a symmetric function f, then the transeunt triangle $T_{\overline{f}}$ for (symmetric function) $\overline{f}$ is obtained from T by replacing the carrier vector at the base of T by its complement. This is because $A \oplus B = \overline{A} \oplus \overline{B}$. In a similar manner, other rows in a transeunt triangle can be complemented.
Complete set of coefficients	Let T be the transeunt triangle of a symmetric Boolean function f. Then, embedded in T are all polynomial coefficient matrices.
Minimal polarity polynomial	Among the polynomial expansions, one with a minimal number of nonzero coefficients can be found.
The total number	The total number of transeunt triangles is the number of different triangles for a given size n (the number of Boolean variables in the represented Boolean function). For large n [10], The number of transeunt triangles $= \dfrac{2^{n+1}}{3}$,

Design example: data structure manipulations

Given the Boolean function of three variables

$$f = \bar{x}_1 \bar{x}_2 \bar{x}_3 \oplus x_1 x_2 x_3.$$

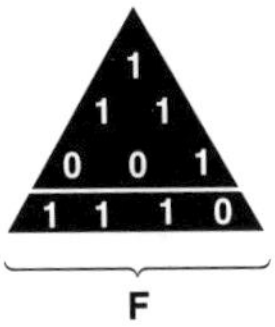

F

Forward procedure: Given the carrier vector $\mathbf{F}_c = [1\ 1\ 1\ 0]^T$ *(the base of the triangle)* the carrier vector of coefficients $\mathbf{RM}(0)_c = [1\ 0\ 0\ 1]$ *is formed as the left side of the triangle (read from bottom to top).*

Inverse procedure: Given the carrier vector of coefficients

$$\mathbf{RM}(0)_c = [1001]^T$$

located at the base of the triangle, the triangle can be restored, so that the carrier vector $\mathbf{F}_c = [1\ 1\ 1\ 1\ 0]$ *can be be formed as the right side of the triangle (read from bottom to top).*

RM(0)

FIGURE 20.10

Forward and inverse procedures for the manipulation of transeunt triangles of the Boolean function $f = \bar{x}_1 \bar{x}_2 \bar{x}_3 \oplus x_1 x_2 x_3$ (Example 20.26).

An algorithm for computing the minimal polynomial expansion

Given: The transeunt triangle.
Step 1: Compute the number of terms of each RM_i.
Step 2: Choose an $\mathbf{RM}(i)_c$ with the fewest product terms.
Result: A minimal fixed-polarity polynomial expansion.

20.5.4 The number of self-similar triangles

The number of generations for transeunt triangles is the parameter related to the following characteristics of the computer structure:

▶ Self-assembly

▶ Class of the implemented Boolean functions

▶ Performance

Given two transeunt triangles, a third can be generated by forming the bit-by-bit modulo two sum. Transeunt triangle T is *self-similar* if and only if a rotation of 120° and 240° leaves T unchanged. The transeunt triangle can be formed from any of its three sides.

Example 20.27 *Figure 20.11 shows all transeunt triangles for symmetric Boolean functions of three variables. Here, the transeunt triangle is not repeated if it is the same as the rotation of another triangle in the figure. Four of these triangles are unchanged by a rotation of $120°$ and $240°$.*

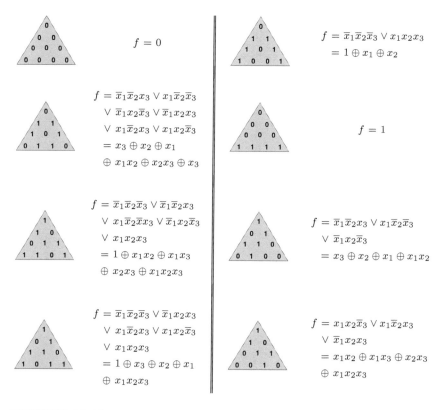

FIGURE 20.11

All transeunt triangles of size 3; they represent all eight symmetric Boolean functions of three variables (Example 20.27).

The number of self-similar transeunt triangles of size n is $2^{\lfloor (n-1)/3 \rfloor + (n-1) \bmod 3}$. For large n, the number of transeunt triangles can be estimated as $2^{n+1}/3$.

Given the size n of the transeunt triangle, all transeunt triangles can be generated from the subset, which includes self-similar transeunt triangles (details are given in [10]). This subset includes $n/3$ *generator* transeunt triangles. In Table 20.8, the following parameters of transeunt triangles are estimated for given number of variables n:

▶ The total number of transeunt triangles

▶ The number of self-similar transeunt triangles

▶ The number of generations

TABLE 20.8

The number of transeunt triangles generated from the subset

n	Total	Self-similar	Number of generation
1	2	1	0
2	4	2	1
3	8	4	2
4	12	2	1
5	24	4	2
6	48	8	3
7	88	4	2
8	176	8	3
9	352	16	4
10	688	8	3
15	21,888	64	6
20	699,136	128	7

20.5.5 Transeunt triangles and decision trees

A transeunt triangle represents all positive polynomial expressions of both fixed and mixed polarity. The corresponding elements of the triangle are components of the carrier vector of polynomial coefficients, and they are formed from the truth vector by modulo-2 addition of its elements. This process, on the other hand, can be represented by a decision tree in which the input (to the root node) is supplied with a sequence of the elements of the truth vector, and this sequence is used to form the components of a vector of any polarity polynomial expression.

Example 20.28 *The Sierpinski and transeunt triangles for the symmetric function $f = \overline{x_1 x_2}$ are interpreted using decision tree and diagram in Figure 20.9d,e.*

Example 20.29 *Figure 20.12 illustrates two different computing paradigms based on the Sierpinski triangle (a) and decision trees for the polynomial forms of a Boolean function f of three variables given its truth vector* $\mathbf{F} = [00101011]^T$*: decision tree (a) computes 0-polarity polynomial and corresponds to the left side of the triangle; decision tree (c) computes 7-polarity polynomial and corresponds to the right side of the triangle.*

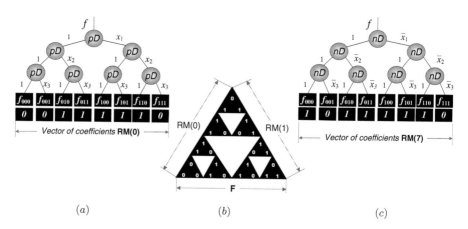

(a)   (b)   (c)

FIGURE 20.12
Representation of a Boolean function $f(x_1, x_2, x_3)$ of three variables given a truth vector $\mathbf{F} = [00101011]^T$ by the Sierpinski triangle (b) and corresponding decision trees for the 0-polarity polynomial (a) and 7-polarity polynomial (c) (Example 20.29).

Example 20.30 *In Figure 20.13, the computation aspects of using the Sierpinski triangle, transeunt triangle, and decision diagrams are compared. The Sierpinski triangle and the corresponding decision diagrams are used for computing a swithching function* $f = \overline{x}_1\overline{x}_2x_3 \vee x_1x_2x_3 \vee x_1\overline{x}_2\overline{x}_3 \vee \overline{x}_1x_2\overline{x}_3$ *given by the truth vector* $\mathbf{F} = [01001010]^T$*. The transeunt triangle and corresponding decision diagrams compute a symmetric swithching function* $f = x_1 \vee x_2 \vee x_3 \vee x_1x_2x_3$ *given by the carrier vector* $\mathbf{F}_c = [0101]$*. Three types of decision diagrams are used in each case: diagrams with pD, nD, and S nodes.*

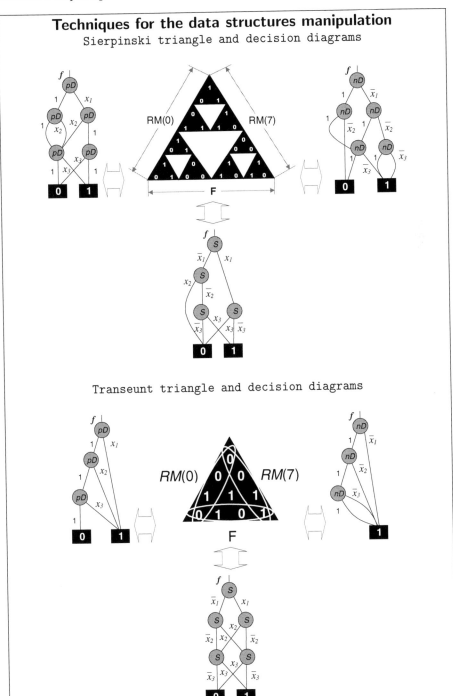

FIGURE 20.13

The relationship between the Sierpinski triangle and the topology of the corresponding decision diagrams given the Boolean function $f = \overline{x}_1\overline{x}_2x_3 \vee x_1\overline{x}_3$ and $f = \overline{x}_1\overline{x}_2x_3 \vee x_1x_2x_3 \vee x_1\overline{x}_2\overline{x}_3 \vee \overline{x}_1x_2\overline{x}_3$ (Example 20.30).

20.5.6   Using transeunt triangles for representation of mixed-polarity polynomials

A mixed polarity of polynomial expressions is a family of 3^n polynomial representations of Boolean functions of n variables. In this family, each binary variable x_i can be represented in one of the following *basic polarities*:

$$\mathbf{c}_0 = [1\ x], \quad \mathbf{c}_1 = [1\ \overline{x}], \quad \mathbf{c}_2 = [\overline{x}\ x]$$

Given $n > 1$, various combinations of terms of variables can be generated by the Kronecker product of basic polarities.

Example 20.31 *The product terms in the $(c_0, c_2) = (0, 2)$-polarity polynomial expression of a Boolean function of two variables x_i and x_j are generated as follows:*

$$\mathbf{c}_0 \otimes \mathbf{c}_2 = [1\ x_i] \otimes [\overline{x}_j\ x_j] = [\overline{x}_j\ x_j\ x_i\overline{x}_j\ x_ix_j].$$

Example 20.32 *Table 20.9 contains various graphical data structures for representation of five elementary Boolean functions. These functions are symmetric, therefore, both the Sierpinski triangle and transeunt triangle for them are shown.*

20.5.7   Transeunt triangles for the representation of multivalued logic functions

In the case of a ternary function, there are 84 bases that form 84^n different mixed-polarity polynomial expressions for a function of n variables [39].

Example 20.33 *In Figure 20.14, a transeunt triangle for a Boolean function of one variable (a) and two variables (b) is pictured. The decision tree (c) corresponds to a polynomial expression $RM(0)$ of zero polarity and elements $RM(0)$ are 0s and 1s on the right side of the triangle. In case of a ternary function, possible polarities of variables are represented by a pyramid (d) in which all possible paths between nodes represent 84 different polarities of polynomial expressions of a function of a single ternary variable.*

TABLE 20.9
Graphical data structures: Sierpenski gasket, transeunt
triangle, decision tree, decision diagram, and
$\mathcal{N}$-hypercube for representation of elementary Boolean
functions of two variables

Sierpinski gasket	Transeunt triangle	Decision tree	Decision diagram

AND-function $\mathbf{F} = [0001]^T$

OR-function $\mathbf{F} = [0001]^T$

EXOR-function $\mathbf{F} = [0001]^T$

NAND-function $\mathbf{F} = [0001]^T$

NOR-function $\mathbf{F} = [0001]^T$

Example 20.34 *One of expansions, $RM(0)$, is represented by the complete ternary decision tree (Figure 20.14e). The product terms of the expression $RM(0)$ are formed as follows:* $[1 \ x_1 \ x_1^2] \otimes [1 \ x_2 \ x_2^2] = [1 \ x_2 \ x_2^2 \ x_1 \ x_1 x_2 \ x_1^2 x_2 \ x_1^2 \ x_1 x_2^2 \ x_1^2 x_2^2].$

Techniques for polynomials of Boolean functions

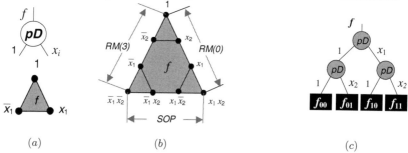

(a) (b) (c)

Techniques for computing polynomials of ternary logic functions

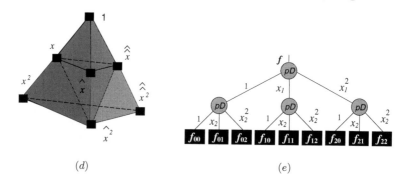

(d) (e)

FIGURE 20.14

Transeunt triangles and decision trees for the representation of a Boolean function (a,b,c) of one and two variables, and a pyramid and ternary decision tree for a ternary function (d,e) of one and two variables (Examples 20.33 and 20.34).

20.5.8 Sierpinski triangle generation using cellular automata

Cellular automata can be defined as a class of computing arrays of spatial topology that are characterized by local interaction and inherently parallel processing. Cellular automata are models for complex systems and processing consisting a large number of identical, simple, and locally interconnected components called *cells*. Cellular automata can be described in terms of dimension (1D, 2D, and 3D lattice of cells), homogeneity (all cells perform the same function), state (each cell can take one of a finite number of possible states), interconnection (each cell can interact only with neighborhoods), and discrete dynamics (each cell updates its current state according to a transition rule at discrete time). Cellular automata are a suitable model of self-assembly.

Example 20.35 *The simplest cellular automaton is a 1D binary cellular automata: the grid corresponds to a row of cells, and each cell can only assume a value of 0 or 1. The state of the i-th cell is updated in discrete time accordingly to the transition rule and depending on the neighborhood states.*

At any time, there is a unique distribution of states of cellular automata. This distribution can be interpreted as a point in the state space. The evolution of a cellular automaton can be studied by observing the the trajectory of states in the state space.

Example 20.36 *Sierspinski triangle can be generated using the following transition rules for the 1D cellular automaton:*

$$0\ \boxed{0}\ 0 \to 0 \qquad\qquad 1\ \boxed{0}\ 0 \to 1$$

$$0\ \boxed{0}\ 1 \to 1 \qquad\qquad 1\ \boxed{0}\ 1 \to 0$$

$$0\ \boxed{1}\ 0 \to 0 \qquad\qquad 1\ \boxed{1}\ 0 \to 1$$

$$0\ \boxed{1}\ 1 \to 1 \qquad\qquad 1\ \boxed{1}\ 1 \to 0$$

where transitions are denoted as follows:

Current states of three cells

$$\underbrace{s_{i-1}(t)}_{Left\ cell}\ \boxed{s_i(t)}\ \underbrace{s_{i+1}(t)}_{Right\ cell} \to \underbrace{s_i(t+1)}_{New\ state}$$

center cell

Example 20.37 *(Continuation of Example 20.36) If the state of the center cell is 0, and the states of both its neighbors (left and right) are 1, then the state of the central cell in the next step will be 1.*

It follows from Example 20.36 that the next state of the cell is formed by the EXOR operation between the states of neighboring cells: $s_i(t+1) = s_{i-1}(t) \oplus s_{i+1}(t)$.

20.6   Evolutionary logic network design

Logic network design using *directed evolution,* also known as *genetic algorithms,* aims at constructing logic networks with desirable characteristics over the library of permissible gates. This approach utilizes properties of biological systems such as the ability to evolve and be optimized under the pressure of natural and artificial selection. The process of directed evolution can to be used as a tool to construct well-behaved logic networks using binary and multivalued gates. Directed evolution is performed by placing random mutations in the sequences of encoded network and restricting mutations to only a specific region.

In this section, the main property of directed evolution, such as stochastic global search is utilized in network design. Despite the simplicity of genetic algorithms, they outperform any random search, because they can exploit information about circuit solution, cumulated during the evolution of the search.

20.6.1   Terminology of evolutionary algorithms

The following terminology is used in evolutionary, or *genetic* algorithms for logic network design:

Chromosome. In the search process, candidate network solutions are encoded in the form of symbolic strings called *chromosomes.* Each chromosome encodes a circuit in the logic network space, that is, a possible network solution. Every symbol in the encoding, called a *gene,* of a possible solution captures some information about the problem.

Population. The simulation of evolutionary process is conducted in a pool of chromosomes called *population* and characterized by *population size,* and performed on an iterative basis. The search algorithm extracts and analyzes the topological and logical information of a search logic network space (each iteration of the search process is called *a generation*), and can, therefore, guide the search to advance along a promising direction.

Fitness function is a kind of *objective* or *cost* function that contains all the information about the logic network synthesis problem. The better the solution encoded by chromosome, the higher the fitness. The genetic algorithm then tries to improve the fitness of the population by combining information contained in high-fitness chromosomes. The objective of solving a logic network synthesis problem in a given design style is to obtain an optimal network by minimizing the total number of gates with respect to their types and levels of implementation. Because the objective is twofold, - representation of initial logic function and its network realization one can propose two ways to describe the objective of this problem.

Selection operator is used to guide the search direction of the genetic search process. It leads to an overall improvement of the chromosomes and cost performance of the search processes. The potential chromosomes with higher fitness values are given higher chances to be selected as parents to breed new chromosomes. The parent chromosomes are placed in a mating pool where crossover and mutation take place. In each generation, the worst $]S \times R[$ chromosomes are removed from the current population in accordance with their fitness value. The best $]S \times R[$ chromosomes are then duplicated and inserted into the same population to replace the rejected chromosomes. Afterwards, the selection operator is applied to select parent chromosomes from the population in accordance with the selection parameter, λ_{select}, of the chromosome. The selection parameter λ_{select} of a chromosome is defined as $\lambda_{select} = \Phi/\Sigma\Phi$, where Φ and $\Sigma\Phi$ are the fitness values of the chromosome, and the sum of fitness values over the entire population, respectively.

Crossover operation transfers a portion of the genetic codes between two parent chromosomes, selected from the mating pool. It leads to an exploitation of the solution space by introducing variations to the parent chromosomes. First, a pair of parent chromosomes is chosen from the mating pool without replacement. The probability of applying the crossover operator to these chromosomes is called *the probability of crossover, P_{cros}*. If the decision is not to cross the chromosomes, they will be cloned to produce a pair of offspring chromosomes, where the offspring chromosomes are identical to their parents. Otherwise, the parent chromosomes will be crossed to produce two offspring chromosomes by using the crossover operator.

Mutation operator is used to safeguard the search process from premature convergence to a local optimum. This operator results in a random walk in the parameter space and introduces new information into the evolution process, which might have been lost with a premature convergence of the genetic algorithm. It is an immediate operation that follows the crossover operation. It attempts to rearrange the structure of a chromosome at random. The probability of mutating a single gene is called *the probability of mutation P_{mutate}*, which is usually a small number. For each gene in a chromosome, an arbitrary choice is made to decide whether the mutation operation is performed or not. If the decision is not to perform the mutation operation, the gene will be kept unchanged. Otherwise, the gene is mutated by swapping its contents randomly with the other gene, on the condition that neither of the gene contents is equal to -1. If the content of a gene is equal to -1, it means that the gene is representing a restricted area, and no mutation should be allowed. The mutation operation is then applied to the next gene, and the entire process is repeated until all genes in the chromosome are tried.

Besides the population and generation sizes, the solution quality can also be affected by the percentage of replication of well-performed chromosomes in each generation R, the probability of crossover P_{cros} and the probability mutation P_{mut}. The effects of these genetic parameters on the solution quality is called *sensitivity analysis*.

20.6.2 Design style

We study evolutionary gate-level combinational logic networks design strategy from the position of design style. The uncertainty of a total search space (the space of all possible logic network solutions) through evolutionary network design is removed faster, if this space is partitioned into subspaces. This idea has been realized through a *parallel window-based scanning* of these subspaces. Such a window is determined by the parameters of a multilevel logic network in a given design style. Moreover, it was shown that information theoretical interpretation of the evolutionary process is useful, in particular, in partitioning of network space and measuring of fitness function.

The idea of evolutionary logic network design is very captivating and attractive for digital design because the designer needs no special software to implement a given architecture and design style. These are provided by evolutionary (adaptive) search for the network solution, and there are many nontrivial network solutions through the evolving process. There main drawback of the gate-level evolutionary design is that the runtime for the evolving process is too long (much longer than in classical heuristics).

Example 20.38 *An extreme example is that an 8-bit counter required about 20 hours of processing to evolve a correct solution. One of the well-known effects is that most of the networks in the logic network space are invalid. For example, authors of [35] reported that correct 3-bit adders evolved twice from about 90,000 generations.*

In this section, a state-of-the-art technique of evolutionary logic network design is introduced. The following features characterize this approach:

(a) For best results, logic network search space must be partitioned into subspaces via decomposition of initial logic function, prior to the evolutionary search.

(*b*) Massive parallel window-based scanning of subspaces of possible logic network solutions is applied after the partitioning of logic network space into subspaces; this provides an opportunity to scan subspaces independently, making this approach the powerful resource (together with the known parallel genetic algorithm) for reduction of executing time.

(*c*) Alternative search strategies can be applied instead of the traditional ones. One of the most powerful is interpretation and estimation of the evolutionary process in the notation of quantity of information. In contrast to the rule-based evolutionary technique, that deals with forms of representation of logic functions (LITERAL, MIN/SUM, AND/EXOR, and so on) and explores algebraic rules, the idea of the evolutionary logic network design is that the network structure is adaptively searched in the space of possible network solutions. The architecture string (chromosome) is mapped onto a network after genetic learning. In this book, we consider the evolutionary logic network design of a combinational networks in terms of a *target design style* that is defined over a fixed library of gates:

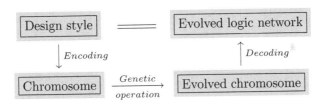

Verification of the evolved logic network consists in testing whether the evolved network performs the given truth table of a function and comparison of an achieved design style with the desired style. One can evaluate the efficiency of the searching process via measure of the fitness function, for example, as the percentage of evolved correct logic network (with the desired functionality and style).

A *target design style* of a multilevel logic network is specified by:

(*i*) Types of logic gates from the library $\mathcal{L}$ of gates

(*ii*) Topology of network (permissible interconnections between gates levels, inputs, and outputs of the network)

(*iii*) The maximal number of levels M and the maximal number of gates N, in every level

(*iv*) Types of gates in every level of the evolved network

The target design style is denoted by $\mathcal{G}_{M \times N}$ and is called a *scanning window*. Suppose that the logic network space is partitioned into R parts. Let us evolve a network, $\mathcal{G}_{R(M \times N)}$, for each part in the same target design style.

Example 20.39 *Given a Boolean function f of n variables, and the number of decomposed variables $d < n$, the logic network space can be divided into $R = 2^d$ parts*

The optimization criteria include (a) the minimal number of gates (sometimes, with given types of gates in each level), and (b) the runtime. There are many factors that influence the final goal, for example, methods to encode a target design style and permissible interconnections to estimate the evolving process, and so on. It is worth mentioning that the realization of the target design style concept is closely related to the well-known concept of permissible functions and perturbations in multilevel logic network optimization in CAD.

20.6.3   Coding of a target design style

The logic network architecture can be encoded by string (of integer numbers) called *chromosome*. Each gene of the chromosome corresponds to a primitive gate. So, first, the basic logic gates are chosen (Table 20.10). Second, these gates are codified along with interconnections between gates, to produce the gene encoding. A *netlist chromosome* is a *configurable string* that codifies the target design style of the multilevel network topology.

TABLE 20.10

Functionality of binary and multivalued gates from the fixed library $\mathcal{L}$ of gates

Binary gates		K-valued gates	
		Code	Function
Code	Function	1	$\text{MAX}(x_1, x_2) = x_1 \vee x_2$
1	$x_1 \wedge x_2$	2	$\text{NOT MAX}(x_1, x_2)$
2	$x_1 \wedge \overline{x}_2$	3	$\text{TSUM}(x_1, x_2) = min(x_1 + x_2, K - 1)$
3	$\overline{x}_1 \wedge x_2$	4	$\text{NOT TSUM}(x_1, x_2)$
4	$x_1 \oplus x_2$	5	$\text{TPROD}(x_1, x_2) = max(x_1 + x_2 - (K - 1), 0)$
5	$x_1 \vee x_2$	6	$\text{NOT TPROD}(x_1, x_2)$
6	$\overline{x}_1 \wedge \overline{x}_2$	3	$\text{MIN}(x_1, x_2) = x_1 \wedge x_2$
7	$\overline{x}_1 \oplus x_2$	4	$\text{NOT MIN}(x_1, x_2)$
8	$\overline{x}_1$	5	$\text{MODSUM}(x_1, x_2) = (x_1 + x_2)_{mod\ K}$
9	$x_1 \vee \overline{x}_2$	6	$\text{NOT MODSUM}(x_1, x_2)$
10	$\overline{x}_2$	7	$\text{MODPROD}(x_1, x_2) = (x_1 \times x_2)_{mod\ K}$
11	$\overline{x}_1 \vee x_2$	8	$\text{NOT MODPROD}(x_1, x_2)$
12	$\overline{x}_1 \vee \overline{x}_2$	9	$\text{NOT}(x) = K - 1 - x$
		10–13	$\text{LITERAL}(x^a) = K - 1,\ a = 0, 1, 2, 3$

A $\mathcal{G}_{R(M \times N)}$ target design style and a netlist chromosome specify an evolutionary design scenario. It includes, in particular, (a) the type of decomposition of the initial function, (b) structure of the network subspaces, (*c*) levels of the logic network realization (for each subspace), and (*d*) a method to estimate the goodness of the evolved logic networks (number and types of gates, power dissipation, delays, testability).

Example 20.40 *A 2-level logic network in $\mathcal{G}_{2\times2}$ target design style is presented in Figure 20.15. There are many scenarios of logic network design in the $\mathcal{G}_{2\times2}$ style. For example, one of the scenarios states that gates MAX, NOTMAX, MODSUM, and MIN can be used without any limitation in a 2-level structure. Regarding the second scenario, gates MAX and NOTMAX can be used in the first level, and gates MODSUM and MIN in the second level of the created network. In the third scenario, a logic network space is partitioned into subspaces and subnets created over different topological limitations.*

Example 20.41 *(Continuation of Example 20.40) The netlist chromosome for the network given in Figure 20.15a is presented in Figure 20.15b. Consider one of the possible scenario in detail. There are 14 genes that correspond to four cells and outputs connections within a network. The first three genes carrier information about the MAX-gate (code 1) of the network with two inputs (labels 2,3). The output of the MAX-gate can be connected to the inputs of MODSUM-gate or MIN-gate. The next three genes represent NOTMAX-gate (code 2) with inputs 4 and 6, and the output connected to the inputs of MODSUM-gate and MIN-gate. In this target design style, no more than two levels and no more than two gates in each level can be used. Under the considered scenario, the genetic algorithm decides which gate is to be used and how to employ it with respect to the given style. It means that the size of the window, that is, the permissible maximal number M of network level and maximal number N of gates in each level, are varying through the window-based scanning process withing the network search space.*

Design example:
2-level $\mathcal{G}_{2\times 2}$ target design style and a netlist chromosome

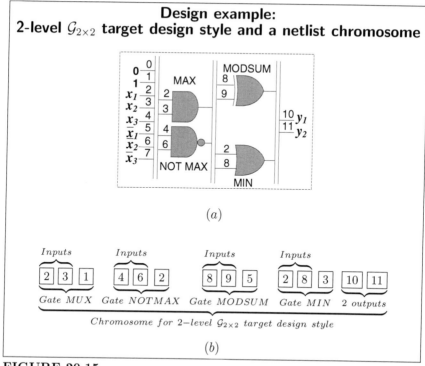

(a)

(b)

FIGURE 20.15
Schematic illustration of 2-level $\mathcal{G}_{2\times 2}$ target design style that provides different
scenarios of the 2-level circuit synthesis (a) (Example 20.40); A typical netlist
chromosome for 2-level $\mathcal{G}_{2\times 2}$ target design style (b) (Example 20.41).

20.6.4 Design example

The following example demonstrates the efficiency of the window-based
scanning process with given target design style. Using the simplest genetic
algorithm, the results were estimated with respect to parameters of M level
target design style $\mathcal{G}_{M\times N}$, that is,

▶ Percentage of correct solutions
▶ Number of experiments
▶ Total number of gates
▶ Total time
▶ Average time to obtain at least one of the correct networks

Here, the percentage of the correct solutions means the percentage of the
runs terminated once a chromosome, which represents a correct solution (the
logic networks of the desired functionality), out of 100 runs. The genetic
algorithm parameters includes:

▶ The population size: $S_{popul} = 60$
▶ The maximal number of generations: $MG = 10^4$
▶ Probability of crossover: $P_{cros} = 0.7$
▶ Level back: $L_{back} = 3$
▶ The *tournament* selection: $T_{select} = 2$ (the number of chromosomes which attended tournament)

Note that the tournament selection selects the best individual from a random subset of a given size and transfers it to the next population μ times; the best individual for each subset is chosen with respect to a condition, for instance, a certain probability. It means that there is 90% chance for a chromosome that is better than the second one, to be selected.

Consider an AND/OR/EXOR 3-level logic network with two-input EXOR gates. This is one of the simplest 3-level topologies. Figure 20.16 gives three network solutions obtained with respect to three different design strategies:

Regular method. The network in Figure 20.16(a) has been synthesized based on the traditional formal method (no genetic algorithm is applied).

Evolutionary strategy. The network in Figure 20.16(b) was evolved via a simple genetic algorithm in 10722 generations (about 4 hours) based on 5-level $\mathcal{G}_{5\times5}$ target design style without limitation on the type of gates in all levels of the network.

Evolutionary strategy. The network in Figure 20.16(c) was evolved in 9947 generations (about 4 hours) but in the AND/OR/EXOR $\mathcal{G}_{5\times5}$ target design style, so that we used a limited set of gate types from the library $\mathcal{G}$ in every level: two AND gates in the 1st and the 2nd levels only, two OR gates in the 3rd and the 4th levels only, two EXOR gates in the 5th level.

It follows from this experiment that the M level $\mathcal{G}_{M\times N}$ target design style enables flexibly manipulation the types of gates and topology of the created logic networks.

20.6.5 Information estimations as fitness function

One of alternative approaches to partitioning the space of possible logic network solutions and to estimation of fitness function is based on the information theoretical criterion, namely, a unified measure of information entropy. This approach aligns well with the fact that evolving process is of stochastic nature itself.

Consider the evolving process for a logic function f in terms of uncertainty. The start state of the process, given the truth table of the function, can be characterized by the entropy $H(f)$ of the function f. Let us consider a conditional entropy

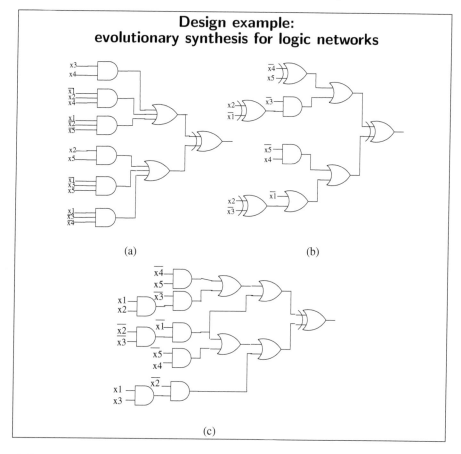

FIGURE 20.16

Comparison of the results of synthesis for a 5-input/single output multilevel network in AND/OR/EXOR design style: (a) 3-level network with EXOR gate and 2,3-input AND/OR gates synthesized via the formal method (b) 4-level evolved network via entropy-based algorithm with 2-input AND/OR/EXOR gates, (c) 5-level evolved network with 2-input EXOR gate and 2-input AND/OR gates.

$$H(f|Network1), \ H(f|Network2), \ ..., \ H(f|Network)$$

Thus, the entropy of the function f with respect to the current solution is used as an information measure for the circuit design process. Figure 20.17 explains this process. Mutual information $I(f; Network)$ between functions f and *Circuit* can be interpret as an amount of the logical work required for the circuit design. The mutual information $I(f; Network)$ and

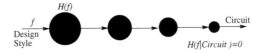

FIGURE 20.17
Information theoretical interpretation of evolving multilevel combinational logic networks: the uncertainty of a total circuit search space is removed via a target design style (window scanning).

the conditional entropy $H(f|Network)$ are related to the entropy $H(f)$ as follows: $I(f; Network) = H(f) - H(f|Network)$.

Let the fitness function be expressed in terms of conditional entropy of a target function f given a current function $Network$:

$$H(f|Network) = - \sum_{j=0}^{K-1} \sum_{i=0}^{K-1} p_{ij} \log \frac{p_{ij}}{p_j}, \qquad (20.1)$$

where f is the given function and $Network$ is the evolved function; p_{ij} is the probability that functions f and g take values i and j, correspondingly; and p_j means the probability that function $Network$ takes value j. Since $p_{ij} = k_{ij}/k$ and $p_j = k_j$, where k_{ij} represents the number of combinations (i, j) of values of the functions f and the $Network$, k_j represents the number of values j in the $Network$; and k is the total number of combinations.

Example 20.42 *Consider design of a ternary adder with carry (Table 20.11). Let function S be evaluated through the evolutionary process as follows. Starting with $s = Network_0 = 2$, the fitness function, in accordance with (20.1), is*

$$H(f|Network_0) = - \sum_{i=0}^{2} \sum_{j=0}^{2} k_{ij}/k \log k_{ij}/k_j$$

$$= -(^6/_{18} \log {}^6/_{18} + {}^6/_{18} \log {}^6/_{18} + {}^6/_{18} \log {}^6/_{18})$$

$$= 1.585 \; bit$$

For instance, in the 380th generation, the truth vector is $S = Network_{380} = [012020202200001012]$, and
$$H(f|Network_{380}) = -(^6/_{18} \log {}^6/_9 + {}^1/_{18} \log {}^1/_9 + {}^3/_{18} \log {}^3/_3$$
$$+ {}^2/_{18} \log {}^2/_6 + {}^1/_{18} \log {}^1/_9 + {}^4/_{18} \log {}^4/_6)$$
$$= 0.853 \; bit.$$

Finally, the correct network solution is obtained at 12610th iteration:

$$H(f|Network_{12610}) = -(^6/_{18} \log {}^6/_6 + {}^6/_{18} \log {}^6/_6 + {}^6/_{18} \log {}^6/_6)$$
$$= 0 \; bit$$

TABLE 20.11

Truth table of a ternary adder with carry

c_i	x	y	c_{i+1}	S	c_i	x	y	c_{i+1}	S
0	0	0	0	0	1	0	0	0	1
0	0	1	0	1	1	0	1	0	2
0	0	2	0	2	1	0	2	1	0
0	1	0	0	1	1	1	0	0	2
0	1	1	0	2	1	1	1	1	0
0	1	2	1	0	1	1	2	1	1
0	2	0	0	2	1	2	0	1	0
0	2	1	1	0	1	2	1	1	1
0	2	2	1	1	1	2	2	1	2

20.6.6 Partitioning of logic network search space

We use decomposition of the initial logic function as a tool to make the network search space partitioned into subspaces. For this, the property of conditional entropy is used: the decomposed variable is chosen with respect to maximum information (minimum entropy criterion) and entropy-based Shannon decomposition. The Shannon decomposition for a Boolean function f is defined as $f = \overline{x}f|_{x=0} \vee xf|_{x=1}$. The entropy of a Boolean function f given the variable x, $H^S(f|x)$, is formed of entropies of subfunctions given $x = 0$ as well as $x = 1$:

$$H^S(f|x) = p_{|x=0}H(f_{|x=0}) + p_{|x=1}H(f_{|x=1}). \tag{20.2}$$

The information measure of Shannon expansion is equal to the conditional entropy $H(f|x)$, that is, $H^S(f|x) = H(f|x)$.

Example 20.43 *The conditional entropy of the function c_{i+1} (Table 20.11) with respect to the variable c_i is computed as follows:*

$$H(c_{i+1}|c_i) = -{}^6/_{18} \log {}^2/_3 - {}^3/_{18} \log {}^1/_3 - {}^3/_{18} \log {}^1/_3$$
$$- {}^6/_{18} \log {}^2/_3 = 0.92 \text{ bit}$$

Similarly,

$$H(c_{i+1}|x) = -2 \cdot {}^5/_{18} \log {}^5/_6 - 2 \cdot {}^1/_{18} \log {}^1/_6 - 2 \cdot {}^3/_{18} \log {}^3/_6$$
$$= 1.12 \text{ bit}$$

$$H(c_{i+1}|y) = -2 \cdot {}^5/_{18} \log {}^5/_6 - 2 \cdot {}^1/_{18} \log {}^1/_6 - 2 \cdot {}^3/_{18} \log {}^3/_6$$
$$= 1.12 \text{ bit}$$

The variable c_i carries more information than variables x and y; therefore, c_i is chosen for the first step of Shannon decomposition. The conditional entropy of the function s with respect to variables c_i, x, y is equal to

$$H(s|c_i) = H(s|x) = H(s|y) = 6 \cdot {}^3/_{18} \log {}^1/_3 = 1.53 \text{ bit}.$$

Algorithm for evolving combinational logic networks in a given design style is introduced in Figure 20.18.

```
       Algorithm for evolving combinational logic networks

 Input Truth table of the initial logic function, target design style,
 genetic algorithm parameters, type of a decomposition
 Output Multilevel combinational network(s)


 {
    ChooseDecompositionType;
    Decompose(TruthTable); // parts are created
    for each part of TruthTable
    {
      SetupDesignScenario(designStyle, SetOfGates);
      SetGAParams(PopulationSize, MaxGenerations, Type of selection,
      Type of mutation);
      CreatePopulation();
      CheckEntropyBasedFitness(newpopulation) (equation (20.1));
      while (CurrentGeneration < MaxGeneration) and (BestFitness > 0)
      {
        Select(chromosome1, chromosome2);
        Crossover(chromosome1, chromosome2);
        Mutation(chromosome1);
        Mutation(chromosome2);
        PutIntoNewPopulation(chromosome1,chromosome2);
        CheckEntropyBasedFitness(newpopulation) (equation (20.1));
        CurrentGeneration++;
      }
    }
 }
```

FIGURE 20.18
Sketch of the algorithm for evolving combinational logic networks via window-based scanning in a given design style of the circuit subspaces.

20.6.7 2-digit ternary adder

The generalization of Shannon expansion for multivalued functions in the Galois filed and its information theory notation can be used for design of multivalued logic networks as well.

Consider a ternary adder with 4 inputs and carry input, and 2 outputs and carry output. The Shannon decomposition with respect to one variable (*carry*) results into six subspaces of network solutions: $C_{00}, C_{01}, C_{02}, C_{10}, C_{11}$, and C_{12}. The window-based scanning of each subspace with 8-level $\mathcal{G}_{8 \times 6}$ target design style and different limitations of types of gates yields six subnetworks with 11, 13, 9, 10, 18, and 15 ternary gates (Table 20.12). These subnetworks are connected automatically by five ternary multiplexers (MUX), and so, the final network contains 76 *gates* + 3 *MUX* (2 *ternary and 1 binary*) = 79 *elements*. Note that the window-based scanning can be realized in independent and parallel ways (six parallel processing), and, therefore, the run time has been reduced from 1297 to 279 minutes, that is, more than in three times.

TABLE 20.12

Effect of parallel scanning logic network space partitioned into six subspaces (ternary adder with carry in 8-level $\mathcal{G}_{6(8\times6)}$ target design style)

Property	Parallel processing						Total
	C_{00}	C_{01}	C_{02}	C_{10}	C_{11}	C_{12}	
Correct solutions (in %)	54	22	62	64	12	24	
Number of generations	500	2400	320	1360	12600	2400	
Total: gates + MUX	11	13	9	10	18	15	76+3
Total CPU-time (in min)	209	254	194	166	279	195	1297
Parallel processing,							
Parallel processing, CPU-time (in min)							279

Parameter setting: $S_{popul} = 60$, $MG = 2 * 10^4$, number of runs is 50, $P_{cros} = 0.7$, $P_{mutat} = 0.8$, $L_{back} = 0.4$.

20.6.8   2-Digit ternary multiplier

One can manipulate design style, that is, use different design styles for independent searching in each network subspace. The results are demonstrated in Table 20.13 for evolving 2-digit (4-inputs) ternary multiplier without carry. The space of network solutions C was partitioned into seven subspaces $C_0, C_{00}, C_{01}, C_{02}, C_{10}, C_{11}$, and C_{12}. The largest subspace C_0 has been created as a result of decomposition of the initial function with respect to the first variable, $a = 0$. The residual subspaces were obtained via decomposition of the initial function with respect to two variables. For instance, the subspace C_{10} has been produced via decomposition with respect to $a = 1, b = 0$.

The window-based scanning of the subspace C_0 with 8-level $\mathcal{G}_{8\times6}$ target design style resulted, in particular, in synthesize subnetworks with 15 ternary gates of the given types. Then, the residual subspaces have been scanned using the 7-level $\mathcal{G}_{7\times4}$ design window (including different limitations of type of gate from the library of gates), and the network solutions with 3, 13, 9, 12, 12, and 9 ternary gates of the assigned types have been obtained. So, the final logic network includes 79 ternary gates (73 gates plus 6 ternary multiplexers), as shown in Figure 20.19.

Design example: a 2-digit ternary multiplier

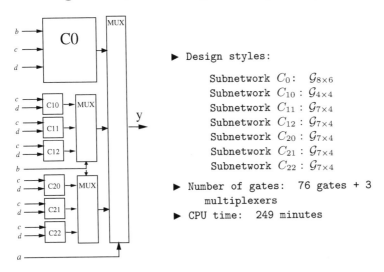

▶ Design styles:

Subnetwork C_0: $\mathcal{G}_{8\times6}$
Subnetwork C_{10} : $\mathcal{G}_{4\times4}$
Subnetwork C_{11} : $\mathcal{G}_{7\times4}$
Subnetwork C_{12} : $\mathcal{G}_{7\times4}$
Subnetwork C_{20} : $\mathcal{G}_{7\times4}$
Subnetwork C_{21} : $\mathcal{G}_{7\times4}$
Subnetwork C_{22} : $\mathcal{G}_{7\times4}$

▶ Number of gates: 76 gates + 3 multiplexers

▶ CPU time: 249 minutes

FIGURE 20.19
An evolved 4-input ternary multiplier in the mixed target design style $\mathcal{G}_{8\times6}/\mathcal{G}_{6(7\times4)}$ with parameters given in Table 20.13.

20.7 Neural-based computing

Neural-based computing is inspired by the nervous systems. It is characterized by a massive parallel and distributed processing. In this section, the focus is on neural-based computing of elementary logic functions.

20.7.1 Computing paradigm

A *Hopfield* computing paradigm (network), also known as *neuromorphic* computing, is based on the distributed architecture principles and can be implemented by various models [21]. Using a Hopfield network, an arbitrary logic function can be computed via a process similar to the gradual cooling process of metal. A value of a Boolean function given an assignment of Boolean variables is computed through the relaxation of neuron cells (Figure 20.20):

$$\text{Distributed network} \rightarrow \text{Relaxation} \rightarrow \text{Solution}$$

Hopfield networks are capable of reliable computing despite imperfect neuron cells and degradation of their performance. This is because degraded neuron cells store information (in weights, thresholds, and topology) in a

TABLE 20.13

Effect of the parallel window-based scanning of network subspaces on evolving a 2-digit ternary multiplier in mixed 4-, 7-, and 8-level $\mathcal{G}_{4\times4}/\mathcal{G}_{7\times4}/\mathcal{G}_{8\times6}$ target design styles

Property	Parallel processing							Total
	C_0	C_{10}	C_{11}	C_{12}	C_{20}	C_{21}	C_{22}	
Design style	$\mathcal{G}_{8\times6}$	$\mathcal{G}_{4\times4}$	$\mathcal{G}_{7\times4}$	$\mathcal{G}_{7\times4}$	$\mathcal{G}_{7\times4}$	$\mathcal{G}_{7\times4}$	$\mathcal{G}_{7\times4}$	
Correct solutions (in %)	8	100	44	72	64	52	76	
Number of generations	8416	10	550	1740	6500	900	720	
Total: gates + MUX	15	3	13	9	12	12	9	**73+3**
Total CPU-time (in min)	249	0.03	57	45	48	56	49	**504**
Parallel processing, CPU-time (in min)								**249**

Parameter setting: (*i.*) $P_{popul} = 60$, $MG = 2*10^4$, $P_{cros} = 0.7$. Number of runs is 50. (*ii.*) For the $\mathcal{G}_{8\times6}$ target design style $P_{mutat} = 0.008$ and $L_{back} = 4$. For the $\mathcal{G}_{6(7\times4)}$ and $\mathcal{G}_{4\times4}$ target design styles, $P_{mutat} = 0.13$, $L_{back} = 3$.

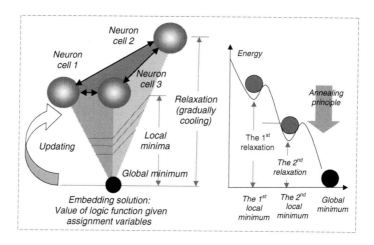

FIGURE 20.20

Computing paradigms based on the Hopfield model.

distributed (or redundant) manner. A single "portion" of information is not encoded in a single neuron cell but is rather spread over many. This contrasts with the traditional computing paradigm (von Neumann architecture), in which data is stored in a specific memory location. The Hopfield computing paradigm is based on the concept of *minimization of energy* in a stochastic system. This concept is implemented using *McCulloch–Pitts'* neuron cells for elementary Boolean functions, and characterized by the *interconnections* between nanocells given a logic function, the *weights* assigned to the links between nanocells; and an *objective* function given the number of neuron cells, their thresholds, interconnections, and weights. The carriers of information in the Hopfield network are a particular topology of connections between neuron cells, the weights of the links, and the neuron thresholds.

20.7.2 Neuron cells

Neuron cells for the computing of elementary logic functions were developed by McCulloch and Pitts in 1943 [27]. They are the most basic component of any artificial neural network, in particular, feedforward networks for computing of logic functions. A McCulloch–Pitts' cell is specified by arithmetic operations and thresholding. An arbitrary logic function can be computed on a network of such cells. Figure 20.21 (top) shows a technique for computing an elementary Boolean function using the McCulloch–Pitts model of logic nanocells, where $\sum$ is an arithmetic sum. Two-input AND, OR, NOR, and NAND elementary logic functions can be generated by a single neural cell. Control over the type of logic function is exercised by the threshold θ and weights $w_i \in \{1, -1\}$ of the arithmetic sum $w_i \times x_1 + w_i \times x_2$, that is, the output is $f = w_1 \times x_1 + w_2 \times x_2 - \theta$.

The McCulloch–Pitts model of a cell is very flexible for functional reconfiguration.

> **Example 20.44** *The threshold $\theta = -2$ for the sum of the weighted inputs $(-1) \times x_1 + (-1) \times x_2$ and assignments $x_1, x_2 \in \{0, 1\}$ results in the output value of the NAND gate. Let $x_1 = 1, x_2 = 0$, then $(-1) \times 1 + (-1) \times 0 - (-2) = 1$. In addition, the weights take only two values, 1 and -1, which is ideal for implementation.*

However, this is a deterministic model, and McCulloch himself tried to overcome this drawback in 1959 [28] before Hopfield proposed his network in 1982 [22].

20.7.3 Relaxation

There are several approaches to increasing the robustness of the McCulloch–Pitts model, in particular, using probabilistic thresholding and weights, and/or extending the set of states of the neuron by adding the third state $\mathbf{X} \notin \{0, 1\}$ that carries uncertain information.

Another approach to increasing the robustness of the McCulloch–Pitts deterministic computing model is a *parallel relaxation* based on the properties of neurons in a distributed system. This model assumes computing in a network of neuron cells of a particular configuration using the random change of states of these cells with respect to the objective function until stable states of cells are achieved. These stable states encode the final result or the solution. In our study, this is the value of an elementary Boolean function. Since the relaxation is nothing more than a search, the various states of the network form a search space. A randomly chosen state will transform itself ultimately

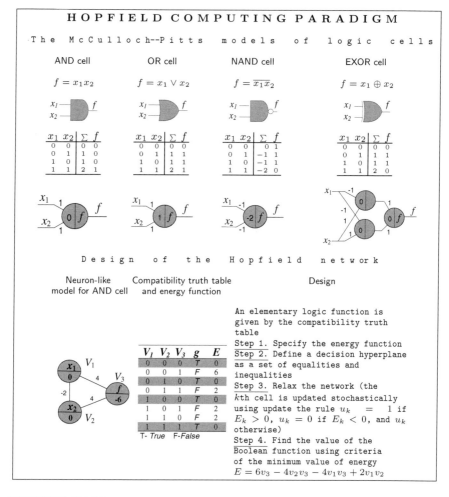

FIGURE 20.21

Design technique based on the Hopfield computing paradigm: the McCulloch–Pitts model of logic nanocells (top) and the Hopfield model of the AND gate, its truth table for compatibility function g and energy function E.

into one of the local minima (the nearest stable state). Even if the initial state contains inconsistencies, a network will settle into a solution that violates the fewest constraints offered by inputs.

The most valuable property of Hopfield networks for logic nanocell design is that given any set of weights and any initial state, (*a*) the relaxation algorithm eventually steers the network of neurons into a stable state, and (*b*) there can be no divergence or oscillation [22]. Each interconnection to cell i from cell j in the Hopfield network has an associated a weight denoted by w_{ij}. In the

Hopfield network, $w_{ij} = w_{ji}$, that is, it is a symmetrical and fully connected network. This provides an ultimate condition for *convergence*, which means that the network attains a stable state after relaxation is exercised.

Using an analogy from physics (the Ising model of spin-glass), the energy is defined as

$$\text{Energy}, E = -\frac{1}{2} \sum_{i} \sum_{i \neq j} w_{ij} v_i v_j, \quad \underbrace{w_{ij} = w_{ji},}_{Symmetry}$$

where $v \in \{0, 1\}$ is a state of a neuron cell. If a neuron cell v_k changes state, the energy change is $\Delta E_k = -\Delta v_k \sum_{j \neq k} w_{kj} v_j$. All allowed changes decrease energy E and gradient descent until a local minimum is reached.

20.7.4 Improved Hopfield network

Successive updates of a Hopfield network enable a convergence, whereby the energy of the overall network gets smaller. In a stable state (local or global) of the network, energy is at a minimum. In the Hopfield network, there is no way to reach a global minimum from a local minimum. Various methods can be used to solve this problem, for example, the *Boltzmann* approach (machine) that uses noise to "shake" the network state out of a local minimum. Our preliminary study shows that the most attractive feature for implementation of logic nanocells based on electrochemomechanical phenomena is that the search strategy is inspired by the annealing principle of quantum physics. This principle describes how heated materials, when their temperature is lowered very slowly in a controlled manner, can cool to very low energy states such as that of a highly ordered crystal lattice (Figure 20.20a). In this technique, local minima are avoided by adding some randomness to the process of energy minimization so that, when the network moves toward a local minimum, it has a chance to escape. Thus, updating of individual neuron cells becomes stochastic rather than deterministic.

The energy function of the updating process is defined by the expression

$$\overbrace{}^{Stochastic\ updaiting}$$

$$E = -\frac{1}{2} \sum_{i=1}^{n} \sum_{j=1}^{n} w_{ij} v_i v_j - \overbrace{\sum_{i=1}^{n} \theta_i v_i + K,}^{} \tag{20.3}$$

$$= -\sum_{t \in I} a_t v_1^{t_1} v_2^{t_2} \cdots v_m^{t_m}, \tag{20.4}$$

where I is a set $0, 1, \ldots, 2^n$, n is the number of neuron cells, $v_1 v_2 \ldots v_m$ is the binary representations of t, a_t is the t-th coefficient (integer number), K is a constant, and

$$v_k^{t_k} = \begin{cases} 1, & t_k = 0, \\ v_k^{t_k}, & t_k = 1, \quad k = 1, \ldots, m \end{cases}$$

Equation (20.4) is a general form derived from the energy function (20.3) using the manipulation of logic functions in arithmetic polynomial form. The results of computations for the Boltzmann machine, that is, the values of a logic function, are decoded in a manner similar to that of the Hopfield network.

In the modeling of nanocells based on electrochemomechanical phenomena, we have utilized the fact that, in the Boltzmann machine, instead of a deterministic threshold θ, neuron cells carry out *probabilistic decision-making*. A neuron cell v_k has an active state 1 with probability $p_k = 1/1 + e^{-\frac{\Delta E_k}{T}}$, where $\Delta E \in \{1, -1, 0\}$ is the energy that corresponds to the state change of the neuron cell, and T is the simulated "temperature" of the network. The output of a neuron cell is always either logical 0 or 1, but its probability distribution is a *sigmoid function* on average. If $T = 0$, this distribution reduces to a *step function*. This allows the system to jump occasionally to a higher energy configuration and to escape from a local minimum. Therefore, while the original Hopfield network uses local minima (which perform the role of the memory of a network), the Boltzman machine, being an improved Hopfield model, uses simulated annealing to reach a global energy minimum. Let the states of a neuron cell be A and B. At thermal equilibrium, the probability of any global state, is constant and obeys the *Boltzmann distribution* $\frac{P_A}{P_B} = e^{\frac{-(E_A - E_B)}{T}}$, that is, the probability ratio of any two states depends on their energy difference, and the lowest energy state is the most probable at any temperature.

20.7.5   Design example

Figure 20.21 shows the main features of the design cycle for a logic nanocell using the Hopfield model. The compatibility function g has a true value, (T), corresponding to a minimum value of the energy function, $E = 6v_3 - 4v_2v_3 - 4v_1v_3 + 2v_1v_2$, (derived Equation 20.3 where $a_0 = 0$, $a_1 = 6$, $a_2 = 0$, $a_3 = -4$, $a_4 = 0$, $a_5 = -4$, $a_6 = 2$, and $a_7 = -4$). This function is calculated based on the constraints of *linear separability*. The minimization of an energy function is referred to as embedding a "correct solution" into the Hopfield network.

The following features of the Hopfield model ensure feasibility of implementation of an elementary logic function based on electrochemomechanical phenomena:

(*a*) It facilitates for implementation that neuron cells always be in one of two states, *active* (encoded by 1) or *inactive* (encoded by 0).

(*b*) The network of neuron cells is fully interconnected and symmetrical, that is, each cell is connected to every other cell, and $w_{ij} = w_{ji}$, while $w_{ii} = 0$. Every pair of cells has a connection in each direction. A positive weighted connection indicates that the two cells tend to activate each other (encode mutually supportive relationships). A negative weight allows an active cell to deactivate a neighboring cell (encode incompatible relationships).

(c) This is a *recursive* model because the outputs of each neuron cell feed into the inputs of other cells. If *asynchronous* control is chosen, each neural cell makes decisions based only on its own local information. All these local actions add up to a global solution and are suitable for parallel asynchronous computing.

(d) An elementary logic function of n Boolean variables can be implemented by a Hopfield network with at least $n+1$ neuron cells.

In the implementation of *parallel relaxation*, the following details are important. Let the neuron cell be chosen at random. If any of its neighbors are active; the cell computes the sum of the weights on the connections țo those active neighbors. If the sum is positive, the cell becomes active, otherwise, it becomes inactive. Let the second cell be chosen at random, and the process repeats until the network reaches a stable state, that is, until no more cells can change state. The state reflects the assignment of truth and falsity to the various hypotheses under the constraints.

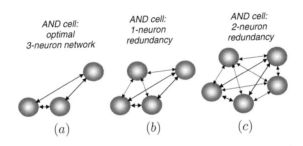

FIGURE 20.22

Fault-tolerance strategy for the AND, NAND, OR, and NOR logic cell design using the redundancy of Hopfield networks at the cell level: optimal 3-neuron Hopfield network for two-input (a); improving fault tolerance using 1-neuron (b) and 2-neuron redundancy (c).

A Hopfield network of logic nanocells is parallel and inherently fault tolerant even in *optimal* configuration. The later is a network configuration such that no other network exists that represents this function using fewer neuron cells. In the fault tolerant design of the network, each logic cell can contain more than three-neuron cells (three is the optimal number for AND, OR, NOR, and NAND, accept for EXOR, which needs more cells) (Figure 20.22). In a redundant network, while a few neuron cells may misbehave or fail completely, the network will still function properly. Given a Hopfield network for a two-input logic cell implementing an AND, NAND, OR, or NOR function, the fault-tolerance of this network can be improved using the following redundance strategy:

(a) At the neuron cell level. additional cells are incorporated into the network (Figure 20.22),

(b) At the cluster level, several copies of Hopfield networks are combined under an appropriate decision profile.

(c) A combination of the above strategies can be utilized.

An alternative approach to improving fault tolerance is as follows. Given an optimal 3-neuron Hopfield network for a two-input AND (also NAND, OR, or NOR) logic cell, redundancy can be implemented by using at least three clusters of cells and another level, decision profile, which perform, for example, a simple voting. In this approach, three electrochemomechanical decoders and at least the simplest electrochemomechanical voter (majority, for example) are needed.

The advantages of a Hopfield network for computing and implementation of an elementary logic function are the following:

▶ A Hopfield network requires a small number of neuron cells

▶ It needs a few iterations to achieve minimum energy and simple interconnect topology.

If an extension of the network is needed, additional inputs are introduced to the logic cell, the total energy function E of the network of logic nanocells $C_1, \ldots, C_m$ is a sum of energies of logic nanocells, $E = E_{C_1} + \ldots, + E_{C_m}$. The following undesirable effects can be observed in such the redundant Hopfield networks:

(a) The interconnect complexity increases (from 3 links in an optimal topology to 9 links in the case of 2-neuron redundancy).

(b) The relaxation time increases (instead of several iterations in an optimal network, it may require many iterations to learn such a network).

(c) The complexity of decoding goes up (in an optimal topology, 2^2 values from 2^3 must be selected; in the case of 2-neuron redundancy, 2^2 values of a logic function must be selected from 2^5).

The same effects can be caused by adding inputs to the logic cells.

The following avenues in modeling of logic nanocells can be formulated:

Interface. Each logic cell is modeled by a Hopfield network. A logic network is assembled by connections of logic cells, that is, by connecting Hopfield networks (clusters). This demands the development of efficient *interfaces* using appropriate physical or chemical phenomena.

Association. The effect of setting a network into local minima is desirable for paradigm based on *association*, but undesirable for calculations using *minimization* of energy function. To compute elementary Boolean functions, the *association* principle can be used.

Robustness can be increased by using three-valued cells, that is, cells with three states: 0,1, and *don't care*. This model can be useful, for example, when physical or chemical phenomena result in a nonstable state. In addition, replacing the deterministic operations of neuron cells by stochastic ones, such as thresholding or stochastic pulse stream encoding, can increase robustness. Hopfield, in particular, attempted to address this issue and proposed an approach to computing based on neuron cells called *spiking neurons* [23].

Learning (training the network by adjusting the weights and thresholds) is not desirable for computing and implementation of the simplest logic functions. This is because of technical difficulties in the implementation and application of redundant techniques. However, the complexity of the problem significantly increases, if additional inputs into logic cells are needed.

3D topology. An arbitrary Hopfield network can be represented by a decision diagram. The valuable feature of decision diagrams is that they can be embedded into a 3D topology. Advantages of decision diagrams can be considered, for example, for the implementation of interfaces in complex networks.

20.8   Further study

Advanced topics of natural computing

Topic 1: **Sierpinski triangles and transeunt triangles** for Boolean function representation. Interest in the Pascal triangle has been induced, perhaps, because of its connection to the expansion of $(x + y)^n$ for the numbers in row n of the triangle given the requisite coefficients. Applications of transeunt triangles for Boolean function manipulation can be found in works by Butler et al. [9, 10], and Yanushkevich et al. [48]. In particular, Butler et al. [9] considered transeunt triangles for calculation of optimal polarity of Reed–Muller expressions. Dueck et al. [17] developed an algorithm to find the minimum mixed-polarity Reed–Muller expression of a given Boolean function. This algorithm runs in $O(n^3)$ time and uses $O(n^3)$ storage space. Popel and Dani [39] used the Sierpinski gaskets to represent 4-valued logic functions, using both Shannon and Davio expansion for manipulation of this topological structure.

Topic 2: **Genetic algorithms for logic network design.** In [35], a technique to apply a genetic algorithm to evolution of simple combinational circuits (i.e. 2-6 bit multiplexers, 2-5 bit adders, 2-bit decoder) and sequential circuits (e.g., 4-state machine, 3-bit counter) has been proposed. Some recent technical reports conveyed results on evolving 16-state machines. Authors of [?] have shown the principal ability of the genetic algorithm technique for FPGA synthesis. Genetic algorithm has been used for multivalued logic network design, decomposition, and testing multivalued networks [46].

Topic 3: Biological sequence processing by HMMs. A Hidden Markov Model (HMM) is a graph of connected states where each state possesses a variety of properties. Processes described by HMMs evolve in some dimension, often parameterized by time stamps. HMMs work under assumption that the current state is not influenced by the entire history of states, but instead only one previous state is considered. Models are interpreted by analyzing possible paths in the structure of states. The structure of model states is built by learning from the data, in which parameters are estimated by applying Bayesian classifiers. Bayesian classifiers provide a framework for obtaining an a posteriori probability (the probability of the model, given the observed sequence) that includes the ability to integrate prior knowledge about the sequences.

For the majority of biological sequences, the time parameterization is replaced by the position in the sequence. HMMs can naturally accommodate variable-length models including operations of mutation, insertion, and deletion. This is generally achieved by having a state which has a transition back to itself. Because most biological data has variable-length properties, machine learning techniques that require a fixed-length input, such as neural networks or support vector machines, are less successful in biological sequence analysis.

Topic 4: DNA computing. The first successful experiment involving the use of DNA molecules and DNA manipulation techniques was reported by L. M. Adelman in 1994 [2]. He proposed the solution of a small instance of the Hamiltonian path problem in a directed graph. The Hamiltonian path problem has been shown to be an NP-complete problem, that is, intractable in the traditional computing paradigm. To solve this problem, Adleman developed the algorithm and encoded this algorithm step by step using DNA molecular structures. He showed that this NP-complete problem can be solved in linear time due to the massive parallelism of DNA computing.

Adleman estimated the potential advantages of DNA computing as follows [2]. DNA computing can be approximately 1,200,000 times faster than the fastest supercomputers; it could permit data storage 10^{12} times more efficiently and it can take 10^{10} times less energy than the existing computers.

Lipton (1995) showed how to use DNA computing to solve the *satisfiability* (SAT) problem for propositional formulas [7]. SAT is a search problem known to be NP-complete. A key aspect of the SAT problem explored by Lipton was the possibility of representing it as a graph search problem. The solution consists of two phases: (a) the generation of all paths in the graph, and (b) the search for a truth assignment satisfying the formula. Note that no method essentially better than *exhaustive search* is known to solve the SAT problem. One has to search through all possible 2^n combinations given a Boolean formula with n binary variables. This makes the problem computationally intractable. For example, it is already computationally infeasible to solve the problem in the case of 200 variable. Lipton's solution to the SAT problem made exhaustive search computationally feasible by the massive parallelism of DNA molecules and manipulation techniques.

Topic 5: Information estimation of chromosome. Consider a chromosome encoded by a string of observed data points Chr_t, $t = 1, 2, \cdots, n$, one after another. The objective is to redescribe the data with a suitably designed code as efficiently as possible (with a short code length), that is, coding of a string such that it can be decoded from its code. This is a typical problem of coding theory [49, 50]. The solution is as follows: (a) select a parametrically defined statistical model $P_\theta(Chr)$ for the data Chr and (b) estimate the vector parameter $\Theta = [\theta_1, \theta_2, \cdots, \theta_m]$ from the observation, where the number of the parameters, m, is also to be estimated.

Consider two chromosomes Chr_1 and Chr_2. The relative information in Chr_1 given the Chr_2 is $I(Chr_1|Chr_2) = Min_{k,\theta}\{-\log P_\theta(Chr_1|Chr_2) + C\}$, where $P_\theta(Chr_1|Chr_2)$ denotes the conditional probability, $C = {}^1/_2 \cdot k \cdot \log n$ parametrized by Θ. Information in the chromosome Chr_1 about the chromosome Chr_2 can be described as mutual information $I(Chr_1; Chr_2) = I(Chr_1) - I(Chr_1|Chr_2)$.

Any chromosome, consisting of some genes (characters of an alphabet) issued by a source with entropy H can be encoded by a string of m symbols, so that the average number l of symbols per gene be approximated to, but no less than, $H/\log m$. For binary coding ($m = 2$), $l \geq H$. An effective encoding means that symbols should take values 0 and 1 with probabilities as equal and independent of the previous one as possible. Such requirements are satisfied by the Huffman (prefix) code.

Topic 5: Reversible computing. In 1976, Charles Bennett proved that, for power not be dissipated in the logic network, it is necessary that this network be build entirely from reversible gates. Reversible logic has been studied at a theoretical level for decades as a counterpoint to conventional switching logic. In a conventional logic gate, several signal lines enter the gate, then the signals are processed and merged, and the output is transmitted on a single output line. Once processed, the inputs cannot be reconstructed from the output; information has been thrown away, and along with it some electrical energy. The logic gates are therefore *irreversible*. Rather than merging a number of inputs into a single output, reversible logic gates have the same number of outputs as inputs. The same amount of information that entered the gate is represented at the output in a transformed state.

Bennet pioneered *logically reversible* computations, leading to the widely known models for reversible computing that admit computations with *energy recovery*. Subsequently, Fredkin and Toffoli demonstrated logical gates that exhibit the same property. In contemporary logic computing, switching is based on *nonrecovering* modes of energy consumption. For example, loss of information for the two input AND gate is $H_{loss} = H_{out} - H_{in} = 1.189$ *bit* for the probability $p_i = 0.25$ of input pattern.

Irreversible computations involve a loss of information that can be equated to a loss in heat. Conventional (irreversible) logic gates leads to energy dissipation. An m-input, m-output Boolean function is *reversible* if it maps each input pattern to a unique output pattern. A reversible function can be realized by a cascade of reversible gates in which there is no fan-out or feedback.

The notion of reversibility arises from the fact that, since each input pattern is associated with a unique output pattern, a reversible function always has an inverse. Frequently, the function is its own inverse, particularly for primitive reversible gates. If one replaces each gate in the cascade with the gate realizing the inverse of the function realized by the original gate (it may be itself), then the cascade applied in reverse order realizes the inverse of the function realized by the original cascade.

The main idea of Fredkin gates were that energy dissipation could be avoided if simple crossover switches were used in place of conventional logic gates. The result, called conservative logic, was that logic levels would be conserved wherever possible. It was proposed in [8] that information should also be preserved, and so the use of feedback was prohibited. The Fredkin gate is shown in Figure 20.23.

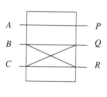

There are three inputs and three outputs in the Fredkin gate. The data path from input A to output P is called the control; its function is to determine the flow of data along the other two paths. If the input $A = 0$, then the data flow straight through the device, so

$$Q = B$$
$$R = C$$

$P = A$

$Q = \overline{A}B \vee AC$

$R = \overline{A}C \vee AB$

Alternatively, if $A = 1$, the paths crossover, so that

$$Q = C$$
$$R = B$$

FIGURE 20.23
Fredkin gate.

The formal definition of a reversible Boolean function is as follows: an m-input, m-output, totally specified Boolean function $f(x_1, x_1, ..., x_m)$ is reversible if it maps each input assignment to a unique output assignment. Note that, for $B = \{0, 1\}$ $f: B^m \to B^m$, and $f^{-1}: B^m \to B^m$.
A reversible function

▶ Can be written as a standard truth table
▶ Can also be viewed as a bijective mapping of the set of integers $\{0, 1, ..., 2^m - 1\}$ onto itself

A reversible function is thus a permutation of its domain, which can be expressed as an ordered list of the integers 0 to $2^m - 1$. For example, a 3-input, 3-output reversible function that specifies cyclic increment is given in Table 20.14. The inverse function is cyclic decrement, as can be seen by reading the truth table from right to left, rather than the normal left to right. This function realizes the permutation $\{1, 2, 3, 4, 5, 6, 7, 0\}$.

An n-input, n-output gate is reversible if it realizes a reversible function. The most commonly used set of reversible gates is the Toffoli family, defined as follows. An $n \times n$ Toffoli gate:

▶ Passes the first $n - 1$ lines (controls) through unchanged.

TABLE 20.14
A 3-input, 3-output reversible function

Inputs			Outputs		
x_1	x_2	x_3	x_1'	x_2'	x_3'
0	0	0	0	0	1
0	0	1	0	1	0
0	1	0	0	1	1
0	1	1	1	0	0
1	0	0	1	0	1
1	0	1	1	1	0
1	1	0	1	1	1
1	1	1	0	0	0

▶ Inverts the nth line (target) if the control lines are all 1; otherwise the target line is passed through unchanged.

Toffoli gates are clearly self-inverse. An $n \times n$ Toffoli gate will be denoted $TOFn(x_1, x_2, ..., x_n)$ where x_n is the target line. Using a prime symbol to denote the value of a line after passing through the gate, we have $x_i' = x_i$, $i < n$, $x_n' = x_1 x_2 ... x_{n-1} \oplus x_n$. $TOF1(x_1)$ is the special case where there are no control inputs, so x_1 is always inverted, that is, it is a NOT gate. $TOF2(x_1, x_2)$ has been termed a Feynman or controlled-NOT (CNOT) gate. $TOF3(x_1, x_2, x_3)$ is often referred to simply as a Toffoli gate, and sometimes as a controlled-controlled-NOT (CCNOT) gate. Toffoli gates are often drawn as shown in Figure 20.24.

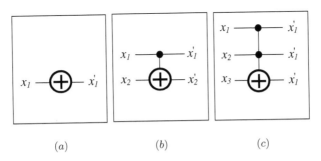

(a)   (b)   (c)

FIGURE 20.24
Toffoli gates: $TOF1(x_1)$ (a), $TOF2(x_1, x_2)$ (b), and $TOF3(x_1, x_2, x_3)$ (c).

A second family of interest is the set of *Fredkin gates*. Each such gate interchanges two target lines if all the control lines have the value 1. Such a

gate with no control lines is a swap gate. Fredkin gates are also self-inverse.
We do not elaborate further on Fredkin gates or other possible reversible gates
since the decision diagram techniques illustrated for Toffoli gates extend to
other reversible gates in a straightforward manner. A reversible function can
be realized by a network that is a cascade of reversible gates. An example is
given in Figure 20.25.

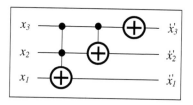

▶ From left to right, this circuit realizes the cyclic increment function.
▶ Since each Toffoli gate is self-inverse, when applied from right to left the circuit realizes cyclic decrement.

FIGURE 20.25
Cyclic increment reversible function.

The specification of a 3-bit full adder given in Table 20.15 and its embedding
into a reversible specification given in Table 20.16 are shown. Input d is a
constant input that is set to 0 to realize the adder function. Output p is a
garbage output, which in this case is the adder propagate signal $a \oplus b$. g is a
garbage output, which in this case is the input a. The logic network in Figure
20.26 was found from specification given in Table 20.16 using the method
described in [29].

TABLE 20.16
Reversible specification.

d	c	b	a	$carry$	sum	p	g
0	0	0	0	0	0	0	0
0	0	0	1	0	1	1	1
0	0	1	0	0	1	1	0
0	0	1	1	1	0	0	1
0	1	0	0	0	1	0	0
0	1	0	1	1	0	1	1
0	1	1	0	1	0	1	0
0	1	1	1	1	1	0	1
1	0	0	0	1	0	0	0
1	0	0	1	1	1	1	1
1	0	1	0	1	1	1	0
1	0	1	1	0	0	0	1
1	1	0	0	1	0	0	0
1	1	0	1	0	0	1	1
1	1	1	0	0	0	1	0
1	1	1	1	0	1	0	1

TABLE 20.15
Irreversible
specification

c	b	a	$carry$	sum
0	0	0	0	0
0	0	1	0	1
0	1	0	0	1
0	1	1	1	0
1	0	0	0	1
1	0	1	1	0
1	1	0	1	0
1	1	1	1	1

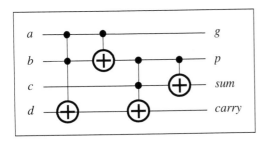

- $carry = ab + ac + bc$
- $sum = a \oplus b \oplus c$
- Constant input d is set to 0 to realize the adder.
- g and p are garbage outputs that in this case realize useful functions. That is not generally the case.

FIGURE 20.26
Reversible logic network for a 3-bit full adder.

Garbage outputs and constant inputs must generally be added in transforming an irreversible specification to a reversible one. The minimum number of garbage outputs required is $\lceil \log_2 q \rceil$, where q is the maximum output pattern multiplicity of the original irreversible function, i.e., the maximum number of times a single output pattern is repeated in the specification. Constant inputs must be added so that the reversible specification has the same number of inputs and outputs.

Further reading

[1] Adamatzky A. Computing with waves in chemical media: Massively parallel reaction-diffusion processors, *IEICE Trans. Electron.*, E87-C(11):1748-1756, 2004.

[2] Adleman LM. Molecular computation of solutions to combinatorial problems. *Science*, 226, November, pp. 1021–1024, 1994.

[3] Aguirre AH and Coello CAC. Evolutionary synthesis of logic circuits using information theory. In Yanushkevich SN, Ed., *Artificial Intelligence in Logic Design*, pp. 285–311, Kluwer, Dordrecht, 2004.

[4] Aoki T, Homma N, and Higuchi T. Evolutionary synthesis of arithmetic circuit structures. In Yanushkevich SN, Ed., *Artificial Intelligence in Logic Design*, pp. 39–72, Kluwer, Dordrecht, 2004.

[5] Al-Rabady A and Perkowski M. Shannon and Davio sets of new lattice structures for logic synthesis in three-dimensional space. In *Proc. 5th Int. Workshop on Applications of the Reed–Muller Expansion in Circuit Design*, Mississippi State University, pp.165–184, 2001.

[6] Ash RB. *Information Theory*. John Wiley & Sons, Hoboken, NJ, 1967.

[7] Lipton RJ. Using DNA to solve NP-complete problems. *Science*, 268, pp. 542–545, 1995

[8] Bennett C. Logical reversibility of computation. *I.B.M. J. of Research and Development*, 17:525–532, 1973.

[9] Butler JT, Dueck GW, Shmerko VP, and Yanushkevich SN. Comments on SYMPATHY: fast exact minimization of fixed polarity Reed-Muller expansion for symmetric functions. *IEEE Trans. Computer-Aided Design of Integrated Circuits and Systems*, 19(11):1386–1388, 2000.

[10] Butler JT, Dueck GW, Yanushkevich SN, and Shmerko VP. On the number of generators of transeunt triangles. *Discrete Applied Mathematics*, 108:309–316, 2001.

[11] Carbone A and Seeman NC. Circuits and programmable self-assembling DNA structures. In *Proceedings of National Academy of Sciences*, 99(20):12577–12582, 2002.

[12] de Castro LN. *Fundamentals of Natural Computing: Basic Concepts, Algorithms, and Applications*. Chapman & Hall/CRC Taylor & Francis Group, Boca Raton, FL, 2006.

[13] Debnath D and Sasao T. Minimization of AND-OR-EXOR three-level networks with AND gate sharing. *IEICE Trans. Information Systems*, E80-D(10):1001-1008, 1997.

[14] Cheng KT and Agrawal VD . An entropy measure for the complexity of multi-output Boolean functions. In *Proc. IEEE Design Automation Conf.*, pp. 302–306, 1991.

[15] Cheushev V, Shmerko V, Simovici D, and Yanushkevich S. Functional entropy and decision trees. In *Proc. IEEE 28th Int. Symp. on Multiple-Valued Logic*, pp. 357–362, 1998.

[16] Drechsler R. *Evolutionary Algorithms for VLSI CAD*. Kluwer, Dordrecht, 1998.

[17] Dueck GW, Maslov D, Butler JT, Shmerko VP, and Yanushkevich SN. A method to find the best mixed polarity Reed–Muller expressions using transeunt triangle. In *Proc. 5th Int. Workshop on Applications of Reed–Muller Expansion in Circuit Design*, Mississipi State University, pp. 82–92, 2002.

[18] Falconer K. *Fractal Geometry*. Wiley, New York, 1990.

[19] Fredkin E and Toffoli T. Conservative logic. *Int. J. Theor. Phys.*, pp. 171–182, 2003.

[20] Green DH. Families of Reed–Muller canonical forms. *Int. J. of Electronics*, 2:259–280, 1991.

[21] Hopfield JJ. Neural networks and physical systems with emergent collective computational abilities. *Proceedings National Academy of Sciences*, USA, 79:2554–2558, 1982.

[22] Hopfield JJ. Neurons with graded response have collective computational properties like those of two-state neurons. *Proceedings National Academy of Sciences*, USA, 81(10):3088–3092, 1984.

[23] Hopfield JJ. Pattern recognition computation using action potential timing for stimulus representation. *Nature*, 376:3–36, 1995.

[24] Landauer R. Irreversibility and heat generation in the computing process. *I.B.M. J. Research and Development*, 5:183-191, 1961.

[25] Lent CS, Tougaw PD, Porod W, and Bernstein GH. Quantum cellular automata. *Nanotechnology*, 4:49–57, 1993.

[26] Lent CS and Isaksen B. Clocked molecular quantum-dot cellular automata. *IEEE Trans. Electron. Devices*, 50(9):1890–1896, 2003.

[27] McCulloch WS and Pitts WH. A logical calculus of ideas immanent in nervous activity. *Bulletin Mathematical Biophysics*, 5:115–137, 1943.

[28] McCulloch WS. Stable, reliable, and flexible nets of unreliable formal neurons. *Quarterly Progress Report*, MIT, RLE, pp. 118–129, 1958.

[29] Miller DM, Maslov D, and Dueck GW. A transformation based algorithm for reversible logic synthesis. In *Proc. IEEE/ACM Design Automation Conference*, pp. 318–323, 2003.

[30] Miller DM, Dueck GW and Maslov D. A synthesis method for MVL reversible logic. In *Proc. 34th IEEE Symp. on Multiple-Valued Logic*, pp. 74–80, 2004.

[31] Hemmi H, Mizoguchi J, and Shimohara K. *Development and evolution of hardware behaviors*, In [41], pp. 250–265, 1996.

[32] Iba H, Iwata M, and Higuchi T. Machine learning approach to gate level evolvable hardware. In *Lecture Notes in Computer Science*, Springer, vol.1259, pp. 327–393, 1997.

[33] Kabakcioglu AM, Varshney PK, and Hartman CRP. Application of information theory to switching function minimization. *IEE Proceedings, Pt E*, 137:389–393, 1990.

[34] Miller JF and Thomson P. Aspects of digital evolution: Geometry and learning. In *Lecture Notes in Computer Science*, Springer, Heidelberg, vol. 1478, pp. 25–35, 1998.

[35] Miller JF, Thomson P, and Fogerty T. Designing electronic circuits using evolutionary algorithms. Arithmetic circuits: A case study. In Quagliarella D, *et. al.*, Eds. *Genetic Algorithm and Evolution Strategies in Engineering and Computer Science. Recent Advances and Industrial Applications.* John Wiley and Sons, pp. 105–131, 1998.

[36] Moraga C and Wang W. Evolutionary methods in the design of quaternary digital circuits. In *Proc. IEEE 28th Int. Symp. on Multiple-Valued Logic*, pp. 89–94, 1998.

[37] Murakawa M, Yoshizawa S, Kajitani I, Furuya T, Iwata M, and Higuchi T. Hardware evolution at functional level. In *Lecture Notes in Computer Science*, Springer, Heidelberg, vol.1141, pp. 62–71, 1996.

[38] Reif JH. Local parallel biomolecular computation. In Rubin H and Wood DH, Eds., *DNA-Based Computers*, American Mathematical Society, 48:217–254, 1999.

[39] Popel DV and Dani A. Sierpinski gaskets for logic function representation. In *Proc. IEEE 32th Int. Symp. Multiple-Valued Logic*, pp. 39–45, 2002.

[40] Rothemund PWK, Papadakis N, and Winfree E. Algorithmic self-assembly of DNA Sierpinski triangles. In *PLOS Biology — www.plosbiology.org* , 2(12):2041–2053, e424, 2004.

[41] Sanchez E and Tomassini M, Eds., *Towards Evolvable Hardware. Lecture Notes in Computer Science*, vol. 1062. Springer, Heidelberg, 1996.

[42] Shmerko VP, Popel DV, Stanković RS, Cheushev VA, and Yanushkevich SN. Information theoretical approach to minimization of AND/EXOR expressions of switching functions. In *Proc. IEEE Int. Conf. Telecommunications in Modern Satellite, Cable and Broadcasting Services*, pages 444–451, 1999.

[43] Shende VV, Prasad AK, Markov IL, and Hayes JP. Synthesis of reversible logic circuits. *IEEE Trans. Computer-Aided Design*, 22(6):710–722, 2003.

[44] Shukla SP and Bahar RI, Eds., *Nano, Quantum and Molecular Computing*. Kluwer, Dordrecht, 2004.

[45] Stanković R. Functional decision diagrams for multiple-valued functions. In *Proc. IEEE Int. Symp. on Multiple-Valued Logic*, pp. 284–289, 1995.

[46] Wang W and Moraga C. Design of multivalued circuits using genetic algorithms. In *Proc. 26th IEEE Int. Symp. on Multiple-Valued Logic*, pages 216–221, 1996.

[47] Yao X and Higuchi T. Promises and challenges of evolutionary hardware. *IEEE Trans. Systems, Man, and Cybernetics, Part C: Applications and Reviews*, 29(1):87–97, 1999.

[48] Yanushkevich SN, Butler JT, Dueck GW, and Shmerko VP. *Experiments on FPRM expressions for partially symmetric logic functions*. In *Proc. 30th Int. Symp. on Multiple-Valued Logic*, pp. 141–146, 2000.

[49] Rissanen J. *Stochastic Complexity in Statistical Inquiry*, World Scientific, New York, 1989.

[50] Rissanen J. *Information and Complexity in Statistical Modeling*, Springer, Heidelberg, 2007.

Index

Printed and bound by CPI Group (UK) Ltd, Croydon, CR0 4YY

25/10/2024

01779234-0001